HOLMES'
PRINCIPLES
OF PHYSICAL
GEOLOGY

HOLMES' PRINCIPLES OF PHYSICAL GEOLOGY

FOURTH EDITION

P. McL. D. DUFF

Sometime Professor of Applied Geology,
University of Strathclyde
Honorary Fellow, Department of Geology and Geophysics,
University of Edinburgh

CHAPMAN & HALL

University and Professional Division

London · Weinheim · New York · Tokyo · Melbourne · Madras

**Published by Chapman & Hall, 2-6 Boundary Row,
London SE1 8HN, UK**

Chapman & Hall, 2-6 Boundary Row, London SE1 8HN, UK

Chapman & Hall GmbH, Pappelallee 3, 69469 Weinheim,
Germany

Chapman & Hall Inc., One Penn Plaza, 41st Floor, New York
NY 10119, USA

Chapman & Hall Japan, Thomson Publishing Japan, Hirakawacho
Nemoto Building, 6F, 1-7-11 Hirakawa-cho, Chiyoda-ku, Tokyo 102,
Japan

Chapman & Hall Australia, Thomas Nelson Australia, 102 Dodds
Street, South Melbourne, Victoria 3205, Australia

Chapman & Hall India, R. Seshadri, 32 Second Main Road, CIT East,
Madras 600 035, India

First edition 1944, reprinted 18 times
Second edition 1965, reprinted 6 times
Third edition 1978, reprinted 1979, 1980, 1983, 1984, 1986, 1987
Fourth edition 1993, reprinted 1994, 1996, 1997

© 1978 Doris L. Holmes, 1993 Chapman & Hall

Typeset in Palatino 10/12pt by Falcon Graphic Art Ltd, Wallington,
Surrey
Printed in China

ISBN 0 412 40320 X

A catalogue record for this book is available from the British Library
Library of Congress Cataloging-in-Publication Data available

CONTENTS

CONTRIBUTORS

G S Boulton
Department of Geology and Geophysics
University of Edinburgh
Chapters 20 and 21

R L Brown
Department of Geology and Geophysics
Carleton University
Ottawa, Canada
Chapter 31

R F Cheeney
Department of Geology and Geophysics
University of Edinburgh
Chapter 8

I W D Dalziel
Institute for Geophysics
University of Texas at Austin
Texas, USA
Chapter 31

J B Dawson
Department of Geology and Geophysics
University of Edinburgh
Chapter 29

P McL D Duff
Department of Geology and Geophysics,
University of Edinburgh
Chapters 1, 9 and 10

J G Fitton
Department of Geology and Geophysics
University of Edinburgh
Chapter 5

K W Glennie
Formerly Shell UK Ltd.
Chapter 22

C Graham
Department of Geology and Geophysics
University of Edinburgh
Chapter 11

A Hall
Fettes College
Edinburgh
Chapters 15-18 and 23

A Halliday
Department of Geological Sciences
University of Michigan
Michigan, USA
Chapter 14

I B Harrison
formerly British Geological Survey
Chapter 19

B Harte
Department of Geology and Geophysics
University of Edinburgh
Chapter 11

I G Main
Department of Geology and Geophysics
University of Edinburgh
Chapter 26

G Nichols
Department of Geology
Royal Holloway and Bedford New College
University of London
Chapter 30

I Parsons
Department of Geology and Geophysics
University of Edinburgh
Chapter 4

R G Pearce
Department of Geology and Geophysics
University of Edinburgh
Chapter 25

N B Price
Department of Geology and Geophysics
University of Edinburgh
Chapter 24

T P Scoffin
Department of Geology and Geophysics
University of Edinburgh
Chapters 6 and 24

R Thompson
Department of Geology and Geophysics
University of Edinburgh
Chapters 27 and 28

B G J Upton
Department of Geology and Geophysics
University of Edinburgh
Chapters 12 and 13

E K Walton
Department of Geography and Geology
University of St Andrews
Chapters 2, 3, 6 and 7

PREFACE TO FOURTH EDITION

Doris Reynolds (Mrs Holmes) approached me in 1977 — as a former student and colleague of Arthur Holmes — with a view to my undertaking a possible 4th edition of *Principles of Physical Geology*, the 3rd edition then being in press.

I felt highly honoured but rather apprehensive and we agreed that as the science was advancing so rapidly no single person could hope to cover such a comprehensive work, in say 10 years time, and that a team would be necessary. On my retirement and return to Edinburgh in 1987 I was delighted to find enthusiasm for the project by many of the staff in the Departments (as they then were) of Geology and Geophysics in the University and they and graduates of the Geology Department make up most of the 'revisers'. The task, as with most books, took longer than anticipated, not only because progress in the science had been so rapid and extensive that some chapters have had to be completely rewritten and not simply revised, but also because the timing coincided with unprecedented changes in the organization and research-funding of British Geology Departments, involving time-consuming bureaucratic procedures, meetings, and cost-cutting exercises that had to take precedence over all else.

Knowing Arthur Holmes, he would have welcomed all the new evidence which supported, modified, or apparently demolished both old and newer hypotheses. I doubt, however, if he would have survived in the recent climate pertaining in the British University and College scene. He was first and foremost a research scientist and teacher. Arthur Holmes taught at Edinburgh when it was still expected and possible (indeed it was the tradition in Scotland) for a Head of Department to deliver 3 lectures a week each academic year to students attending first-year level courses. In the immediate post-war years the 150-seater Geology Lecture Theatre was filled to overflowing and Holmes had no problems in holding the interest of his audience, of whom fewer than a dozen or so, at most, were intending graduates in geology. Those who did become geologists had the unique benefit of his knowledge and advanced ideas in the Honours classes. It would be fair to say that most

of us who graduated in the 1950s wondered why there was so much excitement when 'Plate Tectonics' appeared on the scene in the 1960s; Holmes' suggestions to us to explain movements of large blocks of the Earth's crust basically lacked only the name! He always emphasized the difference between a hypothesis and a theory and was not afraid to discard ideas if the evidence was there that proved them wrong. The abstract of his famous *Revised Geological Time-scale* in 1959, for example, commenced with the words 'The time-scale constructed in 1947 was based on certain assumptions that have recently been shown to be wrong'.

Just as the second edition of his book (1965) differed from the first (1944) and the third, revised by Doris Holmes (1978) varied from the second, so does this edition discard certain themes and introduce newer ideas. Holmes pointed out in the preface to the second edition that so much progress had been made in the previous 20 years that it took him 7 years full-time work to revise the first edition. He doubted if so many new discoveries in geology would ever be made again in another 20-year interval. In the 14 years that have elapsed since the publication of the third edition, however, geologists and geophysicists are having difficulty in keeping up with their own specialities let alone those pursued by others. The reasons are mainly, of course, because the numbers of geologists (I include geophysicists with geologists) have continually increased and more data are continually becoming available in all branches of the science due to the increased use of sophisticated technology to both produce and process it.

We have tried to retain as much of Holmes' writing as possible and also to retain his hope 'that the book would appeal not only to university students and the senior classes in schools and their teachers, but also to the wide range of general readers whose wonder and curiosity are excited by the behaviour of this mysterious world of ours'. In 1928 Holmes in an article in the *Mining Magazine* quoted Paul Termier's words in the *Smithsonian Report* for 1925. 'The least ignorant among us, the most daring, the most restless, ask ourselves questions; we demand when the voyage of humanity began, how long it will last, how the ship goes, why do its decks and hull vibrate, why do sounds sometimes come up from the hold and go out from the hatchway; we ask what secrets do the depths of the strange vessel conceal, and we suffer from never knowing the secrets.' Let us hope we now know at least some of them!

A book of this nature draws inevitably, and quite properly, on the work of an enormous number of geologists. All are acknowledged by name in the text or in figure captions. In view of the need to keep the book to a reasonable size for the intended readership, we have not listed detailed references. In our experience most first-year level students rarely consult primary sources, even if they are fortunate enough to be studying in institutes which have an adequate coverage of specialist journals etc. For the more advanced students the names of the authors and the dates of publication should suffice to allow them to identify the publication in the *Bibliography and Index of Geology*, published by the American Geological Institute.

All contributors are warmly thanked by the Editor, with my additional thanks for editorial advice from E.K. Walton, J.B. Dawson, J.G. Fitton and M.R.W. Johnson. I am particularly grateful also to K.R. Gill for his assistance in the selection of colour plates and half-tones and to Yvonne Cooper for her photographic work. Acknowledgements to the various bodies and organizations which kindly supplied photographs are given with the captions. Thanks too are due to the staff of the Scottish Science Library (the National Library of Scotland) and to R.P. McIntosh, J.M. Dean, N.G.T. Fannin and D. Gould of the British Geological Survey, and J. Underhill of the Department of Geology and Geophysics, University of Edinburgh, for their help in the tracing of particular items of literature.

Una-Jane Winfield of Chapman & Hall carried out with great dedication the virtually impossible task of ensuring that deadlines for the production of manuscripts, proofs etc, were (almost) kept by a group of some twenty or so geologists, any or many of whom, at any given time, if they were not burdened by teaching duties, were on, or flying over, any of the world's continents or oceans in connection with their researches, when they might have been attending to their chapters.

Finally, I would like to acknowledge the help given to me by Catherine Shephard of C.S. Editorial Services, Oxford, and to thank my wife for her constant assistance, support and encouragement during the assembling of this volume.

P.McL.D. Duff
Edinburgh
June 1992

PREFACE TO SECOND EDITION

It would be difficult to find a more attractive and rewarding introduction to the basic concepts of science than physical geology. The activities of our planet may be compared to the combined operations of the four Elements of the ancient Greek philosophers: Fire, Earth, Air and Water, to which we should now add Life. The ever-changing interplay of these operations is responsible for a fascinating variety of natural phenomena, ranging from landscape forms and scenery to the catastrophes brought about by earthquakes, volcanic eruptions, floods and hurricanes, all of which are of daily interest and concern to a high proportion of the world's inhabitants. The effects of these activities, and – so far as current knowledge permits – their causes, are the chief topics of our subject.

In the preface to the first edition I expressed the hope that the book would appeal not only to university students and the senior classes in schools and their teachers, but also to the wide range of general readers whose wonder and curiosity are excited by the behaviour of this mysterious world of ours. My hopes have been surprisingly surpassed. Partly, I expect, this has been because the subject was presented with a minimum of jargon, with constant reference to observational evidence and with copious illustrations. Since I have found this method of treatment to be consistently successful in arousing and developing the interest of students, including those who had no preliminary acquaintance with the elementary principles of science, I have endeavoured to follow it again in this new edition. Partly, too, the timing of the original edition was auspicious, for the book appeared just as the services of geologists and geophysicists began to be required in ever-increasing numbers all over the world.

Writing in 1852 of *The Next Million Years*, Sir Charles Darwin pointed out that 'we are living in the middle of an entirely exceptional period, the age of the scientific revolution'. Throughout this century the rate of scientific publication has been doubling every few years and in many sciences, including the geological group, the rate of increase is itself increasing, like compound interest. It is not generally realised that, of all the geologists who have ever lived, about ninety per cent are now alive and actively at work. For geophysicists the percentage is naturally higher. The curves representing these remarkable increases have been rising much more steeply than those of the world 'population explosion'. Obviously, since the contrast applies to scientists in general, the upward-sloping curves must gradually turn and flatten out, probably with a few fluctuations dependent on future developments in China and Africa. Otherwise there would sooner or later be as many scientists as people. But despite the enormous wealth of geological data still to be collected and the certainty that some major surprises will continue to emerge (e.g. when the various Mohole projects are completed), it is doubtful whether there will ever again be such a profusion of unexpected discoveries concentrated into so short an interval of time as there has been during the last twenty years.

The above considerations serve to explain why the revision on which I started full-time work seven years ago has taken so long; why its completion now instead of a few years ago is more timely; and why it has resulted in what is practically a new book, almost entirely rewritten and of necessity greatly enlarged. Old chapters have been extended and new ones added, to deal with the more significant results of the explorations and researches that have revolutionised geology and geophysics during the last two decades. These include the radiometric dating of minerals and rocks, and the consequent possibility of estimating the rates of some of the earth's long-term activities; the concept of rheidity and its applications to the flow of glaciers, the formation of salt domes and the intrusion of granite and associated rocks; fluidisation as the key to many aspects of volcanic and igneous geology; the host of complex issues raised by the occurrence of ice ages – some involving the destiny of mankind in the relatively near future and others the totally different distribution of climates in the far-off past; the extraordinary surprises, contradicting all expectations, that have rewarded the explorers of the ocean floors, particularly the astonishing thinness of the oceanic crust and its youthful veneer of deep-sea sediments, paradoxically combined in most places with a far higher rate of heat flow than anyone could have foreseen; the growing evidence that the earth is expanding rather than shrinking; the possible energy sources required to keep the earth's internal 'machinery' going; the maintenance of continents above sea-level and the uplift of plateaux; the contorted structures of mountain ranges and the recognition of gravity gliding as a major factor in the folding of rocks; and finally, the magnetism preserved in ancient rocks as a compass-guide to the wandering of continents.

In terms of observational facts alone the behaviour of the inner earth now appears even more fantastic than when we knew much less about it. In this rejuvenated book I have tried to present a balanced view of the new fields of knowledge which have been so explosively opened up. Every chapter contains exciting stories of man's achievements and speculations, tempered by the sobering reflection that while we may often be wrong, Nature cannot be. It may be noticed that inconsistencies have arisen here and there, when certain problems are approached along different lines. Such difficulties are not glossed over. By bringing them into the open, instead of sweeping them under the carpet of outworn doctrines and traditional assumptions, the reader is helped to see where further research is needed and may even be stimulated to take part in. Primarily, however, the book has been designed to meet the needs of the same sort of audience as before; this time enlarged, I hope, by those of my former readers who may be finding it increasingly time-consuming to dig out the essentials of recent progress from the avalanche of publications in which they are recorded.

It is a pleasure to acknowledge the considerable help I have myself received in this way from the publications and pre-prints kindly sent to me by friends and correspondents in many parts of the world. By discussing various problems and reading parts of the book, generous assistance has been given by Professor Tom F. W. Barth, Dr Lucien Cahen, Dr Lauge Koch, Professor L. Egyed,

Professor Maurice Ewing (and some of his colleagues at the Lamont Geological Observatory), Dr R. W. Girdler, Professor W. Nieuwenkamp and Professor C. E. Wegmann. To all of these I owe my grateful thanks for help, suggestions and constructive criticisms. To my wife and fellow geologist, Dr Doris L. Reynolds, I am, as always, more deeply indebted than can be adequately expressed. Not only has her unselfish devotion ensured the completion of this book, but her professional influence and her flair for the utmost rigour of scientific method have been never-failing sources of inspiration and encouragement. It is, however, only fair to add that I remain entirely responsible for any defects of fact, treatment, judgment or style.

As befits the subject, special care has again been taken to illustrate the book as fully and effectively as possible. Nearly all the line diagrams have been specially drawn, many of them being based on diagrams already published by others, with only slight modifications or translations of the lettering. Particular sources are credited in the captions, and a general expression of my indebtedness and thanks is given here to the many authors whose diagrams have been so freely called upon. In addition to the photographs procured from professional photographers and press agencies (all duly acknowledged in the captions) many striking and instructive subjects have been contributed by friends and official organisations.

It will be noticed that the illustrations are not systematically listed in these preliminary pages. When such lists become unduly long, as would here have been the case, they tend to defeat their purpose. As a practical alternative, which it is hoped will facilitate easy reference, all illustrations are indexed, with their page numbers in italics, under each relevant subject or key word.

Finally, I wish to express my thanks to the staff of my publishers for their enthusiastic and helpful co-operation throughout the production of this book, and particularly to Miss Nine T. Yule and her colleagues for their editorial advice and guidance.

Arthur Holmes
London
October 1964

SCIENCE AND THE WORLD WE LIVE IN

1

In the first place, there can be no living science unless there is a widespread instinctive conviction in the existence of an *Order of Things*, and, in particular, of an *Order of Nature*.

Alfred North Whitehead 1927

1.1 INTERPRETATIONS OF NATURE: ANCIENT AND MODERN

The world we live in presents an endless variety of fascinating problems which excite our wonder and curiosity. The scientific worker, like a detective, attempts to formulate these problems in accurate terms and, so far as is humanly possible, to solve them in the light of all the relevant facts that can be collected by observation and experiment. Such questions as What? How? Where? and When? challenge him to find clues that may suggest possible answers. Confronted by the many problems presented by, let us say, an active volcano, we may ask: What are the lavas made of? How does the volcano work and how is the heat generated? Where do the lavas and gases come from? When did the volcano first begin to erupt and when is it likely to erupt again?

Here and in all such queries the question What? commonly refers to the stuff things are made of, and an answer can be given in terms of chemical compounds and elements. Not in terms of the four 'elements' of the Greek philosopher Empedocles (about 490–430 BC), who considered the ultimate ingredients of things to be Fire, Water, Earth and Air, but chemical elements such as hydrogen, oxygen, silicon, iron and aluminium. The

question What? also refers to the names of things, and particularly to their forms and structures: things ranging in size from the elementary particles of atoms to the galaxies of stars that make up the Universe, with our own Earth falling manageably between these extremes of the inconceivably large and small.

The question How? refers to natural processes and events – the way things originate or happen or change – and in human activities to methods and techniques. This question leads to the very root of most natural problems, and satisfactory answers, although amongst the hardest to find, are to the scientist the most rewarding.

Where? refers to everything connected with space, and particularly to the relative positions and distributions of things. The location and distribution of oil is a topical example of current interest.

The question When? raises all the problems connected with the history of things and events. To unfold the history of the Earth and its inhabitants is the most ambitious aim of geological endeavour.

The scientific worker of today differs in one

very important respect from the detective with whom we have compared him. Except when he is concerned with human nature, as the detective invariably is, and perhaps with some types of animal behaviour, the scientific worker never asks the question Why? in its strict sense of implying motive or purpose. It is quite useless, for example, to ask why a volcano erupts, or why the sky is blue, or why there are earthquakes, because there is no possible means of finding answers to such questions. Of course we all commonly ask Why? in the loose sense of meaning 'How does it happen or come about . . . ?' and so long as it is clearly understood that this usage is just scientific slang to avoid the appearance of pedantry, no great harm may be done.

Why? always implies the further question Who? Thus the old legends of Ireland tell us that giants were responsible for many natural phenomena. They were stone throwers or builders. One of them is reputed to have flung the Isle of Man into the Irish Sea, Lough Neagh being the place it was taken from. The Giant's Causeway, which is a terrace carved by the weather and the sea from a lava of columnar basalt (Figure 1.1), was 'explained' as the work of the giant Fionn Mac-Comhal (or Finn MacCoul). This defiant stonemason began to construct a causeway across the sea, the better to attack his hated rival Fingal, who had built himself a stronghold of similar construction

in the Isle of Staffa (Figure 1.2). Complacent satisfaction with such 'explanations' obviously stifles the spirit of eager enquiry which is essential for the birth and development of science.

It is only during the last century or so that the futility of reading motives and purposes into events has been at all widely realized. The ancients in particular, including Sumerians, Hindus, Babylonians and Greeks, all of whom made remarkable discoveries in mathematics and astronomy, went sadly astray as a result of inventing answers to the dangerous question Why? They regarded the phenomena of Nature as manifestations of power by mythical deities who were thought to behave like irresponsible men of highly uncertain temper, but who might nevertheless be propitiated by suitable sacrifices.

In the Mediterranean region, for example, many of the Olympian gods made familiar by the epics ascribed to Homer (about 850 BC) were personifications of various aspects of Nature, including quite a number of geological processes which have been named after them. Poseidon (later identified with the Roman water-god Neptune) was the ruler of the seas and underground waters. As the waters confined below the surface struggled to escape, Poseidon assisted them by shaking the earth and fissuring the ground. He thus became the god of earthquakes. Typhon, the source of destructive tempests, was a 'many-headed monster of

Figure 1.1 Giant's Causeway, Co. Antrim, Northern Ireland. View of the Grand Causeway, showing vertical columns. Steeply inclined columns are seen in the cliff to the right of the house. (*Northern Ireland Tourist Board*)

Figure 1.2 Fingal's Cave, Island of Staffa, west of Mull, Scotland. (*Paul Pepper Limited*)

malignant ferocity' who was eventually vanquished by the thunderbolts of Zeus, the sky and weather god, who imprisoned him in the earth. Hades, corresponding to Pluto of the Romans, was the deity presiding over the nether regions which share his name. He was not, however, the god of subterranean fire. That responsibility was given to Hephaestus, a son of Zeus, who himself controlled fire in the form of lightning. When the Greeks settled in Sicily, Hephaestus was identified with the local volcano-god Vulcan or Volcanus. The eruptions of lava and volcanic bombs from Etna and Stromboli were feared both for their danger and as expressions of the fire-god's wrath.

The epic poems of Homer had biblical authority for the ancient Greeks, and therefore all the more honour is due to Thales of Miletus (about 624–565 BC) for his courage in making a clean break with all these traditional beliefs. He regarded the activities of Nature not as indications of supernatural intervention but as natural and orderly events which could be investigated in the light of observation and reason. He thus became the first Greek prophet of science. Having noticed that deposition of silt from the waters of the Nile led to the outward growth of the delta, he developed the hypothesis that Water was the source of earth and everything else. No doubt crude, but probably little more so than a present-day hypothesis that hydrogen is the primordial element. The

important point is that the natural and observable activities of water took the place of the imaginary and therefore inscrutable activities of Poseidon. Science had replaced superstition.

Later Heraclitus (about 500 BC), celebrated for his philosophy that 'all is perpetual flux and nothing abides', picked on Fire, the most active agent of change, as the fundamental principle behind phenomena. The observable manifestations of fire took the place of the fire-gods. The next step was taken by Empedocles, who saw that both Fire and Water were necessary, for when natural water is heated it evaporates into Air (in this case steam) and leaves a residue of Earth (the material originally in solution). Thus he established the four 'elements' which roughly correspond to our concepts of energy, and the solid, liquid and gaseous states of matter.

Despite this promising start, science languished for two thousand years. The Ionian philosophers who began the scientific quest were well aware that another 'element' was necessary to account for Man. This they called Consciousness or Soul, and some of them regarded it as divine. In concentrating on this aspect of existence, Plato (427–347 BC) and Aristotle (384–322 BC) revived the idea that Earth, Sun and Planets were all deities. So, unwittingly, one effect of their unrivalled authority was to lend support to the old Babylonian cult of astrology, as against the sterner discipline of science. Now what has this to do with geology? Two examples will suffice. Herodotus (484–426 BC) was a great traveller who made many significant geological observations. He speculated about the effect of earthquakes on landscapes, but nevertheless he thought it quite reasonable to ascribe the earthquakes themselves to Poseidon. Much later Pliny (AD 23–79), who, like Empedocles, lost his life while investigating a volcano too closely, 'explained' earthquakes as an expression of the Earth's resentment against those who mutilated and plundered her skin by mining for gold and silver and iron.

With the rise of Christianity there was no longer any incentive to study the ways of Nature. Because men had a complete theory as to **why** things happen, they were not interested in **how** they happen. Moreover, it was widely believed that the Earth had been created ready-made only a few thousand years before, and that it would soon come to an end. Two factors, amongst others, slowly brought this period of stagnation to a close. The Earth inconsiderately failed to come to an end, until at last it began to seem worth while to collect facts about the world instead of merely arguing about ideas. Another factor was a social one. Among the Greeks it had been bad form for a philosopher to become proficient in the craftsmanship and special skills demanded by any form of technology. For all such practical activities slaves were available. However, as the practice of alchemy came increasingly into vogue during the Middle Ages, a philosopher's study was quite likely to become his laboratory. Philosophy and technology – or more generally theory and practice – gradually ceased to be divorced, and with the coming of the Renaissance the occasional and increasing alliance of the two fostered afresh the spirit of science. By his unsurpassed and versatile genius Leonardo da Vinci (1452–1519) firmly established the new pattern. Were he not more renowned as painter and engineer, he would still be famous as a pioneer in many fields of natural science. He recognized that landscapes are sculptured and worn away by erosion; that the fossil shells found in the limestones of the Apennines are the remains of marine organisms that lived on the floor of a long-vanished sea that once extended over Italy; and that it could not have been the Noachian Deluge that swept them into the rocks of which they now form so important a part. But Leonardo was far in advance of his time, and three more centuries had to elapse before the Deluge ceased to obstruct the progress of geology.

Today we think of natural processes as manifestations of energy acting on or through matter. We no longer accept events as the results of arbitrary – and therefore unpredictable – interference by mythological deities. Typhoons and hurricanes are no longer interpreted as the destructive breath of an angry wind-god: they arise from the heating of the air over sun-scorched lands. The source of the energy is heat from the sun. Volcanic eruptions and earthquakes no longer reflect the revengeful behaviour of the gods of underground fire and water: they arise from the stresses and strains of the Earth's unstable interior, and from the action of escaping gases and heat on the outer and crustal parts of the globe. The source of the energy lies mainly in the material of the inner Earth.

In many directions, of course, and particularly where great catastrophes are concerned, our knowledge is still woefully incomplete. The point is not that we now pretend to understand everything – if we did, the task of science would be over – but that we have faith in the orderliness of natural processes. The steadily accelerating researches of the last two or three centuries have unfailingly justified our belief that Nature is understandable: understandable in the sense that if we ask her questions by way of appropriate observations and experiments, she will answer truly and reward us with discoveries that endure. But it must never be forgotten that Nature is the most perfect of expert witnesses. It is often far from easy to find the right questions and to ask them in the right way. For this reason, to say nothing of human fallibility, the discipline of science does not preclude the making of errors. 'What it does preclude,' as Norbert Wiener has expressed it, 'is the retention of an error which has clearly and distinctly betrayed its wrongness.'

1.2 THE MAJOR FIELDS OF SCIENTIFIC STUDY

The questions we ask when faced with a volcano in eruption are typical of the kinds suggested by all natural phenomena. They indicate that – in general terms – scientific investigation is concerned with the manifestations and transformations of matter and energy in space and time. Put more briefly and philosophically, science is concerned with relationships between events.

Of all the sciences **physics** is the most fundamental, for it deals with all the manifestations of energy and with the nature and properties of matter in their most general aspects, and on all scales from sub-atomic particles to the observable limits of the expanding universe. It overlaps to some extent with **chemistry**, which is particularly concerned with the composition and interactions of substances of every kind in terms of atoms and molecules, elements and compounds. **Biology** is the science of living matter. The nature of life still remains an elusive mystery. As examples of a vast range of phenomena that are quite inexplicable in terms of what we know of matter and energy, despite the vaunted powers of electronic robots, it is sufficient to recall our faculties for transforming physical waves into the conscious miracle of colours and sounds. Nevertheless there are innumerable aspects of organisms and their fossil remains that can be investigated by scientific methods with astonishing success.

All the other sciences have their own fields of interest, but these 'fields' overlap in all directions and trespassing is regarded as welcome co-operation. **Astronomy**, with the unfathomed universe of stars and nebulae as its field of study, deals essentially with the distribution and movements of matter in space on a celestial scale. Its interest in the Earth is limited to the purely planetary aspects of our globe. From an astronomical point of view the Earth may seem to be only an insignificant speck in the immensity of space. But space has no downward limit, and the Earth can equally well be seen as a galaxy of atoms, designed to restore our sense of proportion. As the home of mankind and the mother of all the life we know, we naturally regard the Earth as a supremely important field for special investigation. So we come to the science devoted to the Earth, appropriately known as **geology** (from the Greek *gaia* or *ge*, the ancestral Earth-goddess of Greek mythology; and *logy*, a suffix denoting 'knowledge of').

From the earliest days of exploration **geography** has been recognized as the study of the 'home of mankind'. Modern geography focuses attention on man's physical, biological and cultural environment and on the relationships between man and his environment. The study of the physical environment by itself is physical geography, which includes consideration of the surface relief of the continents and ocean floors (geomorphology), of the seas and oceans (oceanography), and of the air (meteorology and climatology), all of which have developed into important members of the commonwealth of sciences on their own account. As a major science, however, geology deals not merely with the land-forms and other surface features of the Earth's rocky crust, but with the structure and behaviour of every part of the Earth, with special reference to the rocks and structures of the visible crust and all that can be learned from them.

1.3 THE SCOPE AND SUBDIVISIONS OF GEOLOGY

Modern geology has for its aim the deciphering of the whole evolution of the Earth and its inhabitants from the time of the earliest records that can be recognized in the rocks right down to the present day. So ambitious a programme requires much subdivision of effort, and in practice it is convenient to divide the subject into a number of branches, as shown in Figure 1.3, which also indicates the chief relationships between geology and the other major sciences. The key words of the main branches are the **materials** of the Earth's rocky framework (mineralogy and petrology), and their **dispositions**, i.e. their forms, structures and interrelationships (structural geology); the

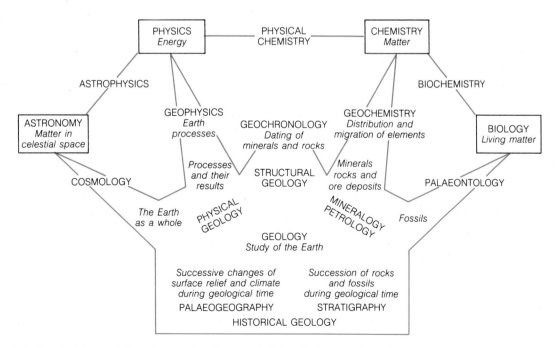

Figure 1.3 Subdivisions of the science of geology and their relation to other sciences.

geological **processes** or machinery of the Earth, by means of which changes of all kinds are brought about (physical geology); and finally the succession of these changes in time, or the **history** of the Earth (historical geology).

The Earth is made up of a great variety of materials, such as air, water, ice and living organisms, as well as minerals and rocks and the useful deposits of metallic ores and fuels which are associated with them. The relative movements of these materials (wind, rain, rivers, waves, currents and glaciers; the growth and movements of plants and animals; and the movements of hot materials inside the Earth, as witnessed by volcanic activity) all bring about changes in the Earth's crust and on its surface. The changes involve the development of new rocks from old; new structures in the crust; and new distributions of land and sea, mountains and plains, and even of climate and weather. The scenery of today is only the latest stage of an everchanging kaleidoscopic series of widely varied landscapes – and seascapes. **Physical geology** is concerned with all the terrestrial agents and processes of change and with the effects brought about by them. This branch of geology is by no means restricted to geomorphology, as we have seen. Its main interest is in the machinery of the Earth, and in the results, past and present, of the various processes concerned, all of which are still actively in operation at or near the Earth's surface or out of sight in the depths. Of these results the changing positions of continents and oceans, of fold mountains, rift valleys and ocean troughs, are important examples. Others are the rock structures – such as folds, Figure 1.4 – that have resulted from movements and deformation of the Earth's crust. **Tectonics**, the study of these structures, is an important part of **structural geology**, which is also concerned with the forms and structures that characterize rocks when they are first formed.

Changes of all kinds have been going on continuously throughout the lifetime of the Earth – that is, for something like 4600 million years. To the geologist a rock is more than an aggregate of minerals; it is a page of the Earth's autobiography with a story to unfold, if only he can read the language in which the record is written. Placed in their proper order from first to last **stratigraphy**, (Figure 1.5) (and dated where possible by determining the ages of radioactive minerals and rocks (**geochronology**)), these pages embody the history of the Earth. Moreover, it is familiar knowledge that many beds of rock contain the remains or impressions of shells or bones or leaves. These objects are called fossils, a term that was first applied by Agricola (1494–1555) to anything of

Figure 1.4 Folded strata of Jurassic age (see Section 7.2). The Stair Hole, west of Lulworth Cove, Dorset. (*K. R. Gill*)

interest dug out of the ground, including minerals. Since the end of the eighteenth century, however, the term has been used only for the relics of animals and plants that inhabited the Earth in former times. **Palaeontology** is the study of the remains of these ancestral forms of life, some of which, resembling certain types of seaweed, can be traced back for at least 3000 million years. Thus we see that **historical geology** deals not only with the nature and sequence of events brought about by the operation of the physical processes, but also with the history of the long procession of life through the ages.

As palaeontological and stratigraphical knowledge grew, it became obvious that similar fauna and flora had developed at similar times in the Earth's history, on continents separated by extensive oceans. Conventional wisdom at the beginning of this century had it that the migration of species was by 'land bridges', long since foundered. (One was considered to have spanned the gap between Brazil and Africa and another between Europe and North America!). Prominent among the dissenters at the beginning of this century (and working independently) were F. B. Taylor in the USA and Alfred Wegener in Germany. The former proposed in 1908 and the latter in 1912 that the opposing continents had once been united and had 'drifted' apart. Scorn was poured on the idea by most geologists and physicists and it was decades before it was taken seriously.

As geochemical data from ancient and modern volcanoes and their sources and from meteorites accumulated, along with more sophisticated interpretations of the ever-increasing geophysical data (from studies of the passage of earthquake waves through the deeper zones of the Earth, for example) opinions changed, as will be seen later.

Geophysicists realized that their fundamental objections to large-scale movements of parts of the Earth's crust over a 'molten' deeper zone were no longer valid. Crucial evidence in the 1960s of changes in the Earth's magnetic field throughout geological time, especially from studies of the ocean floors, made the hypothesis of crustal plates being moved by convection currents in an underlying 'molten' zone tenable. The **Plate Tectonic Theory**, emphasizing that continental and oceanic areas and their relationships with each other and with the Earth's climatic belts have continuously changed throughout geological time – (and will continue to do so) – is now accepted by virtually all geological scientists.

Apart from its intellectual appeal, a knowledge of geology is essential if we of the world's rich

Figure 1.5 A vertical thickness of up to 1600 m of horizontal layers (strata) of sedimentary rocks exposed in the Grand Canyon, Arizona, USA, by the erosive power of the Colorado river. (*P. McL. D. Duff*)

nations are (at the very least) to maintain our standards of living and also help improve the lot of the so-called 'third world' countries. The term 'economic geology' is often used to encompass the search for and extraction of the materials, metals and fuels essential for industry, and indeed our way of life. It is not always appreciated that there is nothing new in this type of exercise which currently can be the subject of much uninformed criticism. Without man's progress through the Stone Age, Bronze Age and Iron Age periods, for example, and his subsequent cultural development you would not be reading this book! Artefacts made of hard materials such as flint and chert, quartz, and glassy volcanic rocks are found along with early traces of man. The firing of clay to make utensils and even ornaments dates from Neolithic times. In fact the extraction and use of clay was probably the first large-scale mineral industry! Since then man has gradually increased his search for and use of minerals and fuels and is now becoming aware of the effect he is having on the world's environments. Geology has an essential part to play in this area, if only to emphasize the changes that have taken place naturally during the Earth's immensely long history and to point out that we are powerless to prevent many geological phenomena such as earthquakes, volcanic activity and many floods. What we can do on occasion is predict them and in other situations highlight those of man's activities such as over-cultivation or deforestation which increase the natural processes of erosion and destruction of a particular environment, and perhaps help to slow them down.

Firstly, of course, we require an immense data base. We must know for example, the natural chemical and physical nature of the rocks and soils of the Earth's surface before we can decide how much contamination and damage is being caused by our actions. The same applies to the sediments and waters of rivers, lakes, seas and oceans, to the nature of our continental ice-sheets and to the Earth's atmosphere.

Government bodies are usually expected to carry out these enormous tasks, though not without continual problems, worldwide, of funding. (Politicians are usually more concerned with short-term rather than long-term projects.)

In Britain environmental research and monitoring is carried out by the Natural Environment Research Council (NERC) of which the largest constituent body is the **British Geological Survey** (founded in 1835, and one of the oldest of its kind). Other NERC bodies concentrate their efforts under the headings of Marine and Atmospheric Sciences, Terrestrial and Freshwater Sciences and Antarctic Science. While most of their work concerns the British Isles, all are involved in international projects. In addition NERC funds University research both in Britain and internationally.

Worldwide, industry uses the basic geological data provided by the government bodies in the shape of maps, reports and advice but in specific areas such as mineral and oil exploration, water supply, large engineering projects (dams, reservoirs, bridges, etc.) industry has its own geologists, who far outnumber those employed in government and academic work.

Today's geologists rely on much more than simply careful observation and recording of what can be seen when walking over the ground, though this remains essential. Mapping of rocks and soils has been speeded up and improved by the use of aerial photography and satellite imagery. Geochemists make use of the most sophisticated analytical apparatus to identify and quantify specific minerals and chemical elements in rocks, ores, soils and natural waters.

Geophysicists investigate the physical properties of rocks such as elasticity, density, magnetization and electrical behaviour. Depending on the property, measurements can be made from ground level, from the sea surface or sea bed, from the air or from boreholes and mines and deductions then made of the nature of the underground rocks and structures present. The final proof of what lies underground, of course, relies on the drilling of boreholes, which in places have recovered rocks from depths measured in kilometres.

The progress made in the past decade of our understanding of the Earth using increasingly sophisticated techniques has enabled geologists, engineers and other scientists to satisfy society's and industry's insatiable demand for raw materials and for power generation. Whether this can be sustained indefinitely without permanent damage to the Earth's atmosphere, and the world can also continue to produce enough food in the light of the continual uncontrolled expansion of the world's population, are matters for debate.

1.4 FURTHER READING

Adams, F.D. (1938) *The Birth and Development of the Geological Sciences*, Dover Publications, London.

Ambrose, E.J. (1990) *The Mirror of Creation*, Scottish Academic Press, Edinburgh.

Farrington, B. (1953) *Greek Science*, Pelican Books, London.

Gould, S.J. (1987) *Times Arrow Times Cycle*, Harvard University Press, Cambridge, Mass. and London.

A VIEW OF THE

EARTH

Let the great world spin for ever down the
ringing grooves of change.

A. Tennyson (1809–1892)

2.1 THE SHAPE OF THE EARTH

The first voyage around the world, begun at
Seville by Magellan in 1519 and completed at
Seville by del Cano in 1522, established beyond
dispute that the Earth is a globe. Today it is
possible to girdle the Earth, like Puck, 'in forty
minutes' and to photograph its surface at heights
from which the curvature of the globe is plainly to
be seen (Plate Ia). Pythagoras (about 530 BC) was
probably the first to consider the possibility that
the Earth might be a sphere. By observing the
approach of ships from beyond the horizon – first
the masts and sails and then the hull – he realized
that the surface of the sea is not flat, but curved.
Three centuries later, when it was already known
that the distance of the sun was so great that the
direction of the sun's rays at any moment could be
regarded as parallel, Eratosthenes (276–196 BC), the
Chief Librarian at Alexandria, devised a simple
and elegant method for estimating the size of the
globe. He had heard that at Syene on the Nile (the
Aswan of today) the sun shines vertically at noon
on Midsummer's Day, so that a vertical stick or
plumb line then throws no shadow. He observed,
however, that at Alexandria, roughly 800 km to the
north of Syene, there were very perceptible shadows at that time. Figure 2.1 illustrates the conditions, but with greatly exaggerated angles and
lengths. At Alexandria a plumb line having a
length AB would throw a shadow of length AC.
These two lengths determine the angle ABC

which, on the simplifying assumptions made,
equals the angle SOA. Eratosthenes made the
necessary measurements and found the angle ABC
to be just over 7°, or almost exactly one-fiftieth of
360°. The approximate length of the whole circumference would therefore be fifty times the distance
between Syene and Alexandria – that is,
$50 \times 800 = 40\,0000$ km. Eratosthenes measured
the distance in **stades**, his result for the circumference being 252 000 stades. It must, however, be
remarked that the result is less accurate than it
appears to be, because Alexandria lies well to the
west of the meridian through Syene, while Syene
itself is several kilometres north of the Tropic of
Cancer, where the midsummer sun shines vertically.

The reason for the spherical shape of the Earth
became clear when Newton formulated his law of
gravitation. All the particles of the Earth are pulled
towards the centre of gravity and the spherical
shape is the natural response to the maximum
possible concentration. Even if a body the size of
the Earth were stronger than steel, it could not
maintain a shape such as, let us say, that of a cube.
The pressure exerted by the weight of the edges
and corners would squeeze out material in depth.
Equilibrium would be reached only when the
faces had bulged out, and the edges and corners
had sunk in, until every part of the surface was
equidistant from the centre.

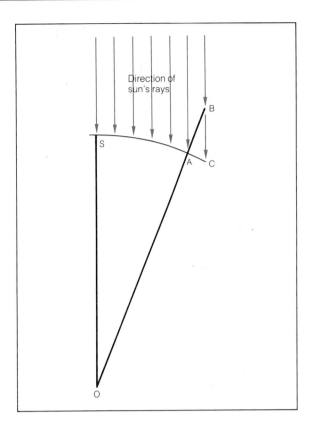

Figure 2.1 Illustrating the method devised by Eratosthenes for measuring the circumference of the Earth. S represents Syene, and A Alexandria. It is assumed that the arc SA lies on a meridan and that O is the centre of the Earth. (Not to scale).

The Earth is not exactly spherical, however. Again it was Newton who first showed that, because of the Earth's daily rotation, its matter is affected not only by inward gravitation, but also by an outward centrifugal force, which reaches its maximum at the equator. He inferred that there should be an equatorial bulge where the apparent value of gravity was reduced, and a complementary polar flattening, where the centrifugal forces becomes vanishingly small. Clearly, if this were so, the length of a degree of latitude along a meridian would be greatest across the poles, where the curvature is flattened, and least across the equator, where the curvature bulges out. It is of interest to notice that Newton's inference was at variance with the few crude measurements that had been made. According to these the Earth was shaped, not like an orange with a short polar axis, but like a lemon with a long polar axis. To settle the matter the French Academy in 1735 dispatched a surveying expedition to the neighbourhood of Chimborazo in the Andes of what is now Ecuador, and followed it up in 1736 by another to Lapland. The results showed that Newton was right. It is, moreover, highly significant that before these expeditions returned, the celebrated mathematician, Clairaut, had calculated what the shape of the Earth would be, assuming the Earth to be a fluid and subject only to the effects of its own rotation and gravitational attraction. The ellipsoid of rotation now internationally adopted for surveying purposes as most closely representing the real shape of the Earth corresponds almost exactly to that calculated by Clairaut.

To sum up: if the surface of the Earth were everywhere at sea-level its shape – the **geoid** or figure of the Earth – would closely approximate to that of an ellipsoid of rotation (i.e. an oblate spheroid) with the equatorial axis 42.8 km longer than the polar axis, as estimated from data derived from satellites. From such data it is now known that the polar axis is slightly longer from the centre of the Earth to the north pole, than from the centre to the south pole, and the Earth is now sometimes described as pear-shaped. The deviation in its form from that of an oblate spheroid, however, is very small.

The Earth continuously tends towards a state of gravitational equilibrium. If there were no rotation and no lateral differences of density of the rocks the Earth would be a sphere. As a result of rotation it becomes a spheroid. As a further result of differences in the density of rocks and their distribution, mountain ranges and ocean basins occur as irregularities superimposed upon the surface of the spheroid (Table 2.1).

2.2 THE OUTER ZONES OF THE EARTH

As it presents itself to direct experience, the Earth can be physically described as a ball of rock (the **crust**), partly covered by water (the hydrosphere) and wrapped in an envelope of air (the atmosphere). To these three physical zones it is necessary to add a biological zone (the biosphere). The system, crust, atmosphere and hydrosphere is usually regarded as being a closed system, i.e. to be in a steady state. This means that losses from any members of the system are balanced by additions to the others. Only hydrogen and helium are light enough to escape from the system.

The **atmosphere** is the layer of gases and vapour which envelops the Earth. In the lower part it is essentially a mixture of nitrogen and oxygen with smaller quantities of water vapour, carbon dioxide and inert gases such as argon. Carbon dioxide in the atmosphere is the main cause of the so-called 'greenhouse' effect in so far as the gas acts in the same way as the glass of a greenhouse which allows the sun's radiation through but prevents part of the reflected light escaping. So energy is retained and the greenhouse heats up. The energy retained in the atmosphere is determined principally by the

Table 2.1 Some numerical facts about the Earth

Land		Oceans and seas	
Greatest known height:	Metres	Greatest known depth:	Metres
Mount Everest	8846	Marianas Trench	11 035
Average height	840	Average depth	3808

Size and shape	Km	Area	Millions sq. km
Equatorial semi-axis, a	6378.2	Land (29.22 per cent)	149
Polar semi-axis, b	6356.8	Ice sheets and glaciers	15.6
Mean radius	6371.0	Oceans and seas (70.78 per cent)	361
Equatorial circumference	40 076	Land plus continental shelf	177.4
Polar (meridian) circumference	40 009	Oceans and seas minus continental shelf	332.6
Ellipticity, $(a-b)/a$	1/298	Total area of the Earth	510.0

Volume, density and mass	Average thickness or radius (km)	Volume ($\times 10^6$ km^3)	Mean density (g/cm^3)	Mass ($\times 10^{24}$ g)
Atmosphere	—	—	—	0.005
Oceans and seas	3.8	1370	1.03	1.41
Ice sheets and glaciers	1.6	25	0.90	0.023
Continental crust*	35	6210	2.8	17.39
Oceanic crust†	8	2660	2.9	7.71
Mantle	2881	898 000	4.53	4068
Core	3473	175 500	10.72	1881
Whole Earth	6371	1 083 230	5.517	5976

*Including continental shelves
†Excluding continental shelves

amount of carbon dioxide. The gas therefore exerts a critical control on the temperature and thereby the climate on Earth, the size of the polar ice-caps and sea-level.

Beyond 100 km above the surface atmospheric pressure is vanishingly small and oxygen becomes more important while from 1000 km outwards helium gradually gives way to hydrogen. Between 40 and 80 km oxygen with two atoms to each molecule is converted to ozone, with three atoms per molecule. This ozone layer absorbs ultra-violet light and shields life on Earth from this destructive radiation. As we shall see in the next chapter, life developed on Earth in the shelter of water where it was protected from harmful radiation. It was only after the ozone layer developed that plants and animals were able to venture on to the land. An additional filter to harmful, high energy short-wave radiation is provided by the ionosphere, a layer above 80 km in which atoms are ionized. The composition and structure of the atmosphere has exerted and continues to exert a control of the first importance on life on Earth and through the medium of the climate and weather determines the nature and intensity of chemical and physical processes acting on the surface of the Earth.

The **hydrosphere** includes all the natural waters of the outer Earth. Oceans, seas, lakes and rivers cover about three-quarters of the surface. But this is not all. Underground, the pore spaces and fissures of the rocks are also filled with water (Chapter 19). This groundwater, as it is called, is tapped in springs and wells, and is sometimes encountered in disastrous quantities in mines. Thus there is a somewhat irregular but nearly continuous mantle of water around the Earth, saturating the rocks, and over the enormous depressions of the ocean floors completely submerging them. If it were uniformly distributed over the Earth's surface it would form an ocean nearly 2750 m deep.

The **biosphere**, the sphere of life, is now a familiar concept. Think of the great forests and prairies with their countless swarms of animals and insects. Think of the tangles of seaweed, of the widespread banks of molluscs, of reefs of coral and shoals of fishes, add to these the inconceivable numbers of bacteria and other microscopic plants and animals. Myriads of these minute organisms are present in every cubic centimetre of air and water and soil. Taken altogether, the diverse forms of life constitute an intricate and ever-changing network, clothing the surface with a tapestry that

Figure 2.2 Two celebrated products of the biosphere. Many millions of years ago the coin-shaped shells of innumerable generations of nummulites (large foraminifera) accumulated on the floor of a vanished sea to form the thick and widespread deposits of Nummulitic Limestone that is now the bedrock of much of the Egyptian and Libyan Desert. From a conspicuous outcrop of this rock the Sphinx was carved, probably during the reign of Chephren, about 2900 BC, whose Pyramid is seen close by, built of gigantic blocks of the same Nummulitic Limestone quarried from the hills on the other (east) side of the Nile. Stone and builders alike were once part of the biosphere. Restoration of the weathered rock is now necessary in places. (*M. R. W. Johnson*)

is nearly continuous. Even high snows and desert sands fail to interrupt it completely, and lava fields fresh from the craters of volcanoes are quickly invaded by the pressure of life outside. Such is the sphere of life, and both geologically and geographically it is of no less importance than the physical zones. Amongst its many products are coal, oil, natural gas, most of the oxygen of the air we breathe, and limestones in great abundance (Figure 2.2).

The **crust** (Figure 2.3) is the outer shell of the solid Earth. It is made up of rocks in great variety. On the lands its uppermost layer is commonly a blanket of soil or other loose deposits, such as desert sands. The idea that the Earth has a crust can be traced back to Descartes (1596–1650), who thought of it as a shell of heavy rock, covered with lighter sands and clays and resting on a metallic interior. Leibnitz (1646–1716) suggested that the Earth had cooled from an incandescent state and that the outside rocky part – the first to cool and consolidate – had formed a crust which covered a still molten interior. This view, that the crust is a relatively thin layer of solid rock on a liquid interior, became widely prevalent until about a century ago, when it was shown to be unsound.

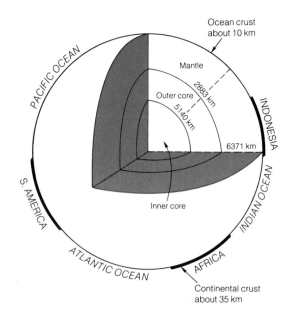

Figure 2.3 Diagrammatic view of the Earth, showing crust (continental and oceanic), mantle and core.

At that time it began to be suspected that if the bulk of the interior were liquid the oceanic tides would rise and fall through a smaller range than they actually do. Tides demonstrate a very conspicuous movement of seawater relative to land. Clearly, if land and sea yielded equally to the attractive forces exerted by the moon and sun – as they would do if the interior were liquid – then there would be no relative movement and no advance and retreat of the sea. Tides do occur in the Earth's crust, but the rise and fall of the land is imperceptible to its inhabitants and can be detected and measured only by special techniques. This elastic response of the Earth to the moon and sun turns out to be just what it would be if the Earth consisted throughout of solid steel. This remarkable conclusion, first reached by Kelvin in 1862, undermined the term 'crust' by depriving it of its original significance. The term was never abandoned, however, and by another route it has now acquired a precise meaning which is generally accepted.

2.3 THE CRUST AND INNER ZONES OF THE EARTH

In the light of modern knowledge the structure of the Earth's interior has turned out to be not unlike the model conceived by Descartes. The deep interior is called the **core**. The surrounding zone is known as the **mantle** and it extends up to a boundary surface above which the rocks are different. This outermost envelope is the **crust** (Figure 2.3).

No direct information is available on the core and this means that its precise nature has been, and still is, the subject of debate. Nevertheless, there are indirect lines of evidence. These include considerations of density, earthquake waves, meteorites and the relative abundances of elements, and the magnetic field. Laboratory experiments under the conditions likely to obtain in the core (where pressure is over a million times atmospheric pressure, 1.3 to 3.5 megabars, and temperature between 4000°C and 5000°C) are also increasingly important.

The size of the core and its physical condition have been deduced from the passage of earthquake waves through the Earth (Chapter 25). The boundary with the mantle lies at a depth of c. 2900 km from the surface and the core therefore occupies 16 per cent of the Earth's volume and over 50 per cent of the radius. The outer portion, down to c. 5000 km responds to earthquake waves like a liquid while the inner core is solid. This contrast between the two parts of the core is important in that the Earth's magnetic field is thought to be maintained by convective movements in the liquid outer zone.

The densities of the crust and the mantle are lower than the mean density of the Earth (Table 2.1). The deficiency in the outer layers therefore requires the compensating presence of a core of higher density. Some meteorites are composed essentially of iron and nickel and a core of this composition would offset the density deficiency. Indeed detailed calculations show that if the core consisted solely of iron or iron with a smaller amount of nickel, the mean density would be too high. Evidently there is some admixture of a lighter element. A number of possibilities, including silicon, sulphur, oxygen, hydrogen and potassium, have been investigated but no consensus has yet emerged.

Earthquake waves and those generated by controlled explosions also make it possible to determine the depth at which the mantle begins in different parts of the world. The boundary surface or **discontinuity** between the crust and the mantle was discovered by a Croatian seismologist, A. Mohorovičić. Since then it has become known as the Mohorovičić discontinuity, the M-discontinuity or more familiarly, the **Moho**. In passing through the rocks immediately above this surface, earthquake waves reach a velocity of about 7.2 km/s, whereas in the rocks below the Moho the velocity suddenly jumps to about 8.1 km/s. Wave velocities together with the materials brought to the surface by volcanic action indicate a mantle made up largely of rocks rich in iron oxides and magnesia and this led Suess (1831–1914) to coin the mnemonic term **sima** (silicon, iron and magnesium). It is expected that the proportion of iron and magnesium increases with depth at the expense of silicon but a change in composition is perhaps not so important as a change with depth to denser silicates. This is likely to be the cause of a discontinuity within the mantle at a depth of 670 km separating an upper from a lower part of the mantle. Nearer the surface, in the upper mantle, at variable depths below the crust down to about 350 km there is a layer in which shock waves have a reduced velocity. This **asthenosphere**, owes its character to the rise in temperature with depth which has caused a small proportion (perhaps only a thin layer of liquid around silicate crystals) of the rocks to melt; in consequence the zone is a layer of weakness which compares with the brittle layer above. The brittle layer which includes the *uppermost part of the mantle and the crust* is termed the **lithosphere**. The Earth, then, may be thought of in terms of composition as crust, mantle and core and in terms of physical properties, as lithosphere, asthenosphere, **mesosphere**, outer core and inner core.

The dominant rocks of the crust fall into two contrasting groups:

Figure 2.4 Basaltic lava-flows of the Antrim plateau looking eastwards from the Giant's Causeway. Above the cliff path three tiers of columns can be seen, corresponding to three successive lava-flows, the lowest of which is that of the Causeway itself. The older lavas below the path are not columnar. (*J. Allan Cash, Northern Ireland Tourist Board*)

(a) a group of light rocks, including granite and related types, and sediments such as sandstones and shales, which form an assemblage having an average specific gravity or density of 2.7. Chemically these rocks, again on average, are very rich in *silica* (65–75 per cent) while *alumina* is the most abundant of the remaining constituents. Suess called these **sial** as distinct from the sima rocks of the mantle.

(b) a group of dark and heavy rocks consisting mainly of basalt and related types with a density about 2.8–3. They are collectively known as **basic** rocks (with about 50 per cent silica) but also include still heavier rocks with densities up to about 3.4 which are distinguished as **ultrabasic** rocks (with about 40–45 per cent silica). These are the sima of Suess, wherein the mantle rocks are mainly ultrabasic and the crustal sima predominantly basic.

The most fundamental feature of the crust is the distinction between continents and oceans. Oceanic crust is made up entirely of basaltic rocks below a variable thin veneer of sediment. The basaltic layer is only 5–6 km thick; then comes the **Moho**. Samples of ultrabasic rocks, thought to represent the underlying mantle, are brought to the surface in many oceanic volcanoes such as those of Hawaii.

Continental crust is much thicker averaging 35 km and reaching 70 km in some places; moreover it is predominantly sialic, at least in its upper part. In the early days of crustal exploration it was thought that the basaltic layer of the oceans continued below the continents: and that there was a worldwide basaltic layer, surmounted here and there by slabs of sial which were the continents. This interpretation has turned out to be too simple, in the sense that sialic and basaltic rocks are inextricably mixed in the continental crust both mechanically and chemically. It is misleading therefore, to picture the continents as having everywhere a basaltic layer as their foundation. Sialic rocks are undoubtedly the dominant materials of the continental crust down to many kilometres; but the continents also have basaltic volcanoes, and in some places, such as the Antrim plateau in Northern Ireland and the Deccan area in India, the sial is covered by thick accumulations of basaltic lava-flows (Figure 2.4). The nature of the lower crust is still uncertain. Seismic studies suggest that the rocks are heterogeneous in composition and structure. In several places, for example in the southern Alps and Calabria in Italy and in Fjordland in New Zealand, sections of the lower crust have been thrust up and exposed at the surface. These confirm a complicated, layered

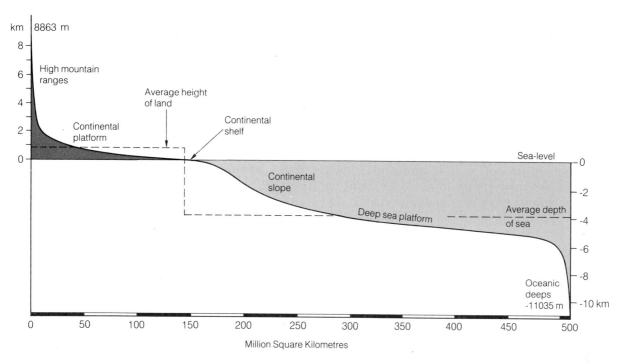

Figure 2.5 Hypsographic curve, showing the areas of the Earth's solid surface between successive levels from the highest mountain peak to the greatest known depth of the oceanic trenches. The curve might suggest that the greatest deeps are farthest away from the lands. In fact they lie close to continental margins (Fig 2.8).

Figure 2.6 View of the Everest group of peaks, seen from a height of about 6000 m. From left to right: Everest, 8846 m; Lhotse, 8511 m; Nuptse, 7827 m. The valley between Everest and Nuptse is occupied by the Khumbu Glacier. (*Royal Geographical Society*)

structure and the presence of basaltic along with strongly metamorphosed rocks of both igneous and sedimentary origin.

2.4 CONTINENTS AND OCEAN FLOORS

The surface of the crust reaches very different

levels in different places. The areas of land and sea floor between successive levels have been estimated, and the results can be graphically represented as shown in Figure 2.5. From this diagram it is clear that there are two dominant levels: the continental platform and the oceanic or deep-sea platform. The slope connecting them, which is actually quite gentle, is called the continental slope.

The continental platform includes a submerged outer border known as the **continental shelf**, which ranges in width up to about 1500 km, but may be absent along mountainous coasts. The older rocks which form the basements of the shelves are overlain by sediments with a thickness up to about 2 km. At one time it was thought that the sediments of the shelves became finer and finer in grain size with increasing distance from the shore line. As a result of investigations begun during the Second World War it has been discovered that it is only where the grain size is sorted by wave action, that the sediments become finer grained away from the shore. The shelves are mostly overlain by coarse sands and the shells of molluscs.

Structurally, the real ocean basins commence, not at the visible shoreline, but at the edge of the shelf. The basins, however, are more than full, and the overflow of seawater inundates nearly 26 million square kilometres of continental shelf. The North Sea, the Baltic and Hudson Bay are examples of shallow seas (epicontinental seas) which lie on the shelf. It is of interest to notice that during the Ice Age, when enormous quantities of water were abstracted from the oceans to form the great ice-sheets that then lay over Europe and North America, much of the continental shelf must have been land. Conversely, if the ice now covering Antarctica and Greenland were to melt, due to natural causes or man's activity in enhancing the "greenhouse effect', the sea-level would rise and the continents would be still further submerged.

Economically, the continental shelves are of prime importance; the epicontinental seas provide fertilizers, and add considerably to the world's food resources, whilst the shelves themselves are a source of oil and gas. At present much of the world's oil and gas comes from continental shelves, and progressively more will be obtained from them in the future as supplies on land are used up.

The continents themselves have a varied relief of plains, plateaus and mountain ranges, the latter rising to a maximum height of 8863 m in Mount Everest (Figure 2.6). The ocean floors, once thought to be monotonous slopes and plains, are characterized by submarine basaltic mountain

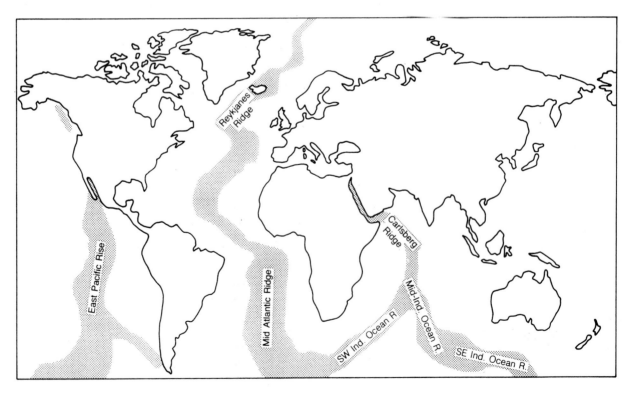

Figure 2.7 Submarine mountain chains extending around the Earth as mid-oceanic ridges that rise to heights of three or four thousand metres above the ocean plains. Unlike continental mountains, submarine mountains are formed of basalt.

ranges which encircle the Earth for more than 40 000 km (Figure 2.7). These underwater mountain chains are known as **mid-ocean ridges** although they are not always precisely in the middle.

Other features of the ocean floors are great numbers of sea-mounts, and ancient volcanic islands (Figures 24.34 and 26.10). Submarine canyons, comparable to the Colorado canyon, are of common occurrence, and deep ocean trenches, the subject of many recent investigations, carry the ocean floor down to more than twice its average depth, the greatest depth so far determined being 11 035 m in the Nero deep in the Marianas Trench (Figure 2.8).

From the figures given above it is clear that the total vertical range of the surface of the crust (continental and oceanic) is around 20 km. This may not be remarkable on the scale of the Earth – for example, on a circle with a radius of 5 cm, representing the size of the Earth, the thickness of the outline of the circle represents 32 km. On this scale, the relief is all contained within the thickness of a pencil line (cf. Figure 2.3). Nevertheless it is curious that mountain chains and ocean trenches, that is regions of greatest relief, often occur not widely separated but close together on the margins of continents and oceans. Is there something in crustal structure and crust–mantle relationships which tends to produce this relationship? Does the situation reflect a system in equilibrium?

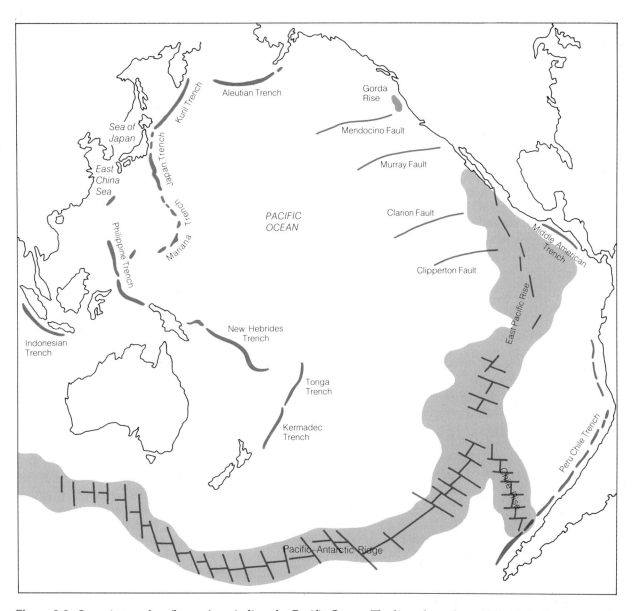

Figure 2.8 Oceanic trenches (brown) encircling the Pacific Ocean. The line along the middle of the East Pacific and Pacific–Antarctic ridges represents the rift along which basaltic magma rises and moves away on either side as new ocean floor. The lines intersecting the rift are a diagrammatic illustration of transform faults (Section 28.3).

2.5 ISOSTASY

For the ideal condition of gravitational equilibrium that controls the heights of continents and ocean floors, in accordance with the densities of their underlying rocks, the term **isostasy** (Gr. *isostasios*, 'in equipoise') was proposed by Dutton, an American geologist, in 1889. The idea may be grasped by thinking of a series of wooden blocks of different heights floating in water (Figure 2.9). The blocks emerge by amounts which are proportional to their respective heights; they are said to be in a state of hydrostatic balance. Isostasy is the corresponding state of balance between extensive blocks of the Earth's crust which rise to different levels and appear at the surface as mountain ranges, plateaus, plains or ocean floors. The idea implies that there is a certain minimum depth below sea-level where the pressure due to the weight of the overlying material in each unit column is everywhere the same. In Figure 2.9 this level of uniform pressure is that of the base of the highest block. The Earth's major relief is said to be **compensated** by the underlying differences of density, and the level where the compensation is estimated to be complete – i.e. the level of uniform pressure – is often referred to as the **level of compensation**. Naturally, individual peaks and valleys are not separately balanced or compensated in this way; the minor relief features of the surface are easily maintained by the strength of the crustal rocks. As indicated in the next chapter perfect isostasy is rarely attained, because of the restlessness of our globe; however, in regions which have remained free from geological disturbances for long periods of time there is generally a remarkably close approach to a state of equilibrium.

If a mountain range were simply a protuberance of rock resting on the continental platform and wholly supported by the strength of the foundation, then a plumb line – such as is used for levelling surveying instruments – would be

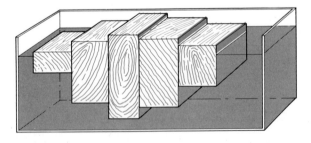

Figure 2.9 Wooden blocks of different heights floating in water (shown in front as a section through the tank), to illustrate the concept of isostatic balance between adjacent columns of the Earth's crust.

deflected from the true vertical by an amount proportional to the gravitational attraction of the mass of the mountain range. The first hint that mountains are not merely masses of rock stuck on an unyielding crust was provided by the Andes expedition of 1735. Pierre Bouguer, the leader of the expedition, made observations both north and south of Chimborazo, and found to his surprise that the deflection of the plumb line towards this towering volcanic peak was very much less than he had estimated. He recorded his suspicion that the gravitational attraction of the Andes 'is much smaller than that to be expected from the mass represented by these mountains!'

Similar discrepancies were met with during the survey of the Indo–Gangetic plain, south of the Himalayas, carried out by Sir George Everest, Surveyor-General of India, over a century ago. The difference in latitude between Kalianpur and Kaliana (603 km due north) was determined astronomically, and also by direct triangulation on the ground. The two results differed by 5.23 seconds of arc, corresponding to a distance on the ground of 168 m. The discrepancy was ascribed to the attraction exerted by the enormous mass of the Himalayas and Tibet (Figure 2.10) on the bob of the plumb line used for levelling the astronomical instruments. The error so introduced does not arise in the triangulation method. A few years later (1855) Archdeacon Pratt made a minimum estimate of the mass of the mountains and calculated what the corresponding gravitational effects would be at the two places on the plains to the south. His estimates for the deflections of the plumb line towards the mountains were

27.853″ at Kaliana, and
11.968″ at Kalianpur.

The difference, 15.885″, was more than three times the observed deflection, 5.23″. Bouguer's suspicion that the mountains were apparently not pulling their weight was now a demonstrated fact. Even more spectacular evidence was provided by French surveyors, who found that in some of the coastal regions of south-western France the plumb line was deflected, not towards the mountains, but away from them, towards the Bay of Biscay.

Clearly the mountains seem to behave almost as if they were hollow, and this apparent gravitational anomaly has since been amply confirmed by the results of innumerable measurements of gravity on and near mountains and high plateaus. The observed results are found to be much lower than those to be expected. Only one physical explanation of these discrepancies or anomalies is available. Since the mountains are **not** hollow, there must be a compensating deficiency of mass in the columns underlying the visible mountain

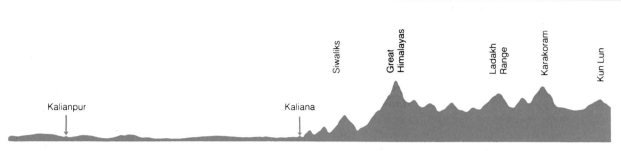

Figure 2.10 Meridional section (about Long. 78°E) through northern India and Tibet, indicating the enormous mass tending to deflect the plumb-line of surveying instruments used on the plains to the south.

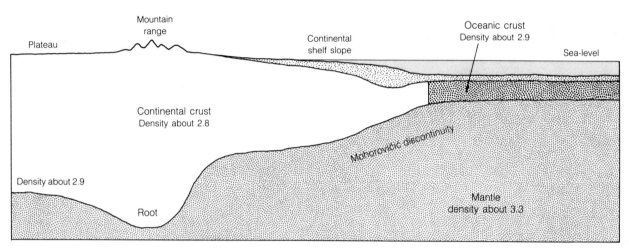

Figure 2.11 Diagrammatic section through the Earth's crust and the upper part of the mantle to illustrate the relationship between surface features and crustal structure. Based on gravity determinations and exploration of the distribution of sial, crustal sima and mantle sima by earthquake waves.

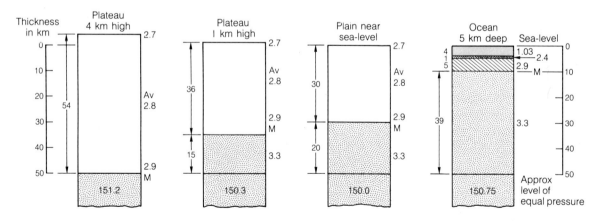

Figure 2.12 Columns of equal cross-section through characteristic parts of the continents and ocean floor. (Shading as in Fig. 2.11). The figures to the right of each column indicate approximate densities. These, multiplied by the corresponding thickness, down to a depth of 50 km, give a total figure proportional to the pressure at that depth; e.g. for the oceanic column: 5 × 2.9 (crustal sima, probably basaltic rock) + 39 × 3.3 (mantle sima) = 150.75. M is the Mohorovičić discontinuity (the 'Moho').

ranges. In simpler language, the density of the rocks must be relatively low down to considerable depths. The possible combinations of density distributions and depths are, of course, theoretically infinite, but fortunately we know something

about the crustal rocks and their densities in most regions, so that in practice the probable combinations are limited. Moreover, as we have seen, exploration of the crust by earthquake waves confirms the inference that mountain ranges have roots, going down to depths of as much as 70 km. Under plains near sea-level the thickness of the crustal rocks is only about 30 km, sometimes less. Beneath all the deeper parts of the oceans sial cannot be detected at all. The suggestion that the crust is supported by underlying denser material and that the weight of mountains is balanced by light materials extending as roots into the denser material – just as icebergs are balanced in water – was first made in 1855 by Sir George Airy, who was then the Astronomer Royal. Figure 2.11 illustrates these relationships between surface relief and crustal structure in a general way.

Figure 2.12 shows characteristic examples of columns, each of which has the same area and extends downwards to the same depth below sea-level, the depth at which the weight of each column exerts approximately the same pressure on the underlying material, irrespective of its surface elevation. For the columns selected, this depth – 50 km – would be the depth of isostatic compensation if the regions concerned had long remained undisturbed by the activities of our restless Earth. When the gravitational effects of the appropriate underlying densities are taken into account in such regions the anomalies referred to above (known as **Bouguer anomalies**) disappear on average to the extent of about 85 per cent. Any residual discrepancy that still remains is called the **isostatic anomaly**.

Where isostatic anomalies occur the crust has not reached equilibrium but the evidence is quite clear that given time balance may be restored. During the last Ice Age thick sheets of ice loaded and therefore depressed the crust of Northern Europe. As the ice melted the load was removed relatively quickly and the land began to rise as isostatic readjustment proceeded. The result is that far above the shores of Finland and Scandinavia there are beaches formerly at sea-level which show that an uplift of about 250 m has occurred (Figure 2.13). Isostatic readjustment is still proceeding and every 28 years another 30 cm are added to surface levels around the northern end of the Gulf of Bothnia. The region is still out of balance and it has been estimated that it has still to rise another 200 m before equilibrium is reached.

Similarly around the northern shores of Hudson Bay new rocky islands have appeared within the memory of older Eskimos, and the land is known to have risen at least 9 m since the Thule eskimos first established themselves there, indicating a rise of about a metre per century.

Isostatic readjustment is very slow and clearly depends on flowage in the mantle. Where increased loading occurs flow will be away from that region towards one where unloading is taking

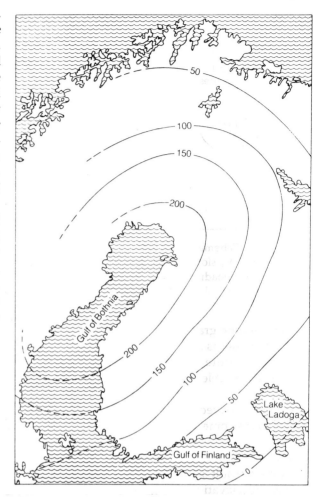

Figure 2.13 Post-glacial uplift of Fenno-scandia (Finland and Scandinavia). The curves are lines of equal uplift, in metres, from 6800 BC to the present day. Around the northern shores of the Gulf of Bothnia the present rate of uplift is 1 cm per year. (*After Niskanen*)

place as illustrated in Figure 2.14. An apparent paradox arises in so far as the crust and mantle are rigid, in that, for example, they transmit earthquake waves, yet isostasy demands flow in the mantle. The rock of the mantle has, as it were, a dual personality; it is, at the same time, an elastic solid and a viscous fluid. To short term stress as in earthquake shocks, mantle rock may act as a solid, while to long-continued stress it acts as a fluid. There are several familiar rocks which display their double personalities. Pitch is perhaps the best known. A lump of it is easily broken into angular fragments by a tap from a hammer, but the same lump, if left to itself at ordinary temperatures, flattens out under its own weight into a thin sheet. To a sharply applied force, it behaves as a brittle solid, while in response to the long-continued but much smaller stress of its own weight it flows as a viscous fluid. Ice is a crystalline solid which fractures when struck but as

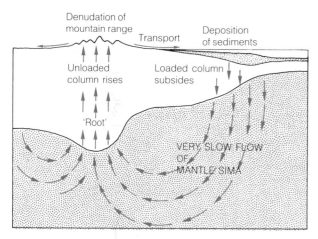

Figure 2.14 Schematic section to illustrate isostatic readjustment, by slow sub-crustal flow in the mantle, in response to unloading by denudation and loading by deposition. Vertical scale greatly exaggerated.

glaciers under gravity it flows downhill (Figure 2.15). Rock salt is another example and crystalline rocks like granite and basalt in the hot depths of the Earth are able to flow as easily as ice or rock salt.

We have arrived at a coherent view of the Earth which gives some account of its size, shape, composition and physical nature. It is, however, essentially a limited, rather static, view. Isostatic responses result in merely up and down movements but elevations and depressions very often cannot be explained as simply as Figures 2.12 and 2.14 would suggest. Examples are deep oceanic trenches and the Colorado Plateau among many others. Gravity measurements over such regions reveal abnormal isostatic anomalies. Moreover, earthquakes and volcanic activity suggest a restlessness within the Earth that requires the operation of processes far more dynamic than those envisaged in isostasy. These processes are explored in the next chapter and, as will appear, they lead to a much more comprehensive and exhilarating view of the Earth.

2.6 FURTHER READING

S 101 4–5 (1979) *Earthquake Waves and the Earth's Interior/The Earth as a Magnet*, Open University, Milton Keynes.

Figure 2.15 The contortions of ice and morainic debris here seen result from the competition for space between glaciers advancing into the same valley at different but comparable rates of flow, Alaska. (*Bradford Washburn*)

S 237 2 (1981) *Earth structure: Earthquakes, Seismology and Gravity*, Open University, Milton Keynes.

Articles in *Scientific American*, September, 1983, **249**, No. 3:

Jeanloz, R. *The Earth's Core*, pp. 40–49.

McKenzie, D.P. *The Earth's Mantle*, pp. 50–62.

Franchetau, J. *The Oceanic Crust*, pp. 68–84.

Burchfiel, B.C. *The Continental Crust*, pp. 86–98.

Broecker, W.S. *The Ocean*, pp. 100–122.

Ingersoll, A.P. *The Atmosphere*, pp. 114–130.

Cloud, P. *The Biosphere*, pp. 132–144.

Leggett, J.K. (1990) *Global Warming; The Greenpeace Report*, Oxford University Press.

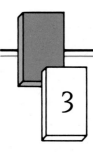

THE DYNAMIC EARTH

3

The same regions do not remain always sea or always land, but all change their condition in the course of time.

Aristotle (384–322 BC)

3.1 WEATHERING, EROSION AND DENUDATION

The circulations of matter that are continually going on in the zones of air and water, and even of life, constitute a very complicated mechanism which is maintained, essentially, by the heat from the sun. A familiar example of such a circulation is that of the winds. Another, more complex, is the circulation of water. Heat from the sun lifts water vapour from the surface of oceans, seas, lakes and rivers; and wind distributes the vapour far and wide through the lower levels of the atmosphere. Clouds are formed, rain and snow are precipitated, and on the land these gather into rivers and glaciers. Finally, most of the water is returned to the oceanic and other reservoirs from which it came (Figure 3.1). These circulations are responsible for an important group of geological processes, for the agents involved – wind (moving air), rain and rivers (moving water) and glaciers (moving ice) – act on the land by breaking up the rocks and so producing rock-waste which is gradually carried away.

Part of every shower of rain sinks into the soil and promotes the work of decay by solution and by loosening the particles. Every frost shatters the rocks with its expanding wedges of freezing water. Water expands on freezing, and through repeated alternations of frost and thaw in water-filled pores and cracks the rocks are relentlessly broken to bits. Fallen fragments dislodged from

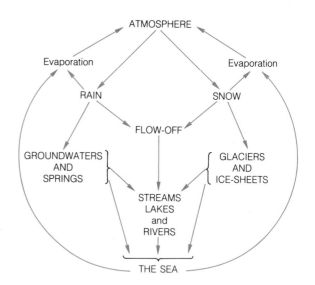

Figure 3.1 The circulation of meteoric water. In addition to the main evaporations here indicated it should be noted that evaporation takes place from all exposed surfaces of water and ice (e.g. lakes, rivers, glaciers and ice-sheets) and also from the soil and from plants and animals. Part of the water that ascends from the depths by way of volcanoes reaches the surface for the first time; such water is called **juvenile water** to distinguish it from the **meteoric water** already present in the hydrosphere and other outer zones of the Earth.

cliffs and precipitous crags accumulate at the base

as screes or talus (Figure 3.2.). Life also co-operates in the work of destruction. The roots of trees grow down into cracks, and assist in splitting up the rocks (Figure 3.3). Worms and other burrowing animals bring up the finer particles of soil to the surface, where they fall a ready prey to wind and rain. The soil is a phase through which much of the rock-waste of the lands must pass before it is ultimately removed. The production of rock-waste by these various agents, partly by mechanical breaking and partly by solution and chemical decay, is described as **weathering**. A familiar example is the fact that inscriptions on exposed slabs of marble – and other rocks – rarely survive the ravages of rain, frost and wind for more than two or three centuries.

Sooner or later the products of weathering are removed from their place of formation. Blowing over the land, the wind becomes armed with dust and sand, carrying them far and wide and often becoming a powerful sand-blasting agent as it sweeps across areas of exposed rock. Glaciers, similarly armed with morainic and other debris, grind down the rocks over which they pass during their slow descent from ice-fields and high mountain valleys. Rainwash, sliding screes and land-slips feed the rivers with fragments, large and small, and these are not only carried away, but are

Figure 3.3 Ice-transported boulder, with crack enlarged by the growth of tree roots, Trefarthen, Anglesey. (*British Geological Survey*)

used by the rivers as tools to excavate their floors and sides. And in addition to their visible burden of mud and sand, the river waters carry an invisible load of dissolved material, extracted from rocks and soils by the solvent action of rain and soil water, and by that of the river water itself. Winds, rivers and glaciers, the agents that carry away the products of rock-waste, are known as transporting agents. All the destructive processes due to the effects of the transporting agents are described as **erosion** (L. *erodere*, to gnaw; *erosus*, eaten away).

It is convenient to regard weathering as rock decay by agents involving little or no transport of the resulting products except by gravity, and erosion as land destruction by agents which simultaneously remove the debris. Both sets of processes co-operate in wearing away the land surface, and their combined effects are described by the term **denudation** (L. *denudare*, to strip bare).

3.2 DEPOSITION OF SEDIMENT

The sediment carried away by the transporting agents is sooner or later deposited again. Sand blown by the wind collects into sand dunes in the desert or bordering the seashore (Figure 3.4). Where glaciers melt away, the debris gathered up

Figure 3.2 Screes of Doe Crag (Borrowdale Volcanic Series), Old Man of Coniston, English Lake District. (*G.P. Abraham Limited, Keswick*)

Figure 3.4 Sand dunes in the Colorado Desert, Imperial County, California. (*W. C. Mendenhall, United States Geological Survey*)

during their journey is dumped down unsorted (Figure 3.5), to be dealt with later by rivers or the sea. When a stream enters a lake the current is checked and the load of sand and mud gradually

Figure 3.5 Öraefajökull Glacier, Iceland, showing its source in an upland ice cap. Its terminus ('snout') is just off the bottom-right of the photograph, and there it melts away amidst the moraines of rock-waste which it transported and deposited before shrinking to its present size. (*Thorvardur R. Jonsson*)

settles to the bottom. Downstream in the open valley, sand and mud are spread over the alluvial flats during floods, while the main stream continues, by way of estuary or delta, to sweep the bulk of the material into the sea. Storm waves thundering against rocky coasts provide still more rock-waste, and the whole supply is sorted out and widely distributed by waves and currents. Smooth and rounded water-worn boulders (Plate IVb) collect beneath the cliffs. Sandy beaches accumulate in quiet bays. Out on the sea floor the finer particles are deposited as broad fringes of sediment, the finest material of all being swept far across the continental shelves, and even over the edge towards the deeper ocean floor, before it finally comes to rest. All these deposits are examples of sedimentary rocks in the making.

We have still to trace what happens to the invisible load of dissolved mineral matter that is removed from the land by rivers. Some rivers flow into lakes that have no outlet save by evaporation into the air above them. The waters of such lakes rapidly become salt because, as the famous astronomer Edmund Halley realized in 1715, 'the saline particles brought in by the rivers remain behind, while the fresh evaporate'. Gradually the lake waters become saturated and rock-salt and other saline deposits, like those on the shores and floor of the Dead Sea, are precipitated. Most rivers, however, reach the sea and pour into it the greater part of the material dissolved from the land. So, as Halley pointed out, 'the ocean itself is become salt from the same cause'. But while, on balance, the salinity of the sea may be slowly increasing, much of the mineral matter contributed to the sea is taken out again by living organisms. Cockles and mussels, sea urchins and corals, and many other sea creatures, make shells for themselves out of calcium carbonate abstracted from the water in which they live. When the creatures die, most of their soft parts are eaten and the rest decays. But their hard parts remain, and these accumulate as, for example, the shell banks of shallow seas, the coral reefs of tropical coasts and islands, and the grey globigerina ooze of the deep-sea floor. All of these are potential limestones. Life as a builder of organic sediments is a geological agent of the first importance. At the close of the twentieth century, when man's activity is threatening to overproduce carbon dioxide and concern is high regarding the 'greenhouse' effect it might be claimed that the more important role of organic sediments lies in locking up the gas in carbonate and reducing its proportion in the atmosphere. Like the glass in a greenhouse the carbon dioxide in the atmosphere allows the sun's radiation through to the Earth's surface but part of the reflected rays are prevented from escaping; heat gradually builds up, the greater the amount of the gas in the atmosphere, the stronger the 'greenhouse' effect.

3.3 THE IMPORTANCE OF TIME

It will now be realized that while the higher regions of the Earth's crust are constantly wasting away, the lower levels are just as steadily being built up. Evidently denudation and deposition are great levelling processes. In the course of a single lifetime their effects may not be everywhere perceptible. Nevertheless they are not too slow to be measured. An average thickness of over one centimetre has already been worn from the outer surface of the Portland stone with which St Paul's Cathedral was built over 250 years ago. Portland stone has justified Wren's confidence in its suitability to withstand the London atmosphere, for the land of Britain as a whole is being worn down rather faster – at an average rate of about 30 cm in three or four thousand years. At this rate a few million years would suffice to reduce the varied landscapes of our country to a monotonous plain. Evidently slowly-acting causes are competent to produce enormous changes if only they continue to operate through sufficiently long periods.

Now, geologically speaking, a million years is a comparatively short time, just as a million miles is a very short distance from an astronomical point of view. One of the modern triumphs of geology and physics is the demonstration that the age of the Earth cannot be less than 4600 million years (Chapter 14). Geological processes act slowly, but geological time is immensely long. The effects of slow processes acting for long periods have been fully adequate to account for all the successive transformations of landscape that the Earth has witnessed.

It was James Hutton (1726–97), the founder of modern geology, who first clearly grasped the full significance and immensity of geological time. In his epoch-making *Theory of the Earth*, communicated to the Royal Society of Edinburgh in 1785, he presented an irrefutable body of evidence to prove that the hills and mountains of the present day are far from being everlasting, but have themselves been sculptured by slow processes of erosion such as those now in operation. He showed that the alluvial sediment continually being removed from the land by rivers is eventually deposited as sand and mud on the sea floor. He observed that the sedimentary rocks of the Earth's crust bear all the hallmarks of having accumulated exactly like those now being deposited. He realized, as no one had done before, that the vast thicknesses of these older sedimentary rocks implied the operation of erosion and sedimentation throughout a period of time that could only be described as inconceivably long.

Hutton was the first to apply to geological problems as a whole that faith in the orderliness of nature which is the basic principle on which all science is founded. The doctrine known as **catastrophism** – the notion of successive destructions of the face of the Earth by violent and supernatural cataclysms, of which the Noachian Flood was the classic example – was widely prevalent in Hutton's day. With such extravagances Hutton would have nothing to do. He realized with the clarity of sheer genius that 'the past history of our globe must be explained by what can be seen to be happening now'. This principle, which gradually replaced the catastrophic conception of Earth history, was eventually called **uniformitarianism** by Sir Charles Lyell – an unhappy word, liable to be taken too literally. Hutton did not exclude temporary and local crises: such natural catastrophes, for example, as the great Lisbon earthquake of 1755 or the eruption of Vesuvius which overwhelmed Pompeii and Herculaneum in AD 79. But Lyell's term seemed to do so, since it inevitably suggests uniformity of **rate**, whereas what is meant is uniformity of **natural law**. Hutton made this quite clear when he wrote: 'No powers are to be employed that are not natural to the globe, no action to be admitted except those of which we know the principle.' He insisted that the ways and means of nature could be discovered only by observation. On the Continent the term **uniformitarianism** never won favour, but was gradually replaced by **actualism**, which conveys much more appropriately the real meaning of Hutton's inspired appeal to 'actual causes': the principle that the same processes and natural laws prevailed in the past as those we can now observe or infer from observations.

Nowadays no scientist would support catastrophism as implying supernatural cataclysms. Rather the notion of catastrophism is applied to far-reaching crises such as an asteroid impact on the Earth which is taken by some to have been the main cause of the extinction of the dinosaurs and many other organisms some 66 million years ago.

Hutton's genius received scant recognition in his lifetime, mainly because of the prevalent belief that the world had been created in the year 4004 BC. Far from being welcomed, Hutton's discoveries were generally regarded with righteous horror. Had he lived in India, however, Hutton would have found a ready-made system of world chronology fully adequate for the needs of geology at that time. According to the Hindu calendar, as recorded in the ancient books of Vedic philosophy, the year AD 1977 corresponds to 1 972 949 078 years since the present world came into existence. This astonishing concept of the Earth's duration has at least the merit of being of the right order.

3.4 EARTH MOVEMENTS

It follows that there has been ample time, since land and sea came into existence, for Britain, and indeed for the highest land areas, to have been worn down to sea-level over and over again. How then does it happen that every continent still has its highlands and mountain peaks? The special creation theory, immortalized in James Thomson's words:

> When Britain first at Heaven's command
> Arose from out of the azure main . . .

is not very helpful, yet it does suggest a possible answer. The lands, together with adjoining parts of the sea floor, may have been uplifted from time to time. Alternatively, as suggested long ago by Xenophanes (c. 500 BC), who noticed that fossil shells in the limestone hills of Malta were like those still being washed up on the beaches down below, the level of the sea may have fallen, leaving the land relatively upraised. In either case there would again be land above the sea, and on its surface the agents of denudation would begin afresh their work of sculpturing the land into hills and valleys. An additional factor is the building of new land – like the volcanic islands that rise from oceanic depths – by the accumulated products of volcanic eruptions. Each of these processes of land renewal has been in continual operation during the course of the Earth's long history.

Relative movements between land and sea are convincingly proved by the presence on Islay and Jura and in many other places around Scotland, of typical sea beaches and wave-cut platforms, now raised far above the reach of the waves (Figure 3.6). Behind the beaches the corresponding cliffs,

Figure 3.6 Late Glacial raised shoreline, west of Rhuvaal Lighthouse, Islay, Inner Hebrides, Scotland. (*British Geological Survey*)

Figure 3.7 Submerged Forest, Leasowe foreshore, Cheshire. (*C. A. Defieux*)

often tunnelled with sea caves, are still preserved. As we have seen these features are mainly due to the isostatic response of the crust after the removal of ice. But there are crustal movements on a far greater scale. The former uplift of old sea floors can be recognized in the Pennines, where the grey limestones contain fossil shells and corals that bear silent witness to the fact that the rocks forming the hills of northern England once lay under the sea. Even greater uplift is demonstrated by the occurrence of marine fossils in the shales that cap the granitic rocks of Mt McKinley in Alaska (Figure 31.1) and so form the highest point in North America. Even more impressive, being the present world record, is the summit of Mt Everest (Figure 2.6), carved out of sediments that were originally deposited on the sea floor of a former age.

Crustal movements have not always and everywhere resulted in emergence of the land. Relative movements between land and sea may also bring about the submergence of the land. Recent submergence of parts of the land surface is proved by the local occurrence around British shores of **submerged forests** (Figure 3.7). These are groups of tree-stumps still preserved with their roots in the original position of growth, but now uncovered only at the lowest tides. In Mount's Bay, near Penzance, what is probably the same submergence is indicated by the occurrence of a Stone Age axe-factory, dating from about 1800–1500 BC, which now lies under the sea. The memory of this and other submerged lands of south-west Britain

may well be enshrined in the legends of the lost land of Lyonesse.

When earth movements take place suddenly they are recognized by the passage of earthquake waves. In certain restless belts of the crust, for example in Japan, there may be several shocks every day, occasionally with terribly disastrous consequences. Exceptionally, as in Yakatut Bay, Alaska, in 1899, an upward jerk of as much as 14 m has been measured, but usually these sudden movements are on a much smaller scale (Figure 3.8).

If crustal movements were all uniformly vertical, then beds of sediment uplifted from their original positions on the sea floor would generally be found lying in nearly horizontal positions. So indeed they often are (Figure 1.5), but in many places they have been corrugated and buckled into folds (Figure 3.9), like a wrinkled-up tablecloth which has been pushed along the table. But from a mechanical point of view this is not a good analogy. A familiar small-scale domestic model that in many ways more closely corresponds to the natural conditions of large-scale folding is provided by the skin that forms on hot milk. If the containing vessel is tilted a little the skin slumps into a series of folds, some of which are likely to overlap their down-slope neighbours. Many an Alpine precipice displays great sheets of rock which have been 'overfolded' in much the same way, so that parts of them now lie upside down (Figure 3.9). In other places layers of rock have been folded tightly together, like the pleats of a closed concertina.

Figure 3.8 Displacement of road and production of a fault scarp by a sudden movement on the White Creek Fault which was responsible for the Murchison earthquake of 1929, South Island, New Zealand. The uplifted block rose about 5 metres relative to the foreground. (*New Zealand Geological Survey*)

Figure 3.9 Recumbent folds in the High Calcareous Alps: the Axenstrasse, near Flüelen, at the south end of Lake Lucerne, Switzerland. (*F. N. Ashcroft*)

Such amazing structures as these suggest that certain parts of the Earth's crust have yielded to pressures of irresistible intensity. All the great mountain ranges of the world have been carved out of rocks that have been severely deformed. Long belts of the crust appear to have been so compressed and thickened that eventually they had no alternative but to rise to mountainous heights.

Although there must have been long periods when much of the land lay under the sea, it is unlikely that all the land was ever submerged at once. Earth movements and volcanic additions to the surface have evidently been fully competent to restore the balance of land and sea whenever that balance has been threatened by the levelling processes of denudation. Most of the sediments originally deposited on the shallow sea floor, hardened and cemented into firm and durable rocks, sometimes bent and twisted into intricate folds, sometimes accompanied by lavas and volcanic ashes, have sooner or later emerged to form new lands, either by upheaval or by the withdrawal of the sea.

3.5 VOLCANIC AND IGNEOUS ACTIVITY

Earth movements are not the only manifestations of the Earth's internal activities. Volcanic eruptions provide a most spectacular proof that the Earth's interior is so hot that locally even the crustal rocks pass into a molten state. A volcano is essentially a fissure or vent, communicating with the interior, from which flows of lava, fountains of incandescent spray, or explosive bursts of gases and volcanic 'ashes' are erupted at the surface. The fragmental materials produced during volcanic eruptions are collectively known as **pyroclasts** (Gr. *pyros*, fire; *klastos*, broken in pieces).

It is convenient to have a general term for the parental materials of these hot volcanic products, as they occur in the depths before eruption at the surface. The term commonly used for this purpose is **magma**, a Greek word originally meaning any kneaded mixture, such as dough or ointment. The only connection with lava seems to be mobility or capacity to flow. In its geological application the property of high temperature is also implied. Magma, then, means any **hot** mobile material within the Earth which is capable of penetrating into or through the crustal rocks. So far as we can judge from their products, magmas may consist of hot liquids, gases and solids in all possible proportions and combinations. The most important point to keep in mind, however, is that magma is not merely molten rock. When confined under pressure it is commonly associated with gases and vapours, sometimes in immense quantities. As a highly gas-charged magma ascends towards the surface and the overhead pressure gradually falls,

Figure 3.10 Pictorial representation of a 16 km stretch of the Laki fissure, Iceland, showing conelets formed towards the end of the 1783 eruption. (*After Helland*)

the gases begin to be liberated. Sooner or later the growing gas pressure overcomes the resistance – the soda-water bottle bursts, so to speak, or blows off its seal-cap – and an explosive eruption breaks out (Figure 13.38).

In the more familiar volcanoes the magma ascends through a central pipe, around which the lavas and ashes commonly accumulate to form a more or less conical volcanic mountain. Magma may also reach the surface through long fissures from which lava-flows spread over the surrounding country, filling up the valleys and forming widespread volcanic plains or plateaus. In such cases the lava is generally basaltic, like that of the Giant's Causeway.

The greatest basalt flood of modern times broke out at Laki, in Iceland, during the summer of 1783. From a fissure 32 km long torrents of gleaming lava, amounting in all to over 12 cubic kilometres, overwhelmed 565 square kilometres of country, extending long fiery arms down the valleys, 64 km to the west and 45 to the east. Fertile farmlands were deeply buried beneath a desert of lava. As the activity diminished in intensity, obstructions choked the long rent. Gases that had been effervescing freely then began to accumulate below the surface until they raised sufficient pressure to overcome the resistance. At the hundreds of points along the fissure where the pent-up gases eventually broke through, miniature cones-with-craters, ranging in height from a metre or so to thirty metres or more, were built up and great quantities of pyroclasts were erupted (Figure 3.10).

Despite the obvious dangers to life and human welfare, of which the Laki eruption is an impressive example, volcanic activity is geologically to be regarded as a constructive process in so far as new materials are brought to the surface and new topographic forms built up.

As we have seen, volcanic activity is only the surface manifestation of the movement through the Earth's crust of magma generated in the mantle or in exceptionally heated regions of the crust itself. Naturally not all the magma reaches the surface, and the new rocks formed in the crust by the consolidation of such magma are the chief

examples of what are called **intrusive rocks**, to distinguish them from lavas, which are called **extrusive rocks**. In many places such intrusive rocks are well exposed to observation, as a result of the removal of the original cover by denudation. Thus the feeders of fissure eruptions, like that of Laki, appear long afterwards as dykes (Figure 3.11). Moreover, there are innumerable dykes that never succeeded in reaching the surface. In favourable circumstances the magma may open up a passageway along a bedding plane, making room for itself by uplifting the overlying rocks. The resulting tabular sheet of rock is called a sill (Figure 3.12). Exceptionally large intrusions, consisting of granite and similar crystalline rocks, are particularly characteristic of the hearts of mountain ranges, including those of former ages as well as those of the present day (Figure 3.13).

It is usual to describe the rocks of intrusions or

Figure 3.11 Dyke of dolerite (a basaltic rock) cutting the Chalk and the overlying Tertiary plateau basalts, Cave Hill, Belfast, Northern Ireland. (*R. Welch Collection, Copyright Ulster Museum*)

volcanic extrusions as **igneous rocks** (from the Latin *ignis*, fire; implying that 'subterranean fire' or the Earth's internal heat was involved in their origin). An exception to this is rock-salt which forms intrusions as salt domes and occasionally reaches the surface and may flow like a lava. Lacking a relatively high temperature – several hundreds of degrees Celsius or more – during emplacement, rock-salt is not referred to as igneous.

The basaltic dyke illustrated in Figure 3.11 cuts through the white chalk over which the earliest lava-flows of Antrim were poured. Instead of being soft and friable, like the chalk of southern England, the Antrim chalk has become hardened by heat from the overlying basalts. And on both sides of the wall-like dyke the chalk has obviously been heated still more, having been recrystallized here and there into a fine-grained marble. The process of 'baking' and recrystallization by which the chalk has been transformed is an example of what is called **contact** or **thermal** metamorphism. Such evidence, if found in the rocks bordering an intrusion, indicates that the latter was capable of heating up its surroundings when it was emplaced, and that it may therefore be properly described as an igneous intrusion.

3.6 METAMORPHISM OF ROCKS

Besides the contact metamorphism referred to above, rocks are subject to many other kinds of transformation to which the term **metamorphism** is applied, and examples of these will be considered later (Chapter 11). Here the term is being introduced to draw attention to the fact that rocks respond to the Earth's internal activities not only by bending and fracturing but also by recrystallization. When crustal rocks are buried and come under the influence of any combination of the following –

1. the intense pressure or stress differences set up by gravity in association with other processes responsible for earth movements;
2. the increased temperature associated with nearby igneous activity, or caused by internal friction or the through-passage of hot gases;
3. the chemical changes stimulated by the through-passage of chemically active hot gases and liquids

they respond by changes in structure and mineral composition and so become transformed into new types of rocks.

It must be emphasized that metamorphism is the very antithesis of weathering. Both processes bring about great changes in pre-existing rocks, but weathering is destructive while metamorphism is constructive. Instead of reducing a pre-existing rock to a decaying mass of rock-waste and soil, metamorphism brings about its transformation, often from a dull and uninteresting-looking stone, into a crystalline rock of bright and shining minerals and attractive appearance.

3.7 THE GEOLOGICAL CYCLE: AN EIGHTEENTH CENTURY PARADIGM

It will now be clear from our rapid survey of the leading geological processes that they fall into two contrasted groups, Table 3.1. The first group – denudation and deposition – includes the processes which act on the crust at or very near its surface, as a result of the movements and chemical activities of air, water, ice and living organisms. Such processes are essentially of external origin. The second group – earth movements, igneous activity and metamorphism – includes the processes which act within or through the crust, as a result of the physical and chemical activities of the

Table 3.1 Classification of geological processes

PROCESSES OF EXTERNAL ORIGIN

1. Denudation (weathering, erosion, transport)
Sculpturing of the land surface and removal of the products of rock decay mechanically and in solution.

2. Deposition
   (a) of the debris transported mechanically (e.g. sand and mud)
   (b) of the materials transported in solution:
        (i) by evaporation and chemical precipitation (e.g. rock-salt)
        (ii) by the intervention of living organisms (e.g coral limestone)
   (c) of organic matter, largely the remains of vegetation (e.g. peat).

PROCESSES OF INTERNAL ORIGIN

1. Earth movements
Moving lithosphere; uplift and depression of land areas and sea floors; mountain building; earthquakes.

2. Metamorphism
Transformation of pre-existing rocks into new types by the action of heat, pressure, stress and hot, chemically active, migrating fluids.

3. Igneous activity
Emplacement of intrusions; emission of lavas and gases and other volcanic products.

Figure 3.12 Dolerite sills intrusive into horizontally bedded sandstones of the Beacon Series, exposed in the cliffs on the southern side of the Taylor Glacier, Southern Victoria Land, Antarctica. The sills range in thickness from 3 to 180 metres. At the lower left an offshoot from a sill can be seen cutting across the bedding in places, with corresponding displacement of the overlying sandstone. (*B. C. McKelvey and P. N. Webb*)

materials of both crust and mantle. Such processes are essentially of internal origin.

Both groups of processes operate under the control of gravitation (including attractions due to the sun and moon), co-operating with all the Earth's bodily movements – of which the chief are rotation about its axis and revolution around the sun. But if these were all, the Earth's surface would soon reach a state of approximate equilibrium from which no further changes of geological significance could develop. Each group of processes, to be kept going, requires some additional source of energy. The processes of external origin are specifically maintained by the radiation of heat from the sun. Those of internal origin are similarly maintained by the liberation of heat from the stores of energy locked within the Earth.

Throughout the ages the face of the Earth has been changing its expression. At times its features have been flat and monotonous. At others – as today – they have been bold and vigorous.

We have already noted that Hutton, in the late eighteenth century demonstrated the importance of erosive processes in wearing away the land surface and producing sediments. But he did more, much more. He produced a framework, a paradigm, which has proved the basis and reference point for geological work since that time. His *Theory of the Earth* links the erosive, destructive processes with the constructive processes which renew the land surface by earth movements and igneous activity. The paradigm is encapsulated in the Geological (sometimes called the Geostrophic, from *trophe*, Gr., change) Cycle, a series of events which has taken place repeatedly throughout the history of the Earth and continues today (Figure 3.14).

Suppose the cycle to begin with a new land surface lifted up into the zone of weathering and erosion. Denudation leads to sedimentation and as the deposits accumulate they are lithified by compaction and cementation. Uplift may occur at this stage exposing the sedimentary rocks to erosion and a further cycle. Or the sediments may be taken down to deeper, hotter levels where increased pressure and temperature may metamorphose, for example, a dull limestone into an attractive, glistening, marble. Uplift at this stage may renew the cycle. Or further subsidence may cause melting of the material to produce a magma. The re-mobilized material may be joined by juvenile material, i.e. material from the Earth's interior which has not yet taken part in the cycle. If this magma works its way up to the surface eventually

Figure 3.13 Granite peaks of the Bergeller Massif across the Swiss–Italian border, north-east of Lake Como. Village of Soglio in the foreground. Between, on each side of the valley, are the metamorphic rocks of the Southern Alps in which the Bergellér granite intrusion is emplaced. (*F. N. Ashcroft*)

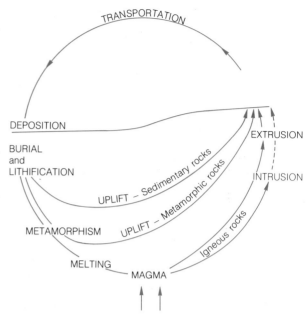

Figure 3.14 Diagrammatic representation of the Geological Cycle. Magma may form from melted continental rocks or from mantle material which may not have taken part in any previous cycle. Magma may reach the surface, lose volatiles and be extruded as lava from volcanoes, and enter the erosive phase of the cycle. Or the magma may crystallize underground as an intrusive igneous rock. The intrusion may be uplifted into the zone of erosion or it may be exposed if denudation reaches down to its level.

to produce a volcano, then the resultant rocks are exposed to erosion right away. On the other hand, if it congeals underground, it will only join a new cycle by erosion of the overburden.

Mountain belts provide the grandest examples of the operation of the geological cycle (Chapter 31). The Caledonian Mountain Belt traceable from Scandinavia through the British Isles across to the Appalachians in North America, may serve as an illustration at this stage. Sedimentation began over 600 m.y. ago and continued, with some phases of uplift, igneous activity and metamorphism for some 200 m.y. when earth movements culminated in the formation of mountains of Himalayan proportions.

3.8 PLATE TECTONICS: A TWENTIETH CENTURY PARADIGM

Despite the elegant and comprehensive nature of Hutton's theory a number of basic features concerning the Earth were imperfectly understood even after almost two centuries of work within that framework. Technology restricted investigations essentially to the land surface, while the oceans, making up two-thirds of the Earth's surface, remained virtually unknown. What was their history? How did they relate to the continents?

The material of the continents (sial) might be regarded as a kind of light slag that floated to the surface early in its history, as it cooled from a molten state. It might have accumulated uniformly and formed a continuous layer. Indeed some geologists supposed that this was so, and that the sialic crust was pulled apart as a result of the gradual expansion of the Earth's interior (Figure 3.15). If, on the other hand, sialic crust was from the start concentrated in continental rafts, a clue as to how such concentration might be brought about is provided by the behaviour of scum on the surface of gently boiling jam (Figure 3.16). A hot current ascends near the middle, and, turning outwards at the surface, it sweeps the scum to the edges, where the current descends. The scum is too light to be carried down, so it accumulates and can be skimmed off. Actual examples of convection occur in the celebrated Pitch Lake of Trinidad. There are reasons, as we shall see later, for suspecting that similar circulations may be going on

Figure 3.15 Schematic illustration to indicate the effects of expansion of the Earth's interior on a sialic crust that may originally have covered the entire globe. Primordial continents (white) first outlined by stretching and fragmentation, are gradually separated as they move outwards in consequence of the earth's increasing volume. Heavy material from the mantle flows into the intervening gaps and forms the floors of the growing ocean basins (black). (*After O. C. Hilgenberg, 1933*) Modern evidence does not support expansion of the order suggested by Hilgenberg nor that the continents were formed by one phase of expansion.

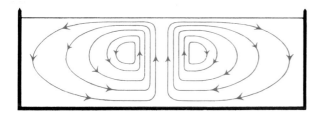

Figure 3.16 Section through a convection 'cell' showing the directions of convection currents in a layer of liquid uniformly heated from below.

inside the 'solid' Earth, though on a much larger scale and at a much slower rate. If at one time such convection currents had been sufficiently vigorous, the horizontal currents spreading out from each ascending column might well have swept some regions clear of sial. These regions would become ocean basins. Where the horizontal current of one convection system met those of a neighbouring system they would be obliged to turn downwards, and the sial, too light to be carried down, would be left behind. Thus regions overlying descending currents would develop into continents.

Since early in the century, a minority of geologists followed Wegener (1912) in thinking that about 300 m.y. ago the continents were joined together as one supercontinent, **Pangaea** (Gr., all earth). The Earth as we see it today, largely from palaeoclimatic evidence, was thought by Wegener to have gradually evolved by the break up of Pangaea, and the drifting apart of the separate pieces. This process is known as continental drift.

It is only since about 1960, however, that new geophysical discoveries have made this conclusion inescapable.

The drifting of continents was envisaged as a movement of sialic material over a substrate. In the new synthesis, which is detailed in Chapter 28, the movement of continents turns out to be a passive rather than an active process. Continents do not move separately, they ride 'piggy-back' on the top of lithospheric plates. In the development of oceans and continents the important boundary is not between crust and mantle but between the rigid, brittle lithosphere and the weak, partly molten asthenosphere.

The Earth is envisaged as being covered by six large carapace-like lithospheric plates and several smaller ones (e.g. Figures 28.2, 28.15). The plates grow at constructive margins where new crust is formed by the injection of basaltic material from below (Figure 3.17). These margins are manifest as ocean ridges (Figure 2.7) which perhaps mark the site of upwelling convection currents and a swelling of the mantle. When upwelling begins below a continent, the crust is split and the continental fragments move apart. The plates slide over the underlying asthenosphere at rates of a few centimetres per year. With the formation of new crust, the ocean formed by the splitting of the continent gradually grows; a process described as **sea-floor spreading**. The Atlantic has formed within the last 150 m.y. as America split from Africa and Europe and spreading took place from the Mid-Atlantic Ridge. The South American plate

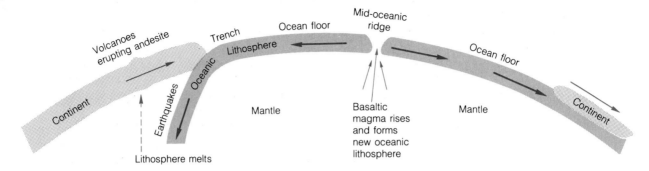

Figure 3.17 Diagrammatic representation of new oceanic crust (brown) forming along a mid-oceanic ridge and, together with the upper part of the mantle moving away as oceanic lithosphere on either side of the ridge. The deep ocean trenches represent the sites of subduction zones, and on their landward side the foci of medium and deep earthquakes outline the surface of oceanic lithosphere descending into the mantle. Movement of oceanic lithosphere away from mid-oceanic ridges is variously attributed to push by basaltic magma emplaced at mid-oceanic ridges, to gravity-gliding away from mid-oceanic ridges, or to downpull exerted on descending oceanic lithosphere as well as convective drag.

consists of ocean crust in the Atlantic and continental crust in South America (Figure 28.15). But some plates such as the Pacific and East Pacific (Nazca) are only oceanic.

From current evidence, any significant growth in size of the Earth's surface over the last 600 m.y. seems unlikely. This being the case growth of plates must be offset by destruction. As plates impinge on one another at least one of them must be bent downwards; a trench at the surface marks the site of down-bending. Ultimately, as the descending plate experiences higher temperatures, it is absorbed into the mantle. This process is referred to as **subduction**. If the two colliding plates are both oceanic, subduction of one leads to melting and the generation of igneous activity; a volcanic island arc is formed. On the other hand if one of the plates is carrying a continent, the lighter sialic crust is usually not subducted, and oceanic crust is forced into the mantle. In addition to the volcanic activity generated by the subducted crust, the continental crust is crumpled by impact and a mountain chain results. It is also thickened and, being pressed down and heated, is partly melted and gives rise to huge intrusions of granitic rocks. The Andes have formed by collision between South America and the East Pacific Plate. In the case of two continent-carrying plates colliding, neither is readily subducted into the mantle. Instead one continent may be forced underneath the other; continental crust may be thickened more than twofold and isostatic response raises mountains and plateaus to excessive heights. The Himalayas and the Tibetan Plateau are the result of India being driven into Asia.

The present array of plates and their movement accounts for innumerable modern, and geologically young, features of the Earth but the arrangement clearly was different in earlier eras. Wegener's Pangaea is evidence of that. The Caledonian Mountain Belt mentioned in connection with Huttonian cycles provides further striking evidence of the growth and decay of oceans and the moulding of the continents. Over 600 m.y. ago, a suture developed running through what is now Scandinavia, across Britain, dividing Scotland from England, and along the eastern coast of America. Sea-floor spreading from the suture produced an ocean which has been called **Iapetus** (after the Greek God, son of Gaia, Goddess of the Earth and Uranus, the personification of Heaven, and father of Atlas). But after spreading, a change in plate movements closed the ocean, caused a continent–continent collision and Europe and America were re-united along the Caledonian mountains (Chapter 31).

The major Huttonian cycles find their explanation in the twentieth century paradigm of Plate Tectonics. The geological history of the Earth is one of constant movement of the lithosphere, birth and death of oceans, reworking and reorganization of continental masses; a chronic restlessness fuelled by sun's radiation and the internal energy of the planet.

3.9   THE EARTH AS A PLANET

Space exploration over the last three decades has boosted interest in, and enormously increased information on, planets and our natural satellite, the Moon. Planetology has developed into a discipline in its own right and beyond the scope of this book. But it is instructive to compare the Earth with its neighbours and discover a surprising range of compositions, structures and histories. It emerges that we inhabit a planet which is unique in many ways. Perhaps most remarkable is the fact that, rather than taking a passive role and accepting, so to speak, a ready-made system, life on Earth has taken an active part in developing and controlling its environment.

The planets are separated into two groups in terms of composition, physical characteristics and distance from the sun (Table 3.2). The inner planets are relatively small, have high densities and are mainly rocky. The outer planets, with the exception of tiny, remote Pluto, are giants by comparison and mainly gaseous. Between the inner and outer planets lies the belt of **asteroids**, small rocky and metallic bodies which vary in size up to the largest with a diameter of 770 km. At the limits of the solar system, orbiting in rather eccentric paths, are the **comets** which consist essentially of ice and dust and have been described as 'dirty icebergs'. Some of the satellites of the giant planets have some similarities with the inner planets and will be brought into our discussion.

In **structure** the rocky portions of the inner planets show considerable similarities of organization; crust, mantle and core being recognizable. There are, however, important differences in the size of these components. The crusts of the Moon (60 km), and Mars (200 km) are much thicker, while the core in Mars is very small, perhaps even

absent. In order to satisfy the smaller overall density of the Moon the core must contain a greater admixture of material lighter than iron (possibly sulphur) compared with the Earth. The present feeble nature of magnetic fields suggests that the cores in the Moon, Mercury, and Mars are not mobile. What magnetism there is, is remnant from fields induced when the cores were mobile, or imposed by the sun's field early in their history. Although its core may yet be molten in part, the magnetic field of Venus has not been maintained perhaps because of its relatively slow rotation compared with the Earth.

Direct evidence of the **composition** of the surface rocks is provided by samples collected from the Moon. They are basalts, very similar to those of the Earth and light-coloured ultrabasic rocks called **anorthosites**. Probes indicate similar compositions from other planets. On Mars the well-known red colour derives from surface materials, lavas or dunes of silt-sized particles, coated with a veneer of ferric oxides. Smooth lava plains on Mercury are separated by cratered areas. Gamma-ray detectors on Venus suggest that granitic as well as basaltic rocks are present, hinting at the possibility of a bi-partite crust like the Earth.

Surface features of the Earth, as we have discovered, are dominated by tectonic movements and volcanic activity as modified by processes of erosion. The landscape of the Earth has been subject to reworking ever since the first crust was formed; there are over 100 probable craters (**astroblemes**) formed by impacting meteorites but most, especially early ones, have been destroyed. Surface features of Venus show considerable volcanic

Table 3.2 The planets

	Mass (E = 1)	Distance from sun ($\times 10^6$ km)	Equatorial diameter (km)	Density	Gases
Mercury	0.06	5790	4880	5.4	none
Venus	0.81	10 820	12 104	5.2	CO_2
Earth	1.0	14 960	12 756	5.5	N_2, O_2
Mars	0.11	22 790	6787	3.9	CO_2
Asteroid Belt					
Jupiter	318	77 830	142 800	1.3	H_2, He
Saturn	95	142 700	120 000	0.7	H_2, He
Uranus	14.6	286 960	51 800	1.2	H_2, He, CH_4
Neptune	17.2	449 660	49 500	1.7	H_2, He, CH_4
Pluto	0.003	590 000	2300	2.0	$?CH_4, N$

activity with some tectonism although recent investigations suggest no de-coupling of surface plates from a substrate, as has happened on Earth.

Apart from the Earth the most common planetary feature to be seen is cratering. Planets are envisaged as forming from the aggregation of gaseous and dust particles in the primordial disc revolving around the sun. The heat produced by accretion radioactivity and impacting bodies would have led to melting. As cooling occurred heavier metallic material sank to form the core and from the silicate portion forming the mantle lighter granitic material floated to the surface around 4.5 billion years ago (Ga). Mercury, Mars and the Moon are now scarred by innumerable craters formed by impacting meteorites, a phase which continued after the formation of the early crusts. On the Moon this phase reached a climax between 3.8 and 3.2 Ga (billion years) when giant impacts were accompanied by intrusion of floods of basalt which now form the Maria, i.e. the dark patches on the face of the Moon. Since that time volcanism has died on the Moon. The satellite has remained inactive, seismic activity is restricted to a few feeble 'quakes' caused by gravitational changes during the Moon's orbit.

The southern hemisphere of Mars is heavily cratered and the northern hemisphere shows smooth plains of basalt. Tectonic features include the 5000-km-long rift – the **Vallis Marineris**. Mercury is also heavily cratered although there are wide areas of intercrater plains of lava. The **Caloris** basin some 1300 km across is the result of such a large impact that shock waves travelled right through the planet and threw the antipodean surface into a chaotic structure. Like the Moon basaltic eruptions were possible till about 3000 m.y. ago but since then the planet has been dormant. Craters in Mars also belong to the early bombardment but on this planet volcanic activity concentrated between 2000 and 1000 m.y. may have continued to as little as 200 million years ago. Volcanism may have been responsible for the puzzling channels which reach 300 km long and are braided like some rivers on Earth. Water may have been released from ice held in the ground but which subsequently dispersed through the atmosphere.

It is when **atmospheres** are considered that significant differences are found. Small bodies like the Moon, Mars and Mercury have insufficient gravitational force to hold the lighter elements which have been dispersed into space. Consequently atmospheres are absent or very thin. Surface temperatures are dependent essentially on distance from the Sun. Polar caps on Mars are of frozen carbon dioxide as 'dry ice'. Venus is larger and has a much denser atmosphere, an atmosphere which has made investigation of the planet very difficult although preliminary mapping of the surface was achieved by the Pioneer Orbiter mission in 1980. The *Magellan* spacecraft began detailed mapping of the whole planet in 1990 and is scheduled to continue until 1995. The carbon content of the planet is about the same as that of the Earth but the absence of any organic activity has meant that none has been locked up in limestones. The atmosphere is dominated by carbon dioxide, the 'greenhouse' effect is rampant and temperatures at the surface are 450°C. The presence of sulphur dioxide and water vapour in small quantities enhances the 'greenhouse' effect, while the former has another bizarre result. A low layer of the atmosphere has clouds but these are not of water; instead they are made up of droplets of sulphuric acid. To be out on Venus, in a rainstorm of concentrated acid, would be very uncomfortable!

The atmospheres of the outer planets bear no comparison with the Earth but may hold clues about our original atmosphere in that they are deficient in oxygen. Some of their satellites are similar in size and density to the inner planets. The Saturn satellite, Titan is larger than Mercury and its density suggests a composition of about half ice and half rock. More relevant is its atmosphere which is made up of nitrogen. This forms dense clouds which obscure the surface but low temperatures (–180°C) suggest oceans of liquid nitrogen. Ethane, acetylene, ethylene and hydrogen cyanide have been detected so that another scenario envisages an ocean of ethane and methane with continents or islands of ice. Another rocky body, the Jupiter satellite Io, is about the size of the Moon and it has lost volatiles including water, but retains the heavier sulphur dioxide. A *Voyager* flypast coincided with an eruption of one of its volcanoes, a spectacular sight rising to 100 km from the surface. The lava was not basalt but sulphur. A molten interior, maintained by heat generated by the tidal effects of nearby Jupiter, results in volcanic activity which constantly reworks the entire surface. Oceans may be of liquid sulphur. The Earth's atmosphere dominated by nitrogen and having 21 per cent oxygen is quite anomalous. The proportion of gases bears no relation to the relative abundances of gases in the solar system, not to the sun, nor any of the giant gaseous planets. Suppose that at the time of its formation the planet had a composition which mirrored that of the solar system, i.e. one dominated by hydrogen and helium with smaller amounts of methane and ammonia. The atmosphere though evolved through time might still retain evidence of its primordial composition. Inert gases for example should have been unaffected. It is found that the proportion of neon is very much smaller on Earth than its overall distribution would suggest. Argon in the sun is mostly the isotopes 36 and 38 whereas on Earth these are poorly represented. Instead argon is mostly ^{40}Ar; an isotope formed by the decay of ^{40}K.

If not derived directly from the solar nebula the atmosphere must have come from the Earth itself; and the probable source is readily identified. Throughout the Earth's history, volcanoes have poured out vast quantities of volatiles, including water vapour, carbon dioxide, chlorine, nitrogen dioxide, sulphur dioxide and hydrogen sulphide. These could have formed the primitive atmosphere and oxygen would be deficient, a condition necessary to allow the earliest organic molecules to survive. In order to arrive at its present composition one or more powerful oxygen-producing processes were necessary.

The earliest evidence of cells preserved as fossils comes from rocks 3.5 Ga old. Life evidently began very early on the planet but the earliest microorganisms lived in an anoxic atmosphere and they produced water in their vital processes, not oxygen. This newly-formed water vapour together with any original water could have been dissociated in the atmosphere by the intense radiation of the young sun into hydrogen and oxygen. Hydrogen would escape, spacewards; rising oxygen would form the three-atomed molecules of ozone and go to form the ozone layer in the upper atmosphere. The amount of oxygen produced in this way is limited by a feed-back process in so far as the growing ozone layer would reduce the radiation available to dissociate the water.

Photosynthesis, by which plants convert the energy of the sun into a form which they can use, produces oxygen from carbon dioxide. It began about 3 Ga ago and oxygen levels began to rise. By 2 Ga ago the oxygen produced was sufficient to oxidize vast amounts of ferrous oxide which had remained soluble in the world's seas. The precipitated ferric oxides are now preserved in the widespread iron ores found today in many continents. Even so the oxygen levels at this time were still of the order of one-hundredth that of today. But the rise continued allowing the development of animals using oxygen in respiration and triggering the vast expansion of forms at about 1 Ga. The present level was achieved about 600 m.y. ago when, in the words of the writer Jacquetta Hawkes, 'life drew a hard line around itself', shelled organisms abounded, and carbon dioxide was fixed in rocks in increasing amounts. Since that time the atmosphere has remained more or less constant, the ozone layer protecting animals and plants, the carbon dioxide content never rising to threatening levels. The runaway greenhouse effect of Venus has been avoided. Life has so affected the physical world as to develop and maintain the conditions that it needs to survive. It could only do this because the Earth is an appropriate size, and far enough away from the burning sun, to retain a significant atmosphere along with vast amounts of water. It did not lose both like Mercury and Mars. It was far enough away from the sun's destructive radiation to allow life to develop while Venus, nearer and hotter, inhibited life and lost the opportunity to take carbon dioxide out of its atmosphere. The satellite Titan may have a composition not unsuitable for life to begin but it is far too cold; while Io was only big enough to hold on to sulphur dioxide.

Earth also has a 'life' that has been lost by the smaller planets. It continues to change, continents move, oceans grow and fail, mountains are gradually worn down but rise again. It is still in a sense youthful; it is inspiringly dynamic.

3.10 FURTHER READING

Baugher, J.F. (1988) *The Space-Age Solar System*, Wiley, New York.

Geological Museum (1977) *Moon, Mars and Meteorites*, HM Stationery Office.

Hallam, A. (1973) *A Revolution in Earth Sciences, from Continental Drift to Plate Tectonics*, Clarendon Press, Oxford.

Lang, K.R. and Whitney, C.A. (1991) *Wanderers in Space*, Cambridge University Press, Cambridge.

Siever, R. (1983) *The Dynamic Earth*, Scientific American, **249**, No. 3, 30–39.

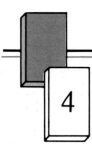

4

MATERIALS OF THE EARTH'S CRUST: ATOMS AND MINERALS

Go, my sons, buy stout shoes, climb the mountains, search the valleys, the deserts, the seas shores, and the deep recesses of the earth. Mark well the various kinds of minerals, note their properties and their mode of origin.

Petrus Severinus (1571)

4.1 ELEMENTS: ATOMS AND ISOTOPES

The vast majority of rocks are aggregates of minerals. Of the remainder some, like obsidian and pumice, are made of volcanic glass, while others, like coal, are composed of the residual products of organic decay. All these materials in turn are made of atoms of the chemical elements, of which 103 are known at the time of writing. Of these only 87 occur naturally in detectable amounts, the others having been synthesized by the modern alchemy of nuclear reactions. With the discovery of radioactivity in 1895 the atom lost its age-long status as the fundamental particle of an element that could not be further subdivided. For centuries the old alchemists tried in vain to change one element into another. They failed because they used only chemical reactions, and chemical reactions affect only the arrangements and associations of atoms; the elements themselves are not changed. Atomic

or nuclear reactions, however, change the actual identities of the atoms concerned. Moreover, as we all know far too well, enormous amounts of energy are released in such reactions. Less alarmingly – indeed quite otherwise – radioactivity and nuclear physics have geological applications to many problems that otherwise could hardly be effectively tackled at all. It is therefore desirable to learn a little about the atom.

An atom cannot really be pictured in any intelligible way, since it behaves as if it were made up of 'particles' which at the same time are also systems of 'waves'. And it would be quite useless – at the moment – to ask what the particles are made of, or what it is that is waving. However, it is convenient for most purposes to visualize the atom as a constellation of **electrons** surrounding an inconceivably small central nucleus. The orbits in which the electrons revolve and spin occupy a relatively vast space – at least a million times as great as that occupied by the nucleus.

It is well known that an electron has (or perhaps is) an electric charge that is regarded as **negative**. This use of 'negative' can be very misleading unless it is clearly realized that the term is only a verbal convention and does not imply that anything is missing. How it came about was purely by chance. *Electron* is the Greek word for amber. When amber is rubbed with wool or fur it acquires the property of attracting little bits of paper; it is then said to be electrified or to have an electric charge. It was recognized that there must be two kinds of electric charges when it was found that certain other materials, e.g. glass rubbed with silk, develop a charge that is in one sense the 'opposite' of that acquired by amber. Two 'amber' charges repel each other, but an 'amber' charge and a 'glass' charge attract each other. Long ago the glass kind of charge came to be distinguished as positive, and the amber kind as negative. If a positively charged point is connected by a wire to a negatively charged point, a momentary current of electricity flows along the wire. Since flow is naturally from high to low, the direction of the current was assumed to be from positive to negative. We now know that this assumption was wrong. The current is a swift flow of electrons along the wire, but the 'high' end, where there is an excess of electrons to begin with, is the negative end. The excess electrons flash along the wire to the 'low' or positive end, where there is a deficiency of electrons. The charge on an electron has therefore to be thought of as 'negative'. This means, of course, that the charge on the nucleus of an atom must be regarded as 'positive', since an atom as a whole is electrically neutral. However, atoms readily become electrically charged by gaining or losing electrons. Such electrified atoms are called **ions**. According to the convention, an atom with more than its normal ration of electrons is a negative ion, whereas one with less than its normal ration is positive.

The simplest atom is that of hydrogen. This has a single electron revolving around a nucleus consisting of a single particle, called a **proton**, the mass of which is 1836 times the mass of an electron. Each particle 'carries' a unit electric charge (e): negative (e^-) in the electron, positive (e^+) in the proton. The nuclei of other elements, however, contain **neutrons** as well as protons. These are particles with nearly the same mass as protons, but without any electric charge. In general the nucleus may be described as a cluster of tightly packed protons and neutrons: a cluster in which nearly the whole mass of the atom is concentrated (Figure 4.1). The fact that the protons

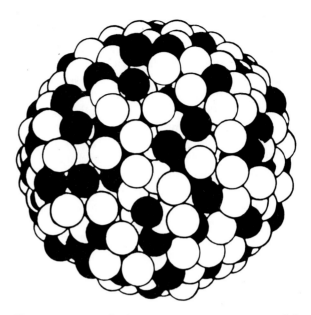

Figure 4.1 A purely diagrammatic representation of the nucleus of an atom of lead (isotope 206); a cluster of 82 protons (black) and 124 neutrons (white). Magnification about ten million million diameters. It is essential to remember that it is merely for convenience that the protons and neutrons are depicted as if they were spherical objects; their real nature is as mysterious as that of electrons and all other 'elementary particles'.

do not fly apart as a result of their electric repulsions proves that they must be held together by inconceivably powerful forces of attraction. These nuclear attractions – by far the most powerful forces known to exist – are at least a million times stronger than the electric attractions that hold electrons and the nucleus together. In turn the electric forces are stronger than the chemical forces that hold atoms together in molecules and crystals, while gravitational forces, capable as they are of raising mountains, are utterly negligible on the atomic scale.

An element is now defined as a substance in which all the atoms have the same nuclear charge, i.e. the same number of protons. The number of protons in the nucleus of a given element is called the **atomic number** (Z), and the elements are numbered accordingly: starting with hydrogen, 1; helium, 2; and so on, through the periodic table of the elements, up to uranium, 92; and now up to the latest of the synthesized elements, lawrencium, 103. The elements were originally arranged in the periodic table according to their atomic weights, and chemists were puzzled by the fact that some elements have atomic weights that depart considerably from whole numbers. Chlorine, for example, has an atomic weight of 35.46. The reason for this and other apparent anomalies became clear when it was discovered that the number of neutrons (N) in the nucleus of a given element may vary, unlike the number of protons, which is invariable. The sum (A) of the number of protons (Z) and neutrons (N) in a nucleus is called the **mass number** (A = Z + N). It follows that an element does not necessarily consist of only one kind of atom. It happens that some of the elements with odd atomic numbers have all their atoms alike, e.g. sodium (Z = 11; N = 12); and gold (Z = 79; N = 118); but the rest, including all the elements with even atomic numbers, are mixtures of nuclear varieties with different numbers of neutrons in their nuclei and therefore with different mass numbers. These varieties are called **isotopes** (Gr. *isos*, equal; *topos*, place) of the given element, because they are identical in their chemical properties and so occupy the same place in the periodic table of the elements. Chlorine (Z = 17) is an example of an odd-numbered element with two isotopes: one with 18 neutrons (making A = 35), the other with 20 neutrons (making A = 37). Approximately three-quarters of the atoms of natural chlorine are the isotope chlorine-35 (or ^{35}Cl) and one-quarter the isotope chlorine-37 (or ^{37}Cl); and accordingly the atomic weight cannot be far from 35.5, thus clearing up the old puzzle. It is of interest that chlorine extracted from meteorites has the same atomic weight and the same mixture of isotopes as terrestrial chlorine, whether the latter is prepared from natural rock-salt (NaCl), from sea-salt or from volcanic gases.

4.2 ELEMENTS: ELECTRONS AND IONS

In a neutral atom – electrically uncharged – the number of electrons swinging in their orbits around the nucleus is the same as the number (Z) of the protons in the nucleus. These electrons do not revolve in a haphazard swarm, but have orbits concentrated in a series of concentric 'shells'. The innermost shell never contains more than two electrons; the second can accommodate up to eight, but no more; the third takes up to eight for some elements, but up to eighteen for others – and so on. The chemical properties of an element depend on the number of electrons in its atoms, and particularly on those in the outermost shell. This can be simply illustrated by considering how sodium and chlorine combine to form sodium chloride (NaCl), familiarly known as common salt or rock-salt (Figure 4.2). The two inner shells of the chlorine atom have their full complement of electrons, but the outer one has only seven of the eight it can take. Now as far as possible, electrons revolve in pairs, one spinning in a clockwise direction as it swings round its orbit, while the other spins in the opposite or anticlockwise

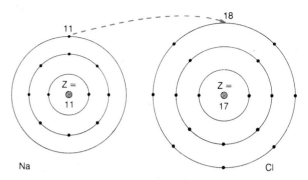

Figure 4.2 The solitary electron No.11 from a sodium atom jumps into the unoccupied orbit No. 18 of a chlorine atom, leaving the sodium as an ion Na$^+$ and transforming the chlorine to an ion Cl$^-$. Na$^+$ and Cl$^-$ then unite to form NaCl in which the electric charges are balanced.

direction. So in chlorine, one of the electrons in the outer shell is left without a dancing partner, so to speak: there is a vacancy to be filled. Sodium is a willing candidate. In sodium the inner shells are also fully occupied, but the outer shell has only one electron and this is so loosely held that, if a chlorine atom comes sufficiently near, it can jump into the vacant place. The chlorine atom (Cl), having gained an electron, has now acquired a negative electric charge and has become a negative ion (Cl$^-$), while the sodium atom (Na), which has been deserted by its outer electron, has acquired a positive charge and has become a positive ion (Na$^+$).

The sodium ion, Na$^+$, and the chlorine ion, Cl$^-$, are now mutually attracted and combine to form the uncharged 'molecule', NaCl, of common salt. In practice, if a bit of sodium is introduced into the poisonous green gas, chlorine, it burns to a white vapour from which tiny crystals of salt settle out. Millions and millions of atoms have turned into ions, all competing for partners. And since there are partners for all, they settle down in an orderly fashion which achieves electric balance and perfect uniformity of composition. Figure 4.3 illustrates the three-dimensional pattern of the ions in a crystal of common salt. Each crystal might be regarded as a sort of giant molecule, but more accurately it is a continuous ionic structure rather than a molecule.

The outer-shell electrons which are responsible for binding together various atoms into the molecules or crystals of a chemical compound are called the valency-electrons. **Valency** is the combining power of an atom (or group of atoms), expressed numerically as the number of electrons which each atom (or group of atoms) has gained, lost, or shared. Hydrogen and sodium ions each lose one electron and their valency is 1+. Some atoms,

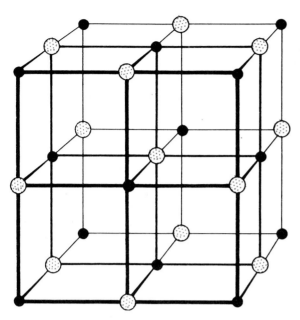

Figure 4.3 Structure of a crystal of common salt, NaCl. Relatively small Na$^+$ ions (black) and larger Cl$^-$ ions (dotted) are arranged alternately at the corners of a set of cubes (here drawn in to show the relationship). Each Na$^+$ ion is surrounded by six Cl$^-$ ions, and each Cl$^-$ ion by six Na$^+$. On the scale of the diagram the ionic 'spheres' are shown – for clarity – as having only about one-third of the radii conventionally ascribed to them. See Fig.4.6.

however, like those of oxygen need two electrons to complete their outer shells. Thus an oxygen atom can combine with two atoms of hydrogen and the result is water, H_2O. But it could also combine with one atom of an element with two valency-electrons in its outer shell, e.g. magnesium, to make magnesia, MgO; or calcium, to make lime, CaO. In these cases the valency is 2+. Where the valency is 3+, as in aluminium, the combination with oxygen requires the proportions represented by Al_2O_3; this oxide is called alumina, and as a gemstone it is familiar as ruby and sapphire. Iron is a more complicated case, as its valency can be either 2+ or 3+. Thus it makes two series of oxides – and of other compounds – which are distinguished as ferrous (Fe^{2+}) and ferric (Fe^{3+}): e.g. ferrous oxide, FeO, and ferric oxide, Fe_2O_3.

In the mineral world the most important element with a valency of 4+ is silicon. Its oxide, silica, represented by SiO_2, is well known as quartz, agate (Plate IIa, b) and flint. Moreover, as indicated below, silica combines with the other common oxides to form a group of crystalline compounds known as **silicates** which, together with quartz, constitute a very large majority of the minerals occurring in rocks. Carbon is another element with a valency of 4, vitally important

because the latter may be either 4– or 4+. Carbon with a negative valency (as in the carbon of methane, CH_4, and of related hydrocarbons in petrol and petroleum) is the essential element in organic compounds and in all living matter. When the valency is positive, however, as in the carbon of carbon dioxide, CO_2, the latter combines with other oxides to form another important group of rock-forming minerals, the **carbonates**, which predominate in limestones and marbles. With this introduction we may now turn to the rocks and minerals themselves.

4.3 CHEMICAL COMPOSITION OF CRUSTAL ROCKS

Although 87 elements occur naturally in minerals, eight of these are so abundant that they make up nearly 99 per cent by weight of all the many thousands of rocks that have been chemically analysed (Table 4.1). Many of the others, such as gold, tin, copper and uranium, though present only in traces in ordinary rocks – for which reason they are referred to as **trace elements** – are locally concentrated in mineral veins and other ore deposits (see Section 4.8) sufficiently to make their extraction profitable. Elements 43, 61, 87 and 89 have not yet been detected at all in minerals, but they have been made artificially by nuclear reactions.

Carrying on from Table 4.1, the elements immediately following are **manganese**, Mn, 0.10; **fluorine**, F, 0.08; **sulphur**, S, 0.05; **chlorine**, Cl, 0.04; and **carbon**, C, 0.03 per cent. The abundances of the rarer or trace elements are more conveniently expressed in parts per million (ppm), which is the same as grams per metric ton (tonne, 10^6g = 2205 lb). Gold and platinum, though well known to everyone as precious metals, are amongst the rarest of elements in ordinary rocks, their average abundances being only about 0.005 ppm.

4.4 MINERALS AND CRYSTALS

Some of the elements, e.g. gold, copper, sulphur and carbon (as diamond and graphite), make minerals by themselves, but most minerals are compounds of two or more elements. Oxygen is by far the most abundant element in rocks. In combination with other elements it forms compounds called oxides, some of which occur as minerals. As silicon is the most abundant element after oxygen, it is not surprising that silica, SiO_2, should be the most abundant of all oxides. Silica is familiar as **quartz** (Plate IVa), a common mineral

Table 4.1 Average composition of crustal rocks (*After V.M. Goldschmidt and Bryan Mason*)

In terms of elements			In terms of oxides		
Name	Symbol and valency	Per cent	Name	Formula	Per cent
Oxygen	O^{2-}	46.60			
Silicon	Si^{4+}	27.72	Silica	SiO_2	59.26
Aluminium	Al^{3+}	8.13	Alumina	Al_2O_3	15.35
Iron	Fe^{3+} Fe^{2+}	5.00	Iron oxides { Ferric { Ferrous	Fe_2O_3 FeO	3.14 3.74
Calcium	Ca^{2+}	3.63	Lime	CaO	5.08
Sodium	Na^+	2.83	Soda	Na_2O	3.81
Potassium	K^+	2.59	Potash	K_2O	3.12
Magnesium	Mg^{2+}	2.09	Magnesia	MgO	3.46
Titanium	Ti^{4+}	0.44	Titania	TiO_2	0.73
Hydrogen	H^+	0.14	Water	H_2O	1.26
Phosphorus	P^{5+}	0.12	Phosphorus pentoxide	P_2O_5	0.28
		99.29			99.23

which is specially characteristic of granites, sandstones and quartz veins. The formula, SiO_2, is a simple way of expressing the fact that for every atom of silicon in quartz there are two atoms of oxygen. Perfectly pure quartz has, therefore, a definite composition. The formulae for other oxides and compounds may be similarly interpreted.

In the cavities of some mineral **veins** (which are usually infilled fault fissures), quartz can be found as clear transparent prisms, each with six sides and each terminated by a pyramid with six faces (Figure 4.4). Thinking of the mineral as a variety of ice that had been permanently frozen, the ancient Greeks gave the name *krystallos* (clear ice) to these beautiful forms, and to this day water-clear quartz is still known as rock crystal (Plate IVa). Most other minerals and a great variety of chemically prepared substances can also develop into symmetrical forms bounded by plane faces, and all of these are now called **crystals**.

The study of crystals by means of X-rays has revealed the fact that their symmetrical forms are simply the outward expression of a perfectly organized internal structure. As already indicated in the case of rock-salt (Figure 4.3), the electrically charged atoms or ions of which a crystal is composed are arranged in an orderly fashion, the different kinds of atoms being built into a definite pattern which is repeated over and over again, as in the design of a wallpaper. In crystals, however,

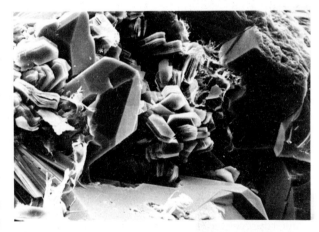

Figure 4.4 Scanning Electron Microscope photograph of crystals that have grown between detrital grains during diagenesis of a sandstone. The entire field is about 0.2 mm across. The largest, well formed hexagonal crystals are quartz, the flat crystals with a marked cleavage, resembling the open leaves of a book, are the clay mineral kaolinite, and the tiny, hair-like fibres are another clay mineral, illite. Clay minerals (Section 4.7) such as these, derived from the decomposition of feldspar, are of great importance in determining the ease with which petroleum may be extracted from pores in sandstone reservoirs. (*Photograph provided by E. K. Walton*)

the repeat pattern is in three dimensions and for this reason it is referred to as a **space lattice**.

The actual arrangement of atoms which repeats

about the points of the space lattice is called the crystal structure. The atomic structure of most minerals has now been obtained by the methods of **X-ray diffraction**, which were largely developed, often using mineral samples, by the Braggs, father and son, in the period 1913–1926. This method relies on two features of the diffraction of X-rays. The regular spacing of identical planes of atoms in the structure causes diffraction to occur at a characteristic angle, and the scattering power of the atoms involved in each diffraction affects the intensity of the diffracted beam. The calculations of the atomic configurations that account for the observed diffraction patterns are lengthy (although much facilitated by modern computers) and the interpretation requires skilful use of morphological, optical and other macroscopic features of the crystals. Although X-ray diffraction is still the routine, unambiguous and most exact way of obtaining crystal structures, in recent years more direct methods of imaging crystal structures have been developed such as **high-resolution electron microscopy**, which relies on the diffraction of

electrons by the crystal structure (Figure 4.5), and lately, the astonishing technique of **atomic force microscopy**, in which the movement of a tiny stylus is amplified as it moves across a crystal surface, leading to a computer-generated image in which individual atoms appear as 'bumps'.

It will have been noticed that diamond and graphite are both crystalline forms of carbon. Corresponding to their strongly contrasted physical properties – one being hard and brilliant, the other soft, opaque and flaky – the crystals of diamond and graphite have very different structures. This contrast in turn reflects the widely different physical conditions under which the two minerals crystallized. Diamond requires a combination of high temperature and high pressure and was first made synthetically in 1955 at pressure above 40 000 atmospheres (40 kbar) and temperature above 750°C. For graphite, however, quite moderate conditions of temperature and pressure suffice. This capacity of certain substances to occur as two or more quite different species of crystals, i.e. to crystallize with lattice structures

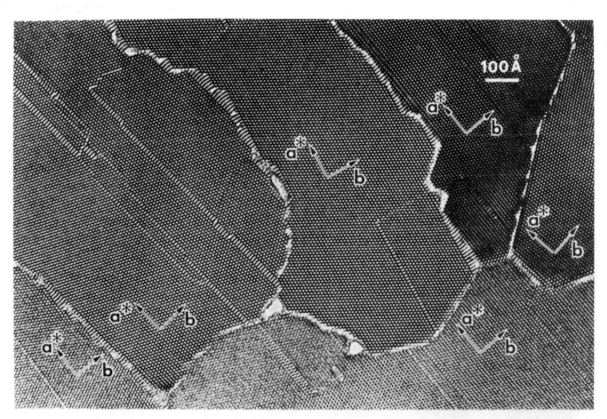

Figure 4.5 A high-resolution electron microscope image showing a bundle of tiny amphibole asbestos fibres viewed end on. Each fibre is about one ten-thousandth of a millimetre across. The regular rows of dots are arrays of points which can be used to define the **lattice** of the crystals. They are not individual atoms but are points marking the ends of pairs of double chains of silicon and oxygen atoms arranged parallel to the long axis of the fibres. There are a few linear defects visible. The boundaries between the individual fibres, each a single amphibole crystal, are curving and irregular, and the orientation of the lattice is quite different in each one. (Note that $1\ \text{Å} = 10^{-10}\ \text{m}$. Photograph kindly provided by Jung Ho Ahn and Peter Buseck, Arizona State University, published in *American Mineralogist*, 1991, vol. 76, 1468–80.)

appropriate to the physical conditions at the time of formation, is a phenomenon described as **polymorphism** (Gr. *polys*, many; *morphe*, shape or form). Other well-known examples are FeS_2, which occurs not only as pyrite, but also as marcasite; and calcium carbonate, $CaCO_3$, which crystallizes mainly as calcite (Plate IVd), but under special conditions as aragonite (e.g. in the shells secreted by certain molluscs and other marine organisms).

Few minerals have the exact chemical composition corresponding to their ideal formulae. The reason is that any ion that happens to be on the spot at the time of crystallization can act as a substitute or proxy for another without seriously disturbing the crystal structure, provided that the ion of the proxy has nearly the same 'size' (Figure 4.6) as that of the ion whose place it takes in the growing crystal. In just the same way bricks of standard size but of different colours could be built into a wall without altering the structure of the wall or its outward shape. In building the crystal edifice such substitution is especially favoured when both ions have the same electric charge or valency. A good example is provided by the green mineral, olivine. The formula is generally written $(Mg,Fe)_2SiO_4$ to express the fact that it is part of a continuous series ranging from Mg_2SiO_4 at one end, through $(Mg,Fe)_2SiO_4$ and

$(Fe,Mg)_2SiO_4$ to Fe_2SiO_4 at the other end, all the members having essentially the same structures and crystal forms. The intermediate members of the olivine series are not to be thought of as mixtures or combinations of the two 'isomorphous' compounds or molecules represented by the formulae of the end-members. It is not the compounds that are mixed, but the ions in the crystal lattice: in this case Mg^{2+} and Fe^{2+}, which are very nearly the same size (see Figure 4.6). Crystals belonging to such series are commonly referred to as solid solutions or, less happily, as 'mixed crystals'.

Equality of valency is, however, by no means essential for effective substitution. All that is necessary, provided that the ionic sizes are not too different, is that electric neutrality is maintained, and this can be achieved by a balancing interchange between the ions of two or more other elements in the structure. Thus, in the plagioclase series of the feldspar group (see Section 4.6) there is a continuous range of composition from albite, $NaAlSi_3O_8$, at one end to anorthite, $CaAl_2Si_2O_8$ at the other. Here Al^{3+} replaces one of the Si^{4+} ions, while at the same time, to balance the resulting deficiency in charge, Ca^{2+} replaces Na^+.

A glance at Figure 4.6 indicates that Na^+ and K^+ are too different in size to be readily interchangeable although, at high temperatures, when

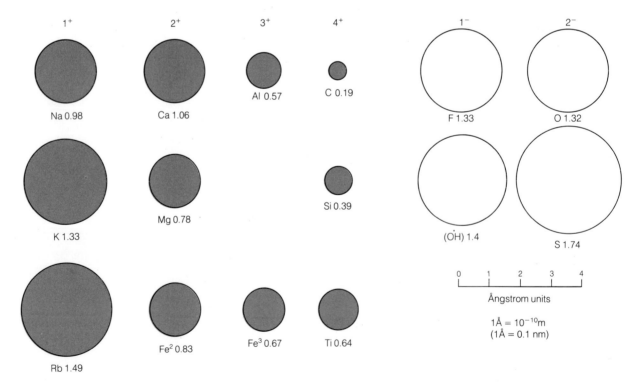

Figure 4.6 Relative sizes of some positive ions (cations) and negative ions (anions) in crystals. The figure at the head of each column refers to the ionic charge or valency. The effective ionic radius of the field of influence of each ion within a crystal, conventionally regarded as spherical, is given in ångstroms (0.1 nanometre).

the whole structure is vibrating, this may be possible. This corresponds to their actual associations in minerals. On the other hand, the relatively rare element, rubidium, Rb^+, is nearly the same size as K^+ and consequently growing potassium minerals provide an equally good home for rubidium. The natural concentrations of rubidium occur almost entirely in potash feldspars and micas (Table 4.3). This is a fortunate circumstance from a geological point of view because both elements have radioactive isotopes, and potassium minerals containing rubidium in sufficient quantity can be dated in two independent ways, as we shall see in Chapter 14.

A few minerals are non-crystalline in the sense that they never develop crystal forms, and for this reason they are said to be **amorphous**. Examples are **opal**, $SiO_2.nH_2O$, which has been described as 'an incompletely dried-out jelly', and **limonite** (Plate IVf), approximately $FeO.OH$, one of the iron hydroxides familiar as iron rust and as the material responsible for the rusty-brown appearance of many weathered rock surfaces. In such materials very tiny particles are arranged haphazardly, like the bricks in a tumbled heap, but investigations with the electron microscope show that within each of the particles the atoms have a regular structural arrangement.

Glass is also a typically amorphous substance. Most glasses are mixtures of silicates that have had insufficient time for the atoms to arrange themselves into the regular patterns of crystals, either because of rapid cooling from the molten state, or because the original melt was extremely viscous from the start. Various spectroscopic techniques have shown, however, that twisted and misshapen chains and networks are present,

indicating that the first steps towards crystallization have been taken. Moreover, this tendency to bring crystalline order out of amorphous chaos is eventually successful. Despite the fact that glass has the mechanical properties of a solid with very high viscosity, it slowly **devitrifies** and turns into an aggregate of minute crystals. The process of devitrification may take a few years or many hundreds in the case of man-made glass at ordinary temperatures. Natural glass, like the highly viscous volcanic lava that solidifies as pumice or obsidian, may require millions of years before it shows visible signs of crystallization. The transformation is speeded up, however, if hot volcanic gases find a way through the glassy material.

Apart from the amorphous materials, minerals are naturally-occurring inorganic crystalline substances, of which each 'species' has its own specific variety of crystal structure. Allowing for the inevitable presence of 'impurities' and trace elements, the chemical composition may be constant (as in quartz) or it may vary (as in the feldspars) within limits that depend on the degree to which the ions of certain elements can substitute for those of other elements without changing the crystal structure.

4.5 ROCK-FORMING MINERALS

Although about 2000 named minerals are known, most common rocks can be adequately described in terms of a dozen series of minerals, as Table 4.2 indicates. It is, therefore, well worthwhile to become familiar with these essential rock-forming minerals and with a few others of special interest, and especially to learn something of their chemical

Table 4.2 Average mineral composition of some common rocks (percentages)

Minerals	Granite	Basalt	Sandstone	Shale	Limestone
Quartz	31.3	–	69.8	31.9	3.7
Feldspars	52.3	46.2	8.4	17.6	2.2
Micas	11.5	–	1.2	18.4	–
Clay minerals	–	–	6.9	10.0	1.0
Chlorite	–	–	1.1	6.4	–
Amphiboles (mainly hornblende)	2.4	–	–	–	–
Pyroxenes (mainly augite)	rare	36.9	–	–	–
Olivine	–	7.6	–	–	–
Calcite and dolomite	–	–	10.6	7.9	92.8
Iron ores	2.0	6.5	1.7	5.4	0.1
Other minerals	0.5	2.8	0.3	2.4	0.3

compositions. An attempt is made here to present this minimum equipment of chemical knowledge as briefly as possible. The student should refer to textbooks for additional information, and above all he should handle typical specimens of minerals and rocks and examine actual rock exposures out of doors whenever opportunity affords.

By far the most abundant of the rock-forming minerals are silicates, but before dealing with these a few other important minerals – oxides, carbonates, etc. – may conveniently be passed in review.

Quartz, SiO_2, has already been referred to as an oxide mineral although it has a framework structure with much in common with other silicates (see Table 4.3), and is described later.

Alumina, Al_2O_3, occurs naturally as **corundum**, the hardest natural abrasive after diamond, and in its rarer transparent form familiar as the gemstones ruby and sapphire.

In their large-scale occurrences the oxides of iron are the chief sources of iron ore (see Chapter 6); as accessories they are notable constituents in a great variety of common rocks. **Haematite**, Fe_2O_3, takes its name from the Greek word for 'blood' in reference to its colour. **Magnetite**, Fe_3O_4, is black and strongly magnetic. **Ilmenite**, $FeO.TiO_2$, is often associated with magnetite, especially in basalts and rocks of similar composition. **Limonite**, averaging about $FeO.OH$, is the rusty alteration product of other iron minerals.

Ice, crystalline water, H_2O, is not commonly thought of as a rock-forming mineral, but glaciers and ice-sheets are rocks on the grand scale composed of granules of ice, and they are none the less rocks because they flow, melt and evaporate before our eyes.

After water, the next oxide in order of abundance (Table 4.1) is that of phosphorus, an element of critical importance in agriculture, and indeed for life generally. Phosphates occur in ordinary rocks as the mineral **apatite**, $Ca_5F(PO_4)_3$, and as a related compound of organic derivation (e.g. from fish bones and teeth and the excreta of birds) which has a similar composition, but with (OH) instead of F. If water charged with traces of fluorine passes or soaks through this rock, known as **phosphorite**, the latter gradually gives up its (OH) in exchange for F, so approaching apatite in composition and becoming more stable. The same process tends to happen during the growth of children's teeth, and perhaps later, and this explains how it is that the addition of traces of fluoride salts (1–3 ppm) to drinking water helps to make teeth more resistant to decay in areas where the water supply is naturally deficient in fluorine. The chief fluoride mineral is **fluorite** CaF_2 (formerly known as fluorspar), which characteristically occurs in association with lead and zinc

ores; it appears in a variety of attractive colours and is mainly used as a flux in making iron and steel.

Among the sulphur minerals, sulphur itself and pyrite (FeS_2) (Plate IIc) is extremely common. Most of the important ores of lead, zinc, copper and nickel are sulphides. As rock-forming minerals, however, two sulphates are outstanding: **anhydrite**, $CaSO_4$, and **gypsum**, $CaSO_4.2H_2O$. Anhydrite, with or without gypsum according to circumstances, occurs mainly in salt deposits (evaporites), such as are left behind when salt lakes dry up, or when enclosed bodies of seawater are strongly evaporated. When the brine becomes sufficiently concentrated, **rock-salt** or **halite**, the chief chloride mineral, begins to be precipitated together with anhydrite.

Evaporation of seawater might be expected to begin with precipitation of carbonates, and so it does, but only on a trifling scale. Most of the carbonate rocks have different modes of origin, as we shall see later (Chapter 6). The chief minerals of these rocks are:

Calcite, $CaCO_3$, the predominant mineral of limestones

Dolomite, $CaMg(CO_3)_2$, which occurs intermixed with calcite in magnesian limestones and, essentially by itself, as the predominant mineral of a carbonate rock that is called dolomite

Siderite, ($FeCO_3$), once an important ore of iron.

Carbonates of lead, zinc and copper are locally prominent minerals in many of the ore deposits of these metals.

4.6 CRYSTAL STRUCTURES OF THE SILICATE MINERALS

Silicate minerals can be classified according to their crystal structure. The fundamental unit of silicate structures consists of a tetrahedral arrangement of four ions of oxygen – one at each corner of the tetrahedron – with an ion of silicon tucked into the interstitial space in the middle (Figure 4.7). Picture three tennis balls (6 cm in diameter) placed at the corners of a triangle and just touching; a marble about 1.6 cm in diameter then just fits into the central dimple (as in Figure 4.7, right); a fourth tennis ball placed on top (as in Figure 4.7, left) completes the structure. In this model the tiny silicon ion and the relatively large oxygen ions are magnified about 230 million times.

The four positive charges of Si^{4+} are balanced by four negative charges, one from each of the four oxygen ions, O^{2-}, thus leaving each tetrahedron with four negative charges. By themselves

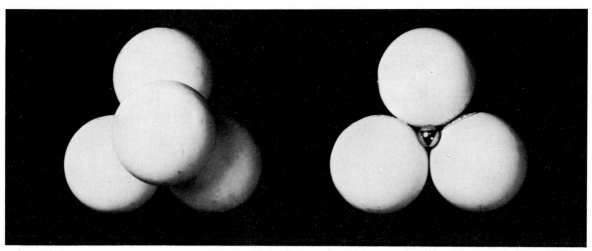

Figure 4.7 Models to illustrate the $(SiO_4)^{4-}$ tetrahedron, the fundamental atomic structural unit of all silicate crystals. A small ion of silicon is enclosed by four much larger ions of oxygen. On this scale the ions are magnified about 65 million times.

these tetrahedral building units would fly apart in consequence of the resulting electrical repulsion. To bind them strongly together, as in crystals, they must be cemented or linked so that the charges, are neutralized. The fascinating architecture of crystals is entirely controlled by the various ways in which this is accomplished. The tetrahedra may be held together by the proper proportion of metal ions (e.g. Mg^{2+} and Fe^{2+}, as in olivine); or they may be linked together by sharing (and thereby neutralizing) all four oxygen ions with their immediate neighbours, as in quartz; or they may share only one, two or three of their oxygens, leaving the remaining charges to be neutralized by metal ions of appropriate kinds. The linkages result in the building up of structures such as separate pairs and rings, single and double chains, and sheets (as illustrated conventionally in Figure 4.8) and finally as the more complex three-dimensional frameworks of quartz and feldspars – patterns spreading out in all directions – so that they cannot be clearly represented except in three dimensions.

The resulting classification is summarized in Table 4.3. As the number of shared oxygens increases, the proportion of O to Si necessarily decreases; this is emphasized in the table by adopting Si_4 as a unit of easy comparison throughout. In explanation of some of the mineral formulae it should be mentioned that Al plays a double role in crystals. It may replace Si inside some of the tetrahedra, as in the feldspars. Si_4 then becomes $(AlSi_3)$ or (Al_2Si_2), in which cases corresponding additions of suitable ions are necessary to balance the altered charges. A further complication arises in minerals that contain ions like hydroxyl, $(OH)^-$, or fluorine, F^-; these

may occupy relatively large 'hexagonal' spaces in the crystal structure, as in mica (Figures 4.8(h) and 4.11), or they may substitute for oxygen. It should be noticed that the ion hydroxyl, $(OH)^-$, does not imply that the 'water' introduced into the crystal structure has lost hydrogen; what happens is that the place of an oxygen ion is taken by two hydroxyl ions: $H_2O + O^{2-} = 2(OH)^-$.

It used to be said that the beauty of a crystal depends on the planeness of its faces. But now it has been realized that the beauty of a crystal is far from being only skin deep, as that old academic quip might suggest. The well-developed faces that are still commonly thought of as the most distinctive features of crystals are only the outward expressions of a pattern of ions cemented by their own electric charges. The internal architecture of the crystal edifice displays a natural beauty that is even more astonishing than the external façade, and one, moreover, that is equally present throughout the most minute fragment, however worn or broken it may be.

4.7 ROCK-FORMING SILICATE MINERALS

Feldspars are the most abundant minerals of the earth's crust, and, as we have seen, they consist of frameworks of SiO_4 – and AlO_4 – tetrahedra, with ions of potassium, sodium or calcium occupying the appropriate places in the structure. These minerals may therefore be considered as solid solutions of three ideal components which, as indicated below, are commonly distinguished from the minerals themselves by the use of the convenient symbols **Or**, **Ab** and **An**.

Table 4.3 Classification of silicate and aluminosilicate structures

Structural relationship of silicon–oxygen tetrahedra	Si : O ratio	Characteristic examples
NESOSILICATES or Island Silicates (Gr. *nesos*, an island) Separate SiO_4 tetrahedra, held together by ions such as Mg^{2+}. No oxygen shared (Fig. 4.8c)	$(SiO_4)^{-4}$	OLIVINE GROUP Forsterite Mg_2SiO_4 – Fayalite Fe_2SiO_4 GARNET GROUP Pyrope $Mg_3Al_2Si_3O_{12}$ ZIRCON $ZrSiO_4$
SOROSILICATES or Group Silicates (Gr. *soros*, a group) Separate pairs of tetrahedra, formed by sharing one oxygen (Fig. 4.8d)	$(Si_2O_7)^{-6}$	MELILITE GROUP Gehlenite $Ca_2Al_2SiO_7$ – Åkermanite $Ca_2MgSi_2O_7$
CYCLOSILICATES or Ring Silicates (Gr. *kyklos*, L. *cyclus*, a circle or ring) Separate closed rings of 3, 4 or 6 tetrahedra, formed by sharing two oxygens (Fig. 4.8e)	$(Si_6O_{18})^{-12}$	BERYL $Be_3Al_2Si_6O_{18}$ TOURMALINE $Na(Mg,Fe,Al)_3Al_6(Si_6O_{18})(BO_3)_3(OH,F)_4$
INOSILICATES or Chain Silicates (Gr. *inos*, a thread or fibre) Continuous single chains of tetrahedra, formed by sharing two oxygens (Fig. 4.8f)	$(SiO_3)^{-2}$	PYROXENE GROUP Hypersthene $(Mg,Fe)SiO_3$ Diopside $CaMgSi_2O_6$ – Hedenbergite $CaFeSi_2O_6$ Jadeite $NaAlSi_2O_6$
Continuous double chains, formed by the lateral coalescence of two single chains. Alternately two and three oxygens are shared (Fig. 4.8g). The resulting hexagonal spaces accommodate ions of $(OH)^-$ or F^-	$(Si_4O_{11})^{-6}$	AMPHIBOLE GROUP Anthophyllite $(Mg,Fe)_7Si_8O_{22}(OH,F)_2$ Hornblende $(Na,K)_{0-1}Ca_2(Mg,Fe^{+2},Fe^{+3},Al)_5(Si_{6-7}Al_{2-1}O_{22})(OH,F)_2$ Riebeckite $Na_2Fe_3^{+2}Fe_2^{+3}Si_8O_{22}(OH)_2$
PHYLLOSILICATES or Sheet Silicates (Gr. *phyllon*, a leaf) Continuous plane sheets of hexagonal networks, like wire-netting, formed by sharing three oxygens (Fig. 4.8h)	$(Si_2O_5)^{-2}$	TALC $Mg_3Si_4O_{10}(OH)_2$ SERPENTINE $Mg_3Si_2O_5(OH)_4$ CLAY MINERALS Kaolinite $Al_2Si_2O_5(OH)_4$
	$AlSi_3O_{10}$	MICA GROUP Muscovite $KAl_2(AlSi_3O_{10})(OH,F)_4$ Biotite $K(Mg,Fe)_3(AlSi_3O_{10})(OH,F)_4$
TECTOSILICATES or Framework Silicates (Gr. *tekton*, a builder or frame-maker) Continuous framework of tetrahedra in three dimensions; all four oxygens shared	SiO_2 $AlSi_3O_8$ $Al_2Si_2O_8$ $AlSi_2O_6$ $AlSiO_4$ $(Al,Si)O_2$	QUARTZ SiO_2 FELDSPAR GROUP Orthoclase $KAlSi_3O_8$ Albite $NaAlSi_3O_8$ Anorthite $CaAl_2Si_2O_8$ FELDSPATHOID GROUP Leucite $KAlSi_2O_6$ Nepheline $NaAlSiO_4$ ZEOLITE GROUP Natrolite $Na_2Al_2Si_3O_{10}.2H_2O$ Heulandite $CaNa_2Al_2Si_7O_{18}.6H_2O$

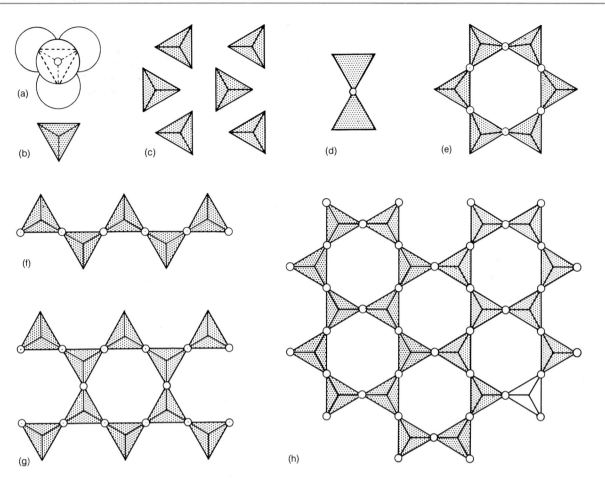

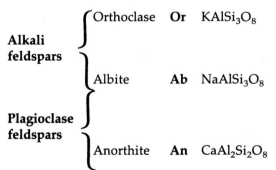

Figure 4.8 Some of the chief structural arrangements of the SiO₄ tetrahedron in crystals:

(a) The SiO₄ tetrahedron, with the ions approximately to scale.

(b) Conventional representations of the SiO₄ tetrahedron, shown by broken lines in (a). In (d) to (h) shared oxygens are indicated by open circles.

(c) No oxygens shared (e.g. olivine).

(d) A pair of tetrahedra sharing one oxygen (e.g. melilite).

(e) A ring of six tetrahedra, each sharing two oxygens (e.g. beryl).

(f) A single chain of tetrahedra, each sharing two oxygens (e.g. pyroxenes).

(g) A double chain of tetrahedra; the outward-pointing tetrahedra share two oxygens, as in (f), while those pointing inwards share three oxygens and so produce a succession of hexagonal 'holes' large enough to accommodate ions of hydroxyl (OH), or fluorine, F (e.g. amphiboles).

(h) A sheet of tetrahedra, each sharing three oxygens and forming a continuous network with hexagonal 'holes' as in (g) (e.g. mica).

Alkali feldspars $\left\{ \begin{array}{lll} \text{Orthoclase} & \textbf{Or} & \text{KAlSi}_3\text{O}_8 \\ \\ \text{Albite} & \textbf{Ab} & \text{NaAlSi}_3\text{O}_8 \end{array} \right.$

Plagioclase feldspars $\left\{ \begin{array}{lll} & & \\ \\ \text{Anorthite} & \textbf{An} & \text{CaAl}_2\text{Si}_2\text{O}_8 \end{array} \right.$

The proportions of these end-member components can be shown in the form of a triangular diagram (Figure 4.9).

In the **alkali feldspars** sodium and potassium can substitute continuously at high temperatures.

Solid solutions (the mineral **sanidine**) intermediate between the **Ab**-rich and **Or**-rich end members are often found in volcanic igneous rocks. At lower temperatures solid solution is limited and in rocks which crystallize at such temperatures separate **Ab**-rich (**plagioclase**) and **Or**-rich (**orthoclase** or a polymorph called **microcline**) feldspars can form. As they cool, solid solutions formed at high temperatures may break down, in the solid state, into slab-like intergrowth of **Ab**- and **Or**-rich feldspar called perthite, visible under the microscope. Some perthites exhibit optical iridescence, caused by the diffraction of light, as seen in the familiar ornamental building stone, larvikite. By looking at such complex mineral microtextures,

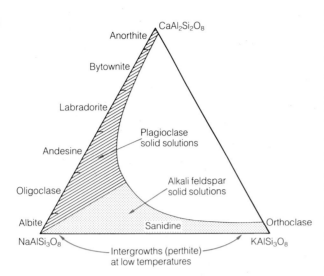

Figure 4.9 Subdivisions of the feldspar group. Feldspars are solid solutions of the three end-member components Ab, An and Or. Only compositions in the shaded areas occur naturally.

modern mineralogists can make interesting deductions concerning the cooling history of minerals.

Ab and **An** also form a continuous solid solution series known as **plagioclase feldspar** (Figure 4.9). The series is divided by convention as indicated in the Figure. Because of its abundance and chemical range, plagioclase is of great importance in classifying rocks. Plagioclase crystals stable at low temperatures probably contain slabs of **Ab**- and **An**-rich structure on a submicroscopic scale and this sometimes leads to iridescence, as in some **labradorite**, (named after Labrador), the characteristic feldspar of basalts. **Andesine** shares its name with that of the volcanic rocks in which it is most abundant: rocks called andesite because of their common occurrence as lavas erupted from the volcanoes of the Andes.

Orthoclase can be easily recognized as the pinkish or cream-coloured mineral in granite. In some granites, like those of Cornwall and Shap Fell (Plate IIe), large slab-like crystals of orthoclase, 2.5 cm or more in length, are sprinkled through the rock. When the crystals are broken across, the surfaces are smooth and glistening. Orthoclase does not break anyhow; it 'cleaves' along parallel planes in the crystal structure across which cohesion is comparatively weak. Just as in many wallpapers the repetition of the unit pattern gives rise to a parallel series of 'open' lanes or bands, so in the atomic pattern of a crystal there may be similar 'open' planes, and it is along these that the crystal splits most readily. Orthoclase has two sets of cleavage planes, and the mineral takes its name from the fact that they are exactly at right angles

(Gr. *orthos*, normal or right; *klastos*, broken). Microcline is another common variety of potash feldspar and its name, meaning 'little slope', refers to the fact that the angle between its cleavages departs by half a degree from a right angle. The plagioclase feldspars also have two cleavages, but here the difference from a right angle is about 4°, varying with the composition from slightly less than 4° at the albite end of the series to slightly more at the anorthite end. Hence the name (*plagios*, oblique). The cloudy, translucent character of many feldspars has recently been shown to result from the presence of myriads of tiny pores. When these are absent feldspars are glassy or dark in colour, as in **larvikite**.

Orthoclase and other varieties of alkali feldspar are extensively used in making glass and glazes. Workable occurrences of these minerals are sometimes found in rocks called **pegmatites** (Gr. *pegma*, thick, coarse) because in them the constituent minerals have generally crystallized to an unusually large size; moreover, both the grain size and the distribution of the minerals may be extremely variable. Most pegmatites occur in the form of irregular dykes, or sheet-like veins, or lenticular bodies, consisting largely of quartz and alkali feldspars, commonly accompanied by mica (Figure 4.10). In bulk, therefore, their composition is granitic. But whereas the minerals are more or less uniformly distributed in granite, there are many pegmatites in which they occur as gigantic crystals or massive aggregates, sometimes in sufficient concentration for profitable quarrying or mining. Single crystals of orthoclase or microcline as big as a house, though exceptional, are not unknown

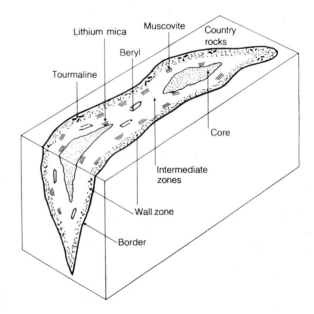

Figure 4.10 Characteristic structure of a lenticular body of zoned pegmatite.

(e.g. Norway and the Urals). Crystals of mica like giant 'books' of hexagonal shape have been found up to 3 and 5 m across (Transvaal and Ontario). Some of these commercially valuable pegmatites are also noteworthy in being veritable museums of beautifully developed rare minerals. Beryls as long as telegraph poles, and much thicker, have been quarried from some of the North American and Indian pegmatites. Sometimes open cavities are encountered in the larger pegmatites during quarrying, and springing from the walls of these some of the finest crystals of quartz, topaz, tourmaline and other showy minerals are found.

When feldspars are decomposed in the course of weathering, or by other processes involving the solvent action of water, practically none of the aluminium is lost by solution, and consequently the residual products become increasingly aluminous. The usual residues are extremely minute flakes consisting of (a) very fine-grained micaceous material called **sericite** or **hydromica**; or (b), when all the potassium (or sodium or calcium) has been lost, a **clay mineral**, of which there are several varieties. Most of the clay minerals are essentially hydrous silicates of aluminium, with formulae such as $Al_2Si_2O_5(OH)_4$, as in the kaolin group. The well-known china clay of Cornwall and Devon consists of kaolin formed by the decomposition of feldspar in granite. Some of the clay minerals, however, like those in fuller's earth and bentonite, are of more complex composition, having a small proportion of their Al^{3+} ions replaced by Mg^{2+}, plus K^+ or Na^+ to balance the valency.

Under appropriate conditions in tropical climates all the silica may be removed from feldspars by weathering. The residue then left is **bauxite**, a mixture of two aluminous minerals with the compositions $Al_2O_3.H_2O$ or $AlO(OH)$, and $Al_2O_3.3H_2O$ or $Al(OH)_3$. Bauxite is of great value as the only workable ore of aluminium.

Quartz is second in abundance to feldspar in granitic rocks, and because of its resistance to weathering is the most important constituent of clastic sediments (Sections 6.1, 6.2), occurring in concentrations as sandstone. Well-crystallized quartz is illustrated in Figure 4.4 and Plate IVa but when it occurs in rocks it is usually relatively shapeless. It is easily distinguished from feldspar by its glassy, colourless appearance and by its lack of cleavage. Natural quartz is chemically very pure SiO_2. It is hard and therefore used as an abrasive. It is also piezoelectric, which means that it deforms slightly when an electric charge is applied across the appropriate axis. Because of the perfect elastic properties of pure quartz, and its ability to vibrate at an exact rate, minute fragments are in everyday use in quartz clocks and watches.

Other important framework silicates include the **zeolite** group of minerals. These, soft, low density minerals occur naturally, often in mineral-lined cavities in basalts. They have the property of being able to take up water and release it when gently heated. They are important in the chemical industry as molecular sieves and as catalysts.

Micas are third in abundance amongst the minerals of granite. There are two leading varieties, one (muscovite) white, silvery and glistening, the other (biotite) dark and often bronze-like. Both are hydrous alumino-silicates of potassium, as the following formulae indicate:

White mica or **muscovite** $KAl_2(AlSi_3O_{10})(OH,F)_2$
Dark mica or **biotite** $K(Mg,Fe)_3(AlSi_3O_{10})(OH,F)_2$

Biotite forms a solid solution with Mg- and Fe-rich end members. The $KMg(AlSi_3O_{10})(OH,F)_2$ component is called **phlogopite**; the iron end member is called **annite**. The expression (OH,F) means that fluorine may replace hydroxyl (OH) to a limited extent in the crystal structure. In the formula for biotite (Mg,Fe) means that Mg and Fe^{2+} are similarly interchangeable; moreover, Fe^{3+} may take the place of some of the Al in biotite, a mineral that can obviously accommodate a wide range of composition within its crystal structure.

Micas all have a perfect cleavage, because their tetrahedral sheets and binding atoms are all arranged in parallel layers. Two layers having the structure represented in Figures 4.8h and 4.11 are tightly bound together like a sandwich, with an infilling of Al ions (in muscovite), or Mg ions (in phlogopite), or Mg and Fe^{2+} ions (in biotite). These 'sandwiches' are held together in turn, though rather more loosely, by layers of K ions, and it is along these 'weak' layers between successive 'sandwiches' that mica cleaves with such

Figure 4.11 Model of part of a continuous sheet of SiO_4 tetrahedra (cf. Fig. 4.8 (h)). Some of the 'top' oxygen ions have been removed to show the silicon ions. (*Trustees of the British Museum, Natural History*)

remarkable facility. The 'books' of mica extracted from certain pegmatites are so called because when the crystals are seen from the side they generally have the appearance of a thick pile of uncut pages. The cleavage is, indeed, so perfect that each crystal can be separated into sheets very much thinner – if so required – than the pages of this book. The cleavage flakes or sheets are both flexible and elastic. Large transparent sheets of muscovite have long been used for lamp chimneys and furnace windows. But the outstanding property that at one time gave mica, and especially muscovite, a high place amongst industrial minerals is its quality as an insulator.

Biotite serves to introduce the silicate minerals that are characterized by an abundance of magnesium and iron, and are therefore commonly known as ferromagnesian or mafic minerals. The leading groups are the **pyroxenes**, the **amphiboles**, and the **olivine** series.

Olivine, as indicated in Section 4.4, is part of a continuous solid-solution series which ranges in composition from Mg_2SiO_4 (forsterite) to Fe_2SiO_4 (fayalite). The common rock-forming variety, found in basalt and gabbro, contains more magnesium than iron, and is called olivine in reference to its usual olive-green colour. The mineral is familiar as the transparent green crystals which are cut as gemstones under the name **peridot**. Golden-yellow varieties used to be known by the old name **chrysolite** (Gr. *chrysos*, gold) but this name has fallen into disuse because it is so readily confused with **chrysotile**, the name of a variety of serpentine (see below) which occurs in the form of flexible fibres and is commercially valuable as a heat-resisting material – asbestos – that can readily be woven into fabrics.

Rocks in which olivine is the most abundant mineral (generally in association with pyroxenes) are called **peridotite**. Peridotite nodules are quite common in certain volcanic rocks and in kimberlite pipes (Section 5.4), and olivine is the commonest mineral in certain types of meteorite. Because meteorites are generally regarded as fragments of a shattered planet or planets, and because of the way in which peridotite occurs on Earth, it is generally accepted to be the main rock of the upper mantle. It also has the property of transmitting earthquake waves at the speeds observed by geophysicists. In the lower mantle, olivine probably changes its structure to a denser form similar to the mineral perovskite. Olivine and 'perovskite olivine' together are therefore the commonest substances on our planet. It is curious that Shakespeare, in a flash of poetic imagery, should have selected the same mineral when, through the mouth of Othello, he speaks of heaven making:

another world
Of one entire and perfect chrysolite.

Serpentine, $Mg_3SiO_5(OH)_4$, is formed from olivine and other magnesium-rich silicates by a process of alteration involving addition of water. Large bodies of peridotite may be partly or wholly replaced by serpentine giving a rock called serpentinite. This is composed of a tough but rather slippery felted mesh or network of minute crystals. The slippery quality is often caused by the presence of the softest of all minerals, **talc**, $Mg_3Si_4O_{10}(OH)_2$. Massive serpentine is generally mottled or variegated in shades of green and brown, like a serpent's skin, and some occurrences (e.g. at The Lizard in Cornwall) find a limited use as ornamental stones.

Pyroxenes are minerals of widespread occurrence in a great variety of rocks. In basalts (including olivine-basalts) and related rocks they are the most abundant constituents after plagioclase. In peridotites they are the most abundant minerals after olivine, and associated masses of rock in which pyroxenes are most abundant are distinguished as **pyroxenites**.

A simple classification scheme for pyroxenes, based on the proportions of Ca, Mg and Fe, is given as Figure 4.13. The simplest pyroxenes, known as **orthopyroxenes**, are **enstatite**, $MgSiO_3$ and **hypersthene**, $(Mg,Fe)SiO_3$. Chemically these are like members of the olivine series, but with more silica. Structurally, however, they are very different. All the pyroxenes are built of innumerable single chains of SiO_4 tetrahedra (Figures 4.8f and 4.12) with some substitution of AlO_4 tetrahedra in the more complex varieties. The chains are parallel and bound together by ions such as Mg^{2+}.

Other common pyroxenes, called **clinopyroxenes**, often contain calcium. **Diopside**, $CaMgSi_2O_6$, forms a solid solution series with **hedenbergite** $CaFe^{2+}Si_2O_6$. Intermediate members, which include the commonest pyroxene, **augite**, usually contain some alumina. This aluminium plays a double role. Where an AlO_4 tetrahedron occurs in a chain an extra valency has to be satisfied. To make up the deficiency and keep the crystal neutral, Al^{3+} or Fe^{3+} can then serve as binding ions between the chains instead of Mg^{2+}. Al^{3+} can also serve to bind together parallel chains of SiO_4 tetrahedra, provided it is associated with a monovalent ion like Na^+. Thus, instead of diopside, $CaMgSi_2O_6$, one can have $NaAlSi_2O_6$. Though very rare, this mineral actually occurs, and because it is beautiful as well as rare it is also familiar. It is, in fact, **jadeite**, the mineral of which jade is composed, i.e. the real jade of Upper Burma which is so greatly prized by

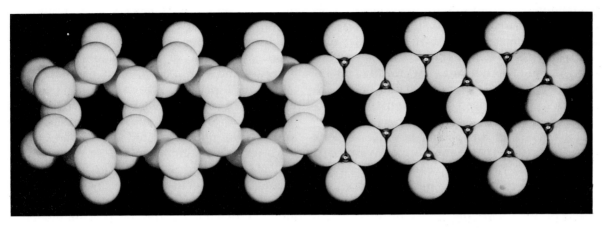

Figure 4.12 Model of part of a continuous double chain of SiO_4 tetrahedra (cf. Fig. 4.8(g)). (*Trustees of the British Museum, Natural History*)

the Chinese; there is also an amphibole, **nephrite**, which is commercially called 'jade'. Another soda pyroxene, known as **acmite** or **aegirine**, $NaFe^{3+}Si_2O_6$, makes a series with augite, the intermediate members of which are known as **aegirine-augite**. These substitutions, of Na^+, Fe^{3+} and Al^{3+}, cannot be shown on a simple diagram like Figure 4.13. Other substitutions include Ti^{4+} (for Si^{4+}, giving **titaniferous augite**) and Cr^{3+}.

Because of the importance of lithium in connection with atomic energy, another alkali-pyroxene has become extremely valuable in recent years. This is **spodumene**, $LiAlSi_2O_6$, now the chief source of lithium. It occurs in certain pegmatites and is one of those minerals which, though rare, is remarkable for the enormous size (e.g. up to 12 m

long) attained by some of its crystals.

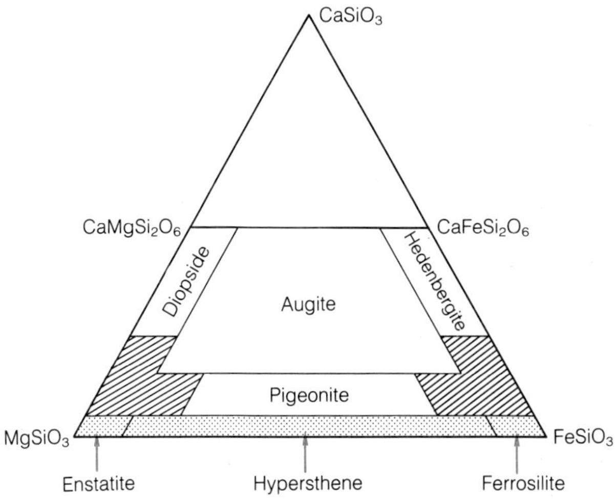

The **amphibole** group, represented mainly by hornblende and its many varieties, is more abundant and widespread than Table 4.2 suggests. Many rocks closely related to granite contain more hornblende than biotite; and amphiboles are especially abundant in the metamorphic rocks known as **amphibolite**. In composition the amphiboles are not unlike the pyroxenes, the essential difference being that the amphiboles all contain ions of hydroxyl, $(OH)^-$, which may be proxied by fluorine, F^-, as in mica. This is made possible by the fact that they are built of long double chains of tetrahedra, illustrated by Figures 4.8 and 4.12, which provide 'holes' (again as in mica) into which these large ions can fit. The resulting complexities of composition are well shown by the following comparisons:

PYROXENES		AMPHIBOLES	
Enstatite	$MgSiO_3$	**Anthophyllite**	$Mg_7Si_8O_{22}(OH)_2$
Diopside	$CaMgSi_2O_6$	**Tremolite**	$Ca_2Mg_5Si_8O_{22}(OH)_2$
Acmite	$NaFe^{+3}Si_2O_6$	**Riebeckite**	$Na_2Fe^{+2}_3Fe^{+3}_2Si_8O_{22}(OH)_2$

These three amphiboles and some of their varieties occur locally as workable masses of asbestos, known commercially as 'amphibole asbestos'. The fibres in asbestos may be extremely small (see Figure 4.5) and it is this feature which makes the substance a health risk. Most asbestos now mined is the mineral chrysotile (see above). The chemical composition of the amphibole group is extremely varied, and there are a large number of possible substitutions. For example, a characteristic composition for hornblende is approximately represented by the formula

$$(Na,K)_{0-1}Ca_2(Mg,Fe^{+2},Fe^{+3},Al)_5(Si_{6-7}Al_{2-1}O_{22})(OH,F)_2.$$

Substitutions such as F for (OH), Fe^3 for Al, and Ti for Si or $2Fe^2$ are general, and there are numerous 'end members', like the three listed above, with

Figure 4.13 Subdivisions of the calcium-magnesium-iron pyroxenes. The end-member molecule $CaSiO_3$ is not a pyroxene. Pyroxenes in the stippled area are known as **orthopyroxenes**, the remainder are **clinopyroxenes**. Compositions in the hatched areas do not occur in nature.

which limited amounts of solid solution are possible. It is a remarkable fact that hornblende contains notable amounts of all the common elements of the earth's crust except potassium.

Augite and hornblende are readily distinguished by their crystal forms and cleavages, both of which features depend on the internal architecture of the crystals. As in the pyroxenes, the amphibole chains are bound together in long parallel bundles. These bundles can be readily split along two particular planes – the cleavage planes – parallel to the length. Thus it happens that both minerals have two well-developed cleavages; but in augite the angle of intersection between them is 87° or 93°, appropriate to the single-chain structure, while in hornblende the angle is 56° or 124°, corresponding to the double-chain structure. Since these angles are nearly 90° and 120° (or 60°) they can be recognized and distinguished at a glance.

Just as olivine and enstatite are altered to serpentine by processes involving addition of water, so the Al- and Fe-bearing ferromagnesian minerals are altered to a group of greenish minerals collectively called **chlorite** (Gr. *chloros*, green, as in chlorine and chlorophyll). These have a double-sheet or 'sandwich' structure like the micas, but the 'sandwiches' are held together not by K, as in the micas, but by (OH), Mg or Fe^2, and Al or Fe^3. The particular composition depends mainly on that of the parent mineral. Because of its structure chlorite has an excellent cleavage, but the resulting flakes lack the flexibility of mica. Moreover, the individual crystals are usually very small. The chlorites bear much the same relation to ferromagnesian minerals as the clay minerals do to feldspars, alkalis and lime being lost in both cases, while the water that is added appears as $(OH)^-$ in the structure of the alteration product.

4.8 ORE MINERALS

Minerals containing metallic elements such as iron, lead, copper, zinc etc. are known as ore minerals and **ore deposits** are aggregates of particular minerals which can be mined and from which metals can then be extracted at a profit.

As noted in Section 4.3 and can be seen from Table 4.1 apart from aluminium and iron most of the metals used in industry occur in tiny amounts on average in our crustal rocks. Consequently there must be certain geological processes which can produce concentrations of metallic elements and the minerals which contain them. The study of these processes is now known as **metallogenesis**. For example, during the crystallization of ultrabasic and basic magmas (Chapter 5) sinking

of the heavier minerals may take place resulting in concentrations, often in distinct bands, in the resulting rocks, for example, of nickel, cobalt and copper sulphides, or of chromium and iron or titanium and iron oxides, or of platinum group alloys. Silica-rich magmas, on the other hand, may produce at a late stage of the solidification process tin and tungsten or copper and molybdenum minerals in quantities in certain parts of the intrusions, in pegmatites and in veins in the invaded country rocks.

Most mineral veins, however, have no obvious igneous source and result from the infilling with minerals of fault fissures by precipitation from **hydrothermal** (hot water) solutions carrying gold or lead and zinc, for example. The metallic minerals are accompanied by so-called **gangue** minerals (i.e. minerals that are not necessarily required commercially) such as quartz and calcite. The hydrothermal solutions may have a magmatic source or they may have been derived from the underground circulation of ground and connate waters (Chapter 19), which 'scavenged' the minute amounts of metals contained in the rocks in which they circulated before being driven upwards by convective processes due to high geothermal gradients. The discovery of hot 'springs' depositing metallic sulphides on the sea and ocean floors indicates that metals can be deposited along with accumulating sediments and this phenomenon probably explains the origin of some of the world's black-shale deposits rich in copper or lead and zinc sulphides, such as those in Germany, Zambia, or Queensland, Australia.

Weathering in hot humid climates can result in silicate minerals being chemically altered, the silica being removed in solution and 'residual deposits' such as bauxite (the ore of aluminium) being left behind (see Chapter 15). In other cases iron can be taken into solution, transported and eventually form various varieties of ironstones (see Chapters 6 and 24).

Placer deposits can be formed from the broken-up debris produced by normal weathering and erosional processes and can result in beach and river sands and gravels having concentrations of minerals such as gold, diamonds, zircon, ilmenite or monazite (a thorium-bearing mineral) due to the physical and chemical resistance of these particular minerals to being worn down or dissolved.

Metallogenetic studies, which incorporate mineralogy with petrology, sedimentology and geochemistry, are essential to ensure optimum exploitation of existing ore deposits and the continuing discovery of the new deposits necessary to meet the increasing demands of the world's industries.

4.9 FURTHER READING

Bragg, W.L., Claringbull, G.F. and Taylor, W.H. (1965) *Crystal Structure of Minerals*, Cornell University Press, New York.

Cox, K.G., Price, B.N. and Harte, B. (1988) *The Practical Study of Crystals, Minerals and Rocks*, McGraw-Hill, London.

Deer, W.A., Howie, R.A. and Zussman, J. (1966) *An Introduction to the Rock Forming Minerals*, Longman, London.

Dixon, C. (1979) *Atlas of Economic Mineral Deposits*, Chapman & Hall, London.

Edwards, R. and Atkinson, K. (1986) *Ore Deposit Geology*, Chapman and Hall, London.

Guilbert, J.M. and Park, C.F. (1986) *The Geology of Ore Deposits*, W.H. Freeman, New York.

Hurlbut, C.S., Jr. (1970) *Minerals and Man*, Thames and Hudson, London.

Hurlbut, C.S., Jr. and Klein, C. (1985) *Manual of Mineralogy*, 20th edn, John Wiley, New York.

Mackenzie, W.S. and Guilford, C. (1980) *Atlas of Rock-forming Minerals in Thin Section*, Longman, London.

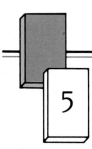

5

IGNEOUS ROCKS: VOLCANIC AND PLUTONIC

Rocks, like everything else, are subject to change and so also are our views on them.

F.Y. *Loewinson-Lessing*, 1936

5.1 DEFINITIONS

Igneous rocks are those which have solidified from hot, molten material called **magma**. Sometimes magma is erupted onto the Earth's surface from volcanoes as lava flows or, if it contains dissolved volatile material, it may be ejected violently as volcanic ash (Chapter 13). Most magma, however, never reaches the surface but solidifies underground to form bodies of igneous rock (**intrusions**) with a large variety of shapes and sizes. At one extreme the magma may solidify in small cracks to form intrusions only a few centimetres wide. At the other it may accumulate in vast subterranean magma chambers measuring tens of kilometres. The mechanisms of emplacement of magma and the forms of the resulting igneous intrusions will be discussed in Chapter 12.

It is convenient to divide igneous rocks into two major types. Those formed by eruption of magma at the Earth's surface, either as lava flows or as ash, are referred to as **volcanic** rocks. When magma solidifies well below the Earth's surface, the resulting igneous rock is said to be **plutonic**. In practice the term plutonic is reserved for the larger igneous bodies (hundreds of metres and larger). Smaller intrusive bodies (centimetres to tens of metres) which have cooled at shallow depths are described as **hypabyssal**.

5.2 RATES OF COOLING AND GRAIN SIZE

Magma cools by losing heat to its surroundings. The rate at which heat is lost from a body of magma depends upon the ratio of its surface area to its volume, and upon the temperature and thermal conductivity of the materials around it. Thin, tabular bodies of magma will clearly have a much larger surface area to volume ratio than will large, roughly spherical or cylindrical plutonic masses, and will cool much more quickly as a result. The tendency for larger bodies to be emplaced in the deeper (and therefore hotter) parts of the crust, and for smaller bodies to be emplaced at shallower levels will enhance this effect. Lava flows and ash particles will tend to cool more rapidly than intrusive bodies. Thus, in general, volcanic rocks will cool faster than hypabyssal rocks which will cool faster than plutonic rocks. Thin lava flows may crystallize completely in a matter of hours whereas plutonic bodies may take tens of thousands of years.

When magma cools it will reach a temperature at which crystals of one or more of the rock-forming minerals will start to grow. The actual mineral species formed will depend upon the bulk composition of the magma. As was shown in Chapter 4, crystals form when atoms join together into regular structures. Liquids (including magmas) consist of atoms which are in constant random motion with respect to each other. For these atoms to form crystals they must move into a more ordered arrangement, and the slower a liquid cools, the more easily its component atoms can arrange themselves into regular crystal structures. Slow cooling will promote the growth of large, well-formed crystals since atoms will have the time to migrate from the regions around the developing crystal to take their place in the structure. With more rapid cooling, atoms will only have time to migrate over short distances before being trapped in a developing structure. Thus rapid cooling will encourage the growth of larger numbers of smaller crystals. At very rapid cooling rates it may not be possible for atoms to form crystal structures at all and they will become 'frozen' in their original random arrangement. The result will be an amorphous glass. Glasses are not stable indefinitely at low temperatures and may, in time, **devitrify** into a mosaic of tiny interlocking crystals.

A plutonic magma body crystallizing over a period of thousands of years will have the opportunity to form large crystals and will therefore cool to a coarse-grained (> 5 mm) igneous rock. At the other extreme, droplets of magma sprayed out of a volcano or lava quenched rapidly by contact with water may cool in a few seconds and form a glassy rock. Magmas rich in silica (such as those which form granite) can only crystallize if cooled very slowly and so even quite large flows of silica-rich lava may solidify to a glassy rock. In general, however, lava flows will tend to solidify to fine-grained (< 1 mm) igneous rocks. Between these extremes, hypabyssal igneous rocks forming small intrusions will tend to have crystals intermediate in size between those in plutonic and volcanic rocks.

If, as frequently happens, magma is stored in a large plutonic body long enough for crystals to form and is then either erupted onto the Earth's surface or emplaced as a hypabyssal intrusion, the result is a fine- or medium-grained igneous rock containing some larger crystals. These larger crystals, which formed while the magma was crystallizing slowly, are called **phenocrysts** and the finer-grained matrix, formed by more rapid cooling, is called the **groundmass**. The rock is said to have a **porphyritic** texture (Plates IIa, IIIb).

The grain size of an igneous rock is one factor used in the naming of rock types and it is common practice to assign different names to fine-, medium-, and coarse-grained varieties of rocks with similar bulk compositions. Taking basaltic rocks (the most common of all igneous rocks) as an example this gives us four rock types:

Tachylyte, basaltic glass often formed when basaltic magma is quenched rapidly on contact with water. **Palagonite** is devitrified and hydrated basaltic glass

Basalt (Plate IIIa), finely crystalline or porphyritic rock forming lava flows and very small intrusions

Dolerite (Plate IIIb) (**diabase** in the United States), medium-grained basaltic rock forming the larger hypabyssal intrusions

Gabbro (Plate IIIc), coarse-grained equivalent occurring in larger, plutonic intrusions.

It must be stressed that specimens of each of these four rock types would have very similar chemical compositions and, with the exception of tachylyte which is non-crystalline, similar mineral assemblages. They differ only in their grain size and hence in their rate of cooling. Rocks with different bulk compositions would have their own sets of names for varieties with different grain sizes. For example the names for glassy, fine-, medium-, and coarse-grained varieties of granitic (silica-rich) igneous rocks are, respectively, **obsidian**, **rhyolite**, **microgranite** and **granite** (Plates IIe, IIId).

The pressures prevailing in deep magma chambers are sufficient to allow the magmas to hold some water and other volatile components (such as carbon dioxide, fluorine and chlorine) in solution. Since these components enter the structure of only a few of the common igneous rock-forming minerals, such as mica and amphibole, and then only in small amounts, their concentration in the magma will tend to increase as crystallization proceeds. The result is that the concentration of volatile components in the last dregs of magma in a crystallizing pluton can exceed their solubility in the magma and an aqueous phase will separate as an immiscible fluid. This will be saturated with dissolved silicate material and will remain fluid long after all the silicate magma has crystallized. It will eventually precipitate its dissolved silicate minerals as veins or irregular bodies either within the pluton or in the surrounding rocks. Since atoms can move through aqueous fluids much more readily than through silicate magmas, the crystals which form can grow to a considerable size. The resulting rock, known as **pegmatite**, has conspicuously larger crystals than is normal for an igneous rock (Figure 4.10), sometimes over a metre in size. Since some trace elements present in the original magma tend to be concentrated along with the volatile components, pegmatites frequently contain rare minerals (Chapter 4). Pegmatites of this type occur mostly in and around granite plutons.

5.3 VESICLES AND AMYGDALES

Volatile components can only dissolve in magmas under pressure and will therefore escape from the magma when the pressure is released during eruption or emplacement in shallow intrusions. The violent release of gases is responsible for such dramatic effects as lava fountains and ash eruptions (Chapter 13). Gases continue to be released after eruption in much the same way as bubbles of carbon dioxide continue to form in a glass of beer after it has been poured. This accounts for the slaggy or frothy appearance of the tops of many lava flows and for the bubbles and irregular cavities usually present in otherwise compact volcanic rocks. Such bubbles are referred to as **vesicles** and frothy lava is described as **vesicular**.

The percolation of heated groundwater (Chapter 19) through a pile of lava flows will, in time, fill any vesicles with minerals. The filled bubbles sometimes look like almonds, and so the name **amygdales** is given to them (*amygdalos* is the Greek word for almond). Volcanic rocks that are studded with numerous amygdales are described as **amygdaloidal**. One of the commonest minerals found in amygdales is agate, banded in concentric layers of different tints (Plate IIb). Agate is a variety of chalcedony, a cryptocrystalline form of silica; and sometimes, inside a lining of agate, crystals of quartz or of amethyst (purple quartz) may be found projecting into the hollow space within. The occurrence of these minerals in a basalt suggests that there was some free silica left over after the crystallization of the rock. One would not expect this to happen in olivine basalts because the presence of olivine implies a deficiency in silica. For this reason olivine basalts generally have calcite and zeolites (Table 4.3) in their amygdales.

5.4 ORIGIN OF MAGMAS

Active volcanoes testify to the existence of bodies of magma at depth but this does not imply that all of the interior of the Earth is molten. Indeed, we know from seismic evidence that the Earth is solid down to the outer core, *c.* 2900 km below the surface (Figure 2.3 and Chapter 25). From this we infer that magma production must be a localized phenomenon and this raises the questions of where and how magma is produced.

Volcanoes are not randomly distributed over the globe but tend to occur in distinct belts, usually associated with earthquake activity. Two of the main environments in which volcanoes occur are along mid-ocean ridges (e.g. Iceland) and in island arcs (e.g. Japan) (Figure 13.13). These are the regions where the great lithospheric plates forming the outer layers of the Earth are respectively created and destroyed (see Chapter 28). Some volcanoes, however, are not related to plate boundaries but form isolated 'hot spots' in the interiors of plates (see Chapter 13). The Hawaiian islands are perhaps the best examples of such intraplate volcanoes. The restricted occurrence of active volcanoes provides us with important clues as to why magmas are produced within the Earth.

We know from boreholes and deep mines that the temperature in the Earth increases with depth. The average rate of temperature increase (the **geothermal gradient**) in regions remote from volcanic activity is about 20 to 30°C per kilometre. If this rate were maintained through the outer layers of the Earth, the melting point of basaltic rocks would be reached at the base of the continental crust. However, heat flow in continental regions is dominated by the decay of radioactive elements in the upper part of the crust so that the geothermal gradient in the lower crust and upper mantle (Figure 2.3) is much less than at the surface.

The likely temperature distribution in the outer parts of the Earth (the **geotherm**) is shown in Figure 5.1. The geotherm has been constructed on the assumption that heat transfer is dominated by conduction through the rigid lithosphere but by convection through the mobile asthenosphere. This sketch also shows the experimentally-determined melting point of **peridotite**, the rock of which the Earth's mantle is thought to be composed. For melting to occur the geotherm must intersect this melting curve but, in the case illustrated, it doesn't. This sketch represents the situation over most of the Earth.

Figures 5.2, 5.3 and 5.4 show the various ways in which the mantle can be induced to melt. In Figure 5.2a the lithosphere has been stretched and thinned and the asthenosphere has been allowed to flow into the space created. In rising nearer to the surface without much loss of temperature, the mobile asthenosphere has crossed the melting curve and started to melt. This process accounts for the production of magma in continental rift valleys and sedimentary basins (Chapter 29). Note that the geotherm only just crosses the melting curve. Magmas produced in this situation will originate at a depth of about 60 km by at most only a few per cent of melting of the mantle. Figure 5.2b shows the situation at mid-ocean ridges (Chapter 26). These can be regarded as an extreme case of lithospheric extension in which the asthenosphere is allowed to rise almost to the surface with the consequent production of the huge amounts of magma needed to create the oceanic crust (about 20 km^3 per year, globally). The geotherm crosses the melting curve to a much

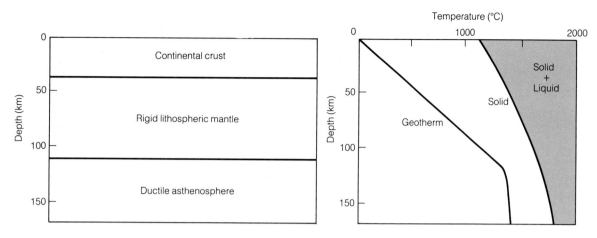

Figure 5.1 The temperature distribution (geotherm) through the outer parts of the Earth in non-volcanic regions. The temperature at which the mantle begins to melt is shown as a brown curve. Note that the temperature does not get high enough to cause the mantle to melt. (*After R.S. White and D.P. McKenzie*)

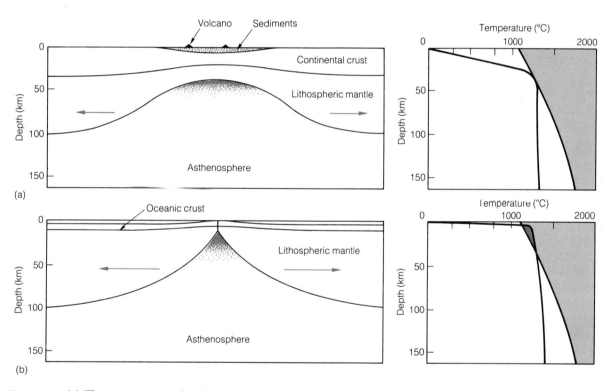

Figure 5.2 (a) The temperature distribution beneath the axis of a sedimentary basin or rift valley in which the lithosphere has been stretched and thinned. The asthenosphere has flowed upwards to fill the space created and consequently the geotherm has been brought into contact with the peridotite melting curve, causing a small amount of melting to occur (shown brown in the left diagram). (b) The situation beneath a mid-ocean ridge where lithospheric plates are being pulled apart and asthenospheric mantle is rising almost to the surface. The geotherm crosses the melting curve to a much greater extent than in (a) with the consequent production of huge volumes of magma. (*After R.S. White and D.P. McKenzie*)

greater extent and at a shallower depth than beneath rift valleys (Section 29.5) with the result that the magmas are produced at only a few kilometres depth by more than 10 per cent melting.

In Figure 5.3 the asthenosphere is hotter than normal. This situation is thought to occur in regions where convective plumes of anomalously hot mantle rise from great depths, as beneath Hawaii. The exceptionally vigorous volcanic activity in Iceland is thought to be due to the superimposition of a mid-ocean ridge on a hot mantle

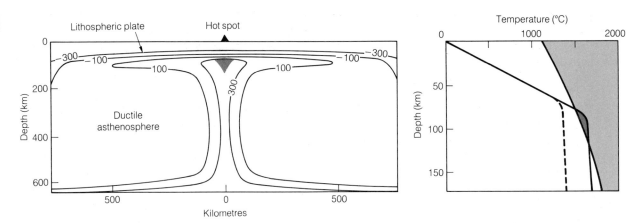

Figure 5.3 The temperature distribution beneath a convecting plume of anomalously hot mantle. The contours show the deviation (in degrees Celsius) from the average upper-mantle temperature of 1340°C. The geotherm oversteps the melting curve and the mantle melts (the normal geotherm is shown as a broken line for comparison). The partially-melted region is shown in brown on the left diagram. Note that the vertical scales on the left and right diagrams are not the same. (*After R.S. White and D.P. McKenzie*)

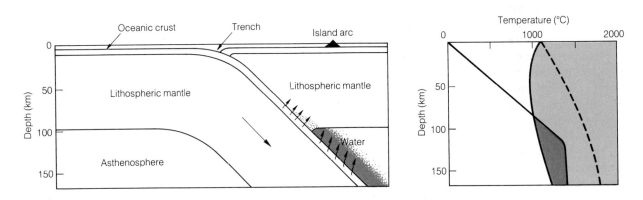

Figure 5.4 A sketch section across a subduction zone where a plate of oceanic lithosphere is returned to the convecting asthenosphere and ultimately consumed. Water released by dehydration of the descending plate lowers the melting point of the overlying mantle and causes melting. The resulting magma is responsible for the formation of a volcanic island arc. The solid line shows the melting behaviour of the mantle in the presence of water. The dry melting curve is shown for comparison as a broken line.

plume; i.e. a combination of Figures 5.2b and 5.3. The superimposition of a continental rift system on a hot mantle plume similarly produces large volumes of basaltic magma (flood basalts; Chapter 13).

Figure 5.4 illustrates the situation in those regions (**subduction zones**) where lithospheric plates are returned to the convecting mantle. These ought to be the least volcanic areas on Earth because subduction takes the cold rocks of the ocean floor down into the mantle. They are, however, some of the most volcanically active areas (the Circum-Pacific belt of volcanoes, for example) (Figure 13.13). Figure 5.4 shows how this apparent paradox is resolved. Along with the sea floor, subduction takes enormous volumes of seawater down into the mantle. This water has the effect of lowering the melting point of the mantle and

causing it to melt. The eventual release of this water, when the resulting magmas reach the surface, accounts for the extraordinary violence of volcanic eruptions above subduction zones (Chapter 13).

5.5 PRIMARY AND EVOLVED MAGMAS

Magmas produced by partial melting of the mantle and whose composition has not been changed by subsequent crystallization or assimilation of crustal rocks are referred to as **primary magmas**. These are rarely erupted at the surface or preserved in intrusions. More commonly, magmas undergo considerable amounts of crystallization on their way through the elaborate system of conduits and magma reservoirs en route to the surface. The crystals produced will tend to be left behind and the magma will therefore change in composition before it reaches the Earth's surface. Crystallization with separation of the crystals from residual liquid is referred to as **fractional crystallization** and the resulting magmas are said to be **evolved**. Fractional crystallization is probably the most effective process of **differentiation**, whereby magmas, and hence igneous rocks, with a wide range of composition can be generated. This process will be discussed in the next section.

We know that the mantle is the site of generation of primary magmas because earthquakes thought to be associated with the generation and transport of magma beneath active volcanoes extend down into the mantle. Magmas frequently carry fragments of rock broken from the walls of the conduits along which they flow. These are referred to as **xenoliths** (derived from the Greek for 'foreign rock') and are most commonly composed of sedimentary, metamorphic and igneous rocks from the crust. Some xenoliths, however, are made of peridotite, whose component minerals (olivine, pyroxene and garnet or spinel) suggests that it originates at the pressures prevailing in the mantle. Peridotite is a good candidate for the composition of the Earth's mantle since it will transmit earthquake waves at the same velocity as actually observed in the mantle (Chapter 26). Peridotite xenoliths, therefore, provide us with direct information on the composition and mineralogy of the Earth's mantle.

Laboratory melting experiments carried out at high pressures and temperatures have shown that the partial melting of peridotite yields silicate liquids similar in composition to basalt. It seems likely, therefore, that basaltic magmas are produced by the partial melting of the mantle. Most basaltic magmas, however, will have cooled and crystallized to some extent before eruption and will not be strictly primary. Because of this uncertainty, basaltic magmas are usually described as **primitive** rather than primary.

Another result of laboratory melting experiments is the demonstration that variation in the amount of melting of mantle peridotite is reflected in compositional variation in the resulting basaltic magma. Large-degree (> 10 per cent) melting of peridotite produces basaltic magma which is richer in silica than that produced by small-degree (< 1 per cent) melting. This explains why the voluminous basaltic lavas erupted at mid-ocean ridges and over vigorous mantle plumes tend to be relatively silica rich (~50 per cent SiO_2) whereas small flows of silica-poor basaltic lava (40 to 45 per cent SiO_2) are found in continental rift valleys where mantle melting is much less extensive. The large-degree melting associated with mid-ocean ridges and mantle plumes produces magmas of an extremely uniform composition. By contrast, small-degree melting of the mantle beneath some continental rift valleys is responsible for the production of an extraordinarily wide range of silica-poor igneous rock types such as nephelinite, melilitite and carbonatite (Chapter 29). **Kimberlite** is perhaps the most extreme product of small-degree melting of the mantle. This rock type is confined in occurrence to cratonic areas (i.e. areas of the Earth's crust which have undergone minimal tectonic deformation since the Precambrian), where the geothermal gradient is extremely low and, as a consequence, any mantle melting must take place at great depth. Kimberlite frequently contains diamond which means that it must originate from at least 150 km depth.

5.6 MAGMATIC DIFFERENTIATION

Primary basaltic magma is less dense than the mantle but denser than average continental crustal rocks. It will therefore tend to rise through the mantle but may become trapped in magma reservoirs or chambers within the crust. Here it will tend to stagnate, lose heat to its surroundings, and crystallize. Because heat is lost by conduction through the surrounding rocks, the first crystals will form in magma close to the walls and roof of the magma chamber. Most of the mineral species likely to crystallize from basaltic magma are denser than the magma itself and will tend to settle to the floor where they will accumulate, often in strikingly layered deposits resembling sedimentary strata. Rocks formed by the accumulation of crystals in a magma chamber are referred to as **cumulates**. Figure 5.5 is a sketch section through a crystallizing magma chamber.

The layered rocks forming Hallival and Askival on the Hebridean island of Rum (Figure 5.6, Plate VIIIb, VIIIf) are examples of cumulates. In this case the layers are made up of alternating olivine- and plagioclase-rich igneous rocks. This tendency for individual mineral species to concentrate in layers is partly due to 'sedimentary' processes operating on crystals settling through the magma. Valuable minerals such as chromite (the principal ore of chromium) are often concentrated in thin layers within sequences of gabbroic igneous cumulates and sometimes form economic ore deposits. A large proportion of our reserves of chromium, nickel, and the platinum-group elements occur in large layered cumulate bodies (Section 12.00).

The mineral species making up a basaltic igneous rock will not crystallize simultaneously from a magma but in a particular order. Returning to our laboratory melting experiments, it is found that

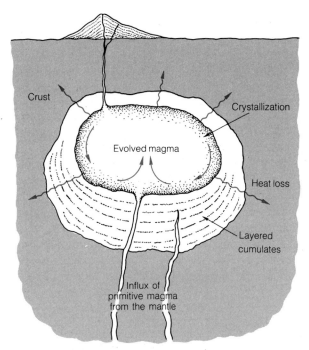

Figure 5.5 A sketch section through a magma chamber. Magma loses heat through the walls and roof of the chamber and begins to crystallize. The crystals accumulate on the floor of the magma chamber to form layered cumulate deposits. The magma left in the chamber will change in composition as crystallization proceeds and may periodically escape to the surface to produce volcanoes. From time to time the chamber may be refilled with primitive magma from the mantle.

olivine is always the first mineral to crystallize when a primitive basalt sample is melted and then cooled. Plagioclase and pyroxene follow at lower temperatures, most often in that order. This is consistent with the observation that basalt lavas frequently carry only olivine phenocrysts but never only plagioclase or pyroxene phenocrysts. These last two always occur with olivine. Thus the common phenocryst assemblages found in basalt lavas is olivine, olivine + plagioclase (or olivine + pyroxene), and olivine + plagioclase + pyroxene. Similarly, gabbros commonly have well-formed crystals of olivine and plagioclase with irregular, interstitial grains of pyroxene, reflecting the same order of crystallization (Plate IIIc).

It is not just the crystallizing mineral assemblage that changes as a basaltic magma cools; the composition of the individual mineral species changes too. With falling temperature the olivine composition will change from magnesium-rich (close to forsterite) to more iron-rich (towards fayalite). At the same time, plagioclase will change in composition from labradorite (calcium-rich) towards albite (sodium-rich). Olivine, pyroxene and plagioclase are all examples of solid solution series (Chapter 4).

At an early stage in the crystallization of a batch of primitive basaltic magma the only crystals forming and sinking to the floor or sticking to the walls of the magma chamber will be Mg-rich olivine. These will have a much higher MgO content than the magma and a slightly lower SiO_2 content. The effect of removing olivine from the magma will be to cause a dramatic decrease in MgO in the remaining magma. SiO_2 and all other major components will increase in concentration. The onset of plagioclase crystallization will be marked by a drop in the CaO content of the residual magma. Once pyroxene (augite; Chapter 4) begins to crystallize, its relatively low SiO_2 and high CaO contents will enhance the increase in SiO_2 in the residual magma while the decrease in CaO content will accelerate. If oxide minerals such as magnetite (Fe_3O_4) begin to crystallize the concentration of SiO_2 in the residual magma will increase more sharply while the FeO and Fe_2O_3 content will begin to fall. As crystallization proceeds the olivine and pyroxene will both become more iron-rich and contribute to the decline in the concentration of FeO in the residual magma.

The result of the crystallization of olivine, plagioclase, pyroxene and magnetite from a basaltic magma is that the contents of MgO, CaO and FeO in the residual magma will fall as temperature drops and crystallization proceeds. The content of Al_2O_3 will remain roughly constant while the Na_2O will rise slightly (both are controlled by plagioclase crystallization) and SiO_2 and K_2O will rise steeply. The result will be a magma rich in SiO_2, Al_2O_3, Na_2O and K_2O but very poor in MgO, CaO and FeO. This will crystallize to an assemblage of orthoclase and albite (feldspars) and quartz and form a granite or rhyolite depending upon whether it crystallizes slowly underground or is erupted.

In this way the entire spectrum of igneous rock types from primitive basalt to evolved rhyolite and granite can be produced by a process of fractional crystallization. Half way through this process the residual magma will have a composition intermediate between the two extremes. If such a magma is allowed to crystallize it will form an intermediate igneous rock such as andesite (fine grained) or diorite (coarse grained).

It was noted earlier that primary basaltic magmas generated by small-degree melting of the mantle are deficient in silica with respect to those generated by large-degree melting. The evolved magmas resulting from the differentiation of silica-deficient basaltic magma may never become sufficiently silica-rich to crystallize quartz. These evolved magmas may crystallize feldspars alone to form a trachyte (volcanic) or syenite (plutonic). If they are particularly deficient in silica they may

Figure 5.6 Layered igneous rocks forming Hallival and Askival on the island Rum. These rocks formed as cumulates on the floor of a magma chamber about 60 million years ago and have subsequently been exposed by erosion. More resistant layers rich in plagioclase alternate with more easily eroded, olivine-rich layers.

also crystallize nepheline to form phonolite (volcanic) or nepheline syenite (plutonic).

Although differentiation through fractional crystallization is the dominant mechanism whereby the compositional diversity of igneous rocks is produced, it is not the only mechanism. Hot magma stored for tens of thousands of years in magma reservoirs is likely to interact with its surroundings in some way. Melting and assimilation of the surrounding rocks can change the composition of a magma body. Since crustal rocks tend to be silica-rich this has the effect of accelerating the evolution of the magma. The influx of new basaltic magma into an evolving magma reservoir would have the opposite effect.

If crustal rocks are heated, either at the margins of a large magma reservoir or in the roots of a developing mountain belt (Chapters 7 and 31), they may melt to produce magma directly. This magma will be produced by the melting out of those components with the lowest melting points and will tend to be granitic in composition. It is therefore possible to produce granitic igneous

rocks directly from the partial melting of crustal rocks rather than from the mantle via basaltic magma and differentiation. Some of the large granite batholiths found in eroded mountain ranges may have been produced in this way.

5.7 CLASSIFICATION OF IGNEOUS ROCKS

Igneous rocks have silica contents ranging from about 40 per cent (basalt) to 75 per cent (rhyolite). The silica content of an igneous rock used to be called its 'acidity' since silicates were regarded as compounds of basic oxides (e.g. MgO, CaO) and silicic acid. This term is no longer used but still forms the basis for a useful classification of igneous rocks. Silica-rich igneous rocks (SiO_2 > 66 per cent) are referred to as '**acid**' and silica-poor rocks (SiO_2 45 to 53 per cent) as '**basic**'. Rocks with silica contents between 53 per cent and 66 per cent are '**intermediate**'. Some igneous rocks are composed largely of olivine and pyroxene and are, therefore, exceptionally rich in 'bases' and poor in silica

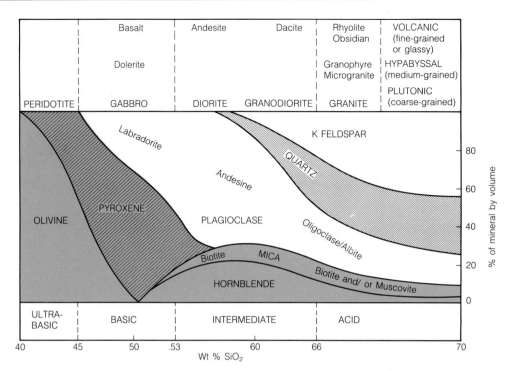

Figure 5.7 A classification scheme for the more common igneous rocks. These are classified on the basis of the proportions of their component minerals and on their grain size. The former reflects the degree of evolution of the parental magma (from basic to acid) and the latter reflects the rate of cooling of the magma. Mafic (dark-coloured) minerals are shaded brown, felsic (pale-coloured) minerals are unshaded. For more unusual rock types, see page 697.

($SiO_2 < 45$ per cent). Such rocks are regarded as **'ultrabasic'**.

These terms are useful as a basis for classification of igneous rocks since they reflect the degree of evolution undergone by the magma. Thus basic rocks are primitive whereas acid rocks are evolved. It is generally possible to recognize ultrabasic, basic, intermediate and acid rocks in hand specimens on the basis of the proportion of dark (ferromagnesian or Fe- and Mg-rich) minerals present. However, it does not necessarily follow that all rocks rich in dark minerals are basic, nor that all pale-coloured rocks are evolved, though this is a useful rule of thumb. To be strictly accurate, a parallel, and looser nomenclature is required when describing hand specimens. Thus igneous rocks rich in ferromagnesian minerals (generally dark coloured) are commonly described as **'mafic'** whereas those rich in feldspars (generally pale coloured) are described as **'felsic'**. The hand specimen terms 'ultramafic', 'mafic' and 'felsic' are broadly equivalent to the more rigorous, compositionally-based, 'ultrabasic', 'basic' and 'acid'. The term 'intermediate' is used in both schemes.

The relationship between composition (SiO_2 content) and proportion of dark and pale minerals is illustrated in the classification scheme shown in Figure 5.7. This diagram shows the proportions of minerals present in some of the more common igneous rocks as a function of their silica contents.

It also classifies the rocks according to grain size, and hence rate of cooling. This classification scheme is by no means exhaustive but covers the more common igneous rocks and serves to illustrate the principles involved. This scheme is appropriate for those igneous rocks derived, through fractional crystallization, from relatively silica-rich primitive basaltic magmas. A similar scheme can be drawn up for the volumetrically less-important group of rocks derived from silica-deficient basaltic magma (e.g. trachyte, phonolite, syenite). This, however, is beyond the scope of this book.

5.8 FURTHER READING

Best, M.G. (1982) *Igneous and Metamorphic Petrology*, W.H. Freeman, New York.

Hatch, F.H., Wells, A.K. & Wells, M.K. (1972) *Petrology of the Igneous Rocks*, Murby, London.

MacKenzie, W.S., Donaldson, C.H. and Guilford, C. (1982) *Atlas of Igneous Rocks and Their Textures*, Longman, Harlow, Essex.

McBirney, A.R. (1984) *Igneous Petrology*, Freeman, Cooper & Co., San Francisco.

Middlemost, E.A.K. (1985) *Magmas and Magmatic Rocks*, Longman, Harlow, Essex.

Thorpe, R. and Brown, G. (1985) *The Field Description of Igneous Rocks*. Geological Society of London Handbook Series, Open University Press, Milton Keynes.

SEDIMENTARY ROCKS

6

Sufficient for us is the testimony of things produced in the salt waters and now found again in the high mountains, sometimes far from the sea.

Leonardo da Vinci (1452–1519)

6.1 SANDSTONES

Sandstone is perhaps the most familiar of all rocks, for it is easily quarried, and it has been used more than any other kind of natural stone for building purposes. Examined closely, using a lens if necessary, a piece of sandstone is seen to consist of grains of sand identical in appearance with those that are churned up by the waves breaking on a beach. Most of the grains consist of more or less rounded grains of quartz, but often there are others of cloudy, weathered-looking feldspar (Figure 6.1 and Plate IIIe), and generally a few shining spangles of mica.

Clearly, sandstone is made of second-hand materials, of worn fragments derived from the disintegration of some older rock, such as granite, which contained the same minerals. It differs from deposits of modern sands only in being coherent instead of loose. Calcite is a common cementing material. Brown sandstones are cemented by limonite and red varieties by haematite. In white, extremely hard sandstones, the cement is silica which has crystallized as quartz. These cementing materials were deposited between the grains by groundwaters which percolated through the sand when it was buried under later sheets of sand or other formations.

White siliceous sandstones in which most of the grains, as well as any cement which may be present, consist of quartz, are often referred to as **orthoquartzite** (to distinguish the sedimentary rocks from the metamorphic, **quartzite**) or **quartz arenite**. However, when an area formed of granites or other feldspar-bearing rocks, is being eroded, climate and topography may be such that all the feldspars are not weathered into clay minerals. When the disintegrated debris is deposited, grains of feldspar, commonly orthoclase or microcline, may be abundant in the resulting sands. The term **arkose** is applied to such sandstones rich in feldspar. Another important type of sandstone is known as **greywacke**, a German name indicating a grey mixed rock and first given to rocks in the Harz Mountains by local miners. It was adopted by Werner and introduced to Britain by Robert Jameson, a professor in Edinburgh in the early nineteenth century. Greywackes differ from other sandstones in a) being less well sorted, i.e. the grains are of widely differing sizes, b) having larger grains held together by a clay matrix rather than introduced mineral cement like calcite or quartz, and c) having a wide variety of constituents as grains and matrix (Plate IIIf). For example, its grains may include not only the common quartz, feldspars and micas, but bits of incompletely weathered ferromagnesian minerals and fragments of rocks from the region that

(a)   (b)

Figure 6.1 Photomicrographs of sandstones (under crossed polars): (a) Quartz arenite – a highly siliceous sandstone consisting of grains of quartz cemented by quartz cement which has grown around original grains to make a very compact quartz mosaic. Carboniferous, Wales (*see also* Plate IIId). (b) Greywacke – a poorly-sorted sandstone with a clay matrix and a mixed composition. Silurian, Scotland (*see* Plate IIIf).

was undergoing erosion, including volcanic types in some areas. The matrix is fine-grained and largely composed of flaky minerals which are alteration products of weathering. Some of these were deposited at the same time as the larger grains but other components of the matrix were formed after deposition by alteration of unstable minerals such as feldspar or the ferromagnesian minerals. In addition the matrix, now mainly micaceous and chloritic, has been formed under conditions of low-grade metamorphism. Even so, it is evident that greywackes are the result of rapid transport and deposition from a region of varied rocks undergoing vigorous erosion.

6.2 OTHER FRAGMENTAL (CLASTIC) SEDIMENTARY ROCKS

The Latin word for rock waste, **detritus**, is used for the products which go to make sandstones and related rocks, as opposed to cement which is introduced later from solution. Detrital sediments are said to have a **clastic** texture (Gr. *klastos*,

broken in pieces) as distinct from the crystalline mosaic texture of chemically deposited sediments, such as rock-salt. Sands can be regarded as the product of medium-energy depositional environments. For example, along the shore, especially near cliffs and exposed areas, boulders and pebbles are heaped up by storm waves. Then, in quieter areas sands occur and finally, in the most sheltered areas and farther out in the deeper water, lie fine deposits of mud. The mud is made up of minute, flaky shreds of clay minerals, finely comminuted grains of quartz and feldspar and often decaying organic material. There is of course every gradation from the coarsest boulder beds down to the finest muds. Conventional size-limits have been set for the various sediments as shown in Table 6.1. It will be seen that an additional category, silt has been placed between sand and clay. All the varieties of modern unconsolidated sediments have their compacted and cemented equivalents amongst the sedimentary strata of all geological periods (Table 6.1).

Just as present-day beaches of boulders and pebbles may be replaced seawards by sands and

Table 6.1 Fragmental sedimentary rocks

	Coarse > 2mm Rudaceous (L.*rudus*, debris)	Medium Arenaceous (L.*arena*, sand)	Fine < 0.06mm Argillaceous (L.*argilla*, clay)
Loose	Boulders, pebbles	Sands	Silt, mud, clay
Indurated	Conglomerate (rounded fragments) Breccia (angular fragments)	Sandstone	Siltstone Claystone including Mudstone (compact) Shale (fissile)

eventually muds, so an ancient sheet of sandstone, as it is traced across country may become thinner and pass into mudstone or shale. Traced in another direction it may become coarser in grain and pass into a conglomerate (Figure 6.2). The term **conglomerate** is applied to cemented fragmental rocks containing rounded fragments such as pebbles and boulders; if the fragments are angular or subangular, the rock is called a **breccia**.

When first deposited, mud is a very weak sediment because of the very large amount of water (sometimes reaching 80 per cent by volume) held between the mineral particles. As more sediment is deposited on top, the growing overburden squeezes out the water and the sediment gradually consolidates into a claystone. Claystones may split very easily, sometimes into paper-thin flakes,

when they are referred to as **shales**; or the rocks may be compact and referred to as **mudstones**. The fissile nature of the shales is often the result of clay and other flaky, sometimes organic, materials occurring as minute films which tend to lie with their flat surfaces parallel to the plane of splitting.

6.3 VARIETIES OF BEDDING

In the steep face of a quarry or a cliff, successive layers can usually be seen, differing from one another by variations in colour and/or coarseness of grain (Figure 6.3). This layering has traditionally been called bedding, but it has been suggested that the general term should be **stratification**; this would include **beds**, (i.e. units

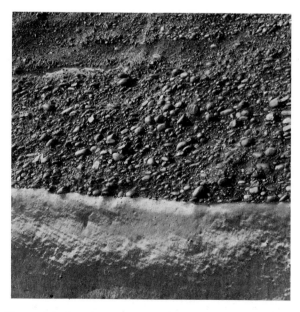

Figure 6.2 Conglomerate overlying cross-bedded sandstone (Triassic System). The pebbles are inclined to the right at a small angle to the bedding. This feature, called imbrication, indicates a current flowing from right to left. The pebbles are arranged with a 'dip' (Chapter 8) up-current. The Cliff, Budleigh Salterton, Devon. (*F.T. Blackburn*)

Figure 6.3 Bedding and jointing in Carboniferous sandstone, Muckross Head, Co. Donegal, Ireland. (*R. Welch Collection, Copyright Ulster Museum*)

thicker than a centimetre) and **laminae** (units less than a centimetre). Conventionally, stratification is said to be massive or thick-bedded (over 1.2 m), well-bedded or blocky (1–0.6 m), thin-bedded or flaggy (below 0.6 m), while lamination is made up with units of less than a centimetre.

The bounding planes between the layers are referred to as **bedding planes**, additionally in thick beds, bedding planes may be picked out by slight changes of grain-size or grain-shape or by planes of easy splitting. Evidently the strata have been formed by successive sheets of sediment. The resulting structure is a primary feature and distinctive of sedimentary rocks (Figures 1.5, 6.4).

Most sediments were originally deposited on nearly flat or very gently inclined surfaces. Stratification closely reflects the original horizontal surface. When sand is deposited from currents in water or by the wind, however, the grains may be heaped up in piles as sandbanks, dunes or small ripples. The front of sand deposition advances, just as a railway or road embankment is built up during construction. The bedding or lamination in a sandbank records the slopes down which the sand rolls as the structure grows, giving a pattern in cross-section like that shown in Figure 6.5. The sand avalanches down the lee side at an angle of rest controlled by the size of the grains and the

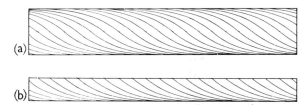

(a)

(b)

Figure 6.5 Sections to illustrate current bedding. In (a) the structure is complete. In (b) the upper part has been eroded down to AB and the current bedding pattern is truncated.

speed of the current. Generally the angle decreases towards the base and is higher towards the top where the current continually sweeps the surface clear.

Later on, another sand unit or, possibly a different kind of sediment may be deposited on the surface thus provided. So we may find that within certain beds, the bedding is oblique and variously inclined to the general 'lie' of the formation as a whole. This structure, which is original and not due to tilting or folding, is referred to as **false**- or **cross**- or, since it is formed under current action, **current-bedding** (Plate IVc).

Another common, primary sedimentary structure in clastic rocks is **graded bedding** (Figure 6.6). A bed in which this occurs displays a change in grain-size from bottom to top. In most sequences, the bottom of the bed lies on a claystone with a sharp boundary often characterized by erosional features (**sole marks**) (Plate IVd) produced by the current bringing in the sand. The base is coarse-grained, sometimes pebbly; above, the grains decrease in size through sand and silt, a fine lamination being developed often before the overlying claystone is reached. Such beds are generally of uniform thickness over considerable areas. They are found at the present day in deep water at the base of continental slopes; in ancient rocks they form thick successions called **flysch** (Ger. *fliessen*, to flow). The name is derived from the Alps where the beds tend to slide down into the valleys when the boundaries between the permeable sands and the impermeable shales become lubricated by rainwater. Flysch sequences consist of hundreds of graded beds alternating with shales. Each graded bed is thought to have formed from a turbidity current on the sea floor. A submarine avalanche of mixed sediment is started on a relatively steep slope, perhaps by an earthquake. The resulting heavy cloud of sediment and water travels swiftly down the slope and spreads out over the deep-sea plains gradually depositing its load in response to decreasing energy (Section 24.4 and Figure 24.11). The individual beds are

Figure 6.4 Well-bedded limestones alternating with shales and mudstones, cliffs west of Harness, Castlemartin, Dyfed, Wales. (*British Geological Survey*)

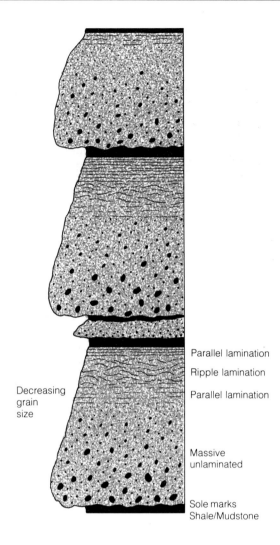

Figure 6.6 Features of graded beds.

Parallel lamination

Ripple lamination

Parallel lamination

Decreasing grain size

Massive unlaminated

Sole marks
Shale/Mudstone

Figure 6.7 The Great Scar Limestone (Carboniferous) above Malham Cove, North Yorkshire. (*J.B. Dawson*)

called **turbidites** and many greywackes as well as individual beds within the flysch have formed in this way.

6.4 LIMESTONES

Limestones are of economic importance for a variety of reasons: they are important aquifers (carriers of water) (Chapter 19) and reservoirs of hydrocarbons (oil and gas) (Chapter 9); they are commonly host to metalliferous deposits (especially those of lead and zinc); they are a source of CaO in the chemical industry; used as a flux for smelting; as a base in agriculture; and as a primary ingredient in cement manufacture (Plate IVe). Hard limestones are crushed for use as concrete aggregate and as roadstone. Thickly-bedded and well-jointed limestones (Figure 6.7) are ideal for building and facing-stone purposes since they can be cut and polished relatively easily. Portland stone, from Dorset, England, for example, has been a favourite choice for many of London's greatest buildings, ever since Wren selected it for the rebuilding of St Paul's Cathedral after the Great Fire of 1666. The towers and steeples of the London churches, the government offices in Whitehall, the front of Buckingham Palace and London University's Senate House also display its use in various styles of architecture.

When we examine limestones in the field we see that they commonly contain the calcified remains of marine organisms – for example stromatolites in the Precambrian are laminated structures created by lithification of soft mucilaginous algae that trapped and bound sediment particles in domical, columnar or branching shapes; Carboniferous-age limestones are in places packed with the skeletal remains of corals and marine shells, and the bead-like relics of the stems of sea lilies – animals like starfish on long stalks (Figures 6.8 and 6.9). The Jurassic-age limestones of the Cotswold Hills of England are also often crowded with fossil shells while belemnites, looking like thick blunt pencils, and the curled forms of the ammonites add further variety. The fine-grained and generally friable limestone known as chalk was shown long ago by Sorby to be composed almost entirely of organic remains, most are very fine grained and many are visible only under the electron microscope. A large proportion of these comminuted shells are of calcareous organisms that lived in the surface waters of the sea. On death they settled to the quiet deep-sea floor where they accumulated as ooze which on partial compaction and lithification became chalk.

Limestones also contain non-skeletal grains in the form of ooliths, faecal pellets, aggregates and calcareous mud. Ooliths (Figure 6.10) have the appearance of very small eggs (hence the name, Gr. *oion*, an egg). They have nuclei, commonly bits

Figure 6.8 Weathered surface of crinoidal limestone (Carboniferous Limestone). (*W.W. Watts*)

of shell, surrounded by concentric layers of $CaCO_3$. Ooliths (or ooids) form today in shallow-marine waters which often have elevated temperatures and salinities, and in tidal shoals of the tropics and subtropics such as the Bahamas and the Arabian Gulf. Warming and evaporation of seawater increase the concentration to a point at which $CaCO_3$ can be precipitated. Where the water is agitated by waves and currents CO_2 is driven off which forces the precipitation of fine crystals of $CaCO_3$ around fine grains which serve as nuclei. It is thought that the initial coating crystals attached with a haphazard or radial orientation and that this fabric is modified to a tangential orientation in high-energy environments where crystals are physically compacted to create a dense concentric fabric. The constant agitation of these ooliths results in excellent sorting, rounding and polishing of the grains. In the south-east coastal area of the Arabian Gulf quiet-water ooliths tend to retain their primary radial structure.

Ooliths also form today in a limited number of non-marine environments, for example in calcareous soils, caliche or calcrete, where they grow with concentric and radial fabrics replacing pre-existing sediment commonly producing a fitted polygonal fabric. Ooliths also grow in splash cups in limestone caves, in hot springs and in saline lakes. The upper size limit of ooliths is 2 mm; above this they are termed pisoliths and the rock they build is a **pisolite** (Gr. *pison*, a pea).

Most modern marine ooliths are aragonite (the orthorhombic form of calcium carbonate). In Baffin Bay, Texas, calcitic ooliths with radial fabric and aragonitic ooliths with tangential fabric occur together but it is noted that grains of identical diameter increasingly consist of aragonite in higher energy areas. It is thought that with increased agitation the greater expulsion of CO_2 from the water produces faster growth which

Figure 6.9 Crinoids. Slab of calcareous shale, River Liddel, south of Penton Bridge, Scottish/English border, showing two specimens of *Woodocrinus*. (*J. Wright*)

Figure 6.10 Ooliths approximately 1 mm diameter, collected from the intertidal zone on the north-eastern margin of the Great Bahama Bank. (*T.P. Scoffin*)

favours aragonite over calcite. Studies of ancient oolites, and other non-skeletal precipitates of calcium carbonate, show there to be a tendency for aragonite forms to have been abundant at certain times whereas calcite forms dominated at others. It is believed that during periods of worldwide low sea-level increased photosynthesis on land depresses atmospheric and oceanic CO_2 pressures, which favours aragonite over calcite in inorganic precipitation in the sea. During high sea-levels calcite is the dominant polymorph of abiotic $CaCO_3$ precipitated. We are presently in a period of relatively low sea-level because of the decrease in the volume of the mid-ocean ridge system (Figure 2.7) (c. 300 m below the Cretaceous sea) and low global temperatures (c. 10°C below Cretaceous average temperatures) so we may expect the present abundance of marine calcium carbonate sediment and the mineralogy of its abiotic constituents to be fundamentally different from those periods in the past when sea-level and global temperatures were relatively high (see also Chapter 21 and Section 24.9).

Lime mudstones are common in the geological record. The original fine particles were either comminuted skeletons, such as those of the weakly-constructed calcareous green algae, or else were precipitated directly from seawaters of elevated temperatures and salinities. Spontaneous precipitation of very fine particles of aragonite from supersaturated seawater is thought to be brought about by the sharp CO_2 uptake during the intense photosynthetic activity of diatoms (microscopic planktonic algae) at the time of blooms.

Present day deposits of calcium carbonate are most abundant in shallow tropical seas where skeletal and abiotic grains form in heated, agitated waters well away from river mouths. Rivers bring quantities of fresh water and terrigenous sediment which inhibit calcium carbonate accumulation. However in places mixtures occur ultimately forming calcareous shales or marls.

Post-depositional modification of carbonate sediments is commonly profound, transforming the deposits in both structure and composition by processes of compaction, dissolution, lithification, recrystallization and replacement. Modern marine carbonate grains consist mainly of aragonite and magnesian calcite which are metastable. During prolonged burial, and also with exposure to the atmosphere, these deposits change to the more stable minerals of calcite and dolomite.

carbonate of calcium and magnesium. In ideal crystal form, layers of cations alternate with CO_3^{2-} groups, and the cation layers themselves are alternatively Ca^{2+} and Mg^{2+}.

Most dolomites initially form by replacing a calcium carbonate precursor. A high fluid flow is necessary which imports magnesium, dissolves the precursor phase, precipitates dolomite and exports calcium (plus trace elements present in the original calcium carbonate mineral). The exchange that occurs can be represented by the following equation:

$$2CaCO_3 + Mg^{2+} = CaMg(CO_3)_2 + Ca^{2+}$$

Calcite | Ion from seawater | Dolomite | Ion into seawater

But this is not the whole story of what happens. In the equation, as in all such chemical reactions, the masses of the reacting substances on one side are equal to the masses of the products on the other. What commonly happens under natural conditions, however, is that a shell or coral is replaced by dolomite without any change of volume, so that every detail of the original organic structure is still perfectly preserved. Such equal-volume replacements – which cannot be completely expressed by chemical equations – are highly characteristic of many mineral and rock transformations. They are described as **metasomatic** replacements (Gr. *meta*, an affix connoting subsequent change; *soma*, body-substance), i.e. replacements involving a change of substance without a change of form.

As it is energetically easier to change from limestone to dolomite than vice versa, as a general rule we find dolomites more common the older the rocks. Though most dolomites develop by the replacement of ancient limestones, there are several examples of dolomites forming today. For example they have been found in playa lake deposits (Section 22.2), in the interstitial waters of deep-sea sediments, in supratidal sediments of arid regions (where the Mg/Ca ratio of seawater is elevated from 3/1 to 10/1 during the removal of Ca^{2+} by the precipitation of aragonite ($CaCO_3$), gypsum ($CaSO_4 2H_2O$) and anhydrite ($CaSO_4$) as salinities rise on evaporation), and beneath tropical islands where fresh and salt water mix in the appropriate amounts for dolomites to precipitate. Most of these contemporaneous dolomites have a poorly-ordered crystal structure that is richer in calcium than magnesium.

6.5 DOLOMITES

Dolomites are rocks made essentially of the mineral dolomite, which is a rhombohedral double

6.6 IRONSTONES

The term **ironstone** is used loosely to describe sedimentary rocks which contain enough iron so that they have been, are or might be mined or quarried as a source of iron for industry. The contained iron-bearing minerals are essentially oxides or hydroxides of iron but can be carbonates or silicates (Chapter 4), and in rare cases sulphides.

Precambrian ironstones, which supply most of the world's iron and steel industries today are widespread and most consist essentially of alternating laminae of haematite or magnetite and chert. They occur in sequences of ordinary clastic sedimentary rocks such as sandstones and limestones/dolomites and are now referred to generally as **banded ironstone formations** (BIFs)but frequently by a variety of names in different countries (e.g. **taconite** in the Lake Superior area of the USA, **jaspilite** in Australia (Plate IVf), **itabirite** in Brazil). Considered to originate as chemical precipitates in seawater (see Section 24.1) their formation has aroused much controversy.

The actual mechanism of the transport in solution of the iron and its precipitation has been the subject of much debate. The problem is that the chemical weathering of rocks at the Earth's surface converts ferrous iron in the individual minerals to the ferric state. Ferric iron compounds are insoluble in normal surface waters and with an oxidizing atmosphere it is difficult to see why the ferric compounds should change into soluble ferrous compounds available for transport in solution. It has been suggested, therefore, that the occurrence of the older Precambrian ironstones supports other lines of evidence of an early Precambrian atmosphere lacking in oxygen. However, the relatively recent discovery of hot water (**hydrothermal**) metal-bearing springs discharging into the sea and ocean floors (Section 13.8, Figure 13.62 and Section 24.1) may support the transport of the iron either by groundwaters rather than surface waters or suggest a juvenile water source, or indeed a mixture of both.

A second important group of sedimentary iron ores are the **oolitic ores**, famous deposits including, for example, the Silurian-age Clinton iron ores of the USA and the Jurassic-age 'minette' iron ores of Europe. The Clinton Ironstones extend sporadically in a belt some 1000 km long from New York State to Alabama in a sequence of thin beds of sandstones, shales and oolitic haematite with occasional limestones. Fossils are abundant (and are often haematized) and the shallow-water marine environment of deposition obvious.

Again the problem of transport of the iron in solution remains, particularly as there seems little doubt that the Earth had an oxidizing atmosphere

in Phanerozoic times. Solution and transport by groundwater circulation is a possibility.

The European oolitic ores are of much younger (mainly Jurassic – Cretaceous) age and occur in England, France, Belgium, Luxembourg and Germany. In England they are worked in a belt stretching from Yorkshire through Lincolnshire. They occur with marine shales and sandstones and have considerable amounts of calcite matrix along with the ooids, which are predominantly of siderite or chamosite (a silicate of iron) or goethite (limonite – a hydroxide of iron). Last worked in Frodingham in 1988 they contained only some 25 per cent iron and in fact were used for their calcium carbonate content as a flux when smelting imported iron ores.

A third group of ironstones, that were once important, (some are forming today and are known as 'bog-iron ores') are the ironstones which occur within most coal-bearing sequences. They consist mainly of siderite and formed when iron-rich waters encountered the organic-rich waters associated with the peat-forming environments. Carboniferous-age ores in Britain were mined along with the coals in the early days of the Industrial Revolution and can occur either as distinct bands or as nodules within shales, precipitation and segregation of the iron having taken place during diagenesis.

6.7 SILICEOUS DEPOSITS: FLINT AND CHERT

Although quartz is practically insoluble in ordinary water and mostly turns up again in sandstones after the journey from its weathered source to its place of deposition, it must not be overlooked that a great deal of soluble silica is liberated during the chemical weathering of most silicate minerals. This can be illustrated by considering the alteration of orthoclase to clay:

$$2KAlSi_3O_8 + 11H_2O = Al_2Si_2O_5(OH)_4$$

Orthoclase　　Water　　Clay (kaolin)
$$+ 2KOH + 4Si(OH)_4$$
Removed in
solution

Similar equations can be written for albite and anorthite, and it should be emphasized that feldspars of the plagioclase series, especially the calcic varieties, weather in this way more readily than orthoclase. The soda liberated from albite helps to hold the hydrous silica in solution, mainly as $Si(OH)_4$. Still more silica is released by the weathering of pyroxenes and amphiboles.

Practically all this dissolved silica must soon be precipitated again, since many river waters contain five to ten times as much soluble silica as seawater. Probably most of it contributes to the finer constituents of alluvium and offshore muds. A little is fixed in iron-silicate minerals and some is used for cementing sandstones. The balance is used by micro-organisms. In freshwater lakes these are mainly diatoms, the remains of which locally accumulate as diatomaceous earth. From the sea both diatoms and radiolarians (Section 24.1) abstract silica to form their tiny opaline shells, while sponges support or protect their ungainly forms with networks of opaline rods or loose frameworks of needle-like spicules, also made of opal. Much of the silica used in this way is soon restored to circulation by being re-dissolved, but the part that survives contributes to the oozes and other deposits that are slowly accumulating on the ocean floor.

Sponges appear to have contributed most of the silica that now appears in the Chalk as **flint**. The relatively soluble opaline silica originally distributed through the Chalk has since been dissolved by percolating waters and redeposited in the insoluble form of flint. The original siliceous chalk has become segregated into flint that is nearly pure SiO_2, and chalk that is practically all $CaCO_3$.

Flint, like chalcedony, is a sort of felt or mosaic of extremely minute but rather imperfect crystals of quartz – imperfect in the sense that there are unfilled gaps here and there in the crystal lattice. For this reason the specific gravity of flint is a trifle lower than that of quartz. Flint occurs as scattered nodules of knobbly and fantastic shapes, commonly concentrated in layers parallel to the bedding planes (Figure 6.11) and in places passing into tabular sheets; occasionally vertical stringers and veins of flint have developed along joint planes.

Remains of sponges are abundant in many parts of the Chalk, but now preserved as calcite, which replaced the original opaline silica. In turn the liberated silica then replaced the finely divided and more vulnerable parts of the Chalk, especially around suitable nuclei such as concentrations of sponge debris within burrows and along bedding planes and joints that served as passageways for migrating solutions. Once deposited as flint, the silica was not again taken into solution.

Such silicification of limestone is not, of course, confined to the formation known as the Chalk, but the related bands and concretions in other lime-

Figure 6.11 Chalk with characteristic bands of flint nodules, Promenade, Rottingdea, East Surrey. (*British Geological Survey*)

stones (and sometimes in calcareous shales and sandstones) are generally referred to as **chert**. There are also radiolarian cherts: these are ancient organic deposits of radiolarians which were cemented by silica into hard, tough or splintery rocks composed essentially of chalcedony or, if coloured by ferruginous impurities, of jasper. Indeed both flint and jasper might be regarded as particular varieties of chert.

Flint, however, has a special interest because one of the most important discoveries made by early man was that flint, of all the natural materials available to him, was unsurpassed for fashioning weapons and tools. Flint, like obsidian, breaks with a perfect conchoidal fracture and sharp cutting edges are easily obtained with a few deft blows. In other parts of the world chert and obsidian were similarly utilized, but wherever the Chalk still remains, as it does over extensive areas of Britain and Europe, multitudes of flint implements have been found in the gravels of river terraces and in cave deposits. Flint, for which the earliest known mines were dug, was the raw material that 'enabled the great creative inventors and the skilful and industrious artisans of the Stone Age to lay the material foundations of modern civilizations' (V. M. Goldschmidt,

Geochemistry, 1954, p. 370).

One of the most remarkable of all replacement phenomena is the transformation of wood into opal, and even into chalcedony or jasper. In **wood opal** the replacement of immense logs is often so perfectly metasomatic that the structures of the wood, down to many details of the original cells, are preserved with astonishing fidelity (Figure 6.12). The process of petrification occurs when waterlogged tree trunks are buried in feldspathic sandy muds, perhaps in the shoals of a river bed. Percolating water becomes charged with additional alkalis and silica as a result of the continuing decay of the feldspar grains. As the solutions soak through the buried wood, some still undiscovered process of natural alchemy slowly operates. The silica so brought in is fixed as opal, taking the place and the form of the wood without disturbing the structure. The wood (cellulose) presumably escapes as marsh gas (methane, CH_4) and carbon dioxide. If chalcedony or jasper should be formed instead of opal there is usually some internal distortion and loss of detail, but the all-over replacement still remains so perfect that a petrified log may be mistaken for a fallen and weathered telegraph pole until the hammer is applied.

Figure 6.12 Wood opal from a silicified log, Petrified Forest National Park, Arizona. (*Jean Tarrant*)

The most celebrated display of silicified logs occurs in the Painted Desert of Arizona, where the number of bark-stripped trunks of tall forest trees, some of them 30 m long, is so great that the region is known as the Petrified Forest. But it is more like the remains of a log-jam than a forest, for few roots or branches remain, and the logs had evidently travelled a long way before being entombed in the sandy muds where they were silicified. At some later time the sands and muds were consolidated into brilliantly coloured sandstones and shales by the deposition of a variety of cementing materials, mainly ferruginous, but also including compounds of manganese, copper and uranium. In the bright sunshine of this semi-arid region the name 'Painted Desert' does no more than justice to a landscape of gorgeously coloured rocks.

6.8 SALT DEPOSITS

The natural history of the potash dissolved from rocks during weathering is very different from that of soda. In average crustal rocks these two constituents are about equally abundant, but while the liberated soda is carried to the oceans where it accumulates for a time, much of the potash remains in the soil, where it serves as an essential nutrient for plants. Eventually, when the soil itself is removed by erosion, the potash goes with it as a constituent of the potash-bearing clay minerals. Fixation of potash by clay continues on the sea floor, together with further fixation in the green mineral glauconite, $(KMg(FeAl)(SiO_3)_6.3H_2O)$. Of the small balance that then remains, much is abstracted by seaweeds, so that the amount left dissolved in the oceans (as shown in Table 6.2) is far less than might have been expected.

The average amount of mineral matter dissolved in seawater is about 3.5 per cent. For convenience this is usually expressed in parts per thousand (‰) and the **salinity** is then said to be about 35. When seawater is evaporated (at, say, 30°C) precipitation begins with quite negligible amounts of carbonates and the salinity has to rise to well over 100 before gypsum, $CaSO_4.2H_2O$, begins to be deposited. When the brine is concentrated to about a tenth of the original volume, rock-salt (halite, NaCl) separates, accompanied by a little anhydrite, $CaSO_4$, and this continues almost to the end. The very soluble salts of potassium and magnesium do not begin to crystallize until the original volume is reduced to about 1/60th.

A geologically recent example of evaporation to dryness, with the resulting salt deposits still left uncovered at the surface, occurs across the boundary of the Ethiopian province of Eritrea, where the Red Sea begins to narrow. This is the **Piano del Sale**, a depressed region below sea-level which was originally part of the Red Sea and is now one of the hottest and driest places on the face of the Earth. The isolation of this great evaporating dish was due to an accumulation of lavas from a series of volcanoes that are still not completely extinct. The depression is now lined with gypsum, representing an early phase of deposition. Within the inner rim of this 'saucer' of gypsum is a wide expanse of rock-salt, 32 km across, with intercalations of anhydrite. Each salt bed has also a saucer-like form, and because the level of the brine continued to fall with continued evaporation each 'saucer' is smaller than the underlying one. Potash salts appear only towards the heart of the depression, which drops to about 120 m below sea-level. Before their commercial exploitation, these were seen at the surface as a broad ring largely composed of carnallite, $KCl.MgCl_2.6H_2O$, thickening to about 60 m in the middle, where it was covered in turn by a metre or so of sylvite, KCl, over an area of 1.5 sq km.

Although the original depth of the basin is not yet known, it could hardly have been more than 300 m. A column of seawater 300 m high would yield no more than about 5 m of salt deposits. In the Piano del Sale the thicknesses revealed by drilling are so great that they can only be accounted for by assuming that for long periods fresh supplies of seawater were continually added to the evaporating basin. Even today there are occasionally inundations from the sea. In the past,

Table 6.2 Constituents and salinity of seawater and natural water

The chief constituents of average sea-water		Average salinities of some natural waters	
	‰		‰
Na	10.56	Seawater	35
Mg	1.30	River water	0.16
Ca	0.40	Baltic Sea	7.2
K	0.38	Caspian Sea	13
Cl	19.00	Gulf of Kara Bogaz	185
SO$_4$	2.65	Black Sea	20
HCO$_3$	0.14	Dead Sea	102
CO$_3$		Great Salt Lake, Utah	203
Br	0.06		

before the growing volcanic barrier reached its present size, it would necessarily go through a stage when it was less effective than now in keeping out the sea.

A process of continuous replenishment is manifestly responsible for the evaporites, mainly gypsum at present, now being precipitated on the floor of the Gulf of Kara Bogaz, a shallow embayment of the Caspian Sea (Figure 6.13). Because of the inflow from the Volga and Ural rivers, the level of the Caspian is always perceptibly higher than that of Kara Bogaz and consequently there is a continuous flow of water of medium salinity into Kara Bogaz through the narrow channel which is its only connection with the Caspian. Evaporation removes this water but leaves the salts behind. The concentration now reached is such that a little rock-salt is precipitated in warm seasons as well as gypsum. If the channel were closed and the Kara Bogaz evaporated to dryness the layer of salts

(mainly rock-salt) would be little more than a metre thick on average, while if the channel remained open until Kara Bogaz filled up with salts the thickness would still be limited to 15 or 18 m, the maximum present depth of the water. To get greater thicknesses the floor of Kara Bogaz would have to subside while supplies of salt water continued to flow in from the Caspian. We can now realize how very extraordinary it is that in some parts of the world beds of rock-salt many hundreds of metres thick have accumulated in the past.

The salt mines of Wieliczka (near Krakow in south-west Poland) have been worked for over a thousand years and have long been celebrated as a tourist attraction because of the underground houses, churches and monuments, roads, railways and restaurants, all excavated far below the surface in a layer of salt over 3500 m thick due to tectonism. Neither the well-known Stassfurt

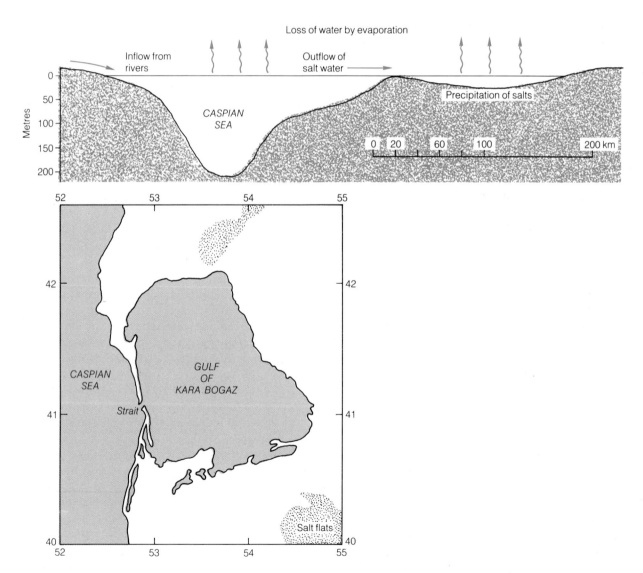

Figure 6.13 Map and section to illustrate the accumulation of saline precipitates ('evaporites') in the Gulf of Kara Bogaz, Caspian Sea.

deposits in Germany, nor those of Alsace, are quite so thick, but they are commercially more important because valuable potash salts have been preserved under a protective cover of impervious clays and marls that prevented their removal and loss by subsequent solution. The thickest known salt deposits are those of Texas and adjoining States. About 3500 m have been proved in New Mexico. Even greater thicknesses are suspected in parts of Texas (see Chapter 10), but since the salt may have a cover of anything up to 9000 m of sediments, it has not yet been penetrated by drilling.

However, 3500 m is enough to raise some tremendous problems, it would require the complete evaporation of a column of seawater 240 km deep! Evidently at several periods in the past the waters of immensely large 'evaporating basins', comparable in size with the Red Sea or the Gulf of Mexico, became all but disconnected from the main body of the oceans, as the Mediterranean nearly is today. A coincidence of special conditions must have persisted for very long periods, the essential conditions being:

1. a hot (but see later) and arid climate to ensure the evaporation necessary to maintain a sufficiently high concentration for salts to crystallize out;
2. an intermittent or continuous supply of salt: in part, perhaps, from rivers and inland seas as in the case of Kara Bogaz, but mainly from seawater, flowing into the basin through one-way channels;
3. sinking of the floor of the basin to make room for the growing thickness of salt deposits;
4. maintenance of the barrier between the sinking basin and the ocean, possibly by volcanic activity, upward growth of coral reefs, or earth movements;
5. a final depression, carrying the barrier with it, to make room for the thick and widespread sediments that eventually covered and preserved the salt deposits.

Modern evaporites are most abundant between latitudes 10° and 40°, which is why fossil deposits are commonly regarded as indicative of past hot climates. Although this is generally the case the occurrence of patches of evaporite deposits on the land surface in the Antarctic oases suggests that it is the aridity of the climate that may be the most crucial factor in their formation.

More widespread than deposits of rock-salt and potash salts are precipitates of carbonates (limestone and dolomite) and anhydrite, which commonly occur as alternating layers. The anhydrite of this association characteristically occurs as nodules dispersed through carbonates, or as tightly packed nodules. Evidence as to how this may happen has been found in the desert region of the Trucial Coast, in the shoal water at the southern end of the Arabian Gulf. Here evaporites of this variety are being precipitated today. Below low-tide level calcareous sediments are being deposited. Mats of calcareous algae characterize the zone between high and low tide levels, and form traps for dolomite and gypsum. Above high-tide level are extensive flats (**sabkhas**) where anhydrite is precipitated as nodules from groundwater of marine origin that has become highly concentrated brine as a result of evaporation and lack of rain. A sabkha model for the formation of thick evaporites may require more sustained hot and dry climates than the deep-basin model. Trenches dug in these flats have exposed mineral sequences, from below upwards, that are respectively the equivalents of those below low-tide level, between high- and low-tide levels, and above high-tide level. The sections revealed in the trenches therefore indicate regression of the shore line. When traced laterally from land to sea, the evaporite of each facies will vary in age; each facies is diachronous.

The evidence from the Trucial Coast may have a bearing on the composition of some of the deposits of rock-salt and potash salts, which differ chemically from evaporites obtained by experimental evaporation of seawater. Within some of these evaporites, for example the Middle Devonian potash deposits of Saskatchewan, not only is sulphate absent, but the proportion of magnesium is also lower than would theoretically be expected to be precipitated after rock-salt, and prior to and concomitantly with potash salt.

Along the Trucial coast, there is early precipitation of gypsum and anhydrite which raises the magnesium to calcium ratio of interstitial waters and this then allows dolomite to precipitate. After this occurs the brines are depleted in sulphate and magnesium. It has been suggested that seawater gaining access to evaporating basins, within which rock-salt and potash salt are deposited, may be conditioned in this way whilst filtering through physical barriers such as barrier reefs.

Thick evaporite sequences have been recovered in cores that penetrated the Late Miocene of the Mediterranean area. Some authors maintain that these represent a complete dessication of the sea when the Strait of Gibraltar was closed during a fall in sea-level. The evidence for this was the predominance of anhydrite, stromatolitic laminations, dessication cracks, barren thick clastic sequences thought to be terrestrial in origin and aeolian cross-laminated silts. However, other authors have compared the anhydrites with those of the Permian of west Texas where similar 'stromatolitic' laminations occur but are thought not to be diagnostic of shallow or subaerial deposition since fine laminae can be traced for miles.

6.9 FURTHER READING

Purser, B.H. (ed) (1973) *The Persian Gulf*, Springer-Verlag, Berlin.

Scoffin, T.P. (1986) *An Introduction to Carbonate Sediments and Rocks*, Blackie and Sons, Glasgow.

Selley, R.C. (1988) *Applied Sedimentology*, Academic Press, London.

Tucker, M.E. (1991) *Sedimentary Petrology*, 2nd edn, Blackwell, Oxford.

PAGES OF EARTH HISTORY

These rocks, these bones, these fossils and shells,
Shall yet be touched with beauty, and reveal
The secrets of the book of earth to man.

Alfred Noyes, 1925

7.1 THE KEY TO THE PAST

So far we have dealt with sedimentary rocks as materials which have been formed from pre-existing materials by the action of geological processes. Rocks are also the pages of the book of Earth history, and the chief object of historical geology is to learn to decipher those pages and place them in their proper historical order. A fundamental principle (called the **Principle of Uniformitarianism**) involved in understanding their significance was first enunciated by James Hutton in 1785 at a meeting of the Royal Society of Edinburgh and in his *Theory of the Earth*, published in 1795, he wrote 'we may look for the transactions of time past, in the present state of things upon the surface of this earth, and read the operations of an ancient date in those which are transacted under our eye'. A century later, Sir Archibald Geikie encapsulated the principle, rather more pithily, in the oft-quoted phrase, 'the present is the key to the past'. Rocks and characteristic associations of rocks, with recognizable peculiarities of composition and structure, are observed to result from processes acting at present in particular kinds of geographical and climatic environments. If similar rocks belonging to a former geological age are found to have the same peculiarities and associations, it is inferred that they were formed by the operation of similar processes in similar environments.

We have already, as a matter of common sense, had occasion to apply this principle. The presence of fossil corals or the shells of other marine organisms in a limestone indicates that it was deposited on the sea floor and that what now is land once lay below the waves. The limestone may pass downwards or laterally into shale, sandstone and conglomerate. The last of these may indicate an old beach and show the old shoreline where the sea and the land came together. Elsewhere, old lava flows represent the eruptions of ancient volcanoes, and in places the vents that were active millions of years ago still figure prominently in the landscape (Figure 7.1). Beds of rock-salt point to the existence of waters that evaporated in the sunshine. Seams of coal which are the compressed remains of peat, suggest former widespread swamps and luxuriant vegetation. Smoothed and striated rock surfaces associated with overlying beds of boulder clay prove the former passage of glaciers or ice-sheets. In every case the characters of older formations are matched with those of rocks now in the making.

Even the weather may be recorded in the structures of the rocks. A brief rainshower falling on the smooth surface of fine-grained sediment spatters it with tiny crater-like pittings known as **rain prints**. **Desiccation cracks** (also called **mud cracks** or **sun cracks**) develop in the mud-flats of tidal

Figure 7.1 Arthur's Seat, Edinburgh, a Lower Carboniferous volcano that was active more than 300 m.y. ago. The bold escarpment of Salisbury Crags is the outcrop of a later sill. At least 14 lava flows can be recognized on Whinny Hill, with Lower Carboniferous sediments above and below. (*Airviews Limited*)

reaches or flood plains when the mud dries up and shrinks (Plate Va and Figure 7.2). As far back as geological methods can be applied to the Earth's history, such relics of 'fossil weather' prove that wind, rain and sunshine have always been much the same as they are today. Nevertheless, as we decipher the climates that have affected land areas through geological time, we find some remarkable variations.

In Great Britain the work of former ice-sheets and glaciers is still written conspicuously in the form of the landscape and in the boulder clays and other deposits left behind when the ice melted (Chapter 20). In striking contrast, the very much older clay through which London's underground railways are tunnelled contains remains of vegetation, shells and reptiles similar to those of the modern tropics. Elsewhere the apparent vicissitudes of climate are equally startling. In India and in central and southern Africa there is clear proof (Figure 20.27 and Plate VId) that while Britain was part of a region of swampy tropical jungles (the time of coal formation) these lands were buried under great ice-sheets like those of Greenland and Antarctica at the present time. In Greenland, however, there are sediments containing the remains of vegetation that could have grown only in a warm climate. Similar discoveries have been made in various parts of Antarctica, beginning with Captain Scott's remarkable discovery of coals near the South Pole in 1912.

Figure 7.2 Desiccation cracks in mudstone of Devonian age, north-east Scotland. The cracks formed when the mud dried out and contracted; they were then filled by sand. (*British Geological Survey*)

Some climatic changes, for example during the last Ice Age, have been worldwide, but as we shall see in later chapters, most of the variations noted above reflect the drifting of continents rather than violent shifts of the climatic belts. The Principle of Uniformitarianism allows us to interpret the environment in which a rock has formed, but how do we plot the changes of conditions through time and space? These questions involving the succession of strata through time and the correlation of strata from one locality to another across the Earth are dealt with in the next sections.

7.2 THE SUCCESSION OF STRATA – STRATIGRAPHY

To place the pages of Earth history in their proper chronological order is by no means an easy task. As a starting point we follow the English civil engineer, William Smith and the French aristocrat, Georges Cuvier who, in the late eighteenth and early nineteenth centuries, applied the **Law of Superposition**. This basic proposition states simply that stratified rocks have accumulated layer upon layer and in a continuous succession of flat-lying beds, the lowest beds are the oldest and those at the top the youngest. We can see this on the slopes of Ingleborough (Figure 7.3) and, more spectacularly in the Grand Canyon of the R. Colorado (Figure 1.5).

In any one vertical section be it valley side, cliff or quarry face where the rocks are lying horizontally, the sequence of rocks which can be examined is limited. When the rocks have been tilted from their original sub-horizontal position, as between the Welsh Borders and London (Figure 7.4) the worn-back edges of layer after layer come to the surface; so long as the rocks are exposed it becomes possible to place a long sequence of beds in their proper order.

Where rocks have been strongly deformed application of the Law of Superposition may not be possible. If we look again at the exposures at Ingleborough (Figure 7.3) we see that while the upper beds are flat-lying and the Law applies directly, there is a lower sequence of strongly-folded slates and hard coarse sandstones ('grits').

In areas such as this, where beds may have been rotated through 90° or more, the Law is inapplicable and other criteria must be used to determine the correct sequence of the beds.

Current bedding provides one of the more easily applied tests, when it is present. As shown in Figure 6.5, the upper surface in sandbanks, dunes and ripples has normally suffered erosion so that the inclined laminae make a high angle with it; by contrast, the current bedding at the base usually curves gently into the main stratification. The truncated top and the original floor of the layer of sandstone can thus be recognized at a glance (Figures 6.5, 7.5). This was first used by Irish geologists more than a century ago to show that parts of the folded sandstones of the Dingle Peninsula in south-west Ireland are upside down.

Graded bedding (Figure 6.6) also provides a useful clue as to whether a particular bed is the right way up or upside down, especially if there are several graded beds all telling the same story. A single instance unsupported by other examples might be otherwise explained and so would remain ambiguous.

Ripple marks like those seen on a beach or in desert sands (Figure 3.4) are often preserved in ancient sandstones (Figures 7.6, 22.23). Where the ripples have been formed under the action of one current moving only in one direction the structures show cross-lamination on a small-scale and can be used in the same way as the larger-scale cross-bedding; the plan view of these ripples is not diagnostic of way-up. On the other hand ripples, formed by the to-and-fro movement in shallow waters, often have sharp crests and rounded troughs and can be used to determine the original top of the bed (Figure 7.6).

Using such criteria allows the correct order of beds to be inferred in the lower sequence at Ingleborough. Between the strongly-folded grits and slates and the flat-lying limestones there is a break in the continuity of the record, a break that may imply a very long interval of time. The

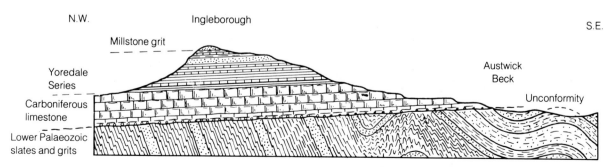

Figure 7.3 Section across Ingleborough and its foundations, Yorkshire, showing the unconformity between the Carboniferous beds above and the intensely folded Lower Palaeozoic strata below. Length of section about 6 km. (*After D. A. Wray*)

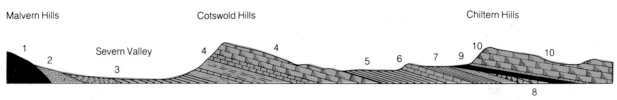

Malvern Hills   Cotswold Hills   Chiltern Hills

Severn Valley

Figure 7.4 Section from the Malvern Hills to the Chiltern Hills: 1. Precambrian and Cambrian; 2. Triassic; 3–8. Jurassic (3. Lias; 4. Lower Oolites; 5. Oxford Clay; 6. Corallian; 7. Kimmeridge Clay; 8. Portland Beds); 9–10. Cretaceous (9. Gault and Upper Greensand; 10. Chalk). Note unconformities (a) between 1 and 2, Palaeozoic to Triassic and (b) between 8 and 9, Upper Jurassic to Lower Cretaceous.

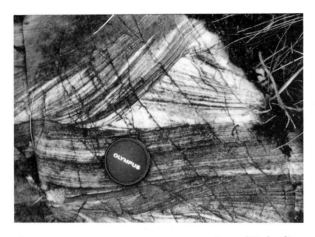

Figure 7.5 Current bedding in quartzites of Dalradian (? Precambrian) age, Kinlochleven, Argyll, Scotland. (*M. R. W. Johnson*)

Figure 7.6 Sharp and double-crested intertidal ripples preserved in a Precambrian sandstone. Madkya Pradesh, India. (*E. K. Walton*)

physical representation of the missing part of the record is evidently an old erosion surface and this feature where lower and upper sequences of rocks represent quite different episodes of sedimentation and deformation is described as an **unconformity** (Figures 7.3 and 7.7). The lowest of the upper **conformable** beds are said to rest **unconformably** on the underlying rocks. After the latter had been deposited on the sea-floor, they were folded and uplifted, deep in the heart of a mountain system which now forms ranges that extend throughout the length of Norway and across much of the British Isles. Because much of Scotland is carved out of its hard, contorted rocks, it is known the world over as the Caledonide mountain system or, more technically, as the Caledonian **orogenic belt** (Greek *oros*, mountain; see Section 7.7 and Chapter 31). By denudation, the folded grits and slates were gradually uncovered and ultimately the land-surface was reduced to an undulating lowland. Then the worn-down surface was submerged below the sea, to become the floor on which the horizontal sheets of Pennine limestones were deposited. Successive stages of the events which occurred during the time gap represented by the unconformity are shown in Figure 7.7.

In general terms, every unconformity is an ero-

sion surface of one kind or another, indicating a lapse of time during which denudation (including erosion by the sea) exceeded deposition at that place. If sediments were deposited there during the interval, they must subsequently have been removed. The time gap is likely to be represented by strata somewhere else and the problem of how to recognize such rocks and fill in that gap involves the question of correlation to which we now turn.

7.3 CORRELATION

In the first instance the rocks are recognized and differentiated on their lithology. This determines their overall appearance in the field – whether they will stand out as ridges and cliffs, or whether they will weather easily and form hollows or be covered by weathered debris. The hard rocks of the Grand Canyon, the sandstones and the limestones, form massive walls in the valley sides and are separated by the softer shales (Figure 1.5). Colour may also be important as for example in the Redwall Limestone in the Grand Canyon. In this exceptional locality almost two kilometres of sedimentary rocks can be seen but most rock

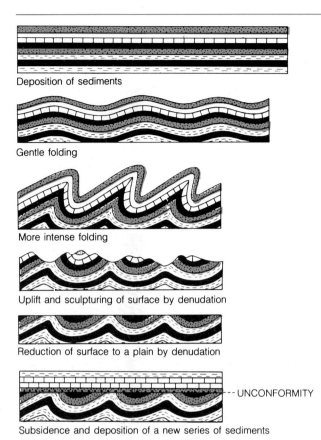

Deposition of sediments

Gentle folding

More intense folding

Uplift and sculpturing of surface by denudation

Reduction of surface to a plain by denudation

-- UNCONFORMITY

Subsidence and deposition of a new series of sediments

Figure 7.7 Successive stages in the development of an unconformity.

(a)   (b)   (c)

Lst. fossil-iferous
Coal
Seat -earth
Sst and shale
Sst
Shale
Anhydrite
Dolomite
Lst with distinctive fossils
Conglomerate
UNCONFORMITY
Slate & greywacke

Figure 7.8 Diagram to illustrate the correlation of short sequences exposed in separate localities (a,b,c) over a wide area allowing the building up of longer successions which record events over extended periods of geological time. Note that lateral variation from one locality to another may complicate correlation; beds commonly change thickness and some may disappear altogether e.g. the sandstone–shale sequence at locality (b) is represented only by shale at locality (c).

outcrops are much more limited in size. Geological history has to be pieced together from one partial succession to another. So long as the rocks are distinctive and widespread, it is possible to trace them from one locality to another. Each short sequence may add to the total number of beds either at the top or the bottom and so fill in more of the pages of Earth history (Figure 7.8). Rocks are traced across country from one place to another by a process of correlating distinctive beds. Since no one locality, not even the Grand Canyon, exposes more than a few of the events of the Earth's past, correlation is of fundamental importance in reconstructing a comprehensive history. As an example of lithological correlation we may refer to the Chalk which forms the White Cliffs of the south coast of England. A very distinctive fine-grained compact limestone, it can be recognized in separate exposures as far afield as Northern Ireland and France.

The question then arises, was a distinctive rock which we have recognized across considerable distances, deposited in the same time interval over the whole area? Reference to the 'present is the key to the past' suggests not. Consider the sediments accumulating at any one time in a marine area (Figure 7.9). As we noted in Chapter 6, inshore beach gravels give way seawards suc-

cessively to sands, muds and perhaps shelly deposits. Now suppose a rising sea-level causes the sea to transgress over the land. The strand-line moves forward carrying with it, as it were, the belts of gravel, sand, clay and carbonate. In time continuous beds of conglomerate, sandstone, shale and limestone are produced but the portions of the beds formed at the end of the **transgression** are much younger than those at the beginning. The Chalk deposited in Antrim 'arrived' much later than the first Chalk on the south coast of England. Each rock picked out by our stratigraphic analysis may cut across time-lines, i.e. it is said to be **diachronous**. This can sometimes be demonstrated by a deposit such as a volcanic ash-fall which scatters debris indiscriminately over a very wide area, in terms of geological time, virtually instantaneously. There is an example, in Fife, Scotland, where mining has revealed a thin volcanic tuff band occurring above a coal seam, within the coal seam and below the seam when traced laterally, so marking one time horizon, an instant in geological time.

But these are rather rare events. Is it possible to improve methods of correlation? Can we refine our analysis to define the time-lines relative to the rock types? The answers to both these questions is positive and take us back again to William Smith.

7.4 THE SIGNIFICANCE OF FOSSILS

To the Law of Superposition Smith added another fundamental principle. Born in Oxford in 1769, Smith became a highly successful consultant in connection with canals, collieries, water supply and coastal erosion. As a boy he had collected fossils from the richly fossiliferous beds near his home and in later years he made separate collections from each of the sedimentary sequences (illustrated in Figure 7.4). He noticed that while some of the fossils in the assemblage collected from any particular bed might be the same as some of those from the beds above or below, others were definitely distinctive. Each deposit had, in fact, a suite of fossils peculiar to itself. Thus he was able to distinguish the different rock formations from the Lias to the Gault by means of the fossils found in them. Smith had discovered that the special assemblage of fossils representing the organisms that lived during a certain interval of time never occurred earlier, and never appeared again. The relative age, or position in time, of a formation could thus be ascertained from its distinctive fossils (Table 7.1).

In France, Cuvier (1769–1832) and Brongniart (1770–1847) made the same discovery by collecting fossils from the rocks around Paris. These beds continue the sequence of strata upwards from the Chalk. By 1808 it became possible to correlate the older rocks of England with those lying below the Chalk in France; and similarly to correlate the younger successions of France with those lying above the Chalk in England.

The principle of identifying the ages of strata by their fossils (**biostratigraphy**) has now been firmly established all over the world. Strata in Europe and Australia, for example, are now known to be practically contemporaneous if they contain certain fossils.

Fossils are not only invaluable aids to correlation but they can indicate time-lines more precisely than lithology. The rate of migration of particular species from one region to another does not involve any practical difficulty because the intervals represented by even the smallest divisions into which we divide geological time may run into hundreds of thousands of years. In comparison the time required even for worldwide migration of many organisms is relatively short. Everywhere the sequence of fossils reveals an unfolding of different forms of life, and thus it

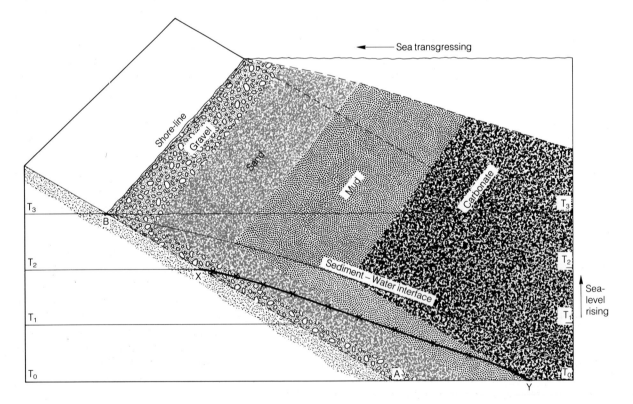

Figure 7.9 Marine sediments developed in relation to a shoreline affected by rising sea-level. At any one time (say T) the sediments form as belts of gravel, sand, mud and carbonate away from the shore. As sea-level rises from T_0 to T_3 the strand-line moves landwards as do the belts of sediment. The lithological boundaries cut across the time lines e.g. the gravel at A was deposited at T_0, many years before the gravel at B (T_3). A particular event, such as a volcanic eruption (XY) may spread ash over the whole area. It is independent of the processes controlling the sediment (facies) belt and forms a time horizon through the sequence when the whole has been lithified to conglomerate, shale and limestone.

becomes possible to divide the whole of the fossil-iferous stratified rocks into appropriate divisions or units each unit having its distinctive fossils and a definite chronological position.

The stratigraphic use of fossils for purposes of correlating rocks from one area to another may proceed without any theoretical interpretation of the nature of evolution. But for the last two centuries the factors controlling biological change have been the subject of a debate marked by a distinct dichotomy of views. Perhaps the simplest interpretation of the extinction of faunal or floral assemblages, and one favoured early in the nineteenth century by French geologists in particular, was that of extinction by catastrophe. On the other hand, with the emergence of evolution as a dominating idea in biological change, **catastrophism** was challenged by notions of gradualism. Diverse forms were thought to arise from ancestral stocks by gradual changes imposed by physical conditions and biological competition; unsuccessful forms died out over variable periods of time. Wherever there were gaps in the record with apparently sudden changes in fossil content, it was maintained that this was due either to lack of preservation or inadequate collecting or both.

Palaeontological researches have proceeded with ever-increasing intensity throughout this century, but in spite of these efforts many of the gaps marking sudden extinctions followed by sudden increases in the number of forms (referred to as radiations) still remain. Thus the major boundaries in the stratigraphic column (see Table 7.1) have been reinforced e.g. the decline of the Cambrian faunas and the Ordovician radiation; the decimation of the Palaeozoic faunas and floras at the end of the Permian period; dinosaurs and other Mesozoic forms superceded by the mammalian faunas of the Tertiary. Linked with modern support for catastrophism is a process which has been called **punctuated evolution**. Long periods of little change in widespread forms are ended by mass extinctions and the vacated ecological niches are then taken over by forms which have diversified in relatively isolated areas.

Much attention has been paid to the Cretaceous/Tertiary (K/T) boundary. At a number of places throughout the world, including the oceans, where sedimentation was continuous through the boundary, examination of clays has shown some chemical and mineralogical peculiarities. In particular the boundary is marked by an enrichment in the element, iridium. The significance of this anomaly lies in the fact that this element is more abundant in planetismals and asteroids than in the Earth. So it is maintained that the Earth was hit by an extra-terrestrial body, the catastrophic and climatic effects being the

reason for much of contemporary life, for example, large foraminifera in the oceans, and on land the dinosaurs, becoming extinct. The topic is still hotly debated; it is pointed out that excessive volcanism, perhaps associated with the outpouring of the lavas of the Deccan Traps in India at that time, would have climatic effects to which some organisms could not adapt. Some palaeontologists while accepting the reality of different rates of evolution would resist catastrophism and instantaneous extinctions. With regard to the dinosaurs and the K/T boundary many would maintain that these reptiles had been gradually dying out for some time during the Cretaceous. On the other hand the case for mass extinctions by catastrophe involving not the survival of the fittest but the survival of the luckiest has been developed recently in a stimulating and provocative account *Wonderful Life* by the American writer S.J. Gould.

7.5 AGE DATING

Superposition and changing fossil content makes it possible to date sedimentary rocks relative to one another without being able to give an actual date in years. Lavas interbedded with sediments can also be placed according to their position in the sequence; intrusive igneous rocks present a different problem but it is still possible to date their formation relative to their enclosing rocks. Dykes and granite batholiths have cross-cutting relationships with the country rocks; they are clearly later (Figure 3.11 and Plate VIIIa). Moreover as hot magma is introduced into the colder surrounds, the igneous rock is chilled, i.e. it has fine-grained margins and the country rock is heated up and metamorphosed. These features are important in conformable intrusions such as sills (Figure 3.12). At first sight these are like lavas and might be interpreted in terms of superposition. This would clearly be wrong because the sill was intruded at some time, perhaps a very long time, after the sediments were formed. The sill however can be recognized as being later (and thereby distinguished from a lava) by possessing chilled margins and having caused some baking of the enclosing rocks. The same principles of cross-cutting and chilling can be used in igneous complexes where there has been successive intrusive phases.

Sills are younger than the country rocks but how much younger? The Whin Sill (Figure 12.16) forms an important topographic feature running across the north of England; Emperor Hadrian used the escarpment to build his wall in an attempt to keep the Picts out of Roman England. The sill intrudes and is therefore later than lower Carboniferous

Table 7.1   Classification of geological Eras and Periods

Distinctive life of the geological periods

CENOZOIC	Recent	Modern man
	Pleistocene	Stone Age man
	Pliocene	Great variety of mammals Elephants widespread
	Miocene	Flowering plants in full development Ancestral dogs and bears
	Oligocene	Ancestral pigs and apes
	Eocene Paleocene	Ancestral horses, cattle and elephants appear
MESOZOIC	Cretaceous	Extinction of dinosaurs and ammonites Mammals and flowering plants slowly appear
	Jurassic	Dinosaurs and ammonites abundant Birds and mammals appear
	Triassic	Flying reptiles and dinosaurs appear First corals of modern types
PALAEOZOIC	Permian	Rise of reptiles and amphibians Conifers and beetles appear
	Carboniferous	Coal forests First reptiles and winged insects
	Devonian	First amphibians and ammonites Earliest trees and spiders Rise of fishes
	Silurian	First spore-bearing land plants Earliest known coral reefs
	Ordovician	First fish-like vertebrates Trilobites and graptolites abundant
	Cambrian	Trilobites, graptolites, brachiopods, molluscs, Crinoids, radiolaria, foraminifera Abundant fossils first appear
	Late Precambrian	Scanty remains of primitive invertebrates, sponges, worms, algae, bacteria
	Earlier Precambrian	Rare algae and bacteria back to at least 3000 m.y. for oldest known traces of life

The Stratigraphical sequence

Eras	Periods and systems	Derivation of names	Age (× 10⁶ years)
CENOZOIC *Kainos or Cenos* = recent *Zoe* = life (Recent life)	QUATERNARY Recent *or* Holocene	*Holos* = complete, whole	
	Glacial *or* Pleistocene	*Pleiston* = most	−2
	TERTIARY Pliocene	*Pleion* = more	
	Miocene	*Meion* = less (i.e. less than in Pliocene)	
	Oligocene	*Oligos* = few	
	Eocene	*Eos* = dawn	
	Paleocene	*Palaios* = old	
	The above comparative terms refer to the proportions of modern marine shells occurring as fossils		
			−66
MESOZOIC *Mesos* = middle (Mediaeval life)	CRETACEOUS	*Creta* = chalk	−144
	JURASSIC	*Jura* Mountains	−208
	TRIASSIC	*Threefold* division in Germany	
	(New Red Sandstone = desert sandstones of the Triassic Period and part of the Permian)		
			−245
PALAEOZOIC *Palaios* = ancient (Ancient life)	**UPPER PALAEOZOIC** PERMIAN	*Permia*, ancient kingdom between the Urals and the Volga	
			−286
	CARBONIFEROUS	*Coal* (carbon)-bearing	−360
	DEVONIAN (Old Red Sandstone = land sediments of the Devonian Period)	*Devon* (marine sediments)	
			−408
	LOWER PALAEOZOIC SILURIAN	*Silures*, Celtic tribe of Welsh Borders	
			−438
	ORDOVICIAN	*Ordovices*, Celtic tribe of North Wales	−505
	CAMBRIAN	*Cambria*, Roman name for Wales	
			−570
	PRECAMBRIAN ERAS *PROTEROZOIC	*Proteros* = earlier	
			−2500
	*ARCHAEOZOIC or ARCHEAN	*Archaeos* = primaeval	
	The time between the formation of the Earth (4600 m.y.) and the oldest dated Archean rocks (3800 m.y.) is referred to as Pre-archean		

* These are sometimes differentiated as EONS together with PHANEROZOIC ('life appearing') which comprises Cambrian to the present.

88

rocks and experience suggested that it belonged to one of the two main phases of igneous activity which have affected the region, either Permo-Carboniferous or Tertiary; the former about 300 m.y. before the latter. The problem of the sill's age was only solved when pebbles of the sill were found in nearby Permian rocks; the sill evidently belonged to the earlier phase. Derived pebbles therefore provide another important criterion of relative age.

These methods of relative dating were successful enough to allow, throughout the last century, the building up of the stratigraphic column (Table 7.1). But there was a serious limitation. The nineteenth-century geologist was in the position of a historian who might have evidence that the Roman Empire existed before the British Empire but was unable to put actual dates in years on either. It was not until the discovery of radioactivity and the spontaneous decay of some elements in naturally-occurring minerals that measurements of ages (referred to sometimes as absolute or, more accurately, radiometric dating) that measurements of ages in years could be attempted (Chapter 14). This century has seen a continuous refinement in methods and allocation of absolute ages to the established divisions of the stratigraphic column. While it is true that radiometric dating is an essential and powerful technique, the traditional methods of relative dating continue to be essential as a means of establishing sequences of events and providing a framework both to check, and be checked by, radiometric measurements.

7.6 THE STRATIGRAPHIC TIME SCALE

As the book of Earth history is immensely long, it has been found convenient to divide and subdivide its contents in much the same way as a long book is divided into volumes, chapters, paragraphs and sentences. If the book is read consecutively, then the order of the sentences, etc. represents a time order. In building up stratigraphy on a firm basis it has been found desirable in the interests of clarity of thought to employ sets of terms for three kinds of unit; one for the time interval, one for the strata which were deposited during that time interval and one for rocks irrespective of their time connotation. Respectively these are **time units, time–rock units** and **rock units** as illustrated in Table 7.2.

Because of the abundance and variety of the fossils that characterize the Jurassic strata, the above scheme has been very successfully applied to the Jurassic Period and System. The System is divided into three Series, Lower, Middle and Upper, each of which in turn is subdivided into three or four Stages. In the example given, the Portlandian Stage is named after the well-known Portland Limestone of southern England. It should be noticed that terms like **Jurassic** are commonly used as if they were nouns instead of adjectives. For example, it is customary to speak of the Portland Limestone as belonging to the Jurassic (System); or of certain ammonites as having flourished during the Jurassic (Period).

The names for the Eras, Periods and Systems (as listed in Table 7.1), and for a few of the smaller units, have won general acceptance internationally; but most of the names given to the smaller units have only limited application and vary from one country to another. The main difficulties in achieving the ideal that is aimed at (i.e. uniform nomenclature and precise correlation), are (a) that the field geologist does not map 'stages' but formations and (b) that the formations being mapped may be lacking in the guide fossils necessary to determine their ages.

A **formation** is a bed or assemblage of beds, of distinctive lithology and with well-marked upper and lower boundaries; it can be mapped over a considerable tract of country. A **member** is part of a formation, perhaps one or a number of beds, which can be traced over a smaller area. A number of formations go to make up a **Group** and a number of Groups may be amalgamated into a **Supergroup**. For example much of the Highlands of Scotland is formed from two Supergroups, the Moine to the north and the Dalradian to the south and the latter comprises three Groups, **Grampian,**

Table 7.2 Stratigraphic terms

Geological time units			
Age	Epoch	Period	Era
Corresponding stratigraphical time–rock units			
Stage	Series	System	
Example			
Portlandian	Upper Jurassic	Jurassic	Mesozoic
Rock units (these do not correspond to the above)			
Member	Formation	Group	Super Group

Appin and **Argyll**. Each one of the groups is made up of sequences of metamorphosed rocks hundreds of metres in thickness. All of these rock units are based on lithological features because fossils are very rare and only, (as we shall see later), when radiometric dates are obtained is it possible to assign any of them with confidence to their precise place in the stratigraphic column.

In Britain it has long been the custom to use the term 'series' as an ordinary English word and therefore to speak of a succession of formations having some features in common, as a 'series'. The more technical use of the word, which then appears as '**Series**', is for the strata deposited anywhere in the world during a particular interval of time to be known as an Epoch.

As examples of the non-technical use of the term 'series' we may consider the strata of the Carboniferous System. At the time when William Smith was first establishing his sequence of strata, the beds containing the chief English coal seams had long been known by the miners as the Coal Measures (see Section 9.5). A coal seam is a formation associated with interbedded sandstone and shales and it was natural to group the whole assemblage together as a series for which the obvious name was the Coal Measures. Underlying the latter comes another 'series' consisting of massive sandstones (often coarse-grained and referred to as 'grits') which had long been famous for the manufacture of millstones and grindstones. In consequence this assemblage was given the quarrymen's name of Millstone Grit. Beneath this in turn are the massive limestones of the Pennines, originally called the Mountain Limestone but now known collectively as the Carboniferous Limestone.

As the last-named was traced northwards through Northumberland into Scotland, it was found to pass into an increasing number of thinner beds of limestone separated by intervening strata which included coal seams (emphasizing the point made earlier about lateral variation). The three 'series' together were therefore styled the Carboniferous System from the Latin meaning 'coal-bearing'. It should be added that in North America the Carboniferous is divided into two Systems and Periods; the upper, **Pennsylvanian**, and the lower, **Mississippian**.

Table 7.1 shows the general scheme of classification by Eras and Periods that was gradually built up by the pioneer workers of the last century. It will be noticed that the Eras have names which broadly express the relations of the life forms then flourishing to those of the present. Underneath the oldest Palaeozoic beds there are enormously thick groups of sedimentary strata on every continent passing down into, or more often resting

Figure 7.10 Carboniferous Limestone lying uncomformably on an erosion surface of steeply inclined Silurian slates, Arco Wood Quarry, 6 km north of Settle, North Yorkshire. (*S. H. Reynolds*)

Figure 7.11 Conglomerate of Old Red Sandstone age, deposited during part of the unrepresented interval of the Silurian/Carboniferous unconformity of Figure 7.3. Gorge of the R. North Esk, Grampian/Tayside boundary. (*British Geological Survey*)

unconformably upon, widespread areas of crystalline rocks, metamorphic and igneous.

Only rare and obscure forms of life, usually restricted to impressions of soft-bodied organisms, have been found in the less altered strata of the ancient rocks, and so far have not proved to be of value for defining worldwide Systems. The only collective name for them all is the **Precambrian**. Since classification into Systems and Periods is based on fossils it cannot be properly extended into the Precambrian. However with the development of radiometric methods of dating rocks (see Chapter 14) it is becoming possible to recognize a few divisions which correspond approximately with Periods and Systems. It should be noted that the duration of time represented by the Precambrian rocks that have already been dated – and probably there are older ones still to be discovered – is more than five times as long as the time that has elapsed since the fossil remains of a great variety of life forms first appeared on the sedimentary rocks of the Cambrian System.

7.7 CRUSTAL MOVEMENTS AND THE GEOLOGICAL TIME SCALE

We can now return to the problem posed in Section 2 of how to recognize the strata corresponding to the time gap implied by the unconformity illustrated in Figure 7.3. From fossil evidence, it is known that around the English Lake District (Figure 18.9), and beneath the Pennines, Carboniferous limestones and associated sediments rest unconformably on the upturned edges of folded beds of Ordovician and Silurian age (Figure 7.10). The time gap is therefore the whole of the Devonian period, plus any part of the lower Carboniferous that may be locally unrepresented and possibly part of the upper Silurian. At various times and places during this long interval the conglomerates (Figure 7.11) and sandstones known as the Old Red Sandstone were being deposited, mainly in a series of deep depressions, sometimes occupied by lakes, that lay between the high mountain ranges of the Caledonian orogenic belt. In various parts of Scotland the 'Old Red' is itself found resting unconformably on an eroded surface of older rocks which had been intensely folded during the Caledonian orogeny (Plate Vb). Obviously in the British area the Caledonian mountain building must have reached a climax at about the end of the Silurian or during the early part of the Devonian.

The celebrated unconformity, exposed at Siccar Point was made classic by Hutton (Figure 7.12). He not only discovered it, but having already inferred its existence, made a special expedition by boat with the confident expectation of finding it. Beneath it he recognized what he called 'the

Figure 7.12 Sir James Hall's 1788 drawing of 'Hutton's Unconformity' (see Plate Vb), Siccar Point, Berwickshire, Scotland. (Reproduced from James Hutton's *Theory of the Earth: The Lost Drawings*, 1978, by kind permission of Scottish Academic Press)

ruins of an earlier world'; a world dating from a time so remote that he could only describe it as 'inconceivably long'; a vanished world that after passing through stages of mountainous land-scapes had been worn down to a surface of upturned rocks such as can now be seen under the Old Red Sandstone at Siccar Point. Hutton had clearly realized that 'one land mass is worn down while the waste products provide the materials for a new one'. He recognized that the destruction of an old land by erosion and the construction of a new land by upheaval of the resulting sediments (hardened, turned into a vertical position and, in places, invaded by granite) implied the existence within the Earth of some agency powerful enough to bring this about. This agency he identified with 'subterranean heat'. To the effects of the Earth's internal heat he ascribed (a) the general uplift (expansion); (b) the hardening and mineral changes suffered by sediments (lithification and metamorphism); (c) the formation of granite and its forcible upward intrusion; and (d) the uptilting and dislocation of sediments on the flanks of the intrusive granites.

The supreme genius of Hutton lay in his dem-onstration that the Earth is a thermally and dynamically active planet, internally as well as externally; and that the history of the Earth can be regarded in terms of an overlapping succession of cycles. The later stages of one cycle necessarily imply the earlier stages of the next cycle. As Hutton expresses it: 'This earth, like the body of an animal, is wasted at the same time as it is repaired . . . It is thus destroyed in one part but is renewed in another'.

The long belts of folded rocks which represent the various cycles referred to, are described as **orogenic belts** to emphasize their relationship to past or present systems of mountain ranges (Fig-ure 7.13). Mountain building by folding and uplift

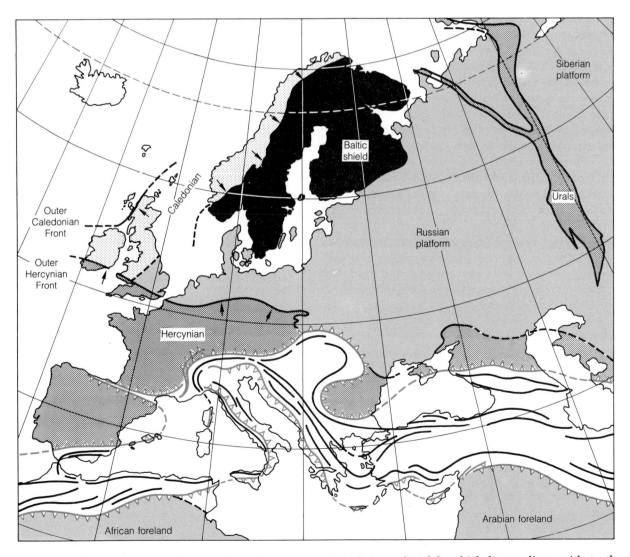

Figure 7.13 Tectonic map of Europe. The Alpine orogenic belt is outlined by thick brown lines with teeth arrowheads indicating the outward thrusts and overfolding towards the forelands.

– as a purely structural or tectonic concept, without reference to the development of high peaks and deep valleys by erosion – is called **orogenesis** (Chapter 31). The sum of the structural changes brought about during the time of mountain building is called an **orogeny** or an **orogenic revolution**.

The Alps and the Himalayas are examples of the last orogenic cycle, in which we are in fact still living (Figure 7.13). The rocks of Cornwall and Devon and their continental continuations are amongst those formed in the immediately preceding European cycle – known as the **Hercynian** (after the Roman name for the forested mountains of Germany, typified by the Harz Mountains) – a cycle that reached its culmination towards the end of the Carboniferous. The rocks of most of Scotland and of the English Lake District, Wales and Norway represent the still earlier **Caledonian** cycle. In Finland and Sweden, on the south-east side of the Caledonian orogenic belt, and in the north-west Highlands of Scotland on the other side, the rocks of very much older cycles appear at the surface.

Orogenic episodes are recorded in mobile belts which have a very complicated and often localized history of events. In order to try and detect events which may have been worldwide, attention has turned to successions in stable areas where sedimentation has responded to widespread earth movements. Those which have upheaved or depressed extensive areas on continents with little folding, if any, apart from broad undulations, are distinguished as **epeirogenesis** (Greek *epeiros*, land, continent). Spectacular epeirogenic uplifts are recorded in plateaus (Chapter 29) but movements often involve relatively small changes in level. The widespread marine sediments of former periods which now blanket much of the continents clearly record fluctuating changes of level between land and sea. These may be due to epeirogenic movements of the crust or to worldwide changes in sea-level – so-called **eustatic** changes. Geological Systems on stable continental areas are characterized by invasions of the land by the sea during which marine beds were deposited. Each invasion produces a sequence of sediments which comprises (a) a phase of advance, referred to as a **transgression** – as the sea overflows the land; culminating in (b) a phase of maximum flooding; which is followed by (c) a phase of retreat, a **regression** – as the sea withdraws or retreats as sediment is built out from the land.

When particular episodes of deposition – **sequences** – followed by phases of non-deposition can be correlated from one continent to another, then it is legitimate to infer that a worldwide event such as a eustatic rise has been responsible. The study now referred to as **Sequence Stratigraphy** is in essence an attempt to detect worldwide events and to document movements affecting smaller regions. In the early 1960s, L. Sloss showed that the Phanerozoic rocks in North America consisted of six sequences of deposition interspersed with periods of non-deposition (Figure 7.14). Later work has shown that these events are similar in time to those in South America and the USSR. Among the mechanisms thought to be responsible for these sequences, eustasy seems the most likely and such changes could be caused by growth or decline of mid-ocean ridges. Ice-ages also provide another means of changing sea-level by locking up water in ice-sheets during glacial periods and returning it to the sea during interglacials.

The concept and application of sequence stratigraphy has received a substantial stimulus from geophysicists interpreting seismic records (see Chapter 25). Picking out the principal boundaries, which mark transgressions and regressions, from seismic logs from all over the world has allowed a picture of changing sea-level to be built up through geological time. Although based on different data the curve is similar to that derived from the Sloss series of sequences. Moreover '**seismic stratigraphy**' allows very detailed analysis so that a hierarchy of

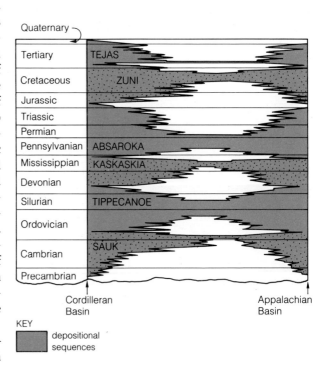

Figure 7.14 The six sequences inferred from the stratigraphic record of North America by L. Sloss, 1963.

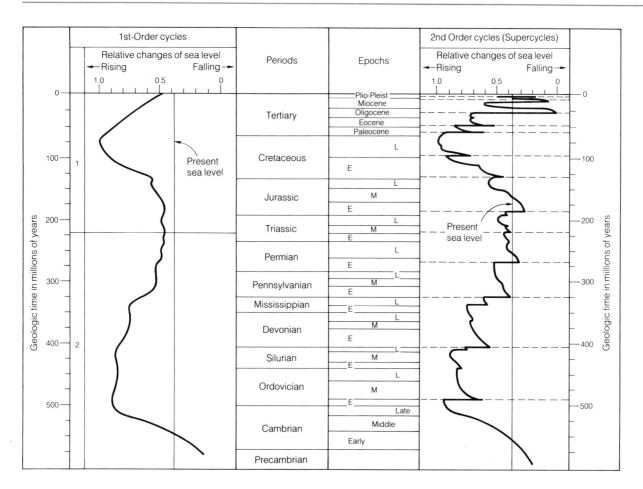

Figure 7.15 First- and second-order cycles deduced from seismic records by P. R. Vail and co-workers in 1977.

episodes or cycles of differing duration may be recognized. Of those marked on Figure 7.15, it seems likely that the long-lasting second order cycles may have been caused by ocean-ridge changes (Chapter 28). Higher-order cycles of shorter duration usually have a restricted area and may be due to local controls rather than world-wide events.

Stratigraphic studies have also been enhanced by the discovery of periodic reversals of the Earth's magnetic field. This has allowed the setting up of a parallel stratigraphic column which is of immense value, especially in the study of Recent, Tertiary and Mesozoic sediments (see Chapter 27).

Since Hutton and his friends first looked back into what Playfair called, the 'abyss of time', a fascinating picture of Earth history has been deciphered. Although new discoveries and powerful new techniques have provided new insights, the principles which he gave us in his enlightened paradigm are just as valid now as they were two hundred years ago.

7.8 FURTHER READING

Alvarez, L.W. *et al.* (1981) Impact theory of mass extinctions and the invertebrate fossil record. *Science*, **233**, 1135–41.

British Geological Survey. *British Regional Geology*, (a series of handbooks), HM Stationery Office, London.

British Geological Survey (1977) *Geological map of the British Isles*, 3rd edn, 10 miles to one inch (2 sheets), Ordnance Survey, Southampton.

Craig, G.Y. (ed.) (1991) *Geology of Scotland*, 3rd edn, Geological Society Publishing House, Bath.

Duff, P.McL.D. and Smith, A.J. (eds) (1992) *Geology of England and Wales*, Geological Society Publishing House, Bath.

Gould, S.J. (1990) *Wonderful Life*, Hutchinson Radius, London.

Holland, C.H. (ed.) (1981) *A Geology of Ireland*, Scottish Academic Press, Edinburgh.

Miall, A.D. (1984) *Principles of Sedimentary Basin Analysis*, Springer Verlag, New York.

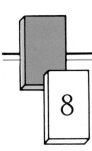

8

TECTONIC FEATURES: FOLDS AND FAULTS

Shall not every rock be removed out of his place?

Book of Job (c. 400 BC)

8.1 SCOPE, OBECTIVES AND VOCABULARY

'Tectonics' is derived from the Greek τεκτου, a builder. Tectonic features arise because the crust of our planet is dynamic: a thin, solid skin concealing a mobile interior. **Structural geology** deals with geometrical forms that develop as a consequence of this dynamism: from microscopic cracks in single crystals within rocks, to crustal blocks of continental dimensions – a scale range in space from 10^{-8} m to 10^7 m. **Tectonic studies** concentrate on the mechanisms by which these crustal structures are formed. Tectonic processes are mostly physical, but we should not overlook important connections with chemical changes. The durations over which tectonic processes operate span a range in time from 10^{-1} s for crack propagation associated with earthquakes to 10^{15} s (30 million years) or greater, for displacement of continental blocks, so some processes are more dynamic than others!

The dynamism of the Earth's crust interrupts the natural tranquillity of our lives from time to time; through destructive earthquakes and through volcanic calamities of sometimes greater proportion. In both cases, enormous amounts of energy are dissipated over very short spans of time. But we have hinted already that the crust is dynamic also on a completely different timescale: slow changes in level of sea with respect to land; slow, gentle warping of land masses; steady transfer, by wind and water, of sediment, and thus crustal load, from one place to another.

There are other mechanical changes, less obvious to us at the surface, at least immediately. This chapter looks at many of the consequences of these changes and the ways in which we can interpret them towards increasing our understanding of the Earth's crust, its evolution, and its mechanisms. Let us begin with two introductory examples before entering into the details.

Iceland (Figure 13.14) is set astride the mid-Atlantic Ridge. The ridge is active from time to time: it separates two crustal plates, one carrying North America, the other, Europe and much of Asia. The plates drift apart and new crust is formed at the ridge: sometimes a little here, or a little there. Near the surface, the mechanism works by the opening of fissures, like that shown in Figure 8.1. It is a simple matter to fix a stake in the ground, one on each side of the fissure, and to measure the distance apart of pegs set at the top of each. If this is done according to a regular timetable, we can calculate the rate of separation. If the

Figure 8.1 One of a swarm of fissures in north-central Iceland, this one with a strain gauge. Two metal pipes are fixed into the ground, one on each side. At the top of each pipe, there is a small metal peg, pointing upwards. The horizontal and vertical components of the distance between the pegs can be checked at regular intervals in time to record the rate and the direction of displacement across the fissure. The fissure helps separate the Eurasian Plate (on the left) from the North American Plate (on the right) (Figure 28.2). One small hop for one small girl from the European Plate to the American Plate – and back again. Great fun! (*R.F. Cheeney*)

rate accelerates, an eruption of lava may be imminent.

Icelandic culture has a strong oral tradition. The Sagas tell, amongst other things, of catastrophic volcanic eruptions: the Eldgja Fissure in AD 950; the Laki Fissure in 1783 (Figures 13.7 and 13.8). The outpourings associated with the latter killed one-third of the population. Clearly, the ridge is active in fits and starts; there is no steady state. So it is with other crustal structures, as we shall conclude. Our tools for such studies need not be elaborate. With the help of an Icelandic plumber to set the pipework in place, a ruler from a school bag can measure the rate of separation of North America from Europe. In addition, we can learn a great deal from careful study and reflection on seemingly long-dead objects. At about the time that William I introduced his Norman culture to England at the Battle of Hastings, Li Kung-Lin in China sketched the willow-pattern picnic scene shown in Figure 8.2, with its backdrop of an arch-like fold in beds of rock. We can follow this path; of travellers with an eye for landscape, with great profit. Simply by looking, sketching, and measuring we can set down invaluable descriptions of the present state of the crust. By induction, we can arrive at an interpretation of these features and, as circumstances may require, we may also be able to predict, or at least offer a limited number of hypotheses regarding short- and long-term patterns of behaviour that may influence our biosphere.

As with any observational science, we shall be involved in description and classification, and thus a vocabulary will emerge. That which we use in this chapter is contained in figures (or their captions), tables, etc, so as to be available for reference later.

Figure 8.2 An exposed anticlinal arch just north of the Yangtse between Hankow and Nanking, China. Painted by Li Kung-Lin about AD 1100. (*From Joseph Needham, 'Science and Civilisation in China'*)

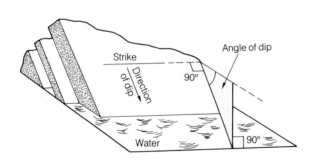

Figure 8.3(a) Sloping beds of rock intersect a horizontal surface in a direction called the **strike**. Perpendicular to the strike is the direction of **dip**, which is the line of maximum slope on the bed. The bed makes an angle with the horizontal called the **angle of dip**, or just 'dip' for brevity if the context is clear.

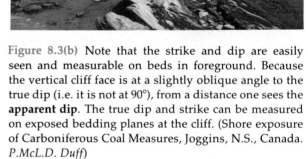

Figure 8.3(b) Note that the strike and dip are easily seen and measurable on beds in foreground. Because the vertical cliff face is at a slightly oblique angle to the true dip (i.e. it is not at 90°), from a distance one sees the **apparent dip**. The true dip and strike can be measured on exposed bedding planes at the cliff. (Shore exposure of Carboniferous Coal Measures, Joggins, N.S., Canada. *P.McL.D. Duff*)

First, we shall look at structures that provide the evidence for crustal dynamism, then see how these structures combine to provide classes of crustal tectonic environments that may be used to categorize different crustal settings. Then we shall see how we can interpret this evidence in terms of learning something of the different ways in which crustal materials behave in response to mechanical and thermal stimuli.

Figure 8.4 Layered metamorphic rocks in a recumbent fold, Kildedalen, east Greenland. The cliff is about 800 m high. (*John Haller, Lauge Koch 1947–54 Expedition*)

Figure 8.5 Air view of plunging anticline in the Parry Island orogenic belt, looking from Peel Inlet to Erskine Inlet, Bathurst Island, Canadian Artic. The core of the anticline is being dissected by a stream and its tributaries, and a small delta is being built in Peel Inlet. (*Royal Canadian Air Force*)

8.2 THE EVIDENCE: the variety and significance of tectonic features.

1. Raised shorelines

We find bench-like features along many continental coastlines in temperate to sub-arctic latitudes. They may occur singly, or at a number of altitudes, both above and below present sea-level (Figure 3.6). To the rear of such benches, caves and stacks may occur and there are frequently accumulations of cobbles reminiscent of storm beaches, but well out of reach of even the most energetic seas recorded in historical times. Nearby, finer-grained deposits sometimes contain shells, demonstrably of Arctic species.

Such raised beaches provide clear evidence of fluctuations in the level of the land relative to the sea. Meticulous surveying of the altitudes of the beaches reveals subtle, systematic variation in their levels: they provide further evidence of very gentle warping of the crust. By inference, this warping is associated with the former presence of Quaternary ice caps and glaciers, the thickness and longevity of which was sufficient to depress

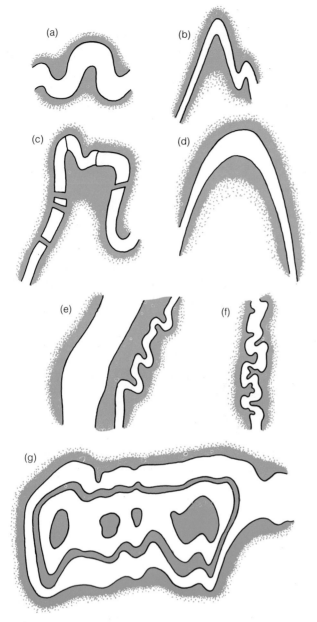

Figure 8.6 (a–f) A selection showing the variety of shapes of fold profiles. There are no scales; these folds occur in all sizes. (g) A more complicated outcrop pattern arising from the interference of two sets of folds belonging to two generations.

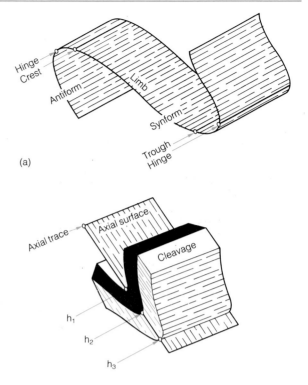

Figure 8.7 Descriptive terms for parts of folds. (a) A single-folded surface ornamented with lines parallel to its generatrix – i.e. the line which, when moved parallel to itself, sweeps out the fold. The direction of the generatrix is called the **axial direction**. There are some special positions of the generatrix: the **hinge** is the line along which curvature of the surface is a maximum; the **crest** and the **trough** are the topographically highest and lowest parts respectively. The **limb** is contained between two adjacent hinges. An **antiform** is convex upwards, a **synform** is concave upwards. (b) Three folded surfaces, bounding two beds. The hinges in the three surfaces are labelled h_1, h_2, and h_3. The three hinges are contained within an imaginary surface called the **axial surface**. The rock in one of the beds is slate; it is cleaved along a family of surfaces that is parallel to the axial surface: the fold has an **axial-surface cleavage**.

If the rocks are sedimentary, and the younger beds are in the **core** (or centre) of the fold, it is called a **syncline**. If the older beds occupy the core, the fold is an **anticline**.

the continental crust on which they developed. Furthermore, the recovery of the crust, its 'rebound', lags behind the melting of the ice by a considerable interval. Unlike a spring balance, the deflection is not recovered immediately the load is removed.

2. Horizontal beds
Amongst the most fundamental of geological observations are those relating to erosion, transportation and deposition of sediment. The last of this triplet, deposition, groups together those processes that result, in most cases, in the accumula-tion of sediment in layers at the bottom of the sea; each new layer deposited in succession on those beneath. Given a basin of sedimentation with a slowly subsiding floor, great thicknesses of beds of rock may be developed, amongst the best known of which is seen in the walls of the Grand Canyon (Figure 1.5), but with other examples readily available, for example, in sedimentary rocks in Donegal (Figure 6.3), and amongst extrusive volcanic rocks, particularly lava flows, as in Antrim (Figure 2.4).

Such successions of horizontally-layered rocks, of great lateral extent, provide a starting point for tectonic studies. Any disturbance of this horizon-

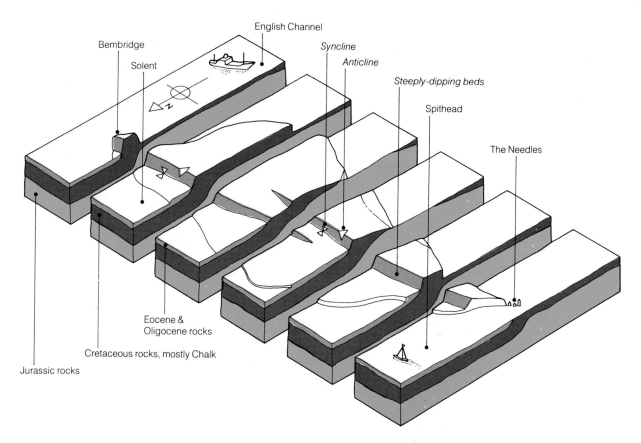

English Channel

Bembridge

Solent

Syncline

Anticline

Steeply-dipping beds

Spithead

The Needles

Eocene &
Oligocene rocks

Cretaceous rocks, mostly Chalk

Jurassic rocks

Figure 8.8 A perspective view of the geological structure of the Isle of Wight. There are some steep beds of rock; they crop out at Alum Bay, near the Needles, at the western extremity, and near Bembridge, at the eastern extremity. The steep beds occupy a fold limb contained between an anticline to the south and a syncline to the north. Beyond these fold hinges, the beds are flat lying, so that overall, the beds form a step-like geological structure (sometimes called a 'monocline') that faces north. However, this step is not continuous across the island, there is an offset at the centre, towards which the height of the step declines, both from the west and from the east. Between these two tapered ends, a **fold bridge**, or **ramp**, slopes gently from the geologically higher levels in the south, to the lower levels in the north.

tal layering, either by marked tilting or by physical disruption, must indicate that tectonic processes have most likely been active.

3. Tilted beds

Although large areas of the globe are underlain by near-horizontal strata, we may find more commonly at surface exposures that the layers are inclined, as shown, for example in Figure 8.3b. Even though the layers may be tilted to a high angle, they may well be otherwise intact, leading to the conclusion that the process involved was one of slow, sustained crustal distortion. Because the more elevated parts of the beds are subject to greater erosion, field observers are presented with exposures of inclined beds which, from place to place, may be orientated at different angles in different directions. To geologists, an important survey activity consists of measuring the inclinations and their directions, for which the terms **dip** and **strike** are used (Figure 8.3a, b). By mapping such variation in orientation over larger areas, we

may reveal a complexity of structure of a different order of magnitude, including the possibility that the beds are smoothly curved; they have been folded.

4. Folded beds

On occasion, an exposure will reveal rock layers, the traces of which are curved, like those of our Chinese anticline (Figure 8.2) and at St. Andrews, Scotland (Figure 1.4, Plate Vc). Not only do these features have great aesthetic appeal, they demonstrate also that the rocks have passed through a stage during which they were sufficiently pliable to be folded. The study of folds is an important branch of structural geology, for not only do they enable us to piece together the impressive architecture of the Earth's crust, we learn also a good deal about the various physical states through which the constituent rocks must have passed.

Despite their apparent simplicity, folds show great variation in their geometrical features; their sizes, shapes and orientations. Amongst all the structures found in crustal rocks, folds show the greatest range of size, from structures of microscopic dimensions, even smaller than those shown in Plate Vd, through exposure-scale structures of the type seen in Plate Vc to examples that can be appreciated only in mountain sides as in Greenland (Figure 8.4) or from the air as in the Canadian arctic (Figure 8.5), or from satellites.

The shapes assumed by folds seem to be a function not only of the amount by which the rocks have been distorted, but also by the types of rocks involved and the thicknesses of the individual layers. Figure 8.6 shows something of the range of profile shapes exhibited and Figure 8.7 introduces some of the terms that we need in order to describe folds and their component parts. In the third dimension, perpendicular to the profile of the folds, and thus parallel to their axial direction, folds can be traced for large, but not infinite distances. They may fade away, through gradually decreasing amplitude, or may suffer mutual cancellation, as in the case of the Isle of Wight folds sketched in Figure 8.8. The under-

standing of the three-dimensional geometry of folds like these is important because some of the layers of rock involved can be important sources of minerals, or may be reservoirs holding water or hydrocarbons.

When surveying folds, we find commonly that their axial surfaces may be inclined at any angle relative to the horizontal. Less commonly, but by no means rarely, we find that the fold hinge is inclined also. Folds that exhibit significant inclination of the hinge are said to **plunge**. With this unrestricted orientation of both axial surface and hinge, it is clear that folds may show a wide range of dispositions. However, in order to provide a manageable terminology, we can use the nomenclature summarized in Figure 8.9.

In the case of folds with very steep angles of plunge of their fold hinges, it may be very difficult to see how they may have developed from initially horizontal layers. However, the situation becomes more easily rationalized if we propose that the beds were initially tilted, and then folded about axial directions oblique to the initial strike.

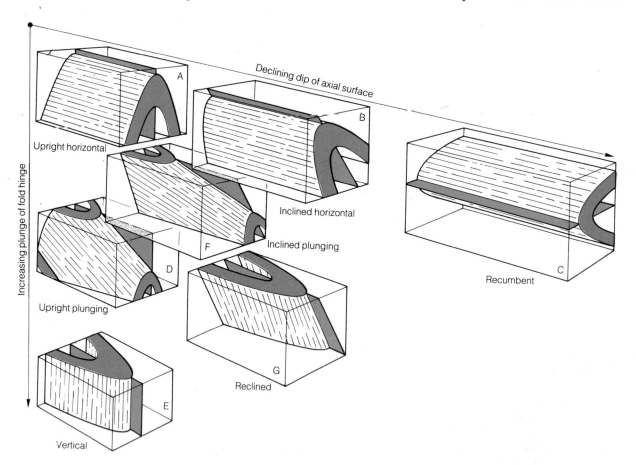

Figure 8.9 Orientations of folds. Two elements determine the orientation of a fold; the angle of dip of its axial surface and the angle of plunge of its hinge. Starting with an **upright horizontal** fold A, we may progressively decrease the angle of dip of the axial surface to define an **inclined horizontal** fold B and a **recumbent** fold C. Alternatively, we may increase the angle of plunge of the fold hinge, via an **upright plunging** fold D to a **vertical** fold E. Decreasing the angle of dip of the axial surface and increasing the angle of plunge of the hinge at the same time generates an **inclined plunging** fold F and a **reclined** fold G. (*After M.J. Fleuty, 1964*)

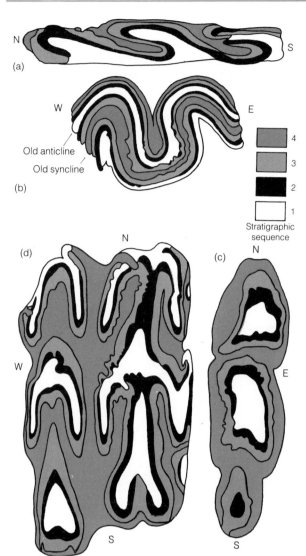

Old anticline
Old syncline

4
3
2
1
Stratigraphic
sequence

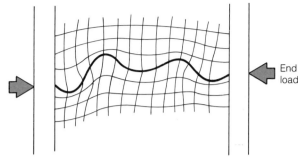

End
load

Figure 8.11 Folds in a rubber model. The model consists of a thin layer of hard rubber, on the opposite faces of which, large blocks of sponge rubber have been glued. Before the model was loaded, a square grid was marked on the sponge rubber. When squeezed laterally between wooden plattens, the end load causes the hard layer to buckle. The resulting sinusoidal strain is transmitted through the softer sponge rubber, but with decreased amplitude, becoming negligible at a distance approximately equal to that of the amplitude of the folds in the harder primary layer.

Indeed, the initial tilting may be a consequence of an earlier period of folding, providing the concept that rocks may be folded more than once. If the earlier folds and the later folds are developed on approximately the same scale, then they may interfere in a way that produces intricate outcrop shapes of the type shown in Figure 8.6 (g) and in a model (Figure 8.10). We find that such structures as these are most frequent in belts of metamorphosed rocks (Chapter 11) that have formed the roots of mountain chains (Chapter 31) in the geological past.

In order to explain the presence of folds in layered rocks, first, we have to invoke the idea that the properties of the rock material must have been substantially different from those that we associate with rocks at the surface today. Second, we must accept that the rock layers have been subject to applied forces. Neither of these requirements should present great difficulty: we know that temperature increases with depth in the crust and that elevated temperatures bring about modification of the properties of materials in general; they tend to become more pliable. Also, we know that parts of the crust are seismogenic: they are in a state of stress that is sufficient to promote rapid failure and it is not unlikely that elsewhere, the rates of deformation are much slower, leading to other modes of crustal distortion, including slow folding.

It remains, therefore, to provide a mechanism for inducing the curvature of rock layers which we describe as 'folding'. One appealing possibility is that due to elastic instability of a layer under end load (Figure 8.11), although others have been proposed.

Figure 8.10 A plasticene model that shows how complex outcrop patterns can arise from the simple superimposition of sets of folds whose axial directions are mutually inclined – at right angles in this case. Roll out three (or four) thin sheets, say about 10 cm × 20 cm, of differently-coloured plasticene to give a stratigraphic sequence as shown in the inset. With the longer side of the sheets running away from you, say North–South, form two or three recumbent isoclinal folds whose axial directions run West–East, as shown in North–South cross-section (a). Then, with care (and possibly some further precursory flattening of the model), re-fold this block about axial directions that run North–South, so that you generate new folds, like those shown in West–East cross-section (b). Now imitate the natural processes of erosion by taking a series of thin horizontal slices off the top of the model: a serrated bread knife proves to be a suitable instrument, but guard against slippages! Initially, you will see a series of loops opening up, as shown in (c), corresponding to the crests of the earlier set of anticlines where they pass over the crests of the later anticlines. As you slice deeper into the structure, you will reveal more complex outcrop forms, as shown in (d), that provide good imitations of those that occur in the centres of orogenic belts. (Beware though, it can be compulsive!) (*After a model by D. Reynolds and A. Holmes, 1954*)

5. Distorted pebbles and fossils

Whatever the mechanism, one important effect of the process of folding is that sections of the crust become shortened and in many circumstances in applied geology, it would be rewarding to be able to answer the question '. . . by how much?' Unfortunately, because of difficulties regarding the details of the mechanisms of folding, it is often impossible to provide an answer with precision or confidence. However, on rare occasions, deformed rocks are found that contain objects whose original shape is known: they can be used as strain gauges.

The flattened objects shown in Figure 8.12 (a) are strained clasts (i.e. cobbles) from conglomeratic rocks in Shetland that have been deformed on a large scale. If we make the assumption, at least to a first approximation, that these objects were initially near-spherical, then their present ellipsoidal shapes record the strains suffered by the rocks in different directions: stretching parallel to the longest axis; compression parallel to the shortest axis. A few measurements and some simple calculations will tell us by how much the rocks have been deformed. Unfortunately, few rocks contain adequate 'strain markers', and in those that do, the computations are more involved. Many such attempts have been based successfully on the study of distorted fossils such as the Welsh trilobites shown in Figure 8.13, including the work of the Rev. Daniel Sharpe, reported in 1847. Examples such as these illustrate very clearly the homogeneous (or 'smooth') way in which rocks may yield to stress under suitable conditions, almost

(a)

(b)

Figure 8.12 Clasts (a) from the Funzie Conglomerate, Fetlar, Shetland and (b) from Upper Devonian conglomerate, Upper Dounans, Aberfoyle, Perthshire, Scotland. The specimens show contrasting responses to deformation. Although the Devonian specimen has developed shear failures, these are now healed. (*R.F. Cheeney*)

Figure 8.13 Three trilobites, originally bilaterally symmetrical, have taken a sheared aspect due to strain. (*Grant Institute collections, University of Edinburgh*)

on an atom by atom basis, in many cases preserving the most intimate detail of fossilized structures.

6. Cleavage

However, despite this preservation of detail in some instances, we find that the most common effects of deformation on the microscopic texture of rocks takes place rather on a crystal by crystal basis. A class of minerals abundant throughout many rocks is the sheet silicates: clay minerals and micas being most common. Under the influence of deformation and a greater or lesser degree of metamorphism (Chapter 11), the tendency is for crystals of these minerals to align themselves in a common direction. Because the individual crystals cleave, or split, readily in this common direction, they impart a ready cleavage to the rock as a whole. If the rock is fine grained, with crystals microscopic or nearly so, it is termed a **slate**; if somewhat coarser grained and more highly metamorphosed, in which case the sheet silicates are almost certain to be micas, it is termed a **schist**. The structure that is imparted to the rock is called **cleavage** or **schistocity**, respectively. Wherever the rock contains an original sedimentary layering, we can see the two structures intersecting each other, as in Cornwall (Figure 11.4). Should the rock contain also strain markers, as mentioned above, then we find that such cleavages develop perpendicular to the axis that corresponds to greatest shortening; the rocks have been flattened in the plane of the cleavage. If the rocks yield simultaneously by folding, then we find the cleavage developed parallel to the axial surfaces of the folds (Figure 8.7).

7. Joints

While slates and schists tend to split with ease and regularity along their cleavage or schistocity surfaces, forming slabs, all rocks, without exception, are pervaded by joints, the most common and universal structure found in lithified earth materials. Joints take the form of plane or curved surfaces of discontinuity across which there is little or no cohesion (Figures 8.14 and 8.15). The surfaces of individual joints may be smooth or decorated; variably or systematically orientated; regularly or irregularly spaced; and with lateral extents that are small (a few millimetres) or large (several hundred metres). In certain zones, joints may be profusely developed (Figure 8.16), giving rise to linear features from which the highly-fractured rock is easily eroded. These 'master joints' may form topographical features that are prominent on aerial photographs and satellite images. Although joints are easily the most abundant structures of tectonic origin, yet paradoxically, they are the least understood and a profusion of mechanisms has been advanced in explanation.

That joints are of importance is self evident. Regularly spaced and orientated joint surfaces allow the ready extraction of stone for building or paving and of coal (Figure 8.14); close-spaced,

Figure 8.14 Rectangular set of joints in 3 m thick Permian coal seam, Blackwater, Queensland, Australia. Jointing in coal is known as cleat and is usually very consistent in directions throughout the coalfield. (*P.McL. D. Duff*)

irregularly orientated joints provide material that is worked with ease as a bulk mineral for aggregates used in the construction industry. Great value has been placed on those rare instances where joints are widely spaced. The Ross of Mull Granite in western Scotland provides an example, yielding joint-free blocks of stone up to 5 m in dimension and formerly used in lighthouses and harbour works worldwide. On the other hand, joint surfaces with critical spacings and/or orientations are crucial agents in weathering and erosion (Figure 15.8) and can render natural or artificial rock slopes unstable, thus hindering construction work on the surface and underground. Systems of joint surfaces are most significant also in relation to transport properties of rocks, determining the way in which fluids can pass through. Such fluid circulation through natural fractures is an important consideration in hydrogeology, hydrocarbon exploitation, in waste containment, reservoir engineering, tunnelling, etc.

There can be little doubt that joints arise

because of the operation of a failure mechanism within the rocks, but the nature of the mechanism can be surmised in a relatively small number of cases, and verified even more rarely. As in many such situations, we are dependent on the chance discovery and recognition of particularly clear instances, the interpretation of which can be argued with confidence.

8. Some fracture systems and their mechanisms

Stone columns that are polygonal in cross-section, like those of the Giant's Causeway and at Fingal's Cave on Staffa, develop in many extrusive igneous rocks and are perhaps amongst the best-known of such joint features whose mechanism seems clear. They are described in Chapter 13. They arise by tensile fracture of the rock caused by thermal stresses induced as the rock shrinks during cooling and have a direct analogy with the polygonal shapes of mud flakes that develop as a lake bed or muddy puddle dries out and the mud surface dessicates.

Rocks also expand when released from long periods of burial, whether by natural erosion or by artificial excavation. If the amount of elastic strain energy that has been stored in this way is dissipated sufficiently rapidly, the rocks will fracture spontaneously by tensile failure as micro-cracks propagate and coalesce. In tunnels and other artificial excavations in 'hard' rocks, such tensile fracture may take place with explosive violence, giving rise to 'rockbursts' that are potentially hazardous.

Natural fractures of the same class may be seen in certain areas of 'hard' rock that have been buried deeply during glacial times beneath thick accumulations of ice, as in Norway (Figure 8.17). In such areas, wastage of ice removes a load quite rapidly, in geological terms, and the stored strain energy is released as fractures develop concordantly with the land surface. The result is that sheets of rock separate via a process sometimes known as exfoliation, although especially in humid climates, the mechanism may also include, or be dominated by chemical alteration of some of the constituent minerals in a way that promotes expansion.

Although strictly speaking not joints, the study of the details of infilled fractures such as those shown in Figures 8.18, 8.19 and 8.20 gives valuable insight into the mechanisms involved. These structures have a symmetry to which the term **conjugate** is applied, meaning that their component members have features in common, but which are opposite or reciprocal in some character. Thus, they form conjugate zones of failure that traverse the rock mass, dividing it up into wedge-shaped blocks. These wedges have advanced upon

Figure 8.15 Jointed dunite, Rum, Inner Hebrides, Scotland. (*R.F. Cheeney*)

Figure 8.16 A 'master' joint in Dalradian quartzite, Kinlochleven, Argyllshire, Scotland. Scale units are in inches. (*R.F. Cheeney*)

each other, or retreated, depending on whether the acute or obtuse angle is subtended. Analogous structures occur in another guise, as conjugate folds (Figure 8.21), and similar features can be induced in the laboratory in cylinders of rock loaded to failure in testing machines (Figure 8.22).

Despite these clear examples, we must, nevertheless, reiterate the note of caution, emphasizing

Figure 8.17 Sheet joints in granite in a Norwegian valley, developed in response to release of residual stresses following glacial unloading. The face is about 100 m high. (*R.F. Cheeney*)

Figure 8.18 Conjugate systems of quartz-filled veinlets in sandstones north-west of Achnashellach, Ross-shire. For an interpretation, see Figure 8.20.

that most fracture systems in rocks, including the majority that appear to have no obvious pattern, remain to be analysed and fully understood.

9. Shear zones

Above, we have seen how rocks fail along discrete surfaces, or along zones of finite thickness. Another class of failure may be seen, most usually in coarser-grained metamorphic rocks, in which displacement between opposing blocks gives rise to a class of schistosity seen in a Greenlandic example in Figure 8.23. Such shear zones develop with widths from a few millimetres up to several kilometres. By careful measurement and calculation, it is possible to estimate the amount of displacement across such zones and it is realistic to argue that they are probably the correlatives at depth in the Earth's crust of certain near-surface faults, treated below.

10. Faults

Next to joints, faults constitute the most abundant structures of tectonic origin in the near-surface

crust. Like most joints, fault surfaces mark discontinuities in rock masses, across which, the transmission of forces is dependent on friction. However, they differ from joints in giving evidence of relative displacement of the opposing blocks of rock: at some stage, the frictional resistance has been overcome.

The displacement across the fault surface, the **net slip**, may be in any direction (Figure 8.24). Depending on the relative values of the components of the net slip and on the angle of dip of the fault surface, we can classify faults as shown in Table 8.1. In most simple examples, this displacement is seen in the disturbance of continuity of beds of rock, as shown in Plate Ve in an example from southern Scotland. In this case, both the beds of rock and the fault surface itself are near-vertical and, as we note in the caption, there is a dextral (right-handed) component of strike slip on the fault that amounts to a few millimetres.

The shapes of fault surfaces are important; they rarely approach true planes. In general, the fault surface is curved, and this gives rise to space

Figure 8.19 Calcite-filled primary and secondary sigmoidal veins in a siltstone, Dent de Morcles, Switzerland. (*R.F. Cheeney*)

problems if the adjacent fault blocks are rigid. This is because when the opposing blocks are displaced, they cannot remain uniformly in contact; voids must develop opposite concavities in one wall of the fault, causing extra loads to be applied where opposed convexities occur. The voids may be filled by broken rock, or may provide sites at which important minerals are subsequently deposited from circulating fluids. Where the fault blocks are not rigid, then the space problems can be accommodated through deformation of one, or both walls. The deformation mechanism can involve subsidiary faulting and/or folding. There are many variations on this theme, but as one example, we note that normal faults developed in basins of active sedimentation are commonly concave upwards and curved in plan. If they could be exhumed, they would resemble the shape of a shovel. The Greek word for shovel is λιστρου, so these structures are called **listric faults**. To accommodate the concavity as the hanging wall slips down, the beds, initially horizontal, in the block resting on the fault must become (gently) convex upwards; they develop a **roll-over anticline**.

The displacements on faults may reach 50 km or more, but we must not assume that such displacements occur in a single event. Like volcanoes, major faults lie dormant for much of the time; their total displacement is accumulated in increments of up to a few metres between these long periods of dormancy (Figure 25.1), even in the crust's most active zones. Because each increment of displacement occurs within a time span of a few seconds, they are immensely destructive, not only in their primary effects, but also by virtue of their ability to trigger huge ocean waves and mass landslides, both above and beneath sea-level. But the social significance of faults extends further. They may provide conduits through the Earth's crust along which fluids and gases may leak out, providing spas and other mineral springs, etc., or down which, fluids may leak away, as water from reservoirs, for example. During the geological past, circulating fluids of dominantly aqueous composition have deposited valuable concentrations of minerals on fault surfaces. In other settings, hydrocarbons have been trapped against the impervious hanging walls of certain faults. Further, the lateral continuity of coal seams, or other valuable stratiform deposits, may be interrupted by virtue of fault displacements, leaving the mining geologist the tantalizing problem of locating the seam's continuation.

As a further consequence of these displacements of one block of rock against another, at relatively shallow crustal depths, it is inevitable that the rocks are damaged, or at least modified in some way adjacent to the fault surfaces. Thus, we recognize a variety of products, the occurrence of which, in some cases, then provides us with valuable clues to the likely presence of faults, which might otherwise be obscured.

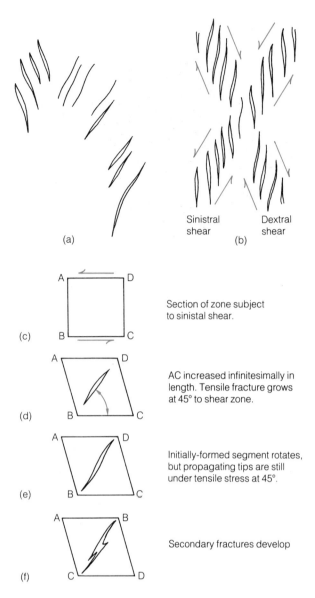

Sinistral shear

Dextral shear

Section of zone subject to sinistal shear.

AC increased infinitesimally in length. Tensile fracture grows at 45° to shear zone.

Initially-formed segment rotates, but propagating tips are still under tensile stress at 45°.

Secondary fractures develop

Figure 8.20 The formation of conjugate arrays of tensile fractures: (a) shows a selection of veins of sigmoidal shape, sketched from Figure 8.18, while (b) is an idealized portrayal of structures of this type. If we take a section (c), initially square, from one of the shear zones, and subject it to an initial small increment of shear (d), then the diagonal AC is stretched slightly and the material will tend to crack along the other diagonal BD. Should shear progress further (e), the section of the crack that was formed first will be rotated, but the growing tips will still propagate at right angles to the direction of tension, at 45° to the shear zone. In a few instances, (f) and Figure 8.19, shear continues to the extent that a second set of cracks is initiated as the first set becomes dormant.

Breccia, from the Italian word referring to road metal, is an aggregate of angular, broken fragments of the rocks that compose the fault walls. The fragments may range in size, but are generally taken to be at least visible to the naked eye. The

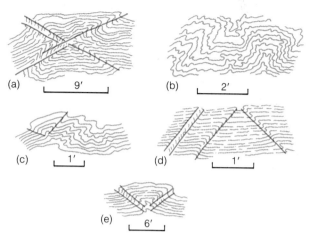

Figure 8.21 Conjugate folds. Typically, these folds are developed in laminated rocks: these examples occurring in the Moine Thrust Zone of Wester Ross, Scotland. They form by a mechanism that involves failure along paired dextral and sinistral shear zones. (*M.R.W. Johnson*)

whole mass is usually held together by some cementing material, commonly calcite or quartz, with lesser amounts of other minerals. On occasion, the infill between the breccia fragments may be of considerable economic importance and many rich veins of metalliferous ores occur in this setting.

The mode of formation of fault breccias seems clear; they arise through fragmentation of the fault walls as a consequence of the fault displacement, with subsequent infilling, or partial infilling of the voids by precipitation from circulating fluids. We can surmise that in certain cases, there is a causal link between the development of master joints, described above, and the subsequent failure of the intensely-jointed zone during faulting.

Fault gouge, or clay gouge, is a clay-like material, yellowish or greyish, that occurs at the interface of the fault blocks and which may contain fragments of breccia on occasion. This association with entrained fragments suggests that fault gouges evolve by further, intense crushing of previously formed breccias. The fine-grain size of the resulting particles, with proportionately increased surface areas, renders them susceptible to accelerated chemical reaction with circulating groundwaters, leading to enhanced production of clay minerals and other products of chemical alteration. It is possible also that the resulting clay gouge becomes impervious, thus effectively sealing the fault surface against further fluid circulation in the long term.

Fault gouges, and many breccias, are weak rocks, highly susceptible to erosion. Because of this, faults are not commonly well-exposed at the

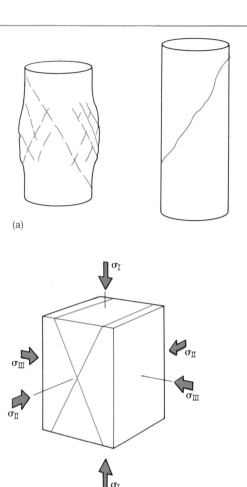

(a)

(b)

Figure 8.23 A shear zone in Kap Hodgson Granodiorite, East Greenland. (*R.F. Cheeney*)

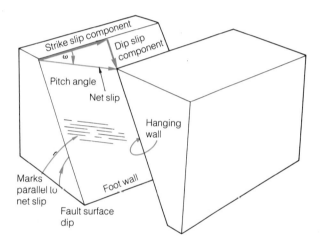

Figure 8.24 Nomenclature for faults. A fault surface separates two blocks of rock, the opposing faces of which are called the **hanging wall** on the upper side of the fault surface, and the **footwall** on the lower side. The displacement across the fault is the **net slip**, the line joining once coincident points. The net slip, a vector quantity, can be resolved into two perpendicular components, the **strike slip** and the **dip slip**, parallel to the strike and to the dip of the fault surface, respectively. The angle between the net slip and the fault surface strike is the **pitch** of the net slip. The angle of dip of the fault surface is measured conventionally. The fault surface may be marked, commonly by slickensides, in a direction parallel to the most recent increment of net slip.

Figure 8.22 Failure and principal stresses. (a) Rock cylinders loaded to failure in laboratory experiments. The specimen on the left has failed on conjugate surfaces orientated in such a way that the applied load bisects the acute angle. The cylinder on the right has failed along a single surface, showing that symmetry is not always maintained, possibly because the structures are sensitive to slight misalignment. Across any such surface in a three-dimensional body, the force that acts may be resolved into two components, acting perpendicular and parallel to the surface and known as the **normal** and **shear** stresses respectively. Mathematically (b), the systems of normal and shear stresses can be summarized and represented by three normal principal stresses acting along mutually perpendicular axes. According to their values, they are designated maximum (σ_I), intermediate (σ_{II}), and minimum (σ_{III}). Should conjugate failure surfaces develop, they intersect along the intermediate principal axis and the acute bisectrix between them is parallel to the maximum principal axis. For other treatments where stresses are involved, we find it convenient to define two further terms. The **mean stress** is simply the arithmetic average of the principal stresses $[\sigma_I + \sigma_{II} + \sigma_{III}]/3$, and is connected with the weight of overburden in the Earth's crust, while the **principal deviatoric stresses** (or 'deviatoric stresses' for brevity) are the amounts by which each of the principal stresses differ from the mean stress. The deviatoric stresses are important in considering any change in **shape** (as opposed to change in **volume**) that a body of rock may undergo.

surface, their traces indeed exploited by the natural drainage system so that many valleys and gorges are controlled by bedrock structures, e.g. the Insubric Fault in southern Switzerland, Plate Vf. Less commonly, a well-cemented fault breccia is more resistant to erosion than wall rocks

Table 8.1 Classification of faults

Name	Dip slip*	Strike slip	Pitch	Dip
Normal	Large, down dip	Small	Large positive	Large
Reverse	Large, up dip	Small	Large negative	Large
Lag	Large, down dip	Small	Large positive	Small
Thrust	Large, up dip	Small	Large negative	Small
Wrench	Small	Large, left or right	Small	Near-vertical

* As seen on footwall.

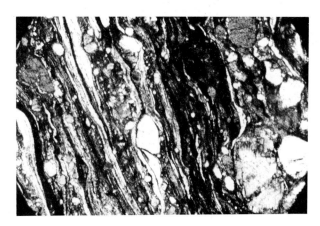

Figure 8.25 Mylonite from the Kap Hodgson Grano-diorite, East Greenland. Field of view is 4 mm wide. (*R.F. Cheeney*)

adjacent and it may stand out as a dyke-like topographical feature.

Mylonite, Figure 8.25, by contrast, is a laminated, fine-grained rock, differing essentially from breccias and gouges in that it is crystalline, and coherent as a consequence. Electron microscope studies suggest that these rocks develop by intense deformation at elevated temperatures, in a mode that allows partial recovery of the crystalline material, thus allowing retention of physical coherence between adjacent grains.

Psuedotachylite is a dark, glassy material, uncommon, but occurring typically in branching veins stemming from fault surfaces in areas of metamorphosed rocks. It appears to represent the products of intense, perhaps shock deformation of the fault walls at greater depths, with production of mobile, liquid material.

We do find situations where, by contrast, constructive rather than destructive processes have taken place adjacent to fault walls. **Slickensides**, Figure 8.26, provide an example where fibrous crystals, commonly of calcite or quartz, have grown in parallel alignment on the fault walls by precipitation from fluids, indeed, thus linking one wall to the other and sealing the fault surface. We interpret slickensides as arising by the slow displacement of one wall relative to the other, at a rate that allows fibrous crystal growth to keep pace. This being the case, the orientation of the slickenside fibres records the direction of the most recent increment of net slip on the fault surface and provides data that may be critical in our structural analysis of the rocks nearby.

In very rare instances, we may find that information of greater detail is available from the wall rocks. Thus, for example, clasts in conglomerate adjacent to the Highland Boundary Fault in Scotland, Figure 8.12, are crossed by annealed fractures that form conjugate sets. From the orientations of these sets, we can infer the distribution of stress at the time of fault displacement.

8.3 STRUCTURAL SETTINGS, STRUCTURES AND LANDFORMS: the associations of structures and their influences on geomorphology

The structures that we have introduced above occur in various combinations with each other in different structural settings in various parts of the world. They also give rise to a wide variety of consequent landforms. Unfortunately, this variety is so great that we can give only a few examples here and have to encourage you to research these examples in depth, or to uncover further examples.

Amongst the simplest of features that arise by erosion are those in areas of horizontal beds or in areas of near-horizontal thrusts. Where beds of rock are horizontal, hill tops preserve outcrops of the younger rocks that are completely surrounded by outcrops of older rocks on the lower slopes (Figure 8.27a). Such an erosional remnant is termed an **outlier**. Similarly, deep valleys may expose older rocks that are completely surrounded by outcrops of younger rocks. These outcrops of older rocks are called **inliers**. If we replace the bedding with a thrust fault, then for these new

Figure 8.26 Fibrous quartz on a fault surface, forming slickensides. Grantshouse, Berwickshire, Scotland. The coin is 20 mm in diameter. (*R. F. Cheeney*)

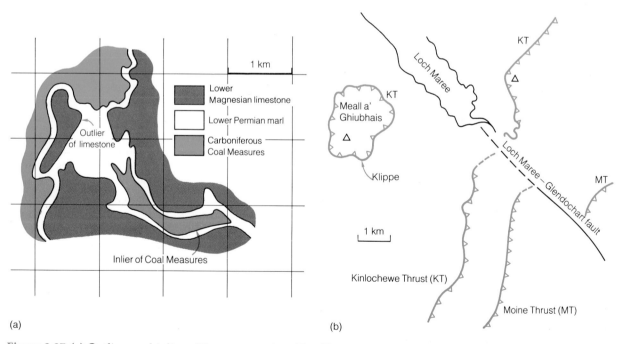

(a)

(b)

Figure 8.27 (a) Outliers and inliers. These are erosional landforms that develop amongst sedimentary rocks whose angle of dip is low. These examples occur north-west of Mansfield, northern England. An **outlier** is an outcrop of younger rock, completely surrounded by older rocks, and typically occurs on higher ground. Conversely, an **inlier** is an outcrop of older rock completely surrounded by younger rocks, and is usually found deep in a river valley. (b) Shows a **klippe**, a structure analogous to an outlier, but rather an outcrop of part of a thrust sheet isolated by erosion. This example is from the Moine Thrust Zone of north-west Scotland, where rocks on top of the Kinlochewe Thrust have been isolated on the mountain Meall a'Ghiubhais. Not illustrated here is a **fenster**, the thrust sheet analogue of an inlier.

Figure 8.28 Looking north-west along the San Andreas fault zone between the southern Temblor Range (visible in upper right) and the Caliente Range (out of sight below the lower left), south-west corner of the Great Valley of California. (*J.S. Shelton and R.C. Frampton, Claremont, California*)

situations, we must introduce **klippe** (Figure 8.27b), (German: cliff) and **fenster** (German: window) respectively.

Faults are such common structures that it is almost natural that we should search for, and find associations of faults that fall into clearly defined patterns. Amongst the classical examples are the **horst** (German: eyrie) and **graben** (German: ditch, trench) structures that contribute to major crustal features such as the African rift system (Figure 29.6). In the USA, the Basin and Range Province (Figure 29.5) provides an example of a terrain divided by faults into rectangular blocks.

Of the major fault zones that have attracted attention, two types dominate: wrench faults and thrust faults. Amongst the former, the San Andreas Fault must be the best-known example (Figure 8.28), but others are well documented: the Great Glen Fault in Scotland (Figure 8.29) and the Alpine Fault in New Zealand. Thrust faults generally feature less prominently, but there are some good examples in the literature: the Glarus Thrust in the Alps (Figure 8.30), probably the first recognized; the Keystone Thrust in Nevada; the Moine Thrust in Scotland.

In the marginal parts of mountain belts, the successions of horizontal beds that accumulate

during the earlier stages of the geological history commonly suffer later folding, ranging from gentle to severe, the latter invariably associated with thrust and wrench faults. There are abundant examples of such zones: the Appalachians; the Canadian Rockies; the Zagros Mountains and the Alpine Jura providing accessible instances.

8.4 THE INTERPRETATION: the physical behaviour of crustal materials

If we take a piece of rock from the Earth's surface, granite for example, we expect that if we exert a sufficiently large force upon it, the rock will fail by fracture; it is **brittle**. We know also that if we heat certain relatively inert materials, their mechanical properties change; hot steel is readily forged; glass is easily blown into intricate shapes. These forged or blown shapes are retained when the material is cooled. During the time of their elevated temperatures, the steel and the glass are **ductile**. In the cases of these materials, the transition between brittle and ductile states is governed by temperature and we anticipate that rocks behave similarly, according to their depth within the crust.

Silicone putty provides an example of a material that is brittle if, suddenly, we exert a sufficiently large force; it will splinter if we hit it with a hammer. Yet the same specimen at the same temperature will spread out slowly over the table top if we leave it at rest under its own weight; it is ductile. Here the transition between brittle and ductile states is governed by the state of stress which in itself determines the rate of deformation. Ice, as in glaciers, exhibits similar behaviours; flowing and/or fracturing as the confines of the valley dictate (Chapter 20).

This behaviour under various loads may be investigated in laboratory experiments termed creep tests. Figure 8.31 shows some results of such tests, in which the deformation of the specimen under a constant load is measured as a function of time. Figure 8.31(a) indicates how the curve can be divided into segments corresponding to **elastic** or **plastic** deformation, depending on whether the specimen returns to its original dimensions when the load is removed and provides a family of curves, each one corresponding to a different load. As the load is increased, so the rate of secondary creep deformation increases too. However, Figure 8.31(b) demonstrates another factor that influences the transition between brittle and ductile behaviour; the chemical composition of the fluid that fills the pore spaces within the specimen. The more chemically reactive is this fluid, the more likely is the material to exhibit ductile rather than brittle properties.

Figure 8.29 The shatter belt of the Great Glen strike-slip fault, viewed from above Loch Oich, with Loch Ness in the distance. (*Robert M. Adam*)

Figure 8.30 Glarus Thrust, Tschlingelhorner, Switzerland. The thrust surface forms the horizontal ledge in the mountain ridge and has carried rocks of Permian age over those of Tertiary age. (*R.F. Cheeney*)

Thus we see from these analogies between common experience and laboratory experiment on one hand, and inductive reasoning based on geological observation on the other hand, that we may provide at least qualitative explanations for the behaviour of rocks deformed in nature. We may refine these interpretations by means of further measurements, some field-based and some laboratory-based, so that our understanding of these tectonic structures advances further into unknown ground, but we must leave these considerations until later.

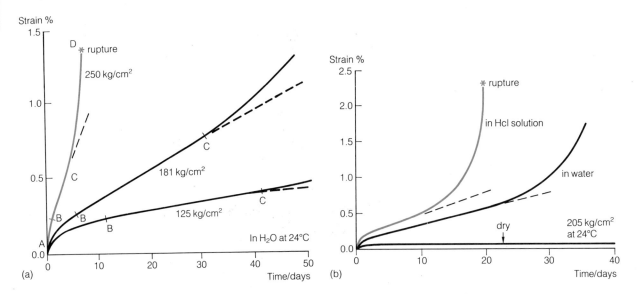

Figure 8.31 Creep tests on alabaster–calcium sulphate. In a creep test, a constant load is applied to the specimen and the deformation (strain) is measured as a function of time. In (a), the members of a family of creep curves each corresponds to a different load, while the temperature and the composition of the pore fluid is held constant. The response of the specimen is measured by its strain, the change in length per unit length, expressed as a percentage. Each curve is divisible into four segments. A short vertical segment OA corresponds to the instantaneous deformation that occurs when the load is applied. This deflection may be recovered immediately if the load is removed promptly: the material behaviour is said to be **elastic**. The curved section AB corresponds to primary creep. If the load is removed within this interval, most of the strain is recovered, although not instantaneously, so that the specimen displays a time-dependent elasticity. Beyond B, at least a part of the strain is permanent: the material has displayed a behaviour known as **plastic**. The segment BC, secondary creep, is approximately linear and corresponds to a more or less uniform flow of the material, rather like a viscous liquid. Note how the gradient of the straight section depends on the applied load. Beyond C, tertiary creep proceeds at an accelerating rate and, given time, the specimen will fail by rupture. Part (b) shows another family of creep curves in which the temperature and load are held constant, but the composition of the pore fluid is changed. The more chemically reactive the fluid, the higher the rate of secondary creep and the sooner rupture occurs. Few natural groundwaters are pure, and at depth in the crust, circulating fluids may be hot, and highly chemically reactive.

8.5 FURTHER READING

Park, R.G. (1989) *Foundations of Structural Geology*, 2nd ed, Blackie, Glasgow and London.

Price, N.J. and Cosgrove, J.W. (1990), *Analysis of Geological Structures*, Cambridge University Press, Cambridge.

Ramsey, J.G. and Huber, M. (1983, 1987)*The Techniques of Modern Structural Geology*, 2 vols, Academic Press, London.

Roberts, J.L. (1982) *Introduction to Geological Maps and Structures*, Pergamon, Oxford.

LIFE AS A FUEL

MAKER: COAL AND

OIL

Organic matter equivalent in quantity to the weight of the earth has been created by living creatures since life originated on this planet.

Philip H. Abelson, 1957

9.1 THE SOURCES OF NATURAL FUELS

Carbon dioxide, water and sunshine are the primary sources of the carbon compounds of all living organisms and of all those that have lived in past ages. Under the influence of the sun's rays green plants, including most of the bacteria, absorb radiant energy and utilize it to synthesize CO_2 and H_2O into carbohydrates, such as cellulose $(C_6H_{10}O_5)_x$, starch $(C_6H_{10}O_5)_y$ and sugar $(C_6H_{13}O_6)$. Since these compounds are equivalent to carbon and water, their formation involves liberation of oxygen. Much research has been undertaken to discover how this **photosynthesis** is done, since many of the urgent problems of mankind – notably the need for ever-increasing supplies of energy – would be eased if it became possible to imitate the methods of nature economically. When the colouring matter of green plants (chlorophyll) is exposed to sunlight in water 'labelled' with traces of $H_2^{18}O$ (instead of the normal $H_2^{16}O$) it is found that most of the oxygen liberated comes from water. In general terms oxygen and hydrogen are separated from water-bearing compounds, and the hydrogen then combines with carbon dioxide to form the carbo-

hydrates mentioned above:

$$2H_2O \rightarrow 2H_2 + O_2; 2H_2 + CO_2 \rightarrow (CH_2O) + H_2O$$
$$\text{liberated} \qquad\qquad \text{carbohydrate}$$

Some of the oxygen so set free combines with the carbon of organic matter to form carbon dioxide; another part is consumed in weathering processes; and the balance passes into the atmosphere or into the sea. The cycle of changes is schematically portrayed in Figure 9.1. If the chlorophyll reaction ceased to maintain the present supply of oxygen, the atmosphere would soon become unbreathable.

If all the decaying remains of dead organisms were completely oxidized there would be no free oxygen left over. Under waterlogged conditions, however, oxidation is not complete. The decomposition products of vegetation, for example, accumulate as humus in the soil and as deposits of peat in bogs and swamps (**mires**). The buried peat deposits of former ages have been transformed into coal seams. In marine sediments a high proportion of the organic material of plant and animal life is either eaten or is lost by oxidation, but some

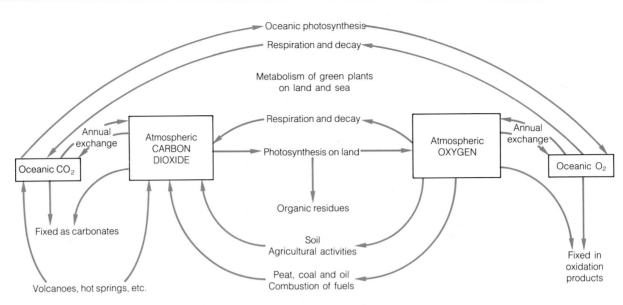

Figure 9.1 Schematic summary of Carbon dioxide – Oxygen cycle.

escapes complete destruction and is entrapped in muddy deposits, ultimately to form minute droplets of oil and bubbles of gas which are the likely source materials of the oil and natural gas of the oilfields. Because of their generally accepted origin from plants or microscopic forms of plants and animals, coal and oil and natural gas are known as the **fossil fuels**.

Life is thus considered responsible for all the natural fuels, including wood, peat, coal, oil and gas, and for the enormously greater amount of carbonaceous and bituminous matter that is dispersed through shales and other sedimentary rocks.

9.2 PEAT

In waterlogged environments, such as swamps, bogs, fens, moors, muskeg, wetlands etc. (non-saline examples are now often collectively known as mires) the normal atmospheric decay of dead material by micro-organisms and fungi is hindered by a lack of oxygen. In addition toxic organic compounds are produced which inhibit and ultimately stop microbiological activity. Under such conditions the softer and finely macerated plant debris changes into a dark brown colloidal liquid. Some of this soaks into the cells of the fragments of wood, bark, roots, twigs, etc. which are also being 'humified' with their cellular structures, in consequence, often being wonderfully preserved. All the humified products, together with a variable proportion of the less destructible materials, such as resin and the waxy pollen cases and spores, accumulate to form deposits of peat.

The process of humification enriches the residue in carbon, as indicated approximately by the following equation:

$$(C_6H_{10}O_5)_n \rightarrow C_8H_{10}O_5 + CO_2 + CH_4 + H_2O$$
cellulose   humified residue   methane

Methane, more familiarly known as marsh gas, is highly inflammable, and its pale flames are responsible for the 'will-o'-the-wisp' which is occasionally seen flickering over the surface of a bog.

The vegetation which contributes to peat formation ranges from mosses (bog peat – Figure 9.2) to trees (forest peat – Figure 9.3), and the environment may be a swampy lowland or a waterlogged upland with imperfect drainage. The climate must therefore be humid and the conditions such that growth exceeds wastage. In the bogs of cool humid regions the rate of decay lags behind because of the low temperature, whereas in the densely forested swamps of tropical regions the phenomenal rapidity of growth more than keeps pace with the high rate of decay.

A special variety of peat accumulates at the bottom of lakes and pools surrounded by marsh vegetation. Wind-blown pollen and leaves fall into the water, and all manner of organic particles drift into it. Year by year these settle down to form layer after layer of organic ooze. Freshwater algae may add further contributions. Locally the spores and algal remains may predominate, giving rise to a deposit that is especially rich in the waxy and oily ingredients of vegetation. If streams are flowing into the water the ooze is likely to be contaminated by a certain amount of muddy sediment.

In thousands of the shallow lakes that formerly occupied depressions in glaciated areas peat has

Figure 9.2 Bog peat, Dunmore Moss, east-south-east of Stirling, Scotland, showing cut peat lying out to dry and stacks of dried peat awaiting transport. (*British Geological Survey*)

Figure 9.3 Forest peat, showing remains of pine trees exposed by removal of peat, near Daless, Findhorn Valley, Highland Region, Scotland. (*British Geological Survey*)

formed due to the steady encroachment of marsh and swamp vegetation, and in others the process of infilling is still in progress. The rushes, reeds and pondweeds gradually advance over the dark gelatinous slime formed from the residues of earlier generations. Floating vegetation sometimes grows out in thick spongy rafts across the surface. Meanwhile the floor is being built up as organic ooze accumulates, and finally the site of the lake becomes a swamp. The treacherous surface may be covered with quaking tussocks of sphagnum moss, as in the bogs of Ireland. Where the drainage conditions are suitable, the plant sequence may culminate in a forest of trees with roots adapted to the precarious foundation through which they spread.

On a more extensive scale swamps are developed from the shallow lagoons and lakes of low-lying coastal plains, flood plains and deltas. The Dismal Swamp of the coastal plain of Virginia and North Carolina is an immense forested area, only a metre or so above sea-level, interspersed with stretches of open water (Figure 23.54). Here over 2400 km^2 have been covered with peat averaging 2 m thick.

Similar thicknesses of peat occur in parts of the Okefenokee 'Swamp' which occupies some 1500 km^2 of the Atlantic Coastal Plain of the southeastern USA on the borders of Georgia and Florida, and in areas within the vast Everglades mangrove swamp–marsh complex in southern Florida.

In Indonesia and Malaysia many scattered swamps support almost impenetrable tropical jungles, the trees of which are often rooted in thick peat while the densely forested swamps of the Ganges and other tropical deltas similarly provide ideal conditions for peat growth. They are thought by some to illustrate the climatic and geographical conditions under which the coal seams of the Carboniferous period originated. Moreover, borings through the Ganges delta, and many others, reveal a succession of buried peat beds with intervening deposits of sand and clay. The sequence points to repeated alternations of sedimentation and plant growth, with the actual surface never far from sea-level.

As raw peat accumulates year after year, water is squeezed out of the lower layers and the maturing peat shrinks and consolidates. It still contains a high proportion of water, however, and before being used as a fuel prolonged air-drying is necessary. In appearance it then ranges from a light brown fibrous or woody material to a dark brown or black amorphous substance.

9.3 COAL AND ITS VARIETIES

Peat becomes still further compacted when it is buried beneath a cover of clays and sands. Microbiological activity ceases and further changes result from physico-chemical processes. As the overhead pressure increases water and gases continue to be driven off and with an accompanying rise in temperature, due to the depth of burial,

coalification commences.

The essential conditions for the development of a coal seam are therefore: long continued growth of peat; subsidence of the area; and burial of the peat beneath a thick accumulation of sediments.

The greater the chemical alteration of the original plant material the greater is the increase in carbon content and the decrease in volatile matter of the resultant coal, key factors in determining its **rank** (Figure 9.4). For guidance as to the potential uses of the coal there are both national and international classifications based mainly on rank. Other factors of economic importance include ash content and sulphur levels. The **volatile matter** referred to includes hydrogen, methane and carbon dioxide, driven off when coal is heated to high temperatures in the absence of oxygen, and important in deciding upon coal's use for coking purposes. Methane occurs in many coal mines as the 'fire-damp' that can cause explosions if various precautions to remove it are not taken. It is now recognized as an important source of natural gas and coal mines in many parts of the world pump it out as a valuable by-product. The methane produced from the southern North Sea gas fields results from the deep burial of the Carboniferous coals below the Permian reservoir rocks (Section 9.5).

While rank increases with depth of burial, due

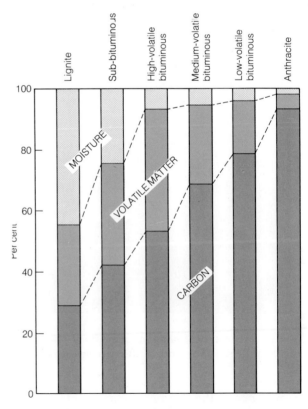

Figure 9.4 Diagrammatic illustration of the variation in composition of coals of different rank. (*After P. Averitt, 1974*)

to the increase in temperature, it is also affected by length of time of burial (i.e. geological age), regional tectonic deformation, and local thermal metamorphism. Anthracites tend to occur in areas where coal-bearing sequences have suffered intense folding. This is clearly seen in West Virginia and Pennsylvania where the flat-lying coals in their western parts are high-volatile bituminous in rank. Eastwards as the coals approach the folded rocks of the Appalachian fold belt (Section 31.3) the rank increases, to semi-anthracite in West Virginia and to anthracite in Pennsylvania. A similar situation may be observed in Canada where flat-lying sub-bituminous Cretaceous coals in the Plains of Alberta are seen to increase in rank towards the Rocky Mountains of Alberta and British Columbia. In the folded rocks of the Rocky Mountain Foothills low-volatile bituminous rank is achieved, while further west in the central Canadian Cordillera (Section 31.10) coals of about the same age, which have been buried to great depths during the folding associated with orogenesis (and hence subjected to greater heat) achieve anthracite rank.

The study of the rank variation over coalfields or an individual coalfield can be carried out by plotting **isovols** (i.e. contours joining points where the coals have equal values of volatile matter). As can be seen from Figure 9.5 the anthracites in South Wales are concentrated in the northwest of the coalfield. (Similar studies can be carried out by measuring the reflectivity of the maceral vitrinite (see later)).

Coals are also referred to as being of a certain **type**, a term which indicates their botanical make-up. Normal coals consist, in the main, of altered wood and bark, indicating derivation from forest peats, and may be referred to as of **humic** type. The lowest rank coals, the **lignites** (or **brown coals**), most closely resemble dark-coloured peat. **Sub-bituminous** coals are generally of a uniform dull black colour, while the **bituminous** coals are very obviously banded, with shiny black and dull black alternating layers (rarely more than a centimetre or so in thickness). The different layers represent differences in the original plant material or its state of alteration. The term **bituminous** does not imply the presence of the material called "bitumen" but has reference to the fact that in the manufacture of coal-gas and coke one of the distillation products, **coal tar**, is of a bituminous nature. Coal of the highest rank, **anthracite**, is shiny black and brittle, does not soil the fingers, burns with a smokeless flame, and has a high heat-producing capacity.

Sapropelic coals are those formed from the organic muds, mentioned earlier, which accumulate in pools, lakes and lagoons from the waxy residues of spores and algal remains etc. As most of the organic debris is fine-grained the coals present a dull, matt black exterior, though sometimes with a slightly waxy sheen. Those consisting mainly of spores, but also containing substantial and varying amounts of resin blebs, algal remains and other waxy-resinous plant debris (e.g. leaf cuticles) are known as **cannel** coals. They occur both as individual seams and as lenticles and bands, usually within or on top of normal seams. The name refers to the fact that splinters of cannel coal can be burnt like a candle, a fact that in turn demonstrates the richness of the cannel in flammable hydrocarbons. Varieties consisting almost

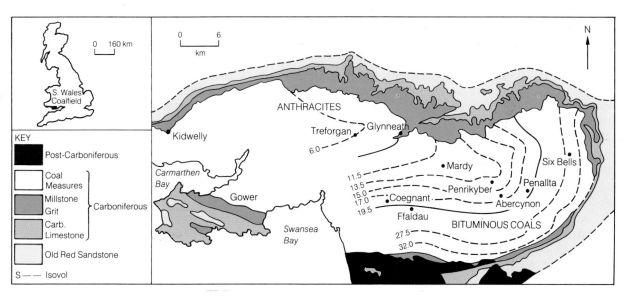

Figure 9.5 Map showing the regional geology of the South Wales Coalfield and the increase in coal rank demonstrated by the isovol contours. (*After W. D. Gill et al, 1979*)

entirely of algal remains are known as **boghead coal** or **torbanite**, named from its discovery at Torbane hill on the Boghead estate, west of Edinburgh. It was the raw material that gave a start in 1851 to the Scottish oil shale industry by James (Paraffin) Young. An increase in inorganic matter (e.g. mud) in the bogheads and cannels results in their being classified as oil shales (which fortuitously occurred in the same area where bogheads were discovered and enabled the Scottish oil shale industry to continue until 1962).

Sapropelic coals were also formed in pre-Carboniferous times as algae, for example, were in existence from Late Precambrian times onwards and an algal cannel ('shungite') of Precambrian age is known from the Lake Onega region of the USSR. In China, Cambrian age 'stone coals' (algal cannels) have been in use locally in many provinces for centuries. Reserves of these types of Lower Palaeozoic coals in China have been calculated to be of the order of 1000 billion tonnes!

9.4 THE CONSTITUTION OF COAL

As mentioned, a piece of bituminous coal can be seen to have a well-marked banded or stratified structure. The commonest bands are composed of soft bright coal which readily breaks into approximately cuboidal pieces with smooth brilliant surfaces. Many of the bands appear to be quite structureless, and since the material is not unlike black glass in appearance it has been called **vitrain**. Other bands are finely laminated and consist of shreds and films of vitrain in a very fine-grained matrix. This type of coal is known as **clarain**. These bright bands are separated by layers of a dull grey-black type of coal which, being relatively hard and tough, is distinguished as **durain**. None of the three types already mentioned soils the fingers. The 'dirtiness' of coal is due to the presence, generally in quite small amounts, of a fourth type of material called **fusain**. It consists of thin flakes or lenticles of extremely friable 'mineral charcoal' which occur at intervals through the seam. Coal naturally splits very easily along these planes of weakness, and as fusain readily crumbles to powder the broken coal becomes dusty.

The various bands of coal, recognizable in hand specimens, are like rocks, made up of different constituents generally recognizable only under the microscope. As they are organic these constituents cannot be classified as minerals so the term **maceral** is used. Most coal petrography is carried out using reflected light microscopes because of the difficulty of making microscope sections of coal thin enough to pass transmitted light. Three main maceral groups are recognized. **Vitrinite** is typical of vitrain and clarain and can show either cellular structures from the woody tissues of plants, or be structureless, denoting solidification from a colloidal gel such as the 'dopplerite' found in peat bogs. **Inertinite** is the group name given to charcoal-like materials where oxidation by microbial or fungal means has occurred or where combustion of similar plant debris to that which forms vitrinite has taken place. (**Fusinite**, for example, is thought to be due to forest fires). The name of the group refers to the relative inertness of macerals of this type during technological processes. Waxy resinous plant material such as spores, pollen grains, leaf cuticles etc. are classified under various names in the **liptinite** (or **exinite**) group. This group dominates in the sapropelic coals.

While vitrain and fusain under the microscope display either humified or carbonized cellular structures of wood or bark, durain is found to be an assemblage of minute particles, like the organic ooze of peat. It contains the resistant coats of spores, more or less crushed and flattened, microscopic shreds of vitrain, lenticles and grains of fusain, and blebs of resin, in an obscure matrix of debris too finely macerated to be identified without an electron microscope. Clarain contains the same ingredients as durain, but in very different proportions; abundant closely-packed strips of vitrain are associated with extremely thin laminae of durain-like material. Cannel coal is essentially durain which is especially rich in spore cases and other waxy and resinous remains.

Lignites and anthracites are found to consist of the same structural types of material as bituminous coals. The variation in properties depends partly on the proportions in which the type of ingredients are present, and partly on the degree of alteration which they have suffered, i.e. on the rank of the coal. Anthracites for example, contain only ghostly carbonized relics of what are probably spores; lignites on the other hand, having undergone much less alteration of the original plant material than the bituminous coals are generally described microscopically by a maceral classification specific to them.

Coal microscopy is now carried out mainly on plane polished surfaces under standardized conditions of reflected light. The **reflectance** of the macerals increases as the rank of the coal increases due to changes in their chemical structures. Vitrinite is an excellent 'reflector' and is used as a measure of the rank of the coal.

9.5 COAL SEAMS AND COALFIELDS

Coal beds (or seams as they are known) do not occur in isolation but are interbedded with sand-

Figure 9.6 Remains of a Carboniferous forest showing fossil stumps and forked roots of *Lepidodendron* rooted in shales, Fossil Grove, Victoria Park, Glasgow; discovered in 1882 and preserved as a natural museum. (*British Geological Survey*)

Figure 9.7 Reconstruction of a Carboniferous forest.

stones and shales, such sequences usually being known as **coal measures**. (The term appears early in coal-mining literature because when sinking shafts and borings the miners measured carefully the vertical distances between the coal seams (see Figure 9.8). It has since been applied to coal-bearing sequences of all ages.) The earliest period of accumulation of coal measures was the Carboniferous – the Period/System being given that name because of the abundant occurrence of coal seams in parts of the stratigraphic sequence.

It was the remarkable evolution and profusion of land plants (more than 3000 species are already known) during the Carboniferous Period, from slow beginnings in late Silurian times, combined with suitable climatic and topographic environments, that resulted in the repeated formation of the varied mires in which peat accumulated. Coals formed during all geological Periods from the Carboniferous onwards though it and probably the Cretaceous Period (by which time the angiosperms dominated the world flora), were the peak coal-forming times. Because of the historic importance of Carboniferous coals in the development of the western world most of the discussion that follows concerns their formation.

The chief coal-makers were tall forest trees (*Lepidodendron* and *Sigillaria*, with widely spreading roots known as *Stigmaria*) which grew to heights of as much as 30 m; and giant reeds called *Calamites* (the ancestors of the little horsetails of today) which flourished in bamboo-like thickets to a height of 15 m or more; together with an under-growth of smaller rushes and ferns, and slender plants of trailing or climbing habits (Figures 9.6 and 9.7). No flowers or birds enlivened these gloomy jungles, but insects – again of extravagant size – were abundant.

The Carboniferous Coal Measures can achieve considerable thickness, indicating prolonged sub-sidence of the area of deposition, the surface mires having been repeatedly drowned and covered by sediments. In South Wales, for example, there are preserved at least 2400 m of measures; in the Ruhr and Saar Coalfields of Germany some 3000 m. Variations in these thicknesses occur within each coalfield and within different parts of the succes-sion, as do the thickness and continuity of the different beds of sandstone, shale and coal seams (Figure 9.8). The latter occupy in aggregate thick-ness perhaps only some 5 per cent of the total successions, and individual seams worked under-ground rarely exceed 2 m in thickness and average perhaps 1.5 m.

Virtually all humic coals of Carboniferous age have certain characteristics from which it is inferred that each seam represents the actual site of the swamp in which the parental vegetation lived and died. In other words the coals are assumed to have formed *in situ*, and are sometimes referred to as being of **autochthonous** (*Gr.* sprung from the soil) in origin.

Beneath each seam there occurs a unique rock-type frequently displaying fossilized rootlets, which is variously named, for example, as a 'seatearth', 'underclay' or 'rootlet bed'. It is con-sidered to represent a fossil soil (a **palaeosol**) and its original composition would be that of the last deposited sediment before plant life was estab-lished on it. This sediment was in most cases of mud grade and in exceptional cases was of such a composition due to chemical alteration that a '**fireclay**' (i.e. a clay rich in kaolinite and quartz used for making refractory bricks) was formed. In other rare instances pure sandstones (consisting of 99 per cent quartz) are found beneath coal seams, often bearing traces of *Stigmaria* (roots), and are called **ganisters** (and used for making silica fire-bricks).

The trunks of the trees growing in the mire either rotted away above water-level (i.e. in aero-bic conditions) or fell into the swamp to contrib-ute to the accumulation of peat. Fine-grained sediments could obviously be brought into the swamps during periodic floodings and they are a main contributor to the 'ash content' of the coal when it is burned. (In many cases so much is deposited that so-called 'dirt-bands' interrupt the bands of the various coal types making up a seam.)

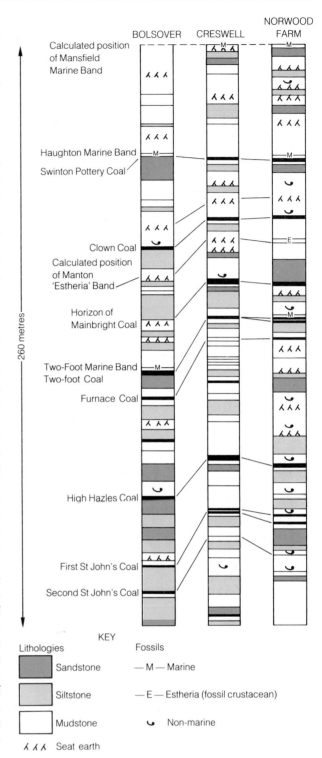

Figure 9.8 Coal Measures sections from Derbyshire, England. Bolsover and Cresswell are colliery shafts. Norwood Farm is site of exploratory borehole. The collieries are about 12 km apart; Norwood Farm is 4 km from Cresswell.

Prolonged periods of peat accumulation were brought to an end when the water deepened sufficiently and permanently so as to prevent plant growth. Sedimentation followed and most coal seams are found to be succeeded by 'roofs' consisting of black mudstones or shales – representing the gentle accumulation of mud as water covered the peats. Sometimes the roof shales preserve impressions of the trunks, barks and leaves of plants, in other cases remains of fossil shells, usually of non-marine species.

In certain instances, however, marine fossils are prevalent and provide important 'marker bands' for correlation purposes, as do the marine limestones which overlie some coals of Carboniferous age (particularly in Illinois, USA for example). As there are no obvious signs of disturbance or erosion of the underlying coals (which might be expected from a sudden invasion by the sea) it is concluded that either a gentle subsidence of the area, or a gradual rise in sea-level, or both, took place to allow the occasional marine encroachment. Deposition of muds, silts, and sands in brackish to fresh water followed until plants established themselves once more. This pattern of events, which was commonly repeated dozens of times, and in some coalfields, over a hundred times, typifies a phenomenon known as **cyclic sedimentation** and much debate has taken place to explain it. In England the average thickness of each cycle of sedimentation between coals is about 10 m, the sedimentary unit being known as a **cyclothem** (see Figure 9.8).

Obviously to understand fully the formation of coal measures we also need to know the climates and the topographic environments in which the peats accumulated. It seems the mires did on occasion extend or encroach over vast flat areas – the Pittsburgh seam in the USA, for instance, is mineable throughout parts of Pennsylvania, West Virginia, Maryland, Ohio and Kentucky, ranging in thickness from about 1 m to 7 m, over an area of some 16 000 km². The Lower Kittanning seam, though in general thinner, extends over an even larger area. This is not to imply, however, that the area now occupied by a correlatable coal was once the site of a mire of that size. As emphasized in Chapters 6 and 7 lithologically correlative beds are not necessarily time-correlative because of, for example, the phenomena of **transgression** and **regression** (Section 7.7).

Important information on peat development has come from studies in present-day 'swamp' areas in the south-eastern USA and from deltas in various parts of the world. Current opinion and botanical and sedimentological evidence favour, for Carboniferous coals, deltaic, alluvial and coastal-plain environments with a tropical or subtropical climate. Many of the sedimentary features are comparable with those of present-day deltas (e.g. those of the Mississippi, Fraser and Niger rivers). Aerial views of deltas show the presence of both large channels and smaller distributaries,

sand banks, swamps and forests, which over a period of time are seen to change in position and extent. Borings to considerable depths through layers of sand, silt, and mud with interbedded peat prove the continual sinking of the area.

Coal measures (with notable exceptions, e.g. certain coal seams and marine fossil horizons) are typified by both lateral and vertical variations (Figures 9.8, 9.9). Evidence from mining, road cuts and boreholes shows, for example, that many sandstones, far from being areally extensive, tabular beds, are actually confined to channels (tens to hundreds of metres wide) which can be traced for many kilometres. Some channels ('**washouts**') cut down into the coal seams and if not foreseen affect both mining and reserve estimates! Sedimentary structures indicate flow directions and other features compatible with delta distributaries, or river sand-bars etc. Coal seams thin out and disappear laterally. Frequently they 'split' into two or more seams, when 'dirt' bands increase in thickness (Figure 9.8).

A problem that arises with the delta model is that many deltaic peats have a high content of sediment – which means that a coal derived from such peat would have a much higher ash content than is usual for most coals (other than those of Permian age). Leaching of the mineral matter in the peat during burial and lithification is of course possible. More likely is that the precursors of coal seams were raised on 'domed' peat bogs. These can occur in both deltaic areas and in the salt flats and lagoons that exist behind the barrier beaches that protect large areas of the world's coastal plains. High rainfall can stimulate plant growth such that the continuing accumulation of peat raises the mire above stream flood-level or high-tide level. Detrital matter becomes negligible. This process can be observed between the various streams flowing through the tropical forests of the deltas in Malaysia, Sarawak, and Brunei. Eventually the growing areas are so far above water-level that they cannot support the growth of forest trees, the vegetation changes and peat accumulation slows down and ceases. A mechanism such as this may, in an area of slow continuous subsidence, provide the means by which the peat is eventually covered by water and sedimentation begins again. Such a process would make unnecessary the problems of stopping peat accumulation by either repeated **eustatic** rises in sea-level or continued changes of the rate of subsidence which have been invoked to explain the cyclicity of coal measures.

There have been occasions during the Carboniferous when major changes in sea-level did take place, as evidenced by the amazing continuity of certain fossiliferous marine horizons ('marine bands') within the Coal Measures over most of

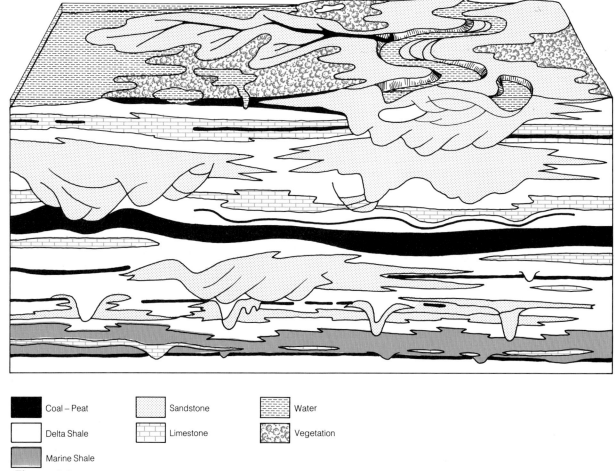

	Coal – Peat		Sandstone		Water
	Delta Shale		Limestone		Vegetation
	Marine Shale				

Figure 9.9 Schematic block diagram illustrating lateral variation in Pennsylvanean-age coal measures near Morgantown, West Virginia, USA, interpreted as having been deposited on a fluvial-dominated deltaic environment. Horizontal scale *c.* 31 m, vertical scale 70 m. (*After A. Donaldson, M. W. Presley, J. J. Renton, 1979*)

Europe. Similar areally widespread beds are found in the USA, Russia and China, perhaps indicating eustatic rises in sea-level due to melting of the polar ice-sheets which existed during the Carboniferous times, but such dramatic changes are surely not necessary for the cessation of every period of peat accumulation recorded in coal measures sequences.

The coalfields of Great Britain are shown in Figure 9.10. Those of central and northern England probably resulted from deposition in an extensive Carboniferous sedimentary basin (now separated into different coalfields by later folding and faulting) which extended eastwards well across what is now Europe. The continuation of the Coal Measures of the East Pennines coalfield under Permian and younger rocks in the North Sea has been proved from oil and gas exploration drilling. It is the deep burial of the coals that has resulted in the formation of the natural gas (methane) now exploited in the southern North Sea gasfields.

While Carboniferous coals now are found in a belt across the Northern Hemisphere from the USA through Europe, the USSR and China it must be emphasized that the distribution of the continents on the then surface of the Earth was very different from that of today. It is thought there was a single supercontinent – **Pangaea** – with very broadly speaking, what are now North America, Europe and Asia forming the northern part **Laurasia**, and South America, Africa, India, Australia and Antarctica forming the southern part – **Gondwanaland**. (The name Gondwana was introduced by H.B. Medlicott (Section 27.7) in an internal report of the Geological Survey of India in 1872 when describing a coal-bearing sequence south of the Narmada Valley in the eastern Deccan area, once the ancient Kingdom of the Gonds, an aboriginal tribe.)

Climatically, the belt of Carboniferous-age coalfields stretching across Laurasia must have been in a tropical or semi-tropical zone (the trees lacking annual growth rings), while the southern part (Gondwanaland) had a considerable portion in South Polar regions. Evidence of this former glaciation is found, for example, in what are now South America, South Africa (Figure 20.27 and

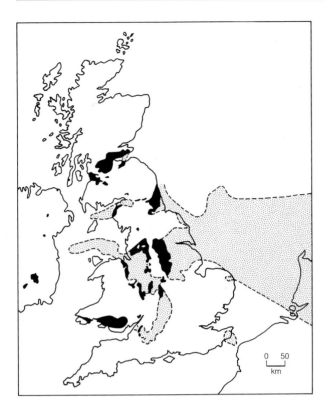

Figure 9.10 Black areas indicate location of coalfields. Stippled areas indicate 'concealed' coalfields where Carboniferous rocks are buried under younger rocks or, as in the case with part of the Northumberland Coalfield offshore and the Firth of Forth Estuary, for example, seabed sediments. (*Based on E. H. Francis, 1979*)

Plate VId), India and Australia. As the climate ameliorated in late Carboniferous times a glaciated landscape with associated glacial deposits (e.g. tillites and varves – see Chapter 20) emerged in southern Gondwanaland, plant life was established and where conditions were suitable peat deposits formed. Tundra conditions changed as the ice-sheets waned and plants, adapted to cold or cool temperate climates, flourished in Permian times.

What became coal measure sequences tended to accumulate in basins within the cratons (Chapter 29) rather than as is the case with most Carboniferous coal measures on the continental margins (Chapter 30). A unique broad-leafed *Glossopteris* flora evolved (with annual growth rings indicating seasonal changes) and because plant degradation in the colder climate was slower the coals are richer in oxidized plant remains (inertinite) than those of Laurasia. The coals also contain much more mineral matter (i.e. ash), which may reflect a wind-blown contribution of loess (Sections 21.7, 22.6) as well as water-borne material.

Another feature is the absence of seatearths (as

recognized in the Carboniferous) which led to the belief that the Gondwana coals were of **drift** (**allochthonous**) origin i.e. the peats were formed from plant material that was washed into the swamps, rather than from the degradation of plants that grew *in situ*. However, remains of standing trees in the position of growth have been found in South Africa, India and Australia. One of the most convincing pieces of evidence can be seen in New South Wales, Australia where a volcanic ash-fall covering a coal seam displays numerous tree stumps in the position of growth (Plate VIa). Nevertheless the absence of obvious seatearths is noteworthy and may be a reflection of the different type of rooting system in the *Glossopterids* and their allies and/or climatic differences affecting the lack of chemical change in the muds underlying the peats.

Sedimentologically the Permian coal measures of Gondwanaland have similarities to those of the Carboniferous in that they consist mainly of alternations of sandstones, siltstones, shales and coals. Depositional environments identified as lacustrine, delta-plains, fluvial and alluvial are common. In Brazil and southern Africa the coal measures are relatively thin (from tens to a few hundred metres) reflecting their origin in cratonic basins (Chapter 29), and contain perhaps only 5–6 laterally persistent mineable seams throughout the various coalfields. These coals tend to be thick – commonly 3–4 metres – though possessing more dirt bands than those of the Carboniferous. Sulphur content is usually low, indicating nonmarine conditions, but ash content ranges up to 25 per cent in South Africa and 50 per cent in Brazil. The coals are generally of bituminous rank, though in South Africa the proximity of dolerite sills has raised certain coals to semi-anthracite rank.

In India important coalfields occur in the east–west Damodar valley, some 200–300 km northwest of Calcutta, the Raniganj coalfield having over 20 workable seams ranging in thickness from 1–30 m in two separate coal measures successions which can attain thicknesses of 600 m and 1000 m respectively. The Jharia coalfield has some 25 seams in the equivalent beds of the lower succession in the Raniganj coalfield. Thicknesses range up to 22 m. Ash content is high in both areas (10–29 per cent) while sulphur content rarely exceeds 0.6 per cent, though coal measures sequences are occasionally interrupted by fossiliferous marine shales.

In Australia, the Permian Sydney and Bowen Basins extend from New South Wales to Queensland and differ from those described so far in that the coal measures can contain thick conglomerates and volcanics suggesting they were formed at the edge of, rather than in the middle, of a craton. Prominent occasional fossiliferous marine sequences interrupt coal measures deposition and support this view. Seams are similarly thicker in general than those of the Carboniferous, have a low sulphur content but a lower ash content than most other 'Gondwanan' coals. An exceptionally thick seam, some 33 m at maximum, occurs in the isolated small coal basin at Blair Athol, Queensland (Plate VIb).

Cretaceous- and Tertiary-age coals have, since the Second World War, come to be recognized as both abundant and important. While, in general, the Cretaceous coals tend to be of sub-bituminous rank this is not always the case. As pointed out in Section 9.3 the Cretaceous coals of western Alberta and British Columbia in Canada attain high ranks suitable for use as coking coals in the steel industry. The coal measures extend in a belt for some 1000 km north of the USA border and are considered to have formed in alluvial/deltaic environments. The Cretaceous coal measures also occur to the south, east of the Cordillera belt, in the USA, intermittently from Montana to New Mexico. Extensive, as yet unworked, Cretaceous deposits are known also in the North Slope of Alaska.

Most Tertiary coals are sub-bituminous or lignitic, often attaining great thicknesses e.g. 100 m in the Miocene deposits near Cologne in Germany. Here, as with the Texas coast lignites, an origin in a coastal-plain environment is suggested. On the other hand, extensive deposits in North Dakota, Colorado, Wyoming and Montana are thought to have formed in alluvial plains in intermontane basins. In Australia, important brown coal deposits occur in the Latrobe Valley, Victoria where there are four seams each averaging about 100 m in thickness, the environment of deposition being considered lacustrine. Rivalling these in thickness is the so far unexploited Hat Creek deposit in British Columbia where three sub-bituminous coals have been proved by drilling to have thicknesses of 160 m, 50 m and 70 m within a 425 m succession!

Few, if any, of the post-Carboniferous coals have the typical seatearth of that period, which as already suggested may be due to differences in plants, climatic conditions and chemical alteration. There also appears to be an increase in thickness of seams, the geologically younger the coal measures are. The thickness of a coal seam is dependent upon the balance between the rate of subsidence and the rate of accumulation of peat. If the subsidence is too fast, for example, the mire is drowned and peat accumulation ceases, if too slow oxidation of the plant material takes place. (In Carboniferous coal measures seatearths commonly occur without an overlying coal, though when traced laterally often prove to have overlying coals).

It is significant that the thicker Tertiary coals (e.g. Lower Rhine area, Germany; Hat Creek, Canada; Latrobe Valley, Australia) often occur in graben-type structures or along the downthrow sides of faults (see Chapters 8 and 29), where increased subsidence has taken place. In Saskatchewan Tertiary lignites thicken above areas where there has been accelerated subsidence due to dissolution of underlying, very much older (Devonian) evaporite beds. That increased rates of subsidence can take place locally during deposition was spectacularly demonstrated some years ago at the open-cast coal site at Westfield, Fife, Scotland. In exposures c. 500 m wide and 200 m deep almost flat-lying seams at the surface gave way to increasingly synclinally-folded seams downwards, the lowest seams having limbs dipping inwards at the steepest angles (Plate VIc).

Coal is now used mainly in electricity generating stations, as coke in the production of iron and steel, to a lesser extent in cement-making and as a feedstock for certain chemical processes. In the western world its use for domestic purposes has fallen dramatically since the Second World War. Various technologies are being developed to reduce the emission of sulphur and nitrogen oxides on combustion so its use may increase in the next century. Intensive research is being undertaken to increase the production of liquid and gaseous fuels from coal – South Africa already produces motor fuels in quantity.

The main coal-producing countries are shown in Table 9.1, while Table 9.2 shows the currently estimated world reserves of coal. There is considerable world trade in bituminous coals because so much outside of Europe is produced by surface mining (which is much cheaper than underground mining) and shipping costs are low.

9.6 PETROLEUM

Petroleum (Gr. *petra*, rock, L. *oleum*, oil) is the general term for all the natural hydrocarbons – whether gaseous, liquid or solid – found in rocks. In common usage, however, it refers more particularly to the liquid oils. Gaseous varieties are distinguished as **natural gas**. Highly viscous to solid varieties are called **bitumen** or **asphalt**, but the latter term is also applied to the bituminous residues left when petroleum is refined, and to natural and artificial paving materials composed of sand, gravel, etc., with a bituminous cement. The most obvious indications that a region is or has been oil-bearing are (a) seepages and springs of gas or oil; and (b) superficial deposits or veins

Table 9.1   Main coal producing countries 1990 (*From BP Statistical Review of World Energy, 1991*)

	Million tonnes	
	Hard coal[1]	Lignite and brown coal[2]
China	916.2	81.5
USA	647.0	305.7
USSR	478.0	151.8
India	184.9	8.5
Africa (mostly S. Africa)	182.2	–
Australia	161.8	48.1
Poland	147.4	47.6
UK	91.8	–
Germany	70.2	357.2
Czechoslovakia	22.9	84.2
Columbia	20.4	–
Yugoslavia	–	76.3
Turkey	–	47.3
Greece	–	46.5
Others	198.8	162.3
Total world	3121.6	1417.0

[1] Anthracite and bituminous coal
[1,2] Commercial solid fuels only
[2] Sub-bituminous coal and lignite

Table 9.2   World coal reserves, 1990 (*From BP Statistical Review of World Energy, 1991*)

	Million tonnes		
	Anthracite and bituminous coal	Sub-bituminous and lignites	World share %
USA	129 543	130 752	24.1
USSR	102 496	136 520	22.1
China	152 831	13 292	15.4
Australia	44 893	45 461	8.4
Germany	23 698	54 964	7.3
India	60 098	1874	5.7
S. Africa	54 811	–	5.1
Poland	28 182	11 487	3.7
Others	40 140	47 692	8.2
Total world	636 692	442 042	100.0

of asphalt and other more or less 'solid' residues of petroleum left behind after loss of the volatile constituents by evaporation. In the early days of the oil industry it was either evidence of these phenomena or the accidental discovery when drilling for water, for example, that determined where oil was produced and explored for.

Historically, the Middle East, particularly in the areas surrounding the Persian Gulf, has been known for seepages of gas, oil and asphalt (pitch) since Biblical times. The Old Testament refers to 'fiery furnaces' into which miscreants were cast and the makers of the Ark and the Tower of Babel (and the mother of Moses) all used pitch for their various purposes. Similarly, there are historical records of gas and oil coming out of the ground on the western shores of the Caspian Sea. Marco Polo recorded the presence of fountains of oil in the thirteenth century, while ignited gas seepages near Baku, were referred to as the 'Eternal Fires' by the thousands of fire-worshippers who made pilgrimages to the area.

At the beginning of this century the Baku region supplied more than half of the world's oil. Occasional gas outbursts overturned boats in the Caspian Sea, while in areas of unconsolidated mud and clayey sediments the explosive escape of high-pressure gas forms so-called 'mud volcanoes' there and in neighbouring areas of the Aspheron Peninsula of the USSR. Mud volcanoes are also known from Burma, Trinidad and California.

Significant discoveries have occasionally been made by accident, as when boreholes put down for water or coal happened to strike oil or gas. Indeed, the modern oil industry began a century ago as a result of a boring for water in Pennsylvania which brought up oil and salt water. This was in 1848 and the oil was sold for medicinal purposes. In 1859 the first well was drilled specifically for oil, and oil was duly found at a depth of only 21 m. The natural gas of western Canada, of which enormous reserves have since been proved, was first tapped in 1883 when it was hoped to find water for the Canadian Pacific Railway, then under construction.

Important seepages of asphalt today include the famous Pitch Lake of Trinidad, which is about 600 m in diameter and some 40 m deep in the centre. The pitch which is 'mined' commercially is continually replenished by asphalt seeping up from below. Bermudez pitch lake in eastern Venezuela is another large deposit, while numerous seepages occur along the Coast Ranges of California.

In north-eastern Alberta, Canada, the Cretaceous McMurray Formation is represented by a sandstone impregnated with an asphaltic oil (or tar). These **tar sands**, the Athabasca Oil Sands, derive their oil from the migration from depth of oil originally trapped within Devonian-age fossil limestone reefs. Exposure at the surface has resulted in loss of most of the volatiles. Because of the low viscosity of the tar, the sands are opencast mined and then treated with steam to separate the oil residue from the rock material. Reserves are enormous – potentially billions of barrels of oil could be extracted, as is the case with other tar sands, in Venezuela, for example. Unfortunately there are both environmental and energy problems associated with their exploitation. The latter demand a careful assessment of the energy used to mine and process the sands compared with the energy produced!

Petroleum consists of an extremely complex mixture of hundreds of different hydrocarbons, generally accompanied by small quantities of related compounds containing nitrogen, sulphur or oxygen. The hydrocarbons fall into several natural series of which the alkane series is the most familiar. Its members, all of which can be expressed by the formula C_nH_{2n+2}, range from light gases (e.g. methane, CH_4 the chief constituent of natural gas), through a long series of liquids (the chief ingredients of successive products of distillation such as petrol or gasoline, paraffin oil and lubricating oil), to paraffin wax (including $C_{20}H_{42}$ and higher members). Crude oils in which these hydrocarbons predominate are generally of pale colour with a yellowish or faintly greenish hue. The darker brown and greenish oils generally contain a high proportion of the cycloalkane series, each member having a composition of the type C_nH_{2n}. These furnish heavy fuel oils and, as they leave a dark asphaltic residue on being refined, they are said to have an asphaltic base. Intermediate varieties have a mixed base of wax and asphalt. In all crude oils there are also smaller proportions of several other series, including the alkynes, C_nH_{2n-2}, and a great variety of aromatic hydrocarbons.

9.7 OIL SHALES

To avoid confusion it should be clearly understood that neither oil shales nor the cannel and boghead coals contain petroleum as such. If they did, it could be dissolved out by carbon disulphide. They do, however, contain **pyrobituminous** substances or kerogens which can be altered into oil and bitumen by heat. **Kerogen** (Gr.–wax; *suffix* gen– that which produces) was the name given to the insoluble organic material that occurred in the Scottish oil shales by A. Crum Brown, Professor of Chemistry, University of Edinburgh (1869–1908). Such deposits can therefore be made to yield a group of petroleum products by destructive distillation. Petrol and related products can be obtained in commercial quantities from suitably powdered coals of ordinary types only by fluidization with hydrogen at high temperatures. Petrol can also be made from the heavier and less valuable oils by a similar but less elaborate process of hydrogenation.

The first significant production of oil from oil shales commenced at Autun in France in 1838, followed in 1851 in Scotland and not long afterwards in Australia, Brazil and New Zealand. Various countries in Europe, the USSR, China and South Africa established plants after the First World War and production continued in many of them until mid-century. In the western world, however, currently, production of oil from oil shale, which involves mining and then pyrolosis, has proved uneconomic because of the availability of cheaper oil from the oilfields of the Middle East and elsewhere. (Figure 9.13 and Table 9.3).

Oil shales were formed during virtually every geological Period since the late Precambrian, the largest presently-known deposits being found in the USA. The Eocene-age Green River Formation extends throughout Colorado, Utah and Wyoming and is considered to represent former lake deposits. Parts of the formation consist of thin alternating laminae of silt and sapropelic material, each 'couplet' being the result of the annual 'blooming' of algae in the lakes when deposition was virtually nil due to annual climatic changes. The oil potential can be measured in hundreds of billions of tonnes, assuming it will sometime be economic to mine and process. In the northern part of the Uinta Basin, Utah, the shales have been buried to depths of 6000 m and are considered to be the source rocks of neighbouring oilfields. The Permian-age Irati Shales in Brazil, and their stratigraphic equivalents in Uruguay and Argentina are another potential resource.

As with the tar sands there are environmental problems concerning the mining of oil shales prior to processing and as said, economic ones. From an energy point of view it is considered that if the average organic content is less than 2.5 per cent by weight more energy is required for the processing than is produced!

9.8   THE ORIGIN OF PETROLEUM

Oil and natural gas accumulations on a scale sufficient to repay the drilling of wells are sometimes referred to as oil or gas pools. The 'pool', however, is merely the part of a sedimentary formation that contains oil or gas in place of, or along with, groundwater. The **reservoir** rocks, as they are called, are porous and permeable (see Chapter 19) and the oil and/or gas occupy spaces between the grains in clastic rocks or in spaces, cracks and fissures in carbonate beds and fossil reefs.

Probably something like 60 per cent of oil occurrences are in sandstones, arkoses, greywackes etc. while about 30 per cent are in limestones and dolomites, though more than 40 per cent of the so-called 'giant' oilfields, such as those in the Middle East, are in carbonate reservoirs. The marine origin of most of the sedimentary sequences in which oil occurs points to a probable marine origin for the oil itself.

Virtually all the individual chemical compounds identified in petroleum are of organic origin so while, unlike coal, no visible evidence of the nature of the material from which it is formed is present, the evidence available points convincingly to an organic origin for petroleum itself. Some of the constituents have the property of altering the direction of vibration of light rays. This 'optical activity' is a characteristic of many substances produced by plants and animals, but it is not shared by similar compounds which are generated by purely chemical reactions. Minor constituents include **porphyrins** (derived, for example, from chlorophylls or from the corresponding colouring substances of animal origin) and other compounds that can be extracted from plant and animal tissues by organic solvents. Significantly, porphyrins are quickly destroyed in the presence of oxygen. Their continued presence in oil points therefore to its origin in a reducing environment.

The existence of oilfields in pre-Carboniferous sediments (as far back as the Ordovician) suggests that land plants were not essential to oil formation. Main production of organic carbon seems to have started some 3 billion years ago when photosynthesis is thought to have commenced on a worldwide scale. Since then there has been an evolutionary development of the different forms of plant and animal life. As has been mentioned algal cannels are found from the Precambrian onwards, pointing to the early arrival of marine phytoplankton which continued to be the main producers of organic matter till the Devonian.

Studies on the organic content of sedimentary rocks indicate that phytoplankton, zooplankton (especially foraminifera), higher plants and microorganisms are the main source material. While at present it has been calculated that phytoplankton and terrestrial plants contribute roughly equal amounts of organic carbon there are basic differences in their chemical make-up. In general terms the terrestrial plants consist mainly of cellulose and lignin, the marine phytoplankton of proteins and lipids.

The deposition of sediments rich in organic matter (>0.5 per cent by weight) depends on the environment of deposition. Obviously in oxidizing conditions little organic matter will be preserved, being destroyed by both chemical and microbial action. If the sedimentation rate is too high then the percentage of organic matter will be lowered. Ideal conditions today are found in quiet-water areas on the continental shelf and slope where there is an abundance of nutrients and a large contribution from both phytoplankton and terrestrial material.

As we have seen the organic material found in cannels and oil shales is known as kerogen. But oil can only be produced from these rocks industrially by **pyrolysis** (involving temperatures of 500°C). Kerogen exists in much smaller quantities in many black shales and it is these which are considered the main **source rocks** of the naturally occurring crude oils and much of the natural gas.

The question then arises as to how this material is changed into oil by a geological process known as **maturation**? The first stage of formation of kerogen commences during **diagenesis** (the physical and chemical changes that occur in sediment undergoing lithification and compaction). Microbial degradation and biochemical alteration are important during the earlier stages of accumulation and burial when proteins and carbohydrates are destroyed.

As sedimentation and subsidence continues, increases in temperature and time become all important and depending upon the nature of the original material different types of kerogen are formed, during the phase referred to as **catagenesis**.

Type-I kerogen contains an abundance of lipid material (e.g. waxes, vegetable oils and animal fats), which may be due either to a predominance of algal remains (similar to those found in present day lacustrine environments) or to the original biomass suffering severe biodegradation by microbial activity or to a combination of these effects. It has a high Hydrogen/Carbon (H/C > 1.5) atomic ratio, and is common in many oil shales.

Type-II kerogen is derived from a mixture of phytoplankton, zooplankton and micro-organisms of the types now found in reducing marine environments. It has a relatively high H/C ratio and a medium to high sulphur content. This is the type that gives rise to important oil occurrences. Crucial factors are the geothermal gradient and the length of time kerogen has been exposed to the appropriate heat. Larger hydrocarbon chains and cyclic compounds are broken down, with the maximum production of crude oil and so-called 'wet gas' taking place at about 100°C (Figure 9.11).

Type-III kerogen has a relatively low H/C ratio (< 1.0) and is mainly derived from terrestrial plant material. The oil potential is low though gas may be generated in quantity at depth.

Increases in temperature to above 150°C result eventually in destruction of the oil and any remnants of the kerogen, 'dry gas' (mainly methane) being all that is produced. **Metagenesis** is the name given to this stage of alteration and if taken far enough, only carbon is left behind in the source rock.

9.9 MIGRATION AND ACCUMULATION OF PETROLEUM AND NATURAL GAS

The exploited reservoir rocks of the world's oilfields contain far more oil and gas than could possibly have originated *in situ*. (Sandstones on average contain only 0.56 per cent of organic matter and limestones 0.5 per cent, compared with an average 1.2 per cent for shales). In addition, limestones commonly carry fossils indicating formation in oxygenated waters. That oil and gas can

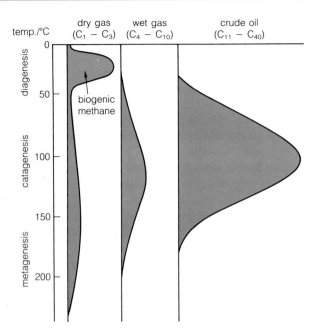

Figure 9.11 The temperatures of the maturation stages and the relative amounts of crude oil, wet gas and dry gas produced. The ranges of number of carbon atoms in each of the maturation products are also shown. (*After G. C. Brown and E. Skipsey, 1986*)

move through reservoir rocks is of course clearly demonstrated by the fact that they flow into commercial wells, either on their own account or by means of pumping. Oil and gas pools, therefore are formed by the **migration** of the petroleum compounds from the source rocks into the reservoir rocks.

The movement of the oil and gas through the capillary-size openings out of the source rocks is referred to as **primary** migration while the movement through the reservoir rocks to what are called **traps** is **secondary** migration. Both migrations occur through water-saturated spaces with the relative porosities and permeabilities (see Chapter 19) of the strata being all important.

Obviously during the early stages of catagenesis the squeezing of water out of the source rocks due to loading of overlying sediments may take such petroleum compounds as have been formed into the overlying rocks of greater porosity and permeability. However, from borehole observations, compaction of source rocks would be virtually completed at depths where temperatures were not high enough for the formation of many petroleum compounds. Squeezing of the oil out of the source rocks cannot therefore account entirely for primary migration. It appears that the continued formation of oil and gas during catagenesis produces such pressures that microfractures occur in the source rocks, thus allowing escape upwards into the reservoir rocks. Secondary migration thereafter takes place upwards and sideways due

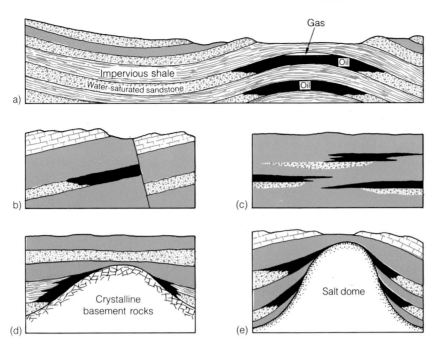

Figure 9.12 Sections to illustrate various types of traps favourable to the accumulation of oil and gas (gas is omitted except in (a)).

to (a) the buoyant rise of the lighter oil and gas through the water-filled reservoir rocks and (b) hydrodynamic conditions (movement of water through the reservoir rocks – see Chapter 19).

In general, then, oil migrates outwards and upwards from the source rocks, passes into porous or fissured reservoir rocks, rises to the highest possible level, and collects into an oil pool wherever the structure provides a trap which impedes further migration. Gas, if present in excess of the amount that the oil can hold in solution, bubbles to the top and forms a **gas cap** over the oil pool. Beneath the pool the pore spaces are occupied by water (usually salt) which is commonly under a very considerable head of pressure. If the pressure and gas content are sufficiently high, the oil gushes out like an effervescent fountain when the pool is tapped by drilling. When the pressure is too low to drive the oil to the surface – or becomes so as the initial pressure falls off – pumping is necessary.

A **trap** is formed when updip movement is stopped by an impervious barrier (the **roof rock**). 'If the roof rock is concave as viewed from below it keeps the oil and gas from escaping laterally and thus forms a trap. This is one of the few generalizations in the geology of petroleum that seems to hold good' as A.I. Levorsen put it in 1954. An anticline is the classic and commonest example (Figure 9.12a) of a structural trap. The great merit of this type of trap is that more than one can occur in the same anticlinal structure. Other types of structural traps can be caused by faulting (Figure

9.12b and Figure 9.14) or by combination of faulting and folding. When the trap is formed primarily by the original depositional environments such as limestone reefs (where they are surrounded by and eventually covered by mud) or the 'pinching out' of beds of sandstone (Figure 9.12c) or the changes in lithology due to the uneven floor of the sedimentary basin (Figure 9.12d) then the term **stratigraphic** trap is used. Many traps are of course formed by a combination of tectonics and stratigraphy. Salt dome traps (Figure 9.12e) are a special type of trap and are discussed in Chapter 10.

9.10 EXPLORATION FOR OIL AND GAS

It is clear from the foregoing that oil and gas fields are situated in areas of sedimentary rocks and therefore the existence, formation and history of former large sedimentary basins (see Chapters 29 and 30) are of prime importance to the oil geologist (Figure 9.13).

Exploration to find new fields still involves fieldwork (as the mapping of surface geological features is known) and the recording of surface 'shows' of oil and gas, their known existence often determining where fieldwork will take place. Much information is also gathered from the study of aerial and satellite photographs, but the existence of possible oil traps at depth requires detailed knowledge of the rock types and the geological structures. Many areas of sedimentary

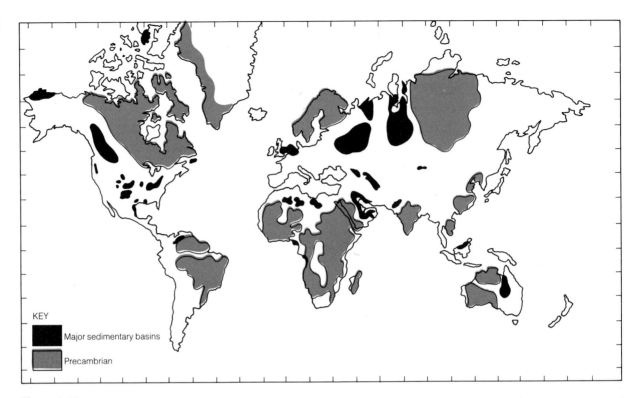

KEY
Major sedimentary basins
Precambrian

Figure 9.13 Distribution of major sedimentary basins containing giant petroleum fields. (*Based on G. C. Brown and E. Skipsey, 1986*)

rocks are covered by, for example, desert sands, tropical jungle or the sea and geophysical exploration methods and the drilling of boreholes are essential to prove the existence of economic deposits of oil and gas. The most effective geophysical method for detailed oil/gas exploration is that known as **seismic reflection profiling** (see Chapter 25). Originally developed on land the technique consists of setting off specifically sited and controlled explosions along surveyed lines, to produce artificial shock (seismic) waves similar to those produced by earthquakes. Some components of the waves are reflected upwards whenever they encounter (e.g. due to changes in stratigraphy or to faulting) rocks of different physical properties from those in which they are travelling. The time taken to reach carefully spaced surface detectors (**geophones**) is recorded and calculations can be made as to the depths at which changes take place and so a picture of the underground structure is eventually revealed.

For a variety of reasons, including obvious environmental objections, this type of prospecting is now carried out on land mainly by the **vibroseis** method. Specially designed 21-tonne trucks, by a system of hydraulics, 'thump' the ground and produce vibrations which penetrate deep into the Earth's crust. The times for the various reflected waves to return are detected by geophones and recorded on magnetic tape with the results being processed by computer. At sea the source of energy is provided mainly by the use of arrays of 'airguns' towed behind the survey ship at a constant depth (usually within the range 6–10 m). The ship carries powerful air-compressors which enable the guns to inject air bubbles of high pressure (14–30 megapascals) into the water at electrically controlled intervals and sound waves penetrate deep into the Earth's crust.

Groups of **hydrophones** (as opposed to geophones) are carried in a 'streamer' which is a tube some 10 cm in diameter, filled with oil to provide neutral density in seawater and perhaps of 3000 m in length. The streamer is towed about 15 m below the surface and can contain up to 120 groups of hydrophones, each group perhaps consisting of 20 hydrophones within a 25 m section of the cable and being separated by a similar interval from the next group. The oil companies generally record two-way reflection times to about 6 seconds only as oil and gas rarely occur at depths of more than a few kilometres. The exact position of the ship and its course is fixed from satellites. The processing and interpretation of the recorded data is complicated and has only reached its present sophisticated level with the use of powerful computing techniques. Figures 10.7 and 10.8 illustrate the type of geological information that can be produced once some detailed knowledge of the rock types of the area have been provided by drilling.

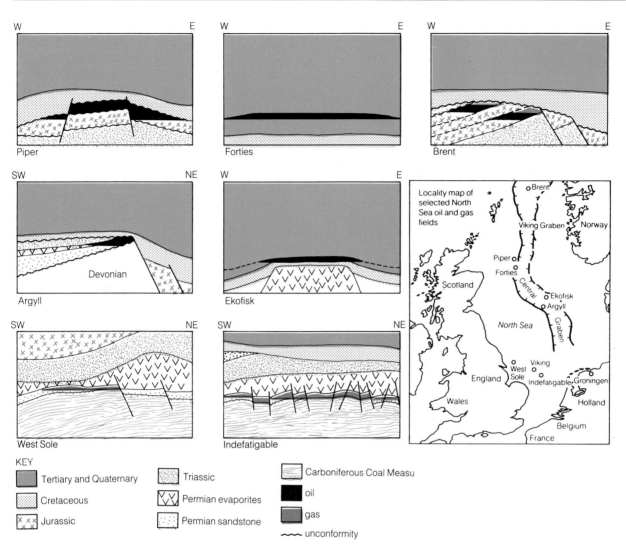

	West Sole	Piper	Ekofisk	Indefatigable	Argyll	Brent	Forties
Location in North Sea	southern North Sea basin	east of Moray Firth basin	Central Graben	southern North Sea basin	Central Graben	Viking Graben	between Viking and Central Graben
Structure of field	faulted low dome	three tilted and folded fault blocks below unconformity	low dome above evaporites (salt plug)	faulted low dome	tilted fault block with unconformities	tilted fault block below unconformity	low dome (drape)
Type of reservoir	sandstone	deltaic sandstones	chalky imestone	sandstone	dolomites and sandstone	deltaic sandstones	deep-water sandstones (submarine fans)
Type of reserve	gas	oil	oil	gas	oil	oil and gas	oil
Age of reservoir	Permian	Jurassic	Cretaceous-Tertiary	Permian	Permian	Jurassic	Tertiary

Figure 9.14 Schematic diagrams of typical North Sea oil and gas fields. West Sole and Indefatigable are in the southern North Sea, the remainder in the north. (*After G. C. Brown and E. Skepsey, 1986*)

Drilling **rigs**, if used at sea can be operated from ships or from floating platforms (**semi-sub-mersible** rigs) in water depths down to 300 m, or in waters less than 100 m deep, from platforms with legs resting on the sea bed (**jack-up** rigs). Essentially drilling rigs consist of a **derrick** from which is suspended a **drillstring** made up of sectional steel tubing (**drillpipes**) at the bottom of which is fixed a specially hardened steel-toothed or diamond-studded **bit**. On rotation the bit grinds its way downwards through the various strata, extra drillpipes being added periodically at the derrick as depth increases. The upper part of a hole is normally lined with steel tubing and may have a diameter of up to 75 cm, but the main exploratory part of the hole, often drilled to depths measured in thousands of metres, is usually of the order of 15 cm in width.

A high-density drilling 'mud' circulates downwards through the pipes and ascends up the sides of the circular drill-hole (the drillstring being narrower than the bit) carrying with it groundup rock fragments for examination. In specific cases an annular diamond-studded bit at the end of a **core-barrel** allows solid cylindrical cores of rocks to be obtained and brought to the surface for examination. (This type of drilling is essential in exploration for ore deposits and coal – Plates IVf, VId.)

When the drilling is completed, geophysical instruments (**sondes**) are lowered down the hole at the end of a wire so that various physical properties of the rocks can be measured, the information being delivered electronically to a recording instrument at the surface as the sonde is drawn slowly up the hole. The resultant **wireline log** indicates changes in such factors as electric potential and resistivity, radioactivity and density of the various strata encountered in the hole, helping to define most accurately thicknesses and properties of the various beds drilled through.

The first exploratory hole in a prospective area is often known as a **wildcat well** and though it may not reveal the presence of oil or gas, it is the essential part of any so-called **play** – a play being the name given to the combination of geological factors in a particular area which together make the accumulation of oil and gas possible. For example, early successful plays in the northern North Sea area favoured drilling in the vicinity of (i) deeply buried rifts (Figure 9.14) where Upper Jurassic black shales (source rocks) had undergone maturation and (ii) the nearby presence of potential sandstone reservoirs. As knowledge of the structures and the stratigraphic history increased during exploration additional plays emerged.

Onshore in Britain oil had been known for some centuries to be seeping into certain coal mines and nearing the end of the First World War drilling commenced in the Derbyshire–Nottinghamshire area. The Carboniferous limestones of the Pennines which outcrop, in generalized terms, in the crest of a large anticlinal structure contain patches of solid bitumen. Flanking the Pennines to the east coal-mining had revealed the presence of anticlines and (by analogy with the large limestone structure of the Middle East) it was decided to follow the 'anticlinal theory' of the day and drill these structures down to the Carboniferous Limestones below the Coal Measures. Eleven holes were drilled between 1918 and 1922 and oil was discovered in some in the Millstone Grit sandstones above the limestones, but not in great quantities. In 1939, however, after seismic surveys indicated an anticlinal structure beneath flat-lying Permian surface rocks at Eakring in Nottinghamshire (cf. Figure 9.12a) oil was discovered in quantity in the Millstone Grit sandstones. Oil production from similar structures in this area of England continues today.

In the south of England, oil seepages were known from Jurassic rocks on the Dorset shore and gas had been found when drilling for water in Cretaceous beds at Heathfield in Sussex. Anticlinal structures were drilled before the Second World War and later in the 1950s with minor success. However Wytch Farm in Dorset commenced production in 1979 and is now the biggest onshore producer of oil (with accompanying gas) in western Europe. The trap is a fault trap (see Figure 9.12b), the reservoir rocks being Triassic sandstones, sourced from younger but structurally lower, Jurassic black shales.

The discovery of methane in an anticlinal structure in the Permian-age Rotliegend Sandstones of Holland in 1959, combined with the knowledge that (a) gas existed in rocks of similar age near the Yorkshire coast of England and (b) that these rocks and the underlying Carboniferous coal measures extended beneath the North Sea (Figure 9.10) led to extensive exploration offshore. Geophysical surveys indicated possible structures and drilling proved methane in 1965 in Permian sandstone in a gentle domal structure (Figure 9.14, West Sole). Since then a variety of structures containing gas have been found, the movement and deformation of Permian halites being influential both in producing structural traps (see next chapter), and acting as a roof rock sealing in the methane rising from the Carboniferous coals below.

9.11 PRODUCTION AND RESERVES OF OIL AND GAS

Tables 9.3–9.6 give details of the world's production and reserves of oil and gas. It is noticeable

Table 9.3 World's oil producing countries, 1990 (*From BP Statistical Review of World Energy, 1991*)

	Million tonnes	World share %
USSR	570.0	18.1
USA	471.6	13.2
Saudi Arabia	327.1	10.4
Iran	155.3	4.9
Mexico	145.3	4.6
China	139.0	4.4
Venezuela	119.4	3.8
Iraq	98.2	3.1
Canada	93.3	3.0
UK	91.6	2.9
Nigeria	89.3	2.8
Abu Dhabi	87.3	2.8
Norway	81.8	2.6
Indonesia	73.0	2.3
Libya	64.7	2.1
Algeria	54.8	1.9
Kuwait	52.7	1.7
Egypt	45.5	1.4
India	33.9	1.1
Others	355.1	12.9
Total world	3148.9	100.0

Table 9.4 Proved reserves of oil 1990 (from B.P. Statistical Review of World Energy. 1991)

	Billion barrels	Billion tonnes*	World share %	
Saudi Arabia	257.5	35.0	25.5	⎫
Iraq	100.0	13.4	9.9	⎪
Kuwait	94.5	13.0	9.4	⎪
Iran	92.9	12.7	9.2	65.6
Abu Dhabi	92.2	12.1	9.1	⎪
Other Middle East countries	25.5	3.3	2.5	⎭
Venezuela	59.0	8.5	5.8	
USSR	57.0	7.8	5.6	
Mexico	52.0	7.3	5.2	
USA	33.9	4.3	3.4	
China	24.0	3.2	2.4	
Libya	22.8	3.0	2.3	
Nigeria	17.1	2.3	1.7	
Indonesia	11.1	1.5	1.1	
Algeria	9.2	1.2	0.9	
Canada	8.1	1.0	0.8	
India	8.0	1.1	0.8	
Norway	7.6	1.0	0.8	
Egypt	4.5	0.6	0.4	
UK	3.8	0.5	0.4	
Others	28.5	3.7	2.8	
Total world	1009.2	136.5	100.0	

* Conversion based on average (Arabian light oil) 33.5⁰ API gravity

that most of the proved reserves occur in the Middle East. Geologically this is due to the fact that the area was part of the southern continental shelf of the Tethys Ocean (Chapter 31), which separated the African and Eurasian plates during Mesozoic and Cenozoic times. In general terms marine shales and carbonate rocks of Mesozoic–Cenozoic age provide the source rocks and reservoirs respectively, most of the traps being of an anticlinal nature.

While Tables 9.2, 9.4 and 9.6 indicate the current figures for the world's reserves of coal, oil and gas it must be emphasized that all such figures are estimates – they are carefully calculated but none-the-less estimates. Many variables affect the calculations, including the world's continually changing political and economic scenes which particularly affect prices and in turn reserves. **Proved reserves** are those quantities which geological and engineering information indicate (with reasonable certainty) can be recovered under existing economic and operating conditions. The latter are another variable, as apart from costs, exploitation reveals new information regarding deposits which may or may not improve the actual reserve figures. In addition, with oil and gas, which eventually have to be forced out of the pores of the reservoir rocks, enhanced recovery methods, such as increasing pressure by pumping water or gas into the reser-

voirs, or increasing the permeability by fracturing the reservoir rocks by explosive or other means, affect the final quantities produced.

The natural gas figures quoted are perhaps the ones which could be drastically affected by future developments. There exist in what are probably enormous quantities, **methane hydrates** in recent marine sediments and in areas of permafrost (Chapter 20). These are ice-like, lattice-type molecular structures formed from methane and water under particular temperature and pressure conditions (between –8°C and 12°C at a pressure greater than 40 atmospheres). They occur in recent marine sediments, where the water depth is between 300 and 600 m, in the upper 300–1000 m of the sediments. (The increased temperatures which occur naturally as one descends the sediment pile destabilizes the hydrates.) The methane is considered to have formed mainly from recent microbial processes, though obviously some is derived from leakages below the sea-floor and some perhaps from abiogenic sources.

No offshore methane hydrates have so far been exploited but already production has taken place on land in western Siberia. The giant Messoyakh

Table 9.5 World's natural gas producing countries, 1990 (*From BP Statistical Review of World Energy, 1991*)

	Million tonnes oil equivalent		World share %	
USSR	655.9 ⎱	694.3	37.2 ⎱	39.4
'Eastern bloc'*	38.4 ⎰		2.2 ⎰	
USA	443.8		25.2	
Canada	88.0		5.0	
Netherlands	54.5		3.1	
Algeria	42.5		2.4	
UK	40.9		2.3	
Indonesia	40.7		2.3	
Norway	25.0		1.4	
Mexico	24.3		1.4	
Australia ⎱ New Zealand ⎰	22.7		1.3	
Argentina	21.8		1.2	
Iran	21.5		1.2	
Venezuela	19.5		1.1	
Malaysia	17.3		1.0	
Others	204.8		11.7	
Total world	1761.6		100.0	

* Albania, Bulgaria, Czechoslovakia, Hungary, Poland, Romania, Yugoslavia

Proved reserves of natural gas, 1990 (*From BP Statistical Review of World Energy, 1991*)

	Trillion cubic metres	World share %
USSR	45.3	38.0
Iran	17.0	14.3
Abu Dhabi	5.2	4.3
Saudi Arabia	5.1	4.3
USA	4.7	4.0
Qatar	4.6	3.9
Algeria	3.2	2.7
Venezuela	3.0	2.5
Canada	2.8	2.3
Iraq	2.7	2.3
Indonesia	2.6	2.2
Nigeria	2.5	2.1
Mexico	2.1	1.7
Malaysia	1.6	1.4
Netherlands	1.7	1.4
Norway	1.7	1.4
Libya	1.2	1.0
Others	12.0	10.2
Total world	119.0	100.0

gasfield consists of a Cretaceous sandstone anticlinal trap covering some 230 km². It is situated in an area of permafrost and temperatures in the top of the deep (800 m) reservoir are low, resulting in the methane and the pore water forming methane hydrate. As the free methane in the lower part of the reservoir was pumped out so the decrease in pressure resulted in the hydrate above dissociating. All production now is from the hydrates 'zone' and while slower than when from the 'free' methane zone, reserve figures have increased by 78 per cent according to J. Krason and colleagues (1989)!

9.12 ENERGY COMPARISONS

Coal reserves are calculated in most countries using relatively rigid rules depending, for example, on the spacing of mine and borehole information and the thicknesses and quality of seams. Unfortunately there is not yet a universally agreed set of criteria but each country attempts to assess their proved reserves in terms of geological and engineering information relevant to their economic situation.

Figure 9.15 shows the world's consumption of primary energy over the past 35 years. For statistical purposes the various ways of producing

energy can be compared in terms of 'oil equivalent'. For example, the calorific value of 1 million tonnes (Mt) of oil is taken as approximately equivalent to 1.5 Mt of coal, 3.0 Mt of lignite and 1.111×10^9 (i.e. 1 trillion) cubic metres of natural gas. Hydroelectric and nuclear power plants are converted to 'oil equivalents' by comparing the amount of oil required to fuel an oil-fired plant in order to generate the same amount of electricity. It will be noted that energy from so-called renewable sources i.e. windpower, wave power, hydroelectric schemes, apart from the last named do not feature. It is unlikely that they will be of significance for some time to come, if at all, and with the slowing down of the development of

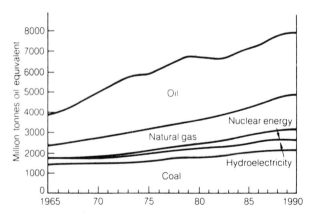

Figure 9.15 (*From BP Statistical Review of World Energy, 1991*).

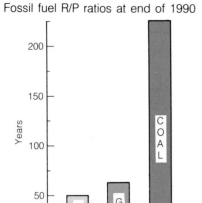

Fossil fuel R/P ratios at end of 1990

Figure 9.16 The Reserves/Production (RP) ratio is calculated by dividing the reserves remaining at the end of a year by the production of that year. The result in years gives the length of time the reserves would last assuming no increase in reserves and production continuing at the level of that year. (*From BP Statistical Review of World Energy 1991*)

nuclear power because of public unease, it appears that the fossil fuels and particularly coal will remain the World's main sources of electrical power generation for the foreseeable future (Figure 9.16).

9.13 FURTHER READING

Brown, G.C. & Skipsey, E. (1986) *Energy Resources,* Open University Press, Milton Keynes.

Ion, D.C. (1980) *Availability of World Energy Resources,* Graham & Trotman, London.

Jenyon, M.K. (1990) *Oil and Gas Traps,* John Wiley and Sons, Chichester & New York.

North, F.K. (1985) *Petroleum Geology,* Allan and Unwin, Boston, Mass.

Scott, A.C. (1987) *Coal and coal-bearing Strata: Recent Advances,* Geol. Soc. London Spec. Publ. 32.

Ward, C.R. (1984) *Coal Geology and Coal Technology,* Blackwell Scientific Publications, Oxford.

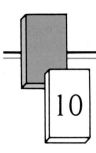

SALT DOMES AND SALT TECTONICS

Many of the world's most prolific oil and gas areas are associated directly or indirectly with evaporite sequences

M.K. Jenyon, 1986

10.1 SALT DEPOSITS

As has been pointed out in Section 6.8 (and will be discussed again in Section 22.8) beds of salt are deposited along with other evaporite minerals when seawater is evaporated. The process occurred as early as Precambrian times (Chapter 22) and extensive salt deposits occur in various parts of the world of Cambrian, Silurian, Devonian, Permian and Jurassic ages, wherever the topography of the then Earth's surface was suitable and the climate was arid.

Rock-salt consists of 98 per cent halite (NaCl), the remaining 2 per cent being made up of variable amounts of anhydrite ($CaSO_4$), potassium minerals and detrital material. Halite is soft (hardness 2.5) and has a low density ($2.17\,gm/cm^3$) which does not increase with depth of burial. Very unusual, however, compared with other rocks (apart from its being edible and soluble) is the ease with which it can become plastic and mobile. With differential pressure on a polycrystalline mass of halite, dislocation and gliding occurs along the same planes of the lattice of each individual crystal. Movement is aided by the presence of free salt water and/or with increase in temperature.

While it has been observed that the world's most productive oil and gas fields occur in areas where halite-bearing evaporite sequences are common, a genetic connection is not obvious. Salt movement, however, can produce both particular types of traps and roof seals, which in turn provide sites for accumulations of oil and natural gas. To understand why it is necessary to consider further aspects of the deformation and folding of rocks.

10.2 PIERCEMENT FOLDS AND DIAPIRS

When a pack of playing cards is folded into, say, an anticlinal form each card slips a little over the underlying card and rises slightly above it towards the crest, so leaving a crescent-shaped opening between them. Similarly when bedded rocks are folded, slipping along the bedding planes takes place. Reference to Figure 8.9 shows that the folded beds are thicker in the hinges of the folds and thinner in the limbs. The idealized form is shown in Figure 10.1. The necessary thickening and thinning is best achieved by weak **incompetent** beds such as clays, mudstones and shales. Under stress they react differently from stronger **competent** beds such as sandstones. Not only do they fill up the spaces beneath the more competent members of, say, a growing anticline, but where they come under sufficient stress they begin to burst out of their confined position by breaking through the overlying beds, taking

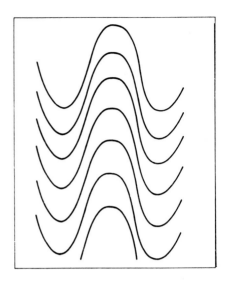

Figure 10.1 Idealized diagram of similar folding in incompetent beds, showing thickening in the crests and thinning in the limbs of the folds.

full advantage of cracks and fissures in the latter. The fold has now become a **piercement** or **diapiric** fold (Figure 10.2). If a structure of this kind were seen only below the level of the break through, it would appear to be no more than a fold; but seen above this level, where the underlying core has pierced its roof, the structure is clearly an incipient intrusion. Shale diapirs have also been

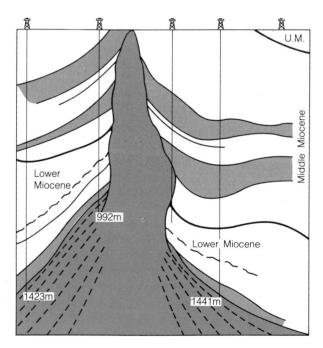

Figure 10.2 Diapir breaking through to the surface from an anticlinal core of Upper Oligocene shale and mudstone. Depths penetrated by some of the oil wells that have revealed the structure are given to indicate scale (vertical = horizontal), Barrackpore Oilfield, Trinidad. (*Simplified after G.E. Higgins, 1955*)

recorded from Alaska associated with the Kaltag Fault (a probable transform (Section 28.3) boundary fault associated with the movement of the North American and Eurasian plates (Chapter 31)), offshore from the Niger delta and in other parts of the world.

The term **diapir** (through-piercing) was proposed in 1910 by the Romanian geologist L. Mrazec (1867–1944) after his discovery in the Carpathians of elongated domes and anticlines with cores of salt that had locally broken through the crests or penetrated along faults. Such a fold becomes a diapiric fold after the mobile core has pierced the stronger mantling rock, but if the salt has continued to develop into a plug-like intrusion the structure is called a diapiric intrusion or, briefly, a diapir. **Diapirism**, a group name for the geological processes involved, is the development of an intrusion from a geological formation which comes under sufficient stress to deform, pierce and break through other rocks. There are many kinds of diapirs, ranging from pingos (made from ice, Section 19.13) and intrusive mudstones (Figure 10.2) to granite plutons in great variety; but, thanks to the exploration activities of oil companies, by far the best known are those that ascend great distances from underlying thick deposits of salt and other evaporites. Salt diapirs are conventionally called **salt domes** in reference to the form of the overlying strata which they have uplifted (Figure 22.34), and which in consequence may reveal their presence underground. They can, however, take various shapes and depending on their three-dimensional form are referred to as **salt plugs, salt pillows, salt stacks** or **salt walls**.

In the Carpathians, and elsewhere in orogenic belts where salt domes were first found, the motive force necessary for their development was ascribed to the compression thought to have been responsible for folding the associated rocks. However, after the significance of the celebrated oilfield around the Spindletop salt dome in southeast Texas was fully realized, hundreds of salt domes (Figure 10.3) were discovered in the thick sediments of the Gulf Coast States and the adjoining sea floor. Throughout this great salt-dome province the strata dip gently seawards, but nowhere do they show any sign of orogenic folding. The only folds are those occurring over the tops or around the flanks of the domes or plugs. Here, then, instead of a diapiric intrusion being a break through resulting from excess of folding, it is the other way about: the folding has resulted from the active ascent of columns of salt. Obviously there are special circumstances in which salt diapirs can develop. What exactly these are is still a puzzle. Presumably the natural 'buoyancy' of salt compared with the denser surrounding and

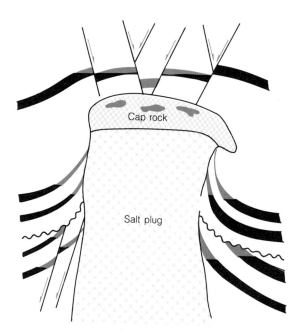

Figure 10.3 Idealized section through a Gulf Coast, Texas, salt-dome field showing various types of traps where oil (black) can accumulate. (*After A.I. Levorsen, 1954*)

overlying rocks is a key factor. But there are many examples where no movement has occurred despite similarities of thickness and depth of burial with areas of extensive piercement. Even in salt-mines, layers of comparable thickness only a few metres apart stratigraphically behave differently, though this may be due to subtle differences in moisture content.

There does, however, appear to be a geographical relationship between salt diapirism and areas susceptible to earthquakes; a spatial relationship to fault trends; and a temporal one to changing rates of movement of lithospheric plates.

10.3 SALT MOVEMENT

As early as 1912 Arrhenius pointed out that if the average density of the surrounding strata is higher than that of the intrusive salt, an upward force of buoyancy would be acting on the latter, tending to lift it towards the surface.

Experimental work in the 1930s by L. L. Nettleton and in the 1950s by T.J. Parker and A.N. McDowell with scale models, using a variety of viscous liquids, did much to support the plastic-flow theory of salt movement. A.I. Levorsen summarized Nettleton's experiments as follows: '[they] consisted in placing a layer of heavy asphaltic oil, with a density of about 1.0, over a layer of heavy syrup with a density of about 1.4, in a glass jar and covering it with a sheet of rubber.

To simulate the salt, the jar is turned upside down and placed on a surface that has irregular projections which distort the rubber and ''trigger'' the upward movement of the lighter oil through the syrup. By photographs taken at regular time intervals, the progress of the vertically moving oil through the syrup could be shown, and the resulting forms look very much like salt plugs.' Parker and McDowell confirmed Nettleton's work using weak muds for the sediment and asphalt for the salt. They were able to reproduce the characteristic patterns that develop around and above growing salt plugs such as those that proliferate in the Gulf Coast of Texas, USA.

The density difference between salt plugs and the intruded rocks has been amply confirmed by the discovery of strongly negative gravity anomalies over many salt domes. Such anomalies imply underlying deficiencies of mass (i.e. the presence of material of relatively low density) extending downwards to great depths, sometimes for many kilometres. However, freshly deposited sediments have a lower density than salt and positive anomalies have been discovered above concealed salt domes, where recent sedimentation has been fast and a salt plug from considerable depth has been rising over a long period of time.

Compaction of sediments during burial increases their density in the process of turning them into rocks and once an overburden thickness of 450 to 600 metres is reached equality of density is achieved with that of a buried salt layer. As subsidence of the area and sedimentation continues the buried sediments become steadily denser than the underlying salt. The process by which the salt rises upwards due to the density difference has been named **halokinesis**.

Like most of the sea floor of today the surface between salt and overburden is likely to have had considerable irregularities here and there, either from the start or at some stage during the deposition of the sedimentary overburden. Underwater erosion, slumping of sediments, folding and faulting could all play their part.

Consider Figure 10.4a where there is a 'high' of salt at B, relative to the surrounding surface AA'. The projection could be a ridge or a mound, for example. Because of the difference in overburden pressure at B (due to the greater thickness of underlying salt) compared with that at A and A', upwards movement could take place at B. This, if it continued, would involve inwards movement of salt towards B and a salt pillow would form (Figure 10.4b). As flowage continues sinking of the overlying sedimentary pile will take place where the salt layer has thinned. What is called a **peripheral sink (PS)** develops, either in annular form if we are dealing with a more or less circular plug, or

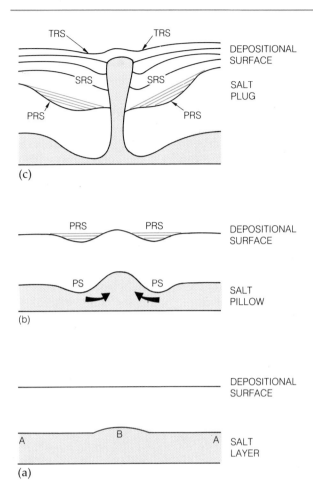

(c)

(b)

(a)

Figure 10.4 Development of salt pillow and plug. (*Modified from M.K. Jenyon 1990*)

as two troughs on either side of a salt wall (Figure 10.4b).

Assuming an original salt layer of the order of one or two hundred metres in thickness and an overburden measured in hundreds or thousands of metres the peripheral sinking would result in subsidence of the contemporary surface (Figure 10.4b). Sedimentation could then take place in the **primary rim syncline** (PRS), one source area being the raised area of the surface immediately above the rising salt mass. Depending upon the rate of the latter's upward movement which depends on continual lateral flow of the original salt layer and the nature and rate of the sedimentation further rim synclines (SRS – secondary; TRS – tertiary) would periodically develop (Figure 10.4c).

During the Permian what are now parts of eastern Britain, the North Sea, The Netherlands, Germany, Poland, and Lithuania were invaded by the Zechstein Sea, probably due to a melting phase of polar ice-sheets causing a worldwide rise in sea-level (Figure 10.5). It flooded an area of desert, now represented by red aeolian sandstones (the Rotliegende).

Oil-companies' drilling, both onshore and offshore, has revealed the extent of this desert area and the sea which flooded it. A ridge of older rocks, now represented by the Mid North Sea– Ringköbing–Fyn High (Figure 10.5) was probably never completely

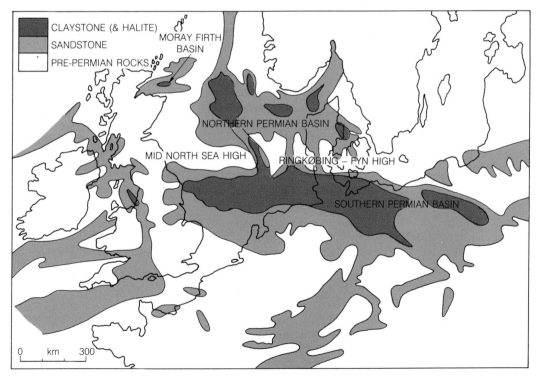

Figure 10.5 Distribution of early Permian aeolian sandstones (Rotliegende), desert lake and sabkha claystones which were subsequently invaded by Zechstein Sea (see also Figure 22.38). (*After K.W. Glennie, 1990*)

submerged so there were two east–west sub-basins north and south of this feature. Evaporation of the seawater and subsequent reflooding of the area took place a number of times, as can be recognized by the occurrence of at least five successive evaporite cycles (Chapter 6) with halite being most prominent in the middle three.

In the North Sea (Figures 9.14, 10.5, 10.6, 10.7, 29.36, 29.37) north-south 'graben' structures (down-faulted blocks between two parallel faults) cut across the 'high' referred to above. The additional presence of many wrench faults and the phenomenon of subsiding depositional areas changing to areas of uplift (a process now included in the term **inversion**) support the view that salt movement is 'triggered-off' and encouraged by earth movements.

J.C.M. Taylor pointed out that diapiric struc-

tures of the Zechstein salts are best developed when the thicknesses of the original salt beds are 130 m or more and the overburden exceeds 900 m. Figure 10.7 is a line drawing across the bed of the North Sea constructed from a seismic reflection survey, while Figure 10.8 shows a detailed seismic section across a salt wall about 3 km thick which lies below sea-bed sediments. A detailed history of the intrusion can only be worked out with the addition of cuttings and cores from boreholes in the area to establish the stratigraphic ages of the rocks. However, rim synclines and/or peripheral sinks are obvious at the sides of the salt wall. The presence of another salt intrusion to the left of the wall suggests that in earlier Cretaceous times the diapir was in the form of a wide pillow. Differences in the stratigraphy on opppposite sides of the wall may indicate a fault, its zone of weakness

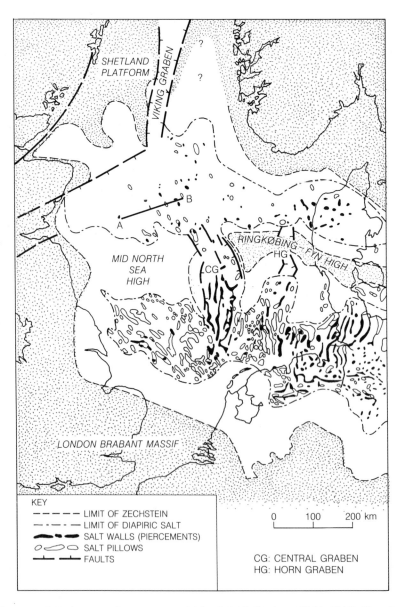

Figure 10.6 Distribution of Zechstein salt intrusions. (*After J.C.M. Taylor, 1986 in K.W. Glennie (ed) 1990*). See Figure 10.7 for cross section A–B.

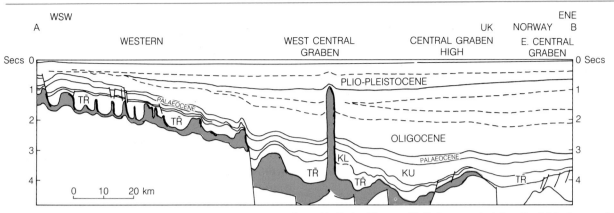

Figure 10.7 Line drawing from Shell-Expro seismic line (A–B in Figure 10.6) interpreted by Paul Veclen and Gerhard Lauer. The diapiric activity on the Western Platform predated a period of Upper Jurassic erosion. In the Central Graben the thickness of overburden deposited during Cretaceous and Tertiary times probably reactivated the diapiric movement, hence the pillar intruding Tertiary sediments. Abbreviations as for Figure 10.8. (*Reproduced with kind permission from K.W. Glennie (ed.), 1990*)

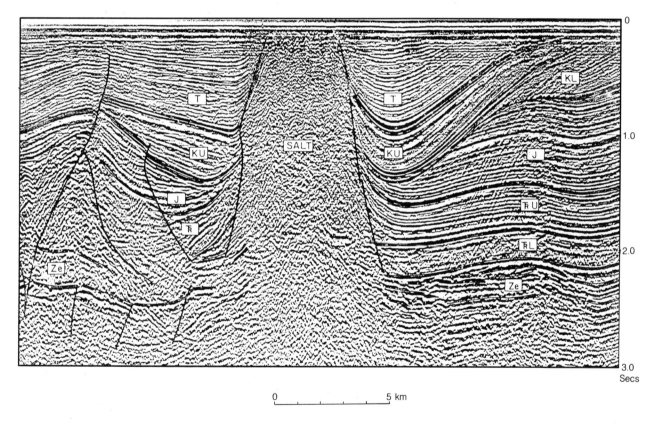

Figure 10.8 Seismic reflection section showing Cretaceous to early Tertiary growth of salt wall. (Ze–Zechstein salt; TR–Triassic; J–Jurassic; K–Cretaceous; T–Tertiary; L–Lower; U–Upper.) (*Reproduced from J.C.M. Taylor, 1990 in K.W. Glennie (ed) (qv.)*)

being selected for further upwards movement which continued into Tertiary times. A comparison with Figure 10.3 will indicate where possible oil traps could occur on the flanks of the wall, assuming that suitable reservoir and roof rocks were present.

If a salt plug rises from so great a depth that its buoyancy or head of pressure has become sufficient to carry it above the surface it will not necessarily do so (but see later). By the time it reaches a level where the density of the surrounding sediments is lower than that of the salt, it may well be easier to spread along the bedding planes and lift the overlying sediments. This will involve less work – expenditure of energy – than forcing its way upwards as a plug. For this reason sills of salt are sometimes encountered, but they do not generally proceed far from the feeding plug. More commonly the plug spreads out laterally to form an overhanging head, like a lop-sided mushroom. Sediments which might have ended steeply or vertically against a cylindrical plug may be turned back on themselves by the overhang (Figure 10.9). Salt movements involving tangential compressive stress are sometimes referred to as illustrating **halotectonism**.

10.4 CAP ROCKS

A feature of many salt plugs, first examined in detail in the Gulf Coast region of Texas, is the presence on top of them of cap rocks (Figure 10.3). They can cover the top of an intrusion and vary between 100 m and 300 m in thickness. Mainly of anhydrite, which tends to occur immediately above the halite, mixed zones of gypsum and anhydrite occur and are sometimes followed by a top layer of calcite (limestone). Free sulphur can also be present as can xenoliths of the country rock through which the salt plug has ascended.

The upper zone of calcite (or limestone) is thought to be due to the reaction on anhydrite/ gypsum by methane and carbon dioxide generated during the maturation of the source rocks of the oil and gas. The free sulphur may result from this process and/or from microbial action.

10.5 EMERGENT SALT PLUGS

Penetration of the surface by a rising salt intrusion is most easily achieved in regions where the more recent overlying lighter sediments have been removed by denudation. Then if the rainfall is not too great, dome-shaped mountains of salt (with or without accompanying gypsum, anhydrite etc.) may be built up high above the surrounding country until the salt is enabled to flow under its own weight as a 'salt glacier'. Conditions of the kind described are characteristic of the salt-dome province of part of the Middle East (Figure 10.10). It appears that during late Precambrian and early Cambrian times extensive beds of evaporites were laid down in subsiding rift basins along the northern edge of Gondwanaland in what are now the Arabian Gulf and Zagros Mountains (the Hormuz Formation), Kerman in central Iran (Ravar Formation) and the Salt Range Province in Pakistan (Salt Range Formation).

The remarkable fact that out of nearly 200 salt domes located round the Arabian Gulf (Figure 10.10) at least 80 still have outflowing 'salt glaciers' implies that the parent bed, here of late Precambrian–early Cambrian age and at a depth of some 2.5 km, must be immensely thick (more than 1000 m). Plugs 4 to 10 km across break through Cretaceous and Tertiary limestones, which are sharply upturned around the margins. The salt carries up with it immense numbers of xenoliths of which some have dimensions which straddle the whole width of the plugs. Both extrusive and intrusive igneous rocks are present, along with many fossiliferous sedimentary blocks. Most are not recognizable in the immediate area, support-

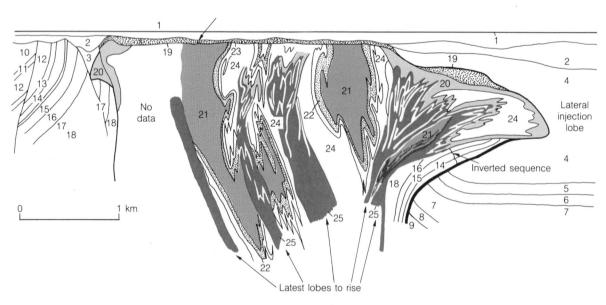

Figure 10.9 Heide salt plug, near Hanover, Germany, showing lateral expansion at its head, with overturning and thrust faulting of the surrounding sediments, and intrusive lobes of older salt (black) into a younger saline series. 1 Recent and Pleistocene, 2 Tertiary, 3–9 Cretaceous, 10–11 Jurassic, 12–17 Triassic, 18 Permian, 19 Residual clay cap-rock, 20 Anhydrite cap-rock, 21–4, *Younger saline series*: Upper rock-salt (21), Red salt clay (22), Potash bed (23), Lower rock-salt (24), 25 Older rock-salt. The highly complex structures have been revealed by borings and mining operations. (*A. Benz, 1949*)

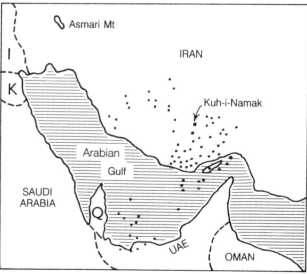

Figure 10.10 A mountain of salt 10 km across and 1310 m high. Kuh-i-Namak salt stock, Iran, showing small outflowing 'salt glaciers'. The contact between salt and intruded Cretaceous limestone is overturned in places. Steep outward dips of 85° are common, lessening to 40° or 30° about 1.5 km from the contact. (*The British Petroleum Company Ltd.*)

Index map of the Arabian Gulf province of salt intrusions which have ascended from Late Precambrian–Early Cambrian evaporite beds. Miocene evaporites are the source of more complex forms of diapirism in the region around Asmari Mountain to the north-west. (I–Iraq; K–Kuwait; Q–Qatar; UAE–United Arab Emirates)

ing the suggestion by P.E. Kent that most or all of the plugs were extruded into rocks or onto surfaces which have long since been eroded away.

10.6 FURTHER READING

Glennie K.W. (ed) (1990) *Introduction to the Petroleum Geology of the North Sea*, 3rd edn, Blackwell Scientific Publications, Oxford.

Husseini, M.I. and S.I. Husseini (1990) Origin of the Infracambrian Salt Basins of the Middle East, in *Classic Petroleum Provinces*, (ed J. Brooks) *Geol. Soc. London, Spec. Publ.* **50**, 279–292.

Jenyon, M.K. (1986) *Salt Tectonics*, Elsevier Applied Science Publications, London.

Jenyon, M.K. (1990) *Oil and Gas Traps*, John Wiley, Chichester, New York.

Kent, P.E. (1979) The emergent Hormuz salt plugs of southern Iran. *Jour. Petrol. Geol.* **2**, 117–144.

Levorsen, A.I. (1954) *Geology of Petroleum*. W.H. Freeman, San Francisco.

Nettleton, L.L. (1955) History of concepts of Gulf Coast salt-dome formation. *Bull. Amer. Assoc. Petrol. Geol.* **39**, 2373–2383.

Parker, T.J. and A.N. McDowell (1955) Model studies of salt-dome tectonics. *Bull. Amer. Assoc. Petrol. Geol.*, **39**, 2384–2470.

METAMORPHISM AND METAMORPHIC ROCKS

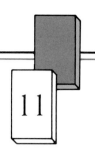

I am not now
That which I have been

Lord Byron (1788–1824)

11.1 INTRODUCTION

That the Earth is an active and dynamic planet is shown throughout this book. Rocks once at the Earth's surface may be buried to great depths, they may be deformed, and their temperatures may be changed as a consequence of burial or by the intrusion of hot magmas. In response to such changes in their surrounding conditions, the characteristics of the rocks commonly become modified – they undergo metamorphism and become metamorphic rocks.

Metamorphic changes are of two principal types: those affecting which mineral species are present, and those affecting the shape and arrangement (microstructure) of mineral grains. The distinction between these types of change may be seen in everyday processes. Thus powdery snow, formed of feathery ice crystals like those shown in Figure 11.1a, is transformed into glassy-looking solid ice with progressive burial under an accumulating snow pile. This transformation involves no change in the mineral constituents and may occur without melting; it results from modification of the complex crystal growth shapes (Figure 11.1a) into simpler shapes (Figure 11.1b)

with much closer contact of crystal grain boundaries as the ice is compacted. Changes in the mineral species in rocks are artificially caused in kilns and blast furnaces. Thus in a lime kiln, high temperatures convert limestone (with the mineral calcite, $CaCO_3$) to lime (CaO) with the evolution of carbon dioxide (CO_2). Similarly, on being subjected to high temperatures in furnaces, rocks containing metal ores (as minerals such as metal oxides or sulphides) undergo a change of mineralogy to produce native metal.

Variations in **temperature** and **pressure** are obvious causes of rock metamorphism, but they are not the only ones. Changes may be induced as a result of changes in **chemical composition**; in such cases the changes are most commonly associated with the movement of fluids (whether they be volatile-rich, that is CO_2-H_2O rich fluids, or melts). Where the rock transformations involve **deformation** (the change in shape of the rock body caused by mechanical stresses), such deformation will have a major impact on the structures in the transformed rocks.

The overall variety of changes that can affect

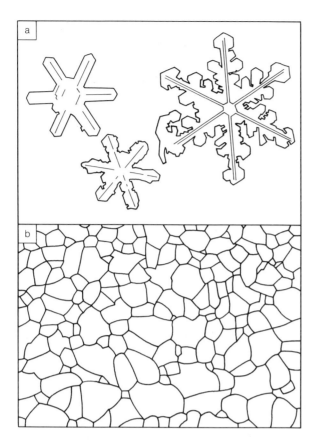

Figure 11.1 (a) Shows the habit of ice crystals forming by crystallization from atmosphere. Note euhedral crystal faces and complex growth forms. (b) Shows ice crystals at depths of 216 m in the Ross Ice Sheet, Antarctica. Note the simple, approximately polygonal, shapes of the crystals fitting together to fill space. (*After A.J. Gow, 1970*)

rocks is extremely large, and for convenience geologists usually limit the **scope of metamorphism** to include only those changes taking place whilst the rocks remain essentially solid bodies. Weathering processes at the Earth's surface involve changing rocks as a result of reaction with the oxygen-rich atmosphere and with rain and groundwaters. These processes in conjunction with erosion and deposition lead to the eventual formation of sedimentary rocks, and are not therefore considered as part of metamorphism. Metamorphic processes are considered to involve temperatures and pressures of transformation which are clearly greater than those usually found at the Earth's surface. The precise boundary used for this distinction is somewhat arbitrary; with increasing depth of burial in the Earth, the conditions which give rise to the consolidation of sediments (e.g. sands and muds) into sedimentary rocks (sandstones and mudstones) have no sharp boundary with conditions of higher temperature and pressure giving rise to metamorphic rocks (meta-sandstones and

meta-mudstones, or psammites and pelites – see summary on metamorphic rock names in Appendix 11.2).

At the other end of the spectrum of metamorphic conditions we may consider high temperature changes. At the highest temperatures, rocks change not simply by developing new minerals but by undergoing melting, and some rocks, called **migmatites**, preserve evidence of the inter-reactions of melts and solids. With sufficient melting (around 30 per cent) the rock ceases to behave like a solid body and becomes more like a fluid in its mechanical behaviour. Once this happens the changes that we are considering belong more to the field of magmatism than metamorphism.

With these considerations in mind, definitions of rock metamorphism do not, therefore, embrace all conditions of rock alteration. A widely-adopted definition of metamorphism is:

> Metamorphism is the mineralogical and structural adjustment of solid rocks to physical or chemical conditions which have been imposed at depths below the surface zones of weathering and cementation, and which differ from the conditions under which the rocks originated.
> (*F.J. Turner and J. Verhoogen, 1960, Igneous and Metamorphic Petrology, McGraw-Hill*)

11.2 CONDITIONS OF METAMORPHISM

Physical conditions; temperature and stress. The variety of circumstances under which metamorphism occurs are very diverse. We may readily contrast the situation adjacent to a gently-emplaced igneous intrusion with that of many rocks in orogenic belts. In the former case, heat from the igneous intrusion changes the **temperature** conditions of the adjacent country rock, and transforms rock mineral assemblages without significant change in **pressure** conditions or **deformation**. In the case of orogenic metamorphism the large-scale major structural changes in active tectonic zones will commonly result in a combination of changes in temperature, **hydrostatic stress** (pressure), and **deviatoric stresses** (which cause deformation; see later).

The imposition of new conditions of temperature and stress may be identified as the principal physical causes of the metamorphism of rocks. Given a constant chemical composition for a particular rock, these physical controls will determine what the mineral assemblages and structures of the rocks should become during metamorphism. Whilst temperature is a familiar concept, the scope of the term stress and its relation to pressure and deformation is worthy of a little more explanation.

The state of stress of a body is a product of the mechanical forces acting upon it. Obviously forces pressing in upon a rock may have different magnitudes in different directions. Conveniently, these variations may always be described in terms of **three principal forces (or stresses)** acting at right angles to one another. Two situations may be readily distinguished. The first is where the forces acting upon a body are the same in all directions and the three principal stresses are therefore equal. In this situation (Figure 11.2a) the forces may change the volume of a rock, but not its shape. In the second case (Figure 11.2b), the forces and the principal stresses vary in different directions with the result that deformation or change of shape will occur. The effects of deformation are seen in many metamorphic rocks. They are seen on all scales as **folds** of the original bedding (Figure 11.3), which shorten the length of a rock body in a particular direction. On a small scale, deformed metamorphic rocks often show crystals of minerals like amphiboles and micas in parallel arrangement, the **cleavage** or **schistose structure** discussed further below.

Where the principal stresses are all equal (in Figure 11.2a, $\sigma^1 = \sigma^2 = \sigma^3$), the rock is said to be

Figure 11.3 Schistose rocks crumpled by small folds.

in a state of **hydrostatic stress or pressure** and the deviatoric stresses are zero. Where the stresses are unequal (in Figure 11.2b, $\sigma^1 > \sigma^2 > \sigma^3$) we can think of the rock being affected by two components:

a) mean stress
$$\bar{s} = \frac{\sigma^1 + \sigma^2 + \sigma^3}{3}$$

b) deviatoric stresses
$$s^1 = \sigma^1 - \bar{s}$$
$$s^2 = \sigma^2 - \bar{s}$$
$$s^3 = \sigma^3 - \bar{s}$$

The mean stress in this situation is sometimes referred to as the hydrostatic component of stress. The presence of deviatoric stresses is accompanied by the presence of shear stresses (see also Figure 8.22).

The natural **slates**, used for roofing, show particularly well the effect of deviatoric stresses and deformation. Their commercial value is due to a structure whereby they can be split regularly into thin slabs or slices, and this fissility or cleavage is often called **slaty cleavage**. It is commonly oblique to bedding (Figure 11.4) and is caused by the orientation of abundant flaky minerals (micas, chlorite, clays) parallel to the cleavage direction. This direction is typically formed at right angles to the direction of greatest compression (maximum principal stress of Figure 11.3) of the rocks, and is aligned with the orientation of the axial planes of folds developed during the same deformation (Figure 11.5). The parallel orientation has been caused both by rotation of mineral grains into the cleavage plane, and by growth of new minerals parallel to the cleavage plane. Slaty cleavage is a

a) System with hydrostatic stress

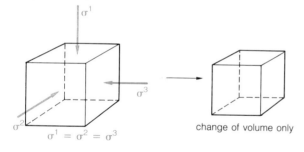

change of volume only

b) System with deviatoric stresses

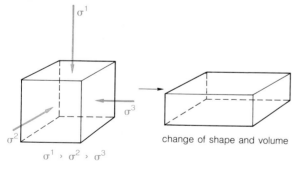

change of shape and volume

Figure 11.2 Effects of stress. (a) Hydrostatic stress or pressure where the principal stresses are equal, the stress acts upon a cube to change its volume but not change its shape. (b) Where principal stresses are not equal, the differences between them (deviatoric stresses) result in change of shape of the cube. The principal stresses in each case are designated σ^1, σ^2, σ^3.

Figure 11.4 Cleavage intersecting bedding at a high angle in grey banded slate. The estuary of the River Camel, west of Wadebridge, Cornwall (*British Geological Survey*).

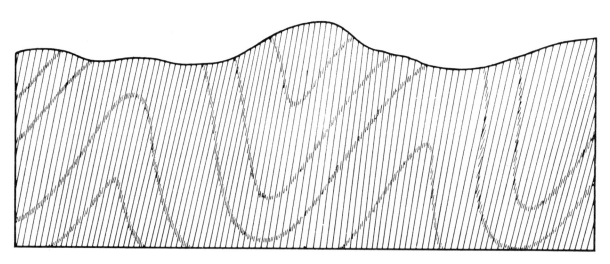

Figure 11.5 To illustrate the relation of slaty cleavage or schistosity to bedding and folding; the cleavage planes are approximately parallel to the axial planes of the folds.

particularly well-developed example of the schistose structure or **schistosity**, which is seen in many metamorphic rocks which have been subjected to deviatoric stress. **Schist** is the general name applied to such rocks, irrespective of grain size, the term slate being kept for fine-grained rocks showing exceptionally good and regular cleavage.

Whilst the relative magnitude of the deviatoric stresses will predominantly determine the extent of deformation and the nature of structures on all scales, the effect of the deviatoric stresses upon the **stability of minerals** in a rock is relatively minor. In general, rocks are not strong enough to support large deviatoric stresses over long periods of geological time at high temperatures, and therefore deviatoric stresses are small in magnitude relative to the pressure exerted by the weight of the overlying rock column. Thus, the stress parameter of greatest importance for mineral stability is pressure. When we talk about pressure, we are either referring to a state of true hydrostatic stress (with deviatoric stresses being zero), or to the mean stress in a system with deviatoric stresses. Pressure may therefore be considered as the overall force tending to decrease (compress) or increase (extend) the volume of the rock. In general, at depth within the Earth's crust, the pressures on rocks are usually approximately equal to the **load stress** caused by the weight of the overlying pile of rocks.

(To assist the reader in the following account Appendix 11.1 is a list of common minerals (with their chemical composition) found in metamorphic rocks. Appendix 11.2 describes a current method of naming these rocks.)

Chemical conditions; protolith composition, metasomatism and fluids. There are two chemical conditions which should be noted for metamorphic rocks.

(a) **Protolith composition.** When one examines metamorphic rocks in the field, one can commonly recognize bedded sequences that clearly represent metamorphosed sedimentary strata. Furthermore, it is evident that the rocks retain the principal chemical features of their parent rocks (**protoliths**). Thus metamorphosed mudstones and shales are rich in Al and K (with micas instead of clay minerals), meta-sandstones are typically rich in SiO_2, and meta-limestones (marbles) have $CaCO_3$ as their main constituent. Similarly, meta-basalts usually retain their distinctive chemical features of moderately high Ca, Al and Mg + Fe.

Such observations lead to the recognition of an important feature of many metamorphic processes, namely that the chemical compositions of the protoliths are usually largely preserved. Therefore the chemical conditions of metamorphism are in many respects very simple – they are defined by the chemical composition of the starting material. Exceptions do occur, and under circumstances where there have clearly been changes in the proportions of elements like Ca, Fe, Mg, Al during metamorphism, the processes are described as **metasomatic**.

(b) **The metamorphic fluid phase.** The **volatile constituents** in any rock undergoing metamorphism, generally behave rather differently to most chemical components. These volatile constituents are dominantly H_2O and CO_2, and under the conditions of pressure and temperature existing during metamorphism, they commonly give rise to a highly compressed fluid – the **metamorphic fluid phase**. There is abundant evidence in metamorphic rocks that crustal metamorphism generally occurs in the presence of such a fluid phase, which is highly mobile and is lost from the rocks prior to their exhumation and exposure at the Earth's surface. Because of the mobility of the fluid phase, the original protolith will usually lose or gain volatile constituents during the course of metamorphism. As a result, changes in H_2O and CO_2 composition are not considered to warrant the use of the term metasomatic, unlike changes in components like Ca, Fe, Mg, and Al.

The volatile content of most metamorphic rocks decreases progressively with increasing temperature of metamorphism. A shale, which initially contains a large proportion of H_2O-rich clay minerals, when metamorphosed at temperature conditions of about 500°C and 0.5 GPa (5 kilobars) to produce a schist, will generate a total of about 12 per cent by volume of H_2O-rich fluid. This fluid will not be released all at one time, but rather continuously or episodically during heating and burial as chemical reactions cause changes in mineral constituents.

The principal volatile components lost from rocks are H_2O and CO_2, which are derived from the breakdown of hydrous and carbonate minerals in **devolatilization reactions**, such as:

$$CaCO_3 + SiO_2 = CaSiO_3 + CO_2$$
calcite   quartz   wollastonite   carbon dioxide

musovite + chlorite = cordierite + biotite
   +quartz   + H_2O
(formulae for minerals given in Appendix 11.1)

Alternatively, it is possible for rocks which are poor in volatiles prior to metamorphism, such as igneous rocks, to take up volatiles in the course of reactions occurring at relatively low temperatures, but these volatiles will subsequently be lost as well if high enough temperatures are attained.

Metamorphic minerals such as quartz or garnet often contain tiny bubbles of fluid called **fluid inclusions** which represent trapped samples of metamorphic fluid present during or after crystallization of the host minerals (Figure 11.6). The fluid trapped after the formation of the host mineral is usually present as trails of inclusions aligned along the sites of the micro-fractures through which fluid passed on its way upwards and out of the system, and may be very different in composition from fluid present during formation of the host mineral.

The most commonly occurring components in trapped fluid inclusions are mixtures of H_2O, CO_2 and dissolved salts such as NaCl, but CH_4 and N_2 have also been reported as major constituents.

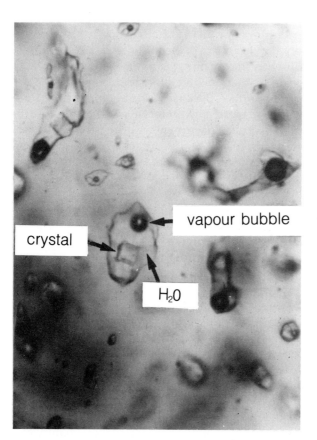

Figure 11.6 Photomicrograph of fluid inclusions in a quartz crystal, from a vein in an Alpine-age shear zone cutting a granodiorite, from the Pyrenees, France. The inclusions contain three phases (see central inclusion): cubic halite (NaCl) daughter, crystals, H_2O-rich liquid, and dark round bubbles of vapour. (*Sample collected by S.A. Tempest, photograph by D.A. Banks*)

pressed at high pressures is trivial compared with that of the loosely-packed, unconsolidated sediments from which it formed. Pore-filling fluid in the latter is gradually expelled as the grains are compressed together, the rock is recrystallized due to rising temperatures, and the pore-space is reduced. Even with such low instantaneous abundance, however, this fluid may be widely distributed and very mobile, and thus gain or loss of the H_2O–CO_2 fluid, or change of fluid composition, may be extensive.

As rocks are heated and devolatilization reactions occur, the fluid-filled porosity may transiently increase until the fluid can escape. However, for most of their history, metamorphic rocks are unlikely to have porosities greater than a fraction of one per cent. The structure and geometry of the fluid-filled pores are nonetheless important, since these will control the **permeability** of the rock to fluid. The fluid in equilibrium with solid mineral grains will be located largely along

H_2O may be derived from devolatilization (dehydration) reactions in the metamorphic rocks themselves, from 'juvenile' water originating in the mantle, or circulation of water such as seawater or meteoric water from the hydrosphere. CO_2 may derive from devolatilization (decarbonation) reactions or from oxidation of organic carbon or graphite within the metamorphic rocks. When graphite remains stable in metasedimentary rocks, then the associated fluid phase may be rich in CH_4. NaCl and other dissolved salts are probably derived from formation waters or seawater trapped in buried sedimentary rocks or from evaporites, and may appear in the fluid inclusions as tiny 'daughter' crystals of the mineral halite (NaCl) as shown in Figure 11.6.

Where does dense, pressurized fluid sit in a rock undergoing metamorphism? Metamorphic rocks do not have obvious pores, and the amount of fluid present in a metamorphic rock at any instant of time is probably very small. The **porosity** of a metamorphosed sedimentary rock com-

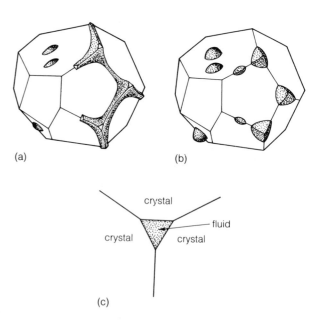

Figure 11.7 Perspective sketches illustrating the distribution of fluid (stippled) around a single grain in a rock. (a) Fluid is mostly present along interconnected tubelike channels along grain-edges. A few isolated fluid-filled pores are found on grain faces. (b) All fluid-filled pores are isolated. (c) Cross-section of tubelike channel at grain-edge. (*After E.B. Watson and J.M. Brenan, 1987*)

grain edges and at grain corners (Figure 11.7), but will not wet the 'faces' between grains unless present in great abundance. If the fluid-filled pores along grain edges are interconnected in a three-dimensional network of tiny tubelike channels (Figure 11.7a), then the ability of fluid to flow through the rock is greatly enhanced. The few data that exist on pore structure at high temperatures and pressures, suggest that interconnection and therefore relatively easy fluid mobility will often be the case, particularly if the fluid is rich in dissolved salts.

However, if the pores are isolated from each other (Figure 11.7b), then fluid flow is only possible along fractures. The most obvious manifestations of such fractures are the mineral-filled **veins** that cut through most metamorphic rocks, the vein-filling having been deposited by the passage of the fluids along the fracture. The formation of the fractures or cracks themselves in rocks under pressure, has usually been caused by pressurized fluid rising rapidly from greater depths, or generated by devolatilization within metamorphosing rock. The presence of the pressurized fluid reduces the bursting (tensile) strength of the surrounding rock, and causes the opening up of fractures perpendicular to the direction of least compressive stress by a process called **hydrofracturing**. This process is probably very important in the escape of fluid from the rocks undergoing metamorphism.

Metamorphic fluids are hot, highly compressed solutions with strong solvent properties, especially above the **critical point**. The critical point of a gas is the temperature above which it cannot be liquefied, however great the pressure. For H_2O, the **critical temperature** is 374°C, and the minimum confining pressure is 0.022 GPa (218 times atmosphere pressure, and corresponding to an overburden of about 835 metres of rock of density 2.7). Above the critical point H_2O passes into the condition of a **supercritical fluid**, which has the density of a liquid but the molecular freedom and mobility of a gas. The low density and high mobility of supercritical fluids results in these fluids being rapidly lost from the metamorphic system, and on the time-scales of crustal metamorphism (see below), fluid may perhaps only be present in a rock for a rather small proportion of time.

Supercritical fluids in metamorphic rocks at shallow levels in the Earth's crust (e.g. in geothermal systems), may be connected to the surface of the Earth through a series of interconnected fractures. In such a system, the fluid is subject only to the pressure exerted by the column of water above, and is said to be **hydrostatically** pressurized. The fluid is free to convect in the manner of a domestic hot water or central heating system if the necessary thermal driving forces are present.

With increasing depth and overburden pressures in the crust, the mechanical strength of a rock will be insufficient to prevent the collapse of any fractures or interconnected pore space which is open to the Earth's surface. Any trapped fluid will then be pressurized at the same pressures as the enclosing rock, which is commonly referred to as **lithostatic** pressure. The transition from hydrostatic to lithostatic pressures has been observed to occur at three to five kilometres' depth in deep holes drilled into oil-bearing sediments. However, this figure may vary widely according to the mechanical properties of the rock.

At high pressures and temperatures, the solubilities of rock-forming minerals in supercritical H_2O-rich metamorphic fluid are significant. For example, the solubility of quartz in aqueous NaCl solution at 0.5 GPa pressure and 350°C is of the order of 1 gram per kilogram of solution. The fluid may transport its dissolved constituents (especially alkalis and silica) over large distances. The absolute concentrations of the various constituents in the supercritical fluid are relatively small, but the total amounts transported through geological time may be very large if the quantities of fluid are large. The passage of metamorphic fluids upwards through rocks undergoing metamorphism produces veins, commonly filled by such minerals as quartz or calcite which are deposited when the fluids become supersaturated in dissolved constituents, such as SiO_2 or $CaCO_3$, during cooling and depressurization. The importance of mass transport in convecting fluids is well illustrated by convecting seawater fluids at hot, newly-formed mid-ocean ridges, where the dissolution of minerals and the resulting chemical alteration of ocean-floor rocks has radically influenced the composition of the oceans over geological time.

The proportions of components such as H_2O, CO_2 and CH_4 in the metamorphic fluid phase can dramatically affect the stabilities of metamorphic mineral assemblages. For example, carbonate minerals are stabilized in fluids containing CO_2 whilst hydrous minerals, such as micas or amphiboles, are stabilized by H_2O-rich fluids. By determining experimentally the quantitative effects of fluid composition on the stabilities of mineral assemblages under metamorphic conditions, we can then use the metamorphic mineral assemblages themselves to deduce the composition of the fluid phase present during metamorphism. It is a common and widespread observation that fluids in regional metamorphic rocks, such as pelites and metabasic rocks, are dominantly H_2O-rich, whereas CO_2 and CH_4 are abundant components of the fluid phase only if the rocks are unusually rich in carbonate or graphite respectively.

We have considered the effects of increasing temperature in promoting reactions amongst the minerals and fluids present in a rock undergoing metamorphism. However, reactions involving minerals and fluids, such as $CaCO_3 + SiO_2 = CaSiO_3 + CO_2$, may equally well occur by changes in the composition of the fluid phase. Thus, if fluid is added to a rock from some external source, such as a nearby intrusion of magma or some adjacent devolatilizing rocks, and the temperature is sufficiently high, then the mineral assemblage of the rock may change in response to the addition or **infiltration** of fluid. The total extent of mineralogical reaction occurring in response to this infiltration is then a measure of the amount of fluid passing through the rock, over the time-period during which the reaction has occurred. This approach allows us to identify the patterns and pathways of fluid movement through a pile of metamorphosing rock. Several such studies have been carried out in recent years. In some of these studies it has been possible to recognize that the metamorphism of some lithologies has been accompanied by the infiltration of quite large quantities of fluid comparable in magnitude to the volume of rock which has been infiltrated, implying that fluid flow through metamorphic rocks tends to be focussed along selected pathways. The relative importance of pore structure and fracture patterns in controlling these pathways has still to be determined in many cases.

11.3 CLASSIFICATION AND SETTINGS OF METAMORPHISM

We have seen above that temperature, pressure and deviatoric stress are the major physical conditions controlling the course of metamorphism, and the mineralogy and microstructure of the metamorphic rocks formed. When we consider the classification of types of metamorphism, it is evident that classification in terms of temperature, pressure and deviatoric stress would be one way of doing this. From this viewpoint, broad descriptions of metamorphism by such terms as **thermal** or **dynamothermal** have been used, but giving precise meanings to such terms is difficult. For example, whilst it is easy to distinguish situations of high and low deviatoric stress, the details of deformation processes are strongly dependent on temperature and pressure conditions. As a result it is difficult to define limits to different magnitudes of deviatoric stress from readily observable features in many rocks.

However, the mineral constituents of rocks are readily determined, and provide a means of distinguishing and classifying temperature–pressure conditions. We shall see in later sections that it is convenient to classify some types of metamorphism according to the **zonal** succession of mineral assemblages encountered in particular areas, or on a wider basis by using mineral assemblages in the **metamorphic facies classification**.

Another approach to distinguishing different types of metamorphism is to consider the settings which have led to metamorphism. For example the circumstances of metamorphism adjacent to a localized igneous intrusion, are very different from those of orogenic metamorphism which results in the production of metamorphic rocks over thousands of km^2 of the Earth's crust. The settings of metamorphism are readily determined by geological associations found by field mapping, and this is therefore a very useful way of defining types of metamorphism.

We therefore distinguish broad categories of local and regional metamorphism according to whether the metamorphic rocks are clearly limited in area and related to a localized event, or are of large areal extent lacking distinct boundaries. Each of these two major categories is subdivided, according to setting, as follows:

Local metamorphism

1. **Contact metamorphism** – where the metamorphic rocks are found adjacent and clearly related to an igneous body.
2. **Dynamic metamorphism** – where the metamorphism is associated with a zone of strong deformation, as adjacent to a fault or in a shear zone.
3. **Impact metamorphism** – where the metamorphism is associated with the impact upon Earth of an extraterrestrial body.

Regional metamorphism

4. **Orogenic metamorphism** – where metamorphic rocks are found associated with orogenic zones of subduction and mountain building.
5. **Oceanic metamorphism** – that found extensively in the oceanic crust, because the repeated injection of basic magmas at oceanic spreading ridges, with accompanying circulation of heated seawater, causes metamorphism of the newly-formed crust before it moves away from the oceanic ridge.
6. **Burial metamorphism** – that found in the lower parts of thick sedimentary basin sequences, where the rocks have been buried to sufficient depths to cause the genesis of minerals which are only stabilized at temperatures and pressures greater than those found near the surface.

Because of the commonness of their occurrence

and the variety of the resulting mineral assemblages, contact metamorphism and orogenic metamorphism have been most extensively studied. Both often show systematic changes in mineral assemblages across the area affected by metamorphism, as a consequence of variations in temperature in the contact case, and temperature and/or pressure in the orogenic case. These features are illustrated by descriptions of specific terrains in the following sections.

11.4 CONTACT METAMORPHISM – AS EXEMPLIFIED BY THE BALLACHULISH AUREOLE

The country rocks around a large igneous intrusion provide a relatively simple setting in which a series of metamorphic changes may be observed. The metamorphic alteration is directly associated with the rise in temperature caused by contact with the hot igneous intrusion, and the zone of altered rock surrounding the igneous intrusion is referred to as the **metamorphic aureole** (Figure 11.8).

In many of these situations the pressure in the contact aureole has remained constant during the

metamorphism, and deviatoric stresses have been insufficient to cause much deformation. Thus the principal cause of metamorphism has been the change in temperature, and this has caused changes in the mineral assemblages. With the formation of new minerals the detailed microstructural relations of the mineral grains also change, but the absence of deviatoric stresses means that larger-scale structures, like bedding and any folds present before the igneous intrusion, are usually preserved in shape even though the mineralogy of the rocks is changed.

In a metamorphic aureole the highest temperatures were reached in the rocks adjacent to the igneous contacts, and the aureole temperatures gradually decreased outwards until at some distance from the contact the rise in temperature was insufficient to induce any metamorphic reconstitution. As a result, the rocks show a progressive series of changes in mineral constitution, becoming more and more modified from their original nature as one moves across the aureole towards the igneous contact. Thus the new mineral assemblages formed in the aureole define a series of concentric zones around the contact.

This is well illustrated by the aureole of the Ballachulish granite–diorite body in the Scottish Highlands. A variety of meta-sedimentary rocks occurs in this aureole, including quartzites (meta-sandstones) and marbles (meta-limestones), but in common with many other locations it is the mineral assemblages in the **pelites** (meta-argillites) see Appendix 11.2), which show the greatest sensitivity to the effect of varying temperature. Figure 11.9 shows the distribution of the mineral assemblage zones in the pelites as it has been mapped in the aureole.

Outside the aureole the pelites, which had previously been affected by regional-orogenic metamorphism, are schists with the principal minerals; quartz, muscovite, chlorite, biotite and sometimes garnet (see Appendix 11.1 for mineral compositions). The first change seen inside the aureole (zone 1, Figure 11.9), is the development of a spotted structure (Figure 11.10a) due to the formation of the mineral cordierite, and as cordierite becomes progressively more abundant chlorite disappears (zone 2). Further into the aureole rising temperature has caused reaction between muscovite, biotite and quartz, so that K-feldspar is extensively developed in addition to cordierite (reaction b, Figure 11.9). At this stage (zone 3) the schistose character of the country rocks is lost, because not only have new minerals formed, but minerals like the micas which persist (though in decreased amounts) have recrystallized (i.e. new grains have formed in place of the old ones). These new micas, having grown in the absence of devia-

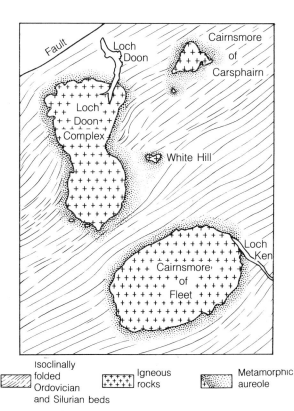

Isoclinally folded Ordovician and Silurian beds

Igneous rocks

Metamorphic aureole

Figure 11.8 Aureoles of contact metamorphism around small plutons (batholiths and stocks), Galloway, southwest Scotland (map is 30 × 40 kms).

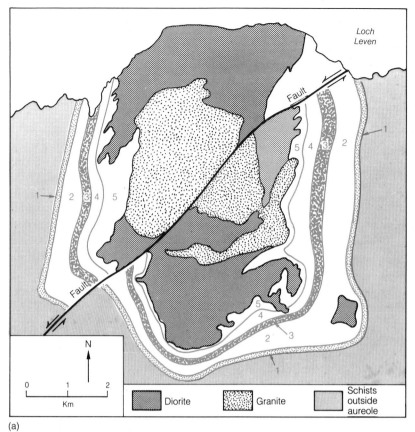

(a)

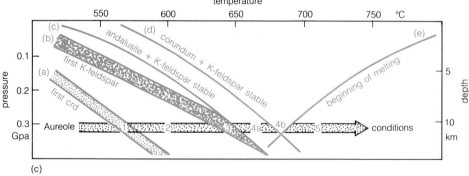

(b)

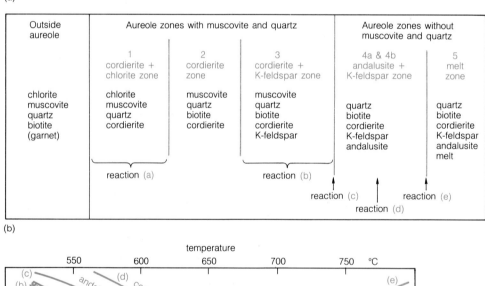

(c)

Figure 11.9 The Ballachulish aureole of contact metamorphism in the western Scottish Highlands. (a) Distribution of metamorphic zones. (b) Characteristic mineral assemblages of zones in pelitic rocks. (c) Temperature–pressure (depth) conditions of zones. Zone names and lines (isograds) separating zones are shown in colour in (a) and (b), and the corresponding reaction lines are shown in colour in (c). (crd – cordierite.) (*After D.R.M. Pattison and B. Harte 1991*)

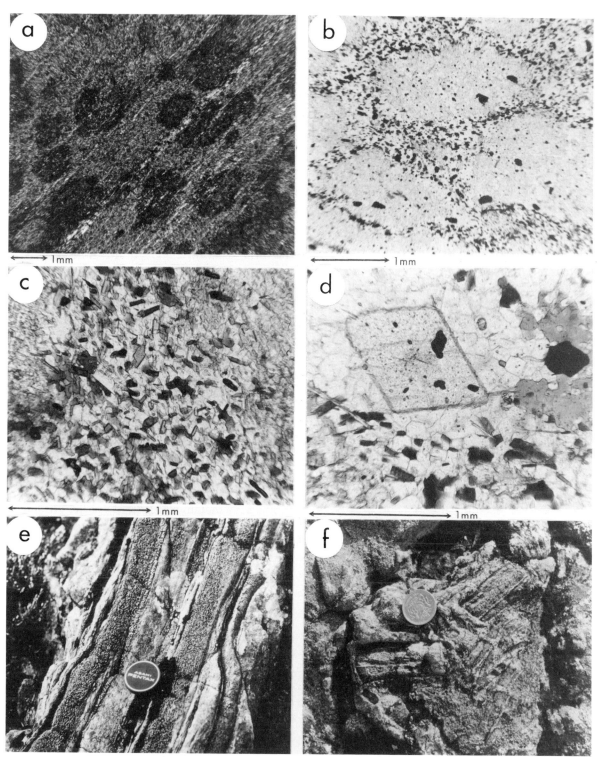

Figure 11.10 Rocks from the Ballachulish aureole. (a) Spotted schist in thin section. Relatively large crystals (porphyroblasts) of cordierite form elongate spots in a fine-grained schistose matrix of muscovite and chlorite. (b) Thin section of typical zone 4 hornfels showing relatively large crystals (porphyroblasts) of cordierite forming ovoid patches in a matrix of quartz, biotite (dark) and K-feldspar. The biotite flakes show random orientation (decussate texture) rather than a schistose structure. (c) Higher magnification of part of (b) showing the decussate texture of biotite. (d) Diamond shaped cross-section of andalusite prism surrounded by K-feldspar together with biotite and cordierite in high grade (zone 4) hornfels. (e) Field photograph of zone 4 hornfelses. The layering of rock types corresponds to original sedimentary beds and is of psammites (meta-sandstones) and pelites (meta-argillites). The pelite layers are the darker ones and show rough surfaces because of upstanding andalusite prisms coupled with pits where cordierite has been weathered away. (f) Small areas of fragmented (brecciated) rock layers, with fragmented blocks of hornfels in leucocratic (granitic) material formed by melting. This is a type of migmatite called an agmatite. (Scale bars in bottom left of corresponding thin section.)

toric stresses, do not usually have a preferred orientation, but grow in a random arrangement known as **decussate** texture (Figure 11.10c). From this zone of the aureole inwards the rocks therefore usually lack schistose character, but are relatively tough and difficult to break – such contact aureole rocks are called **hornfelses**.

Moving closer to the igneous intrusion, the next mineralogical change we encounter, in zone 4 of the Ballachulish aureole, results from the fact that muscovite and quartz have ceased to remain in chemical equilibrium with one another, but have reacted (reaction c, Figure 11.9) to form andalusite and K-feldspar until one of them has been used up. Because of variation in the original rock compositions, which of the minerals muscovite or quartz is used up first varies from rock to rock, and thus nearer the intrusion we may find rocks with muscovite or with quartz but never with both minerals.

The rocks of zone 4 which retain quartz are now hornfelses consisting principally of biotite, cordierite, K-feldspar, andalusite (and/or its Al_2SiO_5 polymorph, sillimanite) together with quartz (Figure 11.10b-e). This assemblage persists up to the igneous contact, although evidence of further reaction is seen in the formation of leucocratic veins and seams, sometimes with fracturing of the rocks (Figure 11.10f) as a result of melting in zone 5 adjacent to the contact. In rocks which lost quartz and retained muscovite at the beginning of zone 4, a further reaction commonly occurs involving the final breakdown of muscovite and production of corundum and yet more K-feldspar. This change from muscovite-bearing to corundum-bearing, may be used to subdivide zone 4 (reaction d, Figure 11.9).

11.5 METAMORPHIC ZONES, GRADE AND METAMORPHIC FIELD GRADIENTS

A sequence of mineral assemblages formed as a result of spatial variation in the conditions of metamorphism, like those described for Ballachulish, defines a particular set of **metamorphic zones**. Under different conditions of temperature and pressure, the mineral assemblages and, therefore, the zones will vary, as will be illustrated in the following section on orogenic metamorphism. Metamorphic zones are commonly defined on the mineral assemblages in pelitic rocks (meta-argillites), but they may also be defined using other rock types (e.g. metabasalts, meta-limestones).

The term metamorphic **grade** is used to indicate **relative variation** in the temperature–pressure conditions of metamorphism within a given area. For example, at Ballachulish, the rocks in the outermost part of the aureole (zone 1) are at lowest grade, whilst those in the innermost part of the aureole (zone 5) are at the highest grade; and proceeding from zone 1 to zone 5 we go **upgrade**. Variations with grade do not have to be purely temperature dependent as at Ballachulish. In other metamorphic areas the variations with grade may result from pressure changes or some combination of temperature and pressure changes. Thus the difference between high-grade and low-grade does not have a single meaning for all metamorphic terrains; it is purely relative within the specific terrain under discussion.

As has been indicated, the change from one zonal mineral assemblage to the next is a result of chemical reaction between the minerals in the rocks, and these chemical reactions have occurred at particular positions which were determined by the maximum temperatures reached at different places in the aureole. In order to show the relation of the mineral assemblages to the temperature (T) and pressure (P) conditions, the positions of the mineral reactions are commonly plotted on a P-T diagram. Such a diagram for the Ballachulish aureole is shown in Figure 11.9c, and it shows the temperature–pressure conditions of the aureole. Note that although the pressure has not changed across the aureole, it has nonetheless had an important bearing on what mineral assemblage has formed at a particular temperature. The set of P-T conditions defined by the rocks on the ground is called the **metamorphic field gradient** (the word 'field' emphasizing that it is the gradient of P-T conditions shown by the rock distribution mapped in the field).

The determination of the P-T conditions defining a metamorphic field gradient, and construction of a P-T diagram like Figure 11.9c, depends on careful determination of the precise mineral reactions occurring in the rocks, and the P-T calibration of these reactions by doing laboratory experiments. The chemical reactions causing the zonal changes in the Ballachulish mineral assemblages (Figure 11.9) are as follows (chemical compositions for the minerals are given in Appendix 11.1):

(a) 1 muscovite + 1 **chlorite** + 2 quartz = 1 **cordierite** + 1 biotite + 1.5 H_2O

(b) 6 muscovite + 2 biotite + 15 quartz = 3 **cordierite** + 8 **K-feldspar** + 6.5 H_2O

(c) 1 muscovite + 1 quartz = 1 **K-feldspar** + 1 **andalusite** (or sillimanite) + 1 H_2O

(d) 1 muscovite = 1 **corundum** + 1 **K-feldspar** + 1 H_2O

(e) K-feldspar + quartz + H_2O + (biotite + cordierite) = **melt** + (andalusite or sillimanite).

The mineral names in bold type are ones whose presence or absence is particularly important in defining the zones, and such minerals are often called **index minerals**. Reaction (a) is involved in the definition of the beginning of zone 1 (by the first appearance of cordierite), and in the definition of the beginning of zone 2 (by the total disappearance of chlorite). Some of the reactions in the P-T diagram (Figure 11.9c) occur over a range of temperatures at a given pressure, because of variations in the precise compositions of the minerals, particularly their Fe : Mg ratios in different rocks. For reactions (a), (b), and (e) this means that the first development of cordierite, K-feldspar and melt, respectively, will not be at identical temperatures in rocks of different composition. In the field the most obvious line to map is usually that where a given mineral and the associated mineral assemblage appears (or completely disappears) for the first time going upgrade, and it is usually this line which is used to define the beginning of each zone. Such lines of first appearance of particular mineral assemblages are called **isograds**, and where applicable the isograd and zone upgrade of it are named after the index mineral (or mineral assemblage) appearing at the isograd, as shown in Figure 11.9 and further illustrated in the following section and Figure 11.11).

11.6 OROGENIC (REGIONAL) METAMORPHISM – EXEMPLIFIED BY THE EASTERN SCOTTISH HIGHLANDS

Regional metamorphic rocks which have been formed in orogenic zones form extensive parts of the Earth's continental crust. Over some extensive areas the metamorphic grade may vary little, but grade variations associated with the occurrence of metamorphic zones have been widely recognized. Indeed it was in the rocks in the eastern part of the orogenic belt in the Scottish Highlands that metamorphic zones were first mapped by Barrow and Horne around the end of the nineteenth century.

Figure 11.11 shows the complex pattern of regional metamorphic zones in the eastern Scottish Highlands. As at Ballachulish the zones are based on mineral assemblages in the pelites (meta-argillites). Low-grade rocks (chlorite and biotite zones) occur in the south,where the metamorphic terrain is truncated by the Highland Boundary Fault, and also in the north of the area (biotite zone). Metamorphic zones of progressively higher grade occur adjacent to both these areas, but they involve quite different mineral assemblages in the two areas. To the north the mineral assemblages involve the relatively low-pressure minerals, cordierite and andalusite; while to the south the higher-pressure minerals garnet and kyanite occur. The zones in the north are referred to as the **Buchan zones**, after the geographic region where they occur, and those in the south are referred to as **Barrow's zones** (or **Barrovian zones**), after the man who first recognized them.

Buchan metamorphic zones. The lowest grade zone is the biotite zone, and around this are a series of higher grade zones (Figure 11.11). To the east of the biotite zone (along Z-Z in Figure 11.11a) the increasing grade sequence of mineral assemblages shows the incoming of the minerals: cordierite, andalusite, and sillimanite, and, following the principles discussed above, the zones are named after these index minerals. In all zones the mineral assemblages in the pelitic rocks include substantial quartz and muscovite, and the changes in mineral assemblage involve the ferromagnesian and aluminous minerals. The diagnostic assemblages are shown in the central chart of Figure 11.11.

There is clearly some similarity in these zones to those of the Ballachulish aureole, but they have formed at slightly higher pressure and the grade of metamorphism does not reach the high temperatures of that aureole. In addition, the Buchan rocks are not hornfelses, but are **schists** with preferred mineral orientations formed as a consequence of the deformation typically associated with orogenic metamorphism. At the highest grades, the schistose structure may be accompanied by some separation of micaceous and quartzo-feldspathic components, resulting in an obvious layering or lenticular structure visible in hand specimen. Such rocks are typically coarse-grained and are described as **gneisses** (see Appendix 11.2).

In places within the sillimanite zone potash feldspar occurs, indicating the beginning of reactions leading to the breakdown of muscovite as seen in the Ballachulish aureole, and there is also some evidence of melting. It is shown by the presence of gneissose rocks called migmatites, which show seams and pods rich in quartz and feldspar, believed to have formed by crystallizations of melts.

The metamorphic zones to the west of the Buchan biotite zone (Y-Y in Figure 11.11) show a different sequence to those of the east, and involve staurolite. This is because the western grade sequence shows increasing pressures from east to west, and shows a transition from the eastern Buchan mineral assemblages to those of Barrow's zones.

Barrow's zones. In this area the pelitic rocks in all zones again contain abundant quartz and musco-

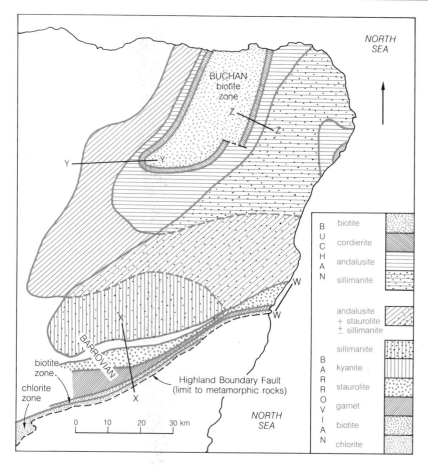

MINERALS PRESENT IN PELITES IN EASTERN BUCHAN ZONES ALONG TRAVERSE Z–Z				
	Biotite	Cordierite	Andalusite	Sillimanite
quartz	———————	———————	———————	———————
muscovite	———————	———————	———————	———————
chlorite	———————	— —		
biotite	———————	———————	———————	———————
cordierite		———————	———————	———————
andalusite			———————	———————
sillimanite				———————

MINERALS PRESENT IN PELITES IN CENTRAL BARROW'S ZONES ALONG TRAVERSE X–X						
	Chlorite	Biotite	Garnet	Staurolite	Kyanite	Sillimanite
quartz	———	———	———	———	———	———
muscovite	———	———	———	———	———	———
chlorite	———	———	———			
biotite		———	———	———	———	———
garnet			———	———	———	———
staurolite				———	———	———
kyanite					———	———
sillimanite						———

Figure 11.11 Map of eastern Scottish Highlands showing distribution of Buchan and Barrovian metamorphic zones, together with charts showing the changes in diagnostic mineral assemblages for the zones. Zone names and lines (isograds) separating zones are shown in colour (*After B. Harte and N.F.C. Hudson, 1979*)

vite, with changes in assemblages being shown by the ferromagnesian and aluminous minerals present. The lowest grade zone here shows chlorite without biotite, and the ferromagnesian and aluminous index minerals appear in the sequence: biotite, garnet, staurolite, kyanite and sillimanite. These minerals are, therefore, used to name the zones. The chart in Figure 11.11 shows the full mineral assemblages which characterize the zones. Mineral compositions are given in Appendix 11.1, and the garnet involved is one predominantly rich in the almandine ($Fe_3Al_2Si_3O_{12}$) molecule.

The rocks of Barrow's zones show structures typical of orogenic belts, with evidence of several phases of deformation, and the rocks are mainly schists. In the lowest grades these schistose pelitic rocks show such fine grain size and good cleavage that they are described as slates (Figure 11.12a). There is a general increase of grain size as one proceeds upgrade. This is shown particularly by the principal minerals, quartz and muscovite, forming the matrix of the rocks. In the lower grade zones (chlorite to garnet) these minerals form grains smaller than 0.1 mm, while in the higher grade zones (staurolite, kyanite and sillimanite) these matrix minerals are 0.3 to 1.0 mm in size (Figure 11.12). Biotite and particularly the minerals garnet, staurolite and kyanite typically occur as larger grains (**porphyroblasts**) with sizes of up to 3 mm (Figure 11.12b, c and d). Sillimanite occurs as mats of fine fibres referred to as fibrolite (Figure 11.12e). The sillimanite-grade rocks show evidence of melt presence by the occurrence of **migmatitic gneisses** (Figure 11.12f), similar to those found in the Buchan high-grade rocks.

Figure 11.11a shows that the sillimanite zone in eastern Scotland partly transgresses the zone boundaries of the other Barrovian and Buchan zones, and the detailed textures of the rocks suggest that the sillimanite tended to grow later than the other index minerals. This is a result of the sillimanite growth occurring in response to the progressive heating of the rocks. Detailed mineral assemblages within the whole sillimanite zone, show that the ferromagnesian and aluminous minerals associated with it change from cordierite + andalusite in the eastern Buchan, through an intermediate andalusite + staurolite bearing assemblage, to a kyanite + staurolite + garnet assemblage in the Barrovian area. Along the coast by W–W in Figure 11.11 a lower pressure variant of the Barrovian sequence is seen in which there is no kyanite zone; it is called the **Stonehavian** sequence after the coastal town of Stonehaven.

In Barrow's zones, as in the Ballachulish and Buchan sequences, increasing temperature during metamorphism is the main cause of the mineral-ogical changes in the given zonal sequence. The control of pressure on the mineral assemblages is shown by the differences between these three sequences. Pressures across Barrow's zones were distinctly higher than those in the eastern Buchan and Ballachulish cases, and correspond to depths of around 20 km (instead of close to 10 km). The P-T conditions (metamorphic field gradients) shown by the eastern Buchan and Barrovian sequences may be seen in Figure 11.13, where they are used as representatives of the **low** and **intermediate pressure baric types** of regional metamorphism (see later discussion). Similar sets of regional metamorphic zones to the Scottish ones described have been recognized in many areas of orogenic metamorphism around the world.

11.7 THE METAMORPHIC FACIES CLASSIFICATION

It has been seen that in areas of both contact and regional metamorphism, such as at Ballachulish and in the eastern Scottish Highlands, the various mineral assemblages are usually arranged in regular zones, providing a ready means of classifying metamorphic rocks and characterizing the metamorphism in a **particular area**. Zonal classifications do not, however, provide a sufficiently general scheme to permit easy comparison of the rocks and conditions of metamorphism in different metamorphic terrains. An alternative, more general, scheme following Eskola (a Finnish geologist) has therefore been developed called the **metamorphic facies classification**, which permits such comparisons to be made.

In the metamorphic facies classification each facies is representative of certain conditions of metamorphism, and metamorphic rocks are allotted to the various facies on the basis of the mineral assemblage they show for their particular rock chemical composition. In any one facies, assuming stable chemical equilibrium, each rock composition will be represented by a particular assemblage of minerals, which is symptomatic of the metamorphic conditions concerned. Within any one facies there will, however, be different mineral assemblages for different rock compositions metamorphosed under the same conditions. The metamorphic facies principle can therefore be stated as:

Rocks from different metamorphic terrains are assigned to the same metamorphic facies if they show the same mineralogy for the same bulk chemical composition.

Since one commonly sees different rock compo-

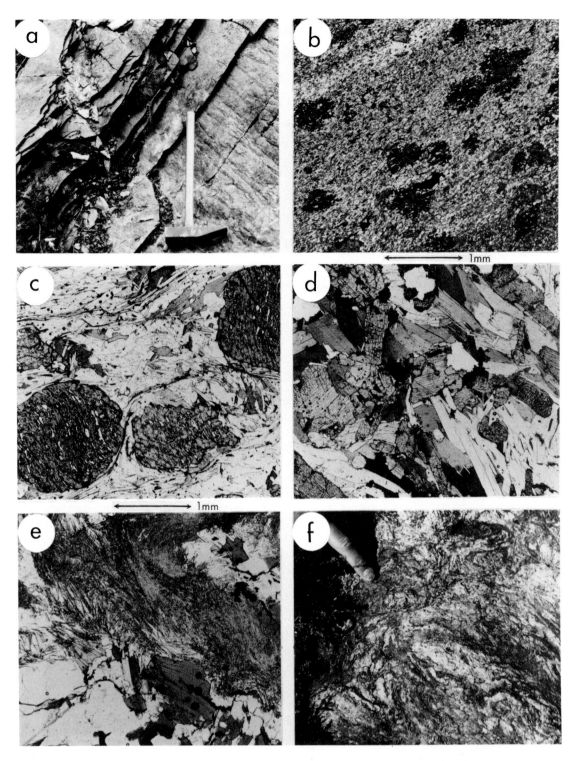

Figure 11.12 Rocks from Barrow's zones. (a) Field photograph of chlorite zone slates. (b) Thin section of biotite zone schist consisting largely of muscovite, chlorite and quartz with well-developed slaty cleavage overgrown by large crystals of biotite (dark). (c) Thin section of staurolite zone schist, showing porphyroblasts of staurolite and garnet in matrix of mica and quartz. (d) Thin section of kyanite zone schist, showing kyanite crystals with high relief and good cleavage in schistose matrix of muscovite, biotite and quartz. (e) Thin section of sillimanite zone gneiss; sillimanite is seen as bundles of fibres (fibrolite). (f) Field photograph of sillimanite zone migmatitic gneiss, showing lenses and layers of pale-coloured, quartzo-feldspathic-rich material. Scales refer to thin sections; note changes in grain size between (b) and (c).

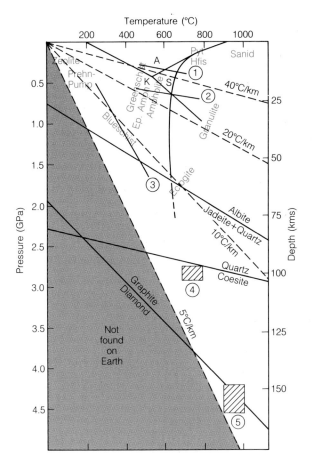

Figure 11.13 Temperature–pressure (depth) diagram showing the conditions appropriate to the ten metamorphic facies listed in Table 11.1. Also shown are: (1) typical low-pressure metamorphic field gradient, eastern Buchan area, North-eastern Scottish Highlands (*Harte and Hudson, 1979*); (2) typical intermediate – pressure metamorphic field gradient, Barrow's Zones, South-east Scottish Highlands (*Harte and Hudson, 1979*); (3) typical high-pressure metamorphic field gradient, New Caledonia (*R.N. Brothers and K. Yokoyama, 1982*); (4) P-T conditions of metamorphism of pyrope-coesite 'quartzite' from the Western Alps (*C. Chopin, 1984*); (5) P-T conditions of metamorphism of diamond-bearing schists and gneisses, Kokcletav Massif, Kazakhstan (*N.V. Sobolev and V.S. Shatsky, 1990*). Useful mineral equilibria for constraining P-T conditions include the Al_2SiO_5 phase diagram (andalusite (A), kyanite (K) and sillimanite (S)) (*Bohlen et al, 1991*), the minimum melting curve for water-saturated granites, the mineral reaction albite = jadeite + quartz, and the polymorphic transformations: quartz = coesite and graphite = diamond. (*Modified after W. Schreyer, 1985*).

sitions (meta-basalts, pelites, meta-limestones) occurring together in the field, it is a straightforward matter of observation to determine which mineral assemblages of differing bulk composition form under the same conditions of metamorphism. For example, by noting the mineral

assemblages developed in pelitic rocks at localities where meta-basaltic rocks also occur, it is possible to determine which mineral assemblages in pelitic rocks form under the same metamorphic conditions as certain mineral assemblages in metabasaltic rocks. Thus, each facies is defined by the field association of a **group** of mineral assemblages.

In principle it is possible to define a very large number of metamorphic facies, by taking into account the possible changes in mineral assemblages of all possible metamorphic rock compositions. However, a practical and general scheme has been adopted which recognizes ten metamorphic facies, each covering a fairly large range of pressure–temperature conditions of metamorphism. This is built around the original facies scheme proposed by Eskola in 1915, and is based to a large extent on the mineral assemblages found in metamorphosed igneous rocks of typical basaltic composition.

The ten widely-recognized metamorphic facies are listed in Table 11.1, and the distinctive mineral assemblages (for the same meta-basaltic bulk composition) of some of them are illustrated in Plate VII. The approximate pressure/depth and temperature conditions to which the ten facies correspond are shown in Figure 11.13. No boundaries are drawn between the various facies in Figure 11.13, since these broad facies are separated by transitional zones. The ten facies principally subdivide the common conditions of metamorphism encountered in crustal rocks. At high pressures the facies field of eclogite extends from conditions of around 1.5 GPa (corresponding to near the base of normal thickness continental crust), to much higher pressures, which are typical of the Earth's mantle but may also arise in thickened crust in orogenic regions.

11.8 METAMORPHIC FACIES SERIES AND BARIC TYPES

It has been shown that contact metamorphic aureoles and regional metamorphic terrains display progressive changes in mineral assemblages with changing metamorphic grade, giving rise to metamorphic zones that define a set of P-T conditions, as illustrated for the Ballachulish terrain in Figure 11.9c and the Buchan and Barrovian terrains in Figure 11.13. In a similar way the sequence of mineral assemblages and P-T conditions for a specific area may be related to a succession of facies or **facies series**. In Figure 9.13, Barrow's zones cover a range of P-T conditions belonging to the greenschist, epidote-amphibolite and

Table 11.1 The ten major metamorphic facies

Facies	General conditions of metamorphism	Characteristic mineral assemblage of meta-basalt
Greenschist facies	Low-grade regional and contact metamorphism (except where glaucophane schist facies rocks occur).	Actinolite–albite–epidote–chlorite.
Amphibolite facies	Middle to high-grade regional and contact metamorphism.	Hornblende–plagioclase (plagioclase more calcic than albite).
Albite-epidote amphibolite facies	Intermediate to those of the greenschist and amphibolite facies in some areas of regional metamorphism.	Hornblende–albite–epidote (-chlorite).
Pyroxene hornfels facies	Highest grade of metamorphism in many contact aureoles.	Clinopyroxene–orthopyroxene–plagioclase (olivine stable with plagioclase).
Sanidinite facies	Exceptionally high-grade (and low-pressure) conditions of contact metamorphism. The relevant mineral assemblages are only seen immediately adjacent to igneous contacts and in xenoliths.	Distinguished from pyroxene hornfels facies by the occurrence of especially high-temperature varieties and polymorphs of minerals.
Zeolite facies	Very low-grade metamorphism during deep burial. Usually regional, but often unaccompanied by deformation. Patchy development of characteristic zeolite minerals.	Zeolites such as laumontite and heulandite, in place of calcic plagioclase, epidote, prehnite, lawsonite of other facies.
Prehnite-pumpellyite facies	Low-grade, usually high-pressure regional metamorphism, commonly intermediate between zeolite facies and greenschist or glaucophane schist facies.	Prehnite–pumpellyite–chlorite–albite (-epidote). Pumpellyite–actinolite–chlorite–albite.
Granulite facies	Typically very high-grade regional metamorphism, but the characteristic mineral assemblages may also develop at lower temperatures in water-deficient environments.	Clinopyroxene–orthopyroxene–plagioclase (olivine not stable with plagioclase).
Glaucophane schist facies	Regional metamorphism where low temperatures have been combined with very high pressures.	Glaucophane–garnet with epidote or lawsonite.
Eclogite facies	This facies embraces temperature and pressure conditions common in the upper mantle. Its characteristic mineral assemblages are only of dry-basic to ultrabasic composition, and form at extremely high pressures at high temperatures.	Clinopyroxene–garnet (olivine stable with garnet in ultrabasic rocks).

amphibolite facies in the scheme of Eskola (Table 11.1), and define a metamorphic facies series of **Barrovian type**. The eastern Buchan zonal sequence, in Figure 11.13, represents a different metamorphic facies series corresponding to the greenschist and amphibolite facies only (the epidote–amphibolite facies being absent). As noted before, this facies series (or zonal sequence or metamorphic field gradient) covers a similar range of temperatures to the Barrovian one but formed at lower pressures (Figure 11.13).

Facies similar to the Barrovian and Buchan have been recognized in many areas of the world, and the commonness of their occurrence and that of other facies series involving the blueschist facies was first pointed out by the Japanese geologist Miyashiro. He also recognized that the differences between the facies series arise principally from the different pressures of regional metamorphism in the different terrains, and introduced the idea of **baric types** of regional metamorphism in recognition of these pressure controls. He recognized three standard types of metamorphism, with various intermediates. These types have been given a variety of names on the basis of the relative pressures involved, or the mineral assemblages formed, or by reference to specific type areas. It is simplest to refer to the three principal types as **low pressure, intermediate pressure** and **high pressure**. Examples of the low-pressure and intermediate-pressure baric types have been shown by the eastern Buchan and Barrovian sequences, and for comparison the P-T conditions of a high-pressure type, illustrated by rocks from New Caledonia, are shown in Figure 11.13. The main characteristics and further examples of all three principal types are listed in Table 11.2.

Drawing on observations of the distribution of different baric types in metamorphic belts around the world, it is found that different baric types sometimes occur within the same orogenic metamorphic belt. The example of the Scottish High-lands (Figure 11.11) has already been cited, where a range of intermediate- and low-pressure baric types of closely similar age are juxtaposed, and the New England area of USA is another well-known area showing a range of similar types. In the extensive Alpine belt of Europe both inter-mediate- and high-pressure types of regional metamorphism repeatedly occur (but without the regularity of the paired metamorphic belts dis-cussed below). The variation in the conditions of metamorphism, reflected by the occurrence of these **metamorphic domains** of varying zonal sequence or baric type is a product of the opera-tion of different thermal and tectonic controls across the metamorphic belt, as discussed in the following section, and it reflects the complex his-tory of the rocks in orogenic belts.

In 1961, Miyashiro observed that in Japan and elsewhere in the circum-Pacific region, linear belts of contrasted high-pressure and low-pressure baric types are commonly juxtaposed, although often separated by major fault zones (Figure 11.14). He termed such associations **paired meta-morphic belts**. Miyashiro also noted that the high-pressure belt in such pairs usually lies on the oceanic side of the adjacent low-pressure belt. Drawing comparison with the thermal and tec-tonic structure of subduction zones and arc mag-matism, he attributed the outer high-pressure (and low-temperature) metamorphism to the influence of subduction in underthrusting cool ocean floor and trench materials to great depths near the trench where the isotherms are depressed (Figure 11.14); whilst the adjacent low-pressure

Table 11.2 Characteristics of baric types of regional metamorphism

Baric type	1. Low pressure	2. Intermediate pressure	3. High pressure
Facies	Greenschist Amphibolite	Greenschist Epidote-amphibolite Amphibolite	Prehnite-pumpellyite Glaucophane schist Eclogite
Distinctive minerals in pelites	Cordierite Andalusite Sillimanite	Garnet Staurolite Kyanite Sillimanite	Garnet Glaucophane Talc Kyanite
Examples	'Buchan', Scottish Highlands. Abukuma, Japan. Maine, USA. Pyrenees, Spain/France.	'Barrovian', Scottish Highlands. Lepontine Alps. Vermont–Massachusetts, USA. Haast schists, South Island, New Zealand.	'Franciscan', California, USA. Cyclades, Greece. Sesia–Lanzo zone, Alps. 'Sanbagawan', Shikoku, Japan.

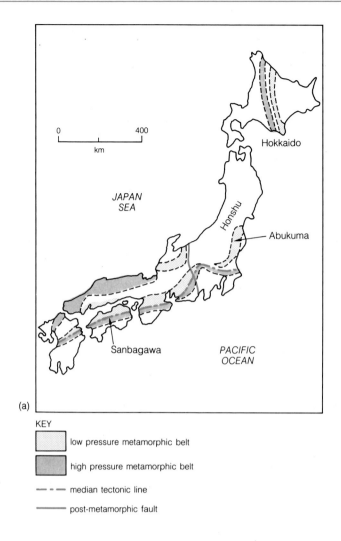

(a)

KEY

▒ low pressure metamorphic belt

▓ high pressure metamorphic belt

— · — median tectonic line

——— post-metamorphic fault

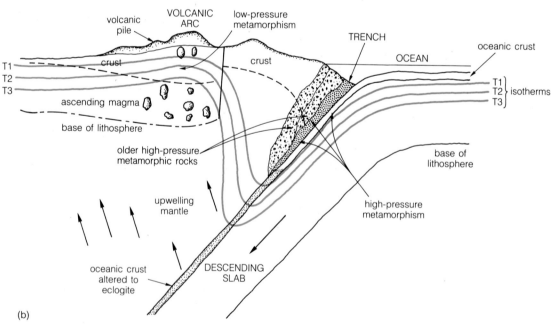

(b)

Figure 11.14 (a) Map of Japanese Islands showing the distribution and pairing of high-pressure and low-pressure metamorphic belts (*After A. Miyashiro, 1961*). (b) Schematic cross-section of subduction zone and associated volcanic arc, to illustrate the tectonic provinces giving rise to adjacent high-pressure and low-pressure paired metamorphic belts (*After Miyashiro, 1973*). Note distribution of isotherms T_1, T_2, T_3.

Table 11.3 Heat production and decay rates of radioactive isotopes

	Half-life/10^9 yr	Heat production mJ/kg/s	Abundance in past relative to present				
			Now	10^9 yr	2×10^9 yr	3×10^9 yr	4.5×10^9 yr
^{235}U	0.71	56.0×10^{-2}	1.0	2.64	6.99	18.50	80.00
^{238}U	4.50	9.6×10^{-2}	1.0	1.17	1.36	1.59	2.00
^{232}Th	13.90	2.6×10^{-2}	1.0	1.05	1.11	1.16	1.25
^{40}K	1.30	2.8×10^{-2}	1.0	1.70	2.89	4.91	10.90

metamorphism was attributed to the effect of high heat flow in the overlying crust caused by under-thrusting and by the influence of upwelling magmas generated above the subduction zone (Figure 11.17).

11.9 THERMAL AND TECTONIC CONTROLS OF METAMORPHISM

Variation in the metamorphic P-T conditions found in contact aureoles can be reasonably attributed to variations in the heat derived from the igneous intrusion and the depth of intrusion. But the causes of orogenic metamorphism, in the form of an obvious heat source, are far less obvious. The importance of tectonic setting and tectonic movements has been indicated in the brief foregoing discussion, but how can a deduction be made of the underlying thermal and tectonic causes of crustal metamorphism from a consideration of metamorphic field gradients and general metamorphic P-T conditions. Conversely, how can the wide variation in regional metamorphic conditions and field gradients (Figure 11.13) in the Earth's crust through geological time be accounted for? To answer these questions it is necessary to develop alongside tectonic considerations, some knowledge of thermal properties (especially the **sources of heat** and means of **heat transport**) in the

rocks which comprise the Earth's crust and upper mantle, and the importance of **time** in metamorphic processes.

Thermal properties of rocks. Heat in metamorphism is required to increase the temperature of the rocks, to drive the heat-absorbing ('endothermic') reactions which occur, and ultimately (in high-temperature metamorphism) to melt the crust. If the rocks undergoing metamorphism are, for example, shales or carbonate rocks, then comparable quantities of heat are required to drive the endothermic devolatilization reactions which remove H_2O and CO_2 from the rocks, as are required to raise the temperature of the rocks. What is the source of this heat?

Most heat is generated within the Earth and lost at the Earth's surface, and hence temperatures increase with depth in the Earth. The Earth's internal heat engine is driven largely by heat from the decay of radioactive isotopes, dominantly ^{235}U, ^{238}U, ^{232}Th and ^{40}K, and the relative heat production amongst these isotopes is shown in Table 11.3. Average abundances of these important elements in crust and upper mantle rock types are shown in Table 11.4. Given that the upper crust has more granitic and andesitic compositions, and the lower crust more basaltic and granulitic compositions, whilst the mantle is largely eclogite and peridotite, it is immediately obvious from Table 11.4 that heat production in the Earth

Table 11.4 Average concentrations (in ppm) of heat-producing elements in various rock-types and equivalent chemical compositions.

	K	U	Th	
Granite	37 900	4.75	18.5 ⎫	typical of
Andesite	18 000	2.00	9.0 ⎭	upper crust
Basalt	8 400	0.60	2.7 ⎫	typical of
Granulite	22 000	0.40	2.1 ⎭	lower crust
Eclogite	360	0.05	0.18 ⎫	typical of
Peridotite	10	0.016	0.05 ⎭	upper mantle

decreases downwards from crust to mantle in the continents. Typical upper crustal rocks have heat productions of the order of $2 \times 10^{-6} \, J \, m^{-3} \, s^{-1}$ (joules per cubic metre per second); this very small figure is nonetheless very significant on the timescale (millions of years) of geological processes affecting the Earth's crust. Regional metamorphic terrains, comprising largely metamorphosed sedimentary rocks derived from the reworking of upper crustal rocks, contain within them a major source of heat which, given time, must make a significant contribution to the overall heat budget of crustal metamorphism.

Depending upon their various half-lives, the heat production from the radioactive isotopes of U, Th and K would have been different in the past to the present. The relevant data is given in Table 11.3, which shows that the abundances of particularly ^{235}U and ^{40}K were very much greater in the early history of the Earth than during the Phanerozoic. The Precambrian would be expected to have had much higher heat production and, therefore, greater temperature increase with depth, than the Phanerozoic. It is notable that the high-pressure baric type of metamorphism (Figure 11.13), which involves the lowest temperatures at the greatest depths, is very rare in Precambrian terrains.

Heat is transported in the Earth's crust by **conduction** or by **advection**. Conduction does not involve motion of the rock body and is the type of heat transport that will occur along the length of a static metal bar; advection does involve motion of whole bodies as, for example, in the form of hot magmas ascending from the mantle. In addition, the processes of uplift, erosion, and tectonic movement in fold-mountain belts move hotter rocks upwards towards the Earth's surface, and constitute a form of heat advection. It is of interest to earth scientists to try to understand the relative importance of these processes in generating the heat necessary for metamorphism.

In contact metamorphism, it is simple to attribute the source of heat to the upward advection of hot magmas ascending from the lower crust or mantle. However, the majority of regional metamorphic terrains cannot be so easily explained, and the control of heat transport by conduction must be considered. Typical thermal conductivities (K) of crustal rocks are of the order of 2 or $3 \, J \, m^{-1} \, K^{-1} \, s^{-1}$, which is very very low, and a fundamental property of **silicate rocks** is that they are **very poor conductors of heat**. Therefore, we can conclude that the processes of heating and cooling the large masses of rock in active orogenic belts, typically some 60 to 70 km thick, require great lengths of time of the order of tens of millions of years.

Geotherms. Temperature almost invariably increases with depth in the Earth, and the rate of increase (dT/dZ, where Z = depth) is called the **geothermal gradient** or **geotherm**, which is related to heat flow (q) through the Earth's crust and the conductivity (K) of the rocks therein, by the relationship:

$$q = K.dT/dZ$$

The heat flow is a function of the heat produced in the vertical column of rocks under consideration. The geothermal gradient varies in both space and time, for example between the continents and the oceans, and from place to place in each, depending on the involvement of the lithosphere in active tectonic or magmatic processes. In a steady-state situation, typical geotherms measured in old and stable areas of continental and oceanic lithosphere unaffected by tectonic or igneous activity in recent geological time are usually in the range 15 to 30°C per kilometre depth. The exact magnitude of the geothermal gradient in any such stable area of lithosphere will depend on the thermal properties (heat production and conductivity) of the various lithologies present in a section through the crust and upper mantle along the geotherm, since the heat production and thermal conductivity will be different in each lithology, and the thermal gradient may vary by a factor of two or more from lithology to lithology.

Steady-state, stable continental and oceanic geotherms are useful reference gradients against which measurements in more active areas of the Earth's lithosphere may be compared. In areas of igneous activity and high surface heat flow, such as mid-ocean ridges or continental rifts, the geotherm will be such that temperature at any depth is greater than that on a reference steady-state geotherm. In areas where cold sediments or lithosphere are sinking, such as rifted sedimentary basins or oceanic trenches, temperature at any given depth will be reduced relative to typical steady-state geotherms. In general, heat flow in the continental crust will be higher than that through old oceanic crust, reflecting the higher radioactive heat production in crustal rocks, but young oceanic crust adjacent to spreading zones and still cooling from magmatic temperatures, has higher heat flow. Regional metamorphism usually occurs in tectonically active areas associated with crustal thickening and mountain building, where uplift and erosion bring hot rocks rapidly towards the Earth's surface, and accordingly young mountain belts are found to have the highest non-magmatic continental heat flow values. The geothermal gradients beneath these areas may be complex and vary systematically through time until the crust is stabilized.

In time, any perturbation of the geotherms will reflect the length and timescale of tectonic or thermal processes responsible for the perturbation. The physical laws of diffusion, in this instance diffusion of heat, require that the timescale (t) of the decay of a thermal perturbation is proportional to the square of the length (l) along which heat is conducted, according to the equation:

$$t = l^2/k$$

where k is a constant for a particular rock called the thermal diffusivity. Recalling that the thermal conductivity and heat production of crustal rocks are very low, thermal perturbations imposed on a pile of rocks will equilibrate through a length of 10 km in about half a million years, and a length of 100 km in about 50 million years, for typical crustal rock conductivities. These lengths and timescales are in the range observed from geological and geochronological data to be typical of contact and regional metamorphism respectively in the Earth's crust. In other words, local heat anomalies or perturbations on the scale of a few kilometres. adjacent to an igneous pluton will be dissipated by conductive processes in less than a million years, whereas crustal-scale thermal perturbations of the large thicknesses of rock found in modern orogenic metamorphic belts, such as the Alps or Himalayas, have very long lifetimes of tens of millions of years. The restoration of an equilibrium heat distribution by conductive processes in these situations is often referred to as **thermal relaxation**.

Even in a contact aureole there will be a cycle of heating and cooling over hundreds of years. Over large lengths of tens of kilometres in the Earth's crust, the timescales required to permit heating and cooling by conductive processes necessitate that metamorphism cannot be considered to be an instantaneous process, even if it often appears 'episodic' on the timescale of Earth's history. Thus, the temperature–pressure relationships recorded by the mineral assemblages which define metamorphic field gradients such as those in the Scottish Highlands, cannot be considered to represent some particular geothermal gradient in the Earth's crust. The **time** of attainment of maximum (peak) metamorphic temperatures may vary significantly along the metamorphic field gradient and these gradients do not represent a 'snapshot' of an ancient geothermal gradient. Thus it is essential to consider the variation of P-T conditions through time (**P-T-time paths**) in evaluating the causes of regional metamorphism.

11.10 P-T -TIME PATHS

(a) **Regional metamorphism based upon con-**

ductive heat transport. To understand the causes of regional metamorphism, it is necessary to look at ways in which a typical 'steady-state' continental geotherm may be perturbed. Looking at the geological record, it is seen that regional metamorphism is usually associated with orogenesis, involving deformation on all scales from the individual mineral grain to the entire crust. The geological record of many orogenic metamorphic belts contains evidence of volcanic activity, accumulation of great thicknesses of sediments, major thrusting and nappe formation, and folding on all scales. Large-scale tectonic processes, probably related to plate collision, accompany regional metamorphism, and these processes are by their nature episodic. As we have seen, by disturbing and redistributing material these tectonic processes will modify the thermal structure of the crust.

Orogenic metamorphism commonly occurs in association with thickening of the continental crust during continent-continent collision, by regional overthrusting and nappe formation. Because these tectonic processes are relatively fast compared to heat conduction, the normal steady-state geotherm is repeated by the thrusting, producing a **sawtooth** gradient as shown in Figure 11.15. As conduction (thermal relaxation) redistributes the heat following this rapid perturbation event, the cool rocks at the top of the lower, over-ridden plate, which have experienced a rapid increase in pressure, heat up (Figure 11.16). This localized heat redistribution, across the 'sawtooth', is moderately fast on geological time scales, but at the same time a slower process of heating up the whole crustal pile commences, because the thickness of that pile and therefore its heat production has been increased. Isostatically-induced uplift and erosion also result from the thickening, and the rocks initially subjected to metamorphic conditions of high pressures and low temperatures heat up whilst undergoing uplift and erosion, resulting in pressure decrease. Thus any given point in the rock column makes a curved pathway through P-T space known as a P-T-time path (Figure 11.17).

In the above model, rocks distributed along the length of a vertical column show different P-T-time paths and therefore have different thermal maxima at different pressures and times (Figure 11.17). Note that during the process of heating and uplift, the geotherm affecting any rock column in the evolving metamorphic belt is continually changing as the heating rock mass rises towards the Earth's surface (Figure 11.17). A consequence of this model is that the maximum temperature attained by any rock occurs during uplift, at pressures (depths) much less than the maximum pres-

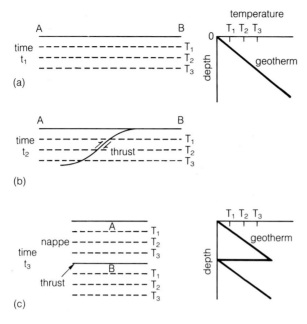

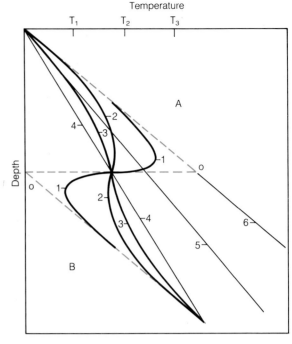

Figure 11.15 (a) Vertical cross-section through undisturbed crust, at time t_1 showing isotherms T_1, T_2, T_3 of increasing temperature with depth. The associated steady-state geotherm at time t_1 is shown to right. (b) Thrusting, at time t_2, starts to carry crust at A over crust at B, with isotherms in A being carried over B. (c) A time t_3, thrusting is completed, with unit A overlaying unit B, separated by the thrust, and the associated geotherm has transiently developed a 'sawtooth' shape. (After E.R. Oxburgh and D.L. Turcotte, 1974)

Figure 11.16 Thermal relaxation of 'sawtooth' geotherm shown in Figure 11.15, with geotherms shown for times from 0 to 6. Note that at times 0 to 2 temperature decreases with increasing depth in the zone at the top of unit B, and that the zone at base of unit A initially loses heat by conduction downwards to unit B as well as upwards to the surface. (After E.R. Oxburgh and D.L. Turcotte, 1974)

sures experienced by the rock. As the column of rock moves upwards, a rock at shallow levels of the column will be cooling when a rock at much greater depth will still be heating, and the times at which these rocks reach their highest temperatures, and ultimately reach the Earth's surface, may be separated by millions of years. Thus in a set of metamorphic zones, rocks formed at different P and T will have experienced their maximum T conditions at different times. Consequently a metamorphic field gradient, like those for the Buchan and Barrovian zones in Figure 11.13, cannot be anything like a single geotherm (Figure 11.17), but will correspond more closely to a line joining the maximum T conditions on the P-T-time paths of all rocks (the piezothermic array in Figure 11.17). A large number and variety of P-T-time paths may be generated in numerical thermal models by variation of the controlling thermal, tectonic, uplift and erosion parameters (England and Thompson, 1985). If one considers lateral variation in these controlling parameters across a small region, the generation of a metamorphic field gradient in an orogenic belt becomes even more complex. Quite often the best developed metamorphic zones (metamorphic field

gradients) occur near the margins of tectonic domains, as suggested in Barrow's zones by their occurrence adjacent to the tectonic boundary of the Highland Boundary Fault (Figure 11.11). This suggests that not only differences in P-T-time path as a function of depth influence metamorphic field gradients, but also lateral differences in P-T history caused by lateral variations in thermal properties and transport.

It is interesting to consider why metamorphic rocks appear to preserve information, in the form of a mineral assemblage, that relates to the thermal maximum on its P-T-time path, or indeed about any high P-T condition. After all, such a mineral assemblage is unstable under the ambient conditions at the Earth's surface. Several factors are involved. Firstly, fluid species are both chemical constituents in low-temperature minerals and catalysts for metamorphic reactions. They act as catalysts by aiding the dissolution of mineral grains and the transport of elements to and from the sites of reaction in a rock. These fluids are released by devolatilization reactions largely during the heating part of the metamorphic cycle. In the absence of fluid during cooling, further reaction is inhibited. A second important factor is deformation, which, although it

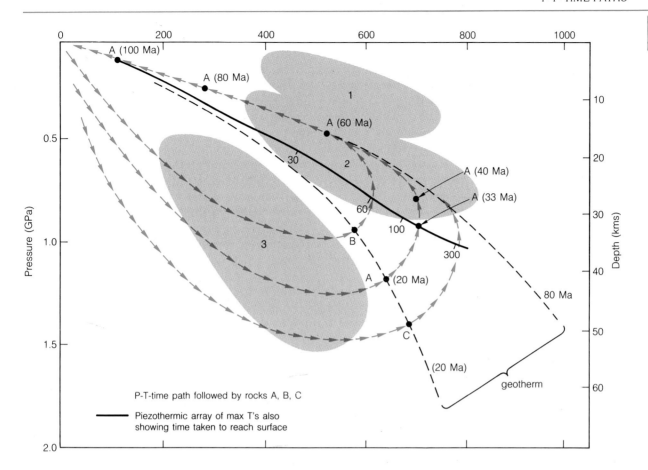

Figure 11.17 Geotherms (dashed), and P-T-time paths (arrows) for rocks A, B and C, during a hypothetic cycle of burial, thrusting, heating and erosion as illustrated in Figures 11.14 and 11.15. The locus of T (maximum) and P (at T maximum) points define a metamorphic gradient (the peizothermic array shown in colour, and which may approximate to the metamorphic field gradient), which is labelled with the times (in Ma) at which the corresponding rocks reach the surface of the Earth. (*After P.C. England and S.W. Richardson, 1977*) Also shown are P-T conditions of low-pressure (1), intermediate-pressure (2) and high- pressure (3) baric types of metamorphism.

does not significantly influence the stability of the mineral assemblage itself, also helps to catalyse metamorphic reactions. Thirdly, the rates of reactions increase exponentially with temperature, and therefore the highest temperature assemblages have the best chance of formation and preservation. If mineral assemblages are changed during decreasing temperature conditions, it is usually because of fluid infiltration into the cooling rock, then new, lower temperature minerals (commonly hydrous minerals or carbonates) will partially replace the higher temperature assemblage in the process of **retrograde metamorphism**.

An interesting consequence of thermal models discussed above is that the P-T-time paths for rocks deep in the metamorphic column may achieve the conditions necessary for partial melting to occur, with the formation of granitic melts. These melts, if sufficient in quantity and mobility (i.e. low viscosity), may coalesce and rise through the crust to intrude lower-grade rocks higher in the column which, in all probability, are already cooling. It is a common observation in regional

metamorphic terrains that granitic intrusions, sometimes in the form of vast 'batholiths', appear as late- or post-metamorphic features relative to the metamorphic rocks they intrude, producing contact-metamorphic overprints on the earlier regional metamorphism. Again, the predictions of calculations based on the thermal properties of thickened continental crust are matched by geological observation. Furthermore the formation and removal of granitic melts serves to deplete the sources of partial melting in the very heat-producing elements which may have been instrumental in producing the metamorphic event in the first place, thereby acting as a buffering mechanism on the metamorphic heating process.

The most readily attained metamorphic conditions compatible with the typical thermal and mechanical properties of overthickened continental crust are those of the intermediate-pressure baric type (Figure 11.13), of which we have seen an example in Barrow's zones in the Scottish Highlands. Such metamorphic conditions are common in regional metamorphic belts throughout the world. Although metamorphic conditions of higher pressures and lower temperatures, corresponding to the high-pressure baric type (Figure 11.13) are quite readily attained in the early stages of P-T-time cycles, the corresponding blueschist facies mineral assemblages may be expected to be overprinted by higher temperature and lower-pressure assemblages later in their histories as the metamorphic pile heats up (Figure 11.17). Thus high pressure, blueschist facies, rocks are relatively uncommon worldwide, and their preservation is thought to require the intervention of tectonic processes capable of exhuming these rocks rapidly before they heat up and form higher-temperature mineral assemblages. The tectonic conditions favourable for the genesis and rapid exhumation of high-pressure rocks are to be found in subduction zones, and linear metamorphic belts of blueschists and related rocks commonly (but not exclusively) mark the locus of former subduction zone metamorphism (Figure 11.14).

Recently, evidence has been found in parts of regional metamorphic terrains in the Alps, Norway and China, of metamorphism at extremely high-pressure conditions for crustal rocks corresponding to 80–100 km depths in the Earth (Figure 11.13). Such conditions are recorded by the stability of such minerals as coesite, the high-pressure polymorph of quartz, and pyrope garnet in metamorphosed sedimentary rocks (Figure 11.13). Even more astonishing is the recent report of the occurrence of diamonds or their graphite pseudomorphs in schists, gneisses and eclogites in Kazakhstan, USSR (Figure 11.13). The tectonic conditions capable of generating and preserving such ultra-high pressure mineral assemblages, far in excess of normal crustal metamorphic pressures (Figure 11.13), are as yet poorly understood.

(b) **Regional metamorphism involving magmatic heat**. The low-pressure baric type (Figure 11.13) of crustal metamorphism, which we have seen exemplified in the eastern Buchan Scottish sequence, cannot be readily explained by tectonic thickening and simple conductive processes alone. P-T-time paths based only on these processes may traverse the P-T field of low-pressure metamorphism, but they require unusually high-heat production and low-conduction properties and only cross the low-pressure facies fields during the cooling stage of their P-T-time trajectory, after previously passing through conditions implying wholesale melting. An alternative way of obtaining such high temperatures at shallow levels of the crust is by major advective heat addition to the crust in the form of magmas. In the Buchan area evidence of such advection of heat is seen in widespread gabbroic rocks (Newer Gabbros),

which were intruded in that area during the period of regional deformation and metamorphism. In some cases of regional metamorphism an association with crustal extension and thinning, and the upwelling of hot mantle material may be important. This has been suggested for some cases in the Pyrenees.

Another important type of regional metamorphism is that found in the large tracts of high-grade metamorphic gneisses that form part of the Precambrian cratons in all the major continents. In this case regional metamorphism at temperatures often much higher than the amphibolite facies of the Phanerozoic orogenic metamorphic belts discussed earlier, belongs to the **granulite facies** (Figure 11.13). In some of the metamorphic terrains concerned, unusually high temperatures of over 900–1000°C were attained at crustal levels of 25–35 km. Such high temperatures at depths typical of normal lower crust cannot be attained by the sorts of heat conduction processes described earlier. A variety of different P-T-time paths have been deduced for occurrences of this type of regional metamorphism, and these require major additional sources of heat, almost certainly from accretion of magmatic material under and within the lower continental crust (the former process is referred to as **underplating**). Thermal models of metamorphism involving thickening of the entire lithosphere (crust plus part of the upper mantle) during continent–continent collision, followed by later detachment and sinking of the lower (mantle) part of the lithosphere and its replacement by hot asthenosphere, may provide the necessary conditions for such high-temperature metamorphic conditions in the lower crust.

On the other hand, if no major thickening of the crust occurs at the time of metamorphism, these hot rocks may not undergo the normal process of rapid uplift and erosion, but remain deep within the crust. Granulite facies metamorphism may thus also occur under conditions of extension and rifting of the continental crust and unusually high heat flow. In this case P-T-time paths involving heating at constant or decreasing pressure during extension, are followed by cooling at constant or increasing pressure when extension stops, and exhumation by much later tectonic events. Thermal histories of these deep crustal rocks contrast sharply with those for regional metamorphic belts described earlier, reflecting different thermal and tectonic processes, and the involvement of convective heat transport in the form of addition of magma to the metamorphic system.

Explanation of the observed patterns of regional metamorphism in the Earth's crust, therefore, involves knowledge of the thermal properties of the crust and lithospheric upper mantle, the input of heat from mantle sources either by conduction or advection (magma addition), and of the tectonic history and crustal structure of orogenic belts. In many instances, the observed P-T conditions can be attained without unusually large input of heat from hot mantle underlying an orogen; the internal heat production of thickened continental crust of low thermal conductivity is sufficient. In other cases, the addition of unusually large amounts of mantle heat, perhaps as magma, to the crustal system is required. The matching of calculated thermal predictions with geological observations and geochronology shows that the causes of most types of crustal metamorphism, and their relationship to large-scale tectonic processes, are beginning to be reasonably well understood.

Appendix 11.1 List of mineral compositions for common diagnostic metamorphic minerals

Minerals	Chemical formulae
Quartz/Coesite	SiO_2
Chlorite	$(Fe,Mg)_5Al_2Si_3O_{10}(OH)_8$
Muscovite	$KAl_3Si_3O_{10}(OH)_2$
Biotite	$K(Fe,Mg)_3AlSi_3O_{10}(OH)_2$
Cordierite	$(Fe,Mg)_2Al_4Si_5O_{18}.0.5H_2O$
K-feldspar	$KAlSi_3O_8$
Andalusite/kyanite/sillimanite	Al_2SiO_5
Corundum	Al_2O_3
Orthopyroxene	$(Fe,Mg)_2Si_2O_6$
Clinopyroxene	$Ca(Fe,Mg)Si_2O_6$
Garnet Almandine	$Fe_3Al_2Si_3O_{12}$
Pyrope	$Mg_3Al_2Si_3O_{12}$
Grossular	$Ca_3Al_2Si_3O_{12}$
Staurolite	$(Fe,Mg)_5Al_{18}Si_{7.5}O_{44}(OH)_4$
Graphite/diamond	C
Albite	$NaAlSi_3O_8$
Plagioclase	$NaAlSi_3O_8–CaAl_2Si_2O_8$
Epidote	$Ca_2(Al,Fe)_3Si_3O_{12}(OH)$
Lawsonite	$CaAl_2Si_2O_8.2H_2O$
Prehnite	$Ca_2Al_3Si_3O_{10}(OH)_2$
Pumpellyite	$Ca_4(Mg,Fe)(Al,Fe)_5Si_6O_{23}(OH)_3.2H_2O$
Jadeite	$NaAlSi_2O_6$
Actinolite	$Ca_2(Fe,Mg)_5Si_8O_{22}(OH)_2$
Glaucophane	$Na_2(Fe,Mg)_3Al_2Si_8O_{22}(OH)_2$
Hornblende	$NaCa_2(Fe,Mg)_4Al_3Si_6O_{22}(OH)_2$
Heulandite	$CaAl_2Si_6O_{16}.5H_2O$
Laumontite	$CaAl_2Si_4O_{12}.4H_2O$
Analcite	$NaAlSi_2O_6.H_2O$

Appendix 11.2 Naming metamorphic rocks

Metamorphic rocks have a huge variety of names. This is because their chemical compositions are very diverse (including all the metamorphic equivalents of sedimentary and igneous rocks) and different conditions of metamorphism can give a wide variety of mineralogy and microstructure to the different starting compositions. Fortunately, from the viewpoint of giving a name that describes both the basic structure and mineralogy, a very simple system of nomenclature exists, for which only a few key terms have to be remembered.

Simple system of descriptive terminology. The features of a rock that can be observed directly are the nature of the minerals present and their structural or textural arrangement. Both these features should be involved in a fully descriptive name, and this may be done by using a small number of terms to characterize the major structural characteristics and adding to them mineral prefixes. When viewed in hand specimen or thin section, important features of the appearance of many metamorphic rocks are the abundance and arrangement of mineral grains which are inequant in shape (that is to say minerals which have elongate or plate-like shapes). Overall the basic structure of a metamorphic rock can be summarized by the use of one of **three structural names**, as follows:

Schist – a rock in which the inequant metamorphic minerals show a preferred orientation, such that the rock tends to split parallel to their alignment. (This type of preferred alignment is called **schistosity** or **schistose structure**.)
Granofels – a rock in which the mineral grains are either largely equant, or are inequant with random rather than preferred orientation. (This type of structure is called **granulose** or **granofelsic structure**.)
Gneiss – a rock of medium- or coarse-grain size in which there may be a limited or a partial development of schistose structure, and in which the arrangement of minerals often gives rise to a layered or lenticular structure; some layers may be schistose whilst others are granulose. (This combination of structural features may be called **gneissosity** or **gneissose structure**.)

Once the appropriate structural term for a metamorphic rock has been decided upon, the essential mineralogy can be indicated by putting mineral names in front of the structural term in a hyphenated series. Only important minerals are put into these hyphenated prefixes, and they are usually put in order of increasing abundance. In a situa-

tion, such as Barrow's zones pelites (Figure 11.11), where all rocks contain abundant quartz and both muscovite and biotite are widespread, it is common to omit quartz as a prefix and to list biotite and muscovite together as 'mica' (especially where they both form the schistosity). Thus rocks from Barrow's zones such as those of Figure 11.12b – e would be respectively called: biotite-chlorite-muscovite-schist, staurolite-garnet-mica-schist, kyanite-mica-schist, and sillimanite-mica-gneiss. Other examples illustrated by Plate VII would be respectively: epidote-amphibole-albite-chlorite-schist, garnet-epidote-plagioclase-amphibole-schist, pyroxene-garnet-epidote-glaucophane-gneiss, pyroxene-plagioclase-granofels, garnet-plagioclase-pyroxene-granofels, garnet-pyroxene-granofels.

Protolith names. In describing metamorphic rocks it is frequently useful to refer to the nature of the rock before metamorphism, as we have done many times in the text of this chapter. Also, in cases where the metamorphism is very weak, the original rock features may still be more prominent than superimposed metamorphic ones.

The general way of naming metamorphic rocks with respect to their original rock type (protolith) is simply to put '**meta**' in front of the protolith name. Thus we may speak of meta-igneous rocks and meta-sandstones etc., or more specifically of meta-granites, meta-basalts, meta-greywackes etc.

With respect to one major rock group, there is, however, an alternative and widely used set of terms used to indicate protoliths. This is for the terrigenous or clastic sediments, and the following terms are widely used:

pelite for metamorphosed argillaceous sediment (i.e. a meta-argillite)
psammite for metamorphosed arenaceous sediment (i.e. meta-arenite)
psephite for metamorphosed rudaceous sediment (i.e. meta-rudite.) The terms pelite, psammite, psephite are in fact the Greek words equivalent to the Latin ones of argillite, arenite and rudite in Table 6.1.

Other Names. There is a very large diversity of other rock names which have been used to characterize either a particular mineralogy or structure (large or small scale). Definitions of the more common of these are given below following proposals by the International Commission on Metamorphic Nomenclature.

Slate, a very fine-grained schist having abundant sheet silicates and showing a very good schistosity (slaty cleavage);
Phyllite, a fine- to medium-grained schist characterized by a lustrous sheen and a well-

developed schistosity resulting from the parallel arrangement of sheet silicates;

Mylonite, a schistose rock characterized by the presence of a fine-scale layering (foliation), which has resulted from strong deformation causing a reduction in grain size of the parent rock;

Cataclasite, a rock containing generally angular fragments in a fine-grained matrix produced by strong deformation but showing little preferred orientation;

Hornfels, a fine- to medium-grained granofels formed in contact aureoles and typically having a tough (not easy to break) character;

Migmatite, a heterogeneous, usually gneissose, rock in which parts largely consisting of felsic minerals are intermingled with parts containing more mafic minerals and which may be schistose;

Amphibolite, a metamorphic rock consisting largely of amphibole and plagioclase (note it is not a rock consisting largely of amphibole alone);

Eclogite, a metamorphic rock composed mainly of pyroxene and garnet and lacking plagioclase;

Marble, a metamorphic rock containing more than 50% carbonate minerals (calcite, dolomite or aragonite);

Calc-silicate rock, a metamorphic rock composed by more than 50% of Ca-rich silicate minerals;

Quartzite, a metamorphic rock containing more than 80% quartz;

Serpentinite, a metamorphic rock composed of more than 80% of minerals of the serpentine group.

Note that these more specialized names may be substituted, if appropriate for the basic three structural terms, in making a descriptive hyphenated name. Thus a hornfels is a type of granofels, and we usually describe the high-grade rocks from contact aureoles as hornfelses rather than granofelses. For example, the rock of Figure 11.10b,c would be described as a biotite-K feldspar-cordierite-hornfels. Notice also that more than one name may apply to a given rock. For example, the rock of Plate VIIb is an amphibolite using the 'specialized' terminology, but a garnet-epidote-plagioclase-amphibole-schist using the systematic descriptive terminology, and from the viewpoint of description alone the latter terminology is more informative and is therefore used in petrographic descriptions. On the other hand the 'specialized' names are more succinct and thus the garnet–pyroxene–granofels of Plate VIIf would usually be described as an eclogite.

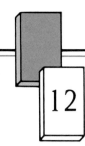

12

IGNEOUS
INTRUSIONS

Nature acts . . . in consolidating bodies by means of heat and fusion, and by moving great masses of fluid matter in the bowels of the earth.

James Hutton (1726 – 1797)

Magmas commonly have densities lower than those of the overlying rocks and consequently tend to ascend through whatever passageways or zones of weakness are available. As they do so, they lose heat through the conduit walls thus bringing about crystallization. The vast bulk of magmas crystallize at depth; only a small proportion reaches the land surface or ocean floors to erupt as volcanic lavas and ashes. The magmatic material congealed at depth is collectively referred to as intrusive. Uplift and erosion leads to the exposure of such intrusive rocks at the surface. Whereas intrusions come in all sizes, and in a wide variety of geometric forms, some forms are sufficiently recurrent to allow a classification on the basis of shape.

12.1 DYKES AND SILLS

Among the commoner forms of intrusions are those referred to as dykes and sills. These are tabular, approximately parallel-sided bodies that are thin in relation to their lateral extent. Dykes are typically vertical or with steeply-inclined contacts and arise roughly perpendicular to the Earth's surface (Figure 12.1). Sills, on the other hand, are horizontal or low-angled. Dykes are particularly common in the deeper parts of the crust whereas the more common habitat for sills is in near-surface bedded sequences, particularly sedimentary strata.

Figure 12.1 A basaltic dyke cutting faster-weathering sandstones, Isle of Bute, Scotland. This early Tertiary dolerite dyke shows joints perpendicular to the contacts and also parallel to the side-walls. (*B.G.J. Upton*)

Figure 12.2 Fair Head Sill, Co. Antrim, N. Ireland. A 75 m thick dolerite intruding Carboniferous strata. Patches of the roof, hornfelsed to a hard brittle porcellaneous rock, veneer the surface.

Dyke walls commonly cut across pre-existing foliations and structures in their host (or 'country') rocks and are thus typically 'discordant' intrusions. Conversely, sills, with their side-walls roughly parallel to the bedding planes of their country rocks, are regarded as 'concordant' intrusions (Figure 12.2). In both cases the space which the magma occupies has been created by the parting of the country rock on either side of an initial fracture plane.

Dyke intrusion is commonly associated with lithosphere that has undergone tensional stress. As a result of tensional 'pull-apart' and emplacement of dykes, the crust can be laterally extended. Dykes rarely occur singly; they are often found composing swarms, and in large swarms many thousands of dykes may be involved. Generally in dyke swarms the dykes are roughly parallel, having been formed by emplacement of magma along fractures brought about by mechanical failure of crust that has been under stress. Radial dispositions of dykes, however, attend some uplifted and eroded volcanic centres (Figures 12.3 and 12.4). Radial swarms, however, are rare in comparison to linear swarms.

There is a great variety in the sizes of both individual dykes and dyke swarms. Dyke widths can vary from the centimetre scale to hundreds of metres and, very occasionally to widths measur-

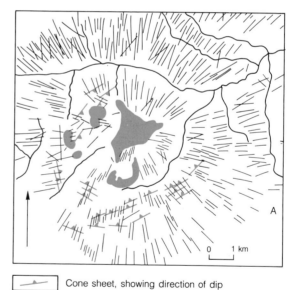

Cone sheet, showing direction of dip

Figure 12.3 Map showing a remarkable swarm of dykes radiating out from a volcanic neck in the Sunlight District, Wyoming. Some cone-sheet intrusions (Section 12.5) are also present. (*After Parsons, I. 1939*)

able in kilometres. Similarly dyke lengths – so far as they can be determined – vary from metres to hundreds of kilometres. Dykes, and dyke swarms, may exhibit sudden sideways jumps in their course to produce an 'en echelon' pattern and may divide into sub-parallel splays (Figure 12.5).

As a broad generalization, the largest swarms,

Figure 12.4 One of the radiating vertical dykes (2-3 m wide) exposed, to form a massive vertical wall extending away from the volcanic neck in the extreme right of the photograph. The volcanic neck, Ship Rock, is some 500 m in height. New Mexico.

and the greatest individual dykes, are to be found in the Precambrian rocks of the continents. Such rocks are to be found in regions that have been subjected to profound uplift and erosion. Huge swarms may have lengths and breadths measurable in hundreds or even thousands of kilometres. The Mackenzie dyke swarm, dating from the Mid-Proterozoic (1270 Ma), has a fan-shaped disposition, stretching from the Arctic Ocean shores of northern Canada southward for some 1500 km (Figure 12.6). The swarm is itself about 1000 km in width. The Muskox dyke, a much studied individual in the swarm is over 200 m wide. Huge Precambrian dykes are known from Canada, Greenland, Sweden, Australia and Africa, (Figure 12.7). The great Abitibi dyke of Ontario is traceable for approximately 600 km, attaining a maximum width of 250 m. The Breven and Halleförs dykes of Sweden have widths of over 1 km; the Jimberlana and Binneringie bodies of Western Australia are dyke-like intrusions in excess of 2 km broad. Finally the Great Dyke of Zimbabwe, traceable for about 600 km, has widths from 3 to 11 km (Figure 12.8). It is accompanied for most of its length by satellite dykes, the Umvimeela and East Dykes, which, while dwarfed by the Great Dyke, are very large dykes in their own right. Whereas small dykes lose heat rapidly to their wall

rocks the larger dykes can have crystallization histories of thousands, or even tens of thousands of years, and can produce coarse-grained rocks of plutonic appearance (Chapter 5). In some instances, e.g. the Jimberlana and Great Dykes, layered cumulates have been produced (Chapter 5).

Among the numerous dyke swarms crossing the British Isles, the best known are those that were emplaced in the Palaeocene (Figure 12.9), at a stage heralding the break-up of northern Europe from Greenland and the initiation of the North Atlantic Ocean. These early Tertiary dykes generally trend north-west–south-east and are concentrated into swarms 10 to 15 km broad such as those traversing the Isles of Mull and Skye in the west of Scotland. For the most part these Tertiary dykes have widths of a few metres, rarely attaining tens of metres across. Among the larger individuals, two or three can be traced from the Hebrides south-east across Scotland and northern England to the North Sea coasts, a distance of about 200 km. Detailed work on one of these, the Cleveland dyke, has shown that magma flow had a very large lateral component as well as an upward component. Magma was evidently supplied from deep-seated sources beneath the Sea of the Hebrides and the fracture plane, and the magma that intruded it, propagated from north-west to south-east.

Figure 12.5 An irregular doleritic dyke cutting Archaean gneisses, south Greenland. In this instance the fracturing accompanying magma intrusion has been complex, allowing ascent of a number of sub-parallel branches, with screens and isolated masses (xenoliths) of gneiss caught within or between the branches. (*B.G.J. Upton; published by permission of the Geological Survey of Greenland*)

The majority of dykes are formed by dilatation of a fracture. As the side-walls move apart magma intrudes. Whereas many dykes appear to have formed by simple dilatation, some lateral and/or vertical movement between their opposing walls may also occur. The magma first admitted into the embryonic dyke comes into contact with cool wall rocks and congeals rapidly to a glass, or a fine-grained crystalline rock. As the dyke fissure continues to dilate, more magma is fed into the axial zone and, being thermally insulated to some extent, by the initial fast-cooled 'chilled margins', cools more slowly to more coarsely-crystalline material (Plate VIIIa). This tendency for the intrusive rocks adjacent to the contact to be fine-grained whereas those closer to the centre are more coarsely-grained is a common feature in all intrusions, and is by no means confined to dykes.

A dyke having been intruded and cooled, the

relative weaknesses provided by the contact zones or even the still-hot axial zone may be exploited by subsequent dykes. Thus, multiple dykes can be generated. Repeated dyke intrusion is regarded as the principal mechanism of sea-floor spreading (Chapter 28). Hot, thin, oceanic lithosphere overlying the axes of the mid-ocean ridges provides favoured sites for dyke emplacement. In an idealized case, a new dyke intrudes the mechanically weak median plane of an earlier dyke chilling against the two halves of its predecessor as it does so, and so a composite dyke is formed. Magma supplying the newly-formed dykes arises from storage chambers beneath the ridge axes. A proportion of the dykes will act as conduits or 'feeder dykes' through which magma is supplied for sea-floor eruptions, typically of 'pillow lavas'. Persistent dyke intrusion along the ridge axes ('spreading centres' or 'constructional plate boundaries') produces the so-called sheeted-complexes, composed of countless numbers of dyke units, thought to be typical of much of the upper oceanic crust. Clearly the youngest dykes, in the idealized situation, would be those beneath the ridge axis, whereas dykes progressively age with increasing distances from the axis. A diagrammatic section of a mid-ocean ridge (or back-arc basin, Table 30.1, Figure 30.3) spreading centre is shown in Figure 12.10. Submarine lavas ('pillow lavas', see Chapter 13) are supplied by dykes contributing to the underlying sheeted complex.

Exposures revealing sheeted complexes, composed of virtually nothing other than dykes cutting earlier dykes, are seen in 'ophiolite complexes' such as those of the Troodos region, Cyprus (Figure 12.11), or the Oman Mountains near the Persian Gulf, where ancient ocean floor remnants have been elevated ('obducted') (Chapter 28) and subjected to subaerial erosion.

With increasing concentration of basaltic dykes, continental crust becomes increasingly dense and begins to acquire many of the characteristics of oceanic crust. A progression from typical continental crust to something closely approximating oceanic crust may be seen along much of the coast of eastern Greenland, between Angmagssalik and Scoresby Sund. Progressing from the interior towards the thinned, stretched and fissured margin of the Greenland continental crust, coast-parallel dykes make a first appearance, increasing in concentration towards the flexed and faulted coastal strip which may closely approximate to the boundary between attenuated continent and true oceanic crust (Figure 12.12). Along this coastal strip, dyke intensity attains approximately 100 per cent, closely resembling the sheeted complexes of ophiolites (Figure 12.13).

The rock strata through which magmas ascend

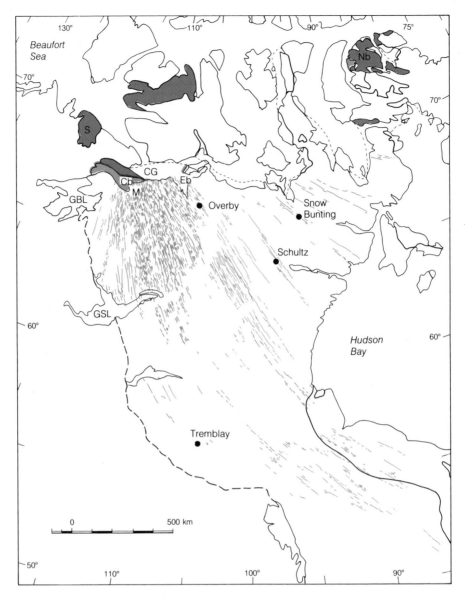

Figure 12.6 Map of the Mackenzie dyke swarm and related igneous rocks in the Beaufort Sea – Hudson Bay area of north-western Canada. This great Precambrian swarm of doleritic dykes fans out from a focus in the Coppermine Bay region south of the Beaufort Sea. The dashed line indicates the edge of Palaeozoic cover. Dyke distribution is based, in part, on aeromagnetic interpretation. In regions with high magnetic background apparent breaks in the dyke pattern may not be significant. M – Muskox intrusion; Cb – Coppermine River basalts; Eb – Ekalulia basalts; Nb – Nauyat basalts; CG – Coronation Gulf; GBL – Great Bear Lake; GSL – Great Slave Lake. (*After W.F. Fahrig, 1987*)

have an overall decrease in density from the lower crust upwards. Magmas commonly encounter materials near the uppermost crust whose density matches, or is less than, that of the magma. At such levels of 'neutral buoyancy' where there is no lithostatic pressure to drive the magmas to higher levels, they may spread sideways, taking advantage of bedding planes or other planes of weakness, to produce sills. Sills are consequently common in volcanic regions where low-density sedimentary strata are prominent in the upper part of the crust. Some of the most extensive suites of sills occur in the Transvaal, in Tasmania and in

Victoria Land in the Antarctic. These resulted from widespread volcanism that attended the Mesozoic break-up of Gondwanaland. Similar extensive sill swarms of early Tertiary age are seen in the coastal region of East Greenland between 70° and 75°N. Much-studied sills include the early Mesozoic Palisade and Lambertville sills of New York and New Jersey, and the late Carboniferous Whin Sill of northern England. The Salisbury Crags sill (Figures 12.14 and 12.15) at Edinburgh, Scotland, has a special place in the history of geology, supposedly providing James Hutton, in the late eighteenth century, with critical evidence

Figure 12.7 Sketch map of part of a Precambrian (Gardar) dyke swarm near Isortoq, south Greenland. Several generations of dyke were involved. The larger dykes attain thicknesses of approximately 0.5 km. (*After D. Bridgwater and K. Coe, 1969*)

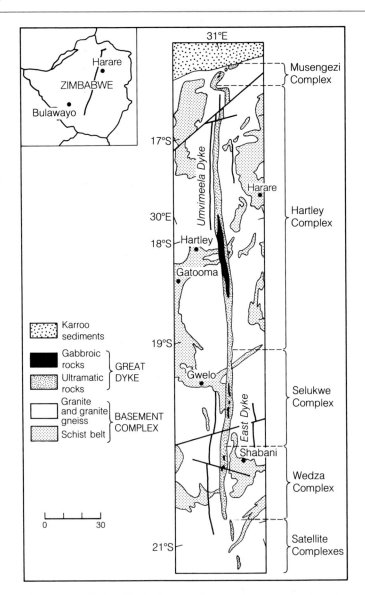

Figure 12.8 Sketch map of the Great Dyke, Zimbabwe and the accompanying Umvimeela and East Dykes. This dyke intrusion was large enough for the cooling history to be very prolonged (probably running to tens of thousands of years). The resulting rocks are coarse-grained layered cumulates (Section 5.6). Most are ultramafic rocks composed mainly of olivine and pyroxene. In some of the highest stratigraphic units of the layered cumulates, the rocks are gabbros, composed of olivine, pyroxene and plagioclase feldspar. (*After R. Bichan, 1970*)

bearing on the 'plutonic' (molten) origin of basaltic rocks, as opposed to an origin by precipitation from seawater ('neptunean').

Igneous intrusions frequently make prominent landscape features, since the rocks of which they are composed are commonly tougher and more resistant to weathering than their country rocks. Dykes owe their name to the fact that they often weather out to form upstanding walls or, conversely, weather in to form troughs. Sills frequently weather to give notable cliffs or escarpments. The Palisade Sill derives its name from the near-vertical, well-jointed cliffs facing the Hudson River, appearing as a natural defence work. The Roman engineers took advantage of the natural defence line provided by the Whin Sill (Figures 12.16 and 12.17) to build Hadrian's Wall to restrain war-like tribes to the north.

Sills may terminate abruptly laterally, (e.g. against a fault or change in the character of the country rock) or may splay into smaller 'leaves', or simply get thinner and wedge out. As with dykes, sills come in a wide size range, large ones having thicknesses of over 100 m. Individual sills may spread over areas of several hundred square kilometres. As sills are followed laterally they commonly step abruptly from one stratigraphic horizon to another (Figure 12.15). Thus whereas sills are characteristically concordant intrusions, they can show local discordance. Figure 12.18 shows diagrammatically how such discordant step-like features may develop.

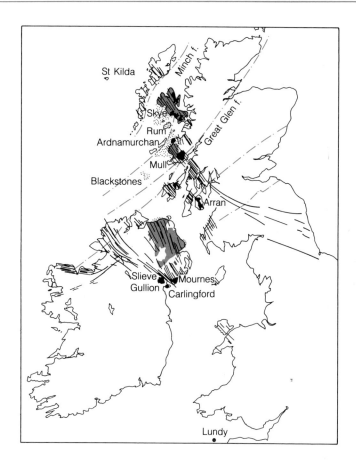

Figure 12.9 Sketch map of the British Tertiary Volcanic Province showing principal on- and off-shore outcrops of early Tertiary lavas (hatched and stippled respectively); main central volcanic complexes (black), and distribution of the dyke swarms (diagrammatic). (*After B.G.J. Upton, 1988*)

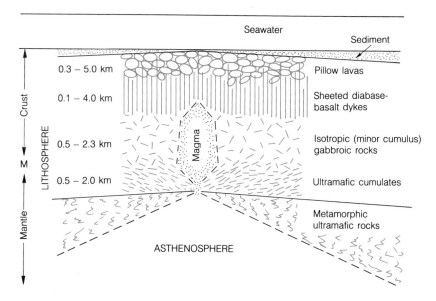

Figure 12.10 Idealized section through a typical ophiolite sequence as envisaged in its original mid-ocean ridge or marginal basin setting. A magma chamber (centre) is supplied by new magma from melt fractions ascending from the underlying asthenosphere. Repeated intrusion of dykes, as sea-floor spreading proceeds, produces the sheeted complex of basaltic or doleritic dykes. Those intrusions that reach the sea-floor erupt basalt in the form of pillow lavas (see Figure 13.33, Plate IXb). (*After M.G. Best, 1982*)

Figure 12.11 Roadside section through a sheeted dyke complex, Troodos ophiolite, Cyprus. The outcrop is composed, almost exclusively, of basaltic or doleritic dykes with individual widths less than 2 m. (*B.G.J. Upton*)

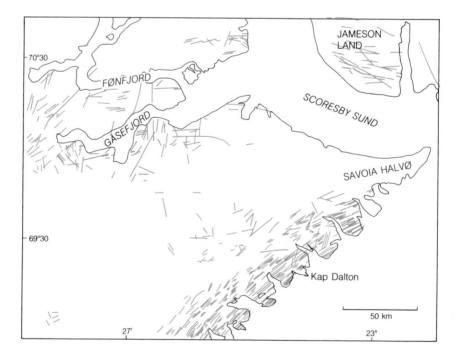

Figure 12.12 Distribution of early Tertiary doleritic dykes in the Scoresby Sund region, East Greenland. Dykes achieve very high concentrations within a narrow zone close to, and paralleling, the eastern coast line. (*After L.M. Larsen, W.S. Watt and M. Watt, 1989*)

Figure 12.13 A sheeted complex at Kap Hammer, East Greenland. The entire outcrop is composed of doleritic dykes forming part of the dyke swarm along the East Greenland coast south of Scoresby Sund (see Fig. 12.12). The dykes were vertical when intruded: the present inclinations are due to tilting resulting from flexuring of the coastal region. (*B.G.J. Upton*)

Figure 12.14 An aerial view of Carboniferous igneous rocks in the Royal Park, Edinburgh, Scotland. The prominent cliff feature (lower right) is formed by the Salisbury Crags dolerite sill, intruded into Lower Carboniferous sediments. The overlying rocks (forming the high ground, centre left) are early Carboniferous lava flows. The highest points (background, centre) are due to basaltic plugs that postdate the lavas. (*Planair, Edinburgh, copyright*)

Figure 12.15 Transgressive lower contact of the Salisbury Crags sill, Edinburgh, cutting and overlying thin bedded sediments.

Figure 12.16 The Great Whin Sill and the Roman Wall, Cuddy's Crag, Northumberland, northern England. (*J. Allan Cash*)

12.2 LACCOLITHS AND LOPOLITHS

Whereas a sill is ideally tabular and approximately parallel-sided over much of its extent, there is gradation to more complex lensoid forms. Sometimes the roofing rocks are arched so that the intrusion has the form of a plano-convex lens, convex upwards, rather like a blister. The term

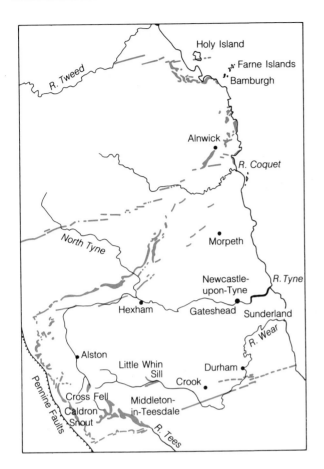

Figure 12.17 Sketch map of north-east England showing outcrop of the Great Whin Sill and associated ENE–WSW trending dykes. A photograph of this late-Carboniferous sill is shown in Fig. 12.16.

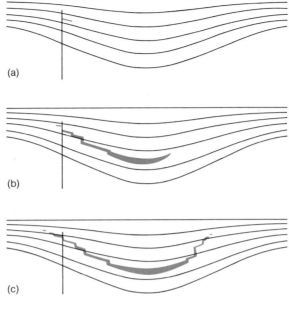

Figure 12.18 A model for emplacement of a sill with discordant step-like features. (a) Magma ascending as a dyke attains a height above the optimum for lateral flow. (b) Lateral sill intrusion follows, with gravitational flow down dip, acquiring discordancies where the magma steps down to accumulate at the bottom of a sedimentary basin. (c) Hydrostatic equilibrium is achieved by ascent of magma from the basin floor by stepping up. (*After Open University, S339, Block 2*)

'laccolith' is commonly applied to such intrusions (Figure 12.19). The Skaergaard intrusion, Greenland, is probably the most intensively studied intrusion on Earth (because it is, in many ways, ideally suited to answer a range of questions relating to the detailed cooling and crystallization history of a large magma body). Whereas it is by no means a typical laccolith, it is now regarded as a thick lensoid body that developed by magmatic inflation at, or close to, the unconformity between gneisses and overlying sediments and lavas (Figure 12.20 and Plate VIIIb).

Very large intrusions that have, at least in part, contacts concordant with those of the country-rock strata were termed 'lopoliths' by Reginald Daly. In the original concept, sagging of the underlying rocks beneath the weight of the intrusive rocks, was regarded as characteristic. The intrusions thus tend to be plano-convex, with the convex surface downwards. There is, no doubt, every intermediate stage between a simple sill and a giant lopolith, whose dimensions may be expressed in

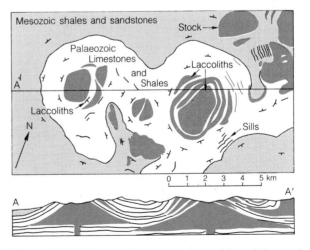

Figure 12.19 Map and cross-section of laccoliths and stocks in the Judith Mountains, Montana. (*After W.H. Weed and L.U. Pirsson*)

hundreds of kilometres lateral extent and several kilometres thickness.

The original 'lopolith' was the Duluth intrusion

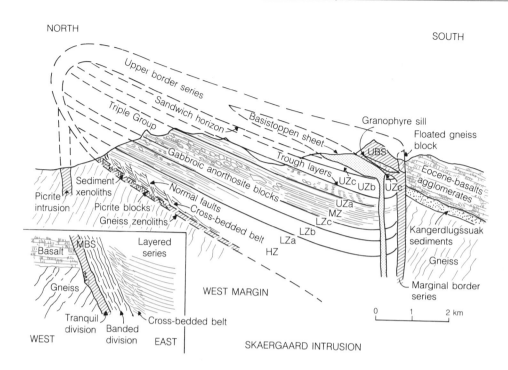

Figure 12.20 Diagrammatic section through the Skaergaard intrusion, East Greenland. This large intrusion of gabbroic rocks was formed by the arrest of rising basaltic magma at a level controlled by the unconformity between Precambrian gneisses and overlying sediments and early Tertiary (Eocene) basaltic lavas. The unconformity plane was approximately horizontal at the time of intrusion. The present dip reflects post-intrusion tilting. Slow cooling of the magma allowed crystallization of a remarkable series of strongly-layered gabbro cumulates (see Plate VIIIb). 'LZ' denotes lower zones a, b and c. 'MZ', middle zone, 'UZ', upper zones, a, b and c. UBS is Upper Border Series – gabbros produced by crystallization inwards from the intrusion roof. (*After T.N. Irvine, 1987*)

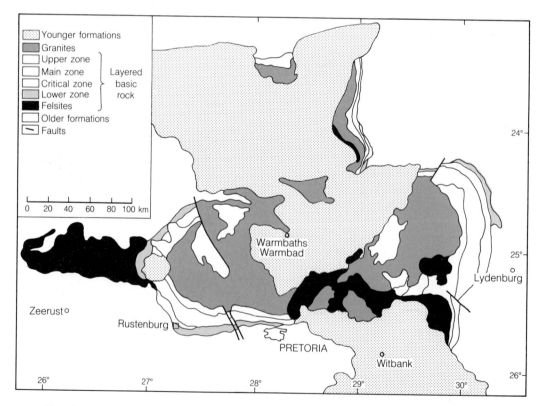

Figure 12.21 Sketch map of the geology of the Bushveld lopolith, South Africa. (*After map by the University of Witwatersrand*)

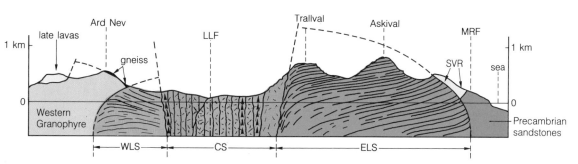

Figure 12.22 Schematic cross-section through the Rum igneous complex, north-west Scotland. WLS, Western Layered Series: CS, Central Series: ELS, Eastern Layered Series: LLF, Long Loch Fault: MRF, main ring fault. The complex probably evolved as a suite of intrusions below a large central-type volcano across the plane of an active fault, the Long Loch Fault. Granophyres and siliceous volcanic rocks – SVR – overlying Precambrian formations were produced early. These were intruded by magnesian basaltic (picritic) magmas from which the Eastern and Western Layered Series formed (Figure 12.23). Subsequent tectonic movements allowed new influx of picritic magma along the Long Loch zone, from which the Central Series formed. Subsidence of the Central Series, while it and its wall rocks were still hot and ductile, gave rise to drag folding of the ELS close to the contact with the CS. (*After C.H. Emeleus, 1987*)

Figure 12.23 Contact zone of the basic–ultrabasic layered cumulates of the early Tertiary volcanic complex on the island of Rum, NW Scotland. The hillside to the left is of rhyolitic ('felsite') volcanic rocks. The snow-filled depression overlies gabbroic rocks forming the margin of the intrusive complex. The stepped topography of the mountain on the right is formed by differential weathering of olivine-rich ultrabasic cumulates and plagioclase feldspar-rich cumulates. The latter form the steeper cliff features whereas the olivine-rich cumulates weather more readily to form the intervening slopes. (*B.G.J. Upton*)

of Minnesota. Supposedly one of the greatest is the vast Bushveld Intrusion of Transvaal (Figure 12.21), some 5 km thick and approximately 300 km across (although recent investigations indicate that the Bushveld may comprise four interconnected lobes or compartments, each of which has a conical form opening upwards). The Dyfek Intrusion of the Antarctic may be even greater. Recent estimates for the Niquelandia intrusion, Brazil, suggested a thickness of about 15 km! These giant

Figure 12.24 The Cão Grande phonolite plug, São Thomé Island, Gulf of Guinea, west Africa. The extrusive cover of a volcano has been removed by erosion leaving only the resistant mass of igneous rock congealed within the main conduit (*J.G. Fitton*).

cousins of the sills (like the largest known dykes) have cooled sufficiently slowly for crystallization to have effected very efficient mineralogical and geochemical sorting (differentiation) with generation of cumulate layering (Chapter 5).

Apart from being highly instructive to the igneous petrologist and geochemist, these great intrusions are of critical economic and political importance as a consequence of the concentrations within them of such elements as chro-

Figure 12.25 El Capitan; a volcanic plug in Arizona. As in Fig. 12.24 this plug has been revealed after erosion of a volcanic edifice. (*B.G.J. Upton*)

Figure 12.26 A volcanic neck composed mainly of agglomerate. Le Puy, Haute Loire, France. (*Burton Holmes*)

mium, nickel, vanadium and platinum-group elements.

12.3 FAULT-CONTROLLED INTRUSIONS

Some less well-defined forms can result when magma ascends along an active fault zone. Lateral motion along faults that are steep or vertical, but with irregular fault planes, can open potential void space into which magma is able to arise. The intrusions so formed can be vertical-sided and tabular, thus resembling dykes but with the open-

ing mechanism involving some lateral shear rather than simple dilatation. An example of a complex intrusion composed of basic (gabbroic) and ultra-basic (peridotitic) rocks whose emplacement may have involved a mechanism of this sort is exposed on the Island of Rum, western Scotland (Figures 12.22 and 12.23).

12.4 PLUGS

Some of the most striking igneous formations, are the crudely cylindrical, steep-sided plugs revealed

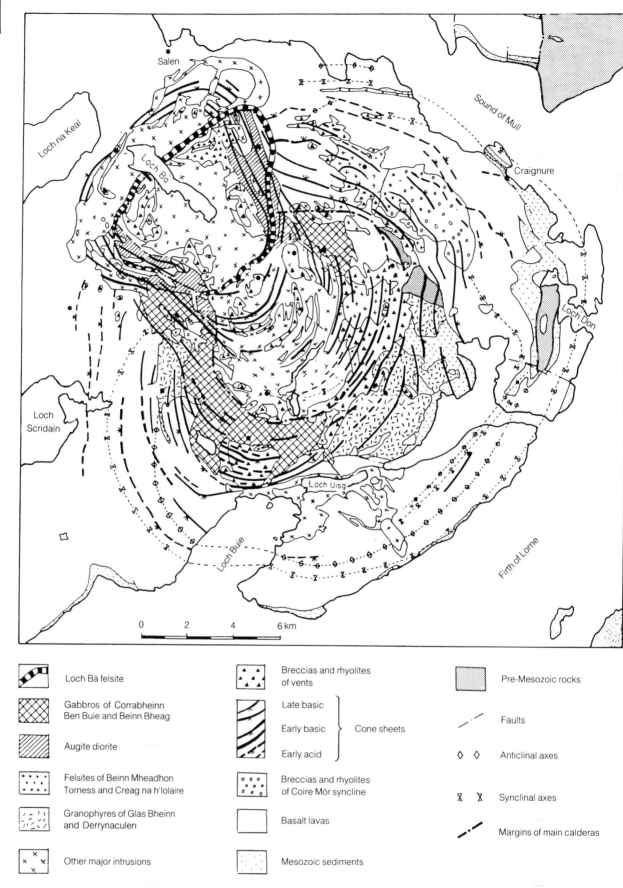

Figure 12.27 Geological sketch map of the Mull central complex, Inner Hebrides, Scotland. This deeply-eroded early-Tertiary volcanic complex displays great numbers of cone sheets – indicated here only diagrammatically – and ring dykes, as exemplified by the intrusion around Loch Bà. (*After Bailey et al, 1924*)

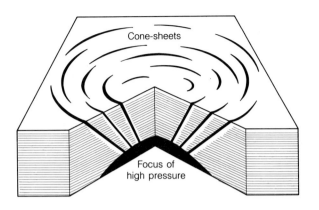

Figure 12.28 Diagram to illustrate the form of a series of concentric cone sheets and their probable relationship to an underlying focus of high magmatic pressure.

by the erosion of central-type volcanoes. These plugs, which represent magma that congealed within the main conduit supplying a former volcanic edifice, commonly are circular or ovoid in plan, with diameters from a few tens of metres to over a kilometre. These intrusions also frequently form positive topographic features. In glaciated terrains they can form steep-sided knolls, Castle Rock, Edinburgh and Mont Royal (from which Montreal takes its name) being well-known examples. More spectacular instances, however, are seen where erosion has occurred in tropical or desert weathering. Figure 12.24 shows a towering plug on the island of São Tomé, Gulf of Guinea. Ship Rock (Figure 12.4) and El Capitan (Figure 12.25) are notable examples in the arid landscapes of the south-western USA.

The space currently occupied by a plug was, of course, formerly occupied by the country rock. Typically the adjacent rocks are not strongly deformed and the magma could not have created the space for its intrusion by simply moving aside the wall rocks. In general, the space was created by removal of the pre-existing rocks, either by engulfing them by subsidence into magma beneath, or by expelling them upwards. Probably, in many cases gases derived from ascending magma rise in advance of the magma, to exploit any potential fissures or fissure intersections. High-velocity passage of escaping gases may provide an agency for blasting out a channel-way through which the magma can escape to the surface. The conduit may be further widened by magmatic erosion caused by the rising molten material. Volcanic plugs are commonly found together with vents choked with assorted fragmental matter (agglomerate) representing infill of gas-drilled conduits (Figure 12.26).

As we have seen, simple tabular or sheet-like intrusions may ideally be subdivided into dykes or sills according to whether they are steep and discordant, or low-angled and concordant. Not surprisingly, in some dissected volcanic terrains, tabular intrusions occur that do not lend themselves to such easy classification. The contacts may dip anywhere from the vertical to the horizontal, and the term 'inclined sheets' is often more appropriate.

Deep erosion has shown that beneath some volcanoes intrusive sheets dip inwards towards a common focus. Attention was first drawn in the 1920s and 1930s to the existence of such intrusions by investigation of the early Tertiary volcanic complexes in the west of Scotland on the Isle of Mull (Figure 12.27) and on the Ardnamurchan peninsula. The Mull sheets, with thicknesses rarely exceeding 3 m, have outcrops concentrically disposed about a particular geometric centre, and dips that (very approximately) converge on a common focus some kilometres below the present erosion level. The term 'cone sheets' was coined to describe such intrusions (Figures 12.28 and 12.29). Cone sheets, or concentrically inclined sheets, are however, relatively uncommon but when present, as on Mull, they may occur in great profusion – in some places, so numerous as to exclude most of the country rock! In the Geitafell volcanic complex, in south-east Iceland, traverses several hundred metres long show only packed, centrally-inclined sheets (Figure 12.30). Such instances, where sheets intrude sheets, which themselves intrude still older sheets, are analogous to the sheeted (dyke) complexes believed to be characteristic of oceanic spreading centres.

The mechanism of cone-sheet intrusion, like that of dyke and sill emplacement, involves opening of a fracture. Since the sheets converge to common foci it is inferred that magma is injected upwards and outwards along fractures propagated in the vicinity of a sub-volcanic magma reservoir. Dilatational injection thus involves the uplift of the rocks above the fracture plane and the process has been compared to the opening of a conical valve. The large amount of mechanical work required for such elevation is probably a factor controlling the thickness of cone sheets. Like those of Mull, they are rarely more than a few metres thick. Most cone sheet-swarms are of basaltic composition. However, some remarkable cone sheet occurrences are known where magma composed principally of molten carbonates, rather than silicates, is involved. Such 'carbonatite' (see Chapter 29) cone sheets have been described for example, from Alnö in the Baltic and from Homa

Figure 12.29 A cone sheet of basaltic composition cutting Mesozoic sediments. Ardnamurchan igneous complex, western Scotland. (*B.G.J. Upton*)

Figure 12.30 A 'sheeted-complex' composed of basaltic cone sheets in an eroded late-Tertiary volcanic complex at Geitafell, south-east Iceland. The cone sheet suite is seen to be cut by four or more dykes. (*B.G.J. Upton*)

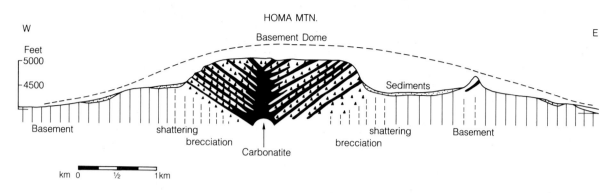

Figure 12.31 A cross-section through the Homa Mountain volcanic complex, Kenya. The complex developed through emplacement of carbonatite magmas (see Chaps 3 and 28), as a remarkable suite of cone sheets, causing updoming of the adjacent country rocks. (*After B.C. King et al, 1972*)

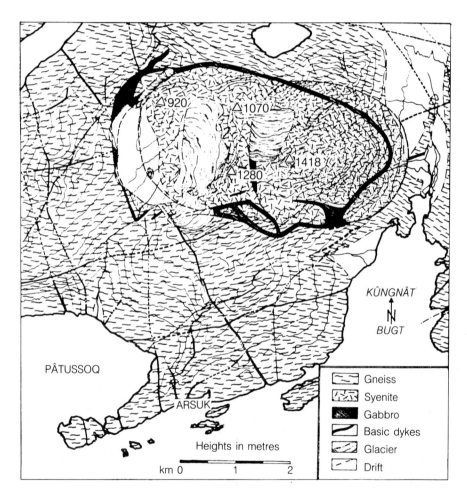

Figure 12.32 Geological sketch map of the Kûngnât Fjeld igneous complex, south Greenland. Two intersecting stocks of syenite are cut by an almost complete 360° (branching) ring dyke of gabbro. The three intrusions are interpreted as representing three successive ring-fault and subsidence events, probably associated with formation of surface calderas (see Chap. 13). (*B.G.J. Upton, 1960*)

Mountain in Kenya (Figure 12.31).

Whereas all of the intrusive forms described above may be developed by magmas of a wide variety of compositions, they are most commonly adopted by basaltic magmas partly because basaltic magmas are abundant and partly because of their low viscosity. Basaltic magma may congeal to fine-grained basalt, more coarsely-grained dolerite or, in the case of the larger dykes and lopoliths, to gabbro and accompanying differentiated rocks.

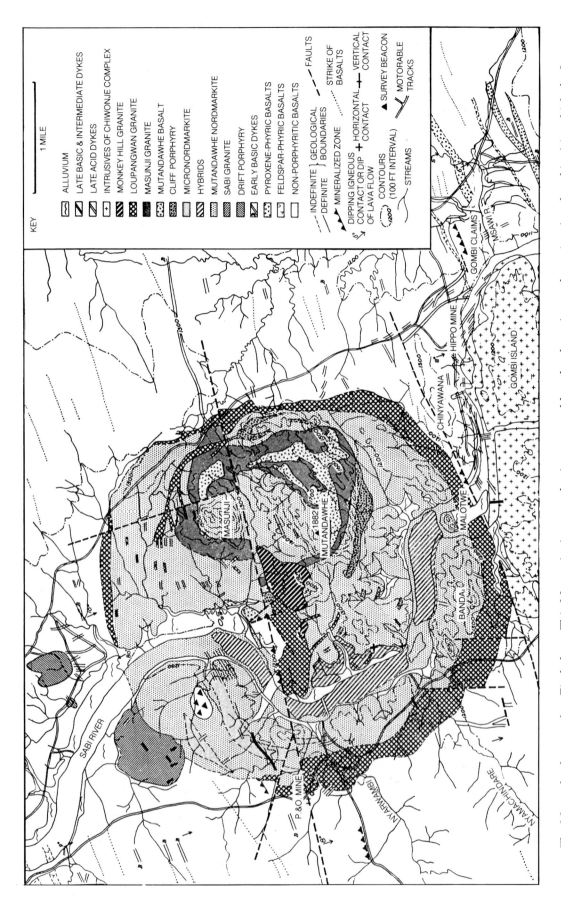

Figure 12.33 The Mutandawhe complex, Zimbabwe. This Mesozoic ring-complex is composed largely of granites and syenites (nordmarkites) in which the Loupangwan granite provides an excellent example of a ring dyke. (*After K.G. Cox et al, 1965*)

Figure 12.34 Aerial photograph of the Ardnamurchan ring-complex, western Scotland. This early Tertiary igneous complex represents a deeply-eroded volcanic centre. It is composed of annular intrusions of gabbro and related rocks, with large numbers of doleritic cone sheets. (The relief of the hills, as shown by the shadows, is best seen by most observers if the photograph is turned upside down. *Ed.*). (*From Dir. Overseas Surv. Mosaic DCS Geol 1042*)

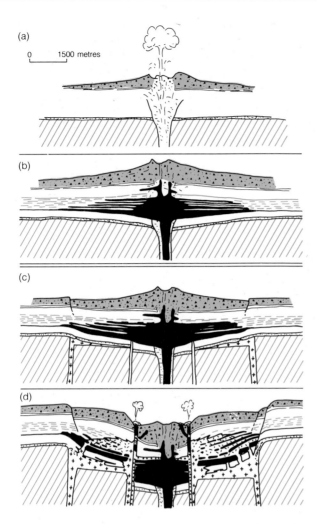

(a)

0   1500 metres

(b)

(c)

(d)

Figure 12.35 Diagrammatic sections illustrating the evolution of the Messum Complex, Namibia. (a) Initial volcanic structure. (b) Emplacement of basaltic magma as conformable sheets (sills), crystallizing to gabbro intrusions. (c) Ring-fracturing and subsidence. (d) Continuation of subsidence with formation of surface caldera (see Chapter 13) and emplacement of ring-dykes. (*After H. Korn and H. Martin, 1954*)

0   200   400 km

N

Coast Range batholith

Idaho batholith

Sierra Nevada batholith

Lower California batholith

Baja California

Figure 12.37 The major granitic batholiths of western North America. These batholiths are great composite bodies, composed of a variety of coarse-grained relatively siliceous igneous rocks. They were intruded during the Mesozoic, mainly in the mid-Cretaceous. (*After W.G. Ernst, 1976*)

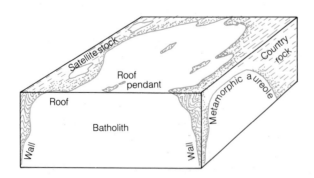

Satellite stock

Roof pendant

Country rock

Roof

Metamorphic aureole

Batholith

Wall

Wall

Figure 12.36 Block diagram illustrating the characteristic features of a batholith according to Daly. Note the steep, but irregular margins with country rock. 'Roof pendants' are relics of the lower portions of the original roof to the intrusion.

However, in the intrusive forms described below, the magmas responsible are more likely to be lower density, more siliceous types of 'intermediate to acid' composition.

12.6   RING DYKES

In large central-type volcanoes, magma withdrawal or expulsion from substantial crustal magma chambers may result in the collapse of the overlying rock to fill the potential void space. The collapse may occur within the limits of roughly cylindrical faults or faults developed concentri-

Figure 12.38 Mt McKinley, Alaska, highest peak of the 800 km long Alaska Range. Folded slates and greywackes of Mesozoic age, with granite behind from the walls of the central cirques up to the summit (see also Figure 31.1). (*Bradford Washburn*)

cally around a volcanic focus that have outwards dipping contacts. Subsidence of the central blocks within such 'ring faults' may permit simultaneous admission of magma up the ring faults. In the case of outward dipping faults, subsidence will clearly produce a progressive widening of the void space between the collapsing block and the surrounding rock. Solidification of magma within such ring fissures brings about the formation of a ring dyke.

Ring fractures, such as those just described, will, if they intersect the surface, generate collapse pits or calderas (see Chapter 13). However, ring fractures may be terminated upwards by low-angled cross fractures and need not necessarily have topographic expression at the surface.

An exposed ring dyke may be continuous through 360° of arc. This, however, is rarely the case and partial, or eccentric, subsidence of the central block will lead to production of an arcuate, often crescentic, partial ring dyke. Ring dykes vary in width from metres to hundreds of metres, and their diameters, by no coincidence, are in the same size range as those of calderas, generally ranging from about 5 to 25 km across. Occasionally much broader collapses occur, resulting in ring dykes such as that at Aïr, in Niger, some 60 km across.

Again it was study of the Mull volcanic centre, that gave the earliest description of a ring dyke. The Loch Bà ring dyke (Figures 12.27 and Plate VIIIc) in Mull is virtually continuous through 360°. Composed mainly of dacitic magma (although imperfectly mixed with more basic varieties), it varies from a few metres to about 100 m thick with the diameter being around 8 km. While many ring dykes have been described around the world in the last 60 years, the Loch Bà ring dyke remains a beautifully exposed, classic example.

There are a number of other fine examples of ring dykes in the British–Irish Tertiary volcanic centres including those at Slieve Gullion in Northern Ireland and on Skye in the Hebrides. The White Mountains of New Hampshire, the Jos Plateau of Nigeria and some of the Mesozoic volcanic complexes of Zimbabwe (Figure 12.33) also present excellent illustrations of the ring-dyke form. Typically ring dykes are associated with continental 'intra-plate' volcanic provinces (Chapter 29). The island of Kerguelen, however, in the Southern Ocean exhibits instances of ring dykes in an oceanic setting.

Ring dykes are commonly produced where relatively high-density country rock collapses into a magma reservoir filled with lower density magma such as rhyolite, dacite or trachyte. Although ring

Figure 12.39 Pucaraju (5346 m) viewed from the south-west Cordillera Blanca of Peru. The steeply-inclined strata seen above and around the snow cirque are Lower Cretaceous sediments, folded in late Cretaceous times, and further deformed and metamorphosed by the transgressive emplacement of a Lower Tertiary batholith of granodiorite (light-coloured outcrops on right and dark slope on left). (*C.G. Egeler and T. De Booy, 1956*)

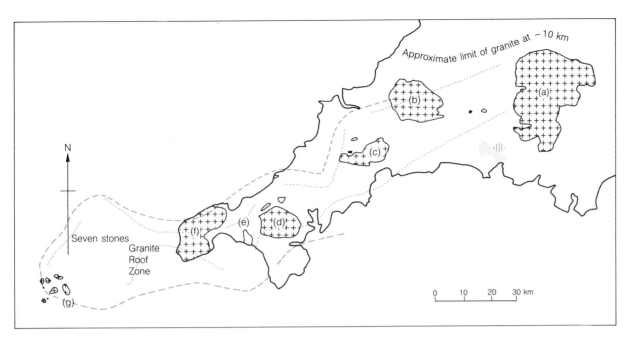

Figure 12.40 Sketch map showing principal outcrops of the Hercynian (latest Carboniferous–earliest Permian) granite batholith of south-west England. (a) Dartmoor, (b) Bodmin Moor, (c) St Austell, (d) Carnmenellis, (e) Tregonning-Godolphin, (f) Land's End, (g) Scilly Isles. Dotted boundary indicates inferred Granite Roof Zone. Dashed line marks underground limit of granite at 10 km depth. (*After C.S. Exley and M. Stone, 1982*)

dykes are most commonly formed from the crystallized products of such comparatively siliceous magmas, basic ring dykes involving basaltic magma are occasionally seen. Even if the bulk density of the central block is less than that of the underlying magma, it may nevertheless still subside for a certain vertical distance just as an ice cube, though less dense than water, will sink substantially in water before equilibrium is attained. A gabbroic ring dyke, continuous through 360°, formed by upwelling of basaltic magma around a subsidiary syenite block at Kûngnât, South Greenland, provides a good example of such a comparatively dense ring dyke (Figure 12.32).

Not infrequently uplift and erosion of central-type volcanoes (Chapter 13) reveals a complex aggregate of intrusions whose borders are more or less concentric around a common focus. These are commonly formed from repeated emplacements of ring dykes, inwardly-dipping cone sheets and/or cylindrical plugs of varying diameters. The whole ensemble is often referred to as a ring complex. The maps shown in Figures 12.27 and 12.33 show representative examples while the air photograph of Ardnamurchan (Figure 12.34) displays another. The evolution of a sub-volcanic ring complex involving sills, laccoliths and ring dykes is shown in Figure 12.35.

12.7 STOCKS AND BATHOLITHS

Relatively small intrusions, whose maximum dimensions are measurable in metres or a few tens of metres, that lose heat sufficiently fast to form rocks with crystal sizes of not more than one or two millimetres, are often referred to as hypabyssal intrusions. The majority are shallow intrusions, congealed within the upper few kilometres of the crust. Most dykes, sills, cone sheets, plugs and many ring dykes fall within this category.

Deeper-seated, large intrusions, whose cooling history is measurable in terms of thousands or tens of thousands of years, produce notably coarse-grained rocks such as gabbros and granites. Such intrusions, normally only revealed by substantial crustal uplift and erosion, are often referred to as 'plutonic' rocks (after Pluto, God of the underworld). The larger dykes, batholiths and more massive ring dykes may show plutonic characteristics.

Among the larger intrusions are the typically steep-sided masses of igneous rock with roughly circular or ovoid plan known as stocks or batholiths (Figure 12.36). Batholith is a term which embraces the largest of all masses of intrusive rock. The common environment for batholiths is deeply eroded fold-mountain belts formed above and behind subduction zones within the cores of volcanic arcs (cf. Chapter 13). Such batholiths are

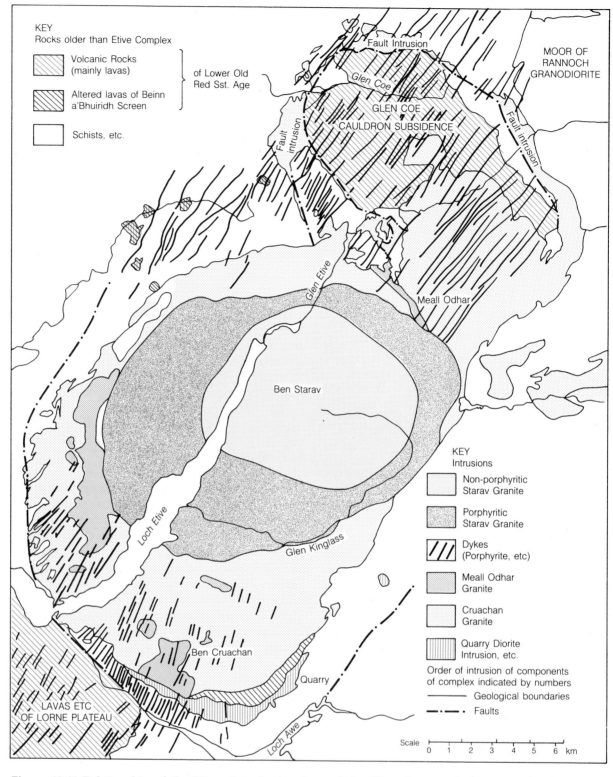

KEY
Rocks older than Etive Complex

Volcanic Rocks
(mainly lavas)

Altered lavas of Beinn
a'Bhuiridh Screen

⎱ of Lower Old
⎰ Red Sst. Age

Schists, etc.

Fault Intrusion

Glen Coe

GLEN COE
CAULDRON SUBSIDENCE

MOOR OF
RANNOCH
GRANODIORITE

Fault Intrusion

Fault Intrusion

Glen Etive

Meall Odhar

Ben Starav

Loch Etive

Glen Kinglass

KEY
Intrusions

Non-porphyritic
Starav Granite

Porphyritic
Starav Granite

Dykes
(Porphyrite, etc)

Meall Odhar
Granite

Cruachan
Granite

Quarry Diorite
Intrusion, etc.

Order of intrusion of components
of complex indicated by numbers

Geological boundaries

Faults

Ben Cruachan

Quarry

LAVAS ETC
OF LORNE PLATEAU

Loch Awe

Scale ⊢————————————⊣
0 1 2 3 4 5 6 km

Figure 12.41 Relationship of the Etive plutonic complex and the Glen Coe complex, western Scotland. These granitic intrusions were emplaced in late Silurian–early Devonian times into late Precambrian–lower Palaeozoic schists overlain by an andesite to rhyolite volcanic cover. Lavas and pyroclastic volcanic rocks in the Glen Coe region were down-faulted by a ring fault and invaded by the Cruachan granite, fault intrusions and Meall Odhar granite. A period of NE–SW dyke intrusions was followed by the intrusion of (a) the porphyritic Starav granite and (b) the non-porphyritic Starav granite. (*After G.S. Johnstone, 1966*)

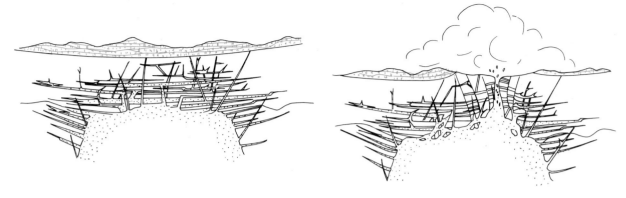

Figure 12.42 Diagrammatic cross-sections illustrating the supposed evolution of the Mio-Pliocene Tatoosh pluton, Mt Ranier National Park, Washington. In an early stage (a) a hypabyssal complex of sills and dykes developed within older volcanic strata overlying the main magma chamber. Subsequently (b), the magma broke through the roofing rocks, venting at the surface in a pyroclastic eruption (see Chapter 13). (*After R.S. Fiske et al, 1963*)

generally elongate in form (Figure 12.37): they may, like the British Columbian and Chilean batholiths, have lengths of over 1000 km and widths measurable in tens or even hundreds of kilometres (Figure 12.38). When mapped in detail, however, these very large batholiths are found to be composite intrusions, formed of large numbers of smaller, coalescing stocks. The well-studied Sierra Nevada batholith of California, for example, has been shown to consist of some 200 separate intrusions and the Peruvian batholith is known to involve many intrusions, emplaced over an interval of about 70 million years from the Cretaceous to the mid-Tertiary period, (Figure 12.39).

The granite masses exposed in south-west England (Figure 12.40) are believed to represent the higher portions of an extensive, partially unroofed batholith, exposed in strongly deformed Palaeozoic strata. Figure 12.41 illustrates a composite stock in Scotland where successive intrusions of progressively smaller size, emplaced about a common centre, produced crudely concentric patterns.

Whereas, once again, the terms stock and batholith do not denote any particular rock compositions, the great majority are composed of relatively siliceous plutonic rocks. Rock types such as tonalite, monzonite, granodiorite and granite (corresponding roughly to magmatic compositions from andesite to rhyolite, with approximately 55–75 per cent SiO_2) are overwhelmingly dominant in the great batholiths, from the Himalayas to the Pyrenees, from Malaysia to Peru. Commonly there is a tendency for successively younger units in the composite stocks and batholiths to become increasingly silica-rich. Despite the wide range in actual compositions, the term 'granite' is commonly employed to embrace them all.

In detail, the margins of such intrusions are commonly ragged and irregular. There is often evidence that the country rocks were considerably pre-heated before the intrusion was emplaced so that temperature gradients were not extreme or abrupt and well-chilled contact zones are characteristically absent. On the other hand, loss of heat through the side-walls, often accompanied by transfer of aqueous fluids and other volatile materials from the intrusion to the country rocks, effects extensive recrystallization so that metamorphic aureoles tens or hundreds of metres broad are commonly developed around such intrusions.

From the 1930s to the 1960s the so-called 'space problem' figured largely in the interpretation of large plutons. A strong school of thought proposed that many granitic bodies had resulted from *in situ* alteration of original country rocks to rocks of granitic texture and composition, through the action of granitizing fluids. Whereas examples of rocks generated in place by such metasomatism (or granitization) undoubtedly exist, they are generally small-scale and irrelevant to the genesis of the world's large granitic plutons. The latter are now acknowledged by virtually all investigators to be of truly igneous (magmatic) origin. Surface volcanism related to the silicic stocks and batholiths (Figure 12.42) is dominantly of pyroclastic type and is discussed in Chapter 13. Room for emplacement of the magmas is made either by engulfment of country rocks, by mechanical deformation (pushing aside) of the country rocks, or by both sets of processes operating concurrently.

Engulfment of country rocks may be largely brought about by differential thermal shattering as the magma comes into contact with them. The angular blocks broken from the roof and walls may settle into the intruding magma and may experience varying degrees of melting or chemical dissolution (Plate VIIId). This process is described as

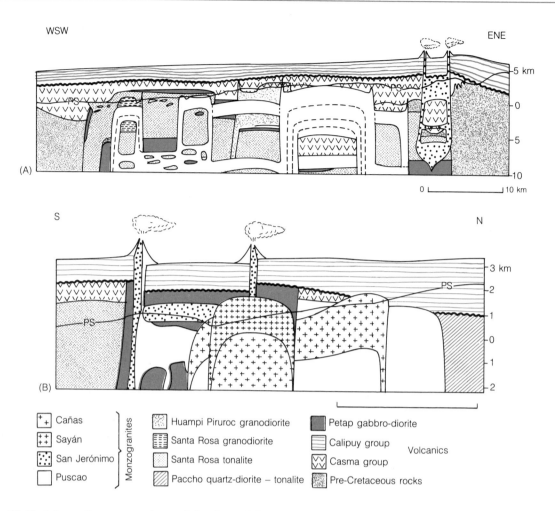

Cañas | Monzogranites
Sayán
San Jerónimo
Puscao

Huampi Piruroc granodiorite
Santa Rosa granodiorite
Santa Rosa tonalite
Paccho quartz-diorite – tonalite

Petap gabbro-diorite
Calipuy group
Casma group | Volcanics
Pre-Cretaceous rocks

Figure 12.43 Schematic cross-sections of the Coastal Batholith of Peru. A: Along northern flank of the Fortaleza Valley (*After Myers, 1975*). B: Along southern flank of the Huavra valley (*After Bussell et al, 1976*). PS – 'present erosion surface'. These Mesozoic batholiths are depicted as having been mainly emplaced by ring-fracturing and subsidence of crustal blocks. (*After W.S. Pitcher, 1978*)

Figure 12.44 Ovoid xenoliths in the Laggan granite, Grampian Highlands, Scotland. (*B.G.J. Upton*)

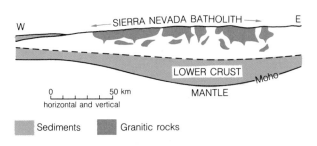

Sediments ▮ Granitic rocks ▮

Figure 12.45 Diagrammatic cross-section of the Sierra Nevada region in western USA indicating probable relatively-shallow crustal roots of the batholithic rocks. (*After Open University, S23, Block 4*)

stoping. Discrete blocks of country rock may not be wholly assimilated by the magma before solidification is completed – such blocks, often seen in eroded stocks and batholiths are referred to as **xenoliths** (literally 'foreign stones'). Stoping, or piecemeal assimilation of country rocks may take place on all scales. Xenoliths may represent small, shattered fragments, tabular slices that may be roughly concordant with the roofs or walls, or great sub-cylindrical blocks (Figure 12.43). In the latter case, the engulfment of large discrete masses of country rock can lead to intrusions with the form of ring dykes.

Ovoid or elliptical xenoliths, ranging from centimetres to tens of centimetres across, composed of coarse- or medium-grained rocks richer in ferromagnesian minerals than the host rock, are commonly found in these large granitic intrusions (Figure 12.44). Their origin has given rise to extensive discussion and there is probably no single answer. Some may represent pieces of incorporated country rock that have been modified texturally and compositionally by reaction with the enclosing magma. Some may represent early, relatively high-temperature, rock types grown on the sides of the intrusion as 'side-wall cumulates' which have been broken off and enveloped in younger, lower temperature magma. Since much of the batholitic magma is thought to have originated from the partial melting of crustal rocks at depth, some of the xenoliths may be portions of the refractory solid residue ('restites') left after the melting process. A further probability, in many cases is that the xenoliths represent imperfect mixing of one magma type with another, much as in the case of the mixed-magma rocks of the Loch Bà ring dyke discussed earlier.

Plutons that arise and displace their country rocks by deformation are said to be diapiric. Fine examples of diapiric granitic plutons where the requisite space has been made by compression and shouldering-aside of the country rocks are provided by the Ardara granite of Donegal (northwest Ireland) and the Flamanville granite in Normandy (France). Figure 12.45 shows the inferred diapiric form of granite units composing the Sierra Nevada batholith of California.

It is increasingly recognized that magmas may be intruded, not into relatively cold, solid country rock, but into pre-existing magma bodies. A common situation may arise where high-temperature basaltic magmas generated in the mantle encounter relatively fusible crustal rocks. Such a situation is likely to be commonplace at or about the mantle–crust boundary (i.e. the 'Mohorovičić Dis-

Figure 12.46 A mixed-magma rock resulting from intrusion of 'pillows' of basaltic magma (dark) into rhyolitic magma (white). Considerable chemical interchange occurred between the two. Both crystallized slowly at depth to coarse-grained rocks. Tugtutôq, South Greenland. (*B.G.J. Upton; published by permission of the Geological Survey of Greenland*)

continuity', Chapter 25) but may be expected also to occur at higher levels within the crust. Melting of the crustal rock, through the agency of heat imparted by the basaltic magma, may generate large bodies of secondary ('anatectic') magmas. These latter will be relatively siliceous with compositions approximating to granite or granodiorite. Further intrusions of basaltic magma may be trapped and congeal within these granitic magmas. The interface between such liquid–liquid intrusive masses is commonly lobate, with the 'basic' magma forming rounded or pillowy blobs within the more siliceous magma (Plate VIIIe and Figure 12.46). Convective stirring may cause the basaltic intrusive blobs to be fragmented, drawn out and stirred into the granitic magma to produce a very heterogeneous rock product.

12.8 FURTHER READING

Best, M.G. (1982) *Igneous and Metamorphic Petrology*. W.H. Freeman, San Francisco.

McBirney, A.R. (1984) *Igneous Petrology*, Freeman, Cooper, San Francisco.

Sutherland, D.S. (ed.) (1982) *Igneous Rocks of the British Isles*, Wiley, Chichester.

VOLCANOES AND THEIR PRODUCTS

13

A blast of burning sand pour out in whirling clouds. Conspiring in their power, the rushing vapours carry up mountain blocks, black ash and dazzling fire.

Lucilius Junior (c. AD 50)

13.1   GENERAL ASPECTS

Volcanic activity includes all the phenomena associated with the surface discharge of magmatic materials, solid, molten and gaseous from conduits communicating with the hot interior of the Earth. In addition to the eruption of hot gases and molten lavas by volcanoes, great quantities of fragmental materials produced by the gas erosion of the conduit walls and by disruption of molten lava, are also discharged.

Vast quantities of gas are emitted during volcanic eruptions. Volatile materials, among which water and carbon dioxide figure prominently, contained in solution under pressure within the magmas, are released as gases when the magmas approach the surface. It is very largely the manner in which gases separate from the liquid fraction, that dictates the nature of the eruption. The manner of separation, e.g. quiet or explosive, depends on such factors as the volatile content, the viscosity of the magma (determining its relative ease of flow), the rates of lava emission and the environment (e.g. subaerial, submarine or subglacial). Volcanism has been, from the dawn of history, one of the principal agencies by which the interior of the planet degasses and by which the oceans and atmosphere have evolved. Volcanic materials at the time of eruption are commonly so hot (frequently over 1000°C) that they are incandes-

cent, particularly by night. The great, dark clouds of gas full of ash particles in suspension, can resemble smoke. At night they may glow as fiery clouds, particularly when reflecting incandescent lavas below. Fires are commonly caused in the

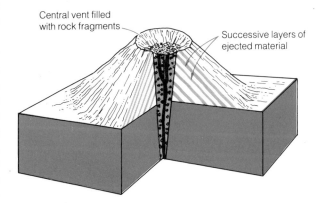

Figure 13.1 A block diagram of a central volcano composed of fragmental (pyroclastic) material. The slopes of the volcano flanks are controlled by the angle of repose of the cinders or ashes ejected from summit eruptions. The form and size of the summit crater are due to explosive gas discharge and gravitational slumping of pyroclasts back into the vent. The conduit (neck) is occupied by coarse (agglomeratic) pyroclastic material and minor intrusions. (*After F. Press and R. Siever, 1982*)

Figure 13.2 Mayon volcano, south-west Luzon, Philippine Islands. This cone, attaining a height of 2421 m, is a typical example of a relatively steep-sided strato-volcano developed above a subducting slab of oceanic lithosphere. The constructional surface is only slightly modified by rain-gullies. The summit crater is about 500 m diameter. (*J.P. Iddings*)

vicinity of eruptions where forests, houses, etc. are ignited by hot gases, particles or lava. These appearances not surprisingly gave rise to the widespread belief that volcanoes are burning mountains. Werner, for example, in Germany in the eighteenth century taught that the heat of volcanoes was due to the subterranean combustion of coal seams. The concept of volcanoes as burning mountains led naturally to the description of the fragments that fell to earth as ashes. We now know, however, that as outlined in Chapter 5, magmas represent molten material ascended from the upper mantle or deep crust and that the heat they contain is in no way related to combustion.

Some 550 volcanoes are known to have been active in historic times; this figure must be regarded as very conservative since many volca-

noes are submarine or are located in unpopulated or sparsely populated regions. Of these, roughly 50 to 60 erupt each year. The majority of these eruptions are small-scale events, causing little or no damage to human settlements. The total volume of lavas erupted subaerially each year is reckoned to be not more than 6 km^3. Submarine productivity, particularly along the mid-ocean ridge constructive plate boundaries, is likely to be very considerably greater, where some 20 km^3 of new oceanic lithosphere is thought to be generated annually.

When eruptions take place through a vertical conduit, the orifice is widened, by explosive gas release and by inward slumping, into a crater with flaring sides. Most volcanoes are positive topographic edifices that grow through repetitive

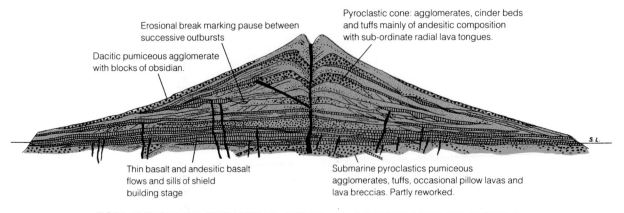

Erosional break marking pause between successive outbursts

Dacitic pumiceous agglomerate with blocks of obsidian.

Pyroclastic cone: agglomerates, cinder beds and tuffs mainly of andesitic composition with sub-ordinate radial lava tongues.

Thin basalt and andesitic basalt flows and sills of shield building stage

Submarine pyroclastics pumiceous agglomerates, tuffs, occasional pillow lavas and lava breccias. Partly reworked.

FORMATION OF SUBAERIAL CONE e.g. EASTERN EPI OR TONGOA

Figure 13.3 A cross-section of a supra-subduction zone strato-volcano, Tongoa Island, New Hebrides. (*A.J. Warden, 1967*)

Figure 13.4 A basaltic shield volcano, central Iceland. (*B.G.J. Upton*)

accumulation of lavas and/or fragmental materials ('**pyroclasts**'). Consequently, around an essentially cylindrical conduit a conical volcano will form, surmounted by a crater (Figure 13.1). Such a volcanic edifice conforms with the general perception of what a volcano should look like, as exemplified by Mayon in the Philippines (Figure 13.2). A volcano of this kind is said to be of 'central type'. Figures 13.1 and 13.2 illustrate central volcanoes composed wholly or largely of pyroclastic deposits. These tend to be relatively steep-sided, with the angle of their flanks (or constructional surfaces) to the horizontal up to 40°. The degree of slope is very largely dictated by the angle of repose of the unconsolidated pyroclasts. Volcanoes composed of crudely alternating layers of pyroclasts and lavas (Figure 13.3) are called **composite volcanoes** or, alternatively, **stratovolcanoes**.

In those volcanoes where explosive gas release is rare and the edifice is mainly composed of lavas,

the constructional surfaces may be at much lower angles. This is especially so for volcanoes built up from emission of very fluid basaltic lavas. In these the flanks may rise at less than 10° and sometimes at as little as 1° or 2° to the horizontal. Such volcanoes are referred to as **shield volcanoes** (Figure 13.4).

The world's greatest central vent volcanoes are of shield type. In this category come the giant oceanic volcanoes such as Mauna Loa in Hawaii, representing huge accumulations of basalt lavas. However, eruption of more volatile-rich, viscous magmas at a late stage in the lifetimes of these volcanoes may construct smaller composite cones atop the great shields. In some volcanoes there is no single central conduit used by successive eruptions and a complex, multi-vent volcano may result with an array of clustered cones or shields. Figure 13.5 shows a section through a relatively complex volcano built sequentially over an array of magmatic conduits.

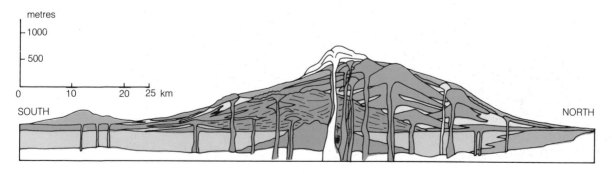

Figure 13.5 Schematic section through a complex rhyolite volcano in the Jemez Mountains, New Mexico, indicating build-up and overlapping of successive cones. (*After R. Bailey and R. Smith, 1968*)

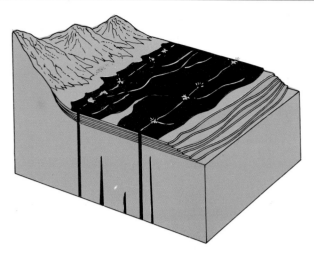

Figure 13.6 Dykes in a volcanic terrain under tension. Some reach surface level and feed fissure volcanoes. The highly mobile basalt lavas spread laterally, building volcanic sequences with nearly horizontal constructional surfaces. (*After R.S. Fiske, 1971*)

Instead of erupting from a roughly cylindrical conduit, magmas (particularly basaltic magmas) in regions where the lithosphere is under tension, erupt through extended splits to produce **fissure volcanoes** (Figure 13.6). Fissure volcanism consequently results where magmatic dykes (Chapter 12) reach the surface. Most volcanic activity along the mid-ocean ridges is likely to be of fissure type. In subaerial fissure volcanoes extended 'curtains of fire' may result from fountaining of incandescent lava along fissures whose lengths may be measurable in hundreds or sometimes thousands of metres. The greatest fissure eruption in recent times took place at Laki in south-east Iceland in 1783. Activity commenced along a fissure zone some 25 km long, with lavas flowing laterally up to 45 km and covering a total area of around 500 km^2 In the latest stages of eruption a series of cinder cones developed along the fissure (Figure 13.7). A sketch map of the fissure zone is shown in Figure 13.8. Some spectacular fissure activity of this kind, but on a much reduced scale, took place recently at Krafla volcano in northern Iceland between 1975 and 1983. Profiles of basaltic fissure volcanoes, transverse to their length, are similar to those of shield volcanoes. With passage of time, eruption along a fissure may become increasingly localized and eventually one (or more) central-type shield volcanoes may evolve over favoured sites of ascent along an otherwise congealed fissure.

Fissure volcanism appears to be not only the dominant process along the constructive plate boundaries of the mid-ocean ridges but is also

Figure 13.7 Looking south-west along the Laki fissure, showing some of the crater-chain lavas and pyroclastic deposits formed during the 1783 eruption. (See also Figure 13.8). (*G. Kjartansson, 1956*)

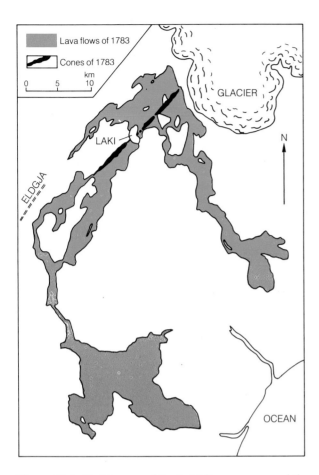

Figure 13.8 Sketch map of the Laki fissure zone and the lavas erupted in the 1783 event. (*After R. Decker and B. Decker, 1983*)

characteristic of continental regions under tensional stress. During the preliminary stages of continental disruption which may ultimately lead to genesis of new oceanic crust, fissure volcanism has, in the past, operated on a huge scale. In the subsiding rifted basins (see Chapter 29) developed over attenuating continental crust in tension, great sequences of fissure-fed basalt lavas accumulated that progressively buried the prevolcanic landscape beneath far-spreading floods of lava. These 'flood basalt' successions can cover vast tracts of country as for example, in the Deccan region of north-west India, where the Deccan basalts have an area of about 500 000 km². Some of the successions can locally attain thicknesses of up to 10 km whereas in the East Greenland flood basalts, (Figures 13.9 and 13.10) typical flows are 15–30 m thick, cover areas of 500 to 4000 km² and have volumes of 10–60 km³. Individual flows can be very much larger, with volumes up to 300 km³. Related fissure volcanism, on a very much more subdued scale, gave rise to lava accumulations in the Hebridean region of Scotland (Figure 13.11). The most recent of the large-scale flood basalt events was that which generated the Columbia basalts, across some 200 000 km² of Oregon, Washington and Idaho in the American north west (Figure 13.12). Among the more prominent flood basalt events of the Mesozoic and early Tertiary were those affecting East Greenland, Ethiopia, the Deccan region of India, large areas of southern Africa (the Karoo volcanics) and substantial areas of Argentina and Brazil. Even greater continental fissure volcano events can be inferred for the Precambrian, even though commonly much or all of the lavas have been lost through erosion, leaving only the underlying dyke swarms that are presumed to have fed them (e.g. the Mackenzie swarm of Canada, Section 12.1, Figure 12.6). One of the largest preserved Precambrian flood basalt sequences is that of the Keweenawan volcanics in the Great Lakes area of North America involving over 1 000 000 km³ of lava.

These dramatic outpourings of fluid basalt lavas appear to have been accomplished within relatively short time intervals. Most appear to have been erupted in less than five million years, with some of the largest having taken place in one to two million years. Within these huge volcanic fields built up from the products of large numbers of fissure eruptions, the constructional surfaces built by the mobile, fast-flowing basalt lavas would generally have been only a few degrees at most from the horizontal.

The heights of volcanoes are controlled by the densities of the ascending magmas and the lithostatic pressure, i.e. the pressure exerted on the magma source by the mass of the overlying column of rock. The highest volcanoes are those of the Earth's continental regions where the lithosphere is thickest. The maximum sustainable heights are in the region of 6000 m above sealevel. Among the continental volcanoes that reach the greatest heights are some of the Andean cones (e.g. Cotopaxi, 5974 m), some in Central Asia and Kilimanjaro in Central Africa. In the oceanic regions, the lithosphere is thinner and the greatest heights are around 4000 m above sealevel. The highest oceanic example is Mauna Kea on Hawaii (4206 m). Other oceanic volcanoes approaching the maximum possible heights are Pico del Teide (Tenerife, Canary Islands), Mawson Peak (Heard Island, south Indian Ocean), Beerenberg (Jan Mayen, north Atlantic) and Mauna Loa, on the main island of Hawaii. When a volcano has reached maturity in terms of height, subsequent magma batches will tend to erupt from vents below the summit crater. In this manner parasitic or 'adventive' cones may develop on the volcano's flanks, together with lateral fissures and chains of vents localized along them.

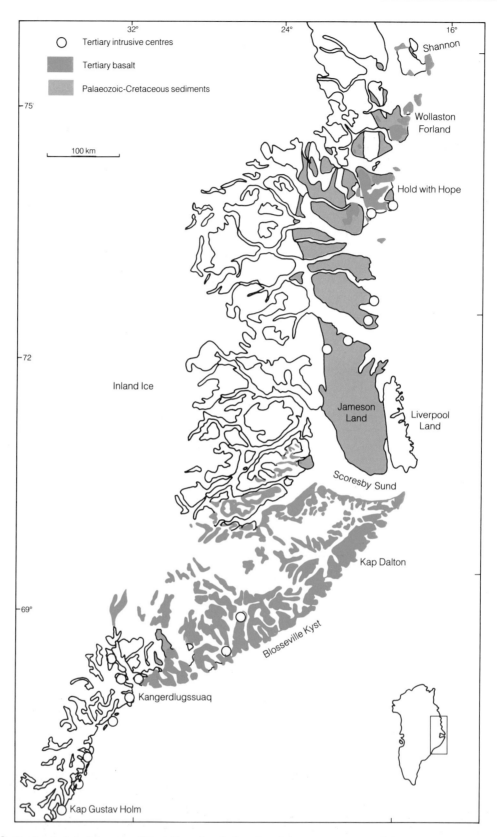

Figure 13.9 Geological sketch map of East Greenland showing the present extent of the early Tertiary flood basalts. (*After L. M. Larsen, W. S. Watt and M. Watt, 1989*)

13.2 DISTRIBUTION OF VOLCANOES

As can be appreciated from the map (Figure 13.13), volcanoes are not randomly distributed globally but tend to be concentrated in linear zones or chains. The distribution is closely related to tectonics and the majority of volcanoes lie along, or in the vicinity of, the margins of the tectonic plates

Figure 13.10 Early Tertiary 'flood basalt' lavas near Scoresby Sund, East Greenland. Note the great lateral continuity of the basalts reflecting the wide lateral spread of the highly fluid lavas. The lavas were originally erupted, in a nearly horizontal attitude, close to sea-level. They have subsequently been tectonically elevated in later Tertiary times (without tilting) and then deeply eroded by glaciers from the Pleistocene to the present. (*W.S. Watt, published by permission of the Geological Survey of Greenland*)

Figure 13.11 Palaeocene fissure-fed basalt lavas on Mull, north-west Scotland. These lavas represent a less-voluminous and smaller-degree melting event than the East Greenland lavas shown in Fig. 13.10. These too were erupted as nearly flat-lying flows, close to sea-level, which have since been uplifted and eroded. (*B.G.J. Upton*)

28). The minority, whose location is remote from the plate margins are said to be intraplate. Some volcanoes occur deep in the continental interiors but most lie within a few hundred kilometres of the sea.

Many volcanoes are situated along the crests of the world-encircling mid-ocean ridge system. The mid-ocean ridges, (as outlined in Chapter 28), coincide with the constructive plate margins, where persistent magmatism provides the mechanism whereby ocean spreading takes place as new lithosphere is generated. The underlying mantle is believed to undergo melting to the extent of some 10–15 per cent in response to pressure-relief as ocean-floor spreading takes place (Chapters 5 and 28). The oceanic spreading rates are generally such that ridge crest volcanism is unable to accumulate extrusive products fast enough for the volcanoes to reach sea-level. Consequently the vast bulk of mid-ocean ridge volcanic activity remains submarine and unseen. Exceptions occur where magmatic productivity is abnormally high and/or where spreading rates are very slow. Such exceptions are found in Iceland (Figure 13.14), where the mid-Atlantic ridge system emerges above sea-level for some 500 km, and in the Afar Depression in northern Ethiopia, where the southern extremity of the Red Sea ridge passes into the continent of Africa.

There are a number of places where other volcanoes occurring on, or relatively close to, the oce-

into which the lithosphere is subdivided (Chapter

Figure 13.12 Sketch map showing the area covered by the Columbia River flood basalts. Basalts erupted from fissure volcanoes were erupted during a brief phase in the Miocene to cover an area of approximately 200 000 km². (*After R.S. Fiske, 1971*)

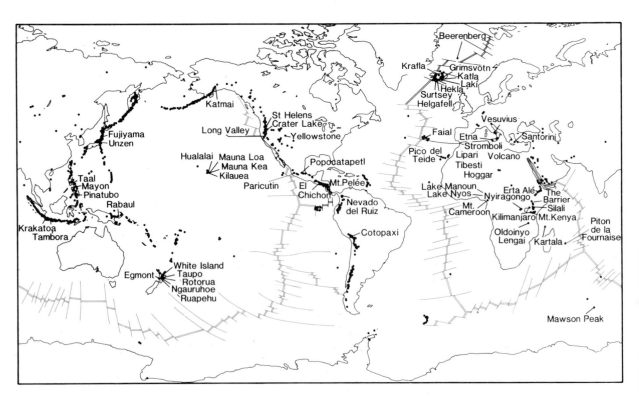

Figure 13.13 Map of the active volcanoes of the world. Tectonic plate boundaries are also indicated.

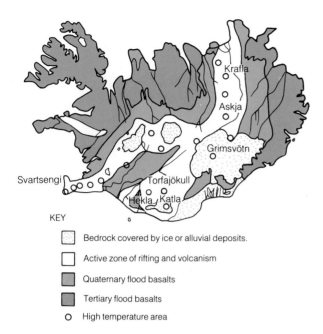

Figure 13.14 Geological sketch map of Iceland. The spreading centres connected with the mid-ocean ridges traverse the island as an active zone of rifting and volcanism. Oriented SW–NE in the south-west of Iceland, the active zone swings into a more nearly N–S orientation in the north. There are two branches in the south-west; an older more north-westerly one (currently inactive), from the Reykjanes Peninsula to central Iceland and a more south-easterly branch (currently active) from the Vestmannaeyjar Islands. An additional zone of recent (post-glacial) activity lies approximately E–W through the Snaefellsnes Peninsula in the west of Iceland. With the exception of this Snaefellsnes zone the volcanic rocks progressively age north-west and south-east away from the recent zones of activity. The principal recent volcanoes are indicated.

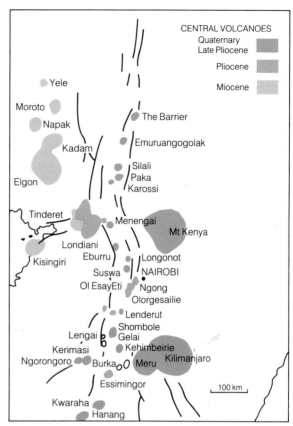

Figure 13.15 Sketch map showing the relationship of late Tertiary to Recent volcanoes to rift faulting in the East African Rift Zone in Uganda, Kenya and Tanzania (*After B.C. King and G.R. Chapman, 1975*).

anic ridges have built up sufficient height to form islands. Prominent examples in the Atlantic Ocean are the volcanoes that form the Azores. In the Pacific, the Galapagos volcanoes arise near the East Pacific Rise and, in the southern Indian Ocean, the remote volcanic islands of St Paul and Amsterdam provide further examples.

Most of the Earth's high volcanoes are associated with destructive plate margins where lithospheric slabs are drawn back down into the mantle. It is believed that addition of water to the mantle rocks immediately above, derived from dehydration processes affecting the down-going slab, promotes melting and the uprise of magma (Chapters 5 and 28). Melting and subsequent volcanism usually occur only when the slab has descended to depths of over 100 km. According to the angle of inclination of the slab, the volcanoes will appear at distances that are generally 200 to 500 km back from the line of intersection between the top of the slab and the Earth's surface. The

chains of volcanoes commonly define curves (volcanic arcs) several hundreds of kilometres long. The volcanoes of these supra-subduction zones are typically cones of lavas and fragmental materials built around a central conduit, with an apical crater. They thus tend to conform to the popular idea of what a typical volcano should look like – celebrated examples include Fujiyama (Japan), Mt. Egmont (New Zealand), Popocatapetl (Mexico) and Mayon (Philippines, Figure 13.2).

Volcanoes overlying subducting slabs near destructive plate margins occur abundantly around the Pacific Basin giving rise to the term 'the Pacific ring of fire'. In the north the volcanoes constitute the Aleutian Islands and continue east into the Alaskan Peninsula. A volcanic chain passes south through British Columbia and into the Cascade Mountains of the American west. Volcanism resumes in Mexico (Figure 23.64), persisting southwards through Central America and on throughout much of the Andean chain. An associated volcanic arc, convex to the east over a westward-diving slab, gives the volcanic islands of the Lesser Antilles.

A somewhat similar arc, convex away from the

Figure 13.16 The Kenya rift east of Silali volcano. In the foreground is an open fissure related to dyke intrusion around Silali. In the background can be seen lava flows and recent cinder cones and, beyond, the eastern escarpment of the rift. (*B.G.J. Upton*)

Pacific, is seen to the south of Cape Horn, forming the volcanic islands of the South Sandwich group. The 'ring of fire' continues in the Antarctic Peninsula and, proceeding clockwise, into New Zealand and the Kermadec–Tonga chain. Around the south-west and western Pacific there are complex patterns of volcanic arcs, mostly convex towards the ocean. With the volcanoes of Japan and the Kurile island arc, the ring is brought full circle back to the Aleutians.

In the vicinity of Indonesia a major chain of supra-subduction volcanoes branches off to the west, through Java and Sumatra passing northwards into more sporadic volcanism off the Burmese coast. This volcanic chain can be traced westwards in a very discontinuous manner. It is, however, manifest in the Elbruz mountains of Iran and the Caucasus mountains of Asia Minor. This intermittent volcanic chain appears still further west in the Aegean arc (e.g. the Santorini (Thera) volcano) and in volcanoes related to subduction processes in southern Italy, e.g. those of the Aeolian Islands (Stromboli, Lipari and Vulcano).

Apart from these principal associations of volcanism with constructive and destructive plate margins, volcanoes occur along, or in the vicinity of, rifted continental zones. Of these, the most striking examples are related to the East African rifts (Chapter 29). The Ethiopian rift, which joins the

constructive (oceanic ridge) plate margins of the Red Sea and Gulf of Aden in a 'triple junction' (Chapter 28) has many recent volcanoes including the active Erta 'Alé volcano. Further south the rift diverges into an eastern and a western system (Chapter 29), both of which are host to Recent volcanoes including some that have been active in historic times. These latter include Nyiragongo in the western rift and Oldoinyo Lengai in the Tanzanian sector of the eastern rift (Chapter 29). Towards the end of the nineteenth century, the Hungarian explorer, Teleki, reported an eruption in northern Kenya, within the Barrier massif at the south end of Lake Turkana (Figure 13.15). Kilimanjaro (5895 m), the highest mountain in Africa is a Recent volcano on the flanks of the eastern rift system. Mt. Kenya (5200 m) is a relatively deeply-eroded rift flank volcano whose former height must have rivalled that of Kilimanjaro. These two volcanoes, and their relationship to the Kenyan rift and to other late Tertiary to Recent volcanoes of the area, are also shown in Figure 13.15. A view of volcanic scenery in the Kenyan rift near Silali volcano, with a fault-scarp and open fissure associated with the volcanism, is presented in Figure 13.16.

Intraplate volcanoes are those occurring either on the continents or the oceans well away from any plate boundary. In Europe, intraplate volcanoes that have been active within the last few thousand years can be found in central France (in the Massif Central, Figure 13.17) and in Germany in the Eifel region.

Some of the most spectacular examples of continental intraplate volcanoes occur in Africa, e.g. in the Tibesti and Hoggar mountains of Chad and Algeria and in the great chain of volcanoes extending from the continental interior south-west out into the Gulf of Guinea, called the Cameroon Line. According to whether one regards the volcanoes associated with the East African Rift System as related to a constructional plate-boundary or as intraplate depends on whether or not the system is to be regarded as a precursor to an oceanic spreading centre (see Section 29.6). The Cameroon Line extends into the Atlantic as a chain of oceanic intraplate volcanoes.

Among other oceanic examples, the greatest are those forming the Hawaiian Islands (Figure 13.18). The main island of Hawaii comprises five large volcanoes (Figure 13.19), three of which are 'active' (although of these, Hualalai, has not erupted since 1801). The other two, Mauna Loa and Kilauea, jointly represent the planet's most productive site of magma generation and emission, with an average eruption rate of about 1.6 km³/year. Kilauea is the most intensively studied active volcano and much of what is known concerning processes in basaltic volcanoes has accrued from research on Kilauea. Figure 13.20 presents a schematic true-scale cross-section of the volcano showing the inferred regions of melting and magma collection, together with the shallow-level magma reservoir from which the surface eruptions are supplied. Hawaii is believed to overlie a rising column of convecting asthenos-pheric mantle (Figure 13.21). The column or plume is thought to ascend at up to 340 mm a year. As the hot, solid, mantle plume ascends to low-pressure zones, partial melting commences, accounting for some 6–7 per cent of the mantle material, with magma then continuing to ascend through the oceanic lithosphere.

Hawaii is a classic example of a **hot-spot** volcano (or more precisely, volcanic group). The oceanic lithosphere through which the magmas rise and upon which the volcanoes grow, is in motion away from the constructional plate-boundary (the East Pacific Rise) whereas the mantle plume has a position that remains essentially static. Consequently, the volcanoes are progressively carried 'down-stream', borne on the spreading ocean floor. Periodically the magma supply from the underlying mantle source ('mantle-plume') is cut off as the volcano is transported away from the melting site and a new volcano begins to grow over the hot-spot (Figure 13.22). Each of the huge basaltic shields intermittently formed above the plume may take approximately 400 000 years to grow.

By this process great chains of volcanoes can develop, the volcanoes passing into extinction and becoming progressively older down-stream from the hot-spot focus. The Hawaiian Islands provide the type example, with the various volcanic islands becoming increasingly old and eroded as traced west-north-west from the active (main) island, i.e. as traced diametrically away from the East Pacific Rise. The earliest activity, that can be attributed to the mantle plume, commenced some 75–80 Ma ago. Embayed, eroded volcanic islands give way to a chain of sea stacks, atolls and seamounts (Figure 13.18). Whereas most of the

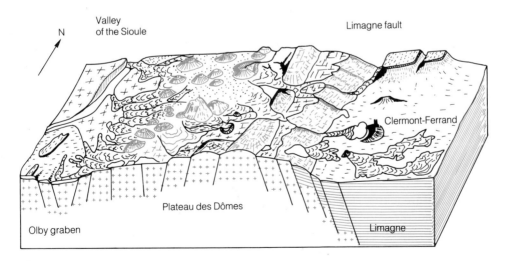

Figure 13.17 Block diagram illustrating the late Tertiary to Recent intraplate volcanoes of the Chaine des Puys in the Auvergne, central France. These volcanoes, including Puy de Dôme, (the highest), are built upon the granitic highlands of the Plateau des Dômes, west of and adjacent to, the Limagne rift basin. (*After G. Camus et al, 1975*)

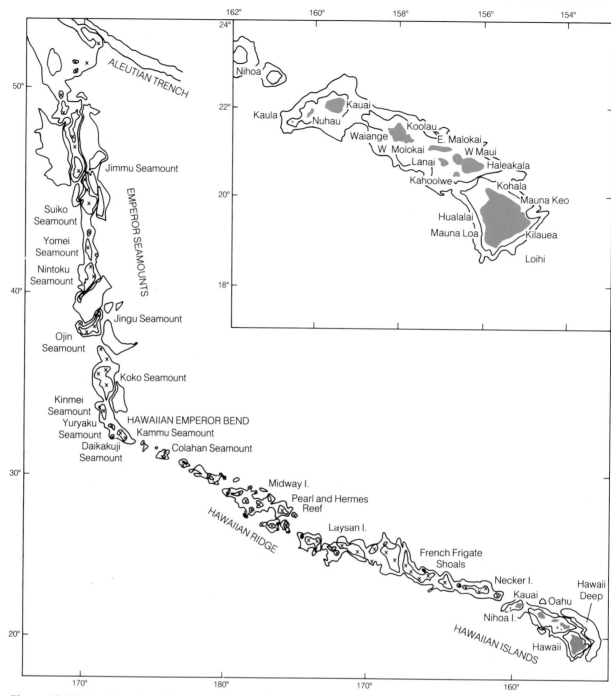

Figure 13.18 The Hawaiian–Emperor volcanic chain in the north-central Pacific Ocean. The bathymetry of the principal Hawaiian Islands is shown in the inset. The distinct bend in the chain, separating the Emperor Seamounts from the Hawaiian Ridge, is attributed to a change in the sea-floor spreading pattern at about 42 Ma ago. Contour interval 1 km. (*After Chase et al, 1970*)

volcanic ridge is now submarine, the component volcanoes are likely to have risen above sea-level during their more mature active stages. Approximately 42 Ma ago, the direction of ocean floor spreading changed, from more nearly north-west to approximately west-north-west. Consequently the great chain of volcanoes represented by the Hawaiian Islands and seamounts shows a deflection or dog's leg; the older part trends north-west towards the north-west Pacific, forming the Emperor seamounts. The whole Hawaiian–Emperor volcanic lineament, which extends for some 6000 km, represents an immense outpouring of over one million km³ of basaltic lavas. Such a feature is referred to as a **hot-spot trace**. The Pacific Ocean contains other notable hot-spot traces such as those of the Line Islands, Tuamoto Archipelago, Marshall and Ellice Island chain and

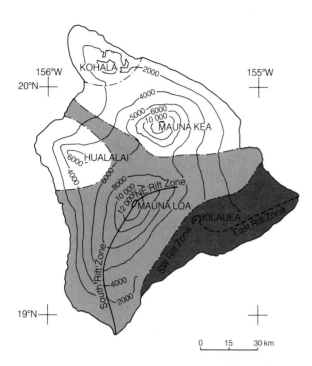

Figure 13.19 Sketch map showing the five volcanoes that make up the island of Hawaii. Contour interval, c. 615 m. Light area, underlain chiefly by lava flows from Mauna Loa; darker area, underlain chiefly by lava flows from Kilauea. (*After T.L. Wright, 1971*)

the Austral Seamount chain, all with the youngest volcanoes at their easterly extremities (Figure 13.23).

In the North Atlantic region another mantle plume is believed to have been initiated in the early Tertiary (at around 55–60 Ma). The early melting products from this plume include the East Greenland flood basalts: Iceland is located over the present site of upwelling. This situation is exceptional and differs from the Pacific examples cited above in that the plume lies more or less directly beneath a mid-ocean ridge axis.

Another prominent hot-spot trace is to be found in the Indian Ocean. During the late Cretaceous a mantle plume was initiated beneath the great southern continent of Gondwana. The extensive lava sequences composing the Deccan region of India resulted from shallow-level decompression melting of this rising mantle material (Chapter 5). The plume is thought to have remained operational during the breakup of Gondwana, creating a hot-spot trace across the western Indian Ocean. Currently the plume axis is believed to lie in the vicinity of Reunion Island, with its melt products feeding the Piton de la Fournaise volcano. While Hawaii, Iceland and Reunion represent some of the most productive volcanic areas thought to be fed by magma from deep mantle plumes, many other intraplate hot-spot volcanoes are also

inferred to mark the sites of rising plumes.

Plume traces are not confined to the oceans. The basaltic lavas of the Snake River plain in Idaho become younger as traced south-eastwards towards the recently active volcanic region of Yellowstone National Park and it has been postulated that a mantle plume, now situated beneath the Yellowstone region, may have been responsible for the Snake River volcanics as the North American plate drifted north-westwards. In some continental areas uplift and erosion has revealed intrusive ring-complexes developed beneath former intraplate volcanoes. Sometimes, as in Namibia and Nigeria these are arranged in chains or elongate zones, with progressive age change from one end to the other. Such phenomena are now recognized as plume traces marking the passage of continental lithosphere across former mantle plumes.

13.3 PRODUCTS OF VOLCANOES

Volcanoes release energy from the Earth's interior. This is overwhelmingly in the form of heat although this may be accompanied to varying extents by light and by seismic tremors and atmospheric shock waves. The more tangible products of volcanoes can conveniently be considered under three headings: (a) gases, (b) lavas and (c) fragmental (pyroclastic) deposits.

13.4 GASES

Gases are invariably emitted during volcanic eruptions. Indeed, the clearest characteristic of many eruptions is the column of ash-laden gas rising above them. The gases can be released with immense energy, driving columns of ash tens of kilometres high. Velocities of up to 600 m/sec have been calculated for the gases leaving the vent in some historic eruptions. Whereas much of the gas represents juvenile matter liberated from the mantle and held in solution in the magma until released at low pressure, there is almost always admixture of potentially gaseous material derived from meteoric and ground waters and/or the atmosphere. Consequently the acquisition of precise and meaningful analyses of magmatic gas is fraught with difficulty. However, the best sampling procedures have shown that H_2O and CO_2 are by far the major species. Whereas for most volcanic emissions, water composes 90 per cent or more of the gas, in some (e.g. some of the African rift volcanoes like Nyiragongo and Oldoinyo Lengai) CO_2 appears to be the dominant component. In continental rift-valley environments (of which

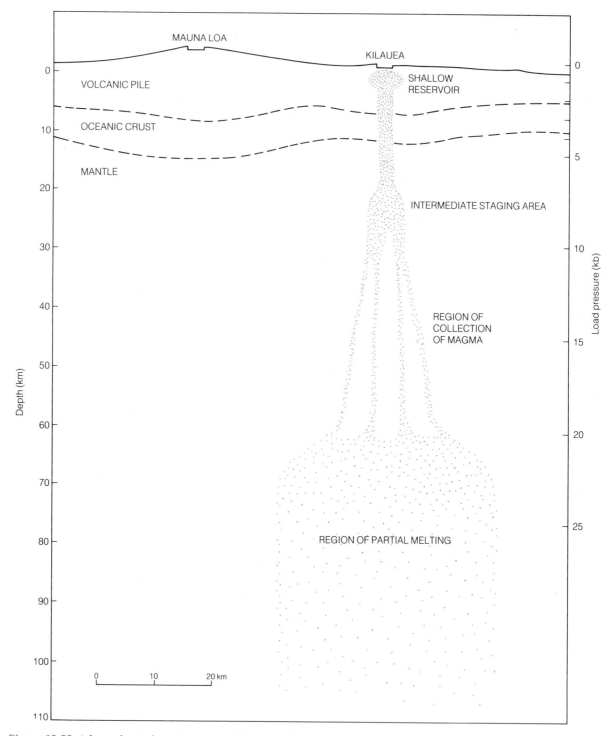

Figure 13.20 A hypothetical section across Kilauea and Mauna Loa volcanoes, Hawaii, outlining regions of melting, and storage of magma prior to eruption. (*After T.N. Wright, 1971*)

more in Chapter 29) CO_2-degassing appears to be a critically important process. Outflow of this gas is so abundant in parts of the Kenyan rift that it can be exploited in commercial wells.

SO_2, HCl, H_2 and H_2S plus He, and other noble gases, together with those metallic halides and sulphides that are volatile at magmatic temperatures, are generally present as subsidiary constituents. The actual gas compositions are, no doubt, highly variable and dependent on the precise geochemistry of the associated magma. The gases from Kilauea, Hawaii, have approximately 80 per cent H_2O with subordinate SO_2, CO_2 and H_2, less than 1 per cent CO, H_2S and HF and trace amounts of Hg, Cu and other metals.

The less volatile components of the gases may be precipitated as sublimates in and around volcanic vents. Yellow crystals of sulphur grown

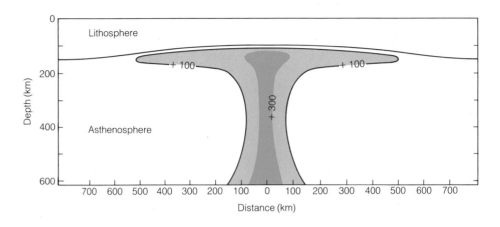

Figure 13.21 Cross-section through a mantle plume of the sort believed to underlie for example Hawaii, Iceland and Réunion, showing isotherms at 100 and 300°C above mean asthenosphere temperature. (*After R.S. White and D.P. McKenzie, 1989*)

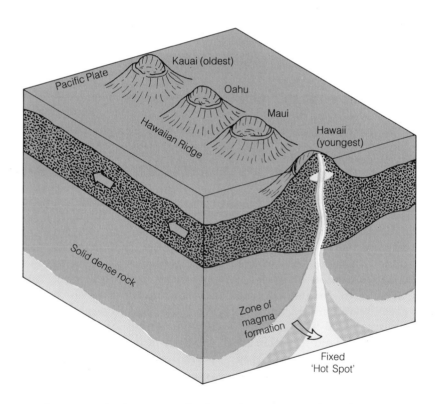

Figure 13.22 Diagram illustrating the formation of volcanic lineaments such as the Hawaii–Emperor chain in the Pacific Ocean. Lithosphere of the Pacific Plate moves north-westwards across an essentially stationary zone of magma formation ('hot-spot') produced by an upwelling mantle plume. An active volcanic centre overlies the 'hot-spot' while older volcanoes are drifted 'downstream', becoming extinct and progressively more eroded. (*After M. Krafft*)

around high-temperature gas vents (**fumaroles**) provide a common example although various chlorides and sulphates are also frequently represented within volcanic sublimate deposits.

Basaltic magmas tend to be relatively poor in volatile components and dissolved H_2O contents are typically less than 1 per cent by weight. At the same time basaltic magmas are relatively fluid and gas exsolution usually takes place in a tranquil fashion compared to gas separation from more

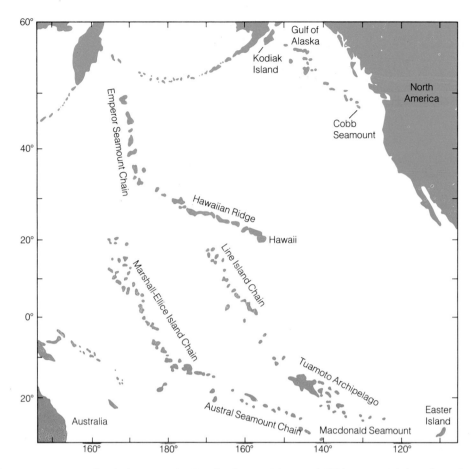

Figure 13.23 Seamount and island chains in the Pacific Ocean. Morgan (1972) proposed that the volcanic chains of the central and south-western Pacific were formed by motion of the Pacific plate over four melting spots, each now located at the south-east extremities of the chains.

viscous magmas. Nonetheless, the formation of gas bubbles (**vesicles**) and their rapid expansion in the magma conduit near the Earth's surface, can be sufficiently vigorous to cause spectacular fountaining. Basalt magma fountains at Kilauea (Hawaii) in the 1959–1960 eruption for instance, attained heights of over 300 m. In historic times huge quantities of gases were emitted relatively quiescently during the great 1783 basaltic fissure eruption at Laki, south-east Iceland to which reference has already been made. The gases produced atmospheric pollution and an acid haze so severe that in the resulting famines following loss of crops and livestock, an estimated 20 per cent of the population of Iceland died.

Some of the components of volcanic gas (e.g. SO_2 and H_2S) are susceptible to oxidation and hydration and, if erupted into the stratosphere, can produce relatively long-lived volcanic aerosols involving sulphuric acid droplets. These can block solar radiation and can be responsible for climatic cooling. Such high-altitude acid particles are eventually precipitated as acid rain or snow. Estimates of the quantities of gas emitted in individual eruptions can now be obtained by study of the acidic layers in the polar ice-sheets that correspond to specific large-scale eruptions that gave global dispersion to such volcanic particles. Despite the comparatively small proportion of sulphur compounds in volcanic gases, the total quantities in absolute terms can be large. Thus the major eruption of El Chichon in Mexico, in early April, 1982 is thought to have sent some 3.3 million tonnes of SO_2 into the stratosphere. El Chichon is thought to have lowered global temperatures significantly for several years. More recently an eruption in the Philippines (Pinatubo volcano) has overshadowed the El Chichon event in terms of SO_2 output, with Pinatubo throwing out some two and a half times as much as El Chichon. It has been speculated that Pinatubo may lower temperatures by around 0.5°C, temporarily counteracting at least a century of global warming due to man-made pollution. The annual flux of SO_2 from volcanoes globally has been estimated at 18.7 million tonnes. It has been suggested that eruptions may be divided into two groups on the basis of their volatile compositions, expressing S and Cl in terms of H_2SO_4:HCl ratios, with wide variations from > 20 to < 3.

By contrast to basaltic magma, highly siliceous (rhyolitic) magma may contain much greater contents of dissolved volatiles (commonly over 10 per cent weight H_2O). The high silica content promotes polymerization and high viscosities. Viscosities increase still further as the gases are lost in near-surface environments. Consequently separation of gas bubbles from the magma tends to be violent and explosive with the viscous silicate liquid being torn to shreds in the process. Whereas the degassed rhyolite may be extruded as a lava flow, it is more commonly erupted in a fragmental state through the explosive loss of gases.

Figure 13.24 shows diagrammatically the succession of events from a magma chamber at depth with dissolved gases, to stages where gas separation occurs to form vesicles, and thence to a stage where gas coalesces to form the dispersed phase and the silicate liquid becomes fragmented.

13.5 LAVAS

Lavas represent magmas that have already lost the greater part of their dissolved volatile materials. The liquids congeal fast (relative to large intrusions) as they lose heat and volatiles, forming rocks that are typically finely crystalline or even glassy. If crystallization was proceeding at depth prior to, or during, the eruptive event, the early grown crystals (**phenocrysts**) will be seen as larger entities, occasionally more than 1 cm across, enveloped in a fine-grained matrix. Such rocks are said to be **porphyritic** (an example is illustrated in Figure 13.25). With increasing phenocryst content the viscosity of the lava rises rapidly.

Because of the viscosity of the lavas a proportion (often a high proportion) of the vesicles will remain trapped as the lava cools, and if gas separation persists until the flow has ceased movement, spherical vesicles will be preserved (Figure 13.26). If, however, vesicle growth terminates before cessation of flow, the vesicles will become distorted into ovoid, pear-shaped or thoroughly irregular cavities. The faster-cooled tops and bottoms of flows are typically more vesicular than the centres which, cooling more slowly, may have lost most of their gas bubbles through flotation. Elongate, tubular pipe-vesicles may occur at the base of flows, extending upwards several centimetres or decimetres. These develop where rapid bubble nucleation occurs over favourable points and where the liquid viscosity is such that the bubbles show confluence into a tube-like cavity. As lavas become buried in accumulating volcanic sequences, new minerals become deposited in the vesicles by circulating aqueous solutions, often at temperatures of several hundred degrees. The filled vesicles are referred to as **amygdales** (Figure 13.27).

Lavas, especially basaltic lavas, are frequently erupted in a highly mobile state and congeal with comparatively smooth, undulatory, upper surfaces (Figure 13.28). The congealing crust of the flow, some centimetres thick, may be deformed by rapidly moving hot lava immediately beneath, causing the stiff crust to be thrown into folds and wrinkles. The resulting surface may resemble folded cloth or coiled rope (Figure 13.29). The term usually applied to such relatively smooth-topped flows, with or without corded or 'ropy' surfaces, is the Hawaiian word, **pahoehoe**.

In pahoehoe flows, the principal lava conduits lie within the interior of the flow, as tubular

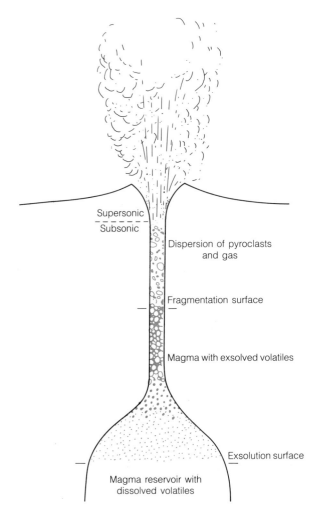

Supersonic
Subsonic

Dispersion of pyroclasts and gas

Fragmentation surface

Magma with exsolved volatiles

Exsolution surface

Magma reservoir with dissolved volatiles

Figure 13.24 Schematic diagram illustrating the release of gas from magma. At the 'exsolution surface' gas begins to come out of solution in the magma, forming bubbles ('vesicles'). At a higher level ('fragmentation surface'), with falling pressure, expanding vesicles coalesce to form a continuous gas matrix containing suspended molten silicate particles. (*After Wilson et al, 1980*)

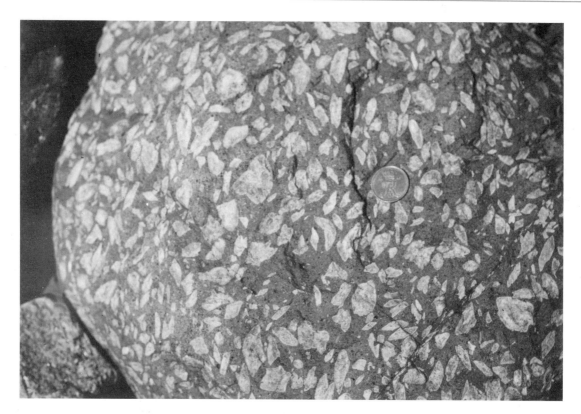

Figure 13.25 A striking porphyritic lava containing rhomboidal phenocrysts of alkali feldspar in a fast-cooled, fine-grained matrix. Surface of a trachytic 'rhomb porphyry' lava from the Permian succession at Tyrijord in south Norway. (*B.G.J. Upton*)

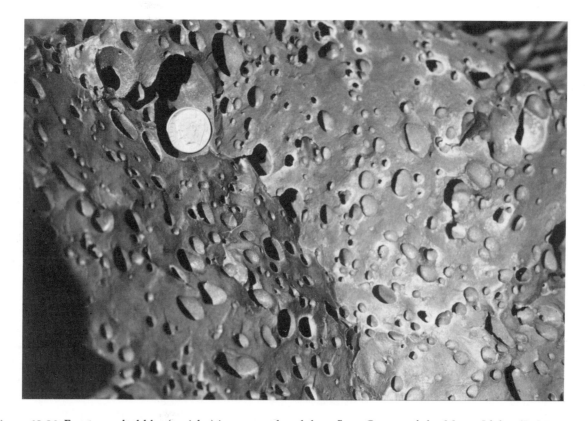

Figure 13.26 Empty gas bubbles (vesicles) in a recent basalt lava flow. Craters-of-the-Moon, Idaho. (*B.G.J. Upton*)

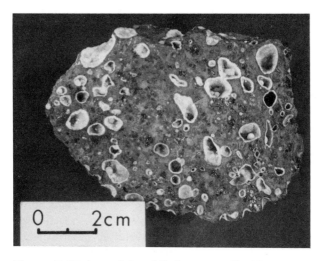

Figure 13.27 Amygdales; filled, or partially filled, vesicles in a basalt lava sample from Réunion, Indian Ocean. The amygdale mineral is chabazite, deposited from hydrothermal solutions. (*B.G.J. Upton*)

as bulbous (toe or finger-like) prolongations. The new-formed crust on these toe-like buds deforms in a ductile manner as they inflate with new lava before cracking or bursting to permit a new pahoehoe toe to form. The lava flow thus advances by continuous production of new inflating toes.

If the supply of new lava to a tube within a pahoehoe flow is stopped, the remaining lava in the tube may be drained off at the distal end, leaving an empty tube. Such lava tubes are common occurrences. Extensive systems of lava tubes are known from many recent volcanic terrains, and can sometimes be followed for many kilometres. The smooth tube walls are often lined with rapidly congealed glass; lava dripping from the roof and sides of the tubes may be preserved as stalactite-like appendages.

Lavas that contrast markedly with pahoehoe are those with ragged, sharp irregular surfaces. These too are generally called by their Hawaiian (bi-syllabic) name **aa**. Aa flows tend to be more viscous and of lower temperature than pahoehoe, to move more slowly and to form thicker flows. Aa flows generally show an open surface 'river' of active lava, confined by ramparts or levees of congealed lava, which supplies the advancing front, sometimes by a series of distributaries, rather in the manner of a delta. Clinker-like blocks

throughways that may be metres or tens of metres in diameter. These tubes supply high temperature (typically in the range 1200–1100°C) incandescent lava to the advancing flow front. The tubular flow reduces heat loss and permits the lava to advance many kilometres or tens of kilometres from the vent. New lava reaching the front extends ahead

Figure 13.28 The ridged and hummocky surface of a pahoehoe basalt flow on Kilauea volcano, Hawaii. (*W.J. Wadsworth*)

of lava tumble forward down the steep frontal slope (Figure 13.30 and Plate IXa). As blocks detach and fall, transient views can be obtained of the incandescent interior. The front of an advancing aa flow can resemble a mass of rubble being bulldozed. Whereas pahoehoe flows tend to be highly vesicular, with roughly spherical vesicles constituting anything from a fifth to a half of the lava, aa flows, particularly the interior parts, are relatively massive and vesicle-poor. Vesicles in aa are more likely to be irregular in shape than those of pahoehoe.

Pahoehoe flows may convert to aa type over a matter of metres when traced away from the vent. The reverse situation, that of aa changing downslope to pahoehoe, has never been observed. The factors controlling whether a flow moves as pahoehoe or aa are complex and by no means do all pahoehoe flows change distally to aa. The factors appear to involve a critical relationship involving both viscosity and the amount of internal disturbance due to flow. For similar rates of emission, pahoehoe flows can travel much further than aa flows.

As lavas cool, contraction causes cracks or joints to appear. Although these may develop in rather irregular patterns, in thick flows (such as those ponded in a crater or blocked valley) which can have cooling histories extending into tens of years, very regular cooling joints can be formed. The joints propagate perpendicular to the cooling surfaces. Low rates of heat loss can lead to regular spacing of joints that intersect roughly at 120° angles and thus define crudely hexagonal columns. Sections through some thick flows show a layered or tiered arrangement with a well-developed suite of regular columns in its lower part (forming a '**colonnade**'), overlain by a much more irregularly-jointed portion referred to as 'the **entablature**' (Figures 13.31 and 13.32).

As has been pointed out, the oceans (and hence the greater part of the Earth's surface) are floored by basaltic lavas, mainly erupted from fissure volcanoes along the mid-ocean ridges but also from growing intraplate seamounts. Most typically the lavas occur as '**pillow lavas**', with rounded, lobate (pillow-like) masses whose size ranges from a few centimetres to a few metres in diameter (Figure 13.33 and Plate IXb). The individual pillows possess shiny black crusts of basaltic glass where they have been abruptly quenched by contact with seawater. These glassy crusts typically surround finely crystalline interiors and thermal contraction leads to growth of radially-disposed joints.

Until recently pillow lavas were widely consid-

Figure 13.29 Pahoehoe flow of basalt lava showing corded or ropy surface features. Krafla volcano, northern Iceland. (*B.G.J. Upton*)

Figure 13.30 The jagged front of an aa lava flow. Hekla, Iceland, 1980. (*B.G.J. Upton*)

Figure 13.31 A sequence of jointed basalt lava flows in the early Tertiary succession of west Greenland. In each flow regular sets of thicker columns characterize the lower parts ('the colonnades'). These are overlain by more chaotically oriented sets of thinner columns, constituting 'the entablature'. These irregular upper portions developed in more complex cooling regimes, probably influenced by surface waters penetrating the tops of the flows. (*A. K. Pedersen, published by permission of the Geological Survey of Greenland*)

Figure 13.32 Polygonal jointing in 'the collonade' portion of a thick basalt lava flow. Giant's Causeway, Antrim. Cracks nucleate at points on surfaces of uniform temperature (isothermal surfaces). 3-pronged cracks at ~ 120° tend to develop, intersecting to form polygonal columns, often approaching regular hexagons. (*B.G.J. Upton*)

Figure 13.33 Pillow lavas, with intervening palagonitic tuff (hyaloclastite). Stappafell, Reykjanes, south-west Iceland. The pen-knife for scale is about 7 cm long. (*B.G.J. Upton*)

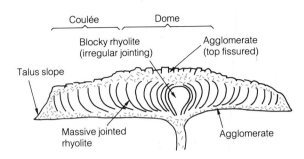

Coulée   Dome

Blocky rhyolite   Agglomerate
(irregular jointing)   (top fissured)

Talus slope

Massive jointed   Agglomerate
rhyolite

Figure 13.34 Generalized cross-section of a rhyolite dome from evidence present on Mount Tarawera, New Zealand. (*After B.N. Thomson et al, 1965*)

ered to consist of discrete separated blobs (pillows!) of lava produced by surface-tension forces interacting between water and molten silicate. However, it now appears probable that most (if not all) pillow lavas represent an extreme case of pahoehoe where progression has been achieved by protrusion of tubular toes, with very pronounced marginal chilling. Thus, apparently separate pillows may really be sections through interconnected accumulations of 'toes'. The lava pillows are often closely associated with accumulations of spalled, broken particles of glassy lava. These '**hyaloclastite**' deposits will be discussed in a later section.

Deep drilling into the ocean floors has shown that not all the sea-bottom lavas are of pillow type and that extensive non-pillowed sheet-like lavas are also present. It is inferred that high rates of emission are required to produce these more extensive (pahoehoe) sheet-flows whereas lower rates give rise to the pillow lavas where the lobate protrusions are individually quenched and consequently preserved as pillows.

Lavas exhibiting features that can be related to modern successions can be found back through the geological column to the oldest formations that still survive. However, in the oldest (Archaean) rocks of the Precambrian, there occur lavas which possess special features that distinguish them from most younger (Proterozoic) Precambrian lavas and from virtually all subsequent (Phanerozoic) lavas. These remarkable ancient lavas were compositionally distinctive in having magnesia (MgO) contents which extended to much higher values than those generally found in younger lavas. Experimental studies have shown that such high-magnesia lavas (called **komatiites** after the Komati River in Swaziland from whose vicinity they were first described) must have been erupted at temperatures in the range 1700–1400°C, i.e. exceptionally high relative to modern basalt eruptions where temperatures rarely exceed 1200°C. In general there is a positive correlation between eruption temperature and the MgO content of the

liquid portion of the lava. Not only were the komatiites distinguished by remarkable compositions and eruption temperatures but they are inferred to have had exceptionally low viscosities (10–100 poises compared to values of over 5000 poises for more 'normal' basalts). With such mobility the manner of flow of these Archaean lavas would have been significantly different from those of less magnesian basalts. The komatiites on eruption would have flowed as Newtonian liquids (in which infinitely small stresses result in deformation) and exhibited 'turbulent flow' in contrast to the 'laminar flow' of cooler and less-magnesian basalts. Turbulent flow would have permitted highly efficient transfer of heat to the flow surfaces and, in consequence may have led to dramatically rapid cooling. Textural evidence that komatiites did lose heat very rapidly is to be found in the sprays and bundles of needle-like and thinly tabular pyroxenes and olivines that grew as the flows congealed. Whereas 'normal' lavas cause little or no melting of the underlying rocks komatiites could melt and thermally erode a wide variety of the rocks over which they flowed, producing incised lava channels.

The discussion of lavas in this section has been principally concerned with basalts (and komatiites), volumetrically the most abundant lavas erupted throughout Earth history. The majority of other magmatic compositions are less magnesian (and are erupted at lower temperatures) and more siliceous (and, largely because of this, more viscous). Additionally they tend to hold higher contents of potentially volatile components in solution so that, on eruption, pyroclastic products are volumetrically important or even dominant. This is particularly so in the case of the most siliceous varieties, the **rhyolites**. **Andesites**, which are compositionally intermediate between basalts and rhyolites, are abundantly represented among the extrusive products of the supra-subduction zone volcanoes. Andesitic pyroclastic deposits are comparatively abundant in association with andesite lavas and the latter are generally more viscous than basalt lavas and are commonly of aa rather than pahoehoe type.

Where the more siliceous categories (which include dacites and, more rarely trachytes and phonolites, as well as rhyolites) are erupted as lavas, they tend to be already extensively degassed, to be of comparatively small volume and to have high viscosities. Thus rhyolite lavas seldom flow for more than a few kilometres from a vent and rarely have volumes greater than 1 km³. There are, however, exceptions to all general rules and the Chao dacite flow in Chile is remarkable for its size; it flowed 12 km from its source and has a volume of about 24 km³. Such stiff, highly viscous, lavas are commonly extruded to form a thick **dome**, i.e. a lava whose length to thickness ratio may be 5 : 1 or even less (Figure 13.34). A lava dome may be extruded more or less symmetrically from a vent so that it has a roughly circular ground plan with little tendency to extend as a typical flow. The lava may even be so viscous that it can be expelled to form an upstanding pinnacle or spine. A remarkable example of such a spine, some 300 m high, was produced as an ephemeral feature, squeezed from the summit vent of Mt. Pelee on the Caribbean Island of Martinique in 1902 following catastrophic pyroclastic eruptions (see below).

Due to their high viscosities and the consequent low rates of diffusion of ions to the potential sites of crystal nucleation, the more siliceous lavas (and hypabyssal intrusions) commonly congeal as black glassy obsidian rather than as crystalline rock. The obsidian however, may subsequently undergo crystallization (devitrification) after flowage has ceased, particularly when this is facilitated by the passage of aqueous vapours or solutions. Devitrification typically takes place around point nucleii, producing radiating clusters of small crystals in roughly spherical bodies. These **spherulites** can occasionally attain sizes of several centimetres. Trachytes and phonolites are relatively siliceous (55–65 per cent by weight SiO_2) magma types that are characterized by high alkali : silica ratios. These alkaline magmas are mainly erupted, in small volume, from large oceanic and also intraplate and rift-related continental volcanoes. Again, lava flows are usually restricted in volume and extent and dome-like emissions are typical (Figures 13.35 and 13.36). However, the high alkali content, exceptionally combined with high content of halogens (especially fluorine and chlorine) confers low viscosity and highly mobile flows can then spread over wide areas. While such eruptions have never been witnessed, the extensive 'flood trachytes and phonolites' of Kenya, erupted in the Pliocene, give testimony to these events in the past. It is possible that they were erupted from fissure rather than central-vent volcanoes similar to the manner of flood basalts. Major dyke swarms involving trachytes and phonolites, in a relatively deeply-eroded continental rift environment in south Greenland may have supplied comparable 'flood trachyte and phonolite' lavas in the mid-Proterozoic.

Among modern lavas by far the most fluid ever observed are the remarkable pahoehoe lavas of Oldoinyo Lengai in Tanzania (Section 29.12). Rather than being silicate-dominated like most magmatic compositions, they are composed mainly of carbonates of sodium and potassium. These lavas, characterized also by high fluorine

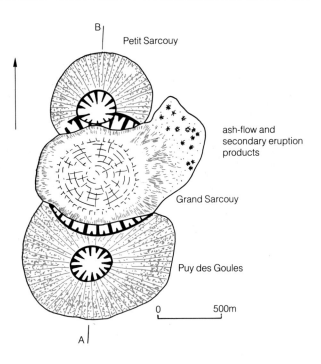

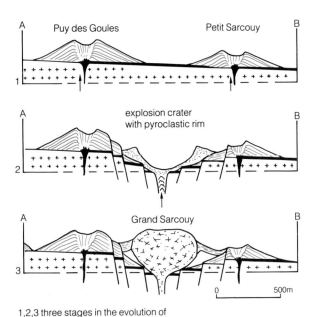

1,2,3 three stages in the evolution of the Sarcouy volcano

Figure 13.35 Geological map and set of cross-sections illustrating volcanic evolution at Sarcouy, Auvergne, France. Early ash cone development was followed by a phreatomagmatic explosive event, followed in turn by formation of a large dome of trachytic lava. (*G. Camus et al, 1975*)

and chlorine contents, are mobile down to temperatures of around 500° (several hundred degrees below the temperature at which most lavas solidify), with viscosities so low that their flowage behaviour resembles that of water more than that of basalt (Figure 13.37).

Figure 13.36 Photograph of the trachytic dome of Puy Grand Sarcouy, Auvergne, France (*see* Fig. 13.35). The morphology at this Recent dome has been essentially unmodified by erosion.

13.6 PYROCLASTIC DEPOSITS

The violent release of gas from volcanic vents is associated with expulsion of fragments of rock and/or disrupted lava (Figure **13.38**). The accumulation of such fragments gives rise to a pyroclastic (literally, 'fire-broken') deposit. Solid fragments torn from the vent walls constitute accidental or lithic clasts and are characteristically angular. Fragments of new lava (juvenile magmatic material), erupted while still molten, will tend to have shapes dictated by surface-tension forces, modified by stresses imposed while airborne, or

acquired on impact or compaction. The new magmatic fragments are consequently commonly spherical, ovoid or lenticular. They are also characteristically more or less vesicular.

Numerous terms are employed to denote different types of pyroclast. '**Cinders**' and '**ashes**' are terms originating from older hypotheses concerning subterranean combustion. 'Scoria' is used for pieces of highly vesicular basalt and andesite. The very low-density and usually white or pale pastel-coloured vesicle-rich fragments derived from dacitic, rhyolitic, trachytic or phonolitic lava, are called **pumice**. Larger volcanic juvenile projectiles,

Figure 13.37 'Rivers' of highly mobile carbonatite lava in the crater of Oldoinyo Lengai volcano, Tanzania. The active lava stream is incised into slightly older pahoehoe flows exhibiting 'ropy' texture. (*J. Keller*) (*See also Plate IXf and Section 29.17*)

Figure 13.38 A tall column of ash-laden gas rising above an eruption of Kliuchevsky volcano, Kamchatka, Russia.

(defined by international convention as those with average size of more than 64 mm), may be termed volcanic bombs (Figure 13.39). The latter often have a spindle-shape acquired while spinning in the air (Figure 13.40). Lapilli denotes smaller (2–64 mm) subspherical particles whereas pyroclasts < 2 mm are designated as ash. **Accretionary lapilli** is the name given to concentrically-zoned pellets formed by sequential adherence of fine ash particles around a nucleus, which may be a water droplet or a solid particle. Commonly these zoned, hailstone-like lapilli form in the steam-rich columns generated when surface waters come into contact with erupting lavas. An outcrop of ancient accretionary lapilli from the Lake District of northern England is shown in Figure 13.41. The term **tephra** (a Greek word meaning ash) can be employed to describe all of the assorted fragments blown out during explosive eruption, regardless of size. Tuff is another useful term referring to all consolidated pyroclastic deposits, again regardless of particle size.

Pyroclastic deposits are normally thickest and composed of the coarsest materials, within or close to, the eruptive vent and to thin and become progressively finer-grained distally. Coarse (particle size average over 64 mm), unsorted proximal accumulations, often within vents, are described as **agglomerates** (Figure 13.42). The feeder pipes to central vent volcanoes can become choked with

Figure 13.39 Radial joint pattern in broken volcanic bomb, Vulcano, (Aeolian Islands). (*B.G.J. Upton*)

Figure 13.40 An ovoid volcanic bomb. The shape was produced while a blob of liquid magma was still air-borne during an explosive eruption, Piton des Neiges, Réunion. (*W.J. Wadsworth*)

Figure 13.41 Accretionary lapilli, Ordovician (Borrowdale) Volcanics, Lake District, England. (*B.G.J. Upton*)

accidental and/or juvenile clasts forming an agglomeratic neck or 'tuff pipe' (see Section 12.4 and Figure 12.26). Such tuff pipes may result from steam-generated explosions following an encounter by rising magma with wet sediments or standing water, or may originate from explosive separation of gas from magma that is inherently volatile-rich. **Kimberlite**, produced by very small-degree mantle melting deep below thick and ancient continental lithosphere (Sections 5.2, 29.12) is one of a number of rare magma compositions that characteristically forms tuff pipes at

Figure 13.42 Coarse, unsorted agglomerate, Pico del Teide volcano, Tenerife, Canary Islands. (*B.G.J. Upton*)

relatively shallow crustal levels. The fragmental fillings of tuff pipes may be complex, incorporating material that has collapsed into the pipe from surface or near-surface levels, as well as juvenile magmatic fragments and wall-rock pieces broken off at deep levels and swept up by the fast-rising magma. Diamonds are among the assorted deep-source produce of kimberlitic tuff pipes, derived from depths of over 120 km (Figure 29.23). A cross-section through a tuff pipe generated by another gas-rich magma type related to kimberlite is shown in Figure 13.43. Such tuff pipes, if followed down to sufficient depth in the crust are found to originate from 'normal' unfragmented intrusions such as dykes or sills.

In the simplest case, vertical discharge of gas above a central vent volcano may be expected to produce a radially-symmetrical sequence of pyroclastic deposits that settled gravitationally from the resultant ash cloud. However reality tends to be more complex: such 'air-fall' deposits are typically asymmetrical, often reflecting the winds prevailing at the time of eruption. Lines connecting points of equal thickness (isopachs) for air-fall deposits thus commonly define ellipsoidal rather than circular patterns, extending downwind from the volcano. An example of such an isopach map determined for air-fall deposits from the AD 79 eruption of Vesuvius is given in Figure 13.44.

Furthermore, explosive discharges may be laterally, rather than vertically directed. Additionally, particulate matter suspended in gas need not be blown high to be dispersed as air-fall but may constitute flows or surges that remain close to the ground.

As pointed out above, basaltic magmas tend to have low gas contents and low viscosities. Separation of gases from basalt magmas can therefore proceed in a relatively tranquil manner and explosive behaviour is uncharacteristic. With more siliceous magmas (e.g. andesites) low-pressure release of higher volatile contents from more viscous silicate liquids leads to explosions and thus to higher proportions of associated pyroclastic deposits. In the case of phonolitic, trachytic and rhyolitic eruptions, the greater part of the extrusive products is likely to be fragmental.

Although there is generally a positive correlation between violence of gas release (and hence of proportion of pyroclastic deposits to lavas) and silica-content, exceptions are to be found in eruptions of volatile-rich, silica-deficient magmas such as nephelinites, lamprophyres and kimberlites, which yield predominantly fragmental deposits.

Whereas basaltic eruptions are not normally accompanied by violent explosions, they may be so if the rising magma encounters surface water such as (shallow) seas, lakes, water-logged sediments etc. In such

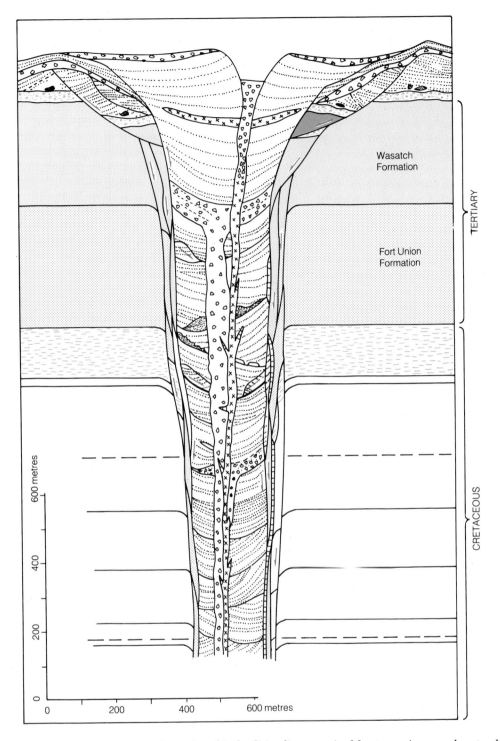

Figure 13.43 A hypothetical section through a kimberlitic diatreme in Montana. Arrows denote direction of movement during post-eruptive subsidence. (*After Hearn, 1968*)

situations meteoric water can be converted to steam and explosions can result. Eruptions involving non-magmatic water are referred to as **phreatic**. Although phreatic eruptions related to basaltic magmatism are commonplace (e.g. the shallow-sea eruption off the Icelandic coast at Surtsey (1963–1965) or that at Faial in the Azores (1957), they are by no means confined to basaltic volcanoes. Phreatic eruptions typically produce small central vent volcanoes with large craters surrounded by relatively low ramparts of pyroclastic deposits. Commonly known as tuff rings, they are also called by the German term, 'maars' (Figure 13.45). An eruption that commences as phreatic may revert to a less violent, non-explosive, mode as the volcanic edifice rises above

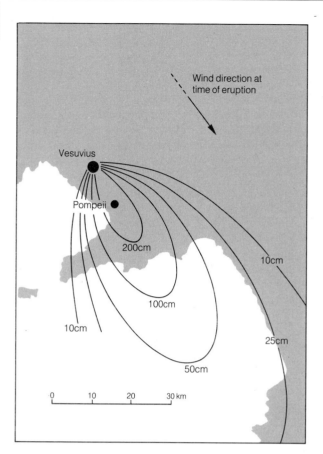

Figure 13.44 An isopach map for air-fall ash (tephra) from the AD 79 eruption of Vesuvius. (*After G.P.L. Walker*)

(or fountains) out from submarine volcanoes rapid supercooling forms glassy rinds (or droplets) and drastic thermal contraction leads to shattering and spalling of the glass into small angular shards. Accumulations of such glassy fragments cemented together by minerals deposited from warm waters in the interstices, give rise to hyaloclastites (literally, glassy broken rock). Hyaloclastites are often found intercalated with, and as infillings of the spaces between, lava pillows. Intimate associations of pillow lavas and hyaloclastites are typical products of submarine (or more generally, subaqueous and subglacial) eruptions. Subsequent hydration and alteration of the glass shards produces yellow, orange or brown material known as palagonite (Figures 13.46 and 13.47). Eruption of lava beneath a cover of ice, such as happened on a large scale in Iceland during the Pleistocene, gave rise to extensive formations of palagonite deposits, intimately associated with pillow lavas. Figure 13.48 illustrates the manner in which these subglacial volcanoes are thought to have grown.

From the description above, it should be clear that major air-fall pyroclastic deposits are related to eruption of the more silicic magma types. In extreme instances violent eruptions may produce a column of ash-laden gas ascending to the upper atmosphere. Such columns are initiated by rapid

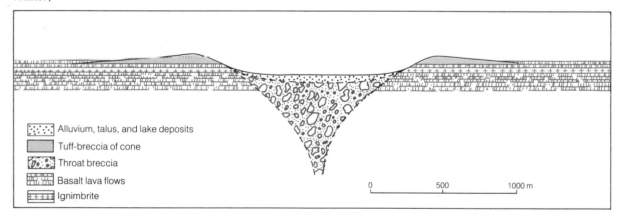

Figure 13.45 Cross-section of Hole-in-the-Ground Crater, central Oregon, showing the low encircling debris cone. (*After G.A. Macdonald, 1972*)

the water level and, with water denied access to the rising magma, quiescent emission of lavas ensues to form a subaerial carapace to the mainly fragmental deposits beneath.

Eruption of magma in deep water (as on the ocean floors) is normally quiescent since the water pressures are sufficient to inhibit separation of volatiles as a gas phase. Thus, basaltic eruptions along the mid-ocean ridges produce mainly lava accumulations. However, as basalt magma flows

decompression but sustained in their upper reaches by thermal convection. The velocities at which these ash-laden gases are discharged from the volcanic conduits can be very high, with several hundred metres per second having been estimated in some instances. As kinetic and thermal energy is lost, the column starts to spread laterally (and downwind) as a cloud.

The earliest description of a very high ash column and culminating cloud was provided by

Figure 13.46 Hyaloclastite (palagonite tuff) with basaltic blocks from disrupted lava pillows (i.e. a pillow breccia). Near Hekla, Iceland. (*B.G.J. Upton*)

Figure 13.47 Large-scale cross-bedding structure in early Tertiary hyaloclastites, west Greenland. The hyaloclastites, formed by eruption of basalt magma into shallow water, are overlain by subaerial basalt lava flows. (*A.K. Pedersen, published by permission of the Geological Survey of Greenland*)

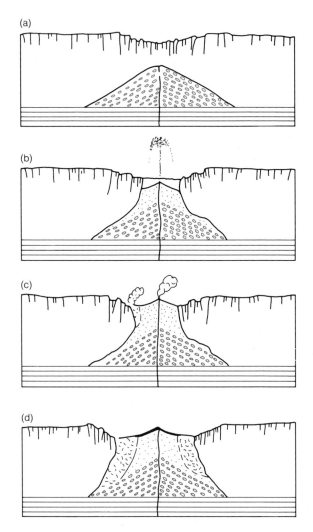

Figure 13.48 Development of subglacial volcanoes. (a). Eruption in meltwater beneath ice forms an accumulation of pillow lavas. (b). At relatively shallow depth (< 200 m) formation of hyaloclastite becomes dominant, with palagonite tuff accumulating above the pillow lavas. (c). The volcanic pile reaches surface level and further sub-aerial eruption produces lavas. (d). Continued eruption sees flows extending laterally into ponded meltwaters and more pillow lava, pillow breccias and hyaloclastites form. Eventually, after complete melting of the ice, the volcano is seen with an essentially flat top and steep, breccia-mantled sides. (*After Jones, 1970*)

Pliny the Younger, writing of the AD 79 eruption of Vesuvius in which his uncle, Pliny the Elder, died along with scores of others in Pompeii and neighbouring towns and villages (see below). Such eruptions are consequently known as 'plinian' (Figure 13.49). Plinian ash-columns have been observed rising to heights of about 45 km and some are thought to have reached heights as much as 80 km. The higher the column, the greater the geographic dispersion of the air-fall deposits. Because of their widespread distribution and rapid deposition (with regard to geological time),

such air-fall deposits can provide critical 'time planes' as marker horizons within sequences of bedded strata.

Ash-fall deposits tend to be very well stratified, (Plate IXd), with variation in particle size reflecting variations in eruptive intensity and wind strength. Phenocrysts may be separated from partly-liquid tephra and crystal-rich concentrates are not uncommon, with glassy matrix material being more widely dispersed. Air-fall deposits also tend to drape the surrounding landscape, much in the manner of snowfalls, so that deposits are of more or less constant thickness over all but the steepest slopes.

Among the greatest pyroclastic eruptions in recent history the huge eruption of Tambora (Sumbawa Island, Indonesia) in 1815 figures prominently. This eruption is estimated to have ejected approximately 100 km^3 of ash. Distributed widely in the upper atmosphere the fine particles are believed to have caused global climatic changes that persisted for several years. A lesser eruption, but still one of great magnitude occurred in the same general region in 1883 when Krakatoa, a volcano lying between Sumatra and Java erupted. Of large pyroclastic eruptions this century, none approaches the ferocity of the nineteenth century Tambora or Krakatoa events but a major eruption took place in southern Mexico in April 1982, at El Chichon volcano. It is possible that globally-dispersed ash (and acidic aerosols) from El Chichon has played some role in the disturbed climatic scene during the 1980s.

Juvenile magmatic ashes, particularly close to the erupting vent, may be sufficiently hot (above, or close to their beginning-of-melting temperature (solidus) for particles to fuse together as they accumulate forming welded air-fall. Fountaining jets of (particularly basaltic) lava may be caught by the wind and the droplets spun out into fast-quenched glass fibres. Deposits of such natural fibreglass, accumulating downwind from the vent are referred to as Pele's Hair, after Pele, the mythical goddess of the Hawaiian volcanoes.

Dense clouds of hot, particle-laden gas are commonly generated during eruptions (particularly those of the more silicic magma types) and flow down the flanks of the volcano in a manner analogous to that of rock or snow avalanches. Such fluidized pyroclastic flows (ash-flows) are preferentially channelled down valleys or gorges in the terrain. The deposits left by ash-flows are typically thickest in (or confined to) the topographic lows and thus do not spread uniformly across the landscape as do air-fall deposits. Furthermore, they differ from typical air-fall in tending to be unstratified, poorly sorted and relatively homogeneous.

Figure 13.49 Paroxysmal emission of ash-charged gas during the 1906 eruption of Vesuvius: the plinian gas discharge continued for 18 hours and visibly cored out and widened the volcanic conduit. (*After F.A. Perret, Carnegie Institute, Washington*)

Ash-flow eruptions are especially dangerous phenomena. Not only are they composed of choking, high-temperature aerosols but they can travel at speeds of over 100 km/hr and, in the case of the larger ash-flows, for long distances. Ash-flows may be sufficiently hot to be incandescent: the French volcanologist, Alfred Lacroix observed ash-flows erupting from Mt. Pelee, Martinique (1902), and, noting the reflected glow on the undersides of overlying ash clouds, coined the name '**nuées ardentes**' (glowing clouds) for this phenomenon. If the ash-flow collapses to a pyroclastic deposit while retaining sufficient heat, with glassy pumice fragments still above their softening points (about 600–500°C), the pumice fragments may flatten and weld together during compaction of the deposit to form a tough, coherent welded ash (Figure 13.50). The name **ignimbrite** is often given to such welded ash-flow tuffs although some workers prefer to apply this name to ash-flow deposits in general, whether welded or not. Thick welded ash-flow deposits, cooling slowly, may develop fine columnar-jointing pat-

terns during thermal contraction (Figure 13.51).

There appear to be several mechanisms by which an ash-flow can be generated (Figure 13.52) Some originate when a plinian column of gas and suspended matter becomes gravitationally unstable as the gas discharge rate slackens. The column, or part of the column, may then collapse, into a dense cloud with a high ratio of particulate matter to gas, which then descends as a ground-hugging flow. Ash-flows may also commence when a viscous, largely degassed lava, extruded as a dome or spine, forms a mass that temporarily plugs a vent. Collapse, or explosive destruction, of such a plug by gas accumulation beneath it, may release a dense pulse of particle-laden gas which behaves as a coherent flow. In a number of instances, ash-flows are thought to have originated from ring fractures propagated as blocks subside in caldera-forming eruptions (Chapter 12).

The geological record preserves many giant ash-flow tuffs that covered great tracts of the landscape. Such deposits testify to pyroclastic eruptions on a scale that vastly exceeds any of those known in historic times. Within the past few million years large-scale ash-flow eruptions occurred, for example, at Yellowstone in Wyoming and at Taupo in New Zealand. In the Sierra Madre Occidentale of Mexico great quantities of volcanic rocks, among which ash-flows figure prominently, were erupted during the interval 34–27 Ma in the mid-Tertiary period. Among the largest individual ash-flow tuffs known is the Fish Canyon tuff in the San Juan mountains of Colorado, with a total volume of over 3000 km³. Eruptions on this scale must involve the catastrophic emptying of great subsurface magma chambers. Surface collapse to fill the potential void space is an inevitable accompaniment producing the volcanic depressions referred to as **calderas**. In the British Isles huge ash-flows, related to caldera formation, attended suprasubduction volcanism in the Ordovician (about 470 Ma ago). Much of the mountainous terrain in the Snowdonia and Lake District National Parks of Wales and England is composed of these resistant igneous rocks.

Ash-flow eruptions are a commonplace phe-

Figure 13.50 Block of welded ash-flow in a Silurian caldera complex, Newfoundland. The pale lenticular components represent deformed and flattened pumice fragments. (*B.G.J. Upton*)

nomenon in volcanic terrains. Probably the largest this century was that produced by Katmai volcano in Alaska in 1912. Whereas there are no witnessed reports of the ash-flow eruption, the products were discovered and investigated four years later by an expedition financed by the National Geographic Society of America. The expedition found that a valley in the vicinity of Katmai had been flooded by at least 10 km³ of friable, unconsolidated ash-flow tuff from which great numbers of steam jets were blowing. The expedition consequently named the area 'The Valley of Ten Thousand Smokes'. It has been calculated that the ash-flow was erupted at a flow rate of around 500 m³/hr during the course of a day. However, neither this nor any other historic eruptions has been large enough or hot enough to generate a welded deposit. Despite its magnitude the Valley of Ten Thousand Smokes deposit is minuscule in comparison with many of the giant pyroclastic flow events known from the prehistoric record.

In addition to air-fall and ash-flows volcanologists recognize a further phenomenon, that of **surges**, distinct from ash-flows (although in all probability there are all the intermediate stages between a flow and a surge.) The concept of a surge is of a very rapidly moving blast of gas with a much lower particle/gas ratio than that characterizing a typical flow. Surges may be thought of as high-velocity pulses travelling close to the ground surface away from the site of the explosion and not as closely confined to valleys as typical ash-flows. Evidence for rapid lateral motion may be gained from studies of surge-borne blocks that have impacted more or less horizontally. Surge deposits are typically highly stratified (laminated), exhibiting dune-type bedding (Figure 13.53). During an eruption at Taal volcano (Philippines) in 1965, water is inferred to have gained access to the rising magma. In the series of phreatic explosions that resulted surges were observed to spread out from the erupting site at velocities of around 100 m/sec. By analogy with the doughnut-shaped pressure waves that propagate, close to ground level, from atmospheric nuclear test sites, these were referred to as 'base surges'. The base surges at Taal flattened all forest out to 1 km and sandblasted obstacles up to a distance of around 8 km. These surges appear to have been wet, mud-laden and relatively cold. Vegetation was not charred and temperatures were deduced to have been less than 100°C. However, there is clear evidence that many surge events are very much hotter than these particular ones at Taal.

Pyroclastic deposits, especially if friable and unwelded, are very prone to reworking and transport by wind and water to produce secondary

Figure 13.51 Columnar jointing in a sheet of ignimbrite, near Waipapa dam, Waikato River, North Island, New Zealand. (*R.H. Clark*)

deposits. In such situations sand grains etc. derived from pre-existing rock formations may also be involved and, with addition of such epiclastic material to pyroclastic materials, all gradations exist between primary pyroclastic deposits and 'normal' sediments.

Water is commonly abundant on active volcanoes, either generated through the melting of snow or ice caps, or released from crater-lakes, e.g. when a new lava ascends as a dome within a crater. Additionally, huge quantities of water vapour in the eruptive columns may accompany an eruption. Consequently pyroclastic deposits, especially on the steeper slopes, may, when water-saturated, become unstable and move as mudslides. The products of these slides, typically chaotic and unsorted with respect to size or composition of components, are called **lahars** (Figure 13.54). Lahars are commonly interbedded with other fragmental deposits and lavas in stratovolcanoes.

Such volcanic mudslides constitute a potentially highly dangerous phenomenon. Of the two major volcanic disasters in the twentieth century, one was caused by ash-flow (nuées ardentes) emission (the destruction of St. Pierre during the 1902 eruption of Mt. Pelee, on Martinique) whereas the other was due to huge mudflows attending the 1987 eruption of Nevado del Ruiz in Columbia. In this latter catastrophe the town of Armero and neighbouring villages, were overwhelmed by these cold slurries and some 20 000 inhabitants are believed to have perished. More recently (1992), in the course of a major eruption at Mt. Pinatubo in the Philippines, lahars constituted one of the principal hazards threatening local communities.

13.7 CALDERAS

A caldera is a topographic depression, commonly circular or ovoid in plan, that has originated from gravitational collapse of part, or all, of a volcano following evacuation or withdrawal of magma from an underlying storage chamber. Examples are shown in Figures 13.55 and 13.56. Caldera diameters are generally in the 5–20 km range but exceptionally large ones, e.g. the Cerro Golan caldera in Argentina, may attain widths of 50–60 km. They may sometimes be distinctly elongate and can grade into fault-bounded troughs

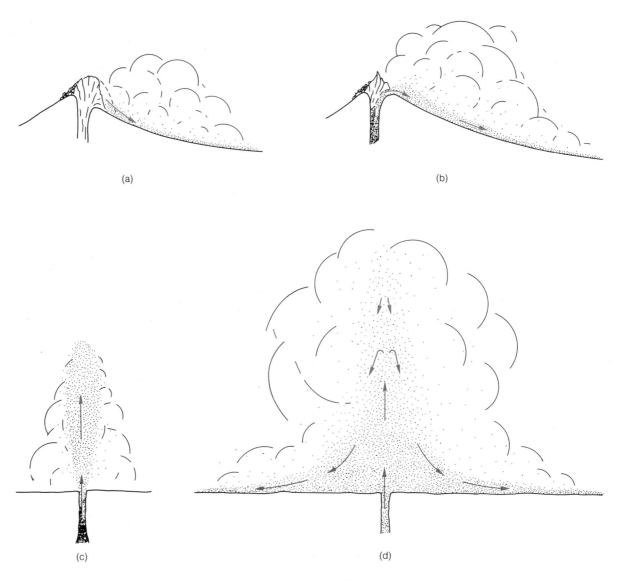

(a)

(b)

(c)

(d)

Figure 13.52 Diagrams illustrating three possible modes of origin for ash-flows. (a) Part of a volcanic dome disintegrates and collapses producing a small hot avalanche; (b) a dome or plug blocking a vent is explosively circumvented by a blast of gas and ash directed laterally; (c) and (d) a tall plinian ash-column becomes gravitationally unstable and collapses, in part or in whole, to produce an ash-flow. (*After M.G. Best, 1982*)

(**volcano-grabens**). Annular fault scarps may surround part or all of the caldera. In the descriptively named **trap-door calderas** collapse is eccentric, the caldera floor hinging about one side as the name implies.

Collapses may be repetitive so that complex calderas develop sometimes with concentrically-nested structures. Caldera walls may be over a hundred metres high, forming striking topographic features: at the other extreme some calderas are merely shallow downwarps where the surrounding faults have little or no surface expression.

A classic caldera is that occupied by Crater Lake in the Cascade range, in Oregon. The lake occupies a caldera about 10 km across that was formed by the collapse of what was once a lofty composite cone rising to an estimated height of around 3700 m. The name Mt. Mazama has been given to this hypothetical cone; the cone must have been there not long ago, for it supported glaciers that have left abundant evidence of their former existence on the outer slopes of the caldera walls. Glacial deposits alternate with beds of pyroclastic deposits: U-shaped valleys partially filled by volcanic debris can be followed up to the caldera rim where they are abruptly truncated. Carbonized wood, the remnants of trees growing at the time of the culminating eruption have been dated by the radiocarbon method at about 8000 years ago. Recent studies indicate that some 50 km³ of magma were ejected in the culminating pyroclastic eruptions. The ashes were very widely distributed and some that have been identified as Mt.

Figure 13.53 Strongly cross-stratified surge deposit within ash-flow tuffs. Santorini. (*B.G.J. Upton*)

Figure 13.54 A lahar or volcanic mud-flow seen in roadside section near Tongariro Volcano, North Island, New Zealand. A figure (lower left) provides scale. The lahar represent a cold slurry of mud and blocks and is a chaotic, unsorted flow, overlain by well-stratified soils and ashes. (*B.G.J. Upton*)

Mazama air-fall have been found in Saskatchewan some 2000 km distant. Collapse and caldera forma-tion attended the evacuation of magma from the subvolcanic chamber. Figure 13.57 shows dia-

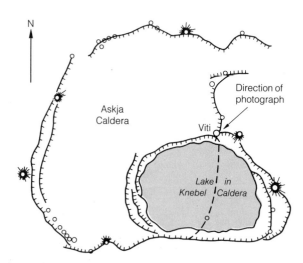

Figure 13.55 Askja volcano (Iceland). The crater of Viti, the most active vent of the eruptions of 1875, which were responsible for the enlargement of the Askja caldera by the formation of the Knebel caldera. Viti itself originated by a sudden explosive eruption in 1724. The Knebel caldera is partly sunk in the Askja floor, as indicated in the accompanying sketch map. See Fig. 13.14 for locality. (*Pall Jóhnsson*)

grammatically the interpreted evolution at Crater Lake.

As we see in this example, loss of structural support, bringing about caldera collapse, may take place because the underlying magma has been evacuated in a cataclysmic eruption in which it is blown out. A similar situation can arise where a subvolcanic chamber has emptied either because magma has leaked sideways, being drained into fissure systems (dykes) related to the volcano, or because it was drained in low-level flank eruptions.

Calderas formed wholly or principally through pyroclastic activity are referred to as **Krakatoan**, with the catastrophic 1883 eruption of Krakatoa, Indonesia, as the type example. Those produced mainly by lateral withdrawal (with or without eruption) are generally called **Kilauean** after the

Figure 13.56 A caldera of Kilauean-type. Silali, Kenyan rift. (*B.G.J. Upton*)

model of Kilauea volcano, Hawaii.

Central-type volcanoes often develop over structurally-favoured sites of easy magma ascent within regimes with well-developed dyke swarms. Where substantial volumes of magma are temporarily stored in shallow reservoirs beneath such volcanoes, renewed distension across the volcanic zone may result in magma draining laterally into propagating dyke fissures, with or without concomitant eruption. Caldera collapse is an inevitable accompaniment. The twin processes of caldera formation and development of associated dyke swarms are frequently intimately related.

Recently it has become more widely recognized that volcanoes, particularly the great oceanic basaltic volcanoes, are prone to gravitational instability where the structure is buttressed on one side by an adjacent volcano, but unsupported on the other, which may slope down to the deep ocean floor. Sliding or slumping towards the unsupported side can generate caldera-like depressions near the volcanic summit. An excellent example of this behaviour is provided by Piton de la Fournaise volcano on Reunion Island, where gravity sliding on a huge scale towards the unsupported eastern flank has created a set of horseshoe-shaped calderas 'closed' to the west but 'open' to the east.

Krakatoan-type calderas are most commonly associated with major pyroclastic eruptions of subduction-related volcanoes. They are most often developed following eruption of andesitic or more silicic magmas such as dacite or rhyolite. Kilauean-style calderas on the other hand, are more characteristically found on basaltic volcanoes.

Ring-dyke emplacement is believed to be closely linked to caldera formation (Chapter 12). Arcuate zones of small cones or domes may grow above caldera ring fractures. Particularly after major pyroclastic eruptions of rhyolitic magma with caldera formation, leakage of largely degassed viscous magma along the arcuate fractures can produce a series of lava domes around the caldera margin. Examples of such behaviour can be seen for example in the Valles caldera, Jemez Mountains, New Mexico, and at Naivasha in the Kenya rift. At the Crater Lake caldera discussed above, a semicircular arc of vents that developed around the northern wall of the caldera suggests ascent of magma along a ring fracture. Ascent of gas-poor magma following a major pyroclastic eruption and caldera growth may cause updoming of the caldera floor, the so-called resurgence phenomenon first described from the Valles caldera (Figure 13.58).

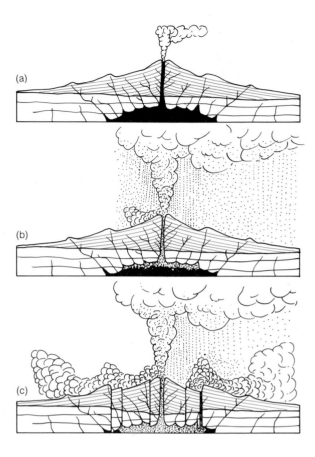

Figure 13.57 Diagram illustrating the formation of the Crater Lake caldera, Oregon. (a) Before the eruption; (b) during an early stage of the caldera-forming eruption, showing a vulcanian ash-cloud and a small ash-flow from the central vent; (c) during the climax of the eruption, with large ash-flows issuing from the central vent and from ring fractures part way down the slope while the summit starts to sink as a series of large blocks; (d) after the eruption; (e) in its present state, with new eruptions on the floor and the caldera partly-filled with water. (*After H. Williams, 1942*)

13.8 HYDROTHERMAL ACTIVITY

After an eruptive cycle, when a volcano falls into dormancy, magma and hot rocks may remain at

relatively shallow crustal depths for hundreds or thousands of years. Groundwaters seeping through faults, joints and porous rocks encounter these high-temperature regions and are themselves heated. The hot waters ascend and more dense, cooler waters take their place. Vigorous convective systems can thereby be established and maintained. The rising waters in such a hydrothermal system may reach the surface as hot springs or, if temperatures are high enough, convert at low pressures to steam (Figure 13.59).

In some subterranean plumbing systems, retained water heats up until the temperature is reached at which it 'flashes' and converts to steam. In the ensuing volume change, cooler water in the conduits above may be violently blown out at the surface as a **geysir** (Figure 13.60). After such an event, time elapses as new waters enter the system, become heated and the cycle is repeated.

Regions where such activity takes place are called **geothermal fields** (Plate IXe) and have, over the past few decades, attracted much attention as potential sources of renewable energy. Some of the better-known geothermal fields with extensive hot springs and geysirs are to be found in Iceland, the USA (particularly the Yellowstone region of Montana–Wyoming) and the Rotorua area of North Island, New Zealand. In a number of geothermal fields high-temperature steam has been harnessed for the generation of electricity. Geothermally operated power-stations have been built, for example in New Zealand (Plate IXe), Iceland (Figure 13.61) and Kenya.

At high temperatures (and relatively high pressures) water becomes a powerful solvent and large quantities of inorganic material can be dissolved and transported by the aqueous fluids, to be reprecipitated elsewhere. Additionally these aqueous solutions can have profound chemical effects on the rocks through which they pass. During their passage through volcanic sequences, the mineral assemblages crystallized at magmatic temperatures ($> 700°C$) are reconstituted to assemblages stable at the lower temperatures (typically $< 400°C$) appropriate to the hydrothermal circulation. Glassy volcanic rocks are devitrified and a wide variety of lower-temperature minerals (including quartz, calcite, epidote and minerals of the chlorite, zeolite and clay groups) may be grown in place of the original minerals or deposited in interstitial cavities, in fissures and joints (to form veins) or in vesicles (to form amygdales).

A major discovery in recent years has been that dramatic hydrothermal circulations occur sporadically along the mid-ocean ridges. Cold seawater, percolating down through jointed and permeable rocks is thought to descend to the lower reaches of the sheeted complexes (Section 12.1), to the vicin-

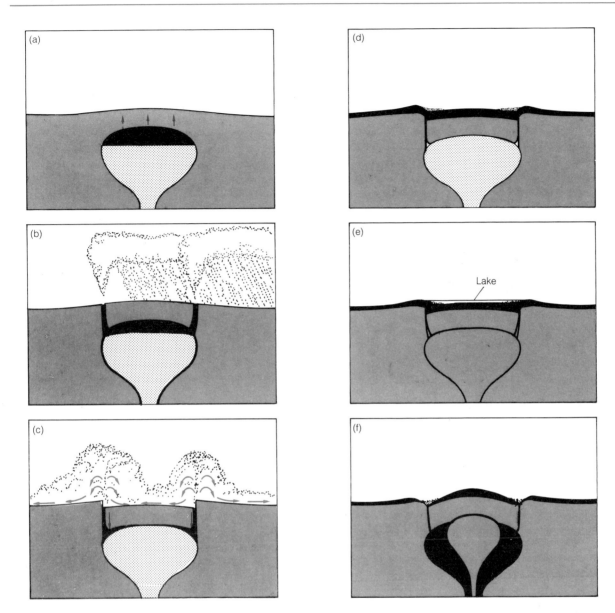

Figure 13.58 Stages in the evolution of a resurgent caldera. (a) Magma fills a chamber a few kilometres below the surface, doming the surface; (b) ring-shaped fractures form around the dome, and the gas-charged magma at the top explodes, erupting columns of incandescent pumice and ash into the atmosphere; (c) the magma chamber emptied, the roof collapses along the fracture, accompanied by the ejection of pyroclastic flows; (d) the caldera and surrounding area are covered by a blanket of ash-flow tuff (ignimbrite); (e) the caldera wall begins to erode, and a lake may form in the depression; (f) some hundreds of thousands of years later, fresh magma enters the magma chamber and the caldera floor begins to dome again. Minor volcanic activity may persist along the ring-fracture for millions of years. (*After P. Francis, 1983*)

ity of the underlying magma bodies. The heated brines ascend and produce pervasive alteration (metamorphism) of the rocks with which they come into contact.

Exciting studies have been conducted, using submersibles, on parts of the mid-ocean ridge system, for example along the East Pacific Rise, close to the Galapagos Islands. Intense jets of hot brines (up to 350°C) arise from the sea floor, in linear zones a few kilometres long by a few hundred metres broad. The jets (in which boiling is prevented by the high hydrostatic pressures of the overlying sea water) squirt out of vents that may be no more than half a metre across, at velocities of up to 5 m/sec. These hot brine jets are complex solutions containing many elements leached from the rock formations that they have traversed. Such sea-floor springs are also astonishing in that they provide a unique sunlight-free environment in which life flourishes. Great colonies of chemosynthetic bacteria, deriving their energy from reactions with the hot brines, pro-

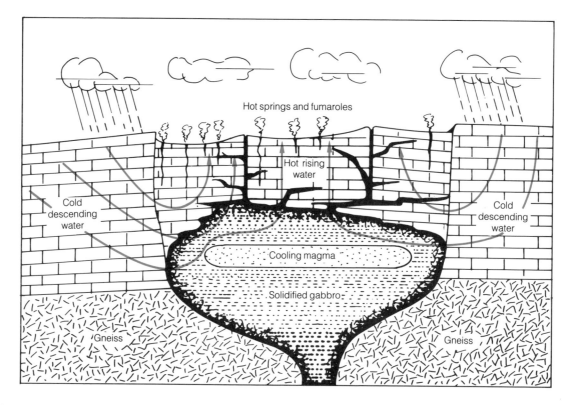

Figure 13.59 Diagrammatic section across the Skaergaard intrusion, East Greenland (Plate VIIIb, Figure 12.20), illustrating the establishment of a convecting hydrothermal system. Meteoric water, flowing through permeable lavas and solidified gabbro, is heated in the vicinity of the residual magma and rises to produce a geothermal field at the surface. (*After A.R. McBirney, 1984*)

mote large-scale precipitation of sulphides, particularly those of zinc, iron and copper. The sulphide precipitates build mounds and chimney-like tubes around the vents, up to 10 m tall (Figure 13.62). The vigorous jet plumes, dark with their content of precipitates, emanating from these vents have been termed 'black-smokers'.

It has been postulated that such hydrothermal systems are regenerated whenever a new fracture in the spreading ocean floor propagates down to the plastic zone surrounding the region of magma. Furthermore it has been calculated that for each 7 km³ of magma solidified, the heat lost to circulating brines leads to the eventual deposition of 1 000 000 tonnes of sulphide ore. In ancient ophiolite bodies, such as that composing the Troodos Mountains complex in Cyprus, huge copper sulphide deposits comprising up to 20 000 000 tonnes, were probably formed from black-smoker-type sea-floor springs.

13.9 VOLCANIC HAZARDS

Some of the dangers associated with active volcanoes have been touched on in the foregoing sections. In comparison with the losses to life and property involved in severe earthquakes, tropical storms or floods, those relating to volcanoes are fairly trivial. However, when they do occur, they can be very dramatic and the possibility of a really major catastrophe affecting heavily populated regions is ever present. This threat has to be taken seriously when the evidence for truly horrendous eruptions in the not-too-distant past is taken into consideration.

Volcanoes and earthquakes tend to be closely linked in popular thought. However, whereas seismic tremors are generated as magma ascends in subvolcanic conduits, with propagation of minor intrusions, the energy of volcanic explosions is largely dissipated in atmospheric shock waves rather than in earth tremors. In brief, while volcano-related earthquakes can, and do, present a risk, they tend to be relatively small-scale and insignificant in comparison to the major earthquakes associated with tectonic processes (see Chapter 25).

Lava flows themselves tend to pose little risk. Lavas move sufficiently slowly to afford minimal danger to life although damage to agricultural land can be substantial and, occasionally, towns and settlements are affected. In recent years houses have been destroyed by lava flows, for example from Mt Etna in Sicily and Helgafell on the Icelandic offshore island of Heimaey. At the time of

Figure 13.60 'Old Faithful' geysir, Yellowstone National Park, Wyoming-Montana, USA. (*B.G.J. Upton*)

writing Kilauea on Hawaii is continuing an eruptive cycle started in the early 1980s and considerable damage is being caused to communities near the south coast of the island. A celebrated example this century of towns being destroyed by lava flows is that of Paricutin village and San Juan Parangaricutiro (Figure 13.63), two small towns enveloped in lava flows erupted from the cone of Paricutin (Mexico), a volcano which first commenced activity in 1943 and continued eruption over the next nine years.

Far more serious are the risks relating to volcanic gases and explosive eruptions. The dangers from asphyxiation by dense gas clouds were highlighted in recent years by two successive events involving small, apparently extinct, volcanoes in Cameroon. On August 15th, 1984, a cloud of gas rich in CO_2 burst from a small crater-lake, Lake Monoun, killing 37 local people. It was deduced that the lake had experienced long-term build-up of dissolved CO_2 derived from carbonated springs supplying the lake. Isotopic study indicated a mantle origin for the carbon. The CO_2 remained in solution in the deeper waters of the lake where retaining pressures were high until a landslide, on the eastern rim of the crater, brought about an

overturning of the lake water. Depressurization at shallow depth allowed the vigorous effervescence of CO_2 to take place.

A remarkably similar but still more serious situation occurred some two years later at another crater-lake (Lake Nyos) in the Cameroon volcanic chain. Lake Nyos occupies a phreatomagmatic crater about 1.6 km wide. In the evening of August 21st, 1986, a sudden and violent degassing appears to have caused the waters to fountain to a height of about 100 m above the lake surface. A dense cloud of CO_2-rich gas, (estimated total mass of CO_2 approximately 1.94×10^6 tonnes) poured down a valley from the crater rim. The power of this density-driven cloud was enough to flatten corn and banana plants in its course. The cloud subdivided into two flows and affected a region of about 63 km^2 before dispersing. The lethal cloud caused death up to a distance of 23 km from the lake killing 1746 people (including virtually the entire population of the village of Nyos) and some 8300 head of livestock.

As at Monoun, it is supposed that CO_2-rich solutions, arising from deep magmatic sources over a long period of time, allowed concentration of gas to rise in the depths of the lake. What precisely triggered the culminating gas release remains unknown. Like the Kenya Rift, the Cameroon Line may well be a zone of weakness characterized by high flux of CO_2 from the underlying mantle.

Reminders of the devastating power of laterally directed gas blasts from volcanoes were provided by violent eruptions at Taal volcano, Philippines (1965) and at Mount St. Helens in the Cascade Mountains of Washington State, USA in 1980, (see below).

Extensive falls of ash can create major problems (Figure 13.64). Apart from the burial of land and buildings, extensive fall-out can seriously disrupt communications and interfere with water supplies. Among historic incidents the most famous is that of the AD 79 Vesuvius eruption when Pompeii was smothered by air-fall tephra. It would appear, however, that in this eruption most of the fatalities were due to asphyxiation. In the huge explosive eruptions of Santorini (Aegean) at around 1470 BC, towns such as Akrotiri (currently being excavated on Santorini) were buried by fallen ash. It was this Santorini event, which affected a wide sector of the eastern Mediterranean, that is generally considered to have been a factor in the demise of the Minoan civilization. In 1944 Vesuvius (Figure 13.65) erupted spreading great quantities of ash across the landscape and materially affecting with the course of hostilities between the German and Allied armies. Vesuvius has not erupted since 1944; in view of the fact that for some hundreds of years previously Vesuvius has generally erupted at intervals of less than 20 years this long quiescence is a cause of concern since the energy of eruptions can be a function of repose time. Since 1944 the city of Naples has grown very considerably, with developments spreading onto the lower flanks of the volcano and a major eruption could give rise to widescale devastation. During violent pyroclastic eruptions in 1991 at Mt. Pinatubo, Philippines, many deaths occurred from collapse of buildings due to the weight of accumulated ash.

A further risk that has only been taken seriously over the past decade or so is that to aircraft encountering ash clouds. Intake of fine ash into jet engines can result in catastrophic loss of power. So far there have been no major air disasters from this cause although there have been incidents where crashes have been narrowly avoided.

Very real risks are attached to floods and mud-flows (lahars) related to volcanic action. Flooding can result from displacement of large volumes of

Figure 13.61 A geothermal power-station at Svartsengi, south-west Iceland. (*B.G.J. Upton*)

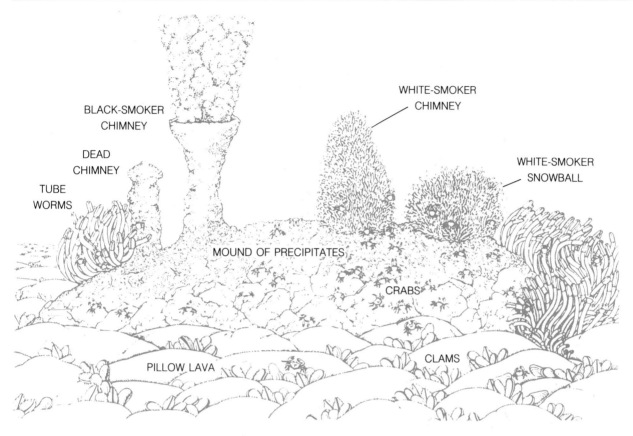

BLACK-SMOKER
CHIMNEY

DEAD
CHIMNEY

TUBE
WORMS

WHITE-SMOKER
CHIMNEY

WHITE-SMOKER
SNOWBALL

MOUND OF PRECIPITATES

CRABS

PILLOW LAVA

CLAMS

Figure 13.62 Idealized scene in the hydrothermal field near the high-temperature vents associated with mid-ocean ridges shows an array of typical vent structures on top of a mound of precipitated minerals and organic debris. The white-smoker chimney is built up out of burrows made by an organism called the Pompeii worm. The solutions emitted by the white-smoker have temperatures of up to 300°C. The hottest water (up to 350°C) comes from the black-smokers, whose chimneys are made up of sulphide precipitates. The mound rises from lava pillows; crabs and bivalves also contribute to the fauna of these sunless environments. (*After Scientific American*)

water held in volcanic crater or caldera lakes when a new eruptive cycle commences or when the retaining walls of such perched reservoirs become breached (Figure 13.66). Alternatively, flooding may occur through damming of water courses by the products of an eruption. Particularly at high latitudes floods can be caused by volcanoes erupting beneath a cover of ice. Two such volcanoes presenting threats to populated areas are Katla and Grimsvotn in Iceland (Figure 13.14). Mudflows can arise when unconsolidated pyroclastic deposits reach gravitational instability through waterlogging such as may be brought about by heavy rain from steam-laden volcanic clouds. Such (usually cold) slurries can have devastating effect as referred to above with regard to the 1987 disaster at Armero (Nevada del Ruiz) in Columbia.

Very violent explosive events in marine environments can cause powerful 'tidal' waves or **tsunami** that can generate ferocious damage to exposed coastal regions. Much of the death toll in 1883 from the Krakatoa eruption was due to drowning by tsunamis hitting the Javanese and Sumatran coasts.

13.10   SOME NOTABLE HISTORIC ERUPTIONS

In the foregoing sections reference has been made to a number of historic eruptions. Brief descriptions will now be given of a few well-documented instances illustrative of some of the varied phenomena that can 'be encountered and of the destructiveness that can be entailed.

The earliest detailed records of an eruption are those concerning the events at Vesuvius in AD 79. We know of this eruption from two letters to Tacitus, written by Pliny the Younger who was a youth of 17 at the time. By AD 79 the volcano had lain dormant for several thousand years. Earthquakes in August of that year were followed by a major explosion and growth of an ash plume several thousand metres high, spreading out at high altitude like a pine tree. Ashes rained down upon Pompeii eventually burying it under about 6 m of deposits. It was not until the nineteenth century that the ruins were discovered and excavations commenced. The nearby town of Herculaneum, on the other hand, became inundated by a massive mudflow that covered it to a depth of some 20 m. At the close of the eruption much of

Figure 13.63 A lava flow from the base of Paricutin which overwhelmed San Juan Parangaricutiro in June 1944. (*Tad Nichols*)

Figure 13.64 Country school buried by showers of volcanic ash erupted from Sakurajima, Japan, in 1914. (*F. Perret, Carnegie Institution of Washington*)

the western slopes of the mountain had been blown away and a new cone had been created. Following this disaster the volcano erupted a further 10 times in the period up to 1037. For the next 600 years the volcano subsided into tranquillity from which it emerged in 1631 with another large-scale eruption. Lava flows, hot lahars, massive explosions and heavy ash falls resulted and it is believed that more than 4000 people died in the Naples area in the course of this eruption.

Figure 13.65 Vesuvius half encircled by the breached caldera known as Monte Somma. (*Fox Photos Ltd*)

One of the largest eruptions of historic times, and the most dramatic within the past 200 years, was at Tambora volcano on the island of Sumbawa in Indonesia. A series of giant explosions in April 1815 blew away the summit area, lowering the mountain by about 1000 m to a height of approximately 3000 m. Much of the ejected tephra was blown into the upper atmosphere and the resultant high-level pollution is thought to have been the cause of the exceptionally cold summers in Europe and North America in 1815 and 1816. Some 10 000 people died on Sumbawa during the eruption, with a further estimated 80 000 subsequently succumbing to famine and disease.

A further eruption of shattering intensity, but of lesser magnitude than that at Tambora, also occurred in Indonesia, at Krakatoa, an island in the Sunda Straits between Java and Sumatra. Krakatoa had had a large eruption in the late seventeenth century but remained quiescent for just over 200 years when earth tremors commenced in the late 1870s. Earthquake activity continued until the 20th of May in 1883 when the volcano began a sequence of explosive eruptions. We may infer that the seismic activity marked faulting and rock-fracturing at depth which was associated with the ascent of new magma from the upper mantle. The violent events that followed almost certainly involved access of seawater to the vicinity of near-surface magma bodies with consequent catastrophic steam generation.

Krakatoa itself at the time was an island some 9 km across, consisting of a number of roughly-aligned cones, Rakata to the north, Danan in the middle and Perboewatn to the south. In the early eruptive phases activity was concentrated at Perboewatn where explosions took place in a 1000 m diameter crater, producing a tall column of steam. The explosions continued, increasing in vigour up to June 19th, by which time much of the upper cone of Perboewatn had been blown away and a new eruptive column was ascending from Danan. Massive detonations and vigorous seismicity persisted and by August 11th, three gas and ash columns were noted, two from the vicinity of Danan and one from Perboewatn. The climax was reached during the last week of August. On the 26th, formidable detonations were heard every 10 minutes; dense volcanic clouds reached a height of 27 km and ashes, converted into stifling mud by the incessant rain, fell over Batavia (now Djakarta) which was plunged into darkness relieved only by vivid flashes of lightning. The morning of August 27th was witness to a series of the largest explosions ever recorded. These were heard over immense distances and were noted, for example, in South Australia and Diego Garcia in the Indian Ocean. The sounds were even heard by the inhabitants of Rodrigues Island in the western Indian Ocean, (who at first mistook it for gunfire), at a distance of over 4800 km. A vast glowing cloud of incandescent pumice and ashes rose to a height of

Figure 13.66 Crater lake on Ruapeho volcano, North Island, New Zealand. (*L. Lyons*)

80 km. The explosions went on until late on the 28th when, after some 3 months of continuous eruptions, the volcano relapsed into silence. The destruction wrought by the events of August 26–27th was so immense that its true scale only began to be fully appreciated two months later with the arrival of a Dutch Scientific Commission. It was found that two-thirds of the island had disappeared. Perboewatn had gone in its entirety and much of the northern part of the island was also missing. Some 20 km² of Krakatoa had been lost, amounting to about 16 km³. Much of this was ejected as dacitic tephra that fell over an area of about 4 million km², the remainder having foundered into a caldera 6–7 km across, that was largely occupied by the sea. Dust particles, distributed through the upper atmosphere, circulated the Earth for many months causing spectacular sunsets, and the total energy of the eruption has been estimated as not less than the equivalent of 5000 megatonnes of TNT. As has been mentioned, the loss of life from the Krakatoa eruption was mainly from drowning: highly destructive tsunami, one of which attained a height of over 36 m, swept over the low coasts of Java and Sumatra, claiming the lives of some 36 000 victims.

Krakatoa remained quiet for about 44 years when renewed, submarine, activity built up a new cone (Anak Krakatoa or 'Child of Krakatoa') within the 1883 caldera. This appeared above sea-level in 1927 and continued to grow by discharge of tephra (basaltic, in contrast to the more siliceous dacites of 1883) up to 1933. Since then Krakatoa has had small eruptions in 1953 and 1959, with further activity on into the 1980s. The products of these more recent eruptions have been dacitic to andesitic.

An eruption which had appalling consequences for those in the vicinity and whose study can, in many ways, be said to have brought about the birth of the science of volcanology, was that on the island of Martinique in the Caribbean in the early part of 1902. Mt. Pelee, which had been dormant for some 50 years, reawoke in February, 1902. In the early stages gas and ash were emitted, with explosive activity that, by late April, was distinctly violent. Matters came to a climax in early May with loud explosions and heavy falls of ash. Hot water resulting from a breach in the wall of the crater lake flooded down the flanks, turning into a mud-slurry as it approached the coast. Explosive detonations increased in intensity during the first week of May and, early in the morning of May 8th, a rapid succession of huge explosions took place, throwing great quantities of ash into the air and blotting out the light of the sun. One dense cloud of superheated gas and suspended particles rolled at high speed down one of the valleys radiating from the mountain, to strike the city of St. Pierre. The force of this volcanic cyclone was such as to flatten masonry buildings and temperatures were high enough to cause widespread fires (Figure 13.67). In this fatal hot wind around 30 000 inhabitants met their deaths from asphyxiation and/or burns; there are records of only two survivors. A further massive explosion took place on 20th May, with a blast that levelled much of St. Pierre not already destroyed on 8th May. The blast created a wedge-shaped area of devastation widening outwards from the summit. The boundary between unaffected land and that within the triangle of destruction was observed to be remarkably abrupt.

Following this St. Pierre catastrophe, scientists were despatched to Martinique from Britain, the USA and France. It was from the reports of these teams that recognition and some understanding of the nature of ash-flow eruptions first came about. Eruptions continued for some time and, on July 9th, two of the observers, Tempest Anderson and John Flett, were able to watch a 'nuée ardente' eruption, from the sea. The ash-flow occurred at night when it appeared as a dull red cloud with a billowy surface, that took only about a minute to sweep down from the crater to the sea. A great surging mass of cloud arose from above the glowing avalanche, illuminated by lightning discharges. A further paroxysm occurred on August 30th, this time with further fatality when some 2000 inhabitants died in the village of Le Morne Rouge. It was from this, and similar eruptions at Mt. Pelée (Figure 13.68) that one of the investigating scientists, Alfred Lacroix, introduced the term 'nuée ardente'. From May 21st a dacitic dome of largely de-volatilized lava grew inside the crater. Gas pressures were able to build up beneath this lava plug and periodically, high-velocity releases of gas with suspended ash shot out laterally from around the base of the dome. Domes or spines of viscous dacite built up several times, tending to be destroyed in subsequent explosive discharges. The most remarkable example came into being in mid-October and was witnessed by Lacroix and his associates. On this occasion, stiff lava was extruded from the volcanic throat to form a spine that grew vertically, sometimes as much as 15 m in a day. It reached its maximum height, of over 300 m some 7 months later (Figures 13.67 and 13.69). This great spine on Mt. Pelee was itself an ephemeral feature and had collapsed by September of the following year.

However, the most closely investigated of all explosive eruptions happened at Mount St. Helens in the Cascade Mountains of the American north-

Figure 13.67 Ruins of St Pierre after the 1902 disaster. Mont Pelée and its spine in the background. (*A. Lacroix*)

Figure 13.68 A nuée ardente eruption of 16 December 1902 approaching the sea down the valley of the Rivière Blanche, Martinique. (*A. Lacroix*)

west in 1980 (Figure 13.70). St. Helens, which had last been in eruption in 1832–1857, was the first of the Cascade volcanoes to erupt since Lassen Peak in California in 1915. Seismic monitoring of the volcano had commenced in 1969 and already, in the early 1970s, there were presentiments about renewed activity. Strong seismic tremors preceded the eruption in the latter half of March, 1980 and, on March 27th, a powerful summit explosion took place. It was deduced that heat from rising magma had melted some of the snow and ice high on the volcano and that interaction of this with the magma provoked a phreatic explosion.

A dramatic change occurred in the middle of the following month, when the north side of the volcano began to bulge ominously. The bulge inflated so rapidly that, by May 12th, it had grown out to more than 150 m from its original surface level, and at this stage the local authorities severely restricted public access to the area around the mountain. Then, after further vigorous earthquake activity, the bulge broke and the entire upper side of the volcano peeled loose as a gigantic landslide. As it did so a great volume of gas and ash shot sideways, followed promptly by another that arose vertically from the summit area. It is

probable that wet rocks within the volcano's superstructure were heated by the magma whose emplacement generated the bulge and that, with release of pressure when the bulge collapsed, superheated waters flashed into steam, propelling the horrific lateral blast. These events may have opened new fractures near the summit allowing the second huge gas and ash column to ascend above the volcano. The laterally directed blast fanned out away from the cone, creating a vast swathe of desolation through the surrounding forests, and the great rush of gas and dacitic ash was hot enough to ignite vegetation. An avalanche, resulting from the breakaway bulge, reached a nearby lake, displacing the waters so that they over-rode the ridge on its far side. In addition, a mudflow lubricated by meltwaters and rain poured down, riding over the avalanche debris.

In this eruption, in a sparsely-populated region, there were 60 fatalities, mainly from asphyxiation. The eruption occurred by day, in good weather and has been closely documented. Despite the size of the St. Helens eruption, it has been calculated that the quantity of ash produced was only about 1/30th of that which was liberated in the prehistoric Mt. Mazama event, further to the south in the Oregon sector of the Cascade Range, which has been discussed in an earlier section.

After these violent happenings degassed magma was erupted as a series of domes within the summit crater, one of which attained a height of over 300 m in May of the following year.

13.11 VOLCANO PREDICTION

Accurate prediction of eruptions remains fraught with difficulty. The best forecasts can be achieved for those volcanoes that erupt frequently and which have been closely studied, so that their eruptive behaviour patterns are reasonably well understood. No volcano however, is entirely regular in its habits. Much can be learned from detailed investigation (volcanological, geochemical, geochronological etc.) of the lavas and pyroclastic deposits produced by its previous eruptive cycles. However, of the 550 or so volcanoes registered as 'active' or 'potentially active' few have had the degree of investigation necessary for even a rudimentary appreciation of average repose times and the nature of previous eruptions.

For volcanoes that erupt frequently and are comparatively well understood (among which Kilauea, Etna and Piton de la Fournaise can be listed), forecasting of eruptions some weeks, or at least several days, in advance of eruption is possible. In the case of Krafla (northern Iceland), which saw a number of small basaltic eruptions between 1975 and 1983, the behaviour patterns were sufficiently well established for prediction of eruption commencement down to days or even hours.

However, most volcanoes remain inadequately studied and largely unpredictable. Such uncertainties present grave problems to civil authorities who may have to organize population evacuations and emergency plans. A false alarm can give rise to severe political and economic consequences whereas failure to make an accurate prediction

Figure 13.70 Steam and gas exploding outwards from the north face of Mt St Helens, May 1980.

could lead to large-scale disaster. Repose periods, i.e. the interludes of quiescence between activity, vary very widely through hours or days (e.g. Stromboli) through years, hundreds of years or even thousands of years. In the case of those with long repose times, any prediction as to future activity on a time-scale long enough to be of use to civil authorities is rather unlikely. To predict events such as those at El Chichon, Nevado del Ruiz or Lake Nyos would require a degree of constant monitoring which, in the world's present political/economic circumstances, is wholly unrealistic and unattainable.

Thorough surveillance of a volcano would certainly require adequate facilities for detecting and interpreting seismic disturbances, since ascent of magma is invariably associated with earth tremors. However, sophisticated seismic monitoring is very expensive and beyond the reach of many economies.

As near-surface conduits fill with new magma (producing shallow sills, dykes etc.) the volcanic edifice inflates. Repeated fillings and emptyings may occur without surface eruption occurring so that a volcano may appear to 'breathe' as it swells and subsides. The changes in elevation and changes in slope are likely to be slow and unde-

Figure 13.69 View of the growing dome of Mont Pelée in 1930, showing small spines in various stages of protusion and collapse. (*After a photograph by F. Perret*)

tectable without precise surveying and use of sensitive tilt-meters. In the case of some volcanoes (e.g. Piton de la Fournaise, Reunion), lunar tidal forces have been shown to be influential in controlling eruptions. Critical clues regarding movement of magma (and its effect on groundwater systems) can be gained from measurement of the volumes, rates, temperatures and composition (including pH) of crater-lakes, springs, geysirs, fumaroles etc. associated with the volcano. The rates of change in all these factors (seismic, ground-elevation, temperatures and compositions etc. of surface waters and gases) can provide critical evidence for surveillance teams. The 1991 eruption at Mt Pinatubo, Philippines has been referred to in preceding sections. During the volcanological monitoring and predictions related to this eruption measurements in the changes in SO_2 concentration in the gases played an important role.

In recent years instrumental surveillance by aircraft and satellites has also proved highly valuable in volcano watching by permitting data on ground temperatures and the compositions of gas plumes to be gathered. By such means information can be collated for volcanoes in remote and/or underdeveloped regions. The preparation of 'hazard maps', based on topographic maps and knowledge of the volcano's previous behaviour can be of major help to local authorities who may have to organize population evacuations. Production of such hazard maps has recently been important in reducing casualties in two contemporary (1991) eruptions (Mt. Unzen, Japan and Mt. Pinatubo, Philippines) where death tolls could have been much greater.

A considerable number of volcanoes (particularly those above subduction zones) present an ever-present threat; many of these are close to population centres in Japan, south-east Asia, New Zealand, Central and South America. In Europe, those in southern Italy merit special attention. There are some volcanoes which have, in their past histories, given rise to major pyroclastic eruptions and which should be regarded as 'sleeping dragons'. Among these might be instanced Rabaul in the Soloman Islands, Long Valley in California and Taupo in New Zealand.

13.12 FURTHER READING

Cas, R.A.F. and Wright, J.V. (1987) *Volcanic Successions, Modern and Ancient.* Allen & Unwin.

Macdonald, G.A. (1972) *Volcanoes.* Prentice Hall, Englewood Cliff, New Jersey.

Volcanoes and the Earth's Interior (1982) Readings from *Scientific American.* W.H. Freeman, San Francisco.

DATING THE PAGES
OF EARTH HISTORY

The terror with which men await the end of
the world decides me to chronicle the years
already passed, that thus one may know
exactly how many have elapsed since the
earth began.

St. Gregory the Great (540–604)

14.1   HUTTON TO KELVIN: THE GREAT CONTROVERSY

Until the discovery of radioactivity, geologists were in the same position as a historian who knew, for example, that the Roman invasion of Britain was followed by the Norman Conquest and that both these events occurred before the Battle of Waterloo, but who could not find any record of the dates of these or any of the other great events of history. The geologist had to be satisfied with putting the scattered pages of Earth history into their proper order, with establishing a purely relative chronology, a chronology without years. A geologist studying the British Isles knew for example that the major deposits of coal had accumulated long before Northern Ireland was flooded with basaltic lavas, but he could no more than vaguely guess at the ages in years of a coal seam or of a column of basalt from the Giant's Causeway.

Hutton himself, in the absence of any guiding data, quite properly made no attempt to estimate the rates of geological processes. He contented himself with having recognized that geological time was unimaginably long. Many of his successors, however, became unduly reckless in their extravagant claims. Kelvin (1824–1907), one of the great pioneers of geophysics, then entered the field with a dramatic counterblast against speculative estimates that assumed a background of almost unlimited time. Penetration of the Earth's crust by bore-holes and mines shows that temperature increases with depth. This fact means that there is a flow of heat from the interior to the surface, and the Earth is therefore losing heat. Kelvin argued that this indicated that the Earth must have been progressively hotter in the past. Whereas Hutton could see no sign of a beginning, Kelvin, as a physicist, saw a beginning corresponding to a time when the Earth was a molten planet, newly born, as was then generally supposed, from the sun. In 1862 he set himself the problem of calculating the time that had elapsed since the Earth's consolidation from a molten state. Because of uncertainty in the data then available he allowed wide limits to cover the possible errors, and within those limits he found that the crust had solidified between 20 and 400 million years ago.

Kelvin's challenge initiated one of the great controversies that enlivened the scientific world in Victorian times. Despite energetic protests from the geologists, however, Kelvin felt justified by 1897 in narrowing his limits to 20 and 40 million years. Here it is of interest to notice that with the far superior data of the present day the solution to Kelvin's problem – as he posed it – is between 25 and 30 million years. Archibald Geikie responded in 1899 by pointing out that the testimony of the rocks emphatically denied Kelvin's inference that geological activities must have been more vigorous in the past than they are today. In this Geikie was perfectly right. Evidently some unconsidered factor – or factors – governing the Earth's behaviour remained to be recognized. And, strangely enough, although one such factor, radioactivity, had already been discovered, its bearing on the Earth's thermal history continued to remain unrecognized for a few more years.

Looking back now, when it is easy to be wise after the event, the general nature of the flaw in Kelvin's argument is quite obvious. Indeed, Kelvin himself was aware of it and pointed it out, but only as a theoretical possibility. The flaw lies in the assumption that the Earth must be cooling because it is losing heat. We do not say an electric fire is cooling because it is losing heat. From the moment when it is first switched on it is losing heat – otherwise it would not begin to heat the room – but it is also getting hotter and it continues to do so until a balance is achieved between the heat electrically generated and the heat lost by radiation, conduction and convection. Only when the current is reduced or switched off does cooling begin. Kelvin's treatment of his problem was concerned only with the limiting case when the 'current' was switched off, or more accurately, when there was no 'current', i.e. no internal source of heat, at all.

The physical nature of at least one of the missing factors in Kelvin's treatment was dramatically disclosed in 1906 when R.J. Strutt (later Lord Rayleigh) detected the presence of radium in a great variety of common rocks from many parts of the globe. This fundamental discovery, that the crustal rocks contain radioactive elements and are consequently endowed with an unfailing source of heat, showed that the Earth is not living on a dwindling capital of internal heat inherited from the sun, as Kelvin assumed, but that it has an independent and regular source of heat-income of its own. Estimates of the age of the Earth based on the outward heat-flow at once became valueless.

14.2 VARVED SEDIMENTS

To measure geological time in years, the first essential is the recognition of a natural process that operates rhythmically or at a known rate from a defined starting-point and brings about a measurable result. The traditional geological method of estimating time from thicknesses of sediment is unreliable because of the fact that rates of sedimentation are extremely variable. However, here and there in the geological record there are rhythmically laminated sediments made up of pairs of laminae (e.g. of silt and clay) which can be recognized as representing a year, like the annual rings of a tree. Each pair is called a **varve** (Swedish, *varv*, a periodic repetition) and sediments characterized by this annual banding are said to be varved. By counting the successive couplets or varves in a given section, the time represented by the latter can be stated in years (Figure 14.1).

The most celebrated example of time measurement by counting varves is that provided by De Geer (1858–1943) and his co-workers, whose detailed work has made possible an exact system of geological chronology covering the period of retreat of the last Scandinavian ice-sheet up to the present day. The receding ice front terminated in a great marginal lake of which the present Baltic Sea is the descendant. Each spring and summer, as the ice thawed, the lake in front received a supply of sand, silt and clay from the streams that flowed into it. The coarser material settled down at once, but the finer particles remained longer in suspension and were not completely deposited until much later in the year. But during the late autumn and winter the glacial streams were frozen and the lake, itself frozen over, received no further sediment. The mud that still remained suspended slowly sank to the bottom, forming a thin layer of dark clay, easily distinguishable from the thicker layer of sandy silt beneath it. The following year the sediment liberated from the ice was again sorted out and deposited in two well-marked seasonal layers, sharply separated from the underlying pair. As this process continued year after year the area of deposit moved northwards with the receding ice, and the varves thus became superimposed after the fashion of wedge-shaped tiles on a roof.

In 1884 De Geer began to count the varves near Stockholm and in later years when he had devised a method of correlation, the record was systematically carried backwards in time to the coast of Scania at the southern end of Sweden, and forwards to Lake Storsjon, 1000 km to the north. At any one locality the thickness of the varved deposit rarely exceeds 10 m, but this may contain several hundreds of varves, each representing a single year (Figure 14.1). Those near the top can then be matched against the lower varves at a neighbouring locality to the north, where the sequence includes the varves of the immediately succeeding years. Thus, by tracing the overlap-

Figure 14.1 Varved sediments in a clay pit, Uppsala, Sweden. The photograph shows 60 varves in a thickness of about a metre. (*A. Reuterskiold, 1920*)

ping varves through sections and borings at more than 1500 localities, and so carrying on the counting bit by bit, De Geer and his students succeeded, after 30 years of laborious work, in establishing an accurate chronology of the retreat stages of the last of the European ice-sheets. The Ice Age was conventionally regarded as having ended about 8700 years ago, when the ice-sheet had retreated to Ragunda and separated into two isolated ice caps. Varves continued to be deposited in Lake Storsjon until 1796, when collapse of the morainic barrier holding back the water catastrophically drained the lake. The varves of the lake-floor sediments made it possible to complete the count up to a date known historically. About 13 500 years have elapsed since the ice front stood along what is now the coastline of southern Sweden.

The only older period of which the duration has been estimated from varve investigations is the Eocene. W.H. Bradley discovered long sequences of varved sediments in the Green River formation, which consists of lake deposits about 650 m thick in Wyoming, Colorado and Utah. Here, the dominant type of varve has two laminae, one being markedly richer in organic matter than its associate, exactly as happens in the annual lamination found in many present-day lake deposits. By counting the varves wherever they occur, Bradley estimated the Green River epoch to have lasted some 5 to 8 m.y. (6.5 ± 1.5 m.y.). Assuming the rate of accumulation of the Eocene lake deposits above and below the Green River Series to have been the same as that of the latter, the duration of the whole Eocene period comes out at about 22 ± 5 m.y. This estimate compares well with the 17 million years allotted to the Eocene in Figure 14.7.

With few exceptions varved sediments record only short intervals of time. Apart from these, geological 'hour-glass' methods are necessarily based on little more than crude guesses.

In 1878 an Irish geologist, Samuel Haughton, introduced a principle that has often been used for lack of a better. In his own words 'the proper relative measure of geological periods is the maximum thickness of the strata formed during these periods'. The figures listed in Figure 14.7 add up to the astonishing total of 137 769 m for the fossiliferous systems alone. However, beyond emphasizing the immensity of geological time these figures are of little help in solving the age problem, because we do not know the rates at which the beds were deposited. To get a preliminary idea of the sort of timescale that is implied we might take the sedimentation of the Nile – a fast worker – as an example. Since the reign of Rameses II, over 3000 years ago, 30 cm of sediment have been added to the neighbourhood of Memphis every 500 years approximately. At this rate the fossiliferous strata of the geological column would represent about 200 million years. But neither this nor any other estimate of its kind can be taken seriously, and it is not surprising that the geologists of last century should have failed to reach agreement in their efforts to assess the time elapsed since the beginning of the Cambrian period. Those who yielded to the cramping influence of Kelvin's authority favoured short estimates, like the 27.6 m.y. of Walcott (1893) and the 18.3 m.y. of Sollas (1900). A few steadfastly refused to be bullied into respectable conformity. Goodchild in particular deserves a salute for his 1897 estimate of 704 million years.

It is easy to find fault with Haughton's principle, but it is only now, about 100 years later, that it can be dispensed with. It should be noticed, moreover, that the enormous maximum thicknesses that have been recorded for the various systems (Figure 14.7) imply that space must have been made available for their accumulation by long-continued down-sagging of the crust. In a very rough-and-ready way the maximum thicknesses are measures of the maximum crustal depressions during the periods concerned. The next task was to tie these maximum thicknesses to actual age determinations based on the phenomena of radioactivity. Sollas was obviously unhappy about his 1900 estimate, for he exclaimed: 'How immeasurable would be the advance of our science could we but bring the chief events which it records into some relation with a standard of time!' Today that hoped-for advance is in full acceleration.

14.3 RADIOACTIVITY

The first of the radioactive methods to be developed was based on the generation of helium and lead from uranium and thorium, and the accumulation of these stable end-products in radioactive minerals. The story begins in 1896 when the French physicist Becquerel most unexpectedly found evidence that uranium salts give out invisible rays. These rays first revealed their existence by their effects on photographic plates wrapped in black paper in a closed drawer. Madame Curie, then a young research student, followed up this epoch-making discovery by testing all the other elements then known. She was soon able to announce that thorium also possesses the astonishing property of spontaneously emitting radiations. For the new phenomenon she proposed the name **radioactivity**. But nothing impressed her more than the discovery that uranium minerals are far more powerfully radioactive than the uranium they contain. This clearly pointed to the presence in the minerals of a previously unknown and highly radioactive element, or possibly of more than one.

Working with pitchblende (a mineral consisting mainly of the oxides of uranium) Madame Curie and her husband, Pierre Curie, detected two new radio-elements before the end of 1898. The second of these was radium. Later, when radium was isolated, it was found to be continuously generating heat. This is true of every radio-element, but in the case of radium, which is about three million times as active as uranium, the heat was easily detectable with an ordinary thermometer. This discovery made it clear that there was a flaw in Lord Kelvin's arguments, since there was now a source of energy capable of keeping the Earth hot for long periods of time.

Ernest Rutherford found that the radiations emitted are of three different kinds which he distinguished as α particles, β particles and γ-rays.

α particles are beams of positively charged particles which were eventually shown to be the nuclei of helium atoms (2 protons + 2 neutrons), ejected from the nuclei of the disintegrating atoms with velocities of thousands of kilometres per second. The resulting atomic collisions are the main source of the heat generated by radio-elements that discharge α particles.

β particles are beams of negatively charged particles, expelled even more swiftly. These were soon identified as electrons, but electrons formed in the nucleus, not from the surrounding orbital swarm. One of the neutrons of a radioactive nucleus spontaneously shoots out an electron (–) and turns into a proton (+).

γ-rays are electromagnetic radiation like X-rays of very short wave length; they travel with the velocity of light.

Now clearly when the nucleus of a radium atom, for example, loses a helium nucleus (i.e. an α particle), the element that is left can no longer be radium. The new, residual, element turns out to be a gas; it is called radon and is also radioactive. Radon in turn loses helium and is transformed into another radio-element. There are several generations of these daughter elements, some losing helium nuclei (α particles), until eventually the series comes to an end. Apart from the helium that has been generated on the way, the stable element that is finally left is an isotope of the familiar element lead.

Obviously, left to itself, radium could not continue to exist indefinitely. It loses helium at such a rate that after 1622 years half of it will have decayed into its daughter elements. This period is described as its **half-life**. Obviously, unless

radium were being continuously generated from something else, the world's supply – however great initially – would have dwindled away to practically nothing long ago. Actually, it was soon discovered that radium is constantly being renewed as a member of one of the families of radio-elements that descend from uranium. The parental element, uranium, as is well known nowadays, consists almost entirely of two isotopes, ^{238}U and ^{235}U. Each is the progenitor of a large family of elements, all of which are radioactive except the end-products, helium and lead. A third family, also terminating in an isotope of lead, has thorium, ^{232}Th, as its parent. From being a father of gods, Uranus has become, with Thor, a father of elements. From being a wielder of thunderbolts, Thor has become, with Uranus, a scintillating generator of high-speed atomic particles (Figure 14.2).

The fact that there is still a good deal of uranium left in the world shows that the half-lives of its isotopes must be enormously longer than that of radium. A near-coincidence worth remembering is that the half-life of ^{238}U is practically the same as the age of the Earth. That is to say there was twice as much ^{238}U in the world 4469 million years ago (the half-life of ^{238}U) as there is today (Figure 14.3).

14.4 THE GEOLOGISTS' TIMEKEEPERS

Omitting the energy that is liberated and appears as heat, the ultimate results of these atomic trans-

 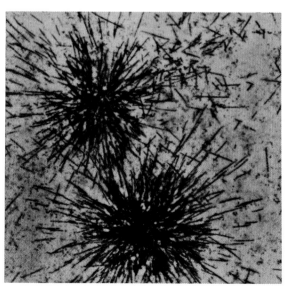

Figure 14.2 Photographic demonstrations of radioactivity. *left* Auto-radiographs of monazite included in feldspar in a thin section of granite from Predazzo, Italian Alps. The section was placed on a photographic plate (unexposed to light) for 36 days. The radiation has blackened the area covered by the monazite ($Ce(PO_4)$), but beyond this the tracks of individual α particles are seen as darkened radial lines. *right* Zircons, containing minute granules of xenotime (YPO_4), included in feldspar in a thin section of Predazzo granite. Exposure 36 days, as in the photograph to the left. (*O. H. Merlin and E. Justin*)

264

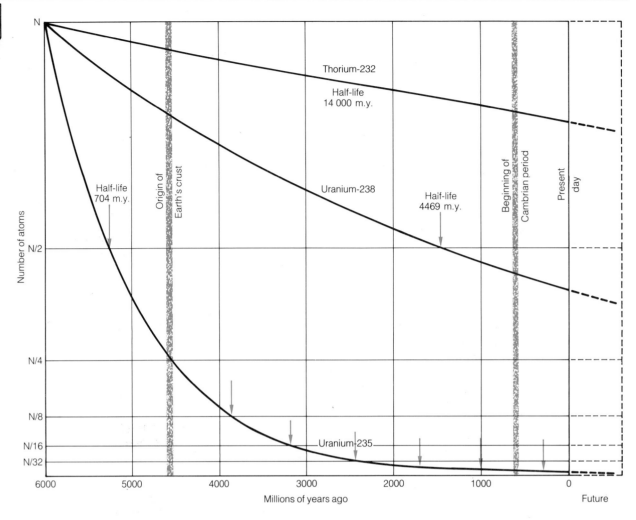

Figure 14.3 Diagram to illustrate the radioactive decay of uranium and thorium, starting with N atoms of each isotope 6 000 m.y. ago. It will be seen that while U-235 has passed through six of its half-lives (and is well into its seventh) during geological time, U-238 has passed through only one, and Th-232 through only a fraction of one. (N.B. Ma instead of m.y. is now used by many authors.)

formations can be summarized as follows:

$$^{238}U \rightarrow \ ^{206}Pb + 8 \ ^{4}He$$
$$^{235}U \rightarrow \ ^{207}Pb + 7 \ ^{4}He$$
$$^{232}Th \rightarrow \ ^{208}Pb + 6 \ ^{4}He$$

The ordinary industrial lead, obtained from common lead minerals like galena, is a mixture of the same three isotopes, together with a fourth, ^{204}Pb, which, unlike the others, is not a product of radioactive decay. Throughout geological time the other isotopes have been steadily increasing in abundance, while ^{204}Pb has remained unchanged in amount. This is highly fortunate, because its amount provides a measure of the proportion of inherited lead which was present as an original constituent of the mineral at the time of its crystallization. Such lead must be subtracted from the total lead in order to arrive at the amount that has been generated from radioactive decay of U and Th within the mineral during its lifetime.

To measure the proportions of atoms of each isotope of an element, requires an instrument that is capable of separating an element into constituent atoms according to their mass. Such a machine is called a mass spectrometer and was developed during this century from an early prototype built by the famous physicist J.J. Thompson at the Cavendish Laboratory in Cambridge University. Thompson had already discovered the electron, and used the first 'mass spectrograph' as it was then called, to discover that the rare gas neon had atoms with more than just one mass, or more than one isotope.

The earliest radioactive-dating measurements utilized uranium-rich minerals such as pitchblende and were soon capable of demonstrating that indeed the Earth was considerably in excess of the 40 million year maximum that Lord Kelvin gave as his last estimate. However uranium minerals are not common constituents of the main spectrum of rocks that most geologists are interested in dating. It was clearly advantageous for techniques to be developed that provided for the direct dating of common rock types. Nowadays there are a number of different decay schemes that are utilized for measuring geological time. Moreover the range of minerals, and hence rock types that can be dated, has expanded considerably over the years. In order for this to happen, the techniques had to be developed for dating minerals that contained only trace amounts of the radionuclide of interest. Nowadays the isotopic compositions of elements present at the parts per million level can be measured accurately, allowing for the direct determination of the age of a wide variety of minerals from common rock types.

The decay schemes that are most commonly deployed for dating rocks are given in Figure 14.4. In addition to the decay schemes of the actinide elements uranium and thorium, radioactive isotopes of the alkali elements potassium (K) and rubidium (Rb) and the rare earth element samarium (Sm) have been extremely useful in deciphering the history of the Earth and Moon.

The radioactive isotope of rubidium, ^{87}Rb, decays in a very simple way. On disintegration an atom of this isotope loses a β-particle from its nucleus, which means that a neutron in the nucleus becomes a proton. The mass number remains the same, but the atomic number increases from 37, that of rubidium, to 38, which is that of strontium. An atom of rubidium has been transformed into one of strontium:

$$^{87}\text{Rb} \rightarrow {}^{87}\text{Sr} + \beta^-$$

The radioactivity of potassium is more complex. The radioactive isotope is ^{40}K, and of the atoms that disintegrate about 89 per cent behave like ^{87}Rb, each losing a β-particle from its nucleus, and so changing to ^{40}Ca, with an atomic number of 20 instead of the original 19:

$$^{40}\text{K} \rightarrow {}^{40}\text{Ca} + \beta^-$$

Since ordinary calcium is a very common element and is itself mostly ^{40}Ca, the minute additions contributed by potassium are generally lost beyond recognition, like raindrops in the sea. This change is therefore of very limited value for dating purposes.

However, the remaining 11 per cent of the ^{40}K atoms disintegrate by an opposite process, so to speak. Each nucleus captures an orbital electron, so that a proton becomes a neutron and the atomic number decreases from 19 to 18, which is that of argon, a noble gas. This transformation can be widely used for dating minerals:

$$^{40}\text{K} + e^- \rightarrow {}^{40}\text{Ar}$$

The radioactive isotope ^{147}Sm decays to ^{143}Nd with the emission of an α particle.

$$^{147}\text{Sm} \rightarrow {}^{143}\text{Nd} + \alpha^{++}$$

Since Sm and Nd are both rare earth elements with very similar properties, the ratio of Sm to Nd is relatively uniform in nature. This means that the differences in Nd isotopic composition brought about by the decay of ^{147}Sm are small in nearly all rocks and minerals. The techniques for making the very precise measurements required with this decay scheme were not developed until the early 1970s.

Various combinations of these techniques have been deployed to study different geological problems. U–Pb dating is nowadays most commonly used for dating the mineral zircon (ZrSiO$_4$), an accessory mineral of widespread occurrence in many granites and gneisses. Zircon has proved most useful because of its resistance to later resetting. (The modification of isotope composition as a result of diffusive losses or changes in mineral chemistry brought about, usually, by heating of the rock). The oldest mineral grains yet identified on Earth (with the exception of samples from

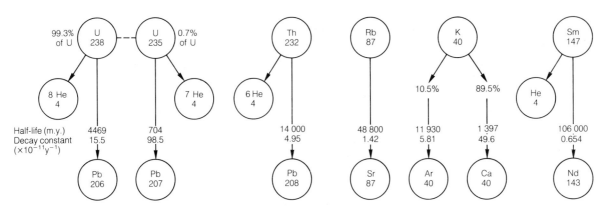

Figure 14.4 Radioactive timekeepers: diagram showing the parent isotopes, their half-lives (in millions of years), decay constituents and end products.

meteorites) are zircons that have been dated using this method. Scientists at the Australian National University have discovered grains that give ages as old as 4.1 to 4.2 gigayears (1 gigayear (Gyr or Ga) = 1 thousand million years). The U–Pb method is also the technique by which the age of the solar system was established at 4.6 gigayears by Clair Patterson at the California Institute of Technology in the 1950s.

The Rb–Sr method has been widely used for dating granitic rocks in particular. This is because Rb is an alkali element with similar ionic charge and ionic radius to potassium. Since granitic rocks tend to be potassium rich, they tend to also have large amounts of rubidium, particularly in micas. While many such age determinations on granitic rocks have been superseded by more precise U–Pb zircon dating in recent years, the Rb–Sr method was the favoured technique for much of the routine geochronology of the continental crust, as well as of many old basaltic lunar samples and meteorites.

The K–Ar method is the most commonly used technique for dating relatively young geological processes. This is for three reasons. First potassium is such an abundant element that it is easy to measure the argon in young rocks. This is particularly true using potassium-rich minerals such as micas and alkali feldspars but even extends to rocks with low potassium such as Cenozoic basalts. For example the only isotopic method in widespread usage for dating the basaltic rocks of the ocean floor is the K–Ar method. Second, because argon is an inert gas it is not chemically bonded and can be released by moderate heating. For this reason rocks that have been disturbed since formation may have lost argon and yield anomalously low ages (more likely in older rocks). This turns out to be an advantage in some respects because the technique is also ideal for dating thermal disturbances. Lastly, the fact that the ^{40}Ar is lost so readily means that there is very little daughter isotope trapped in igneous rocks when they form, hence little correction to be applied for initial argon, even in the case of the small amounts of radiogenic argon released from young rocks.

The samarium–neodymium method is ideal for dating old mafic rocks, indeed the first rocks from Earth to be dated using this technique were Archaean greenstone belt metabasaltic rocks from Africa – rocks that had previously been undateable. Sm–Nd dating is also very relevant to the dating of garnet – a common mineral in metamorphic rocks, permitting the direct dating of rocks such as eclogites, and the determination of thermal histories of metamorphic terranes.

14.5   READING THE RADIOACTIVE TIMEKEEPERS

The rate of decay of all radioactive elements is found to correspond with a very simple law that was discovered and announced by Rutherford and Soddy as long ago as 1902. The number of atoms, that decay during a given unit of time (e.g. a second or a year, as may be convenient) is directly proportional to the number of atoms of the radioactive element present in the sample concerned. Rutherford and Soddy had discovered that radioactive decay was a spontaneous, random process to which the laws of probability could be applied. Unlike human beings, among whom the chance of dying increases with old age, the atoms of a radioactive isotope are just as likely to live or die (i.e. to decay or disintegrate) whatever their age may be. The probability of decay remains constant for any particular radioactive species or isotope; the number of atoms that decay is simply a function of the number of atoms present. The probability is called the decay constant and is represented by the Greek symbol lambda (λ). It is defined as the fraction of atoms of a particular isotope that can be expected to decay in a given period of time (usually one year). Another way of thinking about the decay constant is that it is the probability of an individual atom undergoing decay in a year.

For example the decay constant for ^{87}Rb is 1.42×10^{-11} per year. Therefore, on average, out of a pool of 10^{12} atoms of ^{87}Rb, only 14 will decay with the passage of a year. Clearly as this particular 'stock' of radioactive parent atoms is progressively depleted by decay, the fraction of the remaining radioactive atoms that decay will stay constant, but the absolute number of atoms that decay will decrease with the passage of time. In other words, as Rutherford and Soddy had discovered, the number of atoms that decay is proportional to the number of atoms present. As the atoms decay the number that are left and 'available' for decay decreases and so the number that decay in a given year gradually changes. The decay constant λ, is related to the half-life ($T_{1/2}$), the time taken for the number of atoms of a radioactive isotope to be reduced to a half, as follows:

$$T_{1/2} = (\log_e 2)/\lambda$$

or:

$$T_{1/2} = 0.693/\lambda$$

The decay constant and half-life differ for each radionuclide, because the nucleus of some isotopes has a more stable configuration of protons and neutrons than that of others. The more unstable the radionuclide, the more likely it is to decay, the more radioactive it is and the shorter the half-life. All radioactive isotopes with half-lives of $< 10^6$ years that were originally present on Earth have entirely decayed. Some (such as radium) are re-created in minor amounts by the chain decay of U and Th discussed above. Short-lived radionuclides are also produced in nuclear reactions such as occur in reactors and atomic explosions, or when cosmic rays collide with atoms in the atmosphere. The radioactive isotopes that are useful for dating over a geological timescale need to have very long half-lives otherwise they will have entirely decayed and no longer be detectable. As a rough guide, for a radionuclide to be of use in dating rocks it must have a half-life that is within two orders of magnitude of the age of the Earth (4.6 Gyrs).

It can be seen from Figure 14.4 that the radioactive decay schemes in common use all have such half-lives. Obviously these radionuclides are so mildly radioactive that they are not a serious health threat. However some of the more highly radioactive intermediate daughters of the U decay series, notably the gas radon, can accumulate and form dangerous concentrations that can percolate through shallow crustal acquifers, rocks and soils to pose a serious health risk.

For those who are mathematically minded we may now briefly consider the general formula for calculating isotopic ages.

Let N_P be the number of atoms of a parental radionuclide now present in a given sample of a mineral that crystallized t years ago, t being the age of the mineral.

Let N_D be the number of atoms of the end-product – the stable daughter nuclide – generated in the sample during the time t.

N_P and N_D are known by analysis of the sample, and our problem is to calculate the value of t from the analytical results. For this we need an equation relating t to the decay constant, λ.

From Rutherford and Soddy's fundamental law of radioactive decay, the number of atoms of the parent element originally present in the sample when the mineral crystallized can be expressed as $N_P.e^{\lambda t}$, where e is the base of Napierian logarithms. Thus we arrive at the equation

$$N_P.e^{\lambda t} - N_P = N_D$$

which reduces to

$$e^{\lambda t} = 1 + N_D/N_P$$

Taking logarithms to the base e of each side of this equation

$$t = (1/\lambda) \log_e (1 + N_D/N_P)$$

This is the basic equation used for calculating the ages of a variety of rocks and minerals. An obvious question you might ask is: how do you know that there were no daughter atoms present in the rock when it formed? This can be a problem and limitation to the applicability of the techniques in some instances, but there are two ways around the problem. The first is to analyze minerals that have negligible initial daughter isotopes – this is one of the advantages of the U-Pb zircon method for example. The second is to make a correction based on measurements of the isotopic composition of the daughter element in other minerals and rocks that formed at the same time and from the same, isotopically uniform material (e.g. the same magma).

14.6 THE AGE OF THE EARTH AND THE SOLAR SYSTEM

The uncertainty over the age of the Earth and solar system was finally settled in the 1950s by Clair Patterson from the California Institute of Technology, who, using U–Pb dating of meteorites determined an age of between 4.5 and 4.6 Gyrs. Prior to that time there had been many attempts to arrive at a best estimate using a variety of lead isotopic studies of galenas. Patterson's experiment was impressive in its elegance and quality and fundamental in its significance. He used iron meteorites, which contain measurable amounts of lead but negligible uranium, to determine the initial lead isotopic composition (the initial proportions of the different isotopes of lead) of the solar system. Since iron meteorites were thought to represent fragments of the cores of disrupted planets which are generally thought to form early, the lead isotopic composition of iron meteorites preserves the composition of lead from the earliest history of the solar system. The lead in the Earth's continental and oceanic crust has a very different isotopic composition because of the thousands of millions of years of decay of uranium and thorium that have contributed additional ^{206}Pb, ^{207}Pb and ^{208}Pb to the mantle and crust since formation of the planet and its core. Even though lead from the Earth's surface is isotopically variable because of radioactive decay, Patterson chose some representative samples to determine an approximate average age for the 'silicate Earth'. He also analyzed some silicate meteorites and found that they were even more enriched in ^{206}Pb, ^{207}Pb and ^{208}Pb. The differences in $^{206}Pb/^{204}Pb$ and $^{207}Pb/^{204}Pb$ shown in Figure 14.5 are due to the passage of time and the different amounts of uranium relative to lead (U/Pb ratio) of the iron and silicate meteorites. However the real power of this diagram is that because the differences in U/Pb ratio affect both the $^{238}U–^{206}Pb$ and the $^{235}U–^{207}Pb$ systems,

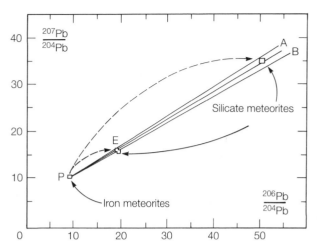

Figure 14.5 Isochrons for 4600 m.y. (A) and 4500 m.y. (B), passing through P, representing the primeval lead of iron meteorites. Broken lines show the changes in isotopic proportions from P to the present day; the silicate meteorites at (E) have isotopic compositions similar to present day average silicate Earth. (*C. Patterson*)

the U/Pb ratio cancels out and the variations in $^{206}Pb/^{204}Pb$ and $^{207}Pb/^{204}Pb$ on their own can be used to determine the age. This is shown in Figure 14.5 where the slopes are calibrated in terms of time. It can be seen that the iron and silicate meteorites are between 4.5 and 4.6 Gyrs in age. The average silicate Earth plots as if it is the same age and formed from the same initial solar system material. In other words Patterson did not directly date the Earth. There are no remains of the original Earth surface that have been identified permitting such a measurement. What he showed was that the lead isotopic composition of the silicate portion of the Earth was consistent with it having formed at the same time as other solar system materials at 4.5 to 4.6 Gyrs.

14.7 THE AGE AND GROWTH OF THE EARTH'S CRUST

The ages of broad sections of the continental crust have been established in a general sense since the 1960s. In more recent times emphasis has been placed on obtaining more accurate ages, finding the oldest traces of the Earth's crust, and deriving an estimate for the average age of the continents. The oldest rocks to be found thus far are the 4 Gyr Acasta Gneisses of Northern Canada, dated by Samuel Bowring. However rocks of very similar age have been found in west Greenland, Labrador, Africa and Australia. In fact it was pointed out in the 1970s by Stephen Moorbath and his colleagues at Oxford University, that the ages of Precambrian gneisses and granites from around the world seemed to define specific episodes, thought to be global-scale crustal growth events. The history of crustal growth has long been a matter of considerable debate and controversy. Figure 14.6 shows a 'geochronological map' of the world with the approximate distribution of rocks

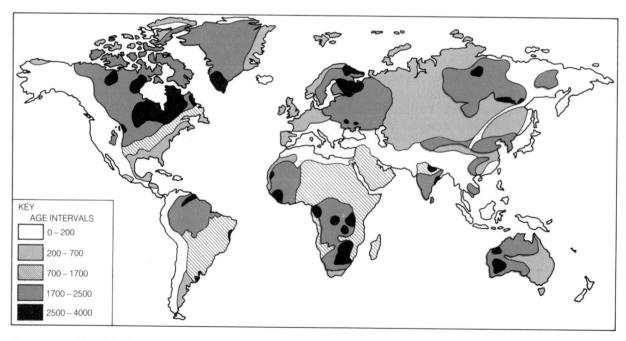

Figure 14.6 Simplified geochronological maps of the world showing the major areas of rocks formed in different time intervals.

of different ages indicated. It can be seen that much of the Earth's continental crust is composed of rocks that formed since the late Precambrian.

This map on its own says nothing about the average age of the continents however. Most rocks that form the continental crust have formed from pre-existing crust. This includes all sedimentary and metamorphic rocks for example. Igneous rocks are a mixture of rocks such as basalts, which are formed by melting of the mantle and represent new additions to the continents, and rocks such as granites, which derive a major portion of their constituents by the remelting of existing continental crust. Clearly then we need some technique for dating the process of formation of continental crust from mantle, rather than simply dating the age of rock formation. The best technique currently available for this is the Sm–Nd method. This is because the continental crust and mantle both have a relatively uniform but distinct ratio of Sm to Nd. The isotopic composition of Nd in an average sample of the continental crust is then a function of the average time that the continental crust grew from the mantle (or the Sm/Nd ratio changed). Obtaining an average sample of the crust is surprisingly simple, although plagued with assumptions regarding representivity. The rivers that carry sediment to the sea and the wind-blown dust that is carried to the middle of oceans through the atmosphere both mix continental crust from broad areas. Keith O'Nions at Cambridge University has made extensive study of such materials and determined an average age for the upper crust of a little less than two gigayears. How representative this is of the total continental crust is unclear at present.

While we know the average age of the continental crust reasonably well, this is still not all we need to know to determine the growth history of the continents. The reason for this is that while crust is being created it is also being destroyed. At subduction zones sediment from the continental crust is transferred back into the Earth's mantle. Until we can find a mechanism for determining the amount of crust that has been transported back into the mantle as a function of time we will be at a loss to find a definite solution to this problem. Several scientists have tried to solve this problem or place limits on it, but the results are so far inconclusive and the debate continues.

What of the period of time between the formation of the Earth at 4.6 Gyrs and the oldest known rocks at 4.0 Gyrs? Was there any continental crust formed at this time and if so how was it destroyed? Funnily enough there are more obvious answers to the latter than the former. From the samples brought back from the Moon by Apollo astronauts, Gerald Wasserburg and his colleagues at the California Institute of Technology were able to determine that the Moon had formed by about 4.4 Gyrs. Unlike the Earth, the Moon became something of a 'dead' planet and ceased to produce volcanism by about 3 Gyrs. This lack of geological activity has preserved the remains of a primitive pre-4.0 Gyr crust. K–Ar dating of lunar samples has been used to provide constraints on the history of lunar bombardment by the huge meteorite impactors that have left its surface so heavily cratered. These studies of argon loss show that the moon received severe bombardment right up until about 4 Gyrs when things appear to have eased off.

This bombardment was probably a tailing off of the accretion of dust and giant impactors that originally resulted in the formation of the Earth–Moon system. The Earth, being of much greater mass, would attract far greater numbers of meteorites and the amount of energy added from this bombardment may have been sufficient to melt the upper portion of the planet to the extent that all of the Earth's 'protocrust' was destroyed. Some have suggested for a variety of reasons that this crust may have been basaltic, rather like the present day oceanic crust and quite different from the continental crust.

There is now some direct evidence for the former existence of an early protocrust. Scientists at the Australian National University, under the direction of William Compston succeeded in building a specially made kind of large mass spectrometer, capable of studying minute samples, called an **ion probe**. They were able to use this instrument to analyze individual microscopic grains of zircon and discovered and dated the oldest mineral grains on Earth yet identified thought to have originated on this planet. A few detrital zircon grains from a metamorphosed sandstone, the Mt. Narryer Quartzite in Australia, have been found to contain grains as old as 4.2 Gyrs. Since zircons are only common in crustal rocks this finding strongly indicates a pre-4.0 Gyr crust.

In contrast to the continental crust, the oceanic crust is entirely young. The age of the ocean crust has been determined by K–Ar dating. Extensive studies have found no sections of the ocean floor that are older than 200 million years. In other words the mean age of the continental crust is far in excess of the mean age of the oceanic crust. This was a very important piece of evidence in the development of our understanding of continental drift and provided the firm confirmation needed that the Atlantic Ocean has been gradually opening.

14.8 THE DEVELOPMENT OF THE ATMOSPHERE AND OCEANS

Determining the age of the crust seems relatively straightforward since it involves the direct dating of rock fragments. But how does one date the formation of the atmosphere and oceans? Until the 1980s there was no obvious solution to this problem. One could certainly say that the atmosphere and oceans were probably early since some of the oldest rocks on Earth include metasediments that must have formed underwater. But trying to determine a more precise chronology would be difficult.

To begin with it was recognized in the 1950s that the Earth's atmosphere had to be secondary. That is, the atmosphere of the Earth was totally different from the sun, even allowing for gravitational loss of hydrogen and helium. For example the Earth's atmosphere is very depleted in rare gases such as neon which should be more abundant if we had the remains of a solar atmosphere. Some components of the atmosphere have clearly formed by degassing of the interior of the Earth. For example the large amounts (1 per cent) of argon in the atmosphere have formed from the degassing of the Earth's interior where ^{40}Ar is produced from decay of ^{40}K. Because ^{40}K is still present on Earth ^{40}Ar is still being produced and added to the atmosphere. In fact if we knew how much potassium there was in the Earth we could place constraints on the average age of atoms of argon in the atmosphere. Unfortunately such information is not readily obtainable although it has long been considered that atmospheric Ar was dominantly produced early in earth history.

However in the 1980s a very important discovery was made. Claude Allègre and his colleagues at the University of Paris extracted the minute amounts of the rare gas xenon trapped in basalts erupting at the present day along the mid-ocean ridges. They then measured the isotopic composition of this xenon and found that it had a greater proportion of ^{129}Xe than is found in xenon from the atmosphere. This was a very exciting discovery because such a difference could only arise if most of the Earth's atmospheric xenon separated from the Earth's interior by 4.4 gigayears ago at the very latest. The reason for this is that excess ^{129}Xe can only be formed by decay of a short-lived radionuclide of iodine, ^{129}I. As pointed out before, the Earth has lost all of its original inventory of short-lived radionuclides, because of radioactive decay. The half-life of ^{129}I is only 17 million years and basically this means that practically all of the ^{129}I would have decayed away within 100 million years of the last time ^{129}I was formed (during nucleosyn-

thesis in stars). Hence there would be no mechanism for producing a difference in xenon isotopic composition after this time. So if the atmosphere formed by degassing of the Earth's mantle, it (or at least, the xenon) had to separate very early, when there was still live ^{129}I in the Earth's interior.

14.9 CALIBRATING THE PHANEROZOIC TIMESCALE

To construct an acceptable timescale it is necessary to have reliable isotopic age measurements on minerals or rocks of known geological age, and also to have determinations which are evenly distributed, through the periods, in time. So far, neither condition is fully realized. Many results have to be discarded either because the evidence as to stratigraphic age is inconclusive, or because the isotopic ages are not completely reliable.

The rocks and minerals used for isotopic age determinations are derived from one of the following:

1. Lava flows and bentonites interbedded with sedimentary rocks whose stratigraphic age is firmly established from fossil evidence. The isotopic ages of such lava flows or pyroclasts will correspond very closely to the actual ages of the strata with which they are interbedded.
2. Sedimentary rocks, accurately dated stratigraphically from fossil evidence, which contain authigenic (generated *in situ*) minerals, which have grown at the time of, or shortly after, the deposition of the rock of which they form part. The isotopic ages are here the minimum age of the sedimentary rock. Glauconite has commonly been used for isotopic dating, particularly glauconite collected from the cores of deep bore-holes, but can yield erroneous K–Ar ages because of loss of argon. A relatively new approach is based on the discovery that carbonate fossils and limestones contain relatively high uranium relative to common lead. Even though the uranium is only present in sub-part per million levels, this permits the direct and reliable dating of sedimentary rocks. This is particularly appropriate for limestones since they undergo lithification shortly after deposition.
3. Intrusive igneous rocks, and minerals derived from them, that intrude sedimentary rocks of established stratigraphic position. The isotopic ages found in this way are minimum ages of the sedimentary rocks, which are necessarily older than the igneous rocks that cut them.

Many timescales have been constructed for the Phanerozoic, that is from the base of the Cambrian upwards. Phanerozoic is a group name, derived from Greek and meaning 'plainly evident life.' When such timescales were first constructed, isotopic age data were extremely scarce. As more data accumulate it frequently becomes necessary for the timescale to be revised by geologists capable of assessing the value of both the stratigraphic and isotopic evidence concerned, so that only reliable data are used. Figure 14.7 is a timescale for the Phanerozoic, constructed from data available up to 1982 (Harland *et al*, 1982).

Although we know more about the Phanerozoic than earlier periods in Earth history, it represents a relatively small portion of the lifespan of our planet. The rocks on Earth that are older than Phanerozoic (>590 million years old) are termed Precambrian. The Precambrian is largely barren of fossils and this renders the correlation of rock strata far less certain. In addition the uncertainty in isotopic ages in absolute millions of years, increases further back in time. With the improvements in isotopic-dating techniques (particularly in U–Pb zircon), over the past decade, significant advances in calibrating the Precambrian timescale have been made. However even our best chronometers are currently limited to precisions of ±0.1 per cent and most isotopic ages have been determined with a precision of ±1 per cent. This translates into uncertainties in the range ±3 to ±30 m.y. in a rock whose age is 3 Gyrs (or 3000 m.y.). Clearly a lot can happen in 30 million years! For these reasons we are still in a far greater state of ignorance over the geology of the Precambrian than over the Phanerozoic.

14.10   UNRAVELLING THE PRECAMBRIAN

While a detailed picture of Precambrian times is only developing slowly, it was recognized early in the history of isotopic dating that it was possible to identify and correlate periods of similar geological activity on different continents and that global subdivisions were possible. The Precambrian is now subdivided into two major episodes – the Archaean and the Proterozoic. The Archaean is used for rocks older than 2500 m.y. (2.5 Gyrs). The term Proterozoic is applied to rocks between 590 (the base of the Phanerozoic) and 2500 m.y. old, i.e. older than the Cambrian and younger than the Archaean.

The Archaean is commonly characterized by gneiss terranes and granitic rocks that have clearly been excavated from great depth in the continental crust (Figure 14.8). Not all Archaean rocks are like this however. There are also the remains of relatively unmetamorphosed sedimentary rocks and even fossils of primitive life forms, the most common of which are bacterial algae termed stromatolites.

The Proterozoic is generally characterized by a greater abundance of unmetamorphosed sedimentary rocks. In addition some of the features of orogenic belts such as occur in rocks of the Phanerozoic are more plentiful in the Proterozoic than in the Archaean.

As well as this simple subdivision there are some general features of the Precambrian that show a progressive change towards the kinds of rocks (hence presumably conditions) that are typical of the Phanerozoic. The most striking of these

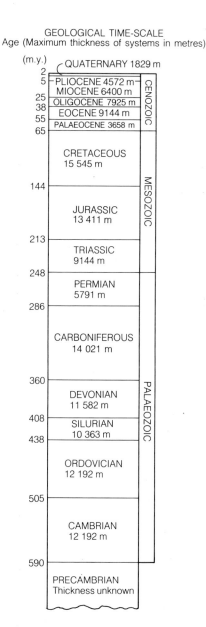

GEOLOGICAL TIME-SCALE
Age (Maximum thickness of systems in metres)

Figure 14.7 The geological radiometric time-scale (1964) with recommended amendments by Harland *et al.* (1982). The radiometric ages, in millions of years, relate to the base of the periods.

Figure 14.8 The peaks of Canisp and Suilven, north-west Scotland, are composed of late Proterozoic Torridonian Sandstone resting unconformably on late Archaean Lewisian Gneiss, which dates from about 2900 million years ago.

is the banded iron formations. These are typically large deposits of iron oxides and silica that are not formed at the present day and of which there are very few examples in the Phanerozoic and late Proterozoic. Nearly all such deposits are Archaean or early Proterozoic in age (pre-2 Gyrs). Scientists think the disappearance of iron formations from the geological record could reflect a gradual increase in the amount of oxygen in the atmosphere and oceans. It is thought that such large deposits of iron could only be formed as precipitates from a large mass of fluid such as the ocean waters. However this requires that seawater built up huge concentrations of iron which in turn means that iron must have been very soluble in seawater in the Archaean and early Proterozoic. At present the amount of oxygen in the oceans prevents the build up of large amounts of iron as it tends to form the insoluble oxidized Fe^{3+} form, rather than the more soluble Fe^{2+}. By inference ancient ocean waters and the atmosphere must have contained less oxygen (see Section 6.6).

A striking parallel change is found in the clastic sedimentary rock record. Red beds such as make up an important part of many Devonian and Triassic sedimentary sequences are virtually absent from the Precambrian. They appear in the late Proterozoic as typified by the Torridonian Sandstone of north-west Scotland (Figure 14.8). Again this is thought to relate to an increase in the amount of oxygen in the atmosphere during the Precambrian.

The most important change related to the buildup of oxygen is the development of life. The sparse Precambrian fossil record indicates that the earliest forms of life were simple cyanobacteria that formed carbonate structures called stromatolites. As time progressed there was the very gradual development of additional, more sophisticated but still relatively simple life forms. Remains of cyanobacteria and bacteria are preserved in sedimentary rocks such as the Gunflint Chert in Canada and the Fig Tree Series of southern Africa. Organisms with some resemblance to Phanerozoic complex life forms are found in rocks dated at 700 m.y. old, that is, at the very end of the Proterozoic. The best examples are in the Ediacara Hills of Australia and include types that resemble worms and jellyfish.

There are also some notable changes in the

kinds of igneous and metamorphic rocks that are found in the Precambrian. Among the basaltic lavas that have been studied are a relatively large proportion with very high magnesium concentrations, called komatiites. Such rocks are virtually absent from the Phanerozoic and may relate to relatively large degrees of partial melting of the mantle. This in turn is explicable in terms of a hotter mantle in the Precambrian, which is hardly surprising given the gradual decay of radioactive nuclides which produce the Earth's heat (Figure 14.3).

The Phanerozoic is also distinguished from the Precambrian in terms of the occurrence of rock assemblages called ophiolites. These are mixtures of pillow basalts, ultramafic rocks, gabbros and dolerite dykes that together represent the kind of assemblages that make up the present ocean floor. In other words they are thought to be fragments of the oceanic crust that were emplaced onto the continental crust. Why they are not so abundant in the Precambrian is far from clear. It is possible that ocean basins or the tectonic processes responsible for emplacing such rocks on the continents were different in earlier times. (See Sections 12.1 and 13.5.)

Lastly there are distinctive metamorphic rocks that appear to be largely absent from the Precambrian. The two outstanding examples are eclogite and blueschist. Both of these are found in Phanerozoic orogenic belts associated with the former presence of subduction zones. Their virtual absence from the Precambrian could again be due to a hotter Earth, since at the present time these rocks form in distinctive geological environments at relatively low temperatures for their depths of burial. Or maybe they have simply not survived later metamorphism.

14.11   RADIOCARBON DATING

The long-lived radioactive isotopes so far considered with half-lives comparable with the age of the Earth, are suitable for measuring long intervals of geological time. For measuring shorter intervals, such as those of historical and archaeological interest, no similar methods were available until 1951, when W.F. Libby discovered that minute amounts of a radioactive isotope of carbon, ^{14}C, exist in air, natural waters and living organisms. Carbon consists mostly of ^{12}C (98.9 per cent) with a little ^{13}C (1.11 per cent), neither of which is radioactive. The proportion of radiocarbon, ^{14}C, would not affect these percentages unless they could be expressed to eight places of decimals.

Radiocarbon is being continuously produced in the atmosphere by reactions caused by cosmic rays. When direct hits are registered on the nuclei of atoms of nitrogen or oxygen, the nuclei are shattered and high-speed neutrons appear among the by-products. In turn some of these collide with other atoms of nitrogen, transforming them into ^{14}C. In due course, each ^{14}C nucleus loses a β-particle and is transformed back into ^{14}N (Figure 14.9). The rate of decay is such that the half-life is 5730 years. There is some evidence that the rate of production of ^{14}C in the atmosphere may not have appreciably varied for many thousands of years, and consequently a natural state of equilibrium should have been achieved, in which the abundance of ^{14}C is such that the gain from new production just balances the loss by decay. The word natural is emphasized here, because human activities have upset the balance in recent years, as pointed out below.

The newly-born atoms of ^{14}C are speedily oxidized to CO_2, which soon becomes uniformly distributed by wind, rivers and ocean currents throughout the circulating carbon dioxide of the world, so that the ratio $^{14}C/^{12}C$ remains theoretically constant. Living creatures, both plants and animals, continually absorb carbon dioxide to replenish their living tissues, and consequently the proportion of ^{14}C in their carbon also remains constant. When the organisms die, however, there is no further replacement of carbon from fresh carbon dioxide and the amount of ^{14}C already present at once begins to diminish as it decays. For example, if the carbon from prehistoric wood is found to contain only half as much ^{14}C as from a living plant, the estimated age of the old wood would theoretically be 5730 years (the half-life of ^{14}C).

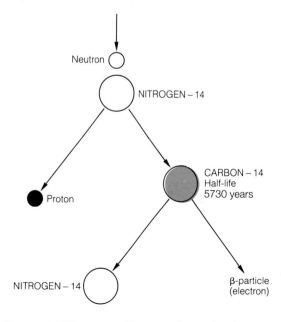

Figure 14.9 Diagram to illustrate the mode of origin and the radioactive decay of radiocarbon (^{14}C).

274

In practice there are sources of error against which special precautions must be taken. The human interference referred to above is a troublesome complication. The ^{14}C content of the atmosphere is being diluted year by year as a result of the combustion of coal and oil. The carbon dioxide so added to the air is made of carbon which long ago lost every detectable trace of ^{14}C. On the other hand, one of the by-products of testing nuclear missiles is ^{14}C, and in a few years this increased the natural abundance of ^{14}C by about one per cent. On balance the carbon from living plants contains less ^{14}C than the carbon from those that lived a century or two ago. Another source of error arises from secular variation in the content of ^{14}C in the atmosphere. This variation was discovered from radiocarbon ages determined from dated annular rings in old wood. For the last 6000 years this error can be allowed for, and prior to this it was small. Errors also arise from the introduction of carbon from various sources carried by groundwater, and the resulting age determination may be too great or too low, dependent on the source of the introduced carbon.

In spite of the possible errors, the radiocarbon method has proved to be of great value for dating events during the later part of the Ice Age, including sediments brought up as cores from borings drilled into the deep-ocean floor. Over the past decade, the techniques have been developed further to allow detection of smaller amounts of ^{14}C using atomic accelerators. This means that older samples and smaller samples can now be dated. Recent studies by scientists from Columbia University, NY, of ^{14}C in corals have shown that radiocarbon ages prior to about 9000 years may be too low, based on comparisons with other new techniques involving the short lived radionuclides of the U and Th decay series. Nevertheless the comparisons that are now possible between ages deduced from tree rings, varve sediments, the new U–Th techniques and even K–Ar dating, if anything, strengthen the usefulness of radiocarbon dating in the future.

14.12 FURTHER READING

Burchfield, J.D. (1975) *Lord Kelvin and the Age of the Earth.* Science History Publications, New York.

De Geer, G. (1940) *Geochronologia suecica Principles (Text and Atlas),* K. Svenska Vetensk, Handl. ser. 3, v **18**, (6), Almquist and Wiksett, Stockholm.

Faure, G. (1986) *Principles of Isotope Geology,* 2nd edn, John Wiley, New York.

Harland, W.B., Cox, A.V., Llewellyn, P.G., *et al.* (1982) *A Geologic Timescale,* Cambridge University Press, Cambridge.

Patterson, C. (1956) Age of meteorites and the earth. *Geochim Cosmochim Acta,* **10**, 230–37.

Snelling, N. J. (ed.) (1986) *The Chronology of the Geological Record.* Geol. Soc. Memoir, **10**, Geological Society, London.

ROCK WEATHERING AND SOILS

The soil, considered as a rock, links
common stones with the atmosphere, and
the dead dust of the earth with the
continuity of life.

Grenville A. J. Cole, 1913

15.1 WEATHERING AND CLIMATE

Weathering is the total effect of all the various subaerial processes that co-operate in bringing about the decay and disintegration of rocks, provided that no large-scale transport of the loosened products is involved. The work of rain-wash and that of wind are both essentially erosional and are thus excluded. The products of weathering, however, are subject to gravity, and there is consequently a universal tendency on the part of the loosened materials to fall or slip downwards, especially when aided by the lubricating action of water (Chapter 16). It is, indeed, only through the removal of the products of weathering that fresh surfaces are exposed to the further action of the weathering processes. No clean-cut distinction between weathering and erosion can therefore be attempted.

The geological work accomplished by weathering is of two kinds:

1. physical or mechanical changes, in which materials are disintegrated by temperature changes, frost action and organisms; and
2. chemical changes, in which minerals are decomposed, dissolved and loosened by water, oxygen and carbon dioxide of the atmosphere and soil waters, and by organisms and the products of their decay.

The physical, chemical and biological agents actively co-operate with one another. Shattering requires stresses powerful enough to overcome the strength of the materials, but the strength is gradually reduced by the progressive action of decomposition. Shattering in turn provides increased opportunities for the further penetration of the chemical agents. Everywhere full advantage is taken of joints and of bedding and foliation planes which, together with the cracks newly formed, admit air, water and rootlets down to quite considerable depths (Figure 15.1). Thus, although the processes of weathering may be considered separately, it must not be forgotten that the actual work done is the resultant effect of several processes acting together in intimate co-operation.

The materials ultimately produced are broken fragments of minerals and rocks; residual decomposition products, such as clay; and soluble decomposition products which are removed in solution. The products of weathering differ widely in different places according to the climatic conditions and the relief and configuration of the surface. In general it may be said that disintegration is favoured by steep slopes and by the conditions characteristic of frost-ridden or desert regions, while decomposition and solution are favoured by low relief and by humid conditions, especially in tropical regions. In the temperate zones the

Figure 15.1 Stac Polly, North-west Highland, Scotland, a peak of Torridonian Sandstone lying unconformably on Lewisian Gneiss. The serrated skyline and the opening-up of joints and other parting planes illustrate the effects of weathering, dominated by frost-wedging. (*L. S. Paterson*)

weather is widely variable, and most of the leading processes are to be found in operation during one part of the year or another.

15.2 PHYSICAL DISINTEGRATION

Frost is a potent rock breaker. When water fills cracks and pores and crevices in rocks and then freezes, it expands by more than 9 per cent of its volume, and exerts a bursting pressure of about 14 kg to the square centimetre. The rocks are ruptured and fragments are wedged apart, to become loose when thaw sets in. Steep mountain slopes and cliffs are particularly prone to destruction in this way, especially where joints are well-developed (Figures 15.1 and 15.2). The frost-shivered fragments fall to lower levels and accumulate there as talus slopes or screes (Figure 15.3). The treacherous nature of these loose aprons of angular debris is implied by the term **scree**, which is an old Norse word meaning rubble that slides away when trodden on. On gentler slopes, fields of angular blocks or **felsenmeer** form. Rocks which hold water in pores break down to their component particles, rather than blocks.

In arid climates the rocks exposed to the blazing sun become intensely heated, and in consequence a thin outer shell expands and tends to pull away from the cooler layer a few centimetres within. Under perfectly dry conditions, however, the stresses so developed are insufficient to fracture fresh, massive rocks. Experiments leave no doubt about this. But experiments also show that in the presence of water repeated alternations of heating and cooling eventually lead to rupture. Under natural conditions massive rocks must first be weakened by chemical weathering before shells and flakes of rock can be set free and broken down into smaller fragments. Even in the desert chemical alteration is continually in progress; slowly, it is true, but time is long. The process may be facilitated by the rapid chilling due to sudden rainstorms, for in the rare downpours of desert regions the rain can be near freezing-point, and even hailstones are not unknown. The resulting contraction opens up any cracks that may have developed at right angles to the surface, and water is thus given a temporary but effective entry into even the most stubborn of stones. Individual minerals swell and shrink and gradually crumble apart, especially in coarse-grained rocks like granite. The disintegration of pebbles on the desert surface is often conspicuous (Figure 15.4).

Figure 15.2 Frost shattering of Ordovician igneous rocks, north peak of Tryfaen, North Wales. (*G. P. Abraham Ltd, Keswick*)

Figure 15.3 Screes from the Tertiary igneous rocks (granophyre and gabbro) of Austerhorn, on the east coast of Iceland. (*L. Hawkes*)

Figure 15.4 Desert surface near the Pyramids outside Cairo, showing 'sun-cracked' pebbles. Shattering by temperature changes follows preliminary weakening by chemical weathering. (*J.J. Harris Teall*)

Figure 15.5 A striking example of the inselberg type of landscape, north of Ribaue, Mozambique. (*Sketch by E.J. Wayland, 1911*)

Where rocks are bedded or jointed parallel to the surface, actual separation of slabs of rock readily takes place. Here the sheet-joints due to pressure-release (Section 8.28) are of importance. However, such separation can occur on a spectacular scale, and in fact generally does, where the rocks affected are either massive or have structural planes of weakness (e.g. foliation) that are not far from being parallel to the existing surface. The scaling or peeling-off of flakes and curved shells of rocks, a phenomenon known as **exfoliation**, gives a highly characteristic appearance to the surfaces and profiles of upstanding hills and peaks in the more torrid lands. The effects of exfoliation are especially well-developed on the bare, steep-sided **inselbergs** (isolated 'island mounts') of Africa (Figure 15.5). Sharp edges, newly broken corners and projecting knobs soon become rounded off, because chemical and temperature changes can penetrate more deeply into the rock in such places than where the surfaces exposed are less irregular. The hills become dome-shaped with increasingly steep sides as time goes on. On the convex slopes (Figure 15.6) successive shells of rock can be seen, overlapping like the tiles on a roof, each ready to fall away as soon as it is liberated by cracks at right angles to the surface.

Exfoliation seems to be the result of a complex

group of processes which include and co-operate with the stresses set up by daily alternations of expansion and contraction, so that the latter eventually became far more effective agents of destruction than they appear to be when dealt with by themselves in the laboratory. This culmination results from the 'fatigue' set up by the many repetitions of expansion and contraction that a shell of rock must undergo before it finally springs loose. The principle involved is that stresses are strongly concentrated, and chemical activity greatly increased, at the inner feather edges of even the tiniest cracks. For this reason preliminary weakening by chemical weathering may be imperceptible. Researches on the growth of cracks in structural materials (e.g. for use in jet planes) have revealed that a microscopic scratch on the surface may be sufficient to initiate a crack that might lead to failure and disaster. At first the incipient crack opens and extends so slowly as to be hardly perceptible, but sooner or later, depending on the stresses involved, a certain critical point is reached and the crack then suddenly develops into a fracture at 10 000 or more times the previous rate.

15.3 THE ROLE OF ANIMALS AND PLANTS

Earthworms and other burrowing animals such as rodents and termites play an important part in preparing material for removal by rain-wash and wind. Worms consume large quantities of soil for the purpose of extracting food, and the indigestible particles are passed out as worm-casts. In an average soil there may be 400 000 worms to the hectare, and

in the course of a year they raise 10 to 15 tonnes of finely comminuted materials to the surface.

The growing rootlets of shrubs and trees exert an almost incredible force as they work down into crevices. It is not usually realized that the cellulose of which all cell walls are made is actually stronger than many metals. Cracks are widened by expansion during growth (Figures 3.3 and 15.7) and wedges of rock are forcibly shouldered aside. Plants of all kinds, including fungi and lichens, also contribute to chemical weathering, since they abstract certain elements from rock materials and consequently liberate others in the process. Moreover, water containing bacteria attacks the minerals of rocks and soils much more vigorously than it could do in their absence. The dead remains of organisms decay in the soil largely as a result of the activities of bacteria and fungi. In this way carbon dioxide and organic acids, together with traces of ammonia and nitric acid, are liberated, all of which increase the solvent power of soil-water. The chief organic product is a 'complex' of brown jelly-like substances collectively known as **humus**. Humus is the characteristic organic constituent of soil, and water charged with humus acids can dissolve certain substances, such as limonite, which are ordinarily insoluble.

Another effect of vegetation, one which is of vital importance in the economy of nature, is its protective action. Rootlets bind the soil into a woven mat so that it remains porous and able to absorb water without being washed away. The destructive effects of rain and wind are thus effectively restrained. Grass roots are particularly effective in this way. Forests break the force of the

Figure 15.6 Exfoliation in a ring-dyke of syenite, rising about 900 metres from the village seen in the foreground. Chambe Plateau Complex, Malawi. (*R.L. Kinsey*)

rain and prevent the rapid melting of snow. Moreover, they regularize the actual rainfall and preclude the sudden floods that afflict more sterile lands. For these reasons the reckless removal of forests may imperil the prosperity of whole communities. Soil erosion is intensified, agricultural lands are impoverished and lost, and barren gullied wastes, 'badlands' (Section 16.8 and Figures 16.21 and 16.22). Except after heavy rainfall, the rivers run clean in forested lands, but after deforestation their waters become continuously muddy. Destruction of the natural vegetation by land clearing and ploughing, and the failure to replace forests cut down for timber or destroyed by fire, have had disastrous economic consequences in many parts of Africa and America. Soil devastation has been one of the major factors in the downfall of past civilizations. Man himself has been, and still is, amongst the most prodigal of the organic agents of destruction. But there is a growing awareness, now almost worldwide, of the urgent necessity for soil conservation and the increase of soil fertility.

15.4 CHEMICAL WEATHERING

The alteration and solution of rock material by chemical processes is largely accomplished by rainwater acting as a carrier of dissolved oxygen and carbon dioxide, together with various acids and organic products derived from the soil. The degree of activity depends on the composition and concentration of the solutions so formed, on the temperature, on the presence of bacteria, and on the substances taken into solution from the minerals decomposed. All natural waters are slightly dissociated into H^+ and $(OH)^-$ ions. The acidity of natural water is measured by the concentration of H^+ ions, pH as it is called. If the pH value is greater than 7 the water is alkaline; if less than 7 it is acid. Any increase in acidity increases the rate of weathering reactions. The pH of rainwater ranges from 5.5 to 6.0. Its acidity comes mainly from dissolved carbon dioxide, which ionizes water:

$$CO_2 + H_2O = H^+ + (HCO_3)^-$$

The lower values of pH may be due to lightning discharge, which produces nitric and other acids in minute amounts. Much lower values (down to 2.8) are also found in certain peat-bog waters, especially where *Sphagnum* and related mosses are actively flourishing.

The chief chemical changes that occur during weathering are solution, oxidation, hydration or hydrolysis, and the formation of carbonates. Only a few common minerals resist decomposition, quartz and muscovite (including sericite) being the chief examples. Others, like the carbonate minerals, can be entirely removed in solution.

Figure 15.7 Tree roots exposed after wedging out the rock masses that originally formed the banks of a stream; cf. Fig.3.3. (*S.H. Reynolds*)

Most silicate minerals break down into relatively insoluble residues, such as the various clay minerals, with liberation of soluble substances which are removed in solution.

Limestone is scarcely affected by pure water, but when CO_2 is also present the $CaCO_3$ of the limestone is slowly dissolved and removed as calcium bicarbonate, $Ca(HCO_3)_2$. Limestone platforms like those around Ingleborough clearly show the effects of solution in their deeply grooved and furrowed surfaces (Figure 15.8). The joints are opened into 'grikes' with intervening ribs or 'clints' of harsh bare rocks. The surface is free from soil except where a little wind-blown dust has collected here and there in crevices. On steep or vertical exposures of limestone, deep grooves, like hemi-cylindrical pipes, may be formed by the solvent action of rainwater that has been concentrated into narrow vertical channels, e.g. at joint intersections. These remarkable flutings are only a special variety of 'grikes', but they are known internationally as **lapiés** or **karren**.

When limestone contains impurities such as quartz and clay, these generally remain more or less undissolved, and so accumulate to form the mineral basis of a soil. The red earth, known as **terra rossa**, is a weathering residue rich in insoluble iron hydroxides derived by long accumulation from the minute traces of iron compounds in the original limestone (Plate Xb). Typically, this material accumulates in enclosed depressions or 'dolines' on the limestone surface, formed by concentrated solution at major joint intersections.

Clay minerals are the chief residual products of the decomposition of feldspars. Under the hydrolysing action of slightly carbonated waters the feldspars break down in the following way:

$$6H_2O + CO_2 + 2KAlSi_3O_8 = \begin{cases} Al_2Si_2O_5(OH)_4 \text{ (kaolinite)} \\ 4SiO(OH)_2 \text{ (silicic 'acid')} \\ K_2CO_3 \text{ (removed in solution)} \end{cases}$$

$$\text{water} \quad \text{orthoclase}$$

From plagioclase the products are similar, except that Na_2CO_3 is formed from albite and $Ca(HCO_3)_2$ from anorthite. Most of the clay is probably at first in colloidal solution, that is to say, it consists of minute particles dispersed through water, the particles being larger than ions but much smaller than any that can be seen with a microscope. The

Figure 15.8 'Clints' and 'grikes' produced by chemical weathering on a shelf of the Great Scar Limestone (Lower Carboniferous), above Malham Cove, North Yorkshire. (cf. Fig.6.7) (*Bertram Unné*)

particles eventually crystallize into tiny scales or flakes. The alkalies easily pass into solution, but whereas soda tends to be carried away, to accumulate in the sea, potash is largely retained in the soil. It is withdrawn from solution by colloidal clay and humus, from which in turn it is extracted by plant roots. When the plants die the potash is returned to the soil. Analyses of river waters show that relatively little ultimately escapes from the lands.

As a result of certain obscure reactions, colloidal forms of clay can break down still further:

$$nH_2O + Al_2Si_2O_5(OH)_4 \rightarrow \begin{cases} Al_2O_3.nH_2O \text{ (colloidal} \\ \text{aluminium hydroxide)} \\ 2SiO(OH)_2 \\ \text{(colloidal silicic 'acid')} \end{cases}$$

water   kaolinite

In the humid conditions of the temperate zone, aluminium hydroxide is not liberated to any important extent, but during the dry season of tropical and monsoon lands it is precipitated in a highly insoluble form, and so accumulates at or near the surface as bauxite (Section 4.8).

The decomposition of the ferromagnesian minerals may be illustrated by reference to a simple type of pyroxene:

water + carbon dioxide + $CaMgSi_2O_6 \rightarrow$

diopside

$2SiO(OH)_2$ (silicic 'acid') +

soluble bicarbonates of Ca and Mg

When Al_2O_3 and Fe_2O_3 are also present (as in biotite, augite and hornblende) clay and chloritic minerals and limonite remain as residual products. In the presence of oxygen, limonite is also precipitated from solutions containing $Fe(HCO_3)_2$. For this reason weathered rock surfaces are commonly stained a rusty brown colour. Ordinary rust is, in fact, the corresponding product of the action of water and air on iron and steel. Rusting is, indeed, the reversion of iron to one of the common ores from which it might have been smelted; it is an economic menace that involves enormous wastage and expense.

The residual products of weathering are naturally more stable in the presence of air and water than the parent minerals from which they were derived. Corresponding to the fact that they are the result of oxidation and hydration, they contain a larger proportion of negative ions, O^{2-} and $(OH)^-$, than the parent minerals. It should be noticed that when a mineral is 'hydrated' it is the ion $(OH)^-$ that is built into the new crystal lattice. The source of the hydroxyl ions is not water alone, but a reaction between water and oxygen ions:

$$H_2O + O^{2-} = 2(OH)^-$$

The resulting increase in the number of negative ions (anions) relative to positive ions (cations) makes for increased stability, i.e. resistance to weathering. If we represent the number of cations in the formula of a mineral by 100, then the corresponding proportion of anions is easily found. For quartz, SiO_2, it is 200. Now quartz is a very stable mineral during weathering, and so is muscovite, with 170 anions. Minerals with less than this number are weathered to more stable residual products with a higher number. The feldspars (160) yield successive products which fall into a series of increasing stability: hydromica (200) → clay (225) → bauxite (250–300). The ferromagnesian minerals, hornblende (160), augite (150), biotite (150) and olivine (133) yield chlorite (180) or serpentine (180) and limonite (200–300). Of these olivine, with the lowest proportion of anions, is the most rapidly weathered of all the common rock-forming minerals. Consistently with these figures, granophyre and most other granitic rocks withstand the attack of chemical weathering more effectively than basalts and dolerites, especially the olivine-bearing varieties.

Chemical weathering contributes to the disintegration of rocks (a) by the general weakening of the coherence between minerals, so that the rock more readily succumbs to the attack of the physical agents; (b) by the formation of solutions which are washed out by the rain, so that the rock becomes porous and ready to crumble (e.g. the liberation of the grains of a sandstone by solution of the cement); and (c) by the formation of alteration products with a greater volume than the original fresh material, so that the outer shell swells and pulls away from the fresh rock within. In exfoliation the only alteration of this kind that has been detected is the appearance of chlorite in place of biotite.

The separation of shells of obviously decayed rock is distinguished as **spheroidal weathering** (Figure 15.9). It is best developed in well-jointed rocks which, like many basalts and dolerites, are readily decomposed. Water penetrates the intersecting joints and thus attacks each separate block from all sides at once. As the depth of decay is greater at corners and edges than along flat surfaces, it follows that the surfaces of rupture become rounded in such positions. As each shell breaks loose, a new surface is presented to the weathering solutions, and the process is repeated again and again, aided, according to the climate, by temperature changes with or without frost.

Each successive wrapping surrounding the core (if any) becomes more nearly spheroidal than its outer neighbour, until the angular block is transformed into an onion-like structure of concentric

Figure 15.9 Spheroidal weathering of dolerite sill, North Queensferry, Fife, Scotland. (*British Geological Survey*)

shells of rusty and rotted residual material. Cores of fresher and more coherent rock may eventually stand out like boulders when their soft outer wrappings have been washed away by rain erosion (Figure 15.10). Residual boulders of granite, sometimes of great size (Figure 15.11) probably result mainly from exfoliation but with the aid of spheroidal weathering.

Figure 15.10 Residual boulders of dolerite produced by spheroidal weathering, North Queensferry sill, Fife, Scotland. (*British Geological Survey*)

Figure 15.11 Residual boulders of granite, produced by combined exfoliation and spheroidal weathering, perched on the summit of an exfoliated surface of granite, Matopo Hills, Zimbabwe. (*C.T. Trechmann*)

15.5 WEATHERING RESIDUES

Weathered rock in place is termed **saprolite**. Its thickness reflects the long-term balance between weathering and transport. Where transport is limited by gentle slopes, weathering can penetrate deeply. In Australia, saprolite thicknesses may exceed 100 m and similar depths are recorded from crystalline rocks in Nigeria and Brazil. Even in formerly glaciated areas, such as north-east Scotland, preglacial disintegration can reach depths of several tens of metres. The penetration of alteration under conditions of low erosion is probably limited only by the depth to which groundwater can circulate.

Saprolite characteristics are largely a product of climate, rock type and time. High surface temperatures accelerate weathering reactions and abundant rainfall allows rapid removal of the soluble products of chemical alteration. Hence the deepest and most advanced chemical weathering of rocks tends to be found in areas of present and former tropical climates (Figure 15.12). Rocks vary in their resistance to chemical attack due to differences in mineralogy, porosity, cementation and fracturing. The resulting contrasts in weathering rates and styles give irregular thicknesses of saprolite in areas of mixed geology. Deep penetration of weathering and advanced alteration requires pro-

longed periods of surface stability and may lead to the development of deep weathering profiles in which surface horizons leached of unstable minerals pass down progressively into less weathered saprolite (Figure 15.13).

In the tropics, weathering profiles occur in association with **duricrusts**. These indurated surface layers take their names from the main crust-forming elements. Most common are **ferricretes**, composed of reddish iron nodules or blocks formed by drying out of the iron-rich clay or saprolite in the upper part of the weathering profile. **Alcretes** occur in tropical rain forest areas and provide most of the world's bauxite reserves. **Silcrete** is a secondary silica deposit, cementing and replacing sediment and weathered rock. **Calcrete** or **caliche** is widespread in arid or semi-arid lands and is a product of the progressive build-up of carbonates in surface horizons from carbonate-saturated surface, or soil, water. Once formed, all duricrusts show considerable resistance to weathering and erosion and act as cap-rocks by protecting underlying loose saprolites (Figure 15.13).

Fossil weathering profiles and duricrusts, buried beneath younger strata, occur widely. Sub-Cretaceous kaolinization occurs in southern Sweden (Plate Xa) and near New York, while Carboniferous bauxites have been worked for refractory bricks in central Scotland. Ancient cal-

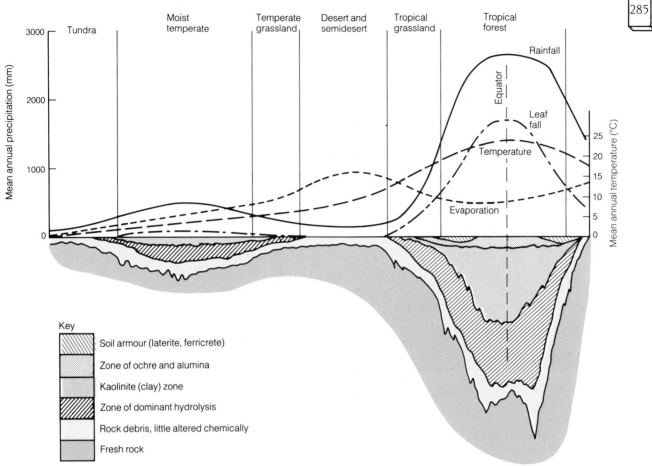

Figure 15.12 Weathering mantles and climate. The effects of climatic change are ignored. (*After N.M. Strakhov, 1967*)

cretes, termed **cornstones** are increasingly recognized in sediments laid down in former arid environments. More important, perhaps, are sediments derived from stripping of thick and extensive weathering mantles, such as the Early Tertiary kaolin clays and quartz gravels of Devon which originated from the removal of deep saprolites which covered the whole of south-west England at the end of the Cretaceous. Such silica-rich sediments have been termed **laterite-derived facies**.

15.6 THE MANTLE OF ROCK-WASTE

The superficial deposits which lie on a foundation of older and more coherent bedrocks form a mantle of rock-waste or **regolith** of very varied characters. In many places the waste-mantle lies directly on the bedrock from which it was formed by weathering. In this case quarry sections and cuttings of all kinds generally show a surface layer of **soil**, passing gradually downwards through a zone of shattered and partly decomposed rock, known as the **subsoil**, to the parental bedrock, still rela-

tively fresh and unbroken by weathering agents. In the soil vegetable mould and humus occur to a varying extent, and under appropriate conditions they accumulate to form beds of peat which it is often convenient to include with the mantle. Soils develop, however, not only on bedrock, but also on a great variety of loose deposits which have been transported into their present positions by gravity, wind, running water or moving ice. These transported deposits will be considered in their appropriate chapters. Here they are summarized (Table 15.1), according to their mode of origin, together with the untransported or sedentary deposits of the waste-mantle. It should perhaps be mentioned that in its engineering usage the word 'soil' refers to all unconsolidated sediments overlying bedrock. 'Soil mechanics' is the study of the behaviour of the ground – whether dry, wet or frozen – under the loads placed upon it by the erection of buildings and the stresses due to engineering structures generally.

15.7 THE GROWTH AND NATURE OF SOILS

The purely mineral matter of the residual or trans-

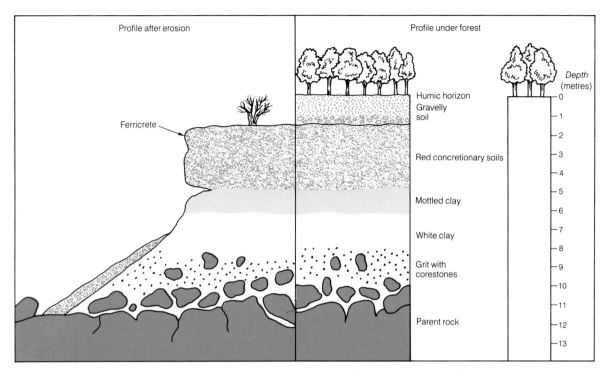

Figure 15.13 Typical lateritic weathering profile in Sudan. (*After A. Faniran and L.K. Jeje, 1983*)

ported deposits is first colonized by bacteria, lichens or mosses. By the partial decay of the dead organisms, mould and humus begin to accumulate; lodgment is afforded for ferns and grasses; berries and winged seeds are brought by birds and the wind; and finally shrubs and trees may gain a footing. The rootlets work down, burrowing animals bring up inorganic particles, and the growing mass becomes porous and spongelike, so that it can retain water and permit the passage of air. Frost and rain play their parts, and ultimately

a mature soil, a complex mixture of mineral and organic products, is formed. But though the soil is a result of decay, it is also the medium of growth. It teems with life, and as the chief source of food it is one of the three indispensable assets of mankind. Soil is the base of the pyramid of life on the lands. Its fertility depends on a proper balance of air and water, minerals and trace elements, humus, and a healthy population of bacteria and other lowly organisms.

From a geological point of view soil may be

Table 15.1 The mantle of rock-waste (Continental deposits)

Mode of origin		Characteristic deposits
Sedentary:		
Essentially inorganic	*Residual*	Gravel, sand, silt and mud, *Terra rossa*, saprolite
Inorganic and organic	*Soils*	Including soils on bedrock and on regolith
Essentially organic	*Cumulose*	Vegetation residues: swamp deposits and peat
Transported by:		
Gravity	*Colluvial*	Screes, landslip and other slope deposits
Wind	*Æolian*	Sand dunes, sand wastes, and loess
Ice	*Glacial*	Till, moraines, and drumlins
Melt-water from ice	*Glacifluvial*	Outwash fans, kames and eskers
Rivers (deposited in lakes)	*Lacustrine*	Alluvium and saline deposits
Rivers	*Fluviatile*	Alluvium, passing seawards by way of estuarine or deltaic deposits into marine deposits

Table 15.2 Soil types and climate

	Climatic regions	Characteristic natural vegetation	Mainly organic	Mixed organic and inorganic
Arctic	glacial	–	–	–
	tundra (short mild summer)	mosses and lichens	peat soils peat	frost-shattered stony soils arctic brown soils
Temperate humid	boreal (cold winter)	coniferous forests	peat soils peat	PODSOL (grey)
	temperate (mild and humid)	coniferous and deciduous forests	peat soils peat	PODSOL (grey) grey and brown forest earths
	Mediterranean (short mild winter)	evergreen shrubs and trees		red and brown Mediterranean soils
Arid and semi-arid	steppes (dry summer)	grasses		CHERNOZEM (black) chestnut-brown soils
	deserts	marginal scrub – marginal scrub		grey marginal soils grey marginal soils ferruginous soils
Tropical rainy	savannahs (wet and dry seasons)	grasses and scattered trees		
	monsoon lands (wet and dry seasons)	forests: mixed, open to sub-tropical evergreen		ferrallitic soils and ferrisols
	equatorial (continuously humid)	evergreen tropical rain forest	swamp soils peat (vertisols)	

defined as the surface layer of the mantle of rock-waste in which the physical and chemical processes of weathering co-operate in intimate association with biological processes. All these processes depend on climate, and in accordance with this fact it is found that the resulting soils also vary with the climate in which they develop (Table 15.2). Other factors are also involved: particularly the nature of the bedrock or other deposit on which the soil has developed, the relief of the land, the age of the soil (that is, the length of time during which soil development has been in progress), and the superimposed effects of cultivation.

The influence of the parental material is easily understood. Sand makes too light a soil for many plants, as it is too porous to hold up water. Clay, on the other hand, is by itself too impervious. A mixture of sand and clay makes a **loam**, which avoids these extremes and provides the basis of an excellent soil. A clay soil may also be lightened by adding limestone, and the natural mixture, known as **marl**, is also a favourable basis. Limestone alone, as we have seen, cannot make a soil unless

it contains impurities. In most climates granite decomposes slowly and yields up its store of plant foods very gradually. Basaltic rocks, on the other hand, break down much more quickly. Volcanic ashes and lavas provide highly productive soils which, even on the flanks of active volcanoes, compensate the agriculturist and vine-grower for the recurrent risk of danger and possible destruction.

These differences are most marked in young soils and in temperate regions. As the soil becomes older, and especially when the climate is of a more extreme type, the influence of long continued weathering and organic growth and decay makes itself felt more and more. Certain ingredients are steadily leached out, while others are concentrated. Humus accumulation depends on the excess of growth over decay, and this in turn depends on climatic factors. The composition of the evolving soil thus gradually approaches a certain characteristic type which is different for each climatic region. The black soil of the Russian steppes, for example, is equally well developed from such different parent rocks as granite, basalt,

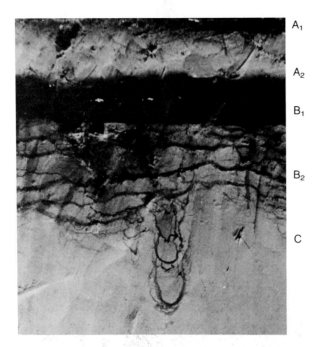

A₁

A₂

B₁

B₂

C

Figure 15.14 Soil profile of podsol type formed from buff-coloured sandy alluvium C by descending soil-water enriched in humus by its passage through A₁ (vegetable mould and humus) here provided by a cover of heather; near Eindhaven, The Netherlands. A₂ is the zone of bleached grey soil from which the podsol type takes its name; iron hydroxides have been leached out as ferro-humus materials and fixed, with washed down clay particles, in the dark and more compact zone B₁, which may eventually develop into hardpan. Continued downward migration leads to periodic precipitation of ferruginous materials in zone B₂ which is generally mottled, but sometimes banded, as here. Where there are favourable passageways, localized tongues of B₂ may invade the parental material C. (*US Department of Agriculture, photograph by Roy W. Simonson*) (See Plate Xc.)

loess and boulder clay. Conversely, a single rock type, like granite, gives grey soils in temperate regions (**podsol**), black soils in the steppes (**chernozem**), and reddish soils in tropical regions of seasonal rainfall (**ferruginous soils**). The colours of soils are almost wholly due to the relative abundance (or paucity) of various iron compounds and humus.

Deeply cultivated soils may be more or less uniform throughout, but this is not the case in purely natural soils. A vertical cutting through an

old natural soil reveals a layered arrangement which is called the **soil profile** (Figure 15.14). The successive layers of one type of profile are clearly developed in the grey soils of the more or less forested north-temperate belt of Canada, northern Europe and Asia. As the drainage is dominantly descending, iron hydroxides, and humus derived from the surface layer of vegetable mould, are carried down in colloidal solution. Thus a bleached zone is developed, and for this reason the soil type is called podsol (Russian, ashy grey soil). By the accumulation of the ferro-humus material at a depth of 5–30 cm, accompanied by particles of silt and clay washed down mechanically, a deep brown or nearly black layer of variable thickness is formed. This may develop into a hard, well-cemented band, impervious to drainage, which is known as **hardpan**. One of the objects of ploughing is to prevent the growth of hardpan. Otherwise, waterlogged conditions may set in, and there will then be a marked tendency for peat to accumulate.

Farther south in the grasslands of the steppes and prairies, summer drought and winter frosts favour the accumulation of humus, largely provided by the grass roots which die each year. During the dry season groundwater is drawn towards the surface, and $CaCO_3$ is precipitated, often in irregular nodules, at a depth of 60–90 cm. Under the influence of the ascending calcareous solutions the humus becomes black and insoluble. Iron hydroxides are therefore not leached out as in the podsol. The upper layer of the soil profile is black, becoming brown in depth where there is less humus. For this reason the soil type is called chernozem (Russian, black earth). The black cotton soils of India and the 'black bottoms' of the Mississippi flood plains are of similar origin.

15.8 FURTHER READING

Millot, G.(1970) *Geology of clays*, Springer-Verlag, New York.

Ollier, C.D. (1984) *Weathering*, Longman, London.

Selby, M.J. (1985) *Earth's changing surface*, Clarendon Press, Oxford.

Small, R.J. and Clark, M.J. (1982) *Slopes and weathering*, Cambridge University Press, Cambridge.

SURFACE EROSION AND LANDSCAPE SLOPES

The surface of the land is made by nature to decay . . . Our fertile plains are formed from the ruins of the mountains.

James Hutton, 1785

16.1   RIVERS AND THEIR VALLEYS

The necessary conditions for the initiation of a river are adequate supply of water and a slope down which to flow. In 1674 Pierre Perrault completed the first quantitative investigation of the relationship between rainfall and stream flow. From the result of his pioneer work in the valley of the upper Seine he justly concluded that 'rain and snow waters are sufficient to make Fountains and Rivers run perpetually'. Rivers are partly fed from groundwaters, and some have their source in the meltwaters from glaciers, but in both cases the water is derived from the meteoric precipitation. In periods of drought rivers may be kept flowing, though on a diminished scale, entirely by supplies from springs and the zone of intermittent saturation (Section 19.2). When these supplies also fail, through the lowering of the water table, as commonly happens in semi-arid regions, rivers dwindle away altogether. Even then water may still be found not far below the river floor, soaking slowly through the alluvium, where it is protected from evaporation.

The initial slopes down which rivers first begin to flow are provided by earth movements or, locally, by volcanic accumulations. Many of the great rivers of the world, e.g. the Amazon, Missis-sippi and Zaïre (Congo), flow through widespread downwarps of the crust which endowed them with ready-made drainage basins from the start. The majority of rivers, however, originated on the slopes of uplifted regions where, often in active competition with their neighbours, they gradually evolved their own drainage areas.

Most rivers drain directly into the sea. But in areas of internal drainage, permanent or intermittent streams terminate in lakes or swamps which spread or contract so that evaporation from the exposed surface just balances the inflow, the conditions being such that the water is unable to accumulate to the level at which it could find an outlet. Notable examples occur in central Asia and Australia.

The development of a river valley depends on the original surface slope; on the climate, which determines the rainfall; and on the underlying geological structure, which determines the varied resistance to erosion offered by the rocks encountered. Where a newly emergent land provides an initial seaward slope, the rivers that flow down the slope, and the valleys they excavate, are said to be **consequent**. The valley sides constitute secondary slopes down which tributaries develop; these streams and their valleys are distinguished as **subsequent**. The resulting slopes and those developed later make possible the addition of further generations of tributaries. The recognition of true consequent and subsequent streams is generally impossible, however, except in young terrains. Rivers soon exploit the weaknesses provided by joints and faults so that their valleys become structurally-aligned and deviate from the original slope. Also, many millions of years may have elapsed since creation of the master surface and all traces of the original slope may have been lost through erosion and earth movements. Hence it is now usual to describe drainage patterns, not by their tenuous relationships to original surfaces, but by their present structural controls, especially to dip and to strike (Chapter 18).

A main river and all its tributaries constitute a river system, and the whole area from which the system derives water and rock-waste is its drainage basin, which has as its boundary, the watershed. Weathering continuously supplies rock-waste, which falls or 'creeps' or is washed by rain into the nearest stream. The latter carries away the debris contributed to it, and at times it acquires still more by eroding its own channel. Valleys develop by the removal of the material carried away by the streams which drain them. The load acquired by the main river is ultimately transported out of the basin altogether or deposited in its lower reaches. Deposits of gravel and alluvium are, of course, dropped on the way at innumerable places, but these are only temporary halts in the journey towards the sea. Rivers are by far the chief agents concerned in the excavation of valleys, not merely because of their own erosive work, but above all because of their enormous powers of transportation.

16.2 PRIMARY AND SECONDARY EROSIONAL SLOPES

The widening of a valley such as that illustrated in cross-profile by Figure 16.1 clearly involves the delivery to the river of an amount of rock-waste that is many times greater than that liberated by the river itself during the down-cutting of its channel floor. If confined to down-cutting, the river would cut a valley in the form of a vertical-sided cleft. Deep gorges approximating to this form are found in mountainous areas and plateaus that have been raised high above sea-level, and where the rocks sawn through by the rivers are chemically resistant and mechanically strong (Figure 16.2). Down-cutting may then proceed a long way before the valley walls are worn back to any great extent by the successive operation of weath-

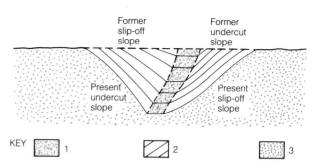

Figure 16.1 Creation of a simple valley by fluvial erosion. (1) Matter removed by fluvial linear erosion. (2) Matter synchronously removed by slope lowering. (3) Bedrock. (*After J. Büdel, 1982*)

Figure 16.2 Gorge of the Bhagirathi, one of the sacred headwaters of the Ganges, Garhwal Himalayas. The river issues from an ice tunnel at the end of the Gangotri glacier and traverses the range through a slit-like gorge about 185 m deep. (*Prem S. Ray*)

ering, migration to the river of the resulting rock-waste under the action of gravity, and its eventual removal by the river. But sooner or later the rate of deepening slows down and widening begins to catch up, with production of a cross-profile that approximates to a V-shape (see Figure 17.20). In less resistant rocks widening on a more conspicuous scale accompanies deepening from the start.

While the valley is being widened by the wasting back of its sides, the river itself begins to widen its valley floor by under-cutting its banks, notably on the outer side of bends (Figure 16.3), where both down-cutting and lateral cutting are taking place simultaneously or alternately from time to time. Instead of the channel floor being sunk vertically into the rocks by river erosion, it digs in obliquely, as indicated by AB in the accompanying diagram. The relatively gentle slope on the inner side of the bend has been fashioned by the stream itself and carpeted with shingle or finer alluvium. This 'slip-off slope' is referred to again in Section 17.7. It is introduced here only in order to emphasize the fact that there are erosional slopes of two very different modes of origin:

1. primary erosional slopes carved by the rivers themselves (e.g. the vertical walls of gorges and the slip-off slope in Figure 16.3 corresponding to the lower side of AB), or by the other chief agents of erosion: glaciers, winds, waves and currents; and

2. secondary erosional slopes developed as a result of weathering and surface erosion of primary slopes, the latter including not only the erosional types, but also those of tectonic origin, such as fault scarps and tilted surfaces due to earth movements.

Here we shall consider the great variety of processes involved in downhill surface erosion and the chief results for which they are responsible. The specific activities of rivers, glaciers, sea waves and currents, and winds are dealt with in later chapters. These agents not only carry out the essential work of transporting the waste-products from the bases of the slopes that provide them; they also make characteristic types of slopes on their own account.

The valley widening that accompanies the excavation of river floors obviously implies the wearing back or **recession** of the slopes that lead down

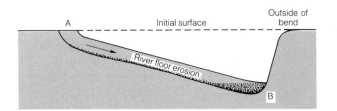

Figure 16.3 The great horseshoe bend of the Rhine, looking eastwards. The town on the right is Boppard, between Koblenz and Bingen. The long gentle slopes on the inside of the bend – the 'slip-off slope' – was left by the river as it cut down from the upland to its present level (e.g. from A to B in the diagram), enlarging its curve and undercutting its banks on the outside of the bend. (*German Tourist Information Bureau*) (See also Figs 17.22 and 17.23.)

Figure 16.4 Escarpment of Carboniferous Limestone, Eglwyseg Mountain, north of Llangollen, Clwyd, Wales. Below the free faces of the outcropping limestone beds the constant slope, determined by the angle of rest of the screes, is seen; this passes into the concave slope that leads down to the river. (*British Geological Survey*)

to the side or sides of the river where erosion and transport are taking place. Coastal cliffs similarly recede, so long as marine erosion undermines them and waves and currents continue to remove the fallen debris. The principles involved in all such cases are essentially the same. The land is consumed sideways, so to speak. For convenience, and to avoid undue repetition, we shall be dealing mainly with the slopes that constitute the sides of widening valleys or the scarps bordering extensive plains.

The slopes between two roughly parallel valleys necessarily approach each other as they recede from their respective valley floors. Eventually they meet and form a divide. The latter is then gradually lowered as the slopes continue to be worn back. Meanwhile, each main divide of this kind is itself being subdivided by the slopes leading down to tributaries of the trunk rivers. When there are several generations of tributaries, slope retreat comes into operation from several directions at once. Residual landforms are then developed in great variety: mountain peaks and hills; escarpments and ridges; inselbergs, tors and isolated pinnacles. Each uplifted area is dissected bit by bit into a slowly changing landscape. Somewhere or other every stage of landscape development is to be seen, from plateaus representing uplifted plains or sea floors, through mountainous or hilly regions of maximum relief where the country is practically all slopes, to those in which the land has again been reduced to a plain.

16.3 VALLEY SIDES AND HILLSIDE SLOPES

A characteristic association of the hillside slopes that constitute the sides of many valleys is illustrated by Figure 16.5. This association forms the basis of an illuminating analysis of slope development presented by Alan Wood in 1942. In addition to the initial upland and the flood plains of broad valley floors, the four slope elements recognized by Wood are shown in idealized form in Figure 16.5. From top to bottom these are listed below, the names given first being those adopted by Wood:

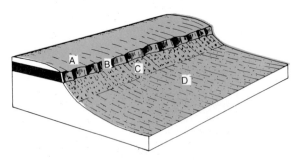

Figure 16.5 Elements of hillside slopes. (*After Alan Wood and Lester King*)

A. The **waxing slope**, the part of the upper surface which tends to become convex by being rounded off at, and towards, its edge with B; more commonly called the **convex slope**, sometimes the **upper wash slope**.

B. The **free face**, any outcrop of bare rock (e.g. a precipitous valley wall, scarp, bluff or cliff) that stands more steeply than the angle of repose of any scree or talus heap that may accumulate from its weathering products; it has also been called a **gravity** or **derivation slope**, because from it is derived the rock-waste that falls or rolls down under gravity.

C. The **constant slope**, which is that of the angle of rest of the scree-debris, whether it is a surface of scree or a bedrock surface on which a sprinkling of fragments may halt for a time. If scree is present its slope is often referred to as the **talus** or **debris slope**. Because scree-debris continues to be weathered, reduced in grain size and washed downhill by rain, the debris slope ceases to be constant towards its lower edge and merges into

D. The **waning slope** (**pediment, valley-floor base-**

Figure 16.6 Grand Canyon of the Colorado River, Arizona. (*J. K. Hilliers, United States Geological Survey*) For location, see Fig. 18.25.

ment or **lower wash slope**), which stretches to the valley floor or other local base level with a diminishing angle, so that it is more or less concave upwards.

According to local circumstances certain of these slope elements may be repeated, or they may be undeveloped or worn out. In Figure 16.4, for example, and far more conspicuously in the pictures of the Grand Canyon of the Colorado (Figures 1.5 and 16.6), it can be seen that B and C are repeated several times, in accordance with the structural fact that the layers of strong limestone or sandstone responsible for the scarps, B, are separated by less resistant beds of shale. The latter have been worn back to a slope, so ensuring that all the debris liberated from the sides of the canyon ultimately reaches the river at the bottom. On the other hand, in areas of low and gentle relief B is likely to be missing and C, if present at all, may be reduced to a short link between the summit convexity, A, and the concave waning slope, D. This case is familiarly illustrated by the smoothly undulating profiles that characterize the more subdued landscapes of parts of the British Isles and New England, USA. The pediment-with-inselberg landscape is a strongly contrasted type in which B and D are the dominant elements. Where parallel down-cutting rivers have V-shaped valley sides that rise up to meet their neighbours in narrow ridges (see Figure 16.28), the constant slope, C, is highly developed. To cross country such as that in Burma between the Chindwin and Assam or between the Upper Irrawaddy and China is to toil laboriously up and down for hundreds of metres over and over again. It should perhaps be added that the valley walls adjoining an actively down-cutting river that has increased its rate of erosion will be

steeper than the higher slopes, thus giving a profile that is convex upwards. However, this is a result of river erosion, to be dealt with in the next chapter. Here we are concerned with the surface erosion brought about over vast areas by the co-operation of gravity with weathering agents and rain.

16.4 PROCESSES OF DOWNSLOPE MOVEMENT

Downslope migration of the mantle of rock-waste and associated displacements of bedrock are commonly referred to as 'mass movement'. The movements range from falls of material and landslides resulting from slips, to soil creep and rain-wash resulting from various types of flow. Significant evacuation of material downslope also takes place due to chemical weathering of soils, regolith and bedrock, leading to loss of solutes in groundwater, although this is not strictly part of mass movement. There are various ways of classifying mass movement. Sharpe based his classification on (a) the dominant kind of movement: slide or flow; (b) the relative rate of movement: fast or slow; (c) the kind of material: rock-waste or bedrock; and (d) the relative content of ice or water in the moving mass (Table 16.1). Sliding requires a slip plane between the moving mass and the underlying stable ground, i.e. an inclined surface on which the component of gravitational force exceeds the resistance due to friction and obstructions. Simple flowage requires no such slip plane, as the movement gradually dies out with depth. But flowage and sliding are often combined and give rise to complex transitional types of movement. Sharpe emphasized the gradation between rivers as trans-

Table 16.1 Classification of mass movement (*After C.F.S. Sharpe 1938*)

	Nature and rate of movement		With increasing ice-content ⟵	─Rock or soil─⟶	With increasing water content	
FLOW	imperceptible	**GLACIAL TRANSPORT**	SOLIFLUCTION ↓	CREEP (ROCK CREEP SOIL CREEP) ↓	SOLIFLUCTION ↓	**FLUVIAL TRANSPORT**
	slow to rapid		DEBRIS AVALANCHE		EARTH FLOW MUD FLOW DEBRIS AVALANCHE	
SLIDE	slow to rapid			SLUMP DEBRIS-SLIDE DEBRIS-FALL ROCKSLIDE ROCK FALL		

porting agents and types of mass movement in which the debris load increases until water acts only as a lubricant or is absent, as in a dry landslide. There is a similar gradation between glaciers and movements of rock-waste in which ice, when present, occurs only in the spaces between the fragments like a sparse cement. In a modified classification scheme, Varnes recognizes the distinctiveness of falls as a mass movement type (Figure 16.7). Types of landslide in rock are illustrated in Figure 16.8.

Falls of newly liberated fragments occur from vertical or overhanging outcrops or, by bouncing

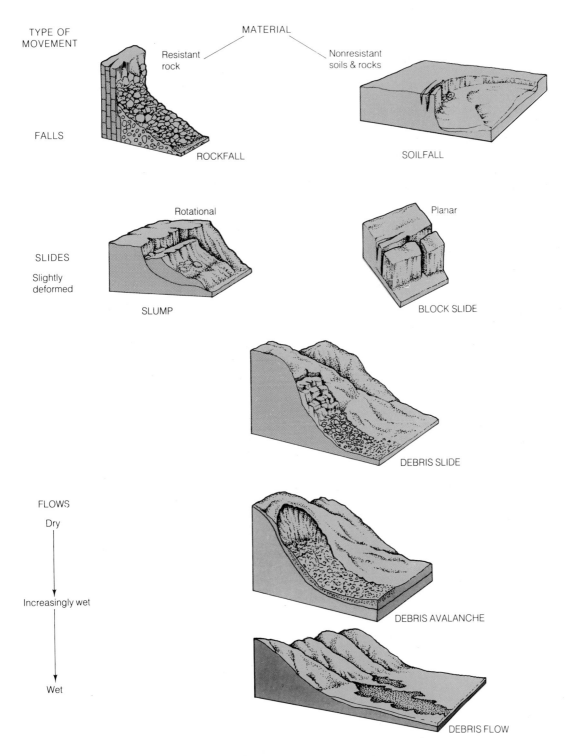

Figure 16.7 Mass wasting types. (*Classification of Varnes (1958) in M. J. Selby (1982)*)

TYPES OF
MOVEMENT

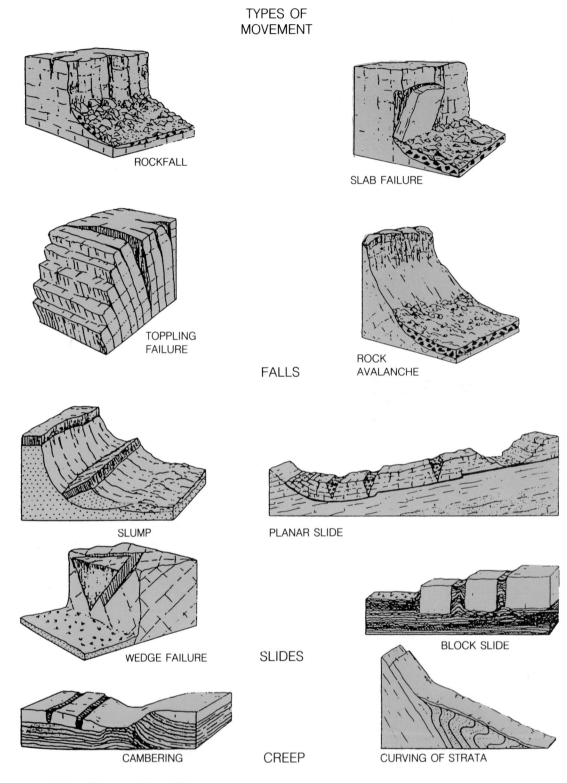

Figure 16.8 A classification of landslides in rock. (*After M. J. Selby, 1982*)

and rolling, from slopes so steep that no slip plane is necessary (Figure 16.8). Such slopes include cliffs and fault scarps as well as the precipitous sides of gorges, glaciated valleys and fjords, and the heads of corries and cirques. Most falls involve individual fragments which may disintegrate on impact, but occasional major slope collapses take place.

The essential conditions for landslides are lack of support in front and a slip surface. Such conditions are liable to occur on the sides of undercut slopes and cliffs, or of road, railway and canal cuttings, particularly where heavy massive rocks (e.g. plateau basalts and basic sills) overlie weak and easily lubricated formations. Sliding (Figure 16.9) occurs when bedding or cleavage planes, master joints or fault fractures dip towards a valley or other depression at a dangerous angle. Slumping (Figure 16.10) takes place on curved shear planes. These are often spoon-shaped and leave an arcuate scar on the defaced cliff or hillside. Because of the rotational slip, backward tilting of the surface and of the dip of the beds often result (Figures 16.11 and 16.12).

Figure 16.13 shows the debris of a landslide that obstructed a tributary of the Ticino valley in 1927. During the preceding years a crack appeared near the top of a peak on the right-hand side and slowly widened until it became a gaping fissure 2 m across. Subsequent movements were carefully measured by setting up a line of stakes and recording their positions every few hours. One day in 1927 an ominous slip of about 3 m alarmed the observers. Warnings were immediately telephoned and the danger zone was evacuated. Two days later the threatened landslide took place, fortunately without loss of life.

The north-western end of the Great Himalaya in Kashmir is a schist-migmatite-granite complex that is still rising and subject to occasional severe earthquakes. In December 1840 an earthquake shook loose part of the western spur of Nanga Parbat (8117 m), where the Indus has cut a gorge 4500 to 5000 m deep through the great mountain range. The gigantic landslide blocked the river and dammed back the water for 65 km. The resulting lake reached a depth of over 300 m before it

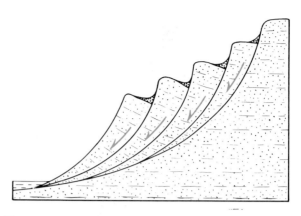

Figure 16.10 Diagram illustrating slumping on curved surfaces in unconsolidated or other weak formations, with production of back-tilted beds that were originally flat (cf. Figs 16.7 and 16.8).

overtopped the obstruction. The water then burst through with such violence that in less than two days the lake had emptied. A devastating flood tore down the valley, sweeping away a Sikh army encamped near Attock and carrying destruction for hundreds of kilometres.

The sediments of the flat terrace seen in the foreground of Figure 16.14 were deposited in a lake that formerly occupied a long stretch of the Upper Rhine and its tributaries. The lake was impounded by a prehistoric landslide that blocked the main valley near Flims, 50 km downstream from the place illustrated. About twelve and a half cubic kilometres of rock plunged down 1000 m or more. The forward momentum of this enormous mass carried it up the slopes on the other side of the valley to form a long tongue that extended 14 km from the deeply scarred mountain side and covered 50 km^2 with a chaos of smashed-up fragments. The sediments deposited in the lake and the great obstruction in front have now been cut through by the Rhine, and it can be seen that the landslide-dam that held up the lake for thousands of years was more than 820 m thick.

Van Bemmelen described a most remarkable landslide which was an important early link in the chain of catastrophes that wiped out the ancient Hindu culture of central Java a thousand years ago. The south-western part of the cone of the Merapi volcano, including the crater and the upper part of the conduit, sheared off in a series of spoon-shaped slices bounded by crescent-like faults that curved down to a basement of soft, Late Tertiary clays. On this weak and wholly inadequate foundation a great volcanic edifice had been built up until, after reaching a height of over 3000 m, parts of it began to slump (Figure 16.15), the movement being probably triggered off by an earthquake. As the colossal weight of displaced

Figure 16.9 Diagram illustrating conditions leading to rock slides on bedding planes after lubrication.

Figure 16.11 A prehistoric landslide at Garron point, Northern Ireland, along the Antrim coast road, where Chalk overlain by Tertiary basaltic lavas has slumped over Jurassic (Lias) clay, resulting in back tilting of the strata as in Fig. 16.10. Differential movement during slumping (see Fig. 16.16) resulted in faulting of the sliding rocks essentially at right angles to the direction of slumping. This caused black basalt, seen on the right overlying white chalk, to be downthrown against chalk on the left. (*P. S. Doughty, Ulster Museum*)

Figure 16.12 Landslide near Exmouth, looking west towards Sidmouth and the estuary of the Exe, south-east coast of Devon. Chalk and Upper Greensand, seen *in situ* on the right, slipped towards the sea (to the left of the illustration) on a surface of Jurassic clay on Christmas Eve, 1839, after a prolonged period of heavy rain. A landslip of large continuous masses of rock was followed by slumping and back-tilting of narrower slices, as seen in the middle of the illustration. (*S. H. Reynolds*)

rock glided forward, it thickened and crumpled up in front. This obstruction created a dam behind which a flourishing countryside, famed for its temples and monuments, was submerged by a deep and extensive lake. But this was not all. The

collapse of the summit of the volcano so reduced the pressure in depth that a cataclysmic eruption burst out through a new conduit and crater. The combination of landslides, earthquakes, floods and fiery ashes ruined the country for hundreds of

Figure 16.13 Landslide of 1927 which blocked the Valle d'Arbedo, near Bellinzona, Ticino Valley, above L. Maggiore, Italy. (*F. N. Ashcroft*)

years. The geological evidence still remaining adds dramatic detail to the brief stone inscriptions which have been found, recording a horrifying calamity that reduced Java to chaos in AD 1006.

The examples described above indicate that major landslides are often responsible for serious floods. Conversely, floods are generally accompanied and followed by a crop of minor slides and slumps. The banks are waterlogged when the streams are swollen; when the water subsides lateral support is withdrawn and the banks tend to fall in. The catastrophically heavy rainfall of August 1952 over Exmoor in south-west England is chiefly remembered for the disaster it brought to the little town of Lynmouth (Section 17.3). It was also the cause of dozens of landslips, mostly slips of earth and debris into the Exe and neighbouring streams draining Exmoor. The wet summer of 1960 again saturated the ground so that in many areas the high rate of runoff accompanying the exceptional rainfalls of the autumn was repeatedly more than the rivers could discharge. Floods were inevitable and these in turn provoked landslips and debris-slides along the river banks of a wide region extending from Dartmoor and the New Forest to the Severn and the Trent.

16.6 EARTH-FLOWS, MUD-FLOWS AND LAHARS

Figure 16.16 illustrates a typical association between earth-flows and certain landslides of the slump variety. Here the earth-flow is a hummocky or crevassed 'toe', composed of materials that slumped down a curved slip plane and thickened over the gentler slopes where the slip plane flattened out or curved upwards. Other earth-flows start independently. When materials such as clay or shale are over-saturated with water and so dilated, any slight shearing movements (which might be due to sea waves, earthquakes or vibrations caused by traffic) tend to bring about a compaction of the 'solids'. The water potentially liberated cannot quickly escape by itself, because of the low permeability. Consequently the sodden material becomes 'quick', i.e. like a quicksand, and readily flows as a whole. Many of the so-called landslips along railway cuttings in the London Clay are of this type.

Mud-flows tend to follow the stream of channels of arid or semi-arid regions, where dry accumulation of debris may be swiftly turned into a sort of 'porridge' by a sudden torrential downpour. If the run-off is high – i.e. if the bedrock near the surface is impermeable – the soaked debris flows down and out of the wadi or canyon (Figure 16.17) and spreads itself over the flatter country beyond. Debris of all sizes, including large blocks of fallen rock, may be transported by mud-flows, like the morainic load of glaciers and the great volcanic blocks carried along by **nuées ardentes** (Section 13.6).

The bog-bursts of Ireland and similar regions of peat bogs are closely related to mud-flows. If the

Figure 16.14 Flat-surfaced deposits marking the site of a prehistoric lake formed by a landslide that dammed the Upper Rhine, Tobel Drun ravine, near Sedrun, Switzerland. The headwaters of the stream illustrate rapid widening by surface erosion. (*F. N. Ashcroft*)

peat is the infilling of a former shallow lake, it may happen that water percolating along or near the underlying rock surface during long-continued rain finds that its previous underground exit has become blocked – probably by peat. It then accumulates at some level within the peat bog or

between it and the bedrock. The bog then swells up until something gives way. If it bursts externally, as occasionally happens, a deluge of muddy peat flows over the surrounding countryside.

Deposits of ash and other pyroclasts on the slopes of a volcanic cone are obviously suitable

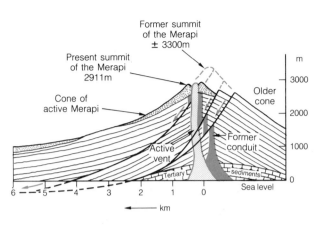

Figure 16.15 Diagram to illustrate the slumping of part of the cone of the active volcano Merapi in central Java on its foundation of weak Tertiary sediments. (*After R. W. van Bemmelen, 1949, Geology of Indonesia, Govt. Printing Office, The Hague*).

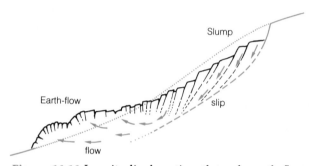

Figure 16.16 Longitudinal section through earth-flow showing typical association of earth-flow with certain landslides of the slump variety. (*C. F. S. Sharpe*)

material, all too favourably placed, for removal as mud-flows. When saturated by rain, melting snow or water from volcanic steam, mud-flows that may be highly destructive are inevitable. Volcanic mud-flows are widely known as **lahars**, this being their name in Java, where they are both frequent and dangerous. Hot and cold lahars are distinguished, the former being one of the special hazards of a volcanic eruption, as Herculaneum, buried in AD 79, still remains a grim witness. Every gradation between a lahar and a flood heavily laden with debris may occur when a crater lake is ejected at the beginning of an eruption. In 1919 Java suffered a disaster of this kind when over a hundred villages and most of their inhabitants were overwhelmed. In Iceland the melting of glacier ice during eruptions and the resulting calamities have led to the abandonment of some of the farms and settlements of earlier centuries.

16.7 SOIL-CREEP AND SOLIFLUCTION

Slow downward movement of soil on hillsides, known as **soil-creep**, is evidenced by tilted fences,

piling up of soil against outward-bulging walls, and the curving of trees near the ground. Creep is the sum of innumerable tiny displacements of grains and particles, with gravity in control at the steering-wheel to ensure that the cumulative effect is a downhill migration. Rain-splashing and rain-wash by surface runoff co-operate; but the effects of rain erosion – and also of wind erosion – are so powerful and distinctive that it is convenient to deal with them separately. Percolation of rainwater through the soil is one of the more important processes in soil-creep proper, and so is frost-heaving in regions subject to frost and thaw. Stones and particles of all sizes are heaved outwards by growing ice crystals. This outward movement at right angles to the slope, followed by a vertical drop when thaw sets in, gives the fragment concerned a slight shift downhill. The return of fine particles to the surface by burrowing animals acts on balance in the same direction. Other less perceptible movements, all of which are dominantly downslope, result from expansion and contraction due to temperature changes; from swelling and shrinking by soaking and drying; from the growth and decay of roots; and from the filling of cracks and other openings, including those made by burrowing animals, and those made by the straining of roots as they respond to the swaying of trees in the wind.

Even the sub-soil and the upper part of the bedrock share in the slow migration downhill. Rock-creep of this kind may lead to 'outcrop curvature', especially where the upper ends of steeply dipping or cleaved beds have been prised apart by frost or rootlets, so that they gradually curve over in the downhill direction (Figures 16.18 and 16.19). The apparent dips of superficial outcrops, exposed in hillside cuttings and gullies, may be very different from those of the undisturbed formations. Care must be taken in mapping not to be misled by such deviations.

The sub-soil activity may also be regarded as a by-product of the processes engaged in the formation of new soil. Where the soil survives, in spite of soil-creep and rain-wash, it means that soil removal is more or less balanced by soil formation in depth.

Many of the agencies responsible for soil-creep also operate in bringing about a similar downward migration of screes. However, the slow creep of screes, also known as **talus-creep**, is liable to sudden accelerations after thorough soaking and lubrication by excessive rain or melting snow. Increase of weight and decrease of friction may locally trigger off the slipping of a mass that was previously in uneasy equilibrium. Some debris-slides and rock-avalanches originate in this way. In permafrost regions screes may be permanently

Figure 16.17 Mud-flow of 1930 at the mouth of Parrish Canyon, on the side of the Lake Bonneville flats, Utah. (*United States Forest Service*)

Figure 16.18 Soil creep, showing the overturning of the cleavage of Cambrian slates, near St Davids, Dyfed, south-west Wales. (*W. Jerome Harrison*)

cemented by ice, except for a shallow spring thaw. With continuous replenishment by frost-shattered fragments from above, the increasing weight eventually forces the interstitial ice to flow. A sluggish **rock-glacier** then creeps down the slope and drapes the flatter ground in front with ridges and mounds of coarse rubble (Figure 16.20).

A rock-glacier has practically no interstitial mud. Where mud or a thin soil is present (with or without stones) on the gentler slopes of the per-

Figure 16.19 Soil creep, showing overturning of Yering-ian beds exposed in a road cutting in Melbourne, Victoria, Australia. (*R. H. Clark*)

Figure 16.20 A rock glacier, Copper River district, Alaska. (*F. H. Moffit, United States Geological Survey*)

mafrost lands, the process corresponding to soil-creep is almost entirely controlled by frost and thaw and is referred to as **solifluction** (soil-flow). The upper part of the waste-mantle creeps over the underlying frozen ground on slopes that may be nearly flat – no more than two or three degrees. Unless the migrating sludge is carried away at the foot of the slope, it accumulates and levels up any depressions that may lie in its path. Solifluction is clearly a slope-reducing process that generally leads to a landscape of low relief.

16.8 RAIN EROSION

Some of the sub-surface effects of percolating rainwater in promoting mass erosion have been indicated above. Interstitial rain-wash contributes to soil-creep; but where the soil has a strong cover of vegetation, and particularly if it is firmly bound together by a mat of interlacing grass roots, it is well-protected against rapid surface erosion. On the other hand, where soil and poorly consolidated mantle deposits lose this reinforcement and are exposed to pelting rain and a copious runoff, the results may be disastrous. Raindrops splashing on wet soil are now known to have far more serious effects than was formerly suspected. From each miniature crater made by the impact hundreds of dislodged particles are scattered in a little

fountain of spray, and on sloping ground most of these fall back a little farther downhill. Recent studies of splash erosion have revealed that during a heavy storm up to 250 tonnes of soil per hectare may be shifted. This will seem less surprising if the energy involved is considered. In such a downpour thousands of millions of raindrops strike a hectare of soil with velocities up to 30 km an hour or so.

The finest material of the top soil or earth is then readily transported to lower levels by rain-wash. The runoff may form a thin sheet of muddy water spread fairly uniformly over the slope (**slope-wash**), but generally it is concentrated into more or less intermittent trickles and rills (**rill-wash**). Observations on smooth sloping beaches show how quickly a thin sheet of water can develop into a network of shallow channels. The bearing on hillside soil erosion is clear. The good topsoil is the first to be carried away. If the process continues unchecked the deeper and inferior soil follows, long before it has had time to mature or to be replenished by the slow weathering of the subsoil. Within a few years the soil may be first impoverished and then lost, either by being washed away down a surface furrowed by rills and gullies (Figure 16.21), or by being blown away, or both.

Present rates of erosion are excessive because of human interference with the natural cover of the

Figure 16.21 Badland type of erosion in the valley of the Ruindi, south of Lake Edward (L. Idi Amin) in the extreme east of Zaïre. Residual buttresses of Pleistocene sands and clays left between the gullies are themselves scored and fluted by rain erosion. (*Félix Inforcongo*)

Figure 16.22 Close up view of the turrets and spires of the Big Bad Lands, South Dakota, USA.

Figure 16.23 Buttress of boulder clay developing into earth pillars, Val d'Herens, tributary of the Upper Rhône. (*G. P. Abraham Ltd. Keswick*)

soil, e.g. by the widespread clearing of forests and the breaking-up of protective turf by ploughing.

Where it was supposed that rain-wash was the major cause of soil erosion on hillsides, contour-terracing like that seen in the background of Figure 16.23 was introduced to minimize the losses. Results often proved disappointing. The reason is that terracing fails to give permanent security against rain-splash erosion. Some additional safeguard, such as a covering of straw-mulch, is necessary to shield the vulnerable soil from the force of the rain.

But man is not always the culprit; he may be the victim. A localized cloudburst may concentrate the runoff into a violent torrent that bites deeply into the turf on sloping ground, sweeping the underlying soil to the foot of the slope, and leaving a long gash in the hillside. The gash is gradually deepened into a gully by recurrent rains, and as soon as the water table is tapped it begins to carry water and becomes a rivulet. This is one of the ways in which tributaries originate or extend their headwaters (Figure 16.14).

In semi-arid regions, where the occasional rains are often exceptionally violent, rain-gashing reaches spectacular proportions on sloping ground underlain by clay or soft earthy deposits. Such land is sculptured into an intricate pattern of gullies and small ravines, separated by sharp spurs and buttresses. The gullies grow backwards into the adjoining upland, and the intervening ridges in turn are further cut up into smaller ribs and trenches (Figure 16.21). Tracts of the almost impassable country so developed are graphically described as **badlands** in North America, where they are widely scattered from Alberta to Arizona (Figure 16.22).

16.9 EARTH PILLARS

The remarkable residual forms known as **earth pillars** are a standing proof of the efficacy of rain erosion. They are developed on the slopes of valleys from spurs and ribs of relatively impermeable material, which contains resistant boulders (or their equivalent) in a more easily eroded matrix: e.g. boulder clay (Figure 16.23), certain conglomerates (Figure 16.24), or tephra (not too porous) containing volcanic bombs or ejected blocks (Figure 16.25). Wherever boulders or other resistant masses, including concretions, large fossils and occasional hard layers, are encountered while the slope is being worn back by rain erosion, they act like umbrellas over the underlying less resistant materials. Where the slope is sheltered from strong winds, earth pillars of surprising height may be etched out before their protective caps topple off.

The valley sides on which earth pillars stand are eroded mainly by rain-wash. Their inclinations are generally between 25° and 35°, the angle being dependent on the properties of the materials of the slope and the nature of its cover of vegetation. The range of inclination is about the same as that of the constant slope (Figure 16.5), which is controlled by the angle of repose of screes. This equivalence implies that the slope of the valley side, eroded mainly by rain-wash, reaches an angle down which the boulders that are continually being liberated can roll or slide, while the finer material is washed down by the runoff. If the slope were less steep the boulders would accumulate. But they do not accumulate, except temporarily in the river bed itself and on small adjoining flats, where the larger ones may rest for a time before being removed by an exceptional flood. This consideration indicates that a slope on which earth pillars are developing must lead down to a river capable (a) of carrying away all the debris fed into it from the slope, and (b) of deepening its floor or widening its channel (or both) in order to make room for the successive slices of material eroded from the slope. Now we have already seen (Section 16.2) that parallel recession of a slope is not geometrically possible unless these conditions are fulfilled. Here, where they are fulfilled, preliminary observations confirm that the slopes on which earth pillars stand do remain parallel, or nearly so, during their recession. And here we also know that the dominant process of surface erosion

Figure 16.24 Earth pillar of soft conglomerate (Old Red Sandstone): one of many isolated stacks, some capped with boulder clay, on the valley side of the Allt Dearg, a tributary of the River Spey, above Fochabers, Grampian Region, Scotland. (*British Geological Survey*)

Figure 16.25 Earth pillars in volcanic tuff erupted from Mt Argaeus (English Dagh), Ghcureme valley, about 240 km south-east of Ankara, Turkey. A lava flow of andesite forms the escarpment seen at the head of the slope. The valley is celebrated for the trogloditic dwellings and monasteries which have been excavated in the tephra. The lava caps have been removed from the earth pillars where practicable, partly for safety and partly to provide stone for the façades of the rock churches. (*Yan, by courtesy of Messrs Thames and Hudson*)

is rain-wash. The boulders are critical in providing the information that the slope of the valley side remains steep enough to ensure their easy migration to the foot, once they are liberated.

The slopes of the earth pillars themselves range from nearly vertical to about 60°; they correspond to the free face of Figure 16.5. These also tend to recede parallel to themselves, since the sides of small pillars have about the same angles as those of the big ones where the materials and conditions are similar. The fact that many pillars develop a convex profile, as may be seen in the illustrations, results from the steepness of the sides, lack of vegetation and high runoff (factors that favour an increasing rate of rain-wash erosion from top to bottom), combined with removal of the debris at the base. However, where the runoff is being continually checked by small obstacles there can be little or no acceleration, and the slopes developed then tend to be straight.

16.10 MEASURED SLOPES AND EROSION RATES

Figure 16.26 shows that two contrasted types of topography have developed under the same climatic conditions on the Brule and Chadron Formations. The latter are of Oligocene age and consist of poorly consolidated clays and other sediments. The Brule, however, is relatively impermeable and dries out to a hard surface down which the runoff from subsequent rainfall is high. Erosion by rain-wash, accompanied by rill-wash, is dominant. As the volume of water increases downhill the flow tends to concentrate into tiny rills, most of which are temporary, as they fill up again with particles of soil or clay when the flow of water abates. Under these conditions small

Figure 16.26 Typical topographic forms developed in the Oligocene strata of Badlands National Monument, South Dakota. The pediment in front rises very gently to the rounded slopes of the Chadron Formation (middle), above and behind which are the steeper, straight slopes of the Brule Formation. Further along the escarpment the pediment locally truncates the Chadron and rises to make a sharp junction with the Brule. (*Stanley A. Schumm*)

Λ-shaped residuals left in front of the Brule escarpment continue to maintain steep, nearly straight slopes. The latter meet in sharp divides except while a more resistant band is exposed at the summit, when the crest becomes rounded or even flat for a time, as seen in Figure 16.27. As material is eroded from the slopes their lengths from pediment to summit are shortened and the divides are lowered. Stanley A. Schumm has measured a variety of these residuals and finds no significant variation of slope angle from the highest and longest to the lowest and shortest, the average being between 44° and 45°. The slopes have remained parallel during their retreat.

The underlying Chadron Formation, exposed by the more rapid retreat of the Brule escarpment, behaves very differently. It dries out to a surface of loose aggregates and, because of its permeability, much of the subsequent rain sinks in, so that the runoff is low. Here soil-creep, accompanied by small slumps and slides of wet mud, accounts for most of the erosion. Schumm's measurements on the Chadron residuals show a consistent decline in the slope angle from 33° for the longest slope, to 8° for the shortest (Figure 16.27). It is evident that convex summits and gently rolling topography are developed where surface creep and related processes predominate over rain-wash.

Measurement of the thickness of the material removed over a period of just over two years, during which 82 cm of rain fell on the area, revealed an exceptionally high rate of erosion, the

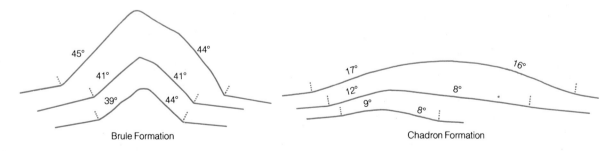

Figure 16.27 Series of slope profiles measured on erosion residuals, Badlands National Monument, South Dakota, to show the changes of angles as the slopes above the pediment become shorter. Dotted lines mark the junctions with the pediment. Brule (low permeability, high runoff): little variation in slope angles. Chadron (high permeability, low runoff): marked reduction in slope angles. (*After Stanley A. Schumm*)

thickness ranging from 2 to 3.8 cm at right angles to the surface. Figure 16.26 shows clearly that the Brule Formation must have been worn back more rapidly than the Chadron, much of which still forms part of the landscape in front of the escarpment. Consistently, Schumm's observations indicate that the rate of present-day slope erosion on the Brule (mainly by rain-wash and rill-wash erosion) is nearly twice that on the Chadron (mainly by surface creep). In most areas, of course, both sets of processes are operative; but generally, as here, in widely different proportions. Parallel retreat of approximately straight slopes would appear to be favoured where rain erosion is dominant. Conditions appropriate to dominant surface creep, which favour slope retreat with declining angles, are of two kinds:

1. solifluction and related processes in the colder regions, and
2. a thick soil cover strongly protected by vegetation.

The latter may be overcome in humid temperate and tropical regions by exceptional rainfall, especially if the underlying bedrock is such that infiltration is low. Figure 16.28 illustrates a case of parallel slope retreat in Sri Lanka, where protection by vegetation looks adequate, but where the

rainfall is very high and the bedrock is a particularly massive quartzite. Schumm tentatively reached the conclusion previously suggested by C. D. Holmes (see Section 16.14) that 'the areas in which creep dominates over rain-wash, and vice versa, are end members of a continuous series ranging through all proportions of both processes' according to the prevailing conditions of vegetation, soils and climate.

16.11 SHEET-WASH AND PEDIMENTS

A curious feature that may have been noticed in Figures 16.26 and 16.27 is the sharp change of angle at the junction of hillside slope and pediment. This implies that all the material that reaches the base of the slope by rain- and rill-wash or creep is removed. Otherwise, as often happens, there would be some deposition at the base of the slope, resulting in a concave surface linking the steeper slope above to the more gently inclined pediment below. In Schumm's area the pediment in turn leads down to an actively eroding river which carries off all the debris supplied to it. This debris must therefore be transported across the pediment. Not only so, but Schumm found that during the period of his investigation the pediments themselves were lowered by nearly 2.5 cm

Figure 16.28 Heavily vegetated hill of quartzite with nearly straight steep slopes, heading in a slightly rounded crest of almost bare rock, near Sigiriya, Sri Lanka. (*Martin Hürlimnan*)

in some places adjacent to a steeper slope. As each slope receded it left behind at its base a small extension of the pediment surface, the width of which ranged between 3.8 and 7.5 cm. Here, then, the pediment is a gently inclined surface that is gradually being extended at the expense of the steeper slopes as they recede. It is, moreover, a transportation surface.

The sharp break in angle between pediment and hill-slopes implies a sudden change in the processes responsible for the erosion and transportation, or in the properties of the materials being transported (or in some combination of both properties and processes). Schumm found that during certain seasons of the year the hillside slopes have a relatively rough and permeable cover of aggregated soil, as compared with the smoother and less permeable surface of the pediment, where the soil aggregates are disintegrated by the greater energy of rain-splash, possibly assisted by changes in vegetation. On the steeper slopes these differences tend to reduce the runoff and its velocity, whereas on the pediments both runoff and velocity tend to be enhanced. If the relative smoothness of the pediment compensates for the gentle slope – given the right conditions of rainfall, runoff and level of the water table – a surface will be provided down which all incoming debris can be efficiently transported in the long run to the nearest stream. If the nature of the surface fails to compensate for the reduction of slope, a certain amount of deposition will occur at the foot of the hillside, thus giving a gradual change of slope down to the pediment (see Section 16.3). But if the conditions lead on balance to over-compensation, the pediment will be not only an efficient surface of transportation but also a surface subject to erosion from time to time by sheet-wash. Such conditions would favour the development and maintenance of an abrupt change of angle between hill-slope and pediment, and active investigations are now in progress to determine more precisely what these conditions may be.

The conditions under which pediments have developed on a large scale require a sudden change of process rather than of material. They have been studied chiefly by Lester King in parts of Africa where violent rainstorms quickly flood the ground over vast areas, so that the whole pediment surface acts like the floor of an extremely wide river channel. The resulting sheetflood may be sufficiently vigorous and turbulent not only to pick up and transport any temporary deposits dropped during the waning stages of earlier floods, but also to accomplish a certain amount of sheetwash erosion by removing rock-waste loosened from the bedrock by weathering. Between the steep sides of an inselberg and the gentle

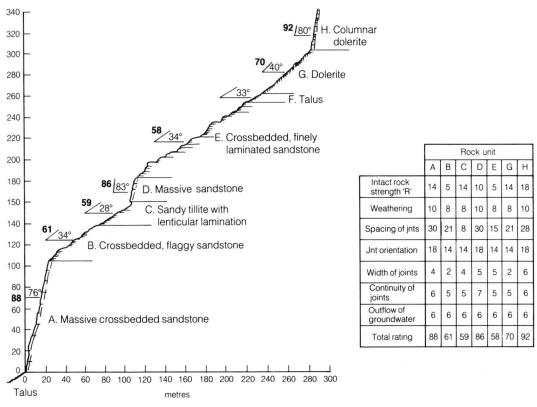

Figure 16.29 A rock slope in Magnis Valley, Transantarctic Mountains, showing the strength ratings for each rock unit and the slope angle. The matrix shows the components of rock mass strength. (*After M. J. Selby, 1982*)

declivity of the surrounding pediment the change of angle is always impressive and often quite abrupt. At that point there is an obvious hydraulic 'break' in the behaviour of running water, such as occurs more familiarly in the plunge-pool at the bottom of a waterfall (Section 17.2). Heavy runoff, streaming rapidly down the smooth steep sides of an inselberg, suffers a sudden check at the bottom, where any accumulated debris is churned up and the fragments reduced in size. Even the transient rills that are largely responsible for the wearing-back of hillsides and escarpments are checked when they reach the bottom. The water then spreads out: perhaps to contribute to a sheetflood if the rainfall and runoff are sufficient; perhaps to dwindle away, leaving a thin veneer of debris as they infiltrate into the soil and begin to co-operate with the processes of surface creep. The possibilities are both numerous and complex.

16.12   EQUILIBRIUM SLOPES

From the preceding example it will be clear that the angle and form of a slope will be controlled by the nature of the slope-forming materials and by the processes operating on them. On rock slopes there are often clear relationships between angle of slope segments and the resistance of the exposed rock types to weathering and erosion (Figure 16.29). The lack of talus on such rock slopes shows that weathered material is quickly removed downslope (Figure 15.6). Hence potential rates of transport exceed actual weathering rates and the term 'weathering-limited' slope is justified. Such slopes are common wherever resistant rocks occur and where rates of weathering are low, such as in arid areas. In contrast, 'transport-limited' slopes occur where rates of weathering are high and the production of new rock-waste exceeds the capacity of mass movement to remove it downslope. Such slopes dominate on weak rocks and unconsolidated sediments and soils and where weathering is especially rapid, as in the permafrost zone. Typically, these slopes carry a thick debris-mantle and are gentle, convexo-concave in profile, lacking the free face associated with the four-unit slope model of Wood, described earlier (Section 16.3). A crucial control on rates of transport and slope form is whether or not debris accumulates at the base of the slope or is removed by rivers or other transporting agents.

16.13   SLOPE EVOLUTION

Historically, a major task of geomorphologists has been to investigate how slopes change through time. This is a difficult problem as most slopes

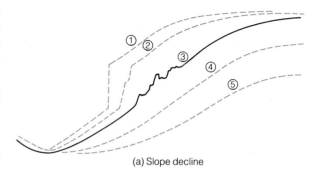

(a) Slope decline

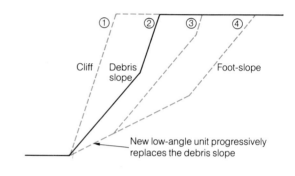

Cliff / Debris slope   Foot-slope

New low-angle unit progressively replaces the debris slope

(b) Slope replacement

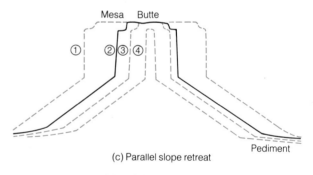

Mesa   Butte

Pediment

(c) Parallel slope retreat

Figure 16.30 Models of slope evolution.

change very slowly. Much of the debate on slope evolution has centred on three models associated with the names of Davis, Penck and the South African geologist, L. C. King. All three models require the prior development of a rock-free face and so strictly do not apply to slopes developed across unresistant materials.

Davis pioneered the slope decline model (Figure 16.30A). Initially, slopes are steep due to stream incision as a result of rapid uplift. Weathering of exposed rock surfaces is rapid and material is transmitted downslope. On the lower part of the slope, however, transport processes are unable to cope with the volume of material generated upslope and accumulation of debris begins at the foot of the slope. Gradually, the angle of the lower slope is reduced and rates of transport across it decline. Weathering and debris removal remain constant on the exposed part of the slope, leading slowly to an overall decline in the slope since its top is lowered more rapidly than its base. Eventually, the free face is buried.

Penck envisaged that lower slope segments progressively replace those higher on the slope (Figure 16.30B). Debris building up at the base of a cliff forms a debris slope which is gradually extended to cover the retreating free face and prevent further weathering. The debris slope is itself consumed from below by the growing foot-slope as the debris is weathered, reduced in size and made susceptible to creep and wash processes.

King observed that the free face may be conserved and retreat without change in angle (Figure 16.30C). Parallel retreat of the cliff is matched on the transport slope below and only the extending concave unit (the pediment) at the slope foot tends to reduce in angle in time.

Classically, slope decline was held to be typical of humid temperate areas and parallel retreat was regarded as the norm in arid and semi-arid regions. In fact, all three types of slope evolution occur worldwide. Slope decline and replacement are models for transport-limited slopes without efficient removal of debris from the foot of the slope. Both apply equally well to certain slopes developed in periglacial areas, as well as the temperate zone. Parallel retreat requires the removal of debris from the slope foot and hence is a form of weathering-limited slope evolution. Parallel retreat appears to reach its finest development in arid areas, where weathering is slow and effective transport is performed by sheet-wash across sparsely-vegetated surfaces. The key factor, however, is the existence of thick and nearly horizontal cap rocks to maintain the free face during slope retreat. Thus, parallel slope retreat can be recognized as an element in the development of stepped slopes in periglacial regions and duricrusted scarps and mesas in the humid tropics (Figure 15.13).

16.14 FURTHER READING

Holmes, C.D. (1956) Geomorphic development in humid and arid regions, *American Journal of Science*, vol. 253, pp. 377-90.

Selby, M.J. (1982) *Hillslope materials and processes*, Oxford University Press, Oxford.

Selby, M.J. (1985) *Earth's changing surface*, Oxford University Press, Oxford.

Sharpe, C.F.S. (1938) *Landslides and related phenomena*, Columbia University Press, New York.

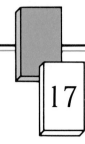

17

THE WORK OF RIVERS

All these varied and wonderful processes, by which water mightily alters the appearance of the earth's surface, have been in operation since the most remote antiquity.

Agricola, 1546

17.1   PROCESSES OF EROSION

Erosion is the cumulative effect of a great variety of processes. In general these can be conveniently divided into four groups: two involving the agent alone (chemical and mechanical activities); one combining the agent with its load of detritus, which arms it, so to speak, with 'tools' or 'teeth'; and one concerning the detritus alone. The four groups are listed in Table 17.1 for each of the chief agents of erosion, with the technical terms often used in describing them. Such a classification unduly emphasizes some of the distinctions, but it serves as a useful analysis provided that the general co-operation and overlapping that inevitably occurs in nature is not overlooked.

17.2   RIVER EROSION

(a) **Corrosion**, as indicated in the table, here includes all the solvent and chemical activities of river water on the materials with which it comes into contact.

(b) Several **hydraulic** processes co-operate in bringing about the mechanical loosening, lifting and removal of materials by flowing water. Loose deposits are readily swept away, the initial lifting force being provided by turbulence, that is by eddies in which local velocities are rapidly changing and are often far higher than the rate of flow of the stream. Except when a river is actively downcutting its floor or undercutting its banks, it may not acquire much new material by erosion of its channel, but the coarser part of its load of debris is likely to be dropped again and again during transit (Figure 17.1), and each time the fragments have to be picked up afresh by the lifting forces before they can be transported farther downstream.

Changes of velocity during rapid turbulent flow can produce some very remarkable effects. For example, a rapid increase of velocity is accompanied by a corresponding decrease of internal pressure. If the pressure falls below a certain critical amount, small bubbles of water vapour form and the water 'foams'. Sooner or later the velocity is decreased by friction against the floor or sides of the channel, the internal pressure increases again and the 'foam' becomes explosively – or rather implosively – unstable. The bubbles then suddenly and violently collapse, with production of shock waves which may shatter the adjoining surface with hammerlike blows, so liberating a

Table 17.1   Agents and processes of erosion

Type of erosion	Active agents	Agent alone		Agent armed with detritus	Detritus alone
		Solvent and chemical action	Mechanical loosening and removal of materials passed over	Wearing of surfaces by transported materials	Mutual wear of materials transported
Rain erosion	Rainwater	Corrosion	Splash Rainwash Sheetwash	Localized corrasion	Attrition slight, if any
River erosion	Rivers	Corrosion	Hydraulic lifting and scouring Cavitation	Corrasion	Attrition
Glacial erosion	Glaciers Ice-sheets	Corrosion limited to sub-glacial streams	Plucking and quarrying	Abrasion (e.g. striated surfaces)	Attrition
Wind (or aeolian) erosion	Wind	–	Deflation (blowing away)	Wind corrasion (sand-blasting)	Attrition
Marine erosion	Sea and ocean waves, tides and currents	Corrosion	Various hydraulic processes	Marine abrasion	Attrition

Note It should be noticed that the term **soil erosion** does not imply erosion by soil, but erosion of the soil, e.g. by wind, rain-wash, etc.

Definitions

Abrasion   Wearing-away of surfaces by mechanical processes such as rubbing, cutting, scratching, grinding, polishing
Attrition   Reduction in size of detrital fragments by friction and impact during transport
Cavitation   Collapse of bubbles of water vapour in highly turbulent eddies of water; such collapse is like a negative explosion and sets up powerful shock waves which tend to disintegrate any adjacent rocks
*In ordinary foams bubbles are occupied by gases of a different composition from the liquid, and cavitation does not normally occur. Sea foam, for example, is mainly air in seawater; its peculiar stability on certain beaches is due to the presence of protein in the films of water.
Corrasion   Cumulative effects of mechanical erosion by running water or wind when charged with detritus and so provided with 'tools' or abrasives
Corrosion   Wearing-away of surfaces and detrial particles and fragments by the solvent and chemical action of natural waters
Deflation   Lifting and removal of dust and sand by wind (Section 22.3)

crop of particles ready to be carried away. Wherever this process of **cavitation*** occurs, the rate of erosion is greatly speeded up. Cavitation accounts for the hollows in stream beds that are later developed into pot-holes (Figures 17.2 and 17.3). It plays an important part in the erosion of plunge-pools below waterfalls (Figure 17.4) and it also helps to account for the extraordinary erosion achieved by turbidity currents on the ocean floor (Section 24.4).

A related application of the above principle is

that water forcibly driven along a joint exerts a very powerful pressure at the feather edge. The opening of the joint is extended and wedging action comes into play. Where the current is swift enough, as when a river is in full spate, the pressure of water driven along joints or bedding planes may be sufficient to liberate slabs of rock from the channel floor, or sides, and so prepare them for transport.

(c) **Corrasion** (an unhappily chosen term, too easily confused with 'corrosion') is the wearing

Figure 17.1 The 'tools' of a river, left stranded in dry weather. Valley cut in boulder clay, Anglezark Moor, east of Chorley, Lancashire. (*British Geological Survey*)

Figure 17.2 Pot-hole drilled in Tertiary plateau basalt, Glenariff, Co. Antrim, Northern Ireland. (*R. Welch Collection, Copyright Ulster Museum*)

Figure 17.3 Pot-hole erosion of Carboniferous sandstone forming the bed of the River Taff, near Pontypridd, South Wales. (*British Geological Survey*)

away of the sides and floor with the aid of the boulders, pebbles, sand and silt which are being transported. With such tools even the hardest bedrocks can be excavated and smoothed. The drilling of **pot-holes** is one of the most potent

methods of downcutting. These develop in the depressions of rocky channels or from hollows

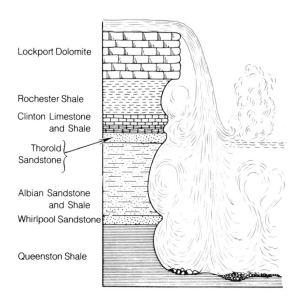

Lockport Dolomite

Rochester Shale

Clinton Limestone
and Shale

Thorold
Sandstone

Albian Sandstone
and Shale

Whirlpool Sandstone

Queenston Shale

Figure 17.4 Section across the Niagara Falls showing the sequence of formations, to illustrate the mechanism of recession by undercutting, and the erosion of the river bed in the 'plunge pool' beneath the falling column of turbulent water.

started by cavitation and scouring. Boulders and pebbles, acting like drilling-tools, are rapidly swirled round by eddies (Figure 17.2). Vertical holes are cut deeply into the rock as the water plunges in and keeps the drilling-tools in action by its spiral motion. As the boulders wear away, and are swept out with the finer materials, new ones take their place and carry on the work. In front of a waterfall very large pot-holes may

develop in the floor of the plunge-pool. This leads to deepening of the channel, and at the same time a combination of hydraulic and scouring activities undermines the ledge of the fall. The eddying spray behind the fall itself is particularly effective in scouring out the less resistant formations that underlie the ledge. Blocks of the overhanging ledge are then left unsupported and break off at intervals, thus causing a migration of the fall in an upstream direction, and leaving a gorge in front (Figures 17.5 and 17.6).

(d) **Attrition** is the wear and tear suffered by the transported materials themselves, whereby they are broken down, smoothed and rounded. The smaller fragments and the finer particles are then more easily carried away.

17.3 DISCHARGE AND TRANSPORTING CAPACITY

The load carried by a river includes the rock-waste supplied to it by rainwash, surface creep, slumping, etc., and by tributaries and external agents such as glaciers and the wind, together with that acquired by its own erosive work, as described above. The debris is transported in various ways. The smaller particles are carried with the stream in suspension, the tendency to settle being counterbalanced by eddies. Larger particles, which settle at intervals and are then swirled up again, skip along in a series of jumps – a process known as **saltation**. Pebbles and boulders roll or slide along the bottom, according to their shapes. Very large

Figure 17.5 General view of the Niagara Falls looking southwards. The America Falls on the left are separated from the Canadian or Horseshoe Falls by Goat Island. (*Photogrophic Survey Corporation Limited, Toronto*)

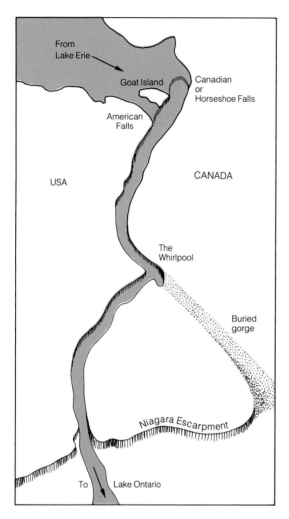

Figure 17.6 Sketch map of the Niagara Falls and the 11 km gorge cut by headward recession since the Niagara River fell over the Niagara Escarpment about 12 000 years ago; looking southward as in Fig. 17.5. The Whirlpool is scoured out of glacial drift which now fills the Buried Gorge of an earlier 'Niagara River'.

blocks may move along on a layer of cobbles which act like ball-bearings.

As the velocity of a river is checked, the bed load is the first to come to rest. With continued slackening of the flow the larger ingredients of the suspended load are dropped, followed successively by finer particles (Figure 17.7). Very fine sediment – clay and silt – behaves rather differently due to the effects of compaction and cohesion. A ball of modelling clay can be washed under the tap without significantly reducing its size because the tiny particles are bonded together by electrostatic charges. Once disaggregated, however, clay and silt particles will only settle in still water. Often these particles must first be lumped together again or **flocculated**, for example, in estuaries due to contact with seawater, before

deposition will occur. When the stream begins to flow more vigorously the finer materials are the first to be moved on again. In consequence a river begins to sort out its burden as soon as it receives it. The proportion of fine to coarse amongst the deposited materials tends on average to increase downstream, but there may be many local interruptions of this tendency because of additions of coarse debris from tributaries or from landslides and slumping of the banks.

It should be noticed that the term **load** does not specifically mean the maximum amount of debris that a stream could carry in a given set of conditions; that amount is referred to as the **transporting power** or **capacity**. When used by itself the term load is technically defined as the total weight of solid detritus transported in unit time past the cross-section of the river at the place of observation. The corresponding quantity of water that passes is called the **discharge** and is usually represented by Q and is expressed in cubic metres per second, or cumecs for short. The load includes all the material being transported in solution and derived from weathering of minerals in soils and rocks and corrosion along the river channel. The latter is termed **solution load** or **dissolved load** to distinguish it, if necessary, from the **solid load**. The solid load consists of the suspended load plus the bed or traction load, which includes the coarser debris travelling by saltation, rolling and gliding. The traction load is difficult to measure and is generally taken to be not more than about a tenth of the suspended load. The following annual estimates for the Mississippi, averaged over a period of many years, are amongst those determined most accurately (opposite):

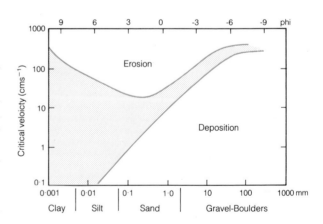

Figure 17.7 Hjulström's diagram, relating critical flow velocity and particle size. At velocities greater than those along the upper line the river will pick up particles of a given size. Where velocities fall within the shaded zone the river will continue to transport particles already picked up. Only where velocities fall below the lower line will particles of a given size be dropped.

	millions of tonnes per annum
Dissolved or solution load	200
Load or solid load { Suspension load	500
Traction or bed load	50
Total load	750

The proportion of solid to dissolved load varies enormously from place to place and from time to time. Taking all the chief rivers of the world over representative periods of years, it appears that on an average about 19 000 million tonnes of the products of rock-waste are transferred to the sea every year and that about 25 per cent of this total is carried in solution.

Both discharge and load depend on the climate and geology (lithology, structure and relief) of the river basin concerned, and both co-operate in carving out the channels down which the water and rock-waste are transmitted to the sea. The size and form of a river channel are determined by adjustments that are continually in progress between a bewildering variety of factors, the chief of which are the discharge, the mean velocity, the slope or gradient, the width, the mean depth, the load and the 'roughness' of the bed. 'Roughness' is a coefficient like that of friction, which expresses the resistance offered to flow; it depends on the varied sizes of the debris or other irregularities of the stream bed, being least when the bed is smooth. Indeed, there are so many factors – or variables – that it is only with the aid of experiments with models of rivers, and with the accumulated experience of the behaviour of real rivers, that river-control engineers can attempt to predict how a change in one factor is likely to affect the others.

It is found, for example, that on the brief occasions, generally seasonal but sometimes years apart, when the channel fills up to the brim (the 'bankfull' stage) and the stream overflows its banks in places, the changes brought about are far greater than those of the intervening months or years. The width in particular is likely to be increased by slumping and caving-in of the channel sides. Thus, in so far as the shape and gradient of a channel are determined by the discharge, it is the bankfull discharge that is the most effective agent, and the bankfull stage that provides the most satisfactory basis of comparison between one river and another, or for one river at different times. Now bankfull discharge and the rarer, more extreme events associated with overbank flow, are the conditions associated with floods. The recurrent floods that afflicted many parts of Britain during the autumn of 1960 naturally provoked a widespread public outcry for systematic river con-

trol. One type of solution would be to try to forestall the possibility of bankfull discharge by increasing the size of the channels concerned. But such measures are extremely expensive. Moreover, if the width and depth are increased, the velocity is decreased; in consequence the new bed of an enlarged channel is liable to be silted up. Continual maintenance of the artificial interference is generally essential.

The Thames provides an instructive example. The mean discharge at Teddington is 2400 cubic feet (67 m³) per second, but the bankfull discharge amounts to 10 700 (300 m³). To protect London from floods the Thames Conservancy, amongst other measures, almost doubled the cross-sectional area of the channel for 12 miles (19 km) above Teddington, completing the work in 1939. This lowered the river by several feet (a metre or so): in fact within 2 or 3 inches (5 to 8 cm) of the predicted level. But to maintain this condition the Conservancy found it necessary to excavate some 250 000 tonnes of sediment from the channel floor every year, and more than twice that amount from tributaries.

Another method of forestalling floods is to provide a new channel through which any dangerous excess of water can be diverted to a depression or an alternative exit. Iraq has a long record of floods dating back to the time of Noah. The melting of snowfields in the mountains to the north has been an almost annual menace to Baghdad. At the time of the great flood of 1954 a channel was already being dug from the Tigris to a large depression, 70 m deep, between the river and the Euphrates. In 1957 Baghdad would have suffered an even greater disaster had not the rising water been drawn aside before the danger level was reached.

The transporting capacity of a stream rises very rapidly as the discharge and the velocity increase. Experiments show that with debris of mixed shapes and sizes the maximum load that can be carried is proportional to something between the third and fourth power of the velocity. But for fragments of a given shape, the **largest size** that can be moved (not the total mass of mixed debris) is proportional to the sixth power of the velocity, provided, of course, that the depth of water is also adequate for the purpose. This explains how it is that enormous boulders, weighing many tonnes and for long periods looking like permanent features of a river bed, can be heaved up by exceptionally swollen torrents and carried a stage further downstream. When this happens (and at correspondingly lower velocities for smaller boulders) the whole bed load grinds and rolls along the channel floor. It is extremely dangerous to attempt to wade across a river with so treacherous a moving bed.

The almost incredible power of rivers in spate, especially when rushing down steep and narrow valleys, was tragically shown by the disaster·that befell the north Devon seaside resort of Lynmouth in 1952. As a result of a downpour of about 22 cm of rain in 24 hours on Exmoor, when much of the ground was already waterlogged, the runoff became far greater than could be carried away through the existing channels. Part of the overflow from the West Lyn river concentrated itself into a new course and became a raging cataract that tore its way down the hillside and burst through the little town before rejoining the original channel (Figure 17.8 and see Figure 17.47). Roads, houses and bridges were demolished; service pipes and cables were destroyed; 40 000 tonnes of boulders, uprooted trees and soil, and great masses of collapsed masonry were churned up together in a bottle-neck entanglement (Figure 17.9). The sea wall was shattered to pieces and swept out to sea. After this grim account of a terrifying night of havoc, it would be unfair not to place on record that Lynmouth, rebuilt and adequately safe-guarded, remained entirely unscathed during the phenomenal floods of 1960, referred to in this section.

17.4 RATES OF DENUDATION

Estimates of the total mass of rock-waste transported to the sea by rivers allow calculation of average rates of global denudation. Corbel has calculated a mean world denudation rate of 30 mm/1000 year. This rate of change appears exceedingly slow on a human time scale but on the

Figure 17.8 Showing on the right the new channel cut by the floodwaters of the West Lyn as they swept down the hillside through Lynmouth in south-west England on the night of 15/16 August 1952. The distribution of boulders and the vanished roads show the extent and violence of this devastating cataract when at its peak discharge only a few hours before the pictures Figs. 17.8 and 17.9 were taken. (*Syndication International Ltd*)

Figure 17.9 The maelstrom of destruction in the bottleneck of Lynmouth. (*Syndication International Ltd*)

geological time scale this rate extrapolates to 30 m in a million years or even 300 m in 10 million years and is clearly sufficient to bring about major landscape change and to produce considerable volumes of sediment. A useful measure of these slow processes is the Bubnoff unit (B): $1\,B = 1\,m/m.y. = 1\,mm/k.y. = 1\,\mu/y$. Corbel's figure then becomes 30 B. This global figure is virtually the same as that arrived at for the whole of North America, and the data for the Mississippi basin yield a result that is little different. For individual drainage basins, however, the average rates of denudation vary enormously, as would be expected. The Irrawady–Chindwin rate is 769 B, whilst for the low-lying areas draining into Hudson Bay the rate falls to about 7 B. One of the highest known sediment-yields per drainage area in Europe is from the Semani River, Albania, with an average annual load of 4593 tonnes/km², equivalent to nearly 3000 B.

In certain regions (see Sections 18.10 and 18.11) remnants of old uplifted erosion plains still survive; and except at their edges these may have remained almost untouched by denudation for millions of years. In the drainage basins themselves the maximum rate of erosion is on the steeper slopes, i.e. on the flanks of the bordering uplands, which may be conspicuous escarpments. Towards the coast, erosion again drops to a minimum and becomes negative at times when deposition of alluvium is the dominant process. It is, however, convenient for purposes of comparison to record rates of denudation as they would be if each were averaged over the whole of a given drainage basin.

Accurate measurements of the sediment accumulated in the great reservoir known as Lake Mead (Figure 17.38) show that during the years 1935–48 it amounted to 2000 million tonnes. To this has to be added another 150 million tonnes for the material removed in solution from the area drained by the Colorado River and its tributaries. In terms of unweathered rock the total corresponds to an average reduction of the whole basin (above the Hoover Dam) by about 1 metre in 7000 years (143 B). At the present rate of infilling the reservoir will last another three centuries.

For South Africa an average rate can be estimated for the period of about 75 million years which has elapsed since the diamond pipes were formed. The pipes originally penetrated some 900 m of higher strata before reaching the surface of that time. The average rate of denudation has therefore been about 1 metre in 80 000 years (12.5 B).

For the Mississippi basin an average rate for a still longer period can be arrived at from the volume of Mesozoic and later sediments deposited in the Gulf Coast Province, USA (referred to in connection with its salt domes (Section 10.2) and oilfields (Section 10.3)). J. T. Wilson has estimated the volume accumulated since the early Jurassic, say during the last 175 m.y., as $5.6 \times 10^6 \text{ km}^3$. This corresponds to the removal of a layer 1.75 km thick from the whole drainage area of $3.2 \times 10^6 \text{ km}^2$; that is 1750 m in 175 m.y., giving an average denudation rate of 1 m in 100 000 years (10 B).

The above figures, which are at least of the right order, leave no doubt that the present rates are far higher than those that can have prevailed during past ages. And this is what we should expect, for a variety of reasons.

(a) The continents are now more elevated and of greater area, relief and climatic contrasts than has been usual. The total energy of rivers available for erosion and transport is therefore abnormally high, and conditions also favour a vigorous circulation of groundwaters. During the long periods when vast areas of the continents were submerged and pre-existing mountains had been reduced to low-lying plains, rivers and groundwaters would be relatively sluggish and ineffective. There must have been times when the rates of both chemical and mechanical processes of weathering and erosion became almost negligible. Geological maps show clearly that, at least since the beginning of the Cambrian, immense areas that are now land must, on balance, have received deposits rather than supplied them.

(b) Extensive regions are now covered with glacial and glacifluvial deposits, most of which offer little resistance to erosion.

(c) Human activities have greatly increased rates of denudation by deforestation and agricultural activities; by excavations and other engineering projects; and by the addition to the atmosphere of carbon dioxide from factories and fires, together with various more corrosive gases.

In the light of these considerations it would seem likely that an average rate of 1 m in 100 000 years (10 B) is more likely to be representative of the geological past than the average based on present rates. It must, of course, not be forgotten that any average of this kind may cover individual rates ranging between the extremes of no erosion at all and the swift cutting of a hillside gully during a sudden cloudburst. But even the adoption of the very low average here suggested leads to some astonishing conclusions. If one assumed, for example, that a rate of 1 m in 100 000 years had applied to the present area of land through the last 3000 m.y., it would mean the transformation into sediments of a slice of rock averaging 30 km thick, with the saline materials dissolved in seawater as by-products. In other words the volume of material denuded during geological time would be about the same as the total volume of the continents, down to the Moho (Sections 2.3 and 25.10). Here we have a most intriguing situation that obviously calls for discussion in connection with the origin of continents and ocean basins. It may be said straight away, however, that the amount of sodium now dissolved in the oceans (plus the smaller amount of former oceanic sodium now imprisoned in evaporites and salt plugs) indicates that the thickness of the slice of rock so transformed – averaged over the present area of land – cannot have been more than 2.5 km. From this result (unless the amount of former oceanic sodium has been grossly underestimated) it is possible to draw either or both of two kinds of inferences:

1. that the present area of land is several times as extensive as its average area since denudation first began; if this inference were taken by itself, 'several' would be at least 12, i.e. 30 km/ 2.5 km,

2. that the same material had been several times re-worked or re-cycled through the processes of denudation and deposition, with varying degrees of metamorphism in successive orogenic belts; here again, if this inference were taken by itself, 'several' would average at least 12.

Geochemical statistics suggest that re-cycling of the same materials has occurred on average about three times, in which case the 'several' of inference 1 would be about four. This combination of 1 and 2 is much more probable than either of the extremes and is consistent with other evidence to which we shall come in later chapters. The conclusion that the present area of land is four times as great as the average area since denudation began may seem to be surprising, but it is well in line with the statement already made that 'since the beginning of the Cambrian, immense areas that are now land must, on balance, have received deposits rather than supplied them'.

Now, however, we can extend our conclusion to a time far beyond the Cambrian; perhaps to a time when no land had yet appeared; to a time when the continents were entirely submerged, as parts of them, like the North Sea and Hudson Bay, are submerged today. We catch a few fleeting glimpses of a fascinating moving picture reproducing the fluctuating 'separation of the dry land from the waters'. At first the lands emerge very slowly, but they become increasingly extensive as time goes on, despite repeated setbacks due to erosion and marine invasions, until today their area falls but little short of that of the continental regions as a whole. And today we see the rivers carrying away the land at such a rate that, had it continued throughout geological time – which of course it could not have done – the whole volume of the continents would have been poured into the oceans three or four times over. An impossible supposition, certainly; but nevertheless an impressive illustration of the immensity of geological time and a dramatic way of indicating the unusual vigour of present-day erosion and of emphasizing the work of rivers in removing the resulting burden of rock-waste.

17.5 BASE-LEVELS AND 'GRADED' PROFILES

Since a river that flows into the sea must have a gradient towards the sea, the deepening of its valley is necessarily limited by sea-level. An imaginary extension of sea-level beneath the land surface is called the **base-level** of river erosion. The long profile of a river (i.e. the line obtained by plotting elevations – generally of the water surface – against their respective distances measured along the stream from, say, the mouth to the source) begins at sea-level, or just below if the profile is that of the river bed, and rises inland. The profile of a river in a region experiencing recent and continuing uplift is likely to be more or less irregular, in conformity with the slopes and undulations of the surface under dissection, and the nature of the rocks being eroded. Characteristic features include lakes and swamps, waterfalls and rapids. However, all but the greatest of these irregularities, such as very deep lakes, are destined to be smoothed out as erosion removes the waterfalls and sediments infill the basins.

In humid regions the discharge of a river increases from the source to the mouth, where

downcutting, if any, is strictly limited. Starting on an initial surface with a general slope towards the sea, downcutting is dominant along the middle reaches of the river. The effect is to steepen the gradient of the stretch leading down from the source, and to decrease the gradient from the middle reaches to the sea. Given ample time and no critical disturbances by earth movements or changes of climate or sea-level, it is not difficult to realize that the profile would be systematically modified until it approximated to a smooth curve, gently concave to the sky, practically flat at the mouth and steepening towards the source. When a river or, more commonly, a particular stretch of a river is found to have such a profile, it is conventionally said to be **graded** (Figure 17.10).

Admittedly, this is only a rough-and-ready definition, but it gives the term graded an objective meaning that in practice has some useful applications. Any attempt at greater precision leads one into a maze of abstraction. The usual interpretation of the graded condition is that the gradient is continually being modified by erosion and deposition so that, allowing for seasonal and other fluctuations of discharge over a representative period of years, the stream is everywhere provided with just the right velocities for the eventual transportation of the rock-waste supplied to it. It must be clearly understood that 'the rock-waste supplied to it' is not restricted to the debris fed into the stream from its drainage basin, but includes all that may be liberated by the stream itself from the bedrock exposed to corrasion on the floor and sides of its channel when and where any temporary cover of alluvium has been swept away. Failure to recognize this part of a graded river's activities has sometimes led to the quite wrong idea that further lowering of a river bed must come to an end when a graded profile has been achieved.

It is now becoming generally realized that a smoothly curved profile cannot be interpreted in terms of velocity and gradient alone. We have already seen (Section 17.3) that there are many other variable factors to be taken into account. For details of the most successful attempts to grapple with some of the complex and difficult problems

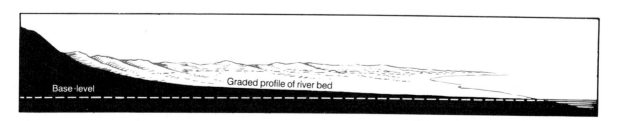

Base-level

Graded profile of river bed

Figure 17.10 Diagram to show the relation between base-level and an idealized graded profile.

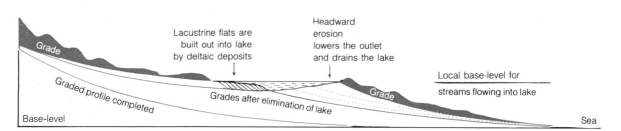

Figure 17.11 Diagram to illustrate the elimination of a lake by sedimentation at the inlets and by headward erosion at the outlet. Successive positions of graded profiles before and after elimination are shown.

Figure 17.12 Infilling of a lake by sedimentation. Head of Derwentwater viewed from the Wathendlath path, English Lake District. (*G. P. Abraham Ltd, Keswick*)

involved in river hydraulics, reference should be made to the invaluable work published by Leopold and his colleagues in 1964. These workers have made it abundantly clear that 'the shape and pattern of the river channel are determined by the simultaneous adjustment of discharge, load, width, depth, velocity, slope, and roughness'. For a given discharge, the velocity depends not on the gradient alone, but also on all these other interrelated variables. The most one can say is that a river oscillates (with seasonal and other fluctuations) about a slowly changing condition of quasi-equilibrium. But this behaviour gives us no reason to suppose that it either requires or guarantees more than an approximation to the sort of idealized profile shown in Figure 17.10.

Some irregularities in the profile there are bound to be, if only because of abrupt changes of discharge and load introduced into a main river by its tributaries. Attempts to define grade in terms of a highly complex and theoretical state of equilibrium, which we could not recognize even if by chance it were temporarily achieved, cannot be correlated with the empirical definition of grade with which we started. The equilibrium concept of grade is a mathematician's dream that no river ever realizes. But if, on the other hand, we accept a profile approximating to a smooth curve as the criterion of grade, we are dealing with something real that can be readily recognized by field work or from the data recorded on accurately contoured maps, even though as yet we do not fully grasp its significance.

It is important to understand, as mentioned above, that erosional lowering of the channel floor does not cease when a river is graded, though it is negligible near the mouth and becomes slower and slower throughout the graded stretch as time goes on. We are still, for convenience, assuming that no major disturbances are introduced by earth movements, or by changes of climate or sea-level. Continued downcutting is then ensured by two associated sets of circumstances:

1. the amount of debris and the sizes of the particles delivered to the river gradually decrease as the whole drainage area is being worn away; while
2. the stream flow or discharge remains about the same.

With reduction of the available load, its finer state of comminution and the resulting reduction of the roughness factor, a considerably reduced velocity would be adequate to do the job of transportation, including that of the bed load. The fact that the actual velocity is more than sufficient for this purpose means that the bed load, which is the active agent of corrasion, can still abrade the bedrock wherever the latter is encountered on the river floor. Energy and tools are thus available for downward erosion and the gradient continues to be lowered. Consequently the graded profile can be slowly flattened out. Base-level, always being approached but never quite attained, is its only downward limit.

The level of the main river at the point where a tributary enters acts as a **local base-level** for that tributary. In the uninterrupted development of a river system, graded tributaries thus become so adjusted to the main stream that they join it tangentially or nearly so. When tributaries fail to behave in this way the absence of adjustment is a clear indication that the progress of erosion has been interrupted by changes of slope or level due, as a rule, to earth movements or to glaciation. The waterfalls from 'hanging valleys' on the sides of deeply glaciated valleys (Figure 20.30) are extreme examples of such lack of adjustment.

Various irregularities in a river channel may postpone the general establishment of grade, although above and below these features individual reaches of the river may be temporarily graded to the local base-levels controlling them. A lake, for example, acts as a local base-level for the streams discharging into it. Lakes that occupy deep depressions have a very long life, but shallow ones are, geologically speaking, soon eliminated. A lake is a trap for sediments, destined to be silted up by deltaic outgrowths from inflowing streams. At the same time the outflowing water erodes the outlet and lowers its level, so that the

Figure 17.13 Alluvial flats occupying the site of a former lake, Borrowdale, looking north towards Derwentwater with Skiddaw in the background, English Lake District. (*G. P. Abraham, Ltd,. Keswick*)

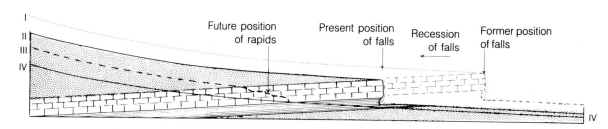

Figure 17.14 Successive stages in the recession and elimination of a waterfall:
I Profile of stream (drawn as graded) above an early position of the falls
II Profile above present falls
III Future profile after degeneration of the falls into rapids
IV Future profile (graded throughout) after elimination of falls and rapids.

lake is partly drained and its area reduced (Figure 17.11). Ultimately the lake is replaced by a broad lacustrine flat through which the river flows (Figures 17.12 and 17.13; see also Figure 16.14). The point where the lower graded profile intersects the upper one is called a **knick-point** (see Figure 18.14). Downcutting through the lake sediments and underlying rock-floor then proceeds, and since this is most rapid where the slope is relatively steep in front of the knick-point, the latter migrates upstream until (in the absence of external interruptions) continuity of grade is established between the upper and lower reaches of the river.

A resistant formation encountered by a river also retards the establishment of grade and acts as a temporary base-level for the stream above until it is cut through by waterfalls and rapids. The latter persist so long as the outcrop of obstructive rock remains out of grade with the graded reaches in the softer rocks exposed above and below.

17.6   WATERFALLS

Where an outcrop of resistant rock is underlain downstream by a weaker formation, the latter is relatively quickly worn down, and the resistant bed begins to be undercut. At the junction, and subsequently above it, the river bed is steepened and in this way **rapids** may be started. If the face of the resistant rock becomes vertical, the stream plunges over the crest as a **waterfall**. A waterfall is a knick-point of the most spectacular kind. The processes that bring about its recession upstream, leaving a gorge in front, have already been described (Section 17.2). Waterfalls eventually degenerate into rapids, which may persist for a long time before they are smoothed out and cease to make a break in the profile (Figure 17.14). A fall that descends in a series of leaps is sometimes referred to as a cascade. An exceptional volume of water is implied by the term **cataract**, which may be applied either to waterfalls or, more commonly,

Figure 17.15 Kaieteur Falls, Potaro River, Guyana. (*John Smart*)

to steep rapids. Rapids are characteristic throughout the wearing-through of an obstructive formation if its dip is downstream or steeply upstream.

Where a bed of strong rock, horizontal or gently inclined upstream, overlies weaker beds, the former is the 'fall-maker' and scouring of the softer beds underneath leads to undermining and recession. In the Yorkshire dales, for example, falls are well-developed under ledges of Carboniferous limestone underlain by shales.

The **Niagara Falls** are the classic example of this

tric stations and the St Lawrence Seaway have greatly reduced the flow of water over the falls. Even so, the falls will recede until Lake Erie is reached and partly drained – an event too many thousands of years ahead to cause any present anxiety.

The **Kaieteur Falls** on the Potaro river in Guyana are also of this type, although in this case the ledge, over which the river makes a sheer leap of 244 m, consists of hard, flat-lying conglomerate and massive sandstones, underlain by softer shales (Figure 17.15).

The **Yellowstone Falls** (Upper, 33 m; and Lower, 94 m) are cutting through an immensely thick succession of ignimbrites, parts of which have been altered and weakened, and at the same time gorgeously coloured, by chemical changes due to thermal waters. Hot springs still emerge along the floor of the canyon, which in places is 780 m deep. The fresh layers of ignimbrite, being more resistant, are the fall-makers (Figure 17.16). Rugged spurs and pinnacles are left as the walls of the canyon are worn back, all vividly splashed with colours of every hue.

Figure 17.16 Great Falls, Yellowstone River, Wyoming. (*T. L. Bierwert, American Museum of Natural History, New York*)

type (Figures 17.4 and 17.5), the ledge being a strong dolomitic limestone of Silurian age. Before the diversion of much of the water to hydroelectric plants, the mean discharge over the falls (c. 5660 m³ per second) was 85 times that of the Thames at Teddington. Most of the water passes over the Horseshoe Falls which, as their name implies, have been receding much more rapidly than the American Falls. After the last great ice-sheet withdrew from this region, and uncovered the pre-glacial Niagara escarpment about 12 000 years ago (Section 17.3), the Niagara river followed a course which descended from Lake Erie (174 m) to Lake Ontario (75 m) by way of a series of rapids and also one big vertical drop, where the river fell over the escarpment. Here the falls began and since then they have receded 11 km, leaving a gorge of which the rim is about 60 m above the river and on average about 110 m above the river floor (Figure 17.6). The rate of recession must have varied a good deal from time to time, but the measured rate for the Horseshoe Falls during the nineteenth century – about 1 m a year – is not far from the average of 0.9 m a year. Now, hydroelec-

Figure 17.17 'Tears of the Glen' waterfalls, Glenariff, Co. Antrim, Northern Ireland. (*Northern Ireland Tourist Board*)

Uplifted areas of plateau basalts may provide the structural conditions for a long series of falls. The strong and compact internal parts of flows make ideal ledges, which are undermined by the more rapid removal of the far less resistant vesicular or amygdaloidal margins. The many falls of Glenariff, where the river is vigorously descending step by step through the Tertiary basalts of the Antrim plateau, would be hard to surpass for their varied and exhilarating scenic attractions (Figure 17.17).

The **Victoria Falls** of the Zambezi River, which have already cut back some 130 km through basalts of Karroo age, owe their unique features to structural controls of a quite different kind. Here the Zambesi drops 110 m, along a frontage of 2 km, from a nearly level basaltic plateau into a narrow gorge which is remarkable in being parallel to the falls (Figure 17.18). A break on the downstream side leads the river into a series of gorges through which it rushes as a powerful torrent with surging rapids at intervals (Figure 17.19). The reason for the astonishingly acute swerves is to be found in the zones of structural weakness that characterize the plateau basalts of this region. Along narrow zones in one direction the rocks are strongly jointed, while along shatter zones following another direction they are fractured and brecciated. Both directions lie athwart the general course

of the river and the falls have receded alternately along one direction and the other, everywhere following the line that furnished the most easily liberated masses of basalt. Occasionally a third direction of special weakness and least resistance has been encountered by the river, like that between the present falls and the railway bridge (Figure 17.19). At the western end of the present falls the river is just beginning to cut back along another shatter zone. If this zone continues obliquely across the river, as it appears to do from the disposition of the islands, it will become the next stretch of the zigzag gorge.

Where rivers pass from uplifted highlands of metamorphic and massive igneous rocks to a coastal plain of weakly resistant sediments, waterfalls are initiated and often gain height as they recede. The rivers flowing from the hard rocks of the Appalachian uplands have thus developed rapids and falls along the 'fall-zone' before they reach the softer sediments of the Atlantic coastal plain. More spectacular examples of this type are the **Aughrabies Falls** (140 m) of South Africa, where the Orange River passes into a grim and desolate gorge of naked granite and gneiss (Figure 18.46); and the **Gersoppa** or **Jog Falls**, on the frontier of Mysore and Bombay in the Western Ghats, which have a sheer drop of 250–260 m, according to the level of the water in the plunge

Figure 17.18 Victoria Falls and zig-zag gorge of the Zambezi River, which here separates Zambia from Zimbabwe. Length of present falls 2 km; height 110-120 metres according to the discharge and depth of the river. While the successive zig-zags of the gorge have been cut by the recession of the Falls, practically no denudation of the surrounding erosion surface has taken place. (*Aircraft Operating Company, Johannesburg*)

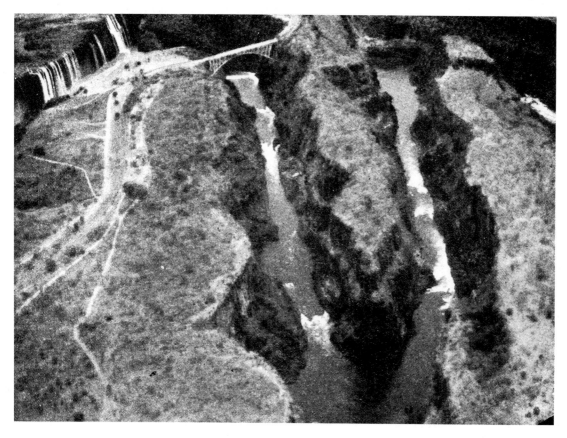

Figure 17.19 Another view of the Zambesi Gorge and Victoria Falls. (*R. U. Light, American Geographical Society*)

pool below. When in flood during the monsoon season the Jog Falls could be claimed as the greatest of the world's waterfalls, because of the combination of exceptional height with exceptional volume. In the dry season, however, the falls dwindle to a few ribbon-like trickles, some of which lose their continuity by passing into spray and mist before reaching the bottom of the rock face.

There are many falls of comparable or even greater height, but most of these result from relatively small tributaries or headwaters falling over precipices already prepared for them by glacial erosion by faulting, or by the long-continued recession of an escarpment at the expense of the uplands of a region that was rising while headward erosion was in progress (Figure 18.44).

17.7 RIVER BENDS AND THE WIDENING OF VALLEY FLOORS

In the last chapter the various processes that bring about the widening of valleys by slope recession were reviewed. Here we are concerned with the work of the rivers themselves in extending their channels sideways and so widening the valley floors at the expense of the pre-existing valley sides. Mountain streams rarely follow a straight course for any considerable distance unless they happen to be flowing rapidly down a steep gradient in a direction favoured by the structures of the rocks through which they are cutting their channels. A stream is likely to encounter belts of least resistance or rocks that are particularly obstructive, and these tend to predispose it to change its direction and so develop a winding course (Figure 17.20). But such obvious physical aids and restraints are not essential for changes of direction, as work with experimental streams in long tanks has clearly revealed.

An experimental stream which begins by flowing down a straight channel on a uniform slope of uniform sediment is always found to modify its channel by erosion and deposition so that a series of roughly symmetrical bends – like a sine curve – is developed (Figure 17.21). This curious behaviour is, in fact, a direct consequence of the general principle that when one medium moves over another the plane of contact tends to be shaped into a wave-like form. To be effective in this way the relative movement must be adequate to overcome such resistances as friction, viscosity and strength. Wind blowing over loose sand builds up sand dunes with rippled surfaces. The backwash of sea waves running down the slope of a beach

Figure 17.20 Steep-sided upland valley with overlapping spurs, Crossdale Beck, Ennerdale, English Lake District. (*British Geological Survey*)

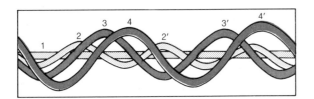

Figure 17.21 Successive stages (1, 2, 3, 4) in the experimental development of incipient meanders in a model stream flowing in a bed of uniform sediment. The meanders migrate downstream and the wavelength (2 to 2'; 3 to 3'; 4 to 4') increases until it is about ten times the width of the bankfull channel.

leaves a characteristic surface of ripple marks. Water falling freely through the air, as from a circular tap, at once begins to change its momentary cylindrical form. It oscillates in and out with rapidly increasing agitation until the pulsating column can no longer hold together, but breaks into drops. This example emphasizes the fact that the phenomena we are concerned with are three-dimensional in space; four-dimensional when time is included.

The oscillations of streams of running water on land are further complicated by the various factors listed in Section 17.5. Amongst these the roughness of the bed and the nature of the load are particularly important in controlling what happens. Suppose we start with an experimental stream flowing down a gently sloping straight channel floored with sand. Like the water from a tap the stream oscillates 'in and out', i.e. up and down and also sideways, but much more slowly and less obviously. However, clear manifestations of wave-like forms soon appear. Down the channel alternating shoals and hollows develop, corresponding to the up-and-down oscillation. Moreover, the shoals are alternately built up a little on one side of the middle line and a little on the other: an indication of the sideways oscillation. This change in the floor affects the flow of the stream, which concurrently begins to swing from side to side. The main current winds round the shoals, first on one side and then crossing over to the other side of the next shoal, and so on. Wherever the current impinges against the banks of the stream, lateral erosion occurs and bends are started. In this way the channel develops a serpentine form as seen in plan (Figure 17.21). The distribution of shoals and hollows is now controlled by the swinging habit of the current. As the curvature of a bend is increased by lateral erosion, the channel on the outside is deepened by vertical erosion, while on the inside it is shallowed by deposition (Figures 17.22 and 17.23).

It will be seen from Figure 17.21 that the successive wavelengths, 2–2', 3–3', 4–4', gradually lengthen, and that the bends themselves migrate downstream. This migration steadily continues, but the wavelength adjusts itself to the width of the channel (bankfull stage) and does not permanently increase beyond a certain average limit, which is about 10 times the width of an average stream, the factor being less for narrower streams and more for wider ones (Figure 17.24). In their studies of natural rivers Luna B. Leopold and his colleagues find that straight reaches have scoured-out hollows in their floors alternating with shoals. Shoals (or hollows) are spaced at distances which

Figure 17.22 Section across the bend of a river to show the asymmetrical profile due, on balance, to erosion on the outside and deposition on the inside. In floods, turbulence and erosion extend to the inner bank.

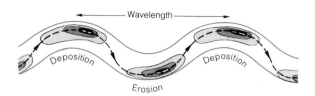

Figure 17.23 Diagram to illustrate the effects brought about by water flowing down a meandering channel. Depth of water is indicated by depth of shading. The broken line indicates the axis of flow under normal conditions. With rising water the axis widens into a belt that may extend to the convex banks and remove some of the alluvium previously deposited there.

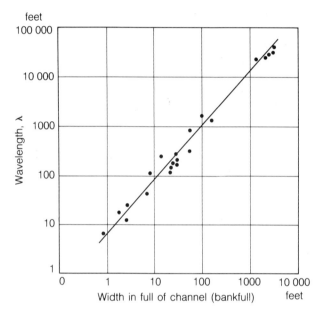

Figure 17.24 Logarithmic graph of the wavelength of meanders, λ, plotted against the corresponding widths, W. The equation of the line is of the general form log λ = log a + b log W (a and b being constants) which is the logarithmic form of λ = aWb = (in this case) 6.5 W^{1.1}. (*After L.B. Leopold and M. G. Wolman, 1964*)

bear the same relationships to widths as the wavelengths of bends described above. In such reaches the rivers are evidently on their way to becoming sinuous. Where they are beginning to develop meanders in neighbouring parts of their courses, the wavelengths correspond very closely with twice the distance between successive shoals (Figure 17.23).

Another discovery relating to the flow of water round a bend is that the position and extent of the belt of high velocity and turbulence depend on whether the discharge is (a) increasing towards bankfull conditions and a state of flood, or (b) decreasing from those conditions towards a state of low water. It is natural to suppose that the main current of the stream is deflected towards the

outer bank, some way beyond the beginning of the bend, and that erosion will take place there while alluvium is being deposited on the inside of the curve. But this is only part of the story. With increasing rate of flow the maximum current impinges farther down the outside of the bend. The velocity also increases on the inside of bends until, during floods, it may be little less, if any, than on the outside. Shingle and sand previously deposited on the inside are then being scoured away. Some bedrock erosion may even occur locally. With decreasing flow the belt of high velocity contracts again towards the outside of the bend, and while the outside or concave bank continues to be eroded, the convex bank receives a new supply of sediment from upstream.

Under the varying conditions indicated above, the resulting changes in the river channel and its valley floor may be summarized in a general way as follows:

1. The channel is deepened towards the outer side of each bend (Figures 17.22 and 17.23), and particularly along the downstream part of the bend.

2. The outer bank is worn back and undercut by lateral erosion, with local development of river cliffs or 'bluffs', which are liable to cave in during or shortly after floods and thaws. At the same time banks of shingle or sand (**point bars**) are built out on the inner sides (Figure 17.25).

3. As the bends are thus widened out by lateral erosion and deepened by downward erosion on the outside, and built out and shallowed on the inside, the river shifts its channel not only towards the outer bank but also downwards (Figure 17.26). Eventually this oblique migration of the channel leaves gently rounded slip-off slopes on the inside of the growing curves; classic examples are those of the Rhine (Figure 16.3) and of the Wye (Figure 17.27). In cross-profile the valley, as well as the channel floor, has become highly asymmetrical at every bend.

4. Since each bend is increasingly enlarged in the downstream direction, it gradually migrates downstream in serpentine fashion. It can now be realized that the changes in a river channel are three-dimensional: downwards (vertical erosion); sideways (lateral erosion); and along (downstream migration of bends). At an early stage, when there are overlapping spurs like those illustrated in Figure 17.20, each spur is eroded and undercut every time it is reached by the outer part of a migrating bend. Eventually, as bend after bend migrates downstream, the spurs are all trimmed off.

Figure 17.25 Meanders of the River Dee, near Braemar, Scotland. Note the point bars on the inside of the meander loops. (*The Scotsman, Edinburgh*)

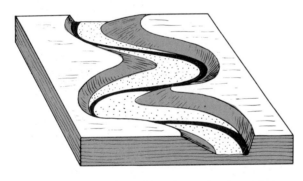

Figure 17.26 Diagram to show the widening and deepening of a valley floor by oblique erosion (vertical and lateral) on the outside of river bends combined with the downstream migration of the bends. The swinging stream eventually trims off the overlapping spurs and slip-off slopes.

Similarly the slip-off slopes are all cut back – apart from a few uplifted relics that have survived from earlier phases of erosion – and a nearly flat trough-like valley floor is developed. The alluvial deposits extend into continuous stretches, with broad embayments bounded by bluffs or hills (Figures 17.25 and 17.28). The flood plain has been established.

17.8 FREE MEANDERS AND MEANDER BELTS

As the river continues to swing from side to side of its valley floor, it undercuts the bordering bluffs wherever a bend impinges on them. So the widening of the valley floor proceeds, while the slopes above are slowly receding. The channel is now mainly in river deposits, bedrock (if any) being exposed only in and below the bluffs. Each part of the material deposited on the growing flood plain is worked over in turn during the downward sweep of the bends, fresh additions from above constantly making good the losses by erosion and transport. The bends, now free to develop in any direction, except where they encounter the valley side, are more quickly modified. Freely developing bends are called **meanders** from their prevalence in the classic river Meander, the Menderes of today, in the south-west of Turkey.

As we have already seen, bends develop into meanders in which 'the wave length is a function of the width' to a degree of approximation that is remarkable. As a direct result of this relationship, a freely meandering river flows in a **meander belt** (Figures 17.25 and 17.28), which is usually about 15 to 18 times the bankfull channel width. There are, as we have seen, many other factors besides width and discharge involved under natural conditions, and these are responsible for local and temporary irregularities. But as straight stretches soon begin to develop meanders (as indicated in Figure 17.21), there is always, barring natural 'accidents' or human interference, an approach to the relationships established experimentally. However,

Figure 17.27 The horseshoe bend of the River Wye at Wyndcliff, near Chepstow, Gwent, Wales, showing well-developed bluffs of Carboniferous Limestone and a gently inclined slip-off slope. (cf. Fig. 16.3.) (*British Geological Survey*)

Figure 17.28 The meander belt of the River Glass; a flat-bottomed, glaciated, trough-like valley floor, veneered with alluvium, Strath Glass, Scotland. (*British Geological Survey*)

there is one type of 'crisis' or emergency that commonly arises in the development of freely meandering rivers. Erosion inevitably picks out the materials, and follows the structures, that offer least resistance. Consequently many meanders continue to swell out into broad loops with gradually narrowing necks, as illustrated in Figure 17.29. If a flood occurs when only a narrow neck of land is left between adjoining loops, the momentum of the increased flow is likely to carry the stream across the neck and thus short-circuit its course. On the side of the 'cut-off' a deserted channel is left, forming an **ox-bow** lake (Figure 17.29), which soon degenerates into a swamp as it is silted up by later floods. By making artificial cut-offs the Mississippi River Commission has straightened long

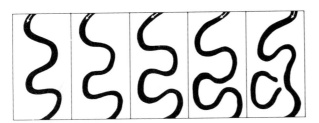

Figure 17.29 Successive stages in the development of meanders, showing the formation of an ox-bow lake by the 'cut-off' of a loop.

stretches of the river and so shortened transport distances, both by road and river, by hundreds of kilometres.

The relics of many old ox-bows, indicating the positions of abandoned meanders, may no longer be obvious topographic features on the ground.

But their outlines can be clearly seen from the air (Figure 17.30). This is because the soil and drainage of an infilled ox-bow lake differs from that of ordinary river alluvium, a difference to which vegetation visibly responds. Ox-bow lakes and their overgrown relics constitute the most convincing evidence of the reality and geological importance of the downstream migration of meanders.

As the bordering slopes of the valley are still cut back every time an individual meander impinges against them, the valley floor continues to be widened. After an immensely long period of time, with no important tectonic or climatic changes, it may become several times as wide as the meander belt. This indicates that as the meander belt becomes 'free', the belt itself proceeds to 'meander' to and fro across the widening flood plain, swinging down the latter towards the sea in

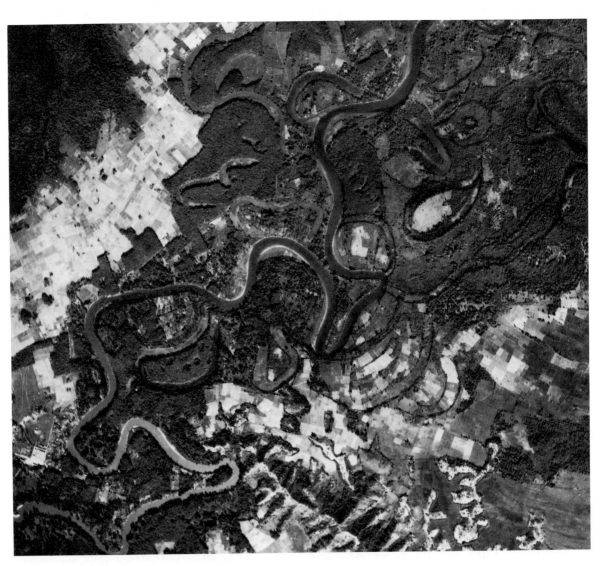

Figure 17.30 Meandering stream with numerous ox-bow lakes and many that have been infilled with sediment and more or less obscured by vegetation; north-east of Kota Kinabalu (Jesselton), Sabah, Malaysia. (R.A.F., *Crown Copyright reserved*)

curves of much longer and more variable wavelengths than those of the actual stream. It should be noticed that while lowering of the channel floor is limited by base-level, there is no such restriction on erosion of the banks. Widening can therefore continue long after significant lowering has ceased.

What appeared to be an interesting example of the meandering of a meander belt was provided by the discovery of the buried ruins of the ancient city of Ur of the Chaldees – the home of Abraham – in what is now Iraq. Five thousand years ago the Euphrates flowed 8 km to the west of Ur; today it is 8 km east of the ruins. But there is another possibility to be considered. It is known that the full-bank width of a meandering river is roughly proportional to the square root of the discharge. If the rainfall were greater at some time in the past, the runoff would also be higher, and the discharge would therefore be very considerably increased. Doubling the rainfall would at least quadruple the discharge, so doubling the width of a river and consequently that of its meander belt. Some exceptionally wide valley floors are now ascribed to the meanderings of ancestral rivers which once had channels several times as wide as those of today. This interpretation, which has been supported by archaeological discoveries in deserts and semi-arid regions, is consistent with other evidence that at various times during the Pleistocene and even in the recent past, the rainfall in these regions was very much greater than it is today.

17.9 BRAIDED RIVERS

Meandering very considerably lengthens the river concerned and so reduces its gradient or slope. This is only one of the ways in which a river can accommodate itself to prevailing conditions so as to achieve an approximation to equilibrium: that is, so as to become graded and remain graded. Another method is for the river to divide into an interlacing network of distributaries with shoals and islands of shingle and sand between. The river is then said to be **braided** (Figures 17.31 and 17.32). The river channel as a whole is characteristically wide and shallow. In keeping with this the floor and outer banks are composed of incoherent sediments, generally consisting wholly or partly of the river's own deposits. Where a river, heavily laden with debris from a mountain range, emerges from a canyon or ravine on to the bordering plain, its velocity is suddenly checked by the abrupt change of gradient and a large part of its load of sediment, including all the coarser debris, is therefore dropped. Along the fronts of scarps and ranges such debris is commonly spread out in the form of alluvial fans (Section 17.12). While thus alleviating the sudden change of slope by deposition, much of which is obstructive, the stream

Figure 17.31 Braided headwaters of the Rangitati River, Canterbury, New Zealand, where they emerge from glacially eroded valleys in the Southern Alps, the floors of which, like that of the broader valley in front, have been built up by heavy deposition. (*New Zealand Geological Survey*)

Figure 17.32 Braided channel of the Rakaia River, incised in the alluvium of the Canterbury Plain (and locally superposed on the underlying bedrocks), between the Southern Alps of New Zealand and the sea. (*V. C. Browne*)

divides into branches which continually separate and reunite.

Another situation which favours braided drainage is in front of melting ice-caps (Figure 17.33) and of the 'snouts' of certain glaciers (Figure 3.5). At the other end of a river, towards its mouth, braiding may be characteristic for long distances, like an inland anticipation of a delta. The Ganges and the Amazon are good examples.

In experiments designed to reproduce braiding the water fed into the channel at the head of a long tank is suitably charged with sediment from the start. This seems to have the effect of intensifying the up-and-down oscillation, while damping down the sideways oscillation. However this may be, elongated sandbanks are deposited at intervals down the middle of the channel and these divert the flow of water to each side. Where the flanking channels become deepened by floor erosion the level of the water drops a little and the adjoining shoal may become an island. But lateral erosion of the outer banks also occurs, thus widening the channels and so providing materials for shallowing them and building up new sandbanks. Each sandbank migrates downstream as a result of erosion at its upper end and deposition at its lower end. The distribution of material changes, and some of it is continually being carried away, but the 'form' – the sandbank or island – may persist for a long time. Because of this migration

the all-over widening process also migrates downstream, since each of the outermost banks continues to be eroded laterally by the stream directed towards it by the nearest sandbank.

Actual braided rivers follow this pattern of behaviour. During low water the channel is mainly sandbanks, as if the river had been choked with excess of sediment. But with high water, and especially during floods, new channels are cut and the islands, generally for the most part submerged, migrate downstream. A braided river becomes graded by depositing part of its load until the width, depth, slope and velocity are modified so that on average, the given discharge and load can be transmitted with a close approach to equality of deposition and erosion. In saying this, however, it must be repeated that in the long run there always remains a slight balance in favour of erosion, unless this tendency is reversed by a change of climate or a rise of base-level. As Leopold remarks: 'Braiding does not necessarily indicate an excess of total load'.

Figure 17.34 embodies a remarkable range of evidence showing that meandering and braided channels can generally be distinguished in terms of only two of the many factors concerned. With a few exceptions, where the general rule is modified by one or more of these other factors, the diagram makes it clear (a) that for a given bankfull discharge braids usually require a higher gradient

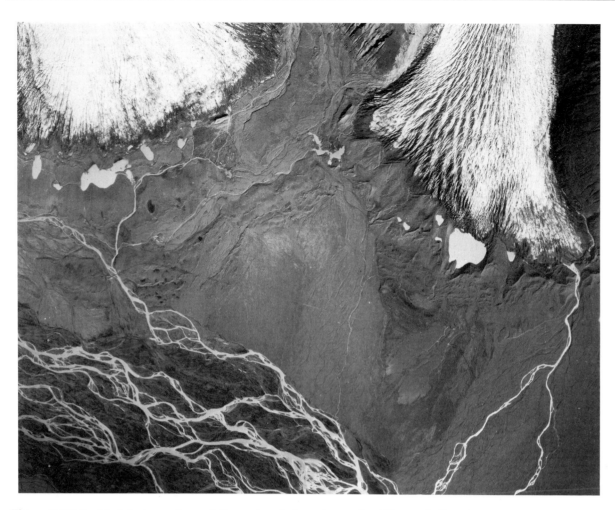

Figure 17.33 Braided drainage through the outwash deposits south of Vatnajökull, Iceland. The two other glaciers are Skaftafells on the west (left) and Svinafells on the east; both are west of Oraefajökull (Fig. 3.5). (*Erlingur Dagsson*)

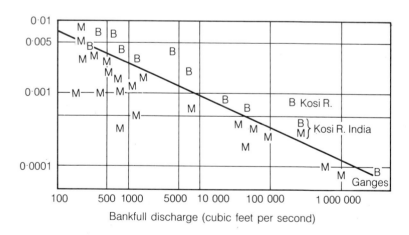

Bankfull discharge (cubic feet per second)

Figure 17.34 Logarithmic graph of the gradient of a wide variety of rivers plotted against the corresponding discharges; M, meandering; B, braided. These types are separated by a straight line, except for one or two extreme examples at each end. For comparison the Thames at Teddington falls near the middle of the diagram in the field of meandering rivers. (*After L.B. Leopold and M. G. Wolman 1957*)

than meanders; and (b) that for a given gradient braids require a higher bankfull discharge than meanders. What is surprising is that there should not be more exceptions.

Amongst the factors not taken into account in Figure 17.34, the more critical ones are roughness of the floor and bed load, and the ratio of all-over width to average depth. Bottom friction is greatly increased when a river is wide and shallow and has a rough floor. These handicaps must be compensated by a higher discharge and/or gradient to ensure the transmission of a given load. It is worthy of notice that a meandering river maintains a fairly steady width because erosion and widening of the channel on one side is balanced by deposition and narrowing on the other. In the development of braiding, however, a river increases its width because erosion and widening occur on both sides at once. The whole régime of sandbanks and inner channels eventually reaches an all-over width that meets the requirements of all but the very greatest floods. This width is likely to remain fairly stable until there is a really catastrophic flood. When this happens it is quite likely to upset the whole of the pre-existing régime by diverting the river into an entirely new course.

The bracketed B and M of the Kosi river in Figure 17.34 illustrate the effect of load and roughness in deciding the pattern of behaviour when gradient and discharge are essentially the same. The Kosi, also known as the Sapt-Kosi, because it is fed by the union of seven rivers, is the most powerful tributary of the Ganga (Ganges) (see Figure 18.31). Four major western Kosi rivers unite with the Sun Kosi, which in turn is joined by the Arun from the north and the Tamur Kosi from the east (Figure 17.35). The greatly enlarged river emerges on the plains near Chatra and flows for 120 km over a gently sloping alluvial fan of its own construction in a broad, conspicuously braided channel. By the time the Kosi has reached the foot of its alluvial fan most of its gravel and coarser sand has been deposited, and from there to its junction with the Ganga the river follows a meandering course. The Kosi is notorious for its frequent and often disastrous floods and some of these have altered the course of the river. Two centuries ago it flowed near Purnea (Figure 17.35), but now its main braided channel is 96 km to the west. By the construction of a barrage and 240 km of marginal embankments, it is hoped to tame the Kosi and utilize its immense water power.

17.10 FLOOD PLAINS

The development of flood plains by the sweeping of meanders to and fro down the valley floor, and later, perhaps, by the swinging of the meander belt itself, has already been mentioned. As the valley floor is widened in this way the slip-off slopes become nearly flat and the flood plain may

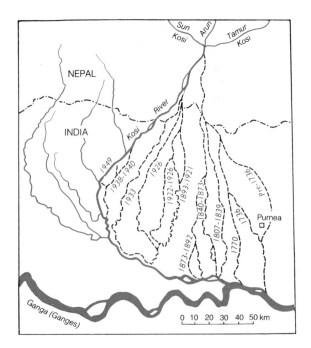

Figure 17.35 Sketch map to illustrate the lateral migration of the Kosi River across its own alluvial deposits. The latter begin near the boundary between Bihar (India) and Nepal, which roughly corresponds to the southern edge of the Siwaliks, the foothills of the Himalayas.

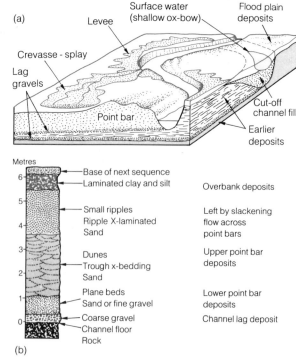

Figure 17.36 Deposition on the floodplain of a meandering river. (a) Facies model and (b) fining-upward sequence. The thickness of overbank, point bar and channel deposits varies and depends, in part, on the volume and grain size of the bed load and suspended load. (*After J. R. L. Allen, 1965*)

then have only a relatively thin veneer of mud and silt, generally with an underlying deposit of sand and gravel, covering a planed-off surface of bedrock. This rock surface results from the slight erosion that occurs each time the bedrock is uncovered at the deepest part of a meander loop which has reached the bordering bluffs (Figure 17.21). Braided streams provide thicker and coarser deposits. But the thickest spreads of alluvium are found where, for some reason, the level of the bedrock floor has been lowered. A common reason is crustal depression in front of mountain ranges, as exemplified by the Indo-Gangetic trough and its infilling of sediment. Another is the lowering of sea-level corresponding to the formation of great ice-sheets. This lowering of base-level stimulated the rivers to deepen their channels. Later, when sea-level rose again with the melting of the ice, these channels filled with sediment. In the Lower Mississippi valley the alluvium is 122 m thick where it fills a channel cut by the river when sea-level was 122 m lower than at present.

The deposits left on the flood plain by a meandering river form a fining-upwards sequence (Figure 17.36). At the base of the channels, resting on bedrock, are gravels left as a lag when coarse bedload is dropped when velocity drops as floods wane (Figure 17.37). Overlying this layer are mainly sandy deposits formed by lateral growth of point bars on the inner bends of the migrating meanders. On top of these channel sediments are deposits left on the flood plain by floodwaters flowing outside the normal channel, termed **overbank deposits**. Each time the river overflows its banks the current is checked at the margin of the channel, and the coarsest part of the suspended load, often a mix of sand and coarse silt, is dropped there. Thus, a low embankment or **levee** is built up on each side (Figure 17.37). Beyond the levees the ground slopes down, and in consequence is liable to be marshy. During floods the levees may be breached and allow 'splays' of sediment to cover part of the floodplain. Behind the levees, the floodwaters become trapped and drop fine suspended sediment, often containing leaves and other organic trash. During floods, levees may grow across the junctions of small

tributaries. The latter are then obliged to follow a meandering course of their own, often for many kilometres, before they find a new entrance into the main river. Depressions occupied on the way become ponds which degenerate into 'back swamps'. The characteristic features of the flood plain of a meandering river include ox-bow lakes and marshes, levees, back swamps and shallow lakes, and a complicated pattern of lateral streams. Levees are not constructed by typically braided rivers, as deposition from the latter tends to be concentrated within the channels.

Levees afford protection from ordinary floods, but it has been found by hard experience that confining the channel raises the water level (for a given discharge). Moreover, in some cases, e.g. the Mississippi, the river begins to silt up its floor with material that would otherwise have been spread across the plain as overbank deposits. This is possibly due to slight tilting of the region towards the sea, since it is known that many deltas are slowly subsiding. However this may be, the bankfull level is raised, and the danger from major floods becomes greater than before. For further protection artificial levees may be built, but these provide only temporary security, since they accentuate the tendency of the river to raise its level. In the flood plains of the Po in Italy and of the Hwang Ho and Yangtze Kiang in China the built-up levees are locally higher than the neighbouring housetops, and the rivers flow at a level well above that of the adjoining land. Such conditions are obviously extremely dangerous, as the swollen waters of a severe flood may breach the defences and bring disaster to agricultural lands over an enormous area.

Along the Mississippi and the Missouri and their tributaries the flood danger has become an increasingly serious menace. Little more than a century ago floods were easily controlled by levees about a metre high. The levels have since had to be raised several times. By 1927 they were three or four times as high, but nevertheless a great flood then broke through in more than 200 places, and devastated 50 000 square kilometres of cultivated land. Stronger and higher levees, up to 6 or 9 m, have now been constructed, but it is clear that by itself this method of flood control is far from

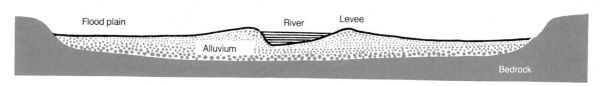

Figure 17.37 Schematic section across the floodplain of a stream bordered by natural levees. Vertical scale greatly exaggerated.

satisfactory. It can, however, be supplemented by reafforesting the upland regions (to reduce the proportion of runoff) and by dredging, cutting through meander loops, and making vast storage reservoirs for floodwaters.

The destructive flood of July 1951, in which over 200 000 people lost their homes, was due to an unprecedented rainfall of 47 mm in one night: like the Lynmouth disaster (Section 17.3), but over a much greater area. Most of the Mississippian floods, however, result from a combination of heavy rainfall with melting of the winter's snow. If, in addition, the snowfall has been abnormal and is followed by a widespread thaw, then the flood crests of all the tributaries may unite with that of the main river (instead of coming one after another in a drawn-out succession, as usually happens). An exceptionally heavy flood is then inevitable.

17.11 DELTAS

The essential condition for the growth of a delta is that the rate of deposition of sediment at or just beyond the mouth of a river should exceed the rate of removal by waves and currents. Where a river flows into a freshwater lake the current is usually checked and much of the load may be quickly deposited (Figure 17.12). Salt lakes provide still more favourable conditions. As river water mixes with saline water the finer particles of clay flocculate into aggregates that are too large to remain long in suspension. Rapid sedimentation is there-

Figure 17.38 Lake Mead 185 km long, held back by the Hoover (Boulder) Dam, the construction of which across the Colorado River was completed in 1935. The river, heavily laden with sediment, enters at the northern end as a turbidity current which can often be clearly seen from the air. By 1960 about 3500 million tonnes of sediment had been deposited. Left to itself the lake would be entirely infilled by the year 2250. (*Fairchild Aerial Surveys, Los Angeles*). For location, see Fig. 18.27.

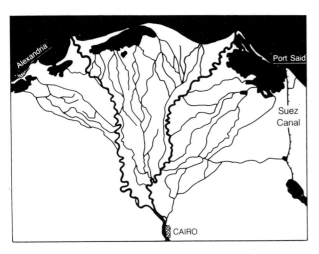

Figure 17.39 The Nile Delta; a classic example of the arcuate type. Since construction of the High Aswan Dam, the reduction in sediment transfer to the Nile Delta has led to erosion and degradation of the delta front, requiring extensive and expensive coastal protection works.

fore promoted. A striking example of this èffect is the Terek delta (between Baku and the Volga), which at one time was visibly advancing into the Caspian Sea at the rate of 1.5 km every 5 or 6 years. As the front has now reached much deeper water its outward growth is naturally slowing down, but even so the rate is still about 10 times that of the Rhône delta at the head of Lake Geneva.

These two examples illustrate another contrast that is of primary importance. Charged with suspended sediment, the Rhône has a milky appearance as it flows into the lake; and because of the higher density its waters visibly dive under the clear lighter water of the lake, forming a **turbidity current** (Section 6.3). This cloudy current flows through an 8 km trench towards the deepest part of the lake. Only a small proportion of the sediment – the coarser part – contributes to the

outgrowth of the visible delta. Turbidity currents are frequently developed where the Colorado river enters Lake Mead (Figure 17.38). The murky inflowing water travels down the floor of the reservoir along the depression marking the submerged channel of the river, and builds a deposit of deltaic structure which extends as far as the Hoover Dam and so is extremely long compared with its width. In these examples and many others the river water, together with its suspended load, is 'heavier' during most parts of the year than the relatively still water into which it flows.

The Caspian water, however, is saline and has about the same density as that of the mud-laden Terek water. Here and in all similar cases the river water fans out on a broad and expanding front; it is rapidly slowed down by mixing with the relatively still body of water through which it is spreading. Sediment is therefore mainly deposited in a broad arc surrounding the mouth of the river. Having blocked its only means of exit, the river breaks through, so making new channels which then proceed to obstruct themselves in the same way. This is the essential mechanism by which deltas of the **arcuate** type are built up (Figure 17.39).

There is still a third possibility: when the river water is less dense than the salt lake or sea into which it is flowing. The fresh water, with a light suspended load of fine mud, then flows over the surface of the salt water, meeting little resistance in front, but rapidly losing velocity sideways. Deposition therefore occurs mainly along the sides of the channel, until low levees are built up. Sooner or later these are breached and the resulting delta grows outwards by way of relatively deep channels resembling outstretched fingers. This is the **bird's-foot** type of delta (Figure 17.40).

The above summary is a very much simplified account of a hydrodynamical treatment by C. C.

Figure 17.40 Fifty years' growth of the Mississippi Delta; a classic example of the bird's-foot type.

Bates who regards the flow of river water into, let us say, the sea, as if it were akin to the discharge of a turbulent jet of fluid into a relatively still body of another fluid. There are then three possibilities:

1. if the river water is the denser (because of its load of sediment) it flows along the bottom as a turbidity current and forms an elongated deposit of the deltaic type;
2. if the river water has about the same density it spreads out fanwise and an arcuate delta is formed;
3. if the river water is the less dense it makes a few confined channels for itself and a bird's-foot delta is formed.

There are, of course, other considerations to be kept in mind. The coagulating effect of salt water has already been mentioned. Waves and currents may sweep away the incoming sediment before there is time for its accumulation above sea-level. The rise of sea-level that accompanied the melting of the last great ice-sheets of Europe and North America must have submerged a great many slow-growing deltas that have not yet had time to reappear. Moreover, the crust itself is sagging beneath many deltas, and not merely because of the weight of sediment. Even when full allowance is made for the thick wedges of light sediment, it is found that the Mississippi, Nile, Indus, Irrawaddy and Ganga–Brahmaputra deltas are all areas of abnormally high gravity. Their tendency is therefore to subside and so to make room for

sediments to accumulate and deltas to thicken – as indeed many deltas have already done to a most astonishing degree.

Large deltas are usually absent from coastlines along colliding ocean–continent plate margins (see Chapter 30). For example, there are no major deltas along the tectonically-active west coasts of North and South America. Here rapid erosion of the mountain chains of the Rockies and the Andes respectively generates abundant sediment, but much of this is lost eastwards to the Atlantic coasts via the extended drainage systems of the Mississippi and Amazon and other large rivers. Sediments that do arrive at the Pacific coasts have no wide, shallow shelf on which to accumulate and are lost to the deep ocean trenches close offshore, and may even be carried down with the descending plate into subduction zones.

The delta of the Nile originally received this name because of the resemblance of its shape to the capital form of the Greek letter delta. It is the prototype of all deltas, but particularly of the arcuate type. It has an arc-like outer edge, modified by fringing sand-spits shaped by marine currents (Figure 17.39). After transversing 1600 km of desert, the Nile has comparatively little water left when it reaches the apex of the delta. Much of the remaining load is deposited before the sea is reached and frontal growth of the delta is correspondingly slow. Such continued accumulation of sediment is possible only because the region is subsiding. Subsidence must, indeed, have been in

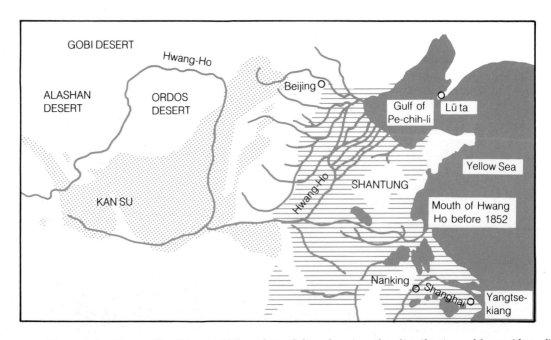

Map of the Huang Ho (Hwang Ho) and its delta, showing the distribution of loess (dotted) and alluvium (horizontal shading) derived from loess. (*After G. B. Cressey*)

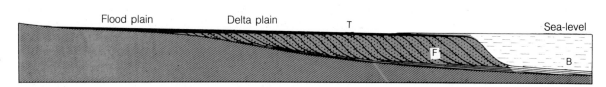

Figure 17.42 Idealized section through the sediments of an arcuate delta; the *foreset* beds, F, grow seawards (like an advancing railway embankment during its construction); finer sediment is deposited farther out on the sea floor as *bottomset* beds, B; and a veneer of *topset* beds, T, forms the seaward continuation of the flood plain. This, now classic, description of delta structure was first made by Gilbert in 1885.

progress here for a very long time, as seismic exploration has revealed a thickness of at least 3000 m of poorly consolidated sediment. Frontal growth is therefore slow. The Po delta extends more rapidly. Adria, now 22 km inland, was a seaport 1850 years ago, the average rate of advance thus indicated being about 1 km in 75 years. Ostia, the seaport of ancient Rome, is now 6 km from the mouth of the Tevere (Tiber). The richly fertile delta of the Huang Ho in north China has grown across what was originally a broad bay of the Po Hai (Yellow Sea). A large island, now the Shantung peninsula, has been half surrounded. Since 1852 the main branch of the river has emptied to the north of Shantung, but before then it flowed across the southern section of the delta and reached the sea about 480 km from its present

mouth (Figure 17.41). In that year and again in 1887 there were calamitous floods in which the loss of life from drowning and famine amounted to many hundreds of thousands. Floods over the Ch'ang Chiang (Yangtze) delta, farther south, have also brought repeated disasters to the inhabitants. In contrast, the annual flood of the Nile was a disaster only when it failed. If, like those mentioned above, a river has a flood plain, the visible part of its arcuate delta is the seaward extension of the plain. Figure 17.42 illustrates the way in which a broad outward-sloping fan of sediments is gradually built up on the sea-floor.

The chief distributaries of the Mississippi delta (Figure 17.43), locally called **passes**, make a typical bird's-foot pattern. The sediment brought down by the river is rich in very fine mud and this

Figure 17.43 Aerial view of part of the Mississippi Delta, showing the chief 'passes' and the levees by which they are confined. (*Photograph by Corps of Engineers, U.S. Army, New Orleans District*)

Figure 17.44 Alluvial fan, eastern side of Death Valley, California. (*J. S. Shelton and R. C. Frampton*)

confines the channels within impervious banks of clay. Because of the great depth of the Gulf of Mexico a good deal of sediment continues to be carried forward, so that slow deposition is in progress over a very wide area. The Mississippi delta is slowly subsiding and here, too, an immense thickness of deltaic sediments has accumulated. Though much of the region is now above sea-level, downwarping of the crust still continues on the seaward flank. Here the Mississippi is adding sediment to the floor of the Gulf of Mexico at the rate of $1\,km^3$ every 4 years or so. Were it not for the submarine downwarping, outward growth of the delta and of the continental shelf would have been on a far greater scale. There have been many interruptions due to changes of sea-level during the Pleistocene, but these can be dealt with more conveniently when the effects of glaciation are being considered.

17.12 ALLUVIAL FANS AND CONES

Many mountain ranges and plateaus descend steeply to the neighbouring lowlands; generally, but not always, because they are bounded by eroded fault scarps. Where a heavily laden stream reaches the plain after flowing swiftly through a ravine or canyon, its velocity is checked, it widens

and much of its load of sediment is dropped. The stream may become braided and keep changing its course, like the notorious Kosi river (Section 17.9); or, as in delta formation, it may divide into branching distributaries; or it may behave in both ways. The deposited sediment is spread out as an **alluvial fan** (Figure 17.44). If the circumstances are such that most of the water sinks into the porous alluvium, while the rest evaporates – as commonly happens in arid or semi-arid regions – then the whole of the load is dropped. The structure gains height and becomes an **alluvial cone** (Figure 17.45). There are all gradations from wide fans 26–260 km across that are usually nearly flat (slopes less than 1°), through fans of moderate width and inclination (4–6°), to relatively small steep-sided cones (up to 15°) built of the coarser debris brought down by short torrential streams.

Where closely spaced streams discharge from a mountainous region across a **piedmont** (a mountain-foot lowland), their deposits may eventually coalesce to form a **piedmont alluvial plain**. Such is the vast crescent of boulder beds, gravel, sand, silt and clay known as the Indo-Gangetic Plain, which extends from the delta of the Indus to that of the Ganga–Brahmaputra. This has long been a 'foredeep' between the loftiest mountains in the world to the north and the Peninsular block to the south. Although the floor has sagged

Figure 17.45 Alluvial cones at the mouths of canyons in southern Utah. (*After C. E. Dutton, United States Geological Survey*)

Figure 17.46 Buttermere and Crummock Water illustrating the division of a lake into two by the growth of a fan-delta. (*G.P. Abraham, Ltd., Keswick*)

transported by the copious drainage from the Himalayas and supplemented on the south by minor contributions from the Peninsula.

Alluvial fans and cones are well displayed in western America: along the eastern base of the Rockies, where the Front Range faces the Great Plains; along the eastern edge of the Sierra Nevada, where the eroded fault scarp slopes down to the Great Basin; and in many other similar situations on the flanks of the block mountains of the Great Basin. They are also developed on a large scale at the foot of the various ranges of the Andes. Where rivers flow along or near the foot of the slope, many examples are to be found of river deflection due to the growth across their channels of alluvial fans deposited by tributaries from the mountains. Where there is a lake in a similar situation, an alluvial fan begun on the land passes into a delta which grows across the lake. Familiar examples which have divided what were originally single lakes into two are the fan deltas of Buttermere in the English Lake District, separating Buttermere from Crummock Water (Figure 17.46); and of Interlaken in Switzerland, separating Lake Thun from Lake Brienz. A fan delta that has spread into the sea at Lynmouth, where the East and West Lyn rivers rapidly descend from the plateau which they drain, is illustrated by Figure 17.47.

17.13 RIVER CHANNELS AND CLIMATIC CHANGE

River channels are continually adjusting their

between 1800 and 3000 m during Tertiary times, the depression has been infilled by alluvium

Figure 17.47 Lynton and Lynmouth, North Devon, showing the plateau drained and eroded by the East and West Lyn rivers, and the fan-delta built into the Bristol Channel at their common outlet. See also Fig. 17.7. (*Aerofilms, Limited*)

form and pattern to changes in discharge and sediment load. Over periods varying from decades to centuries, however, a river channel will become more or less adjusted to evacuate a flow represented by the bankfull stage. This condition of dynamic equilibrium can only persist if there are no long-term changes in the pattern of flow from the drainage basin. In reality, the discharge and sediment load of many rivers has varied greatly in both historic and pre-historic times.

Climatic change is the most obvious cause of changing river discharge. Obviously, an increase in average annual rainfall will give a corresponding increase in discharge. In addition, variations in temperature will also be important in influencing evaporation rates and the contribution of ice- and snow-melt to river regime. Yet the response of a drainage basin to climatic change is often com-

plex. A switch to more seasonal or intense precipitation will enhance flood discharges and require channel adjustment. Extra rainfall may give an increase in vegetation cover (e.g. encourage forest growth) and may lead, paradoxically, to a drop in runoff, due to increased interception and water retention by the vegetation, and a drop in sediment yield, due to the stabilizing effect of the vegetation on soil surfaces. In this way, a change from semi-arid to more humid conditions may bring a drop in flood discharges and reduced sediment transport.

Climatic change affects all aspects of the drainage system, from the size of the drainage network, to channel dimensions and patterns, and the volume and nature of alluvial sediments. In mid-latitude areas, many valleys are simply too large to have been formed by the rivers which now flow in

(a)

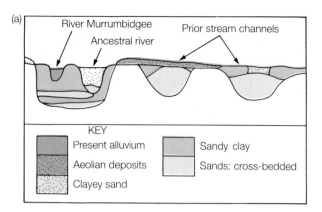

(b)

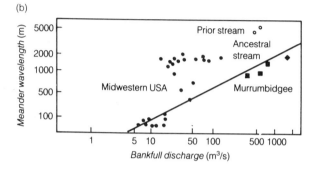

(c)

(d)

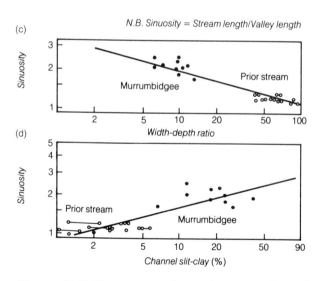

Figure 17.48 Three generations of stream channel beneath the flood plain of the Murrumbidgee River, N.S.W., Australia. (a). Cross-section of the flood plain and the nature of the channel fills. (b)-(d). Logarithmic plots of key characteristics of the contemporary, ancestral and prior channels. (*After Schumm, 1968*)

them. In former glaciated areas, this may simply reflect the occupation of a glacial valley by the post-glacial stream (Figure 17.28). Yet such **underfit** streams are also widespread in parts of North

America and Europe beyond the limits of glaciation. The underfit valleys show meander wavelengths several times greater than those of the present streams. Beneath the flood plain lie buried channels whose width and depth are also far larger than those of the channel that now flows across the flood plain surface. The infilled channels and their valleys were cut under much higher discharges which prevailed at the close of the last glaciation. In the Holocene, discharges have declined, the channels have shrunk and drainage systems have contracted, leaving dry valley heads at the tips of drainage networks.

In the tropics and subtropics, climatic change has had equally dramatic effects on rivers. Schumm identified three generations of channel fill within the alluvial plain of the Murrumbidgee River, New South Wales, Australia (Figure 17.48). The contemporary channel operates under relatively humid conditions and the well-vegetated basin produces only moderate flood discharges and sediment loads. In consequence, the channel width, depth and meander wavelength are low and the channel is filled with a mixture of clay, silt and sand. The Ancestral Channel is much larger, with three times the cross-sectional area and meander wavelength, but is also filled with mainly silt and clay. Schumm suggests that the Ancestral Channel was formed when discharges were higher but sediment input was reduced under wetter climatic conditions, when vegetation cover was more continuous and protective than now. In contrast, the Prior Channel is of similar cross-sectional area to the present channel but wider, shallower, less sinuous and infilled with sand, a product of a drier climate. When mean discharge was reduced flood discharges remained similar in magnitude to now, but sediment loads were larger and coarser due to the reduced vegetation cover.

17.14 FURTHER READING

Leopold, L.B., Wolman, M.G. and Miller, J.P. (1964) *Fluvial processes in geomorphology*, Freeman, San Francisco.

Morisawa, M. (1985) *Rivers*, Longman, London.

Petts, G.E. and Foster, I.D.L. (1985) *Rivers and landscape*, Arnold, London.

Richards, K. (1982) *Rivers: form and process in alluvial channels*, Methuen, London.

Schumm, S.A. (1977) *The fluvial system*, Wiley, New York.

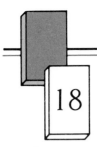

18

DEVELOPMENT OF RIVER SYSTEMS AND ASSOCIATED LANDSCAPES

Straight my eye hath caught new pleasures
Whilst the landscape round it measures

John Milton (1608–74)

18.1 DRAINAGE PATTERNS

Where rocks have no conspicuous grain and offer nearly uniform resistance to erosion, a branching drainage pattern is established, tree-like in plan, and described as **dendritic** (Figure 18.1). Often, however, examination of drainage patterns shows that streams are adjusted to structural trends at both local and regional scales. This correspondence reflects the tendency of streams to exploit lines of weakness provided by joints, faults and beds of unresistant rocks. **Parallel patterns** are found on steep slopes with little vegetation cover. **Rectangular patterns** are often related to the grid-like joint systems found in massive igneous rocks. In **radial drainage** streams radiate from the centre of a volcano or area of domed uplift. **Centripetal drainage** is the opposite of the radial type where drainage is towards a subsiding central basin.

In tilted sequences of sedimentary rocks **trellis patterns** develop. Beds of softer rocks such as clays and shales are eroded more quickly than harder rocks. Trunk valleys are opened along the strike of the rocks. Tributaries enter almost at right angles due to the structural control over surface slopes.

Those which flow down the dip slopes of the resistant beds are termed **dip streams**; those that flow in the opposite direction are called **anti-dip streams** (Figure 18.2). Where sediments are uplifted in a dome-like fashion, exploitation of the curved weaker beds can produce striking **annular** drainage patterns (Figure 18.1).

The valley of a strike stream is widened and deepened between divides formed by the bands of more resistant rock on either side (Figure 18.3a). As the intervening weak bed is gradually worn away, the upper surface of the underlying resistant bed is uncovered, and on this side the valley slope approximates to the dip. On the other side the overlying resistant bed is attacked at its base and soon begins to steepen into the free face of an escarpment. Until it is restrained by approaching base-level, the stream excavates its channel in the weak bed. As a result of combined lateral and downward erosion being specially favoured on one side, the stream shifts its channel obliquely in the direction of dip. The normal recession of the escarpment is thus widened, while a gentler slope – resembling a continuously guided slip-off slope

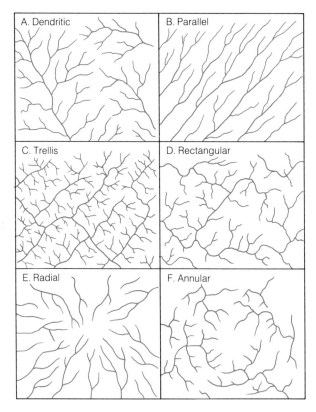

Figure 18.1 Types of drainage pattern. (*After A.D. Howard, 1967*)

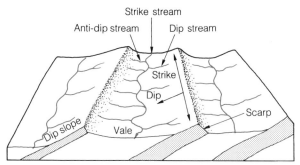

Figure 18.2 Structural relationships and terminology of streams on dipping strata.

– is left on the other side of the valley. This is generally referred to as the dip slope, unless it has been notably steepened or flattened after its exposure to erosion; the term **back slope** is then more suitable. Besides the temporary rills of surface wash, small dip streams flow down the dip slope. Others, known as anti-dip streams, descend the escarpment. Both sets add detail to the trellis pattern of the drainage. As the main strike stream

begins to meander, its valley becomes steadily wider and develops into an **interior lowland**.

Figure 7.4 illustrates the succession of escarpments and interior lowlands between Gloucester and the London Basin. From the Lias clays and marls of the Severn valley the escarpment of the oolitic limestones of the Jurassic rises to the crest of the Cotswolds. The Oxford Clay is responsible for the interior lowland occupied by the Thames above and below Oxford. A minor escarpment, that of the Corallian limestone, is then followed by the interior lowland of the Kimmeridge Clay. Beyond this the Chiltern Hills begin with the Chalk escarpment, the dip slopes of which lead down to the London Basin (Figure 18.3b). On the south side of the basin the Chalk again emerges as the North Downs, with its well-known escarpment overlooking the Weald. The scarp continues through Kent until it reaches the coast near Folkstone. Followed in the other direction it swings round the western border of the Weald, where it makes a less conspicuous feature, and continues on the far side as the South Downs; at Beachy

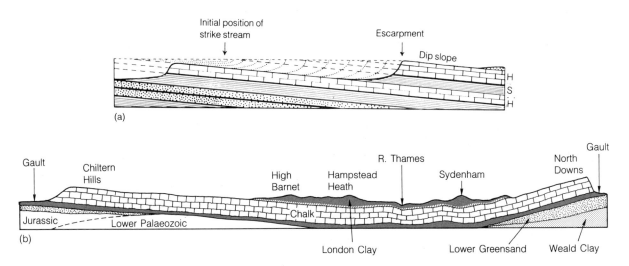

Figure 18.3 (a) Stages in the development and recession of an escarpment. H=relatively resistant formation, S=easily eroded formations. (b) Section 93 km long, through the London Basin.

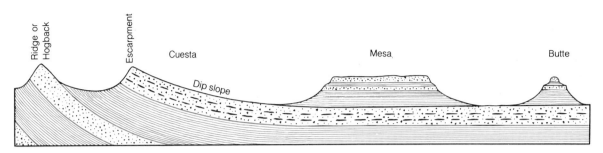

Figure 18.4 Diagram to illustrate the relationship of various erosional land forms to the structure and dip of the strata from which they have been carved.

(a)

(b)

Figure 18.5 (a) Hogback of Dakota Sandstone (Cretaceous) in the foothills east of the Front Range, Rocky Mts, Colorado. (*T.S. Lovering, US Geological Survey*) (b) Mesa topography with marginal 'badland' erosion due to gullying of the 'constant slope', Zion National Park, Utah. (*Grant, US Dept of Interior*)

Head, Sussex, it is again cut off by the sea.

An escarpment together with its dip slope or back slope form a feature for which there is no English name. The Spanish term **cuesta** (pronounced **questa**) for this combination has been internationally adopted (Figure 18.4). If the beds dip at a high angle, so that the dip slope becomes about as steep as the escarpment, the feature corresponding to the cuesta is a ridge or **hogback** (Figure 18.5a). At the other extreme, in horizontal beds, the cuesta becomes a **mesa** (Spanish for table), that is, a tableland capped by a resistant bed and having steep sides all round (Figure 18.5b). By long-continued wearing-back of the sides, a mesa dwindles into an isolated flat-topped hill. In America such a hill is called a **butte**, from its resemblance to the butt or bole of a tree, and the term has been widely adopted. In western America buttes commonly occur where the beds dipping off the mountain flanks flatten out, as illustrated by Figures 18.4 and 18.5(b). Similar residual landforms in South Africa, many of which are capped by outliers of dolerite sills, are called **kopjes**.

18.2 HEADWARD EROSION AND RIVER CAPTURE

A valley may be lengthened by headward erosion (Figure 18.6). This involves the same processes of surface erosion as those concerned in widening (Chapter 16). Nevertheless the rate of recession of a relatively steep valley head may be much more rapid than that of the valley sides. Probably the main single reason for this difference is that the head has one great advantage over the sides. The valley head is concave (downstream) both in plan and in long profile, like the front half of a tilted spoon. Consequently rainwash and rills from a wide area near the top converge towards the point where a definite stream channel begins. This concentration of the runoff towards the lower part of the slope tends to steepen it and so to promote slumping. If the rock is weak and impermeable,

Figure 18.6 Brecon Beacon (886 m), South Wales. Streams on the right drain to the Usk and the sea at Newport; their headward erosion threatens to capture the stream and lake on the left, draining to the Taff and the sea at Cardiff. (*Aerofilms Ltd*)

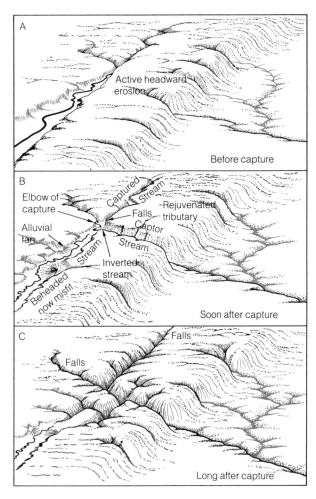

Figure 18.7 Stages in the process of stream capture. (*After A.K. Lobeck, 1939*)

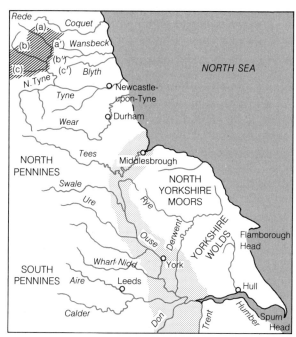

Figure 18.8 River systems of north-east England, to illustrate river capture by the Ouse – along the outcrop of soft Triassic beds (dotted) – and by the North Tyne – along the outcrop of the Scremerston Coal Group (shaded).

such as clay or shale, headward recession is favoured both by the high proportion of runoff and by the ease with which the infant stream can carry away the finely divided rock-waste. Another favourable condition is the emergence of springs or seepages; these tend to undercut the overlying permeable rocks, which fall away at intervals.

The effect of headward recession on a drainage divide is to make it more sinuous, whilst the crest is notched and becomes increasingly varied in height. The speed of headward recession is controlled by rock type and by the gradient of the river. Hence on single rock types the streams with the steepest gradient will move headwards most rapidly. In trellis patterns (Figure 18.7) the anti-dip streams are steepest and tend to breach the ridges before the dip streams. This leads to the diversion or capture of drainage from the dip stream to the anti-dip stream. Continuation of the process then diverts the headwaters of the strike stream to the aggressive anti-dip stream. The rectangular bend at the point of capture is known

as the elbow of capture. The beheaded river, now deprived of much of its drainage, is described as a misfit, since its diminished size is no longer appropriate to the valley in which it flows. The strike valley is deepened as the rivers cut down to a new base-level provided by the encroaching anti-dip stream and the increased gradients of the tributary valleys may allow capture of other gently-flowing strike streams. In this way the anti-dip stream, with the aid of its tributaries, may acquire a very large drainage area at the expense of its neighbours.

The rivers of north-east England illustrate the development of an actual river system by this process of capture (Figure 18.8). The uplift of the Pennines provided the slopes down which a number of streams flowed into the North Sea, which even then occupied a basin with a long history of subsidence and sedimentation. The greater part of modern Britain probably emerged from the late Cretaceous sea with a widespread cover of Chalk, with the exceptions being in the south-east, where subsidence with some fluctuations continued in the London and Hampshire basins; and locally in the north and west, where there were islands representing parts of the 'old land' that had remained unsubmerged. Uplift and tilting of north-east England towards the North Sea in the Early Tertiary led to the development of dip streams which flowed across the chalk covered rocks, soon to be stripped away. Today, only the Aire Humber still maintains an uninterrupted approximation to its original course. Even so, the headwaters of the Aire, including a lengthy strike tributary, were captured by the Ribble, which, flowing to the Irish Sea had the advantage of a shorter run and a steeper gradient. The Wharfe, Calder and Don were probably tributaries of the Aire from an early stage in their history. The Nidd, Ure and Swale, however, have each been captured in turn by the Ouse, a powerful strike stream which lengthened itself northwards by headward erosion along the soft strata of the Trias (Figure 18.8). On the eastern side of the Ouse it is difficult to trace the former courses of the beheaded streams, because of uplift of the Cleveland Hills and the obliteration of the older valleys by glacial deposits.

A more straightforward example is provided by the rivers of Northumberland. The three main streams, a, b and c, of the North Tyne system clearly correspond to the Wansbeck, a', a tributary of the Wansbeck, b', and the Blyth, c'. The headwaters of the forerunners of these were captured by the North Tyne, a subsequent of the Tyne, as it worked back along the soft beds of the Scremerston Coal series.

Tracing the ancestry of rivers and their competition for drainage areas has always had a special fascination for geomorphologists. Like every other branch of the earth sciences it springs from the main trunk of geology, detached from which it cannot flourish. And so we find that the history of rivers cannot be successfully investigated without detailed reference to earth movements and fluctuations of sea-level, and to radical changes of climate, including the highly complex interventions of the Pleistocene glaciations.

18.3 SUPERIMPOSED DRAINAGE

In all the 'old lands' of Britain – representative of innumerable regions of similar history throughout the world – we see ancient folded rocks that were formerly hidden beneath an unconformable cover of later sediments. The Chalk, for example, that originally covered most of the British area has since been worn back to its present escarpment, which extends across England from Yorkshire to Dorset. The underlying Jurassic strata have been pared away almost as extensively. As the cover emerged as land from the late Cretaceous sea, rivers were initiated with a drainage pattern appropriate to the form of its surface. Sooner or later the deepening valleys were incised into the older, more resistant rocks, many of which have

strong structural features trending across the river courses like barriers. As the covering strata were removed, the underlying rocks were gradually exposed over a steadily increasing area and, as uplift continued, were sculptured into a landscape of bold relief, with the rivers maintaining a close approximation to their original courses. The drainage pattern as we see it today has been **superimposed** on the older rocks as an inheritance from the vanished cover.

The clearest example of superimposed drainage in Britain is afforded by the rivers and lakes of the Lake District. As illustrated in Figures 18.9 and 18.10, the Lake District consists of an ovoid area of Lower Palaeozoic rocks (folded during the Caledonian orogenesis (Section 31.2) and having a general trend from ENE to WSW) enclosed in a frame of Carboniferous Limestone and New Red Sandstone, the beds of which everywhere dip outwards. These strata originally covered the older rocks unconformably and were themselves covered in turn by later formations up to the Chalk. During Early Tertiary times the region was uplifted into a slightly elongated dome, with its main axis curving eastwards from the neighbourhood of Scafell towards the Shap Granite. Streams flowed radially down the slopes of the rising dome. After cutting through the cover and removing all the younger strata, they excavated their valleys deeply into the underlying Lower Palaeozoic rocks. There they still persist with only minor adaptations to the very different structures through which they are now flowing. The radial pattern of the valleys and lakes, and of the mountainous ridges centred near Scafell, is particularly striking (Figure 18.10). The attractive contrast of lake and mountain for which the Lake District is renowned, is due to the finishing touches given to the landscape by glaciation.

18.4 EFFECTS OF FOLDING AND FAULTING ON DRAINAGE PATTERNS

The surface expression of active folding is warping. Where river downcutting is weak the drainage pattern will be disrupted by warping (Figure 18.11). The rise of the upwarp beheads the original stream and may locally reverse the direction of flow of its headwaters. Ponding of drainage develops in the downwarped area and overflow may lead to the cutting of a new valley along the axis of the depression. If river downcutting is vigorous then the river may maintain its course during uplift by cutting a gorge through the upwarp. Since the river was in position before warping began, its course is said to be **antecedent**. The case of the Arun gorge is discussed in Section 18.9.

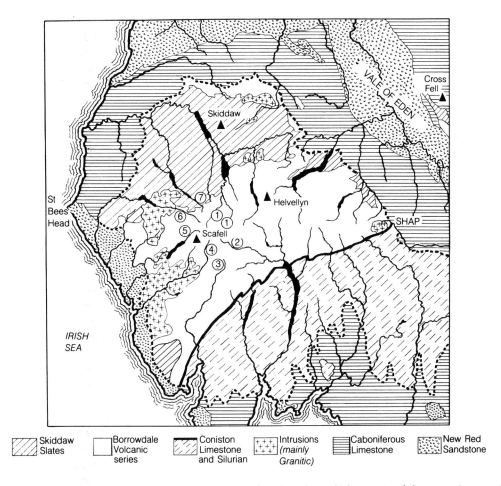

Skiddaw Slates | Borrowdale Volcanic series | Coniston Limestone and Silurian | Intrusions (mainly Granitic) | Caboniferous Limestone | New Red Sandstone

Figure 18.9 Geological map of the English Lake District, showing the radial pattern of the superimposed drainage. Numbers refer to the headwaters of valleys and lakes: (1) Borrowdale and Derwentwater (2) Langdale and Windermere (3) The Duddon (4) Eskdale (5) Wasdale and Wastwater (6) Ennerdale (7) Buttermere and Crummock Water. To the east of (1) the streams leading into Thirlmere, Ullswater, and Haweswater flow off the northern side of the axis.

Active faulting may also modify drainage patterns (Figure 18.12). Above rising fault scarps, stream systems are beheaded and wind gaps are left marking the former stream course on the crest of the scarp. Development of the fault scarps may result in diversion of drainage along the scarp foot, perhaps with formation of a lake. Alternatively, if uplift is slow, river erosion may keep pace with the rise of the fault block or horst and cut an antecedent gorge. Recent lateral fault movements can often be picked out in the landscape by offset drainage lines (Figure 8.28).

18.5 CHANGING BASE-LEVELS

A basic control on river behaviour is the regional

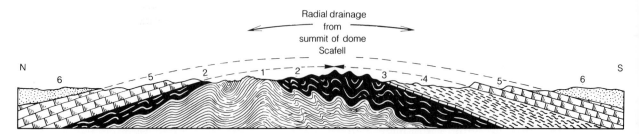

Figure 18.10 Geological section across the Lake District (igneous intrusions omitted): (1) Skiddaw Slates (2) Borrowdale Volcanic Series (3) Coniston Limestone (4) Silurian strata (5) Carboniferous Limestone (6) New Red Sandstone.

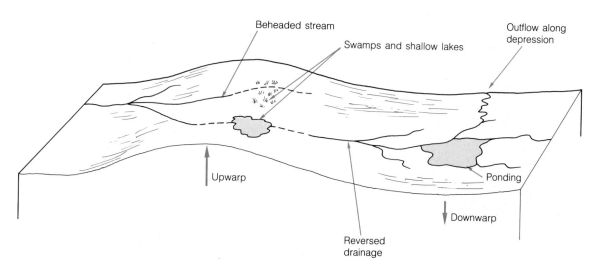

Figure 18.11 Some effects of warping on drainage systems.

base-level provided by sea-level. The level of the sea against the land is not fixed and has varied dramatically in most parts of the world during and before the Quaternary. Sea-level has continued to change significantly during historical time and there are ongoing discussions that it may rise globally by 1–2 m over the next century or so because of the so-called 'greenhouse' effect.

Sea-level fluctuations have three basic components. Worldwide changes in sea-level, whether due:

1. to the growth and decay of ice-sheets;
2. to displacement of seawater by deposition of sediments;
3. to changes in the volume of ocean basins;
4. to expansion of ocean waters due to global warming,

are said to be **eustatic**. These are often involved

with slow crustal movements due to **isostatic** readjustments, in response to the removal of load by the melting of ice-sheets, shallowing of coastal waters and erosion of overlying rocks. Moreover, in many places, independent earth movements may be simultaneously affecting the level of the crust. The term **diastrophism** is used for all movements of the solid crust which result in a relative change of position (including level and altitude), i.e. for the movements commonly referred to as **orogenic** and **epeirogenic** (Section 7.7). Strictly speaking, the term also includes isostatic movements, but it is convenient to consider these separately, so far as it is possible to do so.

If a region is depressed by earth movements its surface is brought nearer to base-level, the work to be done by erosion is diminished and the river may temporarily begin to deposit its load as it adjusts to the change in gradient. When a depres-

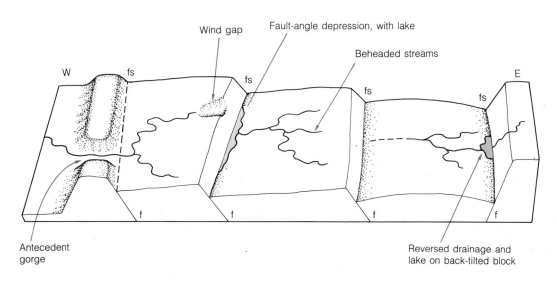

Figure 18.12 Effects of a series of east-facing faults on a west-flowing river. (fs – fault scarp, f – fault).

Figure 18.13 Typical scenery of the Norfolk Broads, near Wroxham 6 km north-east of Norwich, England. River Bure (looking east) with Salhouse Broad on the right (south) and part of Hoveton Great Broad on the left (north). (*Aerofilms Ltd*)

sion is localized across the course of a river a lake is formed. This also forms when a river is ponded back by the up-arching of a fold across its course (Figure 18.11).

When coastal regions subside – unless sedimentation keeps pace by growth of flood plains and deltas – the sea occupies the lower reaches of valleys, and estuaries are formed. Tributaries which entered the valley before it was drowned now flow directly into the tidal waters of the estuary and become **dismembered streams**. Rivers like the Thames and Humber are sufficiently active – with some human assistance – to keep their channels open. More sluggish rivers, however, may be unable to prevent the growth of obstructive bars and spits across their estuaries, and the latter then become silted up and over-

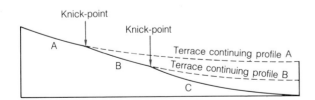

Figure 18.14 A long profile of graded reaches, A, B, and C, with knick-points at their intersections. Profiles A and B may be indicated by terraces preserved on the valley sides.

grown with peat-making vegetation.

The broads of East Anglia (Figure 18.13) were thought to be natural relics of the formerly widespread estuary of the rivers Bure, Yare and Waveney. However, it is now known that the Broads owe their origin not only to changes of level and climate, but also to human interference on a big scale. They are, in fact, the flooded sites of excavations (margining the rivers) dug in peat to depths of 3 or more metres. About 2000 BC the region was widely invaded by the sea, which then withdrew until about 1000 BC, while thick peat deposits were accumulating on the emerging land. Another marine transgression then occurred, marked by a deposit of clay, which was again followed by emergence and further growth of peat until about AD 700. Since then submergence has been almost continuous, with only slight and temporary interruptions. By the year 900 a prosperous peat industry had begun to flourish and this continued to expand for nearly 400 years. During the thirteenth century severe storms and flooding, from the sea as well as from the rivers, not only checked the progress of this industry but began its enforced decline. The worst disaster was in 1287, when over 50 000 lives were lost by drowning in The Netherlands. At that time the coastal region in the neighbourhood of Yarmouth stood 4 m higher above the sea than it does today.

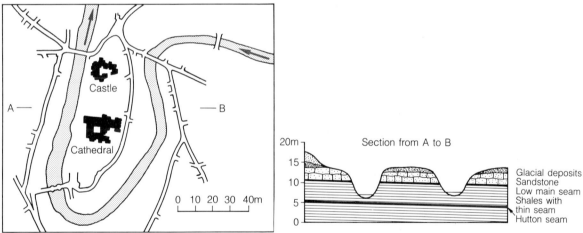

Figure 18.15 The entrenched meander of the River Wear at Durham (*Aerofilms Ltd*) with explanatory map and section.

Even so, some of the pits left by the peat-cutters were flooded, and as they could not be drained they had to be abandoned. Later floods discouraged new enterprise and brought peat-digging practically to an end. By 1350 it had become more profitable to use the flooded pits – the Broads of today – as fisheries. The rivers now leave their flat meadows to flow along the irregular ramparts of peat and sediment which originally separated them from the adjoining peat-cuttings. Full details of this fascinating story, in which geology is linked to living history by archaeology, will be found in the memoir by J. M. Lambert *et al.*, listed in Section 18.16.

Geologically, the Broads may be said to owe the immediate possibility of their existence to a local subsidence, i.e. to a rise of base-level. If, on the contrary, base-level is lowered, the work to be done by erosion is increased and the river is endowed with renewed energy to begin the task of regrading its profile to the new base-level. The river is **rejuvenated**. During the process of regrad-

Figure 18.16 Entrenched meanders of the San Juan River, Monument Valley, Utah. (*Dorien Leigh, Ltd*)

ing there is a more or less marked change of slope at the place of intersection (the **knick-point**) of the newly graded profile with the older one (Figure 18.14). The knick-point is particularly conspicuous when the uplifted region is bounded by a growing fault scarp, or in any other circumstances favourable to the development of rapids or a waterfall. The newly deepened part of the valley may be excavated as a gorge or narrow V in the wide V- or trough-shaped floor of the pre-existing valley. Consequently the cross-profile also shows a steepening of slope below the point where the earlier valley form is intersected by the new. The valley sides tend to become convex as a result of an increased rate of down-cutting.

Headward migration of the knick-point may be retarded by a resistant rock band and it is possible that a succession of migrating knick-points may be held up at such an obstacle. This barrier forms a local base-level for tributary streams and their valley-side slopes and may so delay the advance of rejuvenation that headwater drainage basins begin adjusting to a fall in regional base-level long after this event has occurred. The time lag is obviously greatly extended for long rivers draining from continental interiors. Where the river had formed a flood plain prior to rejuvenation, incision by the stream will leave terraces standing above the river.

18.6 ENTRENCHED MEANDERS

Incised meanders with well-developed slip-off slopes have been described in Sections 16.2 and 17.7. These dig themselves in by oblique erosion while their loops are being enlarged, and are said to be **ingrown**. If, however, erosion is mainly vertical, the existing loops have less opportunity to enlarge themselves and the resulting incised meanders are distinguished as **entrenched**. There is, of necessity, every gradation between these two, entrenchment being favoured by relatively rapid rates of uplift and downcutting. If, for example, at the time of rejuvenation, a stream was freely meandering on a floor of easily eroded deposits underlain by more resistant formations, the deepening channel would soon be etched into the underlying rocks with a winding form inherited from the original meanders. The 'hair-pin gorge' of the River Wear at Durham is a familiar British example (Figure 18.15). A well-protected site within the loop was selected for the Cathedral, which is thus enclosed by the gorge on three sides. The fourth and easily vulnerable side was safeguarded by building a castle which is now the home of the senior college of Durham University.

The change of form of entrenched meanders and the wearing back of the confining walls are relatively slow processes controlled by lateral under-

Figure 18.17 Rainbow Bridge, Bridge Creek, Utah. Water from Lake Powell, impounded by Glen Canyon Dam (1964), now fills and obscures the miniature canyon beneath the arch. See Figs. 18.18 and 18.25. (*Ewing Galloway*)

cutting of the river banks. Localized undercutting on both sides of the narrow neck of a constricted loop sometimes leads to the formation of a natural bridge. On each side a cave is excavated, especially if the rocks at river level are weaker than those above. Eventually the two caves meet, and the stream then flows through the perforation. The

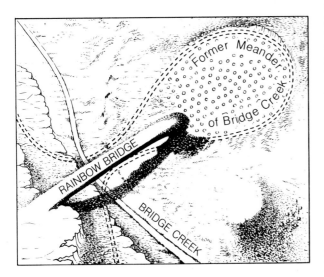

Figure 18.18 Diagram illustrating the mode of origin of Rainbow Bridge.

stronger rocks of the roof remain for a time as an arch spanning the stream, and the loop-shaped gorge at the side is abandoned. In Utah, where recent uplift has made possible the development of many deeply entrenched meanders (Figure 18.16), there are several examples of such arches. The most impressive of these is Rainbow Bridge (Figures 18.17 and 18.18), a graceful arch of sandstone which rises 94 m over Bridge Creek in a span of 85 m.

18.7 RIVER TERRACES

When a river cuts down into its flood plain, the former alluvial surface is no longer flooded and is left as a more or less flat terrace above the new level of the river. In the course of time the new valley is widened and a new flood plain forms within the original one, of which only local remnants may survive. If downcutting resumes, a second pair of terraces may then be left on the valley side (Figure 18.19).

Terraces with surfaces at the same elevation on both sides of the valley are termed paired terraces (Figure 18.20). This pairing generally requires an initial stage of flood plain formation and a later and separate phase of incision. Unpaired terraces indicate that downcutting and lateral cutting by the river are taking place together, so that as the river shifts across the valley it leaves abandoned flood plain fragments at different elevations on opposite sides of the valley (Figure 18.20). Where cycles of channel aggradation and incision occur repeatedly then cut-and-fill terraces form. In stages A and B in the diagram (Figure 18.21), a river begins to cut down into its flood plain and the underlying alluvial fill to form an initial set of paired terraces. At Stage C a phase of aggradation has produced a second but lower flood plain (B). Renewed lateral migration and widening of the channel in Stage D leads to partial removal of material from the earlier flood plains A and B. Later aggradation may even be sufficient to bury older terrace remnants (Stage E). In this way, a highly complex cut-and-fill sequence will develop in the rock channel and the sequence and causes of associated terrace development may be very difficult to unravel.

The onset of downcutting by the river implies an increase in river energy. This can be brought about in many ways. Recent uplift may increase the river gradient and cause rejuvenation and downcutting. The terraces of the Fraser River reflect, in part, Quaternary tectonism in the Rocky Mountains (Figure 18.22). A drop in base-level may also be brought about by a fall in sea-level. During the Quaternary, sea-level has repeatedly

Figure 18.19 The River Findhorn, flowing into the Moray Firth, Scotland, cuts through glacial outwash deposits, the terraces marking successive stages in the erosion of the valley. (*British Geological Survey*)

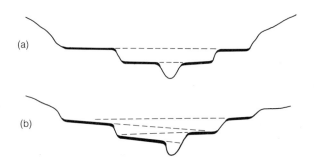

Figure 18.20 (a) Paired terraces, (b) Unpaired terraces. Black shading indicates alluvium.

fallen by 100 m or more during glacial stages. Rapid incision caused terrace formation in the seaward reaches of valleys. These terraces are invisible today, however, as the return of sea-level to close to its present level during interglacials submerged these terraces. Suites of low terraces are found near the mouths of many rivers but these developed during interglacial periods when the drop in gradient caused major aggradation. These interglacial terraces may merge seawards into raised estuarine and beach deposits. Terraces related to eustatic changes in sea-level in this way have been termed **thalassostatic** terraces.

The energy of a river may also increase due to climatic change, with more rainfall giving increased discharge or an increase in vegetation cover leading to a drop in sediment load. The immediate effect of both contrasting changes would be to promote channel incision. A related effect is that of glaciation which has affected the upper reaches of many mid-latitude drainage basins, such as the Thames, Rhine, Danube and Mississippi. During deglaciation, vast quantities of meltwater swell river discharge and the large amounts of debris liberated from the ice add greatly to river load. Aggradation occurs, especially towards the ice margins, and terraces form.

All these factors have interacted in the Thames basin, where the terrace sequence has been intensively studied. Since the early Pleistocene there has been a large, but still unknown number of changes in base-level in the London Basin and a background trend of long-term regional subsidence shared by the whole of the southern North Sea area. These oscillations are recorded by a staircase of terraces and also by buried channels which can be discovered only by borings and excavations. In early Pleistocene times the lower reaches of the Thames followed a more northern course through the Vale of St. Albans (Figure 18.23). During the Anglian glaciation the Thames was diverted to a more southerly course and the present valley alignment through London dates from the time when all earlier exits to the North

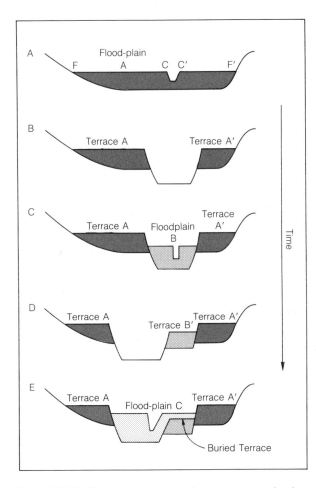

Figure 18.21 The development of river terraces by lateral displacement of the channel and by 'cut and fill' processes. (*After J.J. Lowe and M.J.C. Walker, 1984*)

Sea had been completely blocked by ice and glacial deposits.

As illustrated in Figure 18.24, the Thames valley is shown with three paired terraces that can be most easily recognized in and near London:

1. The **Boyn Hill Terrace** reaches a height of about 70 m, but at Swanscombe (near Gravesend, 80 km nearer the sea) it is much lower and corresponds to a sea-level of about 30 m. At Swanscombe it is particularly well exposed on a platform of Chalk. The gravels, mostly composed of flint, contain the fossil remains of extinct species of elephant, hippopotamus and rhinoceros. These mammalian remains point to a long and rather warm interglacial period, probably the Hoxnian Interglacial. Human skull fragments were found in these gravels in 1938 which represent the oldest known Englishman, Swanscombe man (probably 300 000–400 000 years old).

2. The step down to the **Taplow Terrace**, which would originally have been a fairly steep bluff,

is generally hidden by downwash from the terrace above. This is illustrated by Clapham Common and the slopes in its vicinity. These gravels were graded to a sea-level below that associated with the Boyn Hill Terrace (not more than 10 m above the present level). Bones of lions and bears have been found in the lower gravels, but nearer the surface these are absent and remains of woolly mammoth appear, indicating the oncoming of much colder steppe-like conditions.

3. Less widespread are the fragments of the **Flood plain Terrace**, associated with a sea-level only marginally higher than that of the present day. These deposits include the distinctive hippopotamus-rich fauna found in excavations at Trafalgar Square, London, and formed during the last, Ipswichian interglacial.

4. The **Alluvium** is the most recent of the Thames Terraces and has formed within the last 10 000 years as sea-level has risen to its present level. This rests on deposits which infill the channel cut during the low sea-level at the height of the last (Late Devensian) glaciation. Dredging of the infill of this channel to maintain navigation to the Port of London has produced great numbers of mammoth teeth, reflecting the severe tundra climate which prevailed in the London Basin at that time.

The faunal and floral remains within the terrace deposits around London point clearly to an interglacial age for the main terraces. In the middle and upper course of the Thames, towards Oxford, the situation is very different for here the terrace fragments contain remains of woolly mammoth and woolly rhinoceros and reindeer and are associated with accumulations of periglacial slope deposits (termed locally, Coombe Rock). These terraces obviously formed during glacial stages when ice in the English Midlands poured meltwater and debris into the Thames drainage system, forcing widespread and substantial aggradation upstream from London.

18.8 THE CANYONS OF THE COLORADO RIVER

Bridge Creek is a tributary of the Colorado River, and is thus related to one of the world's most awe-inspiring scenic wonders – the Grand Canyon of the Colorado River (Figures 1.5 and 18.25). At the end of Cretaceous times the region which is now the Colorado Plateau was near sea-level, eastern Utah and Arizona being a coastal plain which extended towards an inlet of the sea in

Figure 18.22 Alluvial terraces of the Fraser River, British Columbia (*P. McL. D. Duff*).

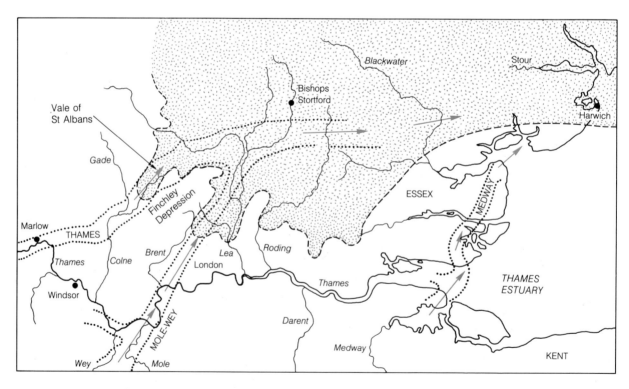

Figure 18.23 Diversion of the River Thames. In the early Pleistocene the Thames entered the sea somewhere near Harwich. During the Anglian glaciation an ice-sheet advanced to the northern outskirts of London (stippled shading) and ice lobes blocked both the Vale of St. Albans and the Finchley Depression, causing the eastward flowing Thames to take a more southerly course. Retreat of the ice left infills of glacial deposits which stopped the Thames and Wey–Mole rivers from re-occupying their former channels. (*After A. J. Sutcliffe, 1986*).

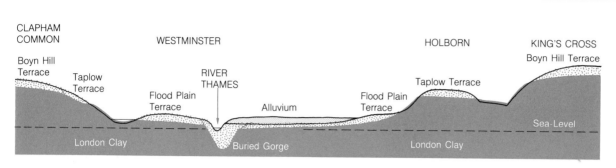

Figure 18.24 Section across London to show the paired terraces and one of the buried 'gorges' of the Thames valley. (*British Geological Survey*)

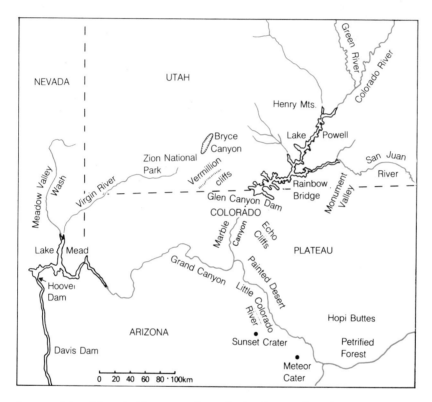

Figure 18.25 Sketch map of the Colorado Plateau to show the localities referred to in text.

Colorado and New Mexico (Figure 18.26). To the west were the early mountains of the Basin and Range province, while to the east the sea was already bordered by the Rockies. After the sea withdrew, the Uinta Mountains arose in the north of Utah, while much of the southern part of the region became the Green River Lake, occupying a downwarp that developed during the Eocene and became the depository of about 3000 m of sediment. Uplift, upwarping, faulting and igneous activity then began more vigorously and have continued at intervals and at varying rates ever since.

Most of the Mesozoic and later formations have been removed from the Grand Canyon region during the Miocene, when the cutting of the present canyons may be said to have begun. J. W.

Powell, the intrepid explorer who was the first (in 1869) to pass through the gorges of the Colorado River, suggested that the river had been able to maintain its course by deepening its valley while uplift was in progress, so becoming permanently entrenched in the rising landscape. A rejuvenated river which thus succeeds in maintaining a downward slope from a source behind the uplands to the plains in front, is called an **antecedent** river, to express the inference that the river must have been in existence **before** the upheaval of the land through which it had to cut its channel.

Although mainly correct, this idea does not cover the whole history of the Colorado River. Of the various upwarps that lie across the course of the river, one was formed early in the Pliocene near the site of the present Lake Mead. Here the

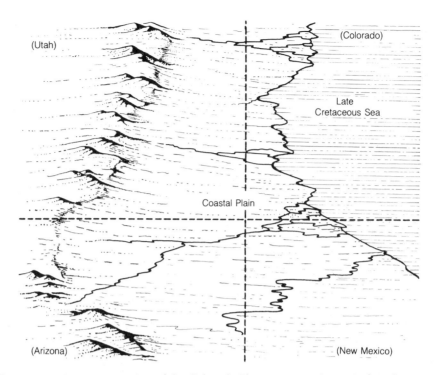

Figure 18.26 Diagrammatic representation of the Colorado Plateau area as it was in late Cretaceous times. (*After C. B. Hunt, 1956*)

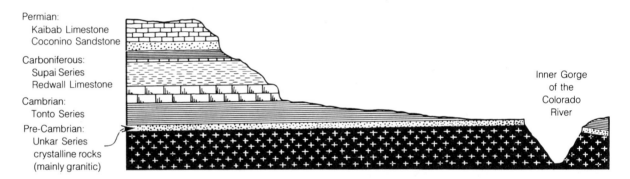

Figure 18.27 Section across the Grand Canyon of the Colorado River at Grand Canyon Station, Arizona. Scale approximately 1 in 63 000 (horizontal and vertical). (*After N. H. Darton, 1933*)

river failed to keep pace with the rate of uparching of the ground beneath it, and in consequence its waters were ponded back, as they are today by the Hoover Dam (Figure 17.38). The Pliocene lake deepened until it could spill over the lowest point on the rim, i.e. the old canyon floor, which thereupon resumed its former function of draining the basin upstream. In due course the lake was eliminated in the usual way by sedimentation from the inlet and headward erosion from the outlet.

The average thickness of strata removed from the Plateau amounts to about 3000 m. At the present rate of denudation (see Section 17.4) this would have taken about 21 million years, which agrees very well with the geological evidence. Canyon cutting was renewed on a big scale during the Pleistocene, in response to further uplift which

raised the surface to heights of about 2000–3000 m. There is some evidence that there were intervals of heavy deposition in the canyons during the Pleistocene, but on balance deepening has predominated. During recent time erosion has been more vigorous than weathering. Much of the region is bare rock, and what little soil there is remains poor and thin. Although the river has to cross hundreds of kilometres of desert country, it receives sufficient water from the Rockies to carry it successfully through.

The Grand Canyon has now reached a maximum depth of about 1900 m. The river drops 457 m in the 322 km stretch above Lake Mead, which is 380 m above sea-level. A narrow inner gorge has been cut through 300 m of crystalline rocks (Figure 18.27). The walls above, carved through a nearly

Figure 18.28 The head of the canyon of one of the Little Colorado tributaries, Arizona. (*US Army Air Force*)

flat Palaeozoic cover of strong sandstones and limestones alternating with weak shales, rise by a succession of steps and slopes of varied colours which add to the architectural grandeur of the scene. As a result of differential erosion the width of the Grand Canyon from rim to rim ranges between 8 and 24 km. During the recession of the walls bold spurs between the bends have been carved into pyramids and isolated pillars. The plateau is trenched by several tributary canyons (Figure 18.28), but otherwise the general surface is but little dissected. Northwards from the Grand Canyon the surface rises by successive cliffs and terraces (Figure 18.29) to over 3350 m in the high plateau of Utah, the edge of which has been sculptured by erosion into such fantastic landscapes as those of Zion and Bryce Canyons. Northeast of Bryce Canyon are the famous stocks and laccoliths of the Henry Mountains. To the southeast of Bryce Canyon, on the other side of the Colorado River, is situated Rainbow Bridge (Figure 18.17); and farther east, south of the San Juan River, is Monument Valley, so called because of its obelisks and towers and other castellated erosion remnants, carved under arid conditions from red Triassic rocks. The celebrated Painted Desert stretches from the Marble Canyon (Figure 18.29) to the Petrified Forest (Figure 6.13) and beyond. All round the edge of the Colorado Plateau there are extensive spreads of lavas and tephra. South of the Grand Canyon the great cone of the San Francisco volcano rises near the edge of the Plateau and is surrounded by a vast volcanic field dotted with hundreds of small vents, many of them so perfectly preserved that they are obviously very youthful features. Some of the eruptions are known to have been quite recent. One that occurred at Sunset Crater in the year 1160 brought disaster to a Pueblo settlement. Not far to the south-east is Meteor Crater, a great depression over 1200 m across and 150 m deep, caused by the explosion which followed the impact of a giant meteorite. This brief outline gives but a faint idea of the variety and interest of the magnificent scenic and geological features for which the region is so justly renowned.

18.9 HIMALAYAN AND PRE-HIMALAYAN RIVERS

The history of drainage patterns in the Himalayas is particularly fascinating, for the courses of several major rivers seem to pre-date uplift of the mountain chain. Upper Cretaceous limestones cover much of the Tibetan Plateau, showing that at that time Tibet lay below sea-level, and drainage flowed southwards to a bay in the Tethys Ocean (Section 9.11 and Chapter 31). By 20 million years

Figure 18.29 Aerial view of part of the Grand Canyon region: looking over the north-western end of the Painted Desert and the Marble Canyon to the Vermillion Cliffs and the Plateau beyond. (*G. A. Grant, US National Park Service*)

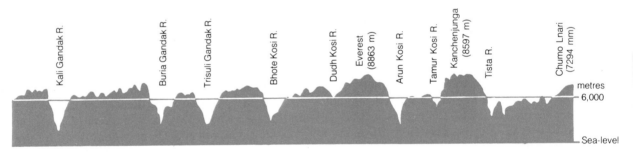

Figure 18.30 Section along the Great Himalaya of Nepal, showing the dissection of the Range by rivers, some of which have their sources on the Tibetan side, nearly 100 km to the north. Length of section about 685 km. (*After E.H. Pascoe, 1950*)

ago, northward movement of the Indian plate had brought collision with Tibet and formation of the first Himalayan peaks. This uplift has continued since and may even have accelerated during the Quaternary.

Rapid uplift has created the most remarkable examples of antecedent rivers in the world, the Brahmaputra, and the Indus and its tributary the Sutlej.

It should be noticed that the term antecedent is not applied to the rivers of drainage basins in areas that have been bodily uplifted. If it were, a great many rivers which have inherited their courses from an earlier cycle would have to be described as 'antecedent' and the term would lose its distinctive value. The significant criteria in the case of the Himalayan rivers are (a) that the river has sawn its way through ranges with peaks that now rise high above the level of the source; and (b) that the deep valleys or gorges traversing the ranges have not resulted from headward erosion (e.g. by waterfall recession), which would have

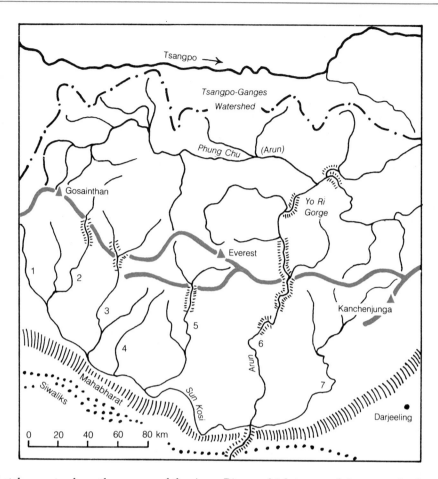

Figure 18.31 Sketch map to show the gorges of the Arun River, which is one of the seven feeders of the Sapt Kosi (see Fig. 17.35): (1) Indrawati (2) Sun Kosi (3) Bhote Kosi (4) Tamba Kosi (5) Dubh Kosi (6) Arun (7) Tamur.

made possible the capture of streams flowing on the far side (Tibet) of what would have been the original watershed. As usual in such cases it may not be easy in practice to dispose of (b) with any confidence.

All the above-mentioned rivers have characteristics favouring a pre-Himalayan origin. They rise in Tibet well to the north of the highest peaks. The Brahmaputra flows eastwards (and there it is called the Tsangpo), while the others flow to the north-west: all three for long distances before they abruptly change direction, turning **towards** the mountain barriers and passing through them by way of stupendous gorges, some of which are deep slit-like gashes cut in the bottom of V-shaped valleys. This latter feature points to an increased rate of rejuvenation during Pleistocene and recent time. As the Indus, for example, leaves Kashmir in the neighbourhood of Nanga Parbat (8117 m), the river itself is about 900 m above its delta, but the precipitous walls by which it is confined rise almost vertically at first, and then by a series of steps, to 6000 m. Like a gigantic saw the river has cut through over 5000 m of rock, keeping pace with an uplift of the same order.

The Himalayan tributaries of the Ganges (or Ganga, as the river is now called in India) nearly all descend from glaciers on the southern side of the Great Himalayan crest line (Figure 18.30), but two or three have their sources in Tibet, i.e. on the northern side, and so may be antecedent. Of these, the Arun, one of the seven rivers that unite to form the Kosi (Section 17.9), is the best known, and yet there is still controversy as to whether it should be regarded as antecedent or not (Figure 18.31). The Arun rises at a height of about 6706 m, and after dropping to about 4267 m it becomes a braided river flowing eastwards through a valley flanked with terraces of gravel. At about 3962 m it abruptly turns and plunges into the heart of the mountains, first through the relatively short Yo Ri gorge and then through a much longer one, between the great mountain groups of Everest and Kanchenjunga, which carries it down to about 1219 m.

Some Indian geologists think that the gorges have been cut by headward erosion, thus enabling the Lower Arun to capture the Tibetan streams in the rear, which are now its headwaters. A good deal of river capture must inevitably have been involved, but there are serious objections to the

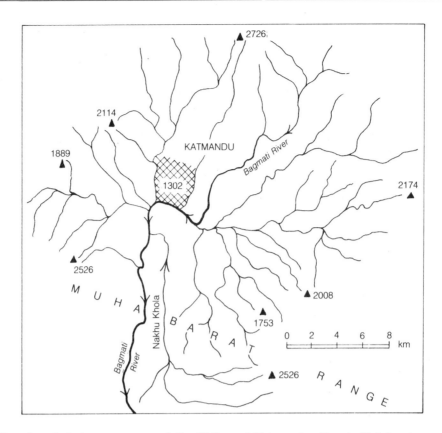

Figure 18.32 Centripetal drainage pattern of the Valley of Katmandu, Nepal. Heights in metres. (*After H.L. Chhibber, 1934*)

headward-erosion interpretation of the gorges. If the Lower Arun had been able to cut back through the crest at its highest point, one might reasonably expect several other Himalayan rivers to have been equally successful where the crest line was not so high. In 1937 L. R. Wager discussed the problem after exploring the critical region during the Everest expedition of 1933. He found that if the Arun had extended itself through the Yo Ri region by headward erosion it could most easily have done so by working along a belt of weak schists. In fact the Yo Ri gorge has walls of highly resistant gneiss. The river has deeply entrenched itself in the hardest rocks of the district and it could have done this only by accidentally coming upon these rocks from above, at a time when its downstream course had already been firmly established. For this and other reasons Wager concluded that the course of the Arun was established long ago on a 'mountain slope which formed the descent from the Tibetan Plateau to the Ganges plain, in those days perhaps an arm of the sea. No such surface exists nowadays because since that time the range of the Himalaya has risen across the middle part of the river's course.' The mountains are still rising against the downcutting Arun at rates of up to 10 mm each year.

On both sides of the valley to the north of the

Great Himalaya the gravel terraces slope up towards the mountains. It has not yet been recorded whether the higher terraces slope more steeply than the lower ones. However, such is the case on the southern side of the mountains, where the upper terraces flanking the Ganges alluvium have been tilted up towards the foothills (the Siwaliks) slightly more than the lower ones. That movements are still in progress is indicated by the occurrence of earthquakes.

Like the Colorado River the Arun has not continuously maintained its course against the rising ground. Locally and at times it has been what Hunt would call an anteposed river. About halfway between the great gorge and the Ganges plain the Arun has cut a channel over 150 m deep through a broad plateau of sediments. These appear to be the infilling of a lake formed at a time when the uparching of one of the Middle Himalaya branches was sufficiently active to pond back the river until it could spill over and resume its downward erosion.

The 'Valley' of Katmandu (the capital of Nepal) has had a similar story. The valley is floored with deltaic lake sediments which everywhere slope inwards towards the city, thus forming a kind of amphitheatre surrounded by spurs of the Middle Himalaya (see Figure 18.32 for heights) and, on the

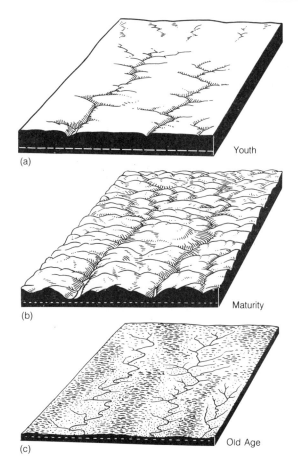

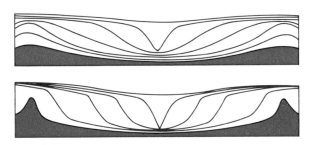

Figure 18.34 Contrasted sequences of valley profiles from youth to old age. *Above:* according to the wearing-down interpretation; *Below:* according to the wearing-back interpretation.

Figure 18.33 Diagram illustrating the three main stages in the wearing down of an uplifted surface according to the Davisian interpretation of the 'normal' cycle of erosion: *Youth*, while parts of the initial surface survive; *Maturity*, after most or all of the initial surface has vanished and the landscape is mainly slopes, apart from valley floors; *Old Age*, when the landscape becomes subdued and gently undulating, rising only to residual hills representing the divides between adjoining drainage basins. Eventually such hills are worn down and the region becomes a peneplain. (*After V. C. Finch and G. T. Trewartha, 1936*)

south, by the Mahabharat range. There is only one exit for the Bagmati River, which drains the amphitheatre, and that is by way of a deep gorge through the Mahabharat, cut after the latter had risen to form a natural dam. Apart from basins of purely internal drainage, the Upper Bagmati and its tributaries make what must be the world's most remarkable example of centripetal drainage. Along the line of exit the Bagmati flows southwards towards the Ganges, while less than 2 km to the east the Nakhu Khola flows due north, to join the Bagmati just before reaching Katmandu.

Wager suggested that the great peaks of the Himalaya owe part of their great elevation to the effects of isostatic readjustment. As the deep valleys have been eroded, an immense load of material is removed. This loss of material must be compensated by an inflow at depth. If this were not so, the crust would thin and eventually be eroded away. In fact, the crust thickens considerably below the mountains. Inflow at depth gives isostatic uplift which serves to greatly prolong the life of the mountain belt, as deep erosion is always partly offset by renewed buoyancy. Isostatic re-adjustment operates, however, in conjunction with tectonic uplift; the exceptional heights of the Everest and Kachenjunga peaks, and of the Kara-koram giants are basically a reflection of the stupendous tectonic forces generated along the boundary of the colliding Indian and Eurasian plates.

18.10 RELIEF EVOLUTION AND CYCLES OF EROSION

In many parts of the world and especially in cratonic areas the relief is dominated by plains cut across varied rocks and structures. The ways in which land surfaces become planate have intrigued theoretical geomorphologists for decades and the debate about surface development is almost inextricably bound up with questions of long-term slope evolution. A highly influential model of relief evolution and eventual planation was introduced to geomorphology early in this century by W. M. Davis. Basically, this model has as its starting point the time when sculpturing of a newly emerged upland begins; rock-waste as well as water starts to stream away from every part of it in turn. As time passes, the rivers, valleys and slopes pass through a sequence of distinctive changes proceeding always towards reduction of slope angles, provided that there has been no significant interference by earth movements, or by changes in sea-level or climate. Davis referred to the whole sequence of changes as a **cycle of erosion**. By analogy with the divisions of a lifetime he divided his evolutionary series into three main stages, metaphorically described as **youth**, **maturity** and **old age** (Figure 18.33).

Figure 18.35 Ayers Rock, 320 km WSW of Alice Springs in the middle of Australia; an inselberg 2.6 km long of nearly vertical Precambrian strata rising in impressive isolation 335 m above the surrounding pediplain. Mt Olga (left background) is 26 km to the west of Ayers Rock. (*Australian National Travel Association*)

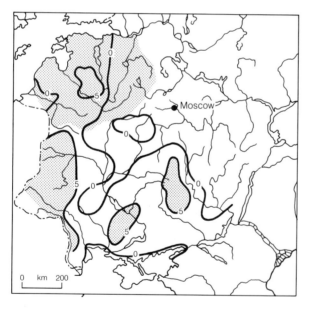

Figure 18.36 Rates of recent vertical movements (in mm/year or kB) in the western Russia. The stippled zone is undergoing subsidence. (*After V. A. Mattskova, 1967*)

Davis considered that as valleys are widened, the bordering slopes tend to become less steep during their retreat. During the stage of maturity, beginning when practically none of the original upland surface remains, the divides between adjoining valleys and drainage basins become rounded crests that gradually lose height. Finally, during a lengthy old age, the region is worn down and reduced to an undulating plain. Relief is faint, apart perhaps from an occasional isolated hill which owes its survival to the superior resistance of its rocks. Such residuals are sometimes called **monadnocks**, after Mt Monadnock in New Hampshire. The low-lying erosion surface which is the ultimate product of old age, Davis called a **peneplain** (L. *pene*, almost). This and the corresponding term **peneplanation** have thus come to be firmly associated with two leading ideas: the decline and flattening-out of hillside slopes during their retreat, and the accompanying down-wearing of divides and residual hills into forms presenting a broad and gentle convexity towards the sky, but merging with equally gentle concavities into the surrounding plain.

This empirical treatment of a highly complex subject did not pass unchallenged, despite its persuasive virtues as a method of description and easy teaching. Penck argued that most hillside slopes should not flatten out as they are worn back, once they have attained an angle that is stable for the type of rock or waste mantle concerned. Instead, he maintained, such slopes would be expected to recede without further change in

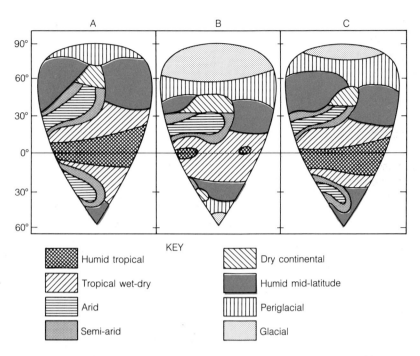

KEY

Humid tropical

Tropical wet-dry

Arid

Semi-arid

Dry continental

Humid mid-latitude

Periglacial

Glacial

Figure 18.37 Models of world climatic zones during the: A, middle Tertiary, B, Pleistocene glacial stages, C, Pleistocene interglacial stages. (*After K. W. Butzer, 1976*)

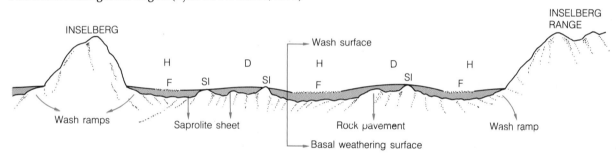

(a) Typical N–S cross section through the Tamilnad Plain (east side of the Deccans), as an example of an active etchplain. H – Wash depressions; D – Wash divides; SI – Shield inselbergs; F – Fine sands in the rainy season riverbed.

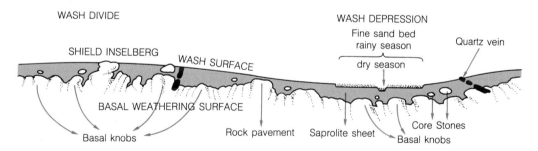

(b) Wash divide and wash depression (detail out of figure above).

Figure 18.38 Formation of etchplains. (*After J. Büdel, 1982*)

declivity, i.e. each one should retreat parallel to itself (Figure 18.34). In a notable contribution in 1930 Davis agreed that this was so for the rocky and stony slopes of arid and semi-arid regions, but not for the forested and soil-mantled slopes of humid regions like New England and western Europe which he had previously been mainly considering.

Thus it came about that a 'cycle of arid erosion' was recognized as well as the original one, which was given a misleading prominence by being called the 'normal cycle of erosion'. The term **normal** in this context implied a standard of reference based on the landscapes now developing in the humid conditions of a temperate climate. It was agreed that in semi-arid regions a

Figure 18.39 The coastal erosion surface in Damaraland, Namibia, surmounted by inselbergs. (*Copyright – Landform Slides*)

gently inclined surface – called a **pediment** – was left in front of the major slopes as a result of their parallel recession. Each pediment led down to the nearest stream or depression from bordering escarpments or from steep-sided isolated hills. In the latter case, well exemplified by the kopjes and inselbergs of Africa and the extraordinary landscape of the heart of Australia (Figure 18.35), it appeared that neighbouring pediments, encroaching from different directions, had coalesced to form a much more extensive **pediplain**.

Architecturally a pediment is a triangular feature crowning a portico of columns in front of a building in the Grecian style. In this sense the term is not an appropriate one for a gently sloping surface leading up to a scarp or butte or inselberg. It is worth remembering that the term in its geomorphological usage could equally well have been derived from the Greek *pedion*, which means flat open country or a piedmont.

The Davisian idea that an uplifted landscape should evolve through a sequence of set stages through time has been highly influential in the teaching of geomorphology due to its simplicity and its apparent worldwide applicability. However, the Davisian concept of cycles of erosion has been severely criticized by researchers.

First, there is now general agreement that the Davisian scheme can no longer be properly described as 'normal'. The landscapes on which it was based are mostly hybrids, or 'fossil' varieties not yet brought up to date. Areas like New England and the British Isles passed through great fluctuations of climate during the Pleistocene. Most of their landscapes are products, not of the temperate conditions regarded as 'normal', but of repeated successions of glacial, periglacial and temperate-humid conditions.

Secondly, present-day landscapes have also been strongly affected by the many changes in base-level that have occurred since the beginning of the Pleistocene. Global sea-levels have fallen by over 100 m during the many glacial stages of the Pleistocene and then risen abruptly to near current levels in interglacials. Consequently, the lower courses of rivers have been repeatedly rejuvenated during periods of low sea-level and their capacity to erode and transport temporarily enhanced and alternately drowned as sea-level has risen again. Similarly, the continuation of significant uplift or subsidence of the land during the Pleistocene is increasingly recognized, even in areas previously thought to be stable, such as the Russian Platform (Figure 18.36). Davis recognized that his cycle could be interrupted by climatic change and fluctuations in base-level but regarded these events as brief and transitory whereas, in fact, change in climate, sea-level and land-level has been continuous over the last few million years of Earth history.

Outside tectonically-active areas many land-

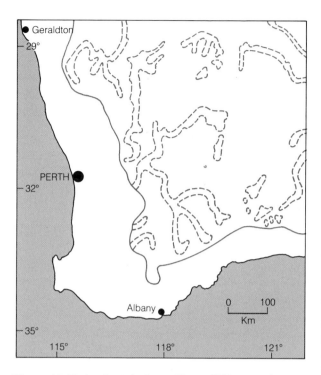

Figure 18.40 Ancient drainage lines of Western Australia. The headwaters of the broad drainage lines appear to have originated to the south before continental drift brought about the separation of Australia and Antarctica. (*After C. D. Ollier, 1981*)

scapes date from before the Pleistocene and developed under very different climates from those which prevail at present (Figure 18.37). Fossilized forests of warmth-loving trees occur on Axel Heiburg Island in the Canadian Arctic and show that the Early Tertiary climate of this area was not only frost-free but sub-tropical. Tertiary tropical deep weathering profiles also occur widely in areas not suited to their formation today, such as in tundra areas of Iceland and the Kola Peninsula, near Murmansk in Russia, in the glaciated terrain of Sweden (Plate Xa) and in the deserts of North America and Australia. Climatic change, rather than stability, has therefore been the norm throughout recent geological time.

The shortcomings of the Davisian cycle have led to a search for new unifying models of long-term landscape development ever since its publication. An American geologist, Hack, developed an alternative – the principle of dynamic equilibrium – partly because of his inability to recognize peneplains or cyclic surfaces in the Appalachians. Differences in slopes and landforms from place to place are explained by differences in bedrock or in processes acting on bedrock, rather than by evolutionary change. Any changes that do take place through time are interpreted as the result of variations in external factors, such as climate or tectonics, or of changes in bedrock character as land

surfaces are lowered by erosion. The concept of dynamic equilibrium is undoubtedly useful in investigations of landscape change in the short- and medium-term, such as the adjustment of river channels to variations in discharge or sediment load, and serves to emphasize the close links which often exist between landforms and geology. The lack of any evolutionary element in the model, however, makes it less useful in the longer term; the landscape cannot remain in a steady state if relief is progressively reduced by erosion through time.

A striking feature of both the Davis and Penck cycles is the lack of attention given to rock weathering and the attendant lowering of rock surfaces. This partly reflects the development of these models for temperate and arid areas, where chemical weathering was thought to be limited. Julius Büdel, a German geomorphologist, had a quite different view based on his experience in the seasonal tropics; denudation required prior deep decomposition of bedrock. Stream incision in such areas is hampered by the lack of available abrasive materials. Hence denudation is dominated by visible wash processes and the more important, but invisible, processes of solution and decomposition. The lack of incision allows the formation of plains which terminate abruptly against escarpments and steep-sided inselbergs. These **etchplains** show a surface of low relief developed across deeply weathered terrain (Figure 18.38). Beneath the surface, weathering effectively etches out differences in bedrock resistance to alteration. An irregular basal surface of weathering forms, with basins of weathering developing over weak or highly-fractured rocks. Only thin saprolites form over resistant rocks. In the rainy season the clayey saprolite is removed by highly effective sheetwash. Over time, slopes are reduced in angle over weathered rocks but zones of resistant rock emerge as inselbergs. These rock slopes shed water easily and weather only slowly once exposed. Gradually, the basal surface of weathering is lowered by the relentless penetration of alteration and the corresponding upper wash surface is lowered with it, conserving the original subdued surface form. At the same time, the inselbergs grow in height and may eventually tower over the adjacent plain. Büdel was able to extend his ideas to the present temperate and arid zones because of the widespread evidence of inheritance of major relief forms from phases of the Tertiary period when humid tropical conditions prevailed in these areas.

18.11 PLANATION SURFACES

Planation surfaces may exist as monotonous flats

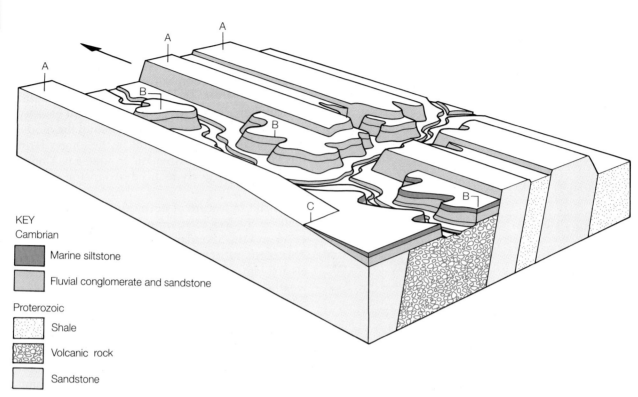

KEY
Cambrian

■ Marine siltstone

▨ Fluvial conglomerate and sandstone

Proterozoic

▨ Shale

▨ Volcanic rock

□ Sandstone

Figure 18.41 Diagrammatic representation of remnants of the Ashburton (Australian) erosion surface (A) with flat-topped terraces and mesas in the valleys (B). Cambrian strata rest on the Ashburton surface at (C).

but often gentle relief is punctuated by spectacular inselbergs and the skyline may show fragments of older plains and residual hill masses at higher levels (Figure 18.39). Planation surfaces are also found not only at the present land surface but also beneath younger cover rocks. Uplifted, tilted and warped planation surfaces have been recognized worldwide in various stages of dissection and destruction, although sometimes the evidence for their original planate character is dubious.

Planation surfaces are never perfectly flat but rather consist of shallow valleys and broad, gentle interfluves with or without residual hills. As they are products of erosion, such surfaces are often difficult to date. Locally, however, planation surfaces may be overlain by sediments and deep weathering profiles which provide a minimum age. Overlying lava flows may also be dated. A final possibility is that the age of the drainage systems on the planation surface can be established (Figure 18.40).

The theoretical models of Davis, Penck and Büdel provide three contrasting methods of planation. The differences between them are important, particularly in terms of dating the surfaces. On a peneplain all parts of the surface are of common age; on a pediplain the surface becomes younger towards the base of the retreating escarpment; on an etchplain the original planate surface is maintained during landscape lowering so that the present planation surface may have been initiated at a higher level long ago. Unfortunately, it is often difficult to apply this genetic classification to actual examples of planation surfaces. This is partly because that the end-product of all three models is, of course, a nearly-flat terrain with little evidence of its origin. Still more problematic, however, is the way which many landscapes show elements of relief development which correspond to all three of the basic theoretical models, in the same way that the various modes of slope evolution can take place within the same area at different times. This is partly a reflection of the major changes in global climate through the Tertiary, when many extensive planation surfaces were forming. Thus etchplain escarpments may retreat if they are formed of strong cap rocks and wash surfaces at first are reduced in angle in the Davisian manner. In Davisian landscapes, weathering must also lower land surfaces through time as slopes decline. Deep weathering profiles formed under humid conditions also survive in many present day arid areas and the ease with which these materials can be eroded, once exposed, will aid slope retreat, as well as the lowering of the land surfaces. The student of landscape evolution must therefore look carefully

Figure 18.42 The Great Escarpment of the Natal Drakensberg, showing gullies in the Jurassic basaltic lavas (over 1200 metres thick) on the precipitous walls leading up to the 3000 metre plateau of Lesotho. (*South African Air Force*)

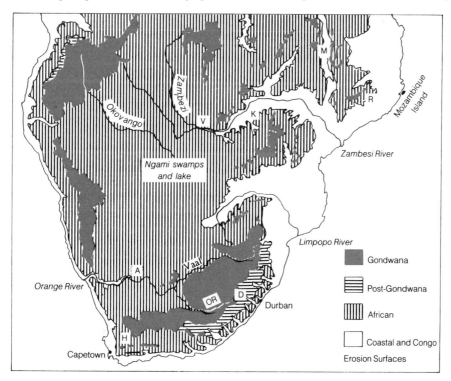

Figure 18.43 Erosion surfaces in southern Africa. A, Aughrabies Falls (Fig. 18.46); D, Drakensberg (Fig. 18.42); K, Kariba Dam; M, Lake Malawi; OR, Orange River headwaters; R, Ribaue Peaks (Fig. 18.49); H, Hex River (Fig. 18.47); V, Victoria Falls (Figs. 17.18 and 17.19).

at the available evidence before piecing together the history of long-term landform development in any region.

18.12 ANCIENT DRAINAGE AND LAND SURFACES IN AUSTRALIA

In areas of rapid uplift and denudation, like the Himalaya, the slopes and peaks seen today have been sculpted almost entirely during the Quaternary period. In stark contrast, the plains and tablelands which dominate the relief of Australia have seen little change during the Quaternary and date back to Tertiary times and earlier. This fundamentally slower pace of landscape change is partly a reflection of the long-term tectonic stability of the Australian craton, large areas of which have remained close to sea-level since the Permian. The preservation of ancient landforms is also a result of the steady drying out of the continental interior since the Early Tertiary, which greatly reduced the effectiveness of the river systems to transport and erode material.

That planation surfaces in many parts of Australia are very old is shown by the Tertiary deposits which rest on them. The plain which surrounds Ayers Rock and the Olgas in central Australia (Figure 18.35) carries a broad depression representing part of a former valley system on the plain which is filled with lignites of late Cretaceous and Tertiary age. In Western Australia drainage systems on an old uplifted planation surface are marked by chains of salt lakes (Figure 18.40). In the south the present watershed between the northward-flowing drainage and the shorter rivers flowing to the south lies at the margin of the uplifted block. Yet at the watershed some of the ancient valleys are 2 to 3 km wide, indicating that the river systems have been beheaded by loss of land to the south. These valleys appear to be have formed before the separation of Australia from Antarctica about 55 million years ago. The deeply weathered and lateritized plateau of western Australia is therefore Mesozoic in age.

The title of oldest landform in the world is difficult to award. Many exhumed ancient land surfaces are known. An Early Tertiary sub-basaltic surface from Northern Ireland is shown in Figure 3.11. In north-west Scotland, hilly landscapes developed in Lewisian gneisses which are buried by Torridonian (late Precambrian) sandstones. In Scandinavia the sub-Cambrian plain forms an extremely flat bedrock surface with a relative relief of less than 20 m which extends over many tens of kilometres. Yet these features have been re-exposed quite recently and lay dormant

beneath cover rocks for long periods. Ancient continuously-exposed landforms are more difficult to find and, once again, the very oldest features come from Australia. In north-central Australia, the highest topographic surface, the Ashburton surface, is dissected by steep-sided valleys up to 100 m deep (Figure 18.41) which contain flat-topped terraces of conglomerate and sandstone. These terraces are remnants of a once extensive sheet of fluvial deposits and are overlain to the south-east by marine deposits which contain Middle Cambrian fossils. The uncovered terraces must therefore be Middle Cambrian in age or older and the higher Ashburton Surface must be older still. Both features appear to have existed as subaerial landforms for much of the last 500 million years. A yet older age has been claimed for fragments of planation surfaces preserved as bevelled cuestas on the Kimberley Plateau in north-western Australia. These flats carry striations from a Precambrian glaciation of the area, about 700 million years ago, and vales between the cuestas contain glacial sediments of similar age. As there is no evidence that the Kimberley Plateau has been covered by deposits younger than the Proterozoic tillites then the tops of the cuestas have been exposed to the elements since that time.

18.13 EROSION SURFACES OF SOUTHERN AFRICA

Vast regions of the southern African craton are dominated by planation surfaces which rise in a series of great steps away from the marginal swells at the coasts of the Atlantic and Indian Oceans (Figure 18.42). They have been made familiar by the writings of Frank Dixey and Lester King, both great geological explorers of the continent in which most of their work has been done. King insisted that such polycyclic landscapes are incompatible with peneplanation, but are a natural consequence of scarp retreat. Almost flat remnants of old erosion surfaces dating from the Jurassic have been preserved almost intact for something like 140 million years, not merely because their bedrocks are highly resistant types or are strongly protected with an armour of laterite, but essentially because they are vulnerable to denudation only from their flanks, downward erosion having been negligible.

At the close of the Triassic the long period of accumulation of sediment in the Karroo basins and troughs was brought to an end by the intrusion of innumerable sills and dykes, and the outpouring of floods of plateau basalts, which continued into the early Jurassic. The basalts form the boldest crags of the Drakensberg (Figure 18.42) and cover large areas around the Victoria Falls (Figure 17.18). Ever since this 'blaze of volcanic fury' the geological history of much of Africa has been characterized by a series of intermittent uplifts on a continental scale. There have been great depressions, too: making room for over 3000 m of Cretaceous sediments in the Mozambique Channel and over its shores, and for over 12 000 m – also marine Cretaceous – in southern Nigeria. But here we are concerned only with the main upheavals and the cycles of erosion that each one initiated.

Sequences of five well-defined erosion surfaces have been recognized in Africa from Madagascar to Angola and Zaïre, and from Uganda to the Cape (Figure 18.43). Detailed work in limited areas has already shown that some of the 'standard' surfaces are composite and that all of them have local features corresponding to minor pulses and warpings, to say nothing of the major interruptions by rift valleys and such giant uplifts as Ruwenzori. Nevertheless the five main surfaces serve well as a provisional basis for discussion. The main difficulty in correlating erosion surfaces over vast distances lies in determining the geological ages or events to which they should be referred. A

Main erosion surfaces	Sequence of main geological events
Gondwana surface	Outpouring of Lower Jurassic plateau basalts Erosion during and since Jurassic

Post-Gondwana surface	Break-up of Gondwanaland: pronounced faulting; down-warping of coastal regions
	    Erosion from early or middle Cretaceous
	    (according to locality)

African surface	Middle to late Cretaceous uplift
	    Erosion from late Cretaceous or early Tertiary
	    (according to locality)

Coastal Plain surface also called Victoria Falls surface	Middle Tertiary uplift Erosion during and since Miocene
	End-Pliocene to recent uplifts (including fluctuations due to oscillating sea-levels)
Zaïre surfaces	Erosion from Plio-Pleistocene to present

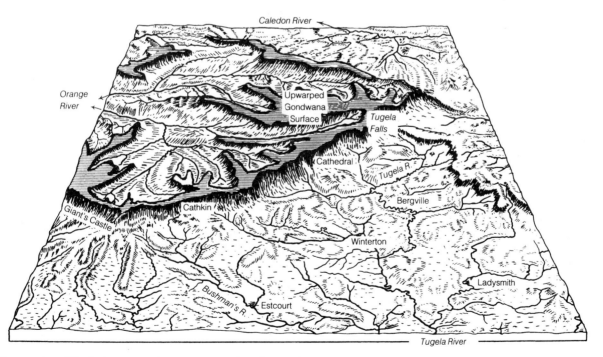

Figure 18.44 Block diagram of western Natal and Lesotho (Basutoland), looking a little south of west showing remnants of the upwarped Gondwana erosion surface above the Great Escarpment. (*After Lester C. King, 1967*)

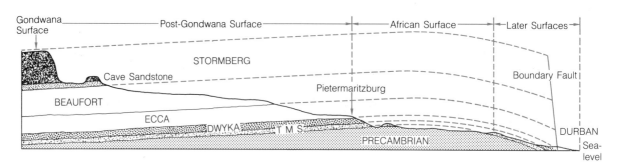

Figure 18.45 Generalized section from the Drakensberg on the west to the coast of Durban on the east, to illustrate the recession of the Great Escarpment since Jurassic time. T.M.S.=Table Mountain Sandstone (late Silurian or early Devonian), resting uncomformably on Precambrian rocks and covered unconformably by late Carboniferous/ Permian tillites and other strata known as the Dwyka Series. Late Karroo sills and dykes, and minor faults and flexures omitted. Length of section 200 km. (*After Lester C. King, 1967*)

Figure 18.46 Aerial view of the Aughrabies Falls and Gorge of the Orange River, South Africa. Here the nearly flat African erosion surface is deeply incised by the Coastal erosion surface, which extends inland as the Falls slowly recede. (*South African Air Force*)

Figure 18.47 Aerial view across the Hex River Valley and Mountains, NE of Cape Town (see Fig. 18.43). The summit level in the background represents a remnant of the Gondwana erosion surface, now largely isolated by the advance into the mountains of the African erosion surface (the valley flats). (*Aircraft Operating Company of Africa*)

surface begins at the coast in Cretaceous times, let us say, and extends itself half-way across Africa, to end at the foot of a scarp or of a waterfall if it has been progressing up a river. There, where it is still working inland, its age is the present. But if it can be traced back towards the coast, it may possibly be found bevelling middle Cretaceous sediments or lavas; and nearer the sea it may be seen emerging from an unconformable cover of, say, Eocene beds. This evidence would enable us to date that spread of the surface as approximately late Cretaceous. In all such cases it is clearly desirable to date a surface from the earliest time when some part of it first became really distinctive. So far as evidence is available, dating on these lines is generally found to be consistent with the geological age of the main uplift responsible for initiating the erosion cycle concerned.

The five surfaces referred to are known by the following names, the uppermost being the oldest:

The **Gondwana** surface is ancient, probably representing fragments of the plains of **Gondwanaland** (Section 9.5) which existed before break up of the super continent in the late Mesozic.

Now much dissected by the headwaters of the Orange River (Figure 18.44) it rises to its greatest height (3292 m) in the plateau of Lesotho, which is separated from Natal by the great castellated wall of the Drakensberg (Figure 18.42). This basalt-capped scarp is the most spectacular part of the Great Escarpment, which faces the stepped landscape leading down to the coast with but little interruption here and there, from the Limpopo southwards and westwards towards the Cape and then northwards across the Orange River into Damaraland and Angola (Figure 18.43). The Drakensberg, which now forms the high western boundary of Natal, affords a magnificent example of scarp recession (Figure 18.45). It began in a small way near the present coast about 140 million years ago and during that time it has been intermittently uplifted and warped, probably to a considerable extent by the isostatic response to unloading by denudation. At the coast a thickness of over 3900 m of rock has been removed, corresponding very nearly to an average rate of 1 cm in 350 years (28 B); but over the whole of Natal – from the coast to the escarpment – this figure is reduced to 1 cm in about 700 years (14 B). The average rate of scarp recession has been 225 km in 140 million years, or 1 cm in 6 years (1700 B). However, a

Figure 18.48 The Sugar Loaf, Rio de Janeiro, Brazil, a bornhardt that has been partly submerged. (*Dorien Leigh Ltd*)

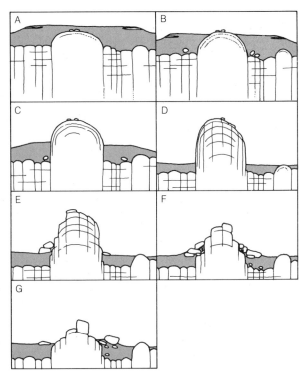

Figure 18.50 The development and decay of a tropical bornhardt. In diagrams A–D, both the landsurface and the basal surface of weathering are lowered around the emerging dome. At stage A, the dome already shows sheet structure and first appears as a low, base rock surface or **ruware** in stage B. Unloading and joint opening in stages D–G, result in the disintegration of the dome to a tower or **kopje**. (*After M. F. Thomas, 1974*)

Figure 18.49 Embayment pediment rising from the African pediplain through the Ribaue group of inselberg peaks, Mozambique. A. Holmes is sitting on a pile of corestones. In the background, is East Peak, over 300 metres above the pediment. (*R. L. Reid, 1911*)

glance at Figure 18.45 shows that this estimate covers an enormous range from place to place, and the rate must also have greatly varied from time to time.

On the plateau side of the escarpment the headwaters of the Vaal and Orange rivers begin their long journey to the Atlantic, while the shorter rivers of Natal begin as mere trickles which descend the escarpment. Not far from its source the Tugela River plunges over the brink of a precipitous amphitheatre in the escarpment. Five clear leaps with intervening cascades make up a total drop of 856 m. Though of insignificant volume, the Tugela Falls are amongst the highest in the world.

The **Post-Gondwana** surface is in many places not yet clearly distinguished from those above and below it. However, it is well established in Natal and the Cape Province (Figure 18.43), and also in Angola, where there is an equally well-developed series of steps rising inland from Benguela. It is probable that the Zaïre and Kalahari Basins began as downwarped depressions of this surface.

The **African** surface, as its name suggests, forms the most extensive plateaus of southern and eastern Africa. Figures 18.39 and 18.46 both show the surface admirably: one at the Victoria Falls, where the Zambezi makes its abrupt descent into the gorge which it has cut deeply back through the African surface; the other at the Aughrabies Falls, where the Orange River has a much more difficult

Figure 18.51 A kopje in northern Nigeria, formed by the collapse of a bornhardt. (*M. F. Thomas, 1974*)

task in sawing through the tough and compact granites and migmatites of a grim and desolate semi-arid land. Along the borders of Uganda and Kenya the African surface supplied the foundation for Elgon and other great volcanic cones. Here an additional surface has been recognized, a good 300 m below the other, and it is a gentle downwarp of this lower surface that is occupied by the shallow Lake Victoria. Rift-valley faulting, however, has brought about changes in level amounting to several hundred metres. One of the 'African' surfaces has been uplifted far above the snow line to form the glaciated summits of Ruwenzori. And one has been depressed far below sea-level, to form the bedrock floor of Lake Tanganyika. Some of the rift-valley fault scarps have been notched by gorges with waterfalls at their heads. Near the south-east end of the Tanganyika rift, for example, the Kalambo Falls have cut back 5 km from the lake, and there a stream flowing placidly across the African surface suddenly takes an unbroken leap of 215 m over the brink. On the western side of the rift the lake finds an outlet into the Congo by way of the Lukuga River, which carries the Coastal surface into the very heart of the continent, where it is represented by the lake surface and its narrow shores. Lake Malawi (Nyasa) is similarly

rimmed by the Coastal surface, which slopes down the valley floor of the Shire River towards the Zambezi. The African surface is itself a broad valley floor in some places, where it is bordered by uplands whose summits represent a higher surface, generally the Gondwana (Figure 18.47).

The **Coastal** surface, already mentioned above, surrounds the basin of the Zaïre. It makes deep inroads into the continent up the Zambezi as far as the Victoria Falls; and up the Limpopo where, in striking contrast to the Zambezi gorges, the surface is wide and monotonous and has all the characteristics of extreme old age. The Limpopo only begins to be vigorous near the Messina copper mines, where it descends by way of a long stretch of cataracts to the Zaïre surfaces which correspond to the real coastal plains. Around south and south-west Africa the coastal plains include the Zaïre and Coastal surfaces, but the latter are too narrow to be separately distinguished on a small-scale map.

The **Zaïre** surface, as its name indicates, covers a vast extent of a country in the basin of the Zaïre. It also makes a broad belt inland from the shores of Mozambique, particularly south of the Zambezi, up which river it has progressed as far as the recently dammed Kariba Gorge. The Zaïre

Figure 18.52 Granite domes in the glaciated terrain of the Yosemite Valley, California. (*R. Cornish, 1983*)

'surface' includes a group of minor surfaces, many of which are represented by river terraces of the kinds already well-illustrated from northern lands.

18.14 INSELBERG LANDSCAPES

The inselberg (German: island mountain) type of landscape was first described in 1900 by W. Bornhardt, one of the early explorers of Tanzania. His name has since become associated with a specific type of domed inselberg, the **bornhardt**. In 1904, after seeing the equally remarkable inselbergs of parts of the Kalahari and Namibia (South-West Africa), S. Passarge convinced himself and many others that they were the typical residual hills of desert regions where resistant crystalline rocks had long been exposed. However, since it is now known that inselbergs and pediments also occur in humid regions, and that the Kalahari and some other deserts have been humid regions in the past, the Passarge hypothesis has become untenable. In 1911 the inselberg landscape was described by J. D. Falconer from Nigeria; and in the same year A. Holmes and his fellow explorers E. J. Wayland and

D. A. Wray independently discovered it – with growing astonishment – in Mozambique. Here is how two of them afterwards described it:

'Proceeding westwards from the coast, the country begins to be diversified with isolated peaks or clusters of hills which rise abruptly from the surface of the plateau and exhibit the most remarkable outlines. Within 30 miles of the coast these island mountains are not numerous, and only rise a few hundred feet above the surrounding country. Further inland, however, they gradually increase in numbers and bulk, until, assembled together in picturesque chains or hurled up, peak upon peak, in towering blocks of rugged gneiss, they cooperate to form fantastic and impressive mountain systems, which arrest the attention by the colossal scale of their daring architecture. The summits of the peaks are of every conceivable shape, varying from gracefully rounded domes of smooth and naked gneiss to irregular knobs and pinnacles. In general, the outlines approximate to smooth curves above, dropping precipitously to the plateau surface below. . . . The most striking feature of these detached groups of hills lies in

Figure 18.53 Inselbergs of position and surrounding pediplain, developed in horizontal sandstones and other sediments, Serovenout, south-east Algeria. (*Copyright – Landform Slides*)

the abruptness of their discontinuity with the plateau.'

(A. Holmes and D. A. Wray, 1913, *The Geographical Journal*, London, vol. 42, pp. 143–52).

Wayland's sketch (Figure 15.5), and Figure 18.48 give a better idea of these fantastic peaks than any description in words. Occasionally summits are seen that approximate to angular or rectangular knobs, with only slightly rounded edges. These result from the fall of rock masses, liberated by fracture along foliation planes or the opening-up of joints, too recently for exfoliation to have restored the normal convexities. At that time (1911) all had to admit that the origin of this type of scenery was not properly understood. But there was general agreement that inselbergs were best developed from tough and massive crystalline rocks, offering more than average resistance to chemical weathering, and so capable of standing up to considerable heights with precipitous sides; many gneisses and granitic rocks fulfil these conditions. In Mozambique it was noticed that on the flanks of some of the inselbergs the gneissic foliation dips radially outwards with increasing angles from the crest, in places being roughly parallel to the surface. This domical structure may continue

(a)

(b)

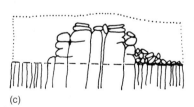
(c)

Figure 18.54 Stages in the evolution of a granite tor. (a) Vertical section of fresh granite with varied spacing of joints. Joint spacing will also vary in the horizontal dimension. (b) After a period of chemical weathering. Decomposed rock is shaded black. Note the irregularity of the rockhead profile or basal surface of weathering. (c) Decomposed rock stripped leaving a tor rising abruptly from the surrounding surface.

Figure 18.55 Staple Tor, Dartmoor, south-west England. A small tor produced by frost weathering, surrounded by a blockfield. (*A. M. Hall*)

Figure 18.56 Small tors and corestones, Katsina, northern Nigeria. (*Dorien Leigh Ltd*)

for some distance over the surrounding pediment, where the dips are steep. In some examples, e.g. those of Figure 15.5, the inselbergs consist of granitic rocks around the crest or farther out,

Figure 18.57 The Barns of Bynack, Cairngorm Mountains, Scotland. A large tor formed on a monolithic granite intrusion, probably initially by differential chemical weathering and erosion with later modification by frost and ice action. (*A. M. Hall*)

passing into migmatite and gneiss on the slopes or on the adjoining pediment. But others are completely independent of this dome-like or 'mantled

dome' structure, the banding of the gneiss being gently inclined and sharply truncated by the steep sides, as seen in Figure 18.49.

Tropical bornhardts are an integral part of many etchplains (Figure 18.38) and their origins have been investigated by Thomas (Figure 18.50). Two elements appear crucial: the deep and rapid penetration of weathering found in humid tropical regions and the contrasting joint density, and hence permeability and resistance to groundwater attack, between a massive rock core forming the bornhardt and the more fractured rock which surrounds it. Over time these contrasts are emphasized as water drains from the emerging rock dome to add to the flow of groundwater through the surrounding regolith. Stripping of the regolith by downwasting or slope retreat leads to an increase in the height of the bornhardt. Differential weathering and stripping over time scales extending to millions of years allow the development of domes hundreds of metres high (Figures 15.6 and 18.48). Eventually, however, opening of joints in the unsupported upper part of the dome will initiate collapse (Figure 18.51).

The first stages of dome formation are illustrated in Plate Xa where removal of kaolin-rich

Figure 18.58 The first stages of tor formation, showing on the left, relatively fresh granite corestones in a zone of well-spaced joints, and on the right, granite weathered to a sandy saprolite in a zone of high-joint density. Two Bridges Quarry, Dartmoor, south-west England. (*A. M. Hall*)

saprolite during quarrying has revealed a hidden gneissic bornhardt. A remarkable feature is the presence of well-developed sheet joints on the surface of the dome even whilst it remained covered by regolith below the ground surface. The example matches many others found in deep cuttings in weathering profiles in the tropics but this dome rests within a saprolite of Mesozoic age and beneath a thin cover of Late Cretaceous sediments on an ice-moulded island in a lake in southern Sweden! The Ivo Klack dome provides an excellent example of the way in which old landforms can be exhumed to provide important inherited elements in the landscape.

Most bornhardts are developed in granite and gneiss, although some, like Mt Olga (Figure 18.35), are formed in massive sandstones. Structural control, specifically, the presence of massive rock cores, is paramount. This accounts for the sporadic occurrence of bornhardts in areas remote from the tropics and where there are no signs of former deep weathering profiles. For example, classic granite bornhardts, with well-developed sheet jointing, dominate the glacial trough of the Yosemite valley in California (Figure 18.52).

Bornhardts are a type of **inselberg of resistance**, where the formation of an isolated hill is largely determined by the existence of a compartment of resistant rock. Other isolated hills develop as surrounding slopes are worn back. The **inselbergs of position** are related features of mesa and buttes produced by scarp retreat in near-horizontal strata. The upper slopes of these inselbergs of position represent the last fragments of old land surfaces, soon to be consumed by erosion (Figure 18.53).

The formation of **tors**, or small, rocky knolls or towers which rise abruptly from their surroundings, are, in some ways, analagous to that of bornhardts. Tors are most common in granite terrain and reflect little more than differential weathering and erosion of unevenly jointed bedrock. In tropical and temperate humid regions, weathering attacks most rapidly the zones of high permeability defined by closely-spaced joint patterns and leads to the development of an irregular basal surface of weathering (Figure 18.54). Stripping of the saprolite then reveals as tors the more

massive and resistant rock zones, which have remained least weathered. Tors also occur widely in periglacial regions. Frost action is more effective where joints are closely spaced and, provided frost-shattered debris is removed by solifluction, zones of widely-spaced joints will gradually emerge as tors. The angular, frost-fretted outline of periglacial tors, and the surrounding scatter of angular frost-riven blocks (Figure 18.55), has been contrasted with the more rounded outline of tropical tors and their corestones (Figure 18.56) but the granular disintegration due to frost and wind action under cold climates can give edge-rounding which makes this distinction unhelpful. In Britain tors occur widely on the granite areas of Dartmoor, in south-west England (Figure 18.55), and the Cairngorms (Figure 18.57), in the Grampian Highlands of Scotland and reflect both modes of origin. During humid temperate preglacial and interglacial periods, chemical weathering led to breakdown of granite into a sandy saprolite of which remnants are preserved in places like Two Bridges, Dartmoor (Figure 18.58). During the Pleistocene, these loose saprolites were removed by periglacial processes and even, in the Cairngorms, by glacial processes, to reveal tor forms. After exposure, these rock outcrops were further shaped by frost action. Many small tors, however, surrounded by blockfields (Figure 18.55), seem to be entirely the product of intense frost action during the many cold stages of the Pleistocene.

18.15 FURTHER READING

Büdel J. (1982) *Climatic Geomorphology*. Princeton University Press, Princeton, New Jersey.

Chorley R.J., Schumm S.A. and Sugden D.E. (1984) *Geomorphology*. Cambridge University Press, Cambridge.

Ollier C.D. (1981) *Tectonics and Landforms*. Longman, London.

Selby M.J. (1985) *Earth's Changing Surface*. Clarendon Press, Oxford.

Summerfield, M. (1991) *Global Geomorphology*. Longman, London.

Twidale C.R. (1976) *Analysis of Landforms*. Wiley, Sydney.

19

UNDERGROUND

WATER

All streams run to the sea, yet the sea never
overflows; back to the place from which the
streams ran they return to run again

Ecclesiastes 1:7 (Revised English Bible)

Most people have an instinctive awareness of the
existence of underground water: caves, springs,
marshy ground and oases are all familiar. Man has
made use of natural emissions of underground
water from ancient time; indeed, ancient civilizations such as the Babylonians, and later the Romans,
placed great emphasis on the supply of water for
their main cities. Many of their engineering structures, such as the underground qanats of Iran or the
Pont du Gard aqueduct at Nimes, can still be seen
and some qanats – horizontal collection tunnels – are
still in use.

19.1 THE OCCURRENCE OF WATER

Free water (i.e. that not bound up in minerals)
comprises only about 0.02 per cent of the Earth's
mass, but without it life could not exist. All
naturally occurring water is a constituent of the
hydrological cycle (Figure 19.1): it is involved in a
perpetual closed loop of movement between the
solid Earth, the seas and the atmosphere; this was
recognized long ago not only by the author of
Ecclesiastes but many other ancient texts. Sometimes, that movement can be swift – as in torren-

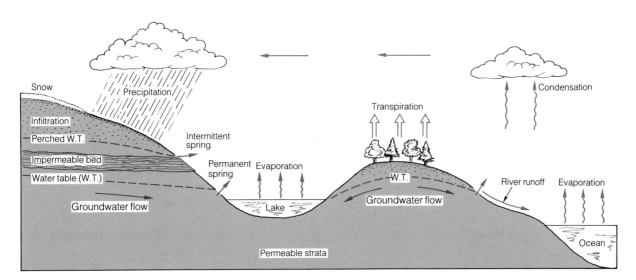

Figure 19.1 The hydrological cycle.

tial rain or a flowing river, but more often it is slow, sometimes imperceptibly slow, as in the case of water trapped in poorly permeable formations at great depths. For supply purposes, Man makes intensive use of a relatively small fraction of the hydrological cycle – surface water in streams, rivers and lakes together with underground water to a depth of about 400 m. Table 19.1 indicates the relative volumes of water on the Earth and emphasizes the importance of groundwater as a repository for terrestrial freshwater.

The study of water movement is known as **hydrology**, the study of underground waters as **hydrogeology**. The modern science of hydrogeology has evolved in several countries in parallel, but the USA has had a dominant impact on developing both the descriptive terminology and the investigative and scientific techniques. A number of terms are in common use which refer to the same parameter or descriptor; those used here are the terms most commonly used today.

Hydrogeology has an important role in many aspects of modern life. Not only can its practitioners produce substantial volumes of potable water from apparently solid rock, but an understanding of the science is essential in civil engineering, mining and minerals extraction, waste disposal and the related sciences of geophysics and geochemistry.

19.2 THE MOVEMENT OF WATER

In order for the hydrological cycle to function, water must change its state at various times. On the surface, or underground, water is usually liquid; but it may be frozen in snow and ice or vaporized in hot springs and geysers. Water vapour in the atmosphere is mainly the result of evaporation from free-water surfaces – the sea,

rivers, lakes etc – and via transpiration from plants.

Precipitation (rainfall plus melted snow and ice) usually infiltrates into the soil and bedrock, and may be discharged as **runoff** via surface watercourses. Water also moves more or less vertically downwards under gravity through shallow soil once the soil has reached its **field capacity** – that volume of water which is required to satisfy the absorbency and adsorbency demand of a unit volume of soil. Water sorbed by soil is available for uptake by plants or may be lost by evaporation. The net precipitation which remains to percolate into the bedrock is known as the **available recharge**.

At some depth, soil gives way to rock; the soil–rock discontinuity may induce lateral movement of infiltrating water into a water course, a more permeable rock stratum or a man-made feature such as a drainage channel. But the bulk of infiltration will continue downwards. Between the ground surface and the level of saturation, three distinct zones can usually be discerned (Figure 19.2):

1. the **unsaturated** or **phreatic** zone where pores and fissures are never completely filled, but through which water migrates: some water is retained in pore spaces
2. the **zone of intermittent saturation** which extends from the highest level reached by the upper surface of the saturated zone to the lowest level to which it falls in drought
3. the **saturated or vadose zone** where pores and fissures are permanently filled with water. The saturated zone extends to a depth – usually several hundred metres below ground level – where the overpressure of rock has closed fissures to the extent that water movement is minimal. Below that level, the rock is effectively dry.

Table 19.1 Quantity and distribution of global water (*After M.I. Lvovich, 1970*)

	Area × 10^6 km^2	Volume × 10^3 km^2	Proportion of total (%)
Oceans	360	1 370 323	93.93
Groundwater – total (involved in active water exchange)	–	64 000 (4 000)	4.39 (0.27)
Polar ice glaciers	16	24 000	1.65
Lakes	510	230	0.016
Rivers		1.2	0.0001
Soil Moisture	–	75	0.05
Atmospheric water	–	14	0.001

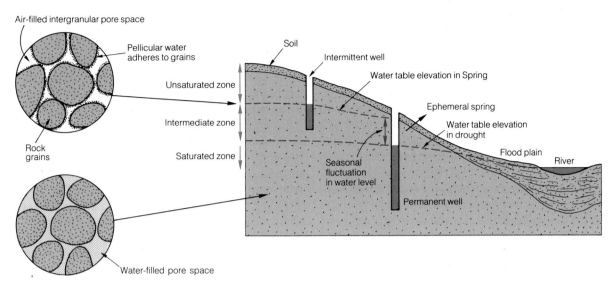

Figure 19.2 Zones of groundwater saturation.

The boundaries between these zones are rarely clear cut; for example, there is a **capillary fringe** above the saturated zone where pore spaces may be partly filled by water drawn upwards by capillary attraction.

Few rock formations are totally impervious to water, even crystalline igneous formations such as granite or a recrystallized metamorphic formation usually contain discontinuities such as fissures, fault or joint planes along which water can migrate. There are many instances where wells tap a fissure or fault system and yield copious quantities of water – sometimes 5000 m³/day or more. Water also moves through granular strata via **pore spaces** or **interstices** between the component minerals or grains of the rock. The diameter of such spaces is very variable; in fine-grained strata such as silts (argillites) they can be less than 0.5 micron, increasing to around 50 microns or more in coarsely-grained formations. A typical sandstone will contain pores of around 10–25 microns. The presence of interstitial cement – calcite, aragonite, silica etc – drastically reduces the pore volume available. The ratio of the pore volume to the total volume of a rock is its **porosity**. A perfectly packed rounded sandstone will have a porosity of 26 per cent (Figure 19.3), but this may be increased to around 35 per cent in loose sand and gravel and reduced to 5–15 per cent in cemented sandstones. Poorly consolidated clay formations with very small intergranular pore spaces may have high porosities of the order of 40–50 per cent because the grains are haphazardly packed, but compaction reduces the porosity of shales and slates to around 3 per cent. Karstic formations – limestones – are a special case (Section 19.9). Usually formed of calcified remains of aquatic organisms, the rock

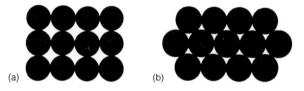

Figure 19.3 Diagrams to illustrate how pore space varies according to the way the grains (here idealized as spheres) are packed together in a porous sediment:

(a) the most open packing: each sphere touches six others (in three dimensions) and the porosity is 47.6%;
(b) the closest packing: each sphere touches twelve others and the porosity is 26%.

mass – largely calcite and magnesian calcite – often has very little intrinsic porosity but is susceptible to dissolution by groundwater. This chemical attack creates enlarged fissures which can lead to the development of caves (see later) and well-defined underground water courses.

19.3 TRANSMISSION OF GROUNDWATER

Rock formations which both store and transmit water are known as **aquifers**; these include most arenaceous formations (sandstones), karstic formations (limestones and chalk) and a wide variety of other compositions. There is some uncertainty regarding the relative importance of intergranular and fissure flow in indurated aquifers: recent evidence suggests that fissure flow, even in arenaceous formations, is very significant as a means of rapid water transmission, but that intergranular flow is important in providing the bulk of aquifer storage. Aquifers in which both fissure and intergranular flow is significant are said to exhibit **dual porosity**: **primary porosity** resulting from intergranular pore spaces and **secondary porosity** due to structural features such as joint and fault fissures. The degree of interaction between water stored in pores and fissures – and the relative magnitudes of fissure and intergranular flow – control the geochemical processes which determine groundwater quality (and also an aquifer's susceptibility to transmitting polluting substances). Britain's most important aquifer – the Chalk – relies almost entirely on fissure flow, even though it has a bulk porosity of over 40 per cent; flow occurs via bedding plane and vertical fissures which range in size from the microscopic to centimetres in width and sometimes several metres in length. Borehole yields are commonly 2000 m^3/day and can rise to 10 000 m^3/day in places.

Formations which cannot store or transmit water are known as **aquicludes**; these include argillites, clayey formations and many metamorphic strata. There are, in fact, very few true aquicludes – most strata will allow some water movement even if it is very slow; the term **aquitard** has been used to describe formations which allow such slow migration. The size and degree of connection of pore spaces determines the ability of a material – rock, soil etc – to allow the passage of water (or other fluid, e.g. oil or gas); this parameter is known as the **permeability** (or permissivity) of the material. The permeability of a given material clearly depends on parameters such as the viscosity and pressure of the fluid passing through it. A rock which is impervious to water at atmospheric pressure may well show higher permeability if the fluid pressure is increased or if the fluid becomes less viscous (e.g. if water is replaced by gas).

Pumice is an example of a rock which, although showing high primary porosity, would commonly be an aquiclude or aquitard because the pores are discrete and would not allow the movement of a fluid unless secondary porosity was also present.

The ability of a formation as a whole to transmit water is a function of its bulk permeability multiplied by the saturated thickness of the formation: this is known as the **transmissibility** (or transmissivity) of the formation.

The rate at which water is transmitted depends also on the local **hydraulic head** – the difference in elevation between the source and the point of measurement. The vertical head difference (effectively the pressure) divided by the distance over which the difference is measured is known as the **hydraulic gradient**; this is the driving force for groundwater movement.

Permeability, transmissibility and hydraulic gradient are analogous to resistance, current and voltage in electrical theory and are related to the **rate of flow in the aquifer** in a similar manner to that whereby under Ohm's law the electrical parameters are related to power. In the case of water flow, the relationship is expressed by **Darcy's law**. This analogous relationship can be used to construct two-dimensional analogue models of aquifer characteristics, substituting electrical inputs, resistances, etc. for the estimated actual values. Measurement of the flow of electrical current through the model gives an indication of the pattern of groundwater flow in the real aquifer. However, with the advent of low cost computers, analogue models have given way to digital modelling of groundwater movement; finite element and finite difference techniques have become widely available in software packages for solving resource and flow problems on both the micro and macro scale – from, for example, calculating the groundwater flow into a civil engineering excavation for a building or opencast mine, up to large-scale regional water resource models for predicting water supply and demand.

19.4 AQUIFERS

When an aquifer crops out at the surface, or immediately underlies the soil profile, it will exhibit water table conditions and is termed an **unconfined** or **water-table** aquifer. The formation above the water table is unsaturated but within it local zones of saturation may occur where a poorly permeable stratum allows storage of water above it, forming a **perched water table**. Broadly, the water table reflects local topography: contours of the elevation of a water table can be plotted on a map in a similar manner to topographic contours and the directions of groundwater flow – orthogonal to the contours – can be inferred. Where the water table intersects the ground, surface springs, seepages, lakes or rivers occur (Figures 19.1 and 19.2). But the water table fluctuates vertically in response to recharge from rainfall, loss of water from storage in drought periods, or by artificial abstraction; this gives rise to intermittent flooding or drying up of watercourses and wetlands. On the Chalk aquifer in England, such **ephemeral streams** are known colloquially as 'bournes'. Groundwater is a constituent of the flow in most rivers. Because it is relatively unaffected by short-term changes in climatic conditions it forms a stable and important part of the river flow and is termed **baseflow**. The surface of a river is essentially a continuation of the water table in the formation which the river is flowing over. River channels are the lowest point of local topography and a natural sink for groundwater flow. River

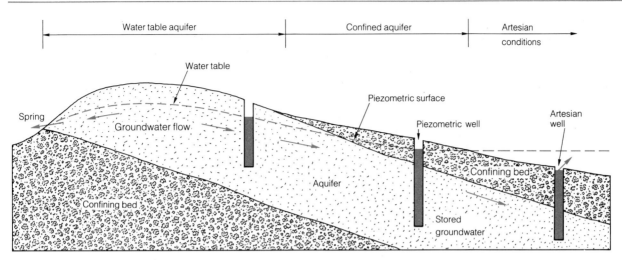

Figure 19.4 An aquifer confined between aquicludes, showing confined and water table (unconfined) conditions. Note that the aquifer becomes artesian where the piezometric surface is higher than ground level.

channel zones frequently comprise unconsolidated strata of heterogenous grain sizes which are highly permeable and allow groundwater to flow quite freely within them as **underflow**, mirroring the visible flow in the river. In drought periods, the local water table may fall below the level of the river bed so that all the discharge becomes underflow, the river appearing to be dry.

Frequently, aquifers occur in alternating sequences of permeable and impermeable strata, e.g. sandstone/shale formations, or a major aquifer may be overlain by an aquiclude. If the local dip and topography combine to take the aquifer to a depth where it is overlain by the poorly permeable formation, it is then said to be **confined** by the overlying **confining bed** or **layer** (Figure 19.4). Water is, however, able to enter the aquifer – either by slow infiltration through the confining layer or by means of **direct recharge** if the aquifer crops out at the surface – where, of course, it becomes unconfined. In Britain, the largest example of such an aquifer is the Cretaceous Chalk which crops out in a sinuous ribbon from Yorkshire, through East Anglia and across the south of England to Wiltshire and Dorset. In the London area, a broad syncline takes the Chalk, from its outcrop in the Chiltern Hills, below the overlying Tertiary formations (the London Clay acting as the confining layer) to re-appear in the North Downs of Kent and Sussex. The Chalk is intensively exploited as an aquifer in both the unconfined and confined zones, particularly for the expanding towns in the area around London.

Other examples of major confined aquifers include the Chalk of the Paris basin in France, the Dakota Sandstone in the USA and the great Jurassic and Cretaceous sandstone basin of Queensland, New South Wales and South Australia. Each of these examples exhibits similar characteristics: an exposed upland unconfined portion of the aquifer receives recharge which is then transmitted into the confined zone. In the case of the Queensland basin, the confined area extends to about 1.5 million km² and is essential for the water supply of the territory.

A confined aquifer is usually fully saturated; the water stored within it is under a pressure greater than atmospheric. If a borehole penetrates the confining layer and enters the aquifer, water will rise up the hole until the water pressure in the aquifer is balanced by the weight of the column of water in the borehole. The water level in the borehole represents the pressure of water in the aquifer; a series of such levels allows the elevation of the **piezometric** (or **pressure**) **surface** to be estimated: it can be plotted on a map in a similar manner to water-table contours. Where an aquifer changes from the unconfined to confined state, the water table coincides with the piezometric surface of the confined aquifer. In some cases the elevation of the piezometric surface is above the local ground level: a borehole in such a location would yield a natural outflow of water from the aquifer. Such conditions are known as **artesian** after the Roman name for the French province of Artois where the condition was first studied; the term is, however, sometimes loosely used to describe any confined aquifer. This loose terminology may have led to the common, but probably erroneous, conception that the Chalk aquifer under London (Figure 18.3b) was artesian, to the extent that the fountains in Trafalgar Square are supposed to have operated under the natural pressure head of the aquifer. Records suggest that the water has, in fact, always been pumped.

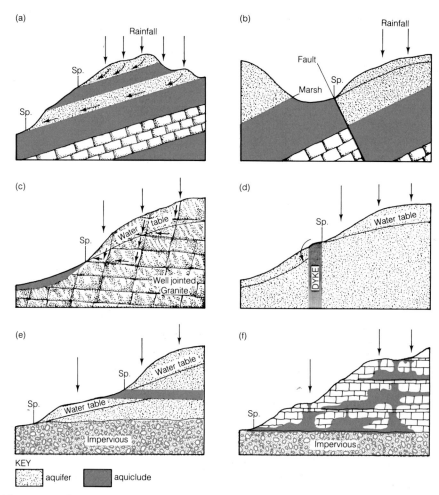

Figure 19.5 (a) Alternating dipping aquifers and aquicludes: springs appear at base of aquifer. (b) Fault throws aquifer against aquiclude: spring appears at fault line and marshy ground on the aquiclude. (c) Spring appears where joints issue water collected from higher ground. (d) Dyke forms a barrier to flow: spring appears where groundwater is impounded but water escapes to lower level. (e) as (a) – spring appears at unconformity, often as a linear feature along the hillside. (f) Similar condition to (e) but in cavernous limestone (this is the situation at Ingleton, Yorkshire).

19.5 EFFECTS OF GEOLOGICAL STRUCTURE

The geological structure and topography of a formation largely controls the movement of water within it. Water infiltrating in high ground will migrate through the formation under gravity until it reaches a point of natural or artificial discharge – a spring, watercourse or borehole (Figure 19.5). Most formations also contain water which is in semi-permanent storage below the level of natural discharge. Such water can only be abstracted by artificial means, but can be replaced if there is an excess of available water in the high-level circulatory system. This water in deep storage is sometimes regarded as a reserve for drought years; the water level in the aquifer may be reduced by overpumping on the assumption that winter rainfall will recharge the loss. Problems can arise if a dry summer is followed by a dry winter and recharge is insufficient to restore water levels to their Spring peak. In such

cases, there is a risk of water levels falling to such an extent that perennial springs dry up, stream flows are reduced and abstractions from shallow wells and boreholes may fail.

Fluctuations of aquifer water levels occur in patterns which can be plotted as a **borehole hydrograph** (elevation of the piezometric surface/time). They are indicative of the characteristics of the aquifer. Typical natural fluctuations in an unconfined aquifer can range from 0.5 m/year to around 20 m/year depending on location, the geological formation and the availability of recharge from infiltration. Natural fluctuations in confined aquifers are not usually so great, ranging from 0.5 m/year to 5 m/year. Where abstraction occurs, however, fluctuations of water level occur over large intervals depending on the volume of water abstracted: wells and boreholes are normally sited to minimize the range of fluctuations expected.

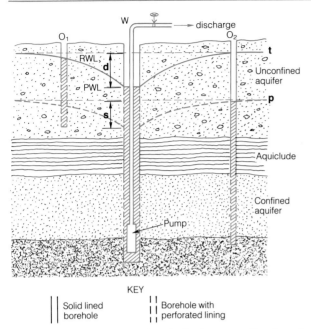

Figure 19.6 A pumping borehole in unconfined and confined aquifers. Pumping borehole, W, draws down the water table, t, in the unconfined aquifer from Rest Water Level to Pumped Water Level (distance d). The effect on the aquifer is monitored in observation borehole O_1. Pumping also depresses the piezometric surface, p, of the confined aquifer by amount s. Effect on the aquifer is monitored in observation borehole O_2. The pumping borehole has solid lining near the surface to prevent surface pollution entering the borehole. It is perforated towards the base of the unconfined aquifer and for the whole of the confined aquifer – which is not dewatered.

19.6 MEASUREMENT AND USE OF GROUNDWATER RESOURCES

Groundwater is abstracted for as many purposes as there are uses for water; public supply companies, large and small industries, farms and individual houses may abstract groundwater. But in order that groundwater can be safely abstracted it is essential to be able to quantify the volumes of water which are moving through an aquifer and being retained in storage. Over-abstraction can lead to a steady lowering of the water table or piezometric surface, which can have profound and sometimes unpredictable effects on the aquifer itself and also on man-made structures.

The water stored in interstitial spaces carries some of the weight of the overlying structure; withdrawal of that water – naturally or artificially – causes the aquifer to be compressed slightly. In normal circumstances this movement is so small as to be inconsequential, but situations have occurred where dewatering has led to spectacular slumping of thick, poorly consolidated aquifers and the collapse of buildings.

Confined aquifers can also respond to imposed loads by the displacement of water when the pore spaces are compressed. Such loads can be man-made, such as by large buildings or heavy vehicles and trains, or they can be natural – such as the imposed load of the tidal fluctuation of the sea on coastal confined aquifers. In the latter case, a borehole hydrograph shows a regular diurnal fluctuation, possibly of the order of 0.25 m, in the water level. These fluctuations lag slightly behind the tides and are damped according to the efficiency of the confining layer and the aquifer as pressure transmitting media. The gravitational attraction of the Moon also produces similar but smaller diurnal fluctuations which can be observed in certain aquifers as the aquifer dilates in response to **Earth tides**. Similar dilations of an aquifer can be caused by compressional or tensional forces in earthquake zones. It has long been recognized that rapidly fluctuating groundwater levels are a precursor of an earthquake; manual observation of water levels was undertaken in countries such as China to provide early warning of earthquakes. As well as seismic methods, telemetric datalogging of borehole hydrographs can help to provide early warning in earthquake-prone areas.

The volumes of groundwater in storage within an aquifer cannot be measured directly; they have to be inferred from measurements of parameters such as the porosity, permeability and transmissivity of the formation, localized variations in formation thickness and petrology, and the effects of geological structures. This rather complex demand has meant that groundwater resources have sometimes been ignored in favour of surface-water storage, which may be less reliable and more prone to contamination, but easier to quantify. Reliable long-term hydrographs and control of abstractions are an important part of maintaining the supply capability – or **safe yield** – of an aquifer.

The collection and abstraction of groundwater for human purposes can often be achieved using very simple techniques; a pit dug in a shallow aquifer will act as a sump from which water can be drawn by hand or a pump. This simplicity, in fact, makes groundwater use attractive in countries where surface water storage would be impracticable or expensive. Groundwater also benefits by its isolation from common forms of contamination such as sewage and industrial wastes; the quality of water percolating through the unsaturated zone is improved by filtration and the adsorption of contaminants by clay minerals. In deeper aquifers, and where sanitary protection is needed, specialized boreholes or wells (large diameter boreholes) are lined with steel or high density plastic lining set in concrete grout. Opposite the aquifer, specially designed 'well screens' – perfo-

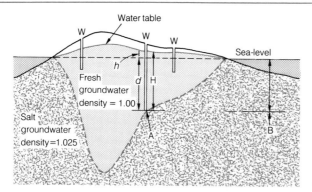

Figure 19.7 Schematic section through an island or peninsula of permeable rock to illustrate the hydrostatic equilibrium between fresh groundwater received from the rainfall and the heavier salt groundwater that seeps in through the sea floor (not to scale). Groundwater continues fresh to a proportionate depth about three times as great as that shown (see text).

rated linings – may be installed which allow the maximum volume of water to enter the borehole, yet prevent the grains of the formation itself from being drawn into the borehole along with the water. Water is usually drawn from a borehole by a high-lift pump, but occasionally an artesian aquifer may overflow sufficiently at the surface to meet the demand. Modern boreholes are rarely more than 70 cm in diameter but before the advent of steel linings, hand-dug brick-lined wells were built, sometimes up to 4 m in diameter and of considerable depth. They may include horizontal adits (small tunnels) to further improve water collection. Many are still in use for public supply or industry. An abstraction borehole should penetrate as much of the aquifer as possible to achieve a predictable inflow and to allow for the **draw-down** or depression of water level which will occur when pumping (Figure 19.6).

By carrying out tests of the yield of a borehole and measuring the drawdown in nearby observation boreholes, the aquifer properties such as transmissivity and permeability – can be calculated and predictions made of the long-term reliability of the borehole source. Groundwater movement and storage can be predicted very satisfactorily in most aquifers by mathematical techniques.

19.7 OVERDRAWN AND SALINE CONDITIONS

In many places, groundwater is a free resource – individuals can sink a well on their land and abstract water without charge. Some nations or federal states have a licensing system to control abstractions and this may involve a fee if the abstraction is large. But even large aquifers can become overdrawn; the Chalk and Triassic Sand-

stone aquifers in England have shown signs of overpumping in places. Commonly, overpumping is exhibited by a decrease in the groundwater contribution to surface-water courses; springs may be dry for long periods, river flows may decrease markedly or even dry up altogether.

In many places around the world, over-abstraction of aquifers has been allowed to occur to such an extent that the aquifer is effectively exhausted or, in coastal zones, saline water has entered the aquifer. Examples of this in the UK are the Mersey estuary area of Cheshire, where the aquifer is Triassic Sandstone, and in localized zones on the Humber and Thames estuaries in the Chalk aquifer. In the US, aquifers in California and Florida have been badly contaminated by saline intrusion. Attempts to correct this have included injecting fresh water into the aquifer to form a localized high water-table barrier to the seawater.

A particular problem in areas where there has been a history of over-abstraction and low water levels is the recovery of groundwater levels which occurs if abstraction ceases. This can lead to widespread inundation of basements and other below-ground structures such as subways and rail tunnels. Emergency pumping arrangements may have to be installed to correct the situation, as happened in the London and Liverpool areas of the UK in the 1980s as traditional industries closed or changed their processes or water supply sources. Cessation of deep coal mining and associated dewatering of workings leads to rapid 'rebound' of the water table with serious consequences for landowners and possible contamination of rivers. Great care must be exercised in abstracting groundwater from coastal aquifers and oceanic islands to prevent seawater from penetrating the aquifer. The controlling parameter is the density difference between fresh water (density, p_f = 1.000 g/cm^3) and seawater (density, p_s = 1.025 g/cm^3) which allows fresh water to 'float' on seawater, provided there is not too much relative movement. A schematic cross-section through an oceanic island or peninsula (Figure 19.7) shows the influence of the low head of fresh water above sea-level within the aquifer in generating a lens of fresh water below ground-level. Hydrostatic pressure, H, at point A must equal that at point B which is at the same depth; the balance is achieved by the additional pressure head due to the water table elevation, i.e. H = h + d. Now

$$d = \frac{p_f}{p_s - p_f} = 40\,h$$

Hence, for every metre of fresh water above sea-level, there is a depth of around 40 m below sea-level – this is known as the **Ghyben-Herzberg**

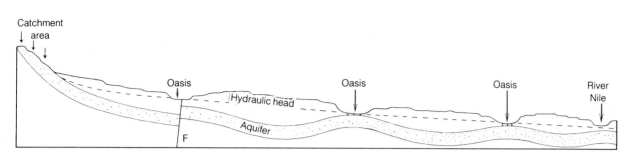

Figure 19.8 Schematic section across the Sahara to illustrate conditions favourable to the development of oases.

relationship. Figure 19.7 assumes a uniform aquifer with no flow; in practice, hydrogeological conditions are much more complex. In particular, it is essential to avoid 'up-coning' of seawater by heavily pumping a single borehole: a large number of shallow boreholes (wells, W, in Figure 19.7) should be used to distribute evenly the demand and maintain equilibrium.

Since the 1960s, experiments have taken place in several parts of the world, for example in the Lea Valley in North London and on the Pacific coast of California, to try to balance over-abstraction by injecting water into aquifers for later abstraction; this 'artificial recharge' has worked quite well in loosely consolidated aquifers, but has been rather less satisfactory in indurated aquifers where bacterial growth around the injection well or borehole has markedly reduced local permeability and demanded a gradually increasing injection pressure. The technique has, nonetheless, been beneficial in preventing contamination of some aquifers which could have been in danger of contamination from seawater or deep saline groundwater being drawn towards the abstraction wells.

19.8 GROUNDWATER IN DESERT CONDITIONS

If often comes as a surprise to learn that abundant water supplies can be found in desert areas. Desertification can be a transient phenomenon – in geological terms. Deserts are often underlain by permeable formations containing groundwater which has been in storage for thousands of years since the area was cooler and wetter than at present (Section 22.2). Groundwater beneath the Libyan desert has been dated by carbon-14 (^{14}C) radiometric methods as being 15 000–35 000 years old. Boreholes drilled to intercept these formations can yield sufficient water to allow substantial irrigation and public supply works to be installed. Clearly, however, the water is not being replaced by modern recharge (or at least not at the rate it is

being abstracted) and so the water is being 'mined'.

More conventionally, groundwater appears at the desert surface in **oases**; these may occur for several reasons: erosion of the land surface to the level of the water table; undulations of a confined aquifer bringing its piezometric surface up to ground level; structural deformities allowing the development of lines of weakness (e.g. faults) which allow water to migrate to the surface (Figure 19.8). Oases may be intermittent, depending on their location and the reason for their existence; depletion of groundwater levels by abstraction could also cause the permanent loss of natural aquatic features such as oases, as has happened in parts of the sub-Saharan region of West Africa.

19.9 GROUNDWATER AS AN EROSIVE AGENT

Water is probably the most active naturally occurring chemical compound known: it can dissolve a huge range of substances; changes its state easily; does not decay; occurs universally and often in abundance, and can demonstrate an infinite range of flow conditions from laminar to turbulent. All these features are important when considering the role of groundwater as an agent of erosion.

Water falling as precipitation dissolves small quantities of carbon dioxide from the atmosphere; the water becomes weakly acidic by the liberation of hydrogen ions:

$$H_2O + CO_2 = (HCO_3)^- + H^+$$

Rainwater is commonly weakly acidic, showing a pH of around 5.5–6.0. As it migrates through soils and bedrock, the water reacts with the rock-forming and cementitious minerals and becomes a complex solution which exhibits chemical characteristics according to the various formations through which it has passed.

Many aquifers are siliceous and granular so groundwater flows through interstitial pore spaces with little erosive effect on the rock matrix.

Figure 19.9 Gaping Ghyll, south-eastern slopes of Ingleborough. (*John Forder, Dent, Cumbria*)

But where the rock-forming minerals or the intergranular cement is calcareous, or otherwise susceptible to solution by water, fissures due to faults, joints and bedding planes can be opened out to create conduits for water movement. Indeed, it is likely that flow through fissures represents a very significant part of all groundwater movement in indurated aquifers. In the Chalk aquifer, virtually all flow is in fissures, albeit many are of microscopic size. At the other extreme, caverns of enormous size may be created in thick limestone formations. In the UK, the caves in the Carboniferous Limestones of North Yorkshire and the Mendip Hills are particularly well-known (and often termed 'sinkholes', 'swallow-holes' or 'pot-holes'). But these are diminutive compared with those in other parts of the world such as the Blue Grass area of Kentucky, USA, the Pyrenees and Dordogne-Lot areas of France, and the Karst of Yugoslavia. Such areas, where soluble rock crops out, are known as 'karst' topographies after the Yugoslavian example, an area about 640 km × 100 km. Around 40 per cent of the Soviet Union and 15 per cent of the USA is karstic, as are large areas of China (Plate Xd) and the Far East, the Mediterranean and North African coasts and Caribbean islands such as Jamaica and Cuba.

Karst topographies are formed under continental conditions. Their unstable and unpredictable underground openings and sinuous ravines can present difficult conditions for civil engineering structures such as buildings, communications and dams. Soils are usually of poor quality and thin, leading to specialized lime-loving flora and cropping patterns.

Underground water in such regions seeks out zones of weakness in the body of the rock, gradually dissolving calcite which may be re-deposited as tufa or travertine. Surface openings become enlarged where there is a concentration of runoff; by continued solution and the falling-in of loosened blocks, cylindrical or conical shafts and chasms are created, often opening into huge chambers, sometimes hundreds of metres below ground level. In Britain, the cavern known as 'Gaping Ghyll' in the Carboniferous Limestone of North Yorkshire (Figure 19.9) includes a chamber 146 m long and 33 m high. The 'Big Room' in the Carlsbad Cavern in New Mexico is nearly 1220 m long, 180 m wide and over 90 m high. Water dripping from a cave roof is subject to slight evaporation and may deposit a tiny quantity of calcite ($CaCO_3$). This deposition gradually builds into the stem of a stalactite, perhaps taking a century to grow a centimetre. Where the dripping water meets the floor, further evaporation leads to the

Figure 19.10 Jenolan Caves, New South Wales, Austalia, have been dissolved out of Silurian limestone by waters from the Jenolan River, which now flows about 80 m below the oldest cave. There are at least 11 caves on four or five levels. (*Australian High Commission, London*)

Figure 19.11 Trow Ghyll, on the slopes of Ingleborough, North Yorkshire. A dry valley due to roof collapse of a former limestone cavern. (*A. Horner and Sons, Settle*)

waterfalls are seen in many caves, including the large show caves such as White Scar near Ingleton in Yorkshire, Wookey Hole in the Mendips, La Cave in the Dordogne, the Carlsbad Caverns in the USA and the Jenolan Caves, New South Wales, Australia (Figure 19.10).

Occasionally the roof of a cave collapses and leaves a corresponding depression at the surface. In the case of a long cavern close to the surface, such collapses can lead to the formation of sinuous ravines, sometimes spanned by arches where the roof has held firm. There may well no longer be any indication of flowing water in such a ravine, which is then termed a **dry valley**. In the UK, Cheddar Gorge in the Mendip Hills and Trow Ghyll near Ingleton (Figures 19.11 and 19.12) are well-known examples of cave collapses. The Karst of Yugoslavia and the ravines of southern Crete are examples on a larger scale.

19.10 GROUNDWATER AS A HEAT SOURCE

Hot underground water is a well-known phenomenon; geysers, sulphurous springs and spas are part of the common awareness of groundwaters. Besides their tourist value, these **geothermal** phe-

accumulation of an upward growth – the stalagmite – which may eventually meet the stalactite to form a column. Complex patterns of these features and screens of tufa and travertine created by

Figure 19.12 Natural Bridge, Virginia. Part of the roof of a former limestone cavern. (*American Museum of Natural History, New York*)

Figure 19.13 Old Faithful in eruption, Yellowstone National Park. A fragment of wood preserved in the siliceous sinter (geyserite) of the mound surrounding the basin has been dated at about 730 BP by the radiocarbon method. (*G. A. Grant, United States Geological Survey*)

nomena are now part of a growing movement to develop groundwater as a general heat source.

Groundwater may be heated by several means: some water is trapped in sedimentary formations as they are laid down and will be retained in the pore spaces as a formation undergoes diagenesis and induration. Tectonic movements may carry this **connate** water to considerable depths where it is heated along with the host rock; the elevated temperature increases the solution potential of the water, allowing it to dissolve minerals such as calcite, dolomite or silica which may be later deposited as interstitial cements or vein minerals on cooling. A circulatory pattern of water movement may be established in response to thermal differences with depth and any local weakness in the rock structure – for example, a fault, or vertically permeable zone – may allow the warm water to migrate to the surface. It is often known as **juvenile** water because it is reaching the surface for the first time. In areas of active volcanicity the water which reaches the surface may be very hot indeed.

More commonly, **meteoric** water – that arising from rain or snow – infiltrates and is brought into contact with a body of relatively hot rock (the average **geothermal gradient** – the rise in temperature with depth – is about 25°C/km). It is heated and rises convectively towards the surface as a

result of density differentials. The water temperature at the surface is usually of the order of 30–50°C. This is the usual form of warm water seen in the hot springs and spas which were the reason for the foundation of 'spa towns' such as Bath, Buxton and Harrogate in the UK. Although 'taking the waters' became socially fashionable in the eighteenth century, the Romans were well-acquainted with the use of geothermal waters for medicinal and heating purposes. Geothermal sources such as these, where the maximum aquifer temperature is less than 150°C, are known as 'low enthalpy' waters.

If the heat source is a relatively shallow active magma body, very high water temperatures may be reached (see Chapter 13). Temperatures in excess of 200°C may be reached under tectonic pressure at a depth of around 1000 m, but at 2000 m the temperature may rise as high as 300–350°C. On ejection, the water temperature will generally fall in the range 150–250°C depending on the volume emitted (lower volumes usually mean higher temperatures).

Water at the lower end of this range is

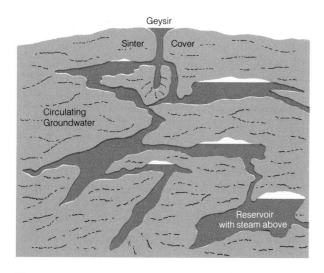

Figure 19.14 Schematic section through *Geysir*, the Great Geyser of Iceland, to illustrate the conditions appropriate to intermittent eruption, showing subterranean reservoirs fed by groundwaters heated from below by the ascent of hot emanations, including superheated steam. (*After T. F. W. Barth*)

abstracted in a number of countries – but notably Japan, New Zealand, Hungary and Iceland – for space heating, horticultural and industrial process uses. Very high temperature water suitable for power generation is less common but is found in countries near active volcanic zones such as Impe-rial Valley in California, Lardarello in Italy, Wair-aki in New Zealand (Plate IXe), Svartsengl, Iceland (Figure 13.61), and in Mexico and the Philippines (Figure 19.13).

Hot rocks are not always associated with hot water; in zones of low permeability 'hot dry rock' it is possible to inject cool water via boreholes and pump it back to the surface after it has been conductively heated by the geothermal source. Characteristically, such sources are relatively deep. Artificial fracture enhancement may be necessary to allow water to flow through the body of the rock mass. In the UK, an experimental system of hot dry rock geothermal energy abstraction has been installed near Redruth in Cornwall, using one of the Cornish granite intrusions as a heat source. As yet, the efficiency of the system has proved only marginal.

Geysers are hot springs from which a column of hot water and steam is explosively discharged at intervals, spouting in some cases to heights of over 100 m. The term comes from *Geysir*, the Icelandic name (meaning 'spouter' or 'gusher') of a spectacular example north-west of Hekla. Other well-known examples include 'Old Faithful' in the Yellowstone National Park in Colorado (Figure 19.13) and those at Wairakei on the North Island of New Zealand. The explosive nature of geysers is thought to be due to the release of dissolved gases at depth (Figure 19.14). The ascending bubbles

Figure 19.15 Terraces of travertine, Mammoth Hot Springs, Yellowstone National Park. (*Dorien Leigh Ltd*)

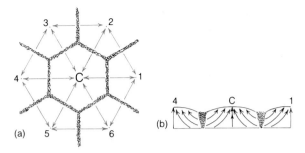

Figure 19.16 Diagram to illustrate the formation of an idealized pattern of hexagonal hummocks.

become saturated with water vapour forming a foam; this relieves the pressure on the water at depth in the geyser pipe which can then erupt in a column of water and steam of vast volume and force. The water carries with it dissolved minerals such as calcium carbonate which is deposited around the neck of the geyser as mounds or terraces of travertine (Figure 19.15); alkaline waters usually carry silica (SiO_2) in solution which is similarly deposited as **siliceous sinter** or **geyserite**.

In addition to these obvious sources of energy, ordinary groundwaters, at ambient temperature, can provide heat. Most groundwaters in temperate latitudes maintain a temperature in the range 8–12°C with a damped seasonal fluctuation within this range; this compares with surface waters whose temperatures fluctuate markedly according to the season, from near freezing in winter to around 15°C in summer. By passing groundwater through a heat pump – rather similar to the arrangement used in a refrigerator – heat can be extracted from the water and transferred in the form of warm air or warm water. The groundwater is reduced in temperature by 2–4°C, but is otherwise unchanged and can be used for supply. Although the cost-effectiveness of this arrangement depends on the price of conventional power, it is a valuable source of non-fossil-fuelled heat which is widely available, particularly where groundwater is already being abstracted for supply purposes.

19.11 FROZEN GROUNDWATER

In boreal latitudes groundwater, like the ground itself, can become permanently frozen, except for a surface layer which thaws during the summer. The frozen ground, known as **permafrost**, occurs

Figure 19.17 Mud polygons of two orders of size near Bruce City, Klass Billen Bay, Spitsbergen. The mountain avens (*Drya Octopetala*) are growing in gutters that outline the large hummocks. The smaller polygons, outlined by mud-cracks, are a seasonal result of summer desiccation and shrinkage. (*J. Walton*)

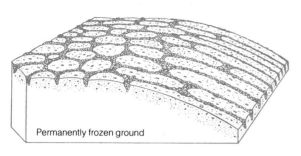

Figure 19.18 Diagram illustrating the merging of stone-rings on a flat surface into stone-stripes on a slope (*C. F. Stewart Sharpe, 1938*)

north of the Arctic Circle in Siberia, Canada and Alaska and is probably a relic from the last Ice Age.

As water freezes, it expands initially, so the density of ice at 0°C is 0.92 g/cm^3 but at –22°C it is 1.03 g/cm^3, corresponding to a decrease in volume of 11 per cent. This behaviour controls most of the physical features associated with permafrost conditions. Over 400 m of permafrost have been penetrated in Canada and over 700 m in Siberia; no significant changes have been observed in the general thickness or extent of permafrost conditions in recent years, the permafrost is thermally stable at depth. In contrast, the summer thaw will affect a thin zone, from a few millimetres to around 3 m according to the climate. Clearly, groundwater cannot be abstracted in permafrost regions, but human activity can cause thawing of

permafrost with unpredictable effects. Towns, railways, airports and other structures built in permafrost zones can create localized freeze–thaw conditions which can cause subsidence and slumping of formerly solid ground. Complex civil engineering is needed to safeguard major structures against collapse.

19.12 MUD AND STONE POLYGONS

Near-surface freeze–thaw conditions set up stresses in the soil/mud which are relieved by the formation of roughly hexagonal patterns (cf. columnar jointing, Section 8.2(8)). Ice wedges form in the cracks between hexagons, the ice gradually expanding and opening the crack downwards so that material is heaved into the centre of the hexagon, forming a shallow mound (Figure 19.16). Ice wedges up to 3 m deep and polygonal mounds up to 12 m across are known. In muddy areas, subject to summer desiccation and shrinkage as well as frost heave, secondary hexagons may be created within larger primary hexagons (Figure 19.17).

Frost causes the movement and migration of stones: the lower surface of a stone in wet ground chills more quickly than the upper; ice begins to form and pushes the stone upwards. The ice spreads rapidly as water in capilliary spaces moves towards the stone and is in turn frozen, so increasing the heave on the stone. In the Spring,

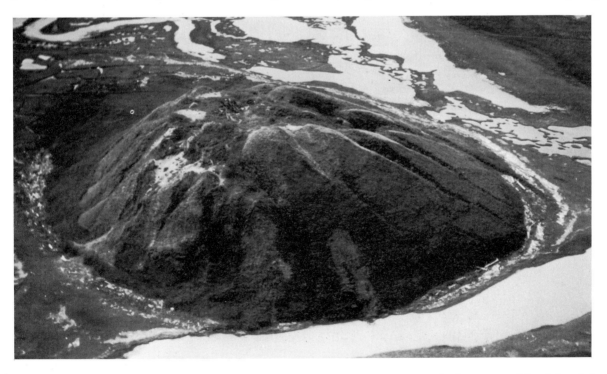

Figure 19.19 'Crater Summit' Pingo, about 40 metres high, and 290 metres across at its base, east of the delta of the Mackenzie River (69°50′N; 133°W) (*National Research Council, Division of Building Research, Ottawa*)

the mud thaws, but still supports the stone in its new position. In this way quite large stones can arrive at the surface supported on a pedestal of ice after several years of upward movement. At the surface, the stone topples off its pedestal and rolls downhill. When associated with the mud mounds described earlier, the localized downhill movement of stones leads to polygonal accumulations of stones, which often provide shelter for alpine flora, thus accentuating the polygonal structures. On steeper hillsides, gradual elongation of the polygons occurs which can result in alternating stony mounds and muddy rills (Figure 19.18).

19.13   PINGOS

In Arctic Canada, Greenland and northern Siberia there are sporadic mounds and cones, with ice in the centre and covered with debris (Figure 19.19). Their dimensions are twenty to one hundred times greater than those of the biggest mud polygons, rising to about 40 m above the surrounding country. These structures are called **pingos** (their Eskimo name), but have also been referred to as 'hydro-laccoliths' or 'cryo-laccoliths'. They appear to form where unfrozen groundwater (a) is trapped between the underlying permafrost and a frozen zone, or (b) is able to migrate into such a position under a hydraulic head (rather akin to a boreal oasis!).

Type (a) pingos are thought to occur where groundwater underlying a frozen lake has migrated to the centre of the lake as the permafrost gradually encroached upon it. The water froze, expanding as it did so, and gradually heaved the lake sediments upwards to form the pingo. Type (b) pingos are found in East Greenland, where snow melt-water in the mountains is able to penetrate the permafrost at lower levels. Where the water approaches the surface, pingo formation occurs as described for type (a).

19.14   FURTHER READING

Burger, A. and Dubertret, L. (eds) (1975), *The hydrogeology of karstic terrains*, International Association of Hydrogeologists, Paris.

Campbell, M.D. and Lehr, J.H. (1973), *Water well technology*, McGraw-Hill, New York.

Chorley, R.J., Schumm, S.A. and Sugden, D.E. (1984), *Geomorphology*, Methuen, London and New York.

Price, M. (1985), *Introducing groundwater*, Allen and Unwin, London.

Rodda, J.C., Downing, R.A. and Law, F.M. (1976), *Systematic hydrology*, Newnes-Butterworths, London.

Todd, D.K. (1980), *Groundwater hydrology*, 2nd edn, Wiley, New York.

(a)

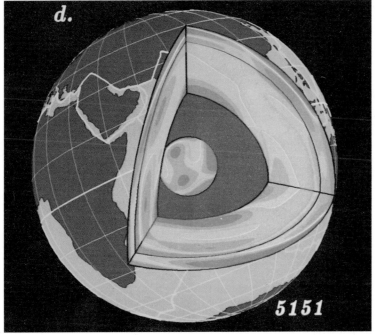

(b)

Plate I
(a) Space shot of Africa, Arabia and the Mediterranean area of the Earth, with adjacent Atlantic and Indian Oceans and north-east South America. The desert areas of North and South Africa and Arabia, together with Spain and the Mediterranean, are free of cloud, whereas the belt of equatorial rain forest is cloud-covered. To the west, Brazil is also only partly covered by cloud. Note the anti-cyclonic swirl of cloud associated with the cold Benguela Current adjacent to the Namib Desert, and the cloud cover building up over western Europe from the west-central Atlantic. Lake Nyasa, at the southern end of the East African rift system, and the Red Sea rift are both clearly visible as are the Arabian Gulf and the Caspian and Black Seas. (*Meteoset image received and processed by the Satellite Applications Centre, Mikomtek, CSIR, PO Box 395, Pretoria 0001, South Africa*)
(b) A composite three-dimensional section of a tomographic model of the Earth's mantle showing a glimpse of the anisotropic properties of the inner core. These subtle anomalies represent small perturbations (of up to a few percent only) from a spherically symmetric Earth model, and may be caused either by lateral variations in composition or temperature. Areas of relatively slow seismic velocities are shown in red, with relatively fast seismic velocities in blue. Plate boundaries are indicated in yellow. (See Section 26.11) (*From J. H. Woodhouse and A. M. Dziewonski, 1989, in* Seismic Tomography and Mantle Circulation. *Eds R. K. Onions and B. Parsons, Royal Society, London*)

Plate II

(a) Quartz is one of the most widely distributed of the rock-forming minerals. However, on occasion, circulatory hydrothermal fluids contain mainly dissolved silica and on cooling in cavities in rocks (e.g. an open fault-fissure) transparent hexagonal prismatic crystals can grow, forming the variety known as rock crystal. (*K. R. Gill*)

(b) Amygdales (Section 5.3) sometimes consist of agate, a microcrystalline variety of quartz with delicate, alternating layers of chalcedony having different colours and porosity. It is named after its first recorded locality on the river Achates, Sicily. (*K. R. Gill*)

(c) Pyrite (FeS_2) is the most common and widespread of the sulphide minerals, occurring in a great variety of environments, e.g. mineral veins and other ore deposits, in black shales, in coal and in the case illustrated in slate. Well-developed cubes are characteristic. (*K. R. Gill*)

(d) A cleavage rhomb of the variety of calcite ($CaCO_3$) known as Iceland Spar, exhibiting its characteristic double refraction. This property was used by William Nicol, an Edinburgh mineralogist, in the late 18th century to construct the 'nicol prisms' for the first polarizing microscopes. The source of the Iceland Spar was a giant cavity in a basalt on the east coast of Iceland discovered in the 17th century. Calcite was named from the Latin 'calx' – burnt lime. (*K. R. Gill*)

(e) Four very different textural varieties of granites, all consisting of similar essential rock-forming minerals, viz. quarts, alkali-feldspars, plagioclase and mica (mainly biotite). From top left clockwise are: (i) Orbicular granite, New Zealand (concentric zoning); (ii) heterogeneous porphyritic granite, Cornwall, England; (iii) Rapakivi granite, Finland; (iv) porphyritic granite with pink-coloured orthoclase phenocrysts, Shap, Cumbria, England. (*K. R. Gill*)

(f) The wide variety of igneous metamorphic rock pebbles found on one of the beaches of the Island of Iona, Inner Hebrides, Scotland. (*K. R. Gill*)

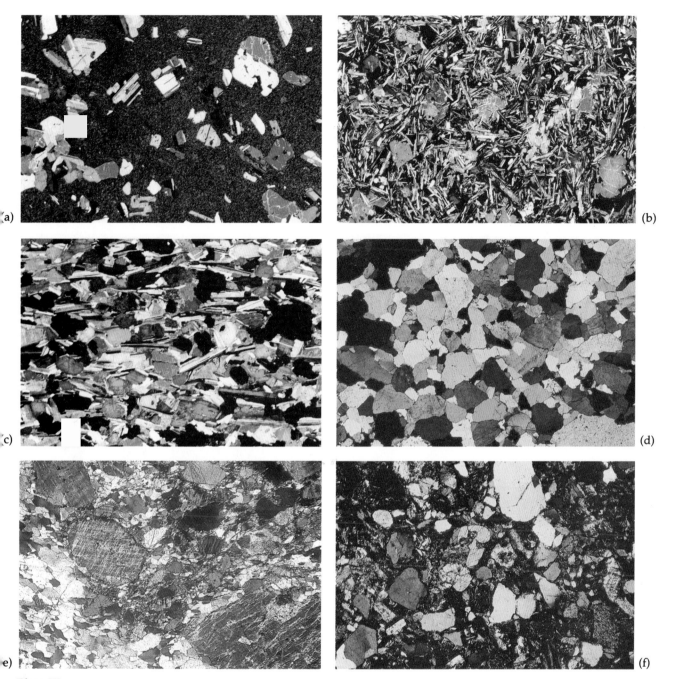

Plate III

Photomicrographs of thin sections of three basic igneous rocks illustrating the variation in grain size and texture resulting from differences in rate of cooling and crystallization. Each photomicrograph shows an area 15 mm by 10 mm and was taken in cross-polarized light. (*J. G. Fitton*)

(a) Porphyritic basalt from an early Tertiary lava flow in east Greenland. Large phenocrysts of olivine (yellow crystal in the centre, for example), plagioclase (grey and white), and augite (mostly showing blue or red polarization colours) are set in a fine-grained groundmass composed largely of these same three minerals. The phenocrysts grew slowly while the magma was stored underground. The magma was then erupted, cooling and crystallizing rapidly (hours) to form the groundmass.

(b) Dolerite from the island of Rodrigues in the Indian Ocean. The rock is composed of the same minerals as the basalt. In this case the magma formed an intrusive body several metres across and cooled relatively slowly (years) forming larger cyrstals (phenocrysts) than those in the basalt groundmass.

(c) Gabbro from the Cuillins in the Isle of Skye, Scotland. Again the sample is composed of olivine, plagioclase and augite. The sample was collected from a large (2 km) intrusion which would have cooled very slowly (thousands of years). Note the lamination due to alignment of the tabular plagioclase crystals. This was caused by settling and compaction as the crystals formed.

Photomicrographs of thin sections in cross-polarized light of three types of sandstone.

(d) Quartz arenite – a highly siliceous sandstone consisting of grains of quartz cemented by quartz cement which has grown around original grains to make a very compact quartz mosaic. Carboniferous, Wales. (*E. K. Walton*)

(e) Arkose – a slightly metamorphosed example, showing quartz and feldspars, the latter displaying 'twinning', the cross-hatched pattern indicating microcline. (?) Torridonian, Tarskavaig, Isle of Skye, Scotland (*R. F. Cheeney*)

(f) Greywacke – a poorly-sorted sandstone with a clay matrix and a mixed composition; white and grey grains are quartz and feldspar, coloured, are hornblende, augite and mica. Silurian, Scotland. (*E. K. Walton*)

Plate IV

(a) The contact of the Peterhead granite with Upper Dalradian psammitic metagreywackes is exposed on this beach at Whinnyfold, Angus, north-east Scotland. Physical and chemical weathering processes give rise to a wide variation in grain size in the beach debris, as seen in (b). (*K. R. Gill*)

(b) Large boulders of granite and metagreywacke are gradually fragmented into sand grains consisting of a single mineral, quartz (the most resistant to physical and chemical attack of the essential rock-forming minerals in this beach environment). (*K. R. Gill*)

(c) An Upper Devonian complex of false-bedded river sandbars which were periodically cut by channelling as stream flow altered in strength and direction. Dunbar, East Lothian, Scotland. (*K. R. Gill*)

(d) Flute-casts exposed on the undersurface (sole) of an upturned block of greywacke. Current-flow direction can be deduced from the V-shaped flutes and is an important palaeoenvironmental feature when the rocks are *in situ*. (In this case the V-shaped flutes indicate the current flow was from bottom to top of the slab.) Ordovician, Cowpeel Bridge, Peebleshire, Scotland. (*K. R. Gill*)

(e) Cement is manufactured by heating a slurry of limestone and mudstone or shale in a ratio of about 3:1. Strip quarrying of interbedded Carboniferous limestones and shales near Dunbar in Scotland which is the site of Scotland's only cement works. Farmland is reclaimed by transferring topsoil from above the working face on the left and depositing it on the quarry backfill to the right. (*K. R. Gill*)

(f) Drill cores of Proterozoic iron ore, Hamersley Range, Western Australia. The original jaspilite (Section 6.6) consisted of alternating bands of chert and magnetite. Prolonged leaching by groundwaters removed the silica leaving a haematite ore of 60–65% iron content (one of the richest ores in the world). Recent hydration of the ore to limonite (red-brown zone) is seen in the cores immediately below the present-day surface. The cylindrical cores (diameter c. 60 mm.), which are laid out in order as they are extracted from the core barrel, have been split vertically for sampling. (*P.McL.D. Duff*)

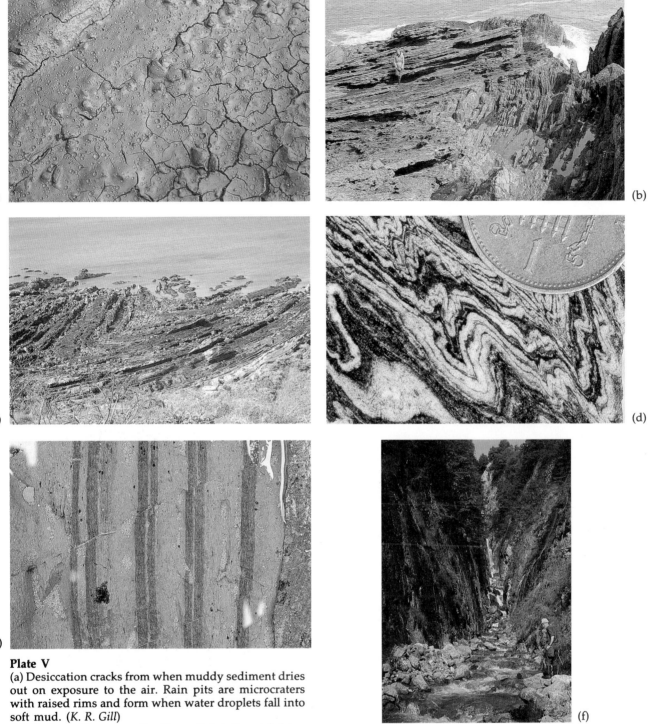

Plate V

(a) Desiccation cracks from when muddy sediment dries out on exposure to the air. Rain pits are microcraters with raised rims and form when water droplets fall into soft mud. (*K. R. Gill*)

(b) Hutton's Unconformity, Siccar Point, Berwickshire, Scotland. Sandstones and breccias of Upper Devonian age overlie unconformably vertically-dipping Silurian greywackes and shales. (*K. R. Gill*)

(c) A plunging synclinal fold in Carboniferous sandstones on the foreshore at St Andrews, Fife, Scotland. The rock platform is about 50 m wide. (*R. F. Cheeney*)

(d) Microfolds in gneiss, Grant Institute Collection, University of Edinburgh. The coin is 20 mm in diameter. (*R. F. Cheeney*)

(e) Plan view of a rock surface with a wave-cut platform at Raeberry, Kircudbrightshire, Scotland. The alternating light and dark layers are siltstones and mudstones respectively, and they are orientated near vertical. They are cut by two faults, also near vertical, and along which veinlets of a pale material have developed. That, with the longer trace, shows a lateral displacement of beds of about 3 mm. To an observer standing on one side of the fault, the beds on the other side appear to have been displaced to the observer's right; the fault has a right-lateral, or dextral component of slip. The exposure under water in a rock pool measures 10 cm from bottom to top of the photograph. (*R. F. Cheeney*)

(f) Gully eroded along line of the Insubric Fault near Gravedona, Tessin, Switzerland. (*R. F. Cheeney*)

(a)

(b)

(c)

(d)

(e)

(f)

Plate VI

(a) Fossilized tree stumps in position of growth showing through a bed of volcanic ash which overlies the Upper Pilot Coal Seam, Newcastle Coal Measures (Permian). Lake Macquarie entrance, south of Newcastle, N.S.W., Australia (*P. McL. D. Duff*)

(b) The Main Seam, which can attain a thickness of 33 m, Blair Athol Coal Measures (Permian), Queensland, Australia. (*P.McL.D. Duff*)

(c) Synclinal structure in Carboniferous (Namurian) Coal Measures where subsidence was occurring during sedimentation resulting in the lowermost seams being more downfolded than the uppermost. A maximum aggregate thickness of 60 m of coal occurs within 150 m of measures in the syncline which covers only 3 km², Westfield Opencast Site, Fife, Scotland. (*P. McL. D. Duff*)

(d) Cores of Dwyka Tillite, the glacial deposit which lies on top of the glaciated surface of the Precambrian Ventersdorp lavas, and marks the base of the Permian-age coal-bearing Karroo Supergroup. Orange Free State Coalfield. (*N. Stavrakis*)

(e) Vein of quartz cutting folded phyllites and mica schists of the Dalradian Series (of garnet-zone grade), Loch Fyne, Argyll, Scotland. Note the fluid-controlled alteration and bleaching of the schists adjacent to the walls of the veins and along selected bands of schist, and the tendency for the quartz to grow inwards from the vein walls. (*K. R. Gill*)

(f) Contact metamorphism of calcareous Jurassic sandstones by a narrow Tertiary basalt dyke. The hornfelsed sediments stand up above the well-jointed (and easily eroded) dyke as two metre high enclosing walls, Bay of Larg, Isle of Elgg, Scotland. (*K. R. Gill*)

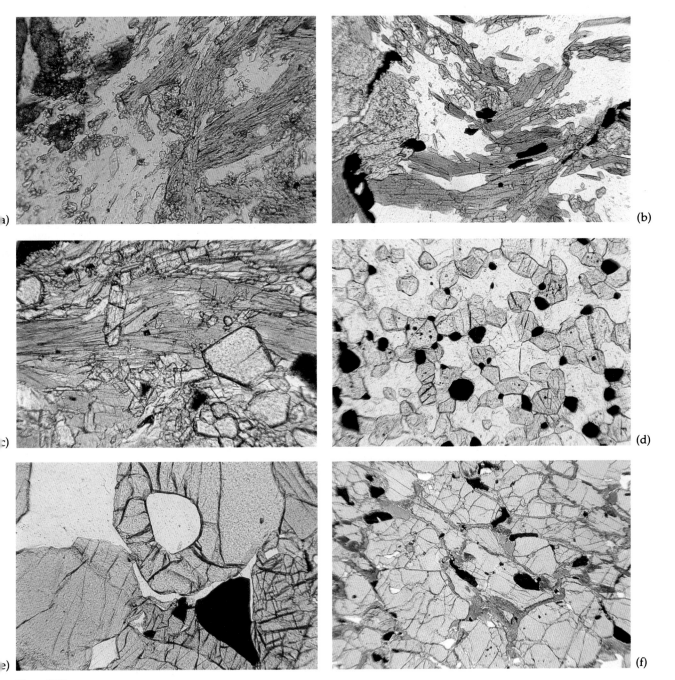

(a)

(b)

(c)

(d)

(e)

(f)

Plate VII

Photomicrographs in cross-polarized light of thin sections of metamorphic rocks of different mineral assemblages. The six rocks illustrated all have similar bulk chemical compositions (equivalent to those of basalts), but have developed widely different mineral assemblages as a consequence of being metamorphosed under different pressure-temperature conditions, corresponding to different metamorphic facies (Chapter 11, Fig.11.13). Note in general that: albite, plagioclase and quartz are colourless-white and have low relief; garnet has high relief and is colourless-brownish; epidote has high relief and is colourless to slightly yellow; actinolite, hornblende and chlorite are green; glaucophane is blue-violet; clinopyroxene is colourless to brown or green. (*B. Harte*)

(a) Greenschist facies with mineral assemblage: albite-actinolite(amphibole)-chlorite-epidote-sphene. The actinolite and chlorite are intergrown with one another; sphene is best seen in the upper left. Field of view is 1.5 mm wide.

(b) Amphibolite facies with mineral assemblage: plagioclase-hornblende (amphibole)-garnet-epidote-ilmenite. The garnet is the large high relief grain on left, whilst epidote forms small high relief grains best seen in upper centre. Field of view is 2.5 mm wide.

(c) Blueschist facies with mineral assemblage: glaucophane-epidote-garnet-sodic pyroxene-rutile. Garnet is abundant in bottom right; epidote in upper centre, and sodic pyroxene at centre left. Field of view is 1.5 mm wide.

(d) Pyroxene hornfels facies with mineral assemblage: plagioclase-clinopyroxene-magnetite. This assemblage forms at pressure-temperature conditions close to those of basalt, and the mineral assemblage is similar to that of basalt, but note that the texture is typically metamorphic with polygonal granular shapes (see Fig.11.1) and not igneous with euhedral rectilinear crystals. Field of view is 1.0 mm wide.

(e) Granulite facies with mineral assemblage: plagioclase-clinopyroxene-orthopyroxene-garnet-magnetite. Orthopyroxene is seen as pinkish-brownish crystals in lower centre; garnet is of high relief in upper centre. Field of view is 4 mm wide.

(f) Eclogite facies with mineral assemblage: largely clinopyroxene-garnet, with some rutile, ilmenite, and quartz. A little alteration to lower temperature conditions has resulted in formation of hornblende (amphibole) along the borders of the clinopyroxene crystals. Field of view is 8 mm wide.

Plate VIII

(a) A basaltic dyke cross-cutting Proterozoic granites, Tugtutoq Island, south Greenland. Note the parallelism of the irregular margins, resulting from dilatational filling of the dyke fissure. The blacker margins to the dyke reflect the finer-grain size of the fast quenched contact facies (representing the first magma to be intruded). (*B. G. J. Upton*)

(b) A view within the Skaergaard gabbro intrusion, East Greenland. Well-developed layering features can be seen in the foreground and near the skyline (upper left). Dark cliff (middle distance) is a younger sill – 'the Basistoppen sheet' of Fig.12.20. High ground in the distance is formed from the Upper Border Series and the Eocene lavas.

(c) A view within the Mull igneous complex, western Scotland. The conspicuous cliff-forming feature in the middle distance is formed by the outer contact wall of the Loch Bá ring-dyke, emplaced during a late-stage ring-fracturing and caldera-forming event (see Fig.12.29). (*B. G. J. Upton*)

(d) Xenoliths of metasedimentary schists at the margin of the Palaeozoic Ross-of-Mull granite, Scotland. Much of the schist has retained its original orientation although some of the xenoliths have been rotated. (*B. G. J. Upton*)

(e) Outcrop of rock in the Østerhorn intrusion, south-east Iceland. Incomplete mixing of basalt and rhyolitic magma has resulted in this very heterogeneous material. The dark areas are products of the basalt: the pale, microgranitic material has resulted from crystallization of the rhyolite. (*B. G. J. Upton*)

(f) Cumulative layering in the ultrabasic rocks of Hallival, Rum, Scotland. Lighter coloured plagioclase cumulates form the cliffs, olivine cumulates the slopes. The Tertiary central complex of the Skye Cuillins is in the background. (*K. R. Gill*)

Plate IX

(a) The jagged front of an aa lava flow, Hekla, Iceland, 1980. (*B. G. J. Upton*)

(b) Basaltic pillow lavas with black glassy rims in a white calcareous mud, South Island, New Zealand. (*K. R. Gill*)

(c) The red layer is the weathered top surface of a lava flow, and is known as a bole (Greek-clod of earth). Chemical weathering after eruption has resulted in the oxidation of the iron-bearing silicates of the lava before the eruption of the succeeding flow. Mt Etna, Sicily. (*K. R. Gill*)

(d) Stratified air-fall pyroclastic deposit, Santonori (Greece). Strata are disturbed by impact of a large volcanic projectile and display 'bomb sag' features. (*B. G. Upton*)

(e) Wairaki geothermal field, North Island, New Zealand. (*K. R. Gill*)

(f) The crater of the active carbonate volcano Oldoinyo Lengai, Tanzania, 23 November 1988. The crater floor is dominated by the breached cone from which new dark-coloured alkali-carbonate lavas are flowing into the depression in the foreground. Older carbonate flows are grey, white or light brown. The crater walls are 45–50 m high and the crater floor 300 m wide. Beyond the crater wall can be seen Lake Natron, an internal-drainage basin lake on the floor of the Rift Valley, with its white evaporite soda-rich crust. (Scale is provided by tents in right foreground.) (*J. B. Dawson*)

Plate X

(a) A kaolinite-rich saprolite which developed on a gneiss-dome before the deposition of Upper Cretaceous Chalk above it, Ivo Klack, Sweden. (*K. Lidmar Bergström*)

(b) Terra rosa soil developed on Cretaceous sandy, shelly, limestone, Carvoeiro, Algarve, Portugal. (*P. McL. D. Duff*)

(c) Podsol, showing zones depicted in Fig. 15.14 (*Courtesy of Professor Gordon Spoor, Silsoe College, Bedford, England*)

(d) Tower karst topography – a residual landform caused by the collapse of caverns, sinkholes and solution cavities which had developed from severe erosion of limestone beds by ground-water. Lijiang River, Guilin, Guangxi-Zhuang Autonomous Region, China. (*Pauline Simpson*)

(e) Braided river crossing the Canterbury Plains, South Island, New Zealand. (*K. R. Gill*)

Plate XI

(a) Corrie on Braeriach, the site of a cirque glacier which fed the valley glacier that flowed down Glen Einnich, Cairngorm mountains, Scotland (cf. Figure 20.30b). (*Mary Allison*)

(b) A valley glacier in Iceland, Breidamerkurjökull, flowing into a lake at its terminus. Crevasses develop as the glacier accelerates into the lake.

(c) Braided streams derived from the ice cap, Vatnajökull, in Iceland, flowing across an outwash plain.

(d) The Lairig Ghru valley in the Cairngorm mountains, Scotland, produced by glacier erosion as ice tongues cut through the mountain mass. (*Angus and Patricia Macdonald*)

(e) The Matterhorn, Switzerland. The classic example of a pyramidal peak (horn) produced by cirque-erosion. (*Courtesy of Petroconsultants, Geneva*)

(f) The Morteratsch Glacier, Bernina Alps, Switzerland, showing the intricate pattern of crevasses developed when valley glaciers change slope and direction. (*Courtesy of Petroconsultants, Geneva*)

Plate XII

(a) Foreground of complex barchanoid dunes with large, coarse-grained ripples in hollow adjacent to flat clay remnant of interdune temporary lake. Beyond scrub-covered middle ground star dunes rise to a height of about 300 m, Namib Desert, Namibia. Note people for scale. (*K. W. Glennie*)

(b) Large compound transverse dunes flanked by the salt marshes of Sandwich Harbour behind the protection of a longshore sandbar on the Atlantic coast of Namibia, south of Walvis Bay. (*K. W. Glennie*)

(c) Sculpted surface of coastal sand dune, indicating prevailing wind direction, Durness, Sutherland, Scotland. (*K. R. Gill*)

(d) Great Barrier Reef, Queensland, Australia. Light-coloured areas are coral and sand just below water-level. Inter-reef areas are 10–30 m deep. The irregular morphology of the patches of coral suggests a pre-Holocene karst foundation below the reefs. (*Courtesy of Australian High Commission*)

(e) Excavated block of modern coral reef from 5 m water depth in Bermuda's North Lagoon, revealing extensive boring by *Lithophaga* bivalves and *Cliona* sponges into coral substrate. [(*T. P. Scoffin*)

(f) Underwater photograph of patch reef with good cover of branching corals (*Acropora* and *Montipora*). Depth 3 m, Pulau Putri, Indonesia. (*T. P. Scoffin*)

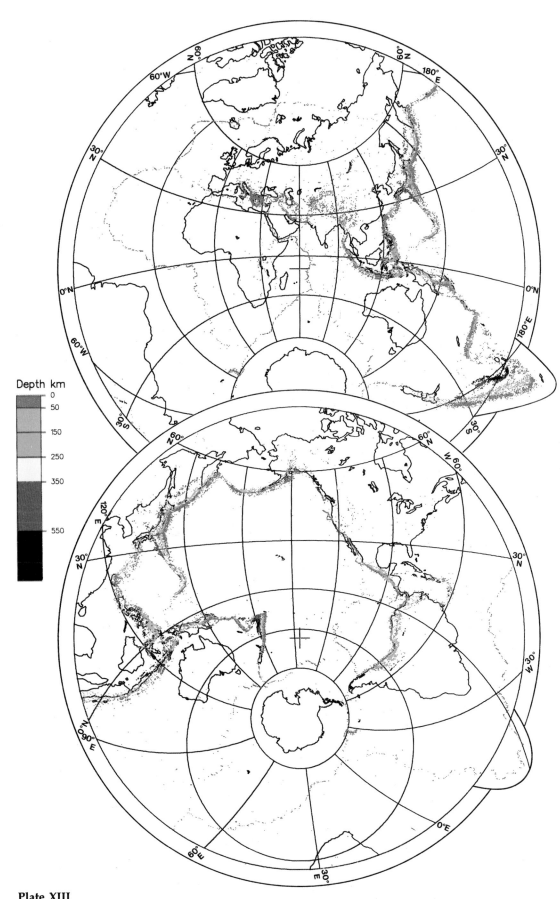

Plate XIII
World distribution of earthquakes greater than body wave magnitude 4.0 from June 1964 to February 1986, shown on two equal angle projections with radius 115°, centred on Lat. 10°S, Long. 60°E and Lat. 37°S, Long. 130°E. This display is designed to show each tectonic plate without gross distortion, and each in one piece except for the North American plate, which is divided between the two plots with the minimum of duplication. Changes in the direction of seismicity (for example at ridge crest/transform intersections) are correctly preserved. Earthquake depth is indicated by a colour scale as shown on the Plate (see Section 25.7). (*Seismicity taken from the Catalogue of the International Seismological Centre, UK*)

Depth km

THE AGE OF THE OCEANS

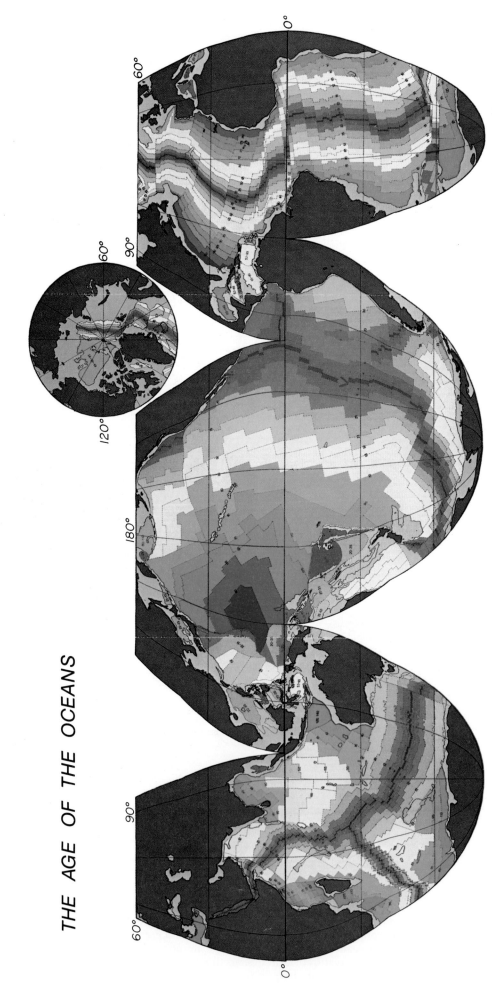

Plate XIV
Ocean floor isochrons (Ma). Spectrum colours from red to blue indicate increasing age. (With acknowledgement to Sclater J.G., Parsons, B. and Jaupart, C. 1981, *J. Geophysic Res.*, **86**, pp. 11541–1154 (Plate 1), and kind permission

Plate XV
(a) The Andean volcano of Osorno in the southern volcanic zone. (*W. Dickson Cunningham*)
(b) Granite intrusion of Cerro Paine, southernmost Chile; the dark-coloured rocks are Upper Cretaceous turbidites that formed the roof of the Miocene pluton. (*I. W. D. Dalziel*)
(c) Upper Cretaceous turbidites of the Magallanes basin foredeep, southernmost Andes. (*I. W. D. Dalziel*)
(d) The Cordillera Darwin metamorphic complex, Chilean Tierra del Fuego. (*I. W. D. Dalziel*)
(e) The active volcano of Mount Erebus, Ross Island, Antarctica. (*I. W. D. Dalziel*)
(f) The Himalayan mountain range where India is colliding with Asia – photographed from a NASA space shuttle.

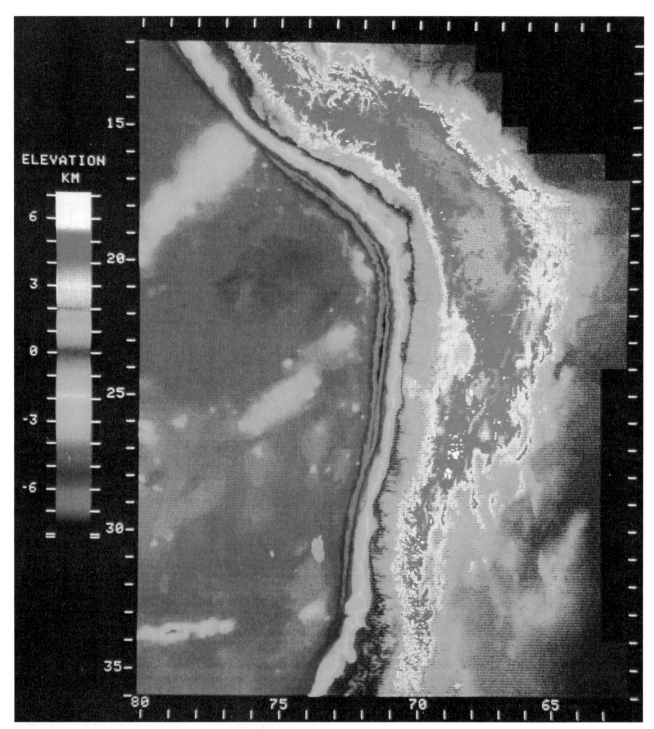

Plate XVI
Computer display of digitized topography of western South America. (*Bryan L. Isacks, 1988, J. of Geophysical Research, vol. 93. Copyright by American Geophysical Union*)

GLACIERS AND GLACIATION

The ground of Europe . . . became suddenly
buried under a vast expanse of ice covering
plains, lakes, seas and plateaux alike.

Louis Agassiz 1840

20.1 ICE ON EARTH

About 14 per cent of the Earth's land area is
covered by ice or underlain by ice-cemented rock,
and about 4 per cent of the ocean surface is
covered by a thin, seasonally fluctuating ice layer
(Figure 20.1). During the glacial periods which
have characterized a large part of the last million

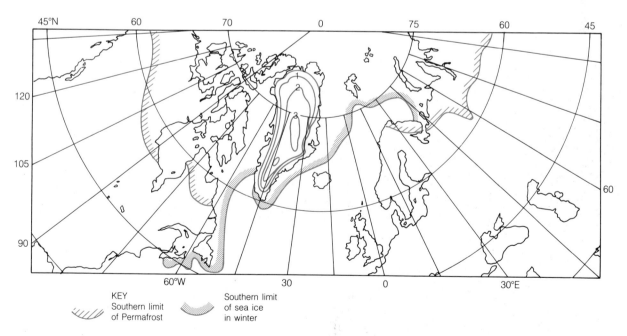

KEY
Southern limit
of Permafrost

Southern limit
of sea ice
in winter

Figure 20.1 Major components of the cryosphere in the Atlantic sector of the northern hemisphere, showing the
southern limits of permafrost and sea ice (in winter) and the form of the Greenland ice-sheet. (Contours are shown
at 1 km, 2 km and 3 km.) Notice the great embayment in the sea ice margin in the north-east Atlantic, produced by
the storms which drive to the north-east past Iceland, and the associated warm water surface currents derived from
the sub-tropics.

Figure 20.2 Sea ice plates in the Ross Sea, Antarctica. In the background is a tabular iceberg which has carved from the Ross ice shelf. The flat top of the iceberg is 37 m above sea-level, suggesting that its keel lies at about 340 m below sea-level (keel depth below sea level ~9 × height above sea-level). See Figure 20.48 for drag marks produced by iceberg keels on the sea bed.

years of Earth history, up to 25 per cent of the land area and up to 6 per cent of the ocean area have been covered or underlain by ice.

This global ice mass (the **cryosphere**) is made up of three principal components

> sea ice
> permafrost
> glaciers

Sea ice (Figure 20.2) is rarely more than 10 m thick and floats as a permanent cover over the Arctic Ocean, the oceans surrounding the Antarctic continent and much of the sea area in the Canadian arctic archipelago. It extends in winter into lower latitudes by as much as several hundred kilometres, and many shallow sea areas of low salinity, such as the Baltic, develop winter sea-ice cover. It leaves little evidence in the geological record.

Permafrost. In polar and high mountain regions where annual average temperatures are lower than –2°C, the ground is **perennially** frozen to depths which can be as great as several hundred metres. In the summer time, the surface of the permafrost melts to form a water-saturated layer up to 1–2 m thick, known as the **active layer**. In summer, soils or sediments in the active layer tend to flow down

Figure 20.3 Tundra polygon on a raised beach in northern Spitsbergen. They are about 200 m in diameter.

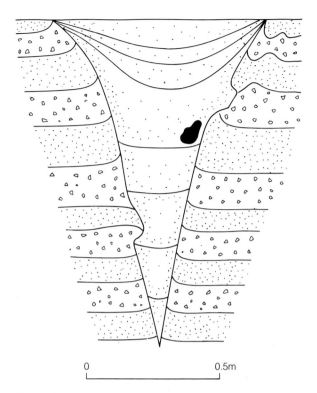

Figure 20.4 Diagram of an ice wedge cast. This represents a tension crack in the ground produced by strong cooling at temperatures below −6°C. The crack is first infilled with ice as it grows, but as the climate warms, the ice melts, and its place is taken by sediment which infills the crack.

slopes as low as 1°, in a process known as **soliflux-ion**. Strong winter cooling, and the contraction associated with it, produces polygonal crack structures known as **tundra polygons** which may be several hundred metres in diameter (Figure 20.3). In summer, the cracks fill in with material from the active layer to produce **frost wedges** (Figure 20.4). Polygons, frost wedges and extensive solifluxion deposits are diagnostic of permafrost in the geological record (see also Sections 19.11 to 19.13).

Glaciers occur widely in polar and sub-polar regions and on mountains in lower latitudes. They consist of moving masses of snow and ice which flow from high altitude where snow and ice accumulate progressively on their surfaces to low altitudes where it progressively melts. At the present day they cover an area of 15.3×10^6 km², 10 per cent of the Earth's land area, and have a total volume of about 30×10^6 km³. During the coldest part of the last glacial period, about 18 000 years BP (Before Present), they covered an area of 40×10^6 km² with a volume of $75–85 \times 10^6$ km³. In contrast to sea ice and permafrost, they leave abundant evidence of their activity in the geological record and, as a consequence, are the subject of the rest of this chapter.

Glaciers occur in two basic forms:

1. those which submerge the topography tend to be dome-shaped, and flow radially outwards from the dome's centre;
2. those which flow along valley channels, being prevented from spreading laterally by flanking mountain ridges.

These two classes can be subdivided as follows:

Dome-shaped glaciers which submerge topography

Ice-sheets. These are continental-size glaciers, represented at the present day by the Antarctic and Greenland ice-sheets, with volumes of 27×10^6 km³ and 2.6×10^6 km³ respectively (Figures 20.5 and 20.6). During the last glacial period, large ice-sheets also existed over northern North America and north-west Europe, with a combined volume of about $37–42 \times 10^6$ km³ (Figure 20.5). They finally disappeared about 7000 years ago.

Ice caps. These are smaller ice domes, such as that of Vatnajökoll in Iceland, and the Penny and Barnes ice caps in Baffin Island. These have volumes of several thousand cubic kilometres.

Summit ice caps. The summits of some flat-topped mountains have small glacier domes. A famous example is that on the Kibo summit of Kilimanjaro in Tanzania, lying at an altitude of 5895 m, and within 3° of the Equator (Figure 20.7).

Glaciers which flow in channels

Valley glaciers. These are long compared with their width. They may be derived from cirques at the head of the valley (Figure 2.6) or from an ice-sheet or ice cap (Figure 3.5) which flows into a valley system and becomes channelled. Confluent valley systems often contain complex confluent glaciers which ultimately feed a single major valley glacier (Figure 20.8). If valley glaciers flow out from between the confining mountain walls onto a flat plain, the ice flows radially outwards to form a semi-circular **piedmont lobe** (Figure 20.9).

Cirque glaciers. These are relatively small ice bodies which occupy cirque hollows high on mountain sides (Figure 20.10). They occur in places where the climate is only just cold enough to maintain a permanent snowfield, and where there is inadequate accumulation of snow and ice to feed a valley glacier. The width and length of the glacier are of similar magnitude. There are many mountain areas which are not currently glaciated (Plate XIa) and where cirques have a moraine at their lip, left by the cirque glacier which occupied it during the last glacial period.

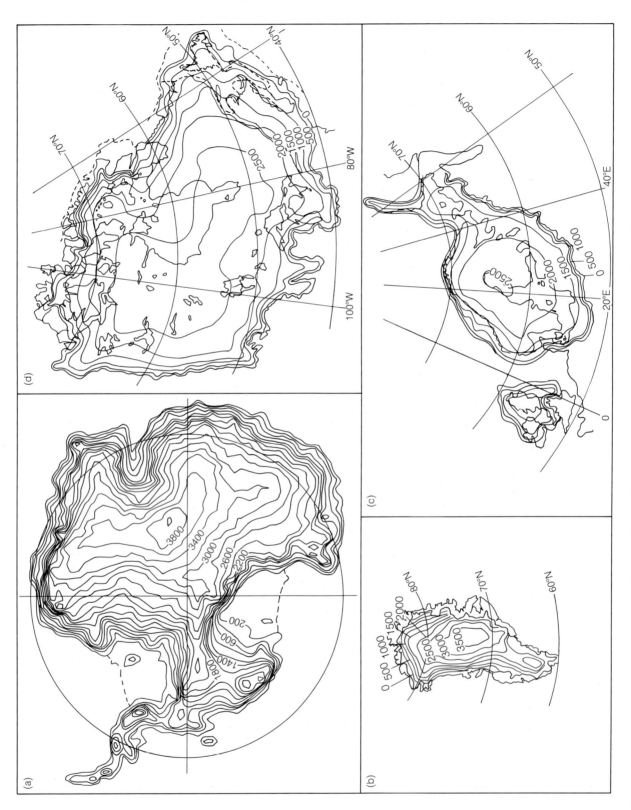

Figure 20.5 Modern and recent ice-sheets on Earth. Figures (a) and (b) show the form of the modern Antarctic and Greenland ice-sheets. Figures (c) and (d) show the reconstructed form of the ice-sheets which existed over Europe and North America during the last glacial period according to Denton and Hughes (1981).

(a)

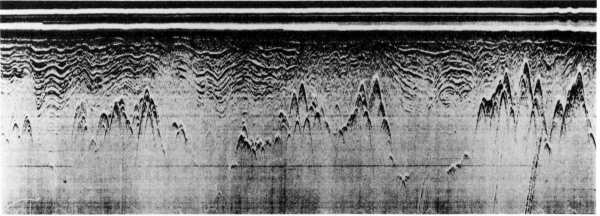

(b)

Figure 20.6 (a) Flow of the Antarctic ice-sheet over a large submerged mountain range. (*British Antarctic Survey*) (b) A radio-echo record from the Antarctic ice-sheet. It shows the ice-sheet surface at the top, and the irregular mountainous bed of the ice-sheet. Internal layering in the ice-sheet can also be seen clearly, and is deformed as the ice-sheet flows over mountainous irregularities. (*D. Drewry*)

Although the plan form of these types of glaciers is very different, they all show several repeated features:

1. The long profile of their surface is parabolic in form, steep at the margin and flattening at or towards their head (at the extreme head of a cirque glacier or valley glacier emanating from a cirque headwall, the ice surface steepens and is tangential to the headwall, figure 20.12).

2. There is a distinctive line on their surface

Figure 20.7 A small ice cap resting on the flat Kibo summit of Kilimanjaro in Tanzania. The summit is at 5895 m and lies 300 km from the Equator. The ice cap is 5 km in width.

separating a zone of new snow above and old ice below.

3. They all show evidence of movement.

20.3 MASS BUDGET OF A GLACIER

Although many glaciers have an extremely long history (the Antarctic ice-sheet has probably existed for more than 30 m.y.), they are in a continual state of flux. The overall mass may change as the climate becomes warmer or colder; they may show a net loss of mass during summer and a net gain during winter; they may be losing mass by melting at a low altitude at the same time as gaining it by snowfall at a high elevation. It is thus very important to know the glacier's **mass budget** – the relationship between the mass gained and mass lost. The budget may be negative (a net mass loss) or positive (a net mass gain) or roughly in balance. Mass is gained by **accumulation** of snow and ice on the glacier surface and lost by **ablation** (the loss of mass by any process, whether it be melting, sublimation or calving of icebergs into the sea or lakes). Almost all the surface of the Antarctic ice-sheet lies in the accumulation area, ablation takes place almost entirely through the loss of icebergs from its margin.

A typical mass budget through a whole year is illustrated in Figure 20.11 for an ice-sheet similar to that of Greenland, by following the pattern of

accumulation and ablation at three points for a year from the end of the summer season. One point is on the ice-sheet summit, one halfway between the summit and margin and a third just above the margin. On the ice-sheet summit (Figure 20.11b) snow accumulates throughout the year, though most rapidly during winter. Temperatures do not rise to 0°C, and thus there is no melting. This is the dry snow zone. Stratification in the snow is produced by the action of wind in sorting the snow crystals as it does desert sand. Further down the ice-sheet surface (Figure 20.11c), temperatures rise above zero during summer, and though some mass is lost by runoff and evaporation, it does not equal the accumulation of snow and ice during winter. There is thus still net accumulation and a positive budget through the year. Some surface meltwater percolates down through the underlying snow to refreeze as a band of ice, often immediately above the old surface from the previous year. At a lower elevation still (Figure 20.11d), the melting season extends from spring to autumn. Not only does all the winter snowfall melt during this period, but melting also cuts down into the snow and ice which formed the surface at the end of the preceding summer.

The first two examples (Figures 20.11b and c) lie in the **accumulation zone** of the glacier where the mass gained during the accumulation season exceeds the mass lost during the ablation season. The third example (Figure 20.11d) lies in the

Figure 20.8 A number of confluent valley glaciers, derived from snowfields in cirque hollows beneath different mountains, which converge to feed a single major valley glacier, the Yentna Glacier, Alaska. The dark ridges are medial moraines (Section 20.14). They stand above the flanking clean ice because the debris cover reduces the rate of melting of the underlying ice.

ablation zone, where mass gained in the accumulation season is exceeded by mass lost in the ablation season. Between the accumulation zone and the ablation zone, lies the **equilibrium line** (Figure 20.11a), where accumulation and ablation balance. It is important to remember (Figure 20.11d) that total accumulation during the accumulation season is greater near to the ice-sheet margin than over the summit dome, as the moisture content of the atmosphere progressively decreases while it moves towards the centre of the ice-sheet, because of precipitation from the air mass. However, because ablation season melting is so strong near to the glacier terminus, ablation in this zone is still greater than winter accumulation, resulting in net ablation and a negative mass budget.

The accumulation zone is a zone of sedimenta- tion where snow crystals are the equivalent of grains of sand in other sedimentary sequences. Year after year, new increments of snow and ice strata are progressively laid down in a sedimentary sequence which obeys the normal laws of stratigraphy and contains a unique record of atmospheric change in the layers of frozen, fossil atmosphere which comprise it. In the Antarctic ice-sheet, the lower parts of such a layered sequence are over 200 000 years old, and the overlying stratigraphy gives vital evidence of global change since that period (Section 21.1).

20.4 THE FLOW OF GLACIERS

If the processes and patterns of accumulation and ablation described above and shown in Figure

Figure 20.9 A small valley glacier fed by upland snowfields emerging from a deep gorge to spread out as a piedmont lobe in the adjacent valley, Ellesmere Island, Canada.

20.11 were the only mass-exchange processes in a glacier, we would expect the accumulation area to increase in elevation and the ablation area to decrease, producing a glacier of diminishing width and increasing height. We know however that glaciers such as the Greenland ice-sheet have only undergone slightly proportional changes in width or summit elevation during the last few thousand years, and yet accumulation and ablation have not ceased during that period.

This is explained by flow within the ice-sheet from the accumulation area to the ablation area. The flow rate has been sufficient to carry away from the accumulation area a mass of ice roughly equivalent to that gained by accumulation and to introduce a mass of ice into the ablation area roughly equivalent to that lost by ablation. If the

glacier is neither to grow nor diminish in volume, the following must be true:

Total accumulation in the accumulation area	= Flow from accumulation to ablation area	= Total ablation in the ablation area

(equation 20.1)

If accumulation exceeds flow, the glacier will grow in size; and if ablation exceeds flow, the glacier will diminish.

The pattern of flow in a cross-section of a glacier parallel to flow is shown in Figure 20.12. Flow occurs in the direction of surface slope. The summit of the ice-sheet/ice cap shown in Figure 20.12a is also a flow divide, called the **ice divide**, from

Figure 20.10 A small glacier lying in a cirque in Spitsbergen. (Cirques are known as 'corries' in Scotland and 'cwms' in Wales.)

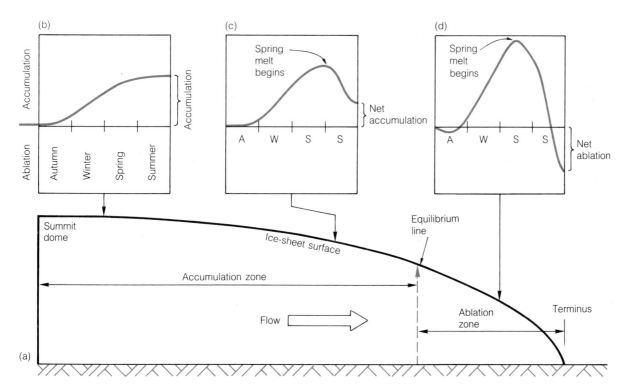

Figure 20.11 Mass budget of an ice-sheet. An ice-sheet (a), or any other glacier, has a zone of accumulation (net mass gain) and a zone of ablation (net mass loss). Ice flows from the accumulation area to the ablation area. High in the glacier or ice-sheet accumulation area (b), the temperature may never rise above freezing and there is continuous accumulation. At a lower elevation (c), some of the snow and ice accumulated during late autumn, winter and spring will be melted during the summer, thus reducing the net accumulation. Below the equilibrium line, where summer melting balances winter accumulation, total summer melting exceeds total winter accumulation (d), to produce net ablation.

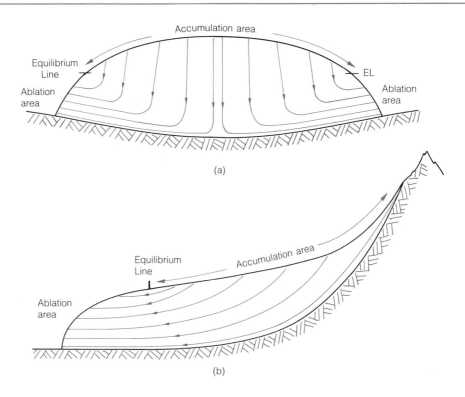

Figure 20.12 Ice flow pattern in a section across an ice-sheet or ice cap (a) and a valley glacier/cirque glacier (b). Flow is downwards, away from the surface in the accumulation area, where, if the glacier shape is constant, downward flow must balance the upward growth of the surface due to net accumulation. Flow is upwards, towards the surface, in the ablation zone, where it must balance the net loss of ice at the surface.

which ice flows in two opposite directions. In the accumulation area, flow takes place downwards into the ice, thus counteracting the upward growth of the surface by snow and ice accumulation. In the ablation area, flow is upwards towards the surface, thus counteracting the lowering of the surface by ablation.

Understanding the rate at which glaciers flow (and many of their characteristics, such as their long profiles as shown in Figure 20.14) requires some knowledge of the flow properties of ice. The force which drives a glacier forward is a gravitational force derived from the fact that the glacier surface slopes towards the terminus. If there were no surface slope, glaciers would not flow. This force is responsible for a **shear stress** (force per unit area) acting horizontally in the ice. This is given by:

Shear	= sine	× ice	× ice	× gravit-
stress	of the	thick-	dens-	ational
	ice	ness	ity	acceler-
	surface			ation
	slope			

$$S = \sin \alpha \times h \times \rho \times g \qquad \text{(equation 20.2)}$$

If we apply a shear stress to a cube of glacier ice as shown in Figure 20.13a it will deform as shown in Figure 20.13b. We measure the deformation rate (equivalent to the flow rate) by measuring the angular distortion of the cube per unit time:

$$\text{Deformation rate} = \frac{a}{b} \times \frac{1}{time} \qquad \text{(equation 20.3)}$$

The results of such experiments for ice are plotted in Figure 20.13c. It shows that if the shear stress is raised from about 0.5 kg per square centimetre (50 kilopascals) to 1 kg per square centimetre, the flow rate will increase not by two, but by about eight times.

20.5 THE SURFACE PROFILE OF GLACIERS

A glacier acts like a conveyor belt whose rôle is to transport mass from a high elevation, the accumulation area, to a low elevation, the ablation area. If the rate of accumulation is very large, the conveyor belt will require large flow rates, whereas if the accumulation rate is small, only small flow rates will be required (equation 20.1). It can be seen in Figure 20.13 that in order to sustain relatively small flow rates, the shear stress needs

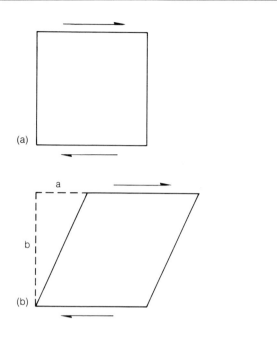

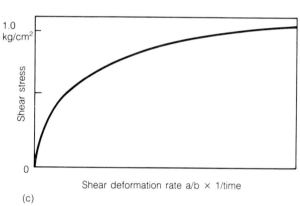

(a)

(b)

1.0
kg/cm²

Shear stress

0

Shear deformation rate a/b × 1/time

(c)

Figure 20.13 How ice deforms. If a shear force is applied to a cube of ice (a) it deforms as shown in (b). a/b × 1/ time is the deformation rate. An experiment in which the deformation rate is measured for increasing values of shear stress gives the result shown in (c), where a doubling of the shear stress gives an eight-fold increase in deformation rate.

possible to reconstruct the form of Quaternary ice-sheets which have long disappeared (Figure 20.5).

Another important consequence is reflected by the fact that, for instance, although the pressure at the bottom of the Antarctic ice-sheet in its central zone, where it is 3–4 km thick, is between 300 and 400 kg per cm², the shear stress responsible for its motion is only about 1 kg per cm².

20.6 THE VELOCITY OF A GLACIER

The relationship in equation 20.1 is a relatively simple but important relationship between the mass budget of a glacier and its average velocity. If we assume a constant average velocity through the whole thickness of the glacier, then we can deduce the change in velocity from the glacier summit to its terminus (Figure 20.14). At a point a distance x from the ice divide, horizontal flow at a velocity u, must be sufficient to remove by flow all the accumulation which occurs up-glacier of that point. The mass flux past this point will be:

Mass flux = average velocity (u) × ice thickness (h)

This must equal the total accumulation rate up-glacier of that point which will be:

distance from the point (x) × average
to the ice divide   accumulation (a)

Thus $u = \dfrac{ax}{h}$ (Figure 20.14b) (equation 20.4)

This will increase from zero at the ice divide (x = 0) to a maximum at the equilibrium line, but from there to the terminus must decrease as ablation continually reduces the mass flux (Figure 20.11c).

20.7 FLOW STRUCTURES IN GLACIERS

The accumulation area of a glacier is like a sedimentary basin in which strata of snow and ice accumulate annually in winter, and contain a stratigraphic record of environmental change. However, this mass also moves horizontally, deforming as it does so. As a consequence of down-glacier velocity increase in the accumulation area, individual ice strata stretch longitudinally, and therefore become thinner vertically. At the same time, recrystallization of the ice tends to occur. As the ice stretches and thins, it comes closer to the glacier bed, which can be extremely irregular (Figure 20.6b), and which sets up large stresses and strains within the ice, resulting in

to be near to 1 kg per cm², whereas to sustain very large flow rates, the shear stress still need not be much larger than 1 km per cm². Thus, for most glacial conditions, the shear stress at the base of a glacier can be assumed to be constant at about 1 km per cm². This has a very important consequence. As explained above (equation 20.2), the shear stress at the base of the glacier is the product of ice thickness and surface slope. If the shear stress is constant, a large ice thickness must be associated with a small surface slope, and a small ice thickness with a large surface slope. The longitudinal profile of a glacier will therefore have a parabolic form, with high slopes at the margin and low slopes in the accumulation area (Figure 20.14). This repetitiveness of ice-surface profile makes it

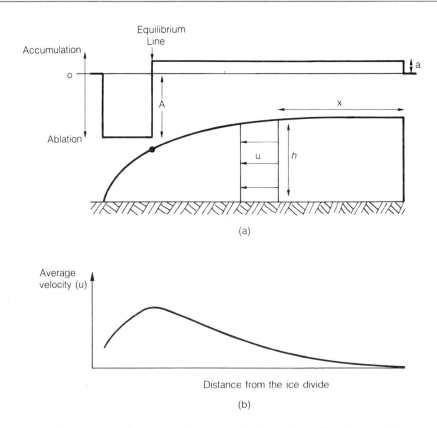

(a)

(b)

Figure 20.14 (a) A simplified pattern of accumulation on a glacier surface. For flow to balance accumulation at a point x, the accumulation up-glacier of that point (a × x) must be balanced by flow through x (h × u). The velocity (u) at x is therefore given by $u = ax/h$. The velocity distribution to which this will give rise is shown in (b). Velocity will rise from zero at the ice divide to a maximum near to the equilibrium line, from where it will decrease to the terminus.

acute folding (Figure 20.15). The ice has, by now, made the transition from a sedimentary to a metamorphic material.

In the accumulation area, the glacier surface is that of the last sedimentary layer, and tends to appear relatively smooth and structureless. In the ablation area, however, these sedimentary layers are progressively stripped away by melting, to reveal older, recrystallized and folded metamorphic ice beneath.

The rate of horizontal stretching resulting from the longitudinal velocity increase due to the glacier's mass balance is relatively small. If, however, the glacier flows down a steep section of bedrock, the stretching rate can be so large as to cause the ice to fracture in the form of **crevasses** (Plates XIb, f). Any process which causes strong longitudinal accelerations will produce crevasses. For instance, strong crevassing is often produced where ice accelerates as it flows into the lakes or sea (Figure 20.16). These crevasses are the natural fractures which permit icebergs to break away from the glacier.

20.8 GLACIER–CLIMATE RELATIONSHIPS

The rate of mass exchange between the accumula-

tion area and the ablation area of a glacier, represented by the average velocity through the equilibrium line, is a useful indication of the dynamic activity of a glacier. (It is important however that this should only be used to compare glaciers of similar length, as velocity does depend on the length of the flowline – equation 20.4). Thus, glaciers with a large accumulation rate also have a large equilibrium line velocity, and those with a small accumulation rate, a small equilibrium line velocity.

Very large accumulation rates tend to occur in areas of maritime climate which are relatively warm and wet, with large accumulation rates and large ablation rates. Very low accumulation rates occur in areas of continental climate which are relatively cold and dry, with low accumulation rates and low ablation rates. The distribution of these climate types is largely determined by the location of atmospheric storm tracks. In the North Atlantic and Norwegian Sea areas for example (Figure 20.17), major storms tend to be derived from the sub-tropical ocean. The glaciated areas which lie along the axes of the storm tracks in Iceland, western Norway, western Spitsbergen, Jan Mayen and south-east Greenland, have glaciers with large winter accumulation rates (> 2 m

Figure 20.15 Aerial photograph of folding in a glacier due to its flow over an irregular bed, Breidamerkurjökull, Iceland. The individual folds are several hundred metres across and are picked out by bands of volcanic ash which fall onto the glacier surface from time to time from the eruptions of Icelandic volcanoes. When the ice flows into the ablation area, melting reveals the ash stratigraphy. The sequence of layers shown here can be thought of as a history of recent volcanic eruptions in Iceland.

per year), high summer ablation rates and very high equilibrium line velocities (Figure 20.14). In more continental areas, glaciers such as those of Sweden, on the eastern side of the Scandinavian mountain chain, and of north Greenland, have low accumulation rates, low ablation rates and low equilibrium line velocities.

The dynamic activity of a glacier, and therefore its capacity to erode is primarily determined by the energy of the regional atmospheric circulation system.

Glacier temperatures. Climate also determines the temperature regimes of glaciers. Some examples of temperature measurements which have been taken down boreholes in glaciers are shown in Figure 20.18. At Camp Century, site of a borehole through the northern Greenland ice-sheet, temperatures near surface are about −25°C, and rise to −12°C at the glacier bed. At Byrd Station, site of a borehole through the west Antarctic ice-sheet, temperatures near surface are about −30°C and rise to the melting point at the glacier

Figure 20.16 The front of the Miles Glacier, Alaska, where crevasses are produced by strong longitudinal stretching as the glacier accelerates to calve into the sea. The photograph was taken a few seconds after the iceberg calved from the 46 m high ice cliff. It is shown largely submerged due to the momentum of its fall, but shortly rose buoyantly to float away from the glacier. Note the beginning of a small shock wave in the sea. (*Bradford Washburn*)

bed where meltwater is present. These temperature gradients, from warm at the base to cold at the surface, reflect (i) that the atmosphere is cold, (ii) that the heat sources are at the base of the glacier (geothermal heat and the heat produced by the friction of the glacier moving over its bed) and (iii) that ice is a relatively good insulator (with a thermal conductivity similar to granite). In contrast, most mountain glaciers in temperate regions and many glaciers in sub-polar regions have temperatures at the melting point throughout. This is

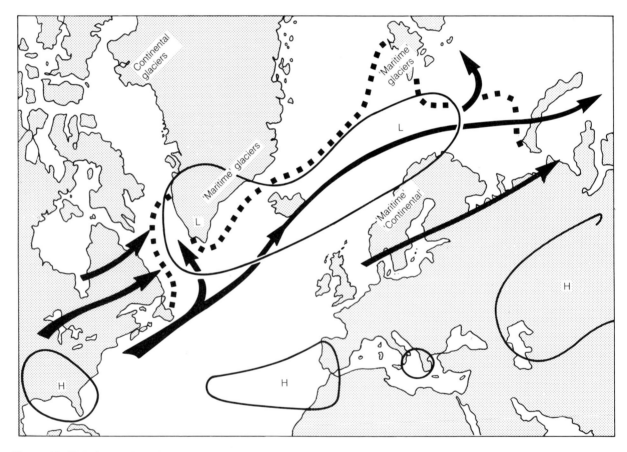

Figure 20.17 Relationships between storminess and glacier activity. Moisture- and heat-transporting storms in the North Atlantic and Norwegian sea produce high accumulation rates and relatively high temperatures on the southern part of the Greenland ice-sheet, on the glaciers of Iceland and those of western Norway and western Spitsbergen. Away from this maritime zone, the glaciers of Sweden and northern Greenland are more continental; much colder and dryer. The 'maritime' glaciers therefore have high accumulation and ablation rates and high flow velocities. More 'continental' glaciers have low accumulation and ablation rates and low flow velocites (see Figure 20.14). Glacier ice in warm maritime zones tends to be at the melting point, whereas in cold, continental areas, tends to be below the melting point (Figure 20.18).

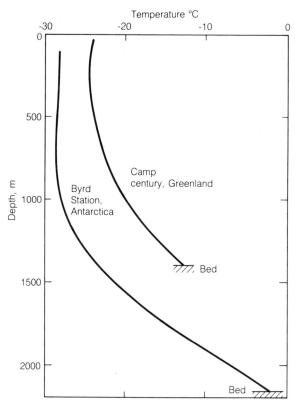

Figure 20.18 Measured temperatures in boreholes at Camp Century, on the north-west flank of the Greenland ice-sheet, and Byrd Station on the West Antarctic ice-sheet. The upward temperature gradient reflects the insulating effect of an ice-sheet. The atmosphere is very cold, but geothermal heat warms the base of the ice-sheet. At Byrd Station the ice is sufficiently thick that basal ice is actively melting.

due to relatively warm surface temperatures and the fact that much of the winter snow in the accumulation area of temperate-zone glaciers is saturated with summer meltwater, which re-freezes in winter, thereby warming up the near-surface layers due to the release of latent heat.

Figure 20.19 shows a reconstruction of ice temperature in an east–west section across the Greenland ice-sheet. The coldest area lies beneath the summit dome, where basal temperatures are far below the melting point. This cold ice is also carried outwards by flow, cooling the outer part of the ice-sheet. Nearer to the margin, basal temperatures rise to the melting point and basal meltwa-

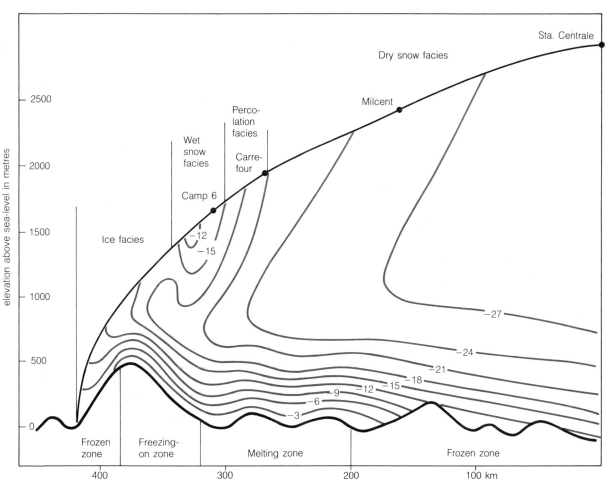

Figure 20.19 The internal temperatures in a cross-section through the Greenland ice-sheet at 70°N. The nature of snow and ice on the surface are also shown as are conditions of melting/freezing on the bed.

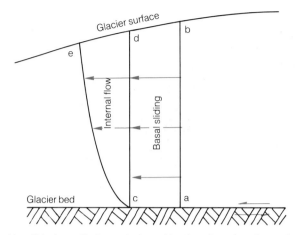

(a) Velocity profile for a glacier melting basally and resting on bedrock

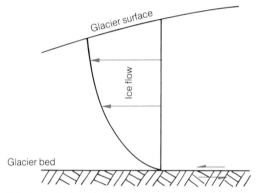

(b) Velocity profile for a glacier resting on a frozen rock or sediment surface

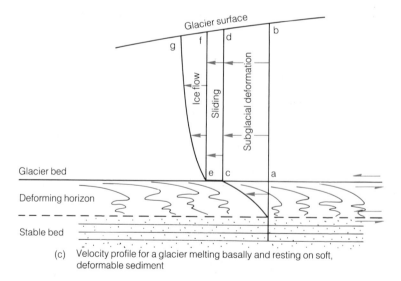

(c) Velocity profile for a glacier melting basally and resting on soft, deformable sediment

Figure 20.20 (a) The velocity distribution within a glacier sliding over its bed, where the bed is at the melting point. If we insert a vertical line within a glacier (ab) at a given time, this line will be displaced because of sliding at the base to position cd. However, internal flow also occurs, which will displace the line further in the same time to position ce. At the base, the total movement during this time will be a–c, produced entirely by sliding. The flow within the ice however, will ensure that the total surface movement will be greater, b–e, the product of both basal sliding and internal flow. (b) Glacier movement where the bed is below the melting point and no sliding occurs. Movement takes place by internal flow alone. (c) Glacier movement where the glacier is at the melting point, and there are soft, water-saturated sediments. There is an additional flow component due to soft-sediment deformation.

Figure 20.21 A panoramic view of a cavity on the down-glacier side of a bedrock hummock where the glacier has lost contact with its bed. The glacier sole loses contact with bedrock on the left and regains it on the right. The far wall of the cavity consists of fractured ice. The distance from cavity opening to closure is 24 m. The maximum height of the cavity is 4 m. The thickness of overlying ice is 200 m.

Figure 20.22 Soft sediments which have lain beneath a glacier and which have undergone shear deformation beneath it. The glacier beneath which they lay probably had a velocity profile such as that shown in Figure 20.20c.

ter is produced. This meltwater is pressed outwards towards regions of lower pressure at the margin. In the extreme marginal zones, the ice is so thin as to be a poor insulator of the bed against the cold atmosphere, so that the bed of the glacier again lies at the melting point.

Glaciers can also be classified according to their thermal conditions:

1. **Polar glaciers** are those whose internal temperatures lie entirely below the melting point of ice. Examples are the north Greenland ice-

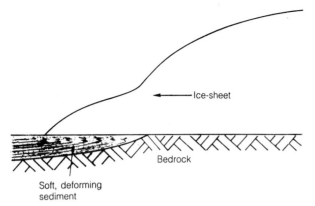

Figure 20.23 Reduction of the surface slope of a glacier in its terminal zone where it has flowed over soft, easily-deformed sediments. Deformation of the type shown in Figure 20.22 may be produced in the outer zone.

Figure 20.24 The highly-crevassed surface of an ice stream in the West Antarctic ice-sheet, where the glacier flow velocity is very much greater than in the flanking more sluggish ice. The ice stream is flowing towards the viewer and is several km in width.

sheet, the east Antarctic ice-sheet and many glaciers in the High arctic, such as the Devon Island ice cap.

2. **Sub-polar glaciers** are those whose internal temperatures are partly below the melting point and partly (generally at the base) at the melting point (Figure 20.19). Examples include the south Greenland ice-sheet, the west Antarctic ice-sheet and the glaciers of Spitsbergen.

3. **Temperate glaciers** are those whose internal temperatures are at the melting point. They include most European Alpine glaciers, Norwegian glaciers, most Andean glaciers, Icelan-

Figure 20.25 Looped and folded moraines produced by the surge of the Sutsina Glacier, Alaska, in 1952. (*Bradford Washburn*)

Figure 20.26 A bedrock surface polished and scratched by the movement over it of the glacier sole which was heavily charged with rock debris. Ice movement was from right to left. Islay, Scotland. (*British Geological Survey*)

Figure 20.27 Striated glaciated pavement of Precambrian basalt marking the Permian glaciation of South Africa. The Dwyke Tillite overlies this surface, Nooitgedacht, South Africa. (See Plate IXd and Section 9.5.) (*A. M. Duggan-Cronen*)

dic glaciers, and New Zealand glaciers. They melt basally throughout, and the meltwater is squeezed from beneath them in large volumetric discharges.

20.9 THE BASAL MOTION OF GLACIERS

The basal plane of the glacier is called the **glacier sole**. The sole of a temperate glacier, or those basal parts of a sub-polar glacier which are at the melting point, is continuously melting. Where the glacier sole moves over a rock bed, meltwater forms a thin film between the glacier and its bed, reducing friction so that sliding of ice over bedrock can occur (Figure 20.20). In these cases, the forward movement of the glacier is achieved by a combination of two processes – internal flow, which has already been discussed, and basal sliding.

Where sliding occurs and a glacier moves over a bedrock knob, the glacier can lose contact with its bed on the lee of the knob to form a natural cavity (Figure 20.21). If the sliding velocity of the glacier increases, the cavity will tend to increase in size as the ice travels further in the lee of the knob before coming to ground. If the ice thickness increases, the cavity will become smaller as ice is pressed down to close the cavity nearer to the knob. Where ice is in contact with bedrock, the bedrock tends to

become smoothed and striated, whilst it tends to remain irregular in the cavity.

Where the glacier sole is below the melting point, increased ice–rock friction in the absence of a water film, and adhesion between cold ice and rock, make the **shear strength** of the ice–rock interface much larger than the typical shear stress at the base of a glacier of about 1 kg/cm². In this case, sliding does not occur, and forward motion of the glacier takes place by internal flow alone (Figure 20.20).

Many glaciers do not flow over rock beds but over beds of soft sediment such as sand, silt and clay. Where the glacier sole is melting, meltwater flows into the interstices of the sediment, producing a soft, easily-deformed slurry-like mass. In response to the shearing force imposed by the overriding glacier it forms a continuously deforming carpet on which the glacier moves. A large part of the forward movement of the glacier may be produced by flow of this soft sediment layer (Figures 20.20c, 20.22, 20.23). In such cases movement of the glacier may not primarily be produced by flow of ice and controlled by the mechanical properties of ice, but by the properties of soft, wet subglacial sediments. As these tend to be weaker

Figure 20.28 Glacially-smoothed and striated bedrock hummocks; (roches moutonnées). Ice movement was from left to right. Ardnamurchan, Scotland. (*British Geological Survey*)

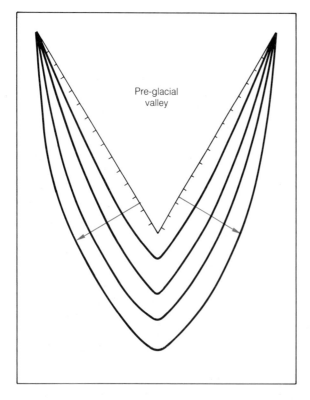

Pre-glacial valley

Figure 20.29 Theoretical pattern of glacier erosion which causes the original cross-profile of a V-shaped valley to evolve into a U-shaped profile because of glacier erosion across the whole cross-section instead of concentration of erosion on the centre-line of the valley as is achieved by rivers.

than ice, the glacier conveyor belt can operate at the required speeds at shear stresses less than the 1 kg per cm^2 typical of glaciers flowing over rock beds, and thus, from equation 20.2, the surface slope of the glacier can be smaller than in rock-floored glaciers. Figure 20.23 shows how the surface slope of a glacier might be reduced as it flows from an area of bedrock to one of soft sediments. Such a change may have occurred in the ice-sheets which built up in mid-latitudes of the northern hemisphere during the glacial periods. They flowed out from both the Scandinavian mountains onto the soft sediments of the Baltic, North Sea and north European plain, and from the hard rocks of the Canadian Shield onto the soft sediments of Hudson Bay, the Great Lakes and the St Lawrence valley.

20.10 FAST ICE STREAMS AND GLACIER SURGES

Ice-sheets are not entirely smooth, sluggishly flowing domes of ice. They are occasionally marked by relatively narrow (5–50 km), highly crevassed streams of fast flowing ice, where velocities may be over ten times greater than in the sluggish flanking ice (Figure 20.24). In the West Antarctic ice-sheet, although these **ice streams** comprise no more than 10–15 per cent of the ice-sheet area, they discharge by flow into the ocean up to 80 per cent of the annually discharged

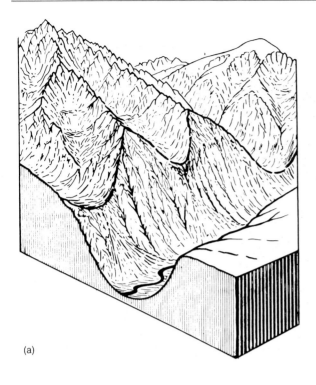

(a)

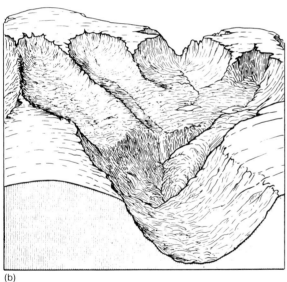

(b)

Figure 20.30 Glaciated landscape in a mountainous region. (a) Shows the contrast between the overdeepened valleys of fast-flowing **trunk glaciers**, into which hanging valleys drain having been eroded less deeply by more sluggish tributary glaciers. The trunk glacier erodes the ends of spurs between tributary glaciers producing **truncated spurs**, whilst adjacent tributary glaciers erode into intervening ridges to produce *arêtes*. (b) Shows the product of cirque glacier and valley glacier erosion in an otherwise unglaciated upland (or one where more extensive ice-sheets have been frozen basally). **Trough-ends** occur at the heads of valleys. (*Modified after W.M. Davis*)

mass. They are thus of great dynamic importance. They may occur because of long troughs in the bed

(cf. Figure 20.33) which channel fast flow, or because an area is underlain by soft deformable sediment. In the Antarctic ice-sheet, many ice streams feed **ice shelves**, which are floating extensions of the ice-sheet.

Ice streams are stable features, rather like valley glaciers with ice walls; but from time to time hitherto relatively sluggishly-flowing glaciers will accelerate into a fast flow for a short period (1–2 yrs) before becoming stagnant. This unstable phenomenon is known as a **surge**. During a surge, the glacier becomes highly crevassed and intensely folded (Figure 20.25). Most surges probably result from the build up of subglacial meltwater, which detaches the glacier from its bed, thereby reducing basal friction and permitting it to accelerate to fast flow, when velocities in excess of 1 km per year are common. Valley-glacier surges occur, as do surges which have affected sectors of ice caps. Ice-sheet surges have not been observed, although there may be evidence for them in the geological record. Potential Antarctic ice-sheet surges are a matter for concern, because if large masses of Antarctic ice were to flow into the ocean there would be a consequent rapid rise in global sea-level.

20.11 PROCESSES AND RATES OF GLACIER EROSION

When glaciers flow over rock 'knobs', there is a tendency for them to **quarry** or **pluck** blocks from the lee side of the knob. This probably occurs because of the pressure release on the lee side of the knob which causes rock fracturing, permitting loosened, fractured blocks to be plucked away by the ice as it flows around the knob.

These plucked blocks are then carried along by the ice and embedded in the glacier sole, where they act as grinding tools by cutting **striations** into bedrock. They themselves are further fractured by this process, producing a wide range of grain sizes from fine **rock flour** to boulders. Whilst the larger blocks cut striations, the **rock flour** polishes the bedrock surface, thereby producing the smoothed and striated bedrock surfaces so typical of glaciated terrain (Figures 20.26, 20.27). The combination of **lee** side plucking and up-glacier (**stoss**) side-smoothing and striation of bedrock knobs produces the widespread forms known as **roches moutonées** (Figure 20.28). Roches moutonées often show a well-defined line dividing the smoothed and striated stoss side from the plucked lee-side cavity opened (Figure 20.21). Not only is the bed smoothed and striated, but so are many of the smoothed and striated, but so are many of the

Figure 20.31 A glacially-overdeepened fjord, Naeröyfjord, on the south side of Sognefjord, Norway. (*Mittet Foto, Norway*)

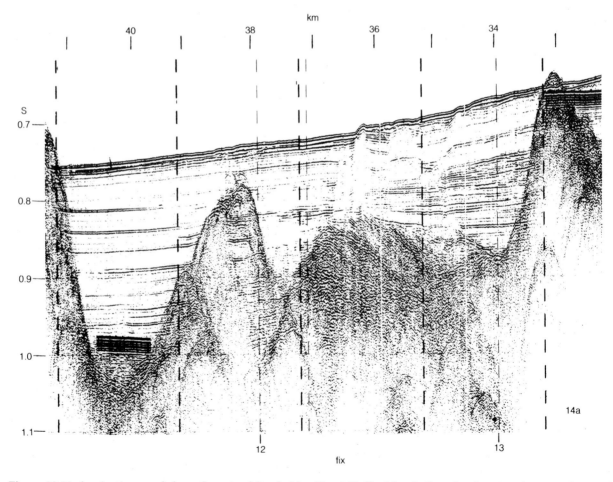

Figure 20.32 A seismic record along the axis of Cambridge Fiord, Baffin Island, Canada, showing the irregularity of the bedrock surface (the strong reflector at the base) and the depth of sediments which fill in overdeepened hollows (the sediments show approximately horizontal seismic reflectors). The horizontal scale is in kilometres. The vertical scale is in two-way travel time (seconds) for the seismic waves. A 0.1 second interval is equivalent to above 75 m of sediment. The maximum depth of sediment shown is about 230 m.

Figure 20.33 Large-scale glacially-deepend valley in Labrador. A fault zone of pre-existing weakness has been accentuated by glacier erosion.

larger blocks which act as cutting tools in the glacier sole.

In ice-sheets and ice caps, which submerge bedrock topography, almost all erosion takes place **subglacially** and erosional debris is derived from the glacier bed. In valley and cirque glaciers erosion also takes place on the flanking mountain walls. Much of the debris prised free from these hillsides by the splitting action of frost or by chemical weathering, is carried down onto the glacier surface by avalanches, by rock-falls, in scree runnels or in mass flows. Rather than building up at the foot of the slope in an avalanche cone, alluvial fan or scree cone, as would happen

in an unglaciated valley, the glacier continuously carries away this debris, thereby keeping the lower mountain slopes clear of superficial detritus and permitting subglacial erosional processes to operate on fresh bedrock.

Measured rates of bedrock erosion by glaciers tend to be in the region of 1 mm per year. Whilst this may appear to be a small rate, on a geological timescale it is very large, being equivalent to 1 km per million years; quite enough to erode many of the more dramatic glacially-eroded features produced during the recent ice age. Detailed studies of glacier erosion suggest that the erosion rate (E) is proportional to between the second or third

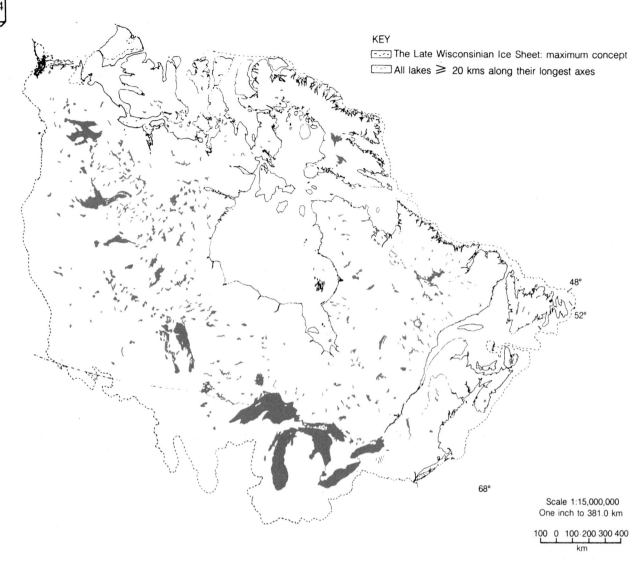

KEY

⌐∙∙∙⌐ The Late Wisconsinian Ice Sheet: maximum concept

⌐‗‗⌐ All lakes ≥ 20 kms along their longest axes

48°

52°

68°

Scale 1:15,000,000
One inch to 381.0 km

100 0 100 200 300 400
km

Figure 20.34 The pattern of large lakes around the Canadian Shield produced in a zone of enhanced erosion near to the margin of the ice-sheets which have occupied the northern part of the North American continent from time to time in the recent geological past.

power of the basal ice velocity (U):

$$E \propto U^{2-3} \qquad \text{(equation 20.5)}$$

20.12 LARGE-SCALE EROSIONAL FEATURES PRODUCED BY VALLEY AND CIRQUE GLACIERS

Valley glaciers and cirque glaciers, which are contained between high, flanking valley walls in mountain regions, are responsible for the most obvious of large-scale features produced by glacier erosion. At the beginning of an ice age, when cirque and valley glaciers form for the first time in a hitherto unglaciated mountain region, glaciers flow down typical river-eroded valleys. Because river channels and flood plains are relatively narrow compared with the total width of the valley, erosion tends to be concentrated in or near the river channel. Erosional lowering of this central zone of the valley tends to produce a V-shaped valley cross-profile. When a glacier occupies the valley for the first time, erosion occurs across the whole width of the valley covered by the glacier. Although ice velocities along the centre-line of the valley are greater than those near the margin, producing greater centre-line erosion (see equation 20.5), general lowering of the whole valley perimeter will tend to produce the *U-shaped cross profile* typical of glaciated valleys (Figure 20.29).

In complicated valley-glacier systems, where relatively minor **tributary glaciers** feed more rapidly flowing **trunk glaciers**, the trunk glacier tends to erode more rapidly than the tributary glacier (see equation 20.5). As a result, the trunk valley tends to be deeper than the tributary valley, which forms a **hanging valley** separated from the main

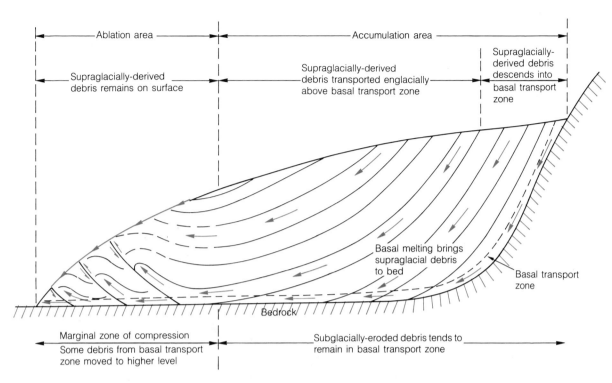

Figure 20.35 The transport pathways of debris through a valley or cirque glacier. Debris eroded from the glacier bed is transported along a basal flowline, from which it may be deposited beneath the glacier, or exposed on its surface. From here it may slide off the frontal slope of the glacier, or be lowered onto the bed as the glacier melts back. Debris falling onto the glacier surface from the glacier headwall also finds its way into a basal flowline. Debris falling onto the glacier surface from a nunatak or valley wall in the accumulation area, travels within the ice before being exposed on the glacier surface as lateral or medial moraine. From there, it is let down onto the glacier bed by melting, or slumps down the terminal slope. Debris which falls on the surface in the ablation area remains on the glacier surface until being deposited.

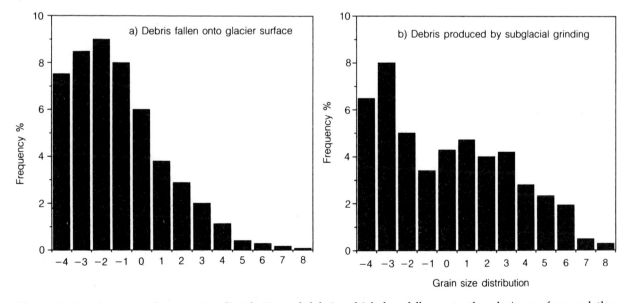

Figure 20.36 (a) A typical grain-size distribution of debris which has fallen onto the glacier surface and then travelled within the ice. (b) A typical grain-size distribution of debris which has travelled in the basal part of the ice, and in which fine sand and silt is produced by grinding against the bed. Grain sizes are shown in ϕ units, -4ϕ is 16 mm, 0ϕ is 1 mm, 4ϕ is 0.06 mm.

valley by a conspicuous rock step (Figure 20.30a). The erosional vigour of the trunk glacier also tends to erode away spurs which descend into it from flanking mountains, producing **truncated spurs** (Figure 20.30a). Erosion on the margins of glaciers in adjacent valleys tends to cut into the interven-

Figure 20.37 Lateral moraines produced by dumping of debris which has been transported within the ice or on its surface in the way shown in Figure 20.35, Ladd Glacier, Cascade Range, USA. (*E. R. La Chapelle and A. Post*)

ing ridges producing narrow **arêtes** (Figure 20.30) and in some cases **horns** (Plate XIe). The cirques from which glaciers originate in valley-glacier systems often have steep back-walls, or **trough-ends** (Figure 20.30b). All these features are characteristic of glaciated mountain regions, and are often very strikingly seen in fjords, which are glaciated valleys which have been inundated by the sea (Figure 20.31).

Another important feature produced by many valley glaciers and cirque glaciers, is the **overdeepening** of valleys at some distance behind a long-term terminal position of the glacier. This can be understood by considering the longitudinal distribution of velocity in a glacier (Figure 20.14b), which reaches its maximum value near to the equilibrium line. Because of the relationship between erosion rate and ice velocity (equation 20.5), greatest erosion will occur in the area of greatest velocity, and thus overdeepening will

tend to occur at some distance back from long-held terminal positions. Overdeepened rock basins in fjords (Figure 20.32) and behind the lips of cirques which have held glaciers can be explained in this way.

Large-scale patterns of erosion produced by ice-sheets may be well-defined troughs and linear patterns in **zones of linear erosion**. These occur where erosion rates beneath the ice-sheet have been relatively high and pre-existing valleys or zones of low rock hardness have been overdeepened and accentuated by glacier action (Figure 20.33). A general regional overdeepening on a very large scale indeed may also occur as **zones of areal erosion**. These are too large to be seen as obvious erosional features and can only be reconstructed by continental-scale mapping. Figure 20.34 shows the pattern of large lakes around the Canadian shield, parallel to, and some distance behind the margin of the east North American ice-sheet.

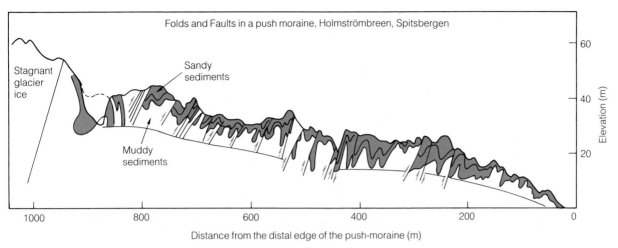

Figure 20.38 A section through a large push-moraine in Spitsbergen. The glacier, Holmströmbreem, surged in the last century and pushed up a series of sands and muds which lay on a planar surface beyond it, and rucked them up into a push moraine some 1 km in width. Its internal structure shows a series of overturned folds and thrust faults similar to those formed in mountain belts. The glacier advanced from left to right.

These are believed to have been produced in a zone of enhanced erosion by successive ice-sheets which have had a similar extent. They probably reflect the location of the zone of high ice velocity and therefore high erosion rate, which lies at some distance behind the margin of an ice-sheet (Figure 20.14). Zones of linear erosion also develop best in these areas of high velocity.

There is very little erosion in the ice-divide region of an ice-sheet because of the low velocities (Figure 20.14), although the tendency for the ice-sheet bed to be frozen in this divide region also inhibits erosion (Figure 20.19). In the Cairngorm Mountains of Scotland, for instance, Tertiary soils survive on the flat mountain-top plateaux at over 1000 m elevation, probably because of the fact that the ice-sheet was relatively thin over the mountain tops, making it a less efficient insulator and permitting very low basal temperatures, and strong basal freezing to inhibit sliding and erosion at the glacier sole.

In ice-sheets where the ice divide lay in mountain areas, such as the Scandinavian ice-sheet, whose initial source was the Scandinavian mountain chain, we probably see evidence of two types of strong erosion. One was during early and late phases of a glacial period when the mountains were occupied by valley and cirque glaciers, producing the assemblage of features illustrated in Figure 20.30 (see also Plates XId, e). The second during glacial maxima, when the ice divide lay to the east of the mountains, but when strong ice flow to the west over and through the mountains took advantage of convenient valleys and enlarged

them as **through troughs** cutting through the mountain chain.

20.13 GLACIAL DEPOSITION

Debris transported by glaciers is termed **till** when it is deposited. The constituent debris is derived, transported and deposited by a variety of pathways.

Ice-sheets and ice caps, except where they are pierced by many **nunataks** (mountain peaks protruding above the ice), only transport subglacially eroded **debris**, which they do in a relatively thin layer of basal ice, up to a few metres in thickness. Valley and cirque glaciers however, also transport debris from flanking mountain walls (Figure 20.8). When this debris falls onto the glacier surface in the accumulation area, it is buried by accumulating snow and ice, and moves along the flow lines shown in Figure 20.35, until it emerges in the ablation area as snow and ice is progressively stripped from the surface by melting, to accumulate on the surface near the margins of the glacier as **lateral moraine** (Figure 20.8). Where two valley glaciers join, their lateral moraines at the point of confluence also join to form a **medial moraine** (Figure 20.8). Each of the multiple medial moraines shown in Figure 20.8 represents a single confluence. The minimum number of confluent glaciers which have merged to produce the trunk glacier can be counted from the number of threads of ice seen between medial moraines. It is a minimum of 28!

Figure 20.39 A series of small push-moraines formed in front of the glacier Fjallsjökull in Iceland. The glacier has been retreating since the beginning of this century, but in winter, it readvances, to produce small push moraines about 1–2 m high. Their spacing indicates the net retreat of the glacier from year to year. (*Landmaelingar Islands*)

Where a subglacial rock bed has been eroded, the processes of crushing and abrasion during transport produce debris with a wide range of grain sizes, from silt through sand to gravel and large boulders. This may be deposited subglacially, as till, or it may be transported to the surface of the glacier in the terminal zone, from where it is dumped at the glacier front. Debris derived from flanking mountains in a valley or cirque glacier, or from nunataks in an ice-sheet or ice cap, can travel along either a basal transport path (Figure 20.35) to be deposited subglacially, or in medial or lateral moraine to be dumped from the glacier surface at the terminus. The process of frost-shattering on mountainsides also gives a wide range of grain sizes in medial and lateral moraine debris, although it tends to lack a silt fraction (Figure 20.36a). Debris travelling at the ice–rock interface, particularly at the sole of a glacier, is reduced in size by a grinding process (Section 20.11). Consequently basal till is finer-grained than the debris travelling within the ice at a higher level. (See Figures 20.35, 20.36b.)

If, however, a glacier flows over a soft sediment bed which then deforms, the deforming layer may itself be transported for very considerable distances beneath the glacier. The process produces

erosion because it transports sediment from its original location. Because clay-rich sediments deform more readily than coarser sediments, the deforming layer will tend to be enriched in clay. However, the deformation process is also a mixing process, which mixes together many different sorts of sediment to produce a mixed-grain deposit which is also a till, although it will tend to be clay-rich compared with bedrock-derived tills (Figure 20.36).

Moraines are ridges which mark present or former positions of ice margins. In their simplest form, moraines are of two fundamental types:

1. **Dump moraines** where material carried on the ice or coming to the glacier surface in the terminal zone (Figure 20.35) is dumped at the terminus or lateral margin. If a glacier is retreating at a relatively steady rate, the amount of debris dumped at the terminus will be constant for each unit distance of retreat, leaving a relatively continuous cover, but no moraine. A dump moraine forms when the glacier terminus is stationary for a while, thus permitting a ridge to build up at the margin. Dump moraines are most important at the margins of valley glaciers where very large

Figure 20.40 A series of drumlins from Victoria Island in Canada. The ice-sheet which produced them was flowing towards the camera. The drumlins are several hundred metres, to kilometres in length, and several tens of metres high. The streamlining process reflects moulding of till and other sediment beneath the moving sole of the glacier. (Credited to *The Ministry of Energy, Mines & Resources, Canada*)

moraines can form (Figure 20.37).

2. **Push moraines** tend to mark the extremities of advances in contrast to dump moraines which tend to mark halts during retreat. When a glacier advances over soft sediments, it acts as a bulldozer, pushing up moraine ridges. Major glacier advances may produce large push moraines, in which pre-existing strati-fied sediments may be folded and faulted, or till may be pushed up into ridges which appear structureless. Large and complex push moraines, often several hundred metres across, are frequently produced by glacier surges (Figure 20.38). Push moraines can also be formed annually. Many very active glaciers undergoing retreat, readvance in winter because ablation ceases and the ice is still flowing forward. For instance, a glacier retreating at an average rate of 50 m per year, may readvance by 10 m in winter, and thereby produce a small push moraine. A record of the net annual retreat of a glacier can be recon-structed by studying the frequency of winter push moraines (Figure 20.39).

Figure 20.41 Flutes and eskers exposed beyond a retreating glacier, the Woodworth Glacier in Alaska. The flutes are at the bottom right of the photograph. They are made of till, and accurately reflect the direction of flow of the basal ice when the glacier terminus lay at the left-hand side of the photograph. The flutes formed as till was squeezed into the low pressure zones on the down-glacier sides of boulders deeply-embedded in the till surface. The flutes are up to 0.5 m in height, and a few metres apart. The two eskers lie in the centre and to the left of the photograph. They are made of gravel and reflect meltwater tunnels at the base of the glacier. At the left-hand extremity of the main esker there is an ice-front gravel fan. This formed at the margin of the glacier as water emerging from the esker tunnel spread out to form a fan at the margin. The esker itself formed when the tunnel became blocked with sediment.

Streamlined features. When glaciers flow over soft sediment they tend to mould them into stream-lined ridges with long axes parallel to ice flow. Large, elliptical streamlined ridges made of soft sediment, with lengths in excess of 10–100 m are termed **drumlins** (Figure 20.40). They may be made of till, or sediment deposited in water, and may have a bedrock core. They tend to have a steep up-glacier nose and a gentler, tapering, down-glacier tail.

Very long, narrow, straight ridges called **flutes** (Figure 20.41) are smaller, almost universal features of subglacially-deposited till surfaces exposed beyond modern glaciers, where they also cover drumlin surfaces. They are parallel-sided, normally less than 1 m in height and 1 m wide, they may be several kilometres in length and generally start at the lee side of large boulders embedded in the till. As ice flows over a boulder embedded in a till surface a cavity begins to form on its lee, which immediately fills with soft till which is extruded into it. As the glacier flows on, this till-filled cavity simply extends by addition of till at its extremity.

20.14 WATER DISCHARGE FROM GLACIERS AND GLACIOFLUVIAL DEPOSITION

The widespread melting of the glacier sole in sub-polar and temperate glaciers produces large volumes of subglacial meltwater. In the terminal zone of the glacier, surface meltwater can find its way to the bed via crevasses or through tunnel systems called **moulins**. This meltwater will flow towards the glacier terminus, from zones of high ice pressure to low pressure. It may flow as a thin layer (> 2 mm thick) between a glacier and its rock bed, or as groundwater, or in subglacial tunnels. An ice-sheet growing during glacial periods in mid-latitudes may completely reorganize ground-water reservoirs into very large systems controlled

Figure 20.42 Outwash from the Vatnajökull ice cap in southern Iceland. Large amounts of summer meltwater from the glaciers flow across the coastal plain as braided river systems. They transport very large amounts of sediment, and are known in Iceland as sandurs.

by the continental-scale ice-sheet pressure gradient determined by ice-sheet thickness, compared with the small reservoirs of non-glacial periods which are roughly coincident with surface-water catchment areas.

The glacier bed is most efficiently drained by tunnels whose floor is the glacier bed and whose arched roof is of glacier ice. Very large, highly pressurized discharges of water can be carried by these tunnels which also tend to meander, as do most river channels cut in homogeneous materials. Eventually they become blocked by sediment transported by the channelled stream, and when the ice which encases them finally melts, on glacier retreat, the channel-fill sediment is left behind as an **esker**, which retains its sinuous, meandering form (Figure 20.41).

When meltwater reaches the glacier terminus and flows out beyond it in meltwater rivers, it is termed **outwash**. In narrow valleys, outwash rivers and the sediments they transport are often confined to deep single channels. In broader valleys and on wide plains outwash rivers tend to spread outwards from **glaciofluvial fans** at the ice margin, and thence pass into broad, **braided outwash** systems (Figure 20.42). Two strong cycles of water discharge are characteristic of glacial rivers. The first, a 24-hour cycle in summer reflects enhanced daytime melting of the glacier surface, especially during sunny periods, which reaches a peak in the late afternoon, and a much reduced

melting rate at night. The second, an annual cycle, simply reflects cessation of surface melting through much of the winter, and enhanced summer melting. These discharge cycles may show contrasts of over 10 to 100 times between discharge troughs and peaks, a unique and very strong rhythmic cyclicity which is characteristic of glacier rivers.

Glaciers are also prone to extreme, catastrophic water discharges, which are often associated with rapid drainage of ice-dammed lakes, or the escape of subglacially-trapped water bodies. Catastrophic floods termed **Jökulhlaups** or **glacier bursts** in Iceland have achieved water discharge rates of the order of 10 km^3 per day. Such catastrophic events are also known from former ice-sheets, associated with, in the case of the last North American ice-sheet, for instance, rapid drainage of large ice-dammed lakes.

The intrinsic variability of meltwater discharge ensures that sediment grain size and sedimentary structures which reflect river channel form and location vary dramatically in glaciofluvial outwash sediments.

20.15 STAGNATION FEATURES AT THE GLACIER MARGIN

When debris melts out onto the glacier surface,

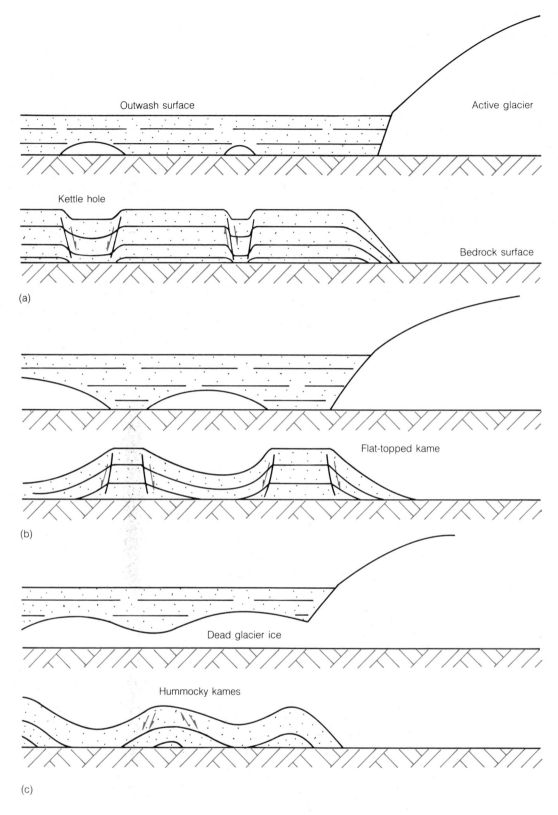

Figure 20.43 (a) Where isolated masses of stagnant ice are left behind and buried beneath accumulating outwash, kettle holes develop on an otherwise planar outwash surface (see Figure 20.44). (b) Where most of the outwash surface is underlain by stagnant ice, melting of the ice produces a hummocky stagnation surface, with occasional flat-topped hummocks where there was no buried ice. The sediment is locally deposited on a solid surface and thus does not collapse beneath the level of the original planar outwash surface. (c) Where the whole surface is underlain by ice, a hummocky stagnation deposit develops. In all cases a prominent ice contact scarp has developed at the right-hand side after retreat of the active glacier front. See Figure 20.45 for example of faults which formed as a consequence of melting of the buried ice.

Figure 20.44 Kettle holes on an Icelandic sandur surface. See Figure 21.43a for an explanation of their origin. The Kettle holes are about 100 m in diameter.

Figure 20.45 Small faults in glacial outwash deposits consisting of river-lain sands overlain by debris-flow deposits. The faults were produced when underlying buried ice collapsed. This collapse also produced kettle holes (see Figures 20.43, 20.44). Lamstedt, Germany.

(a)

Figure 20.46 (a) A lake in a tributary valley blocked by ice flowing down the main valley. The Marjelen Sea and the Aletsch Glacier, Switzerland. (b) The 'Parallel Roads' of Glen Roy, Scotland, formed 10 000 years ago, when the main valley, Glen Spean, was occupied by a glacier. A lake was dammed up in a tributary valley, Glen Roy, and the Parallel Roads represent former shorelines of a lake.

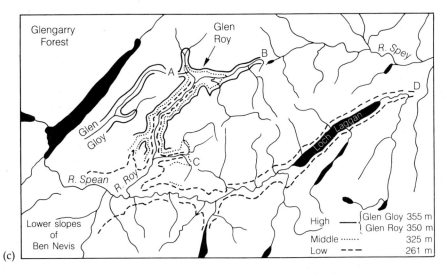

Figure 20.46 (c) Map of the Parallel roads of Glen Roy, Inverness-shire, Scotland.

either in medial moraines or in the terminal zone where basal flow lines come to the surface (Figure 20.35), the melting rate of underlying ice will be changed. When the **supraglacial** debris layer is greater than a few centimetres in thickness it protects the underlying ice from direct radiation and slows down its melting rate compared with that of clean ice. As a result, the clean ice melts down below the level of the debris-covered ice, which therefore forms ridges. Medial moraines almost invariably form ridges on the glacier surface (Figure 20.8).

Even when the clean ice of the active glacier has receded, it may leave stagnant, debris-covered ice ridges, or very extensive buried ice masses behind. Outwash sediments being flushed out from beneath the glacier and building up in thick accumulations at the glacier margin tend to bury this stagnant ice. When the buried ice eventually melts, the sediment underlain by stagnant ice collapses. There are many and complicated variants of the forms of **stagnation terrain** which can develop as a result. Where isolated stagnant ice masses are left behind, isolated **kettle holes** develop beneath the level of the relatively smooth glaciofluvial depositional surface (Figures 20.43a and 20.44). Where much larger areas of outwash sediment are underlain by stagnant ice, but some small areas are not, a hummocky stagnation terrain develops with occasional flat-topped hummocks occurring in places where there was no buried ice (Figure 20.43b). Where the whole area is underlain by stagnant ice, a very disorderly, hummocky stagnation terrain develops (Figure 20.43c). The collapsed sediment often shows simple anticlinal and synclinal folds and numerous faults (Figure 20.45). Till-like **debris flow** sediments derived from the debris covered stagnant ice

masses tend to be mixed with the glaciofluvial sediments (Figure 20.45).

It is very common for belts of stagnation terrain to occur in areas just beyond places where the glacier margin has been stationary for a long period, and for it to occur in a belt parallel to the margin. The line of the former stationary ice margin against which these deposits have built up is often marked by a counter scarp, called an **ice contact slope** (Figure 20.43, Plate XI (c)). Many large so-called moraine belts are the product of such ice marginal stagnation. Mounds of fluvioglacial sediment produced by ice stagnation are called **kames**.

20.16 GLACIAL LAKES

Glaciers preferentially flow towards low ground and tend to block normal river drainage pathways. They may thus block drainage in tributary valleys to produce **valley-side lakes** (Figure 20.46a). Depending on the location of the ice divide, they may advance up valleys or against the slope, thereby producing **ice-front lakes**. Their capacity to produce large-scale kettle holes in stagnation terrain, and ice contact scarps, also favours lake formation whilst lakes also frequently form on top of buried ice.

During the recession of the Pleistocene ice sheets ice-dammed lakes came into temporary existence on an extensive scale in the middle latitudes of Europe and North America. Some of these marginal lakes overflowed at successively lower levels before they eventually disappeared, each stand of the lake being determined by the height of the lowest outlet available at any given stage during the wasting away of the ice barrier. The features and deposits left behind by such

Figure 20.47 Varves deposited in an ice-dammed lake in Kongsforden, Spitsbergen. The light layers are sand deposited during summer from rivers which drain directly from the glacier into the lake. In winter, the lake freezes over, and meltwater rivers also freeze up. The fine sediments suspended in the milky-coloured lake waters in summer, then fall slowly to the floor of the stagnant winter lake to produce the dark silt and clay layers. These varve pairs are therefore deposited annually, and their varying thickness reflects the length and warmth of the summer season. They can be used as climate indicators and also as reliable measures of the passage of time. (The sample is 12 cm in length.)

lakes (thereby providing evidence of their former existence) include: (a) overflow channels or spillways at the outlets, often eroded into conspicuous valleys and gorges (now dry) situated at the heads of the valleys from which the lake waters escaped, or across the ridges and spurs between neighbouring valleys; (b) shore-line deposits and terraces, formed by the action of waves and currents; (c) deltas deposited by streams flowing into the lake; and (d) lake-floor sediments.

A classic example of former lake terraces are the Parallel Roads of Glen Roy (Figure 20.46b,c). These are beaches about 12 to 15 m wide which follow

the contours at the levels shown in Figure 20.46c. Each beach can be traced to the head of a valley, where a spillway is found corresponding to the level at which the lake overflowed while the beach was being formed. Ice extending from Ben Nevis across the valleys to the north blocked the entrances to Glen Roy and Glen Gloy. The highest lake (355 m), that of Glen Gloy, discharged across the watershed at A into the Glen Roy lake (350 m) which overflowed into the river Spey at B. Later an outlet into a tributary of Glen Spean was uncovered at C (325 m), and the lake rapidly drained to the level so determined. This stage lasted until further withdrawal of the ice allowed the lake to extend along Glen Spean, whence it overflowed through D (261 m), the outlet at its head. In due course the lobe of ice that blocked the lower part of the valley dwindled away sufficiently to allow the lake to drain towards the sea (Loch Linnhe) and so, finally, to disappear.

Two important processes in glacial lakes produce characteristic sedimentary features. In summer, large discharges of glaciofluvial sediment into the lake tend to produce a relatively thick, coarse-grain sediment layer on the lake floor. In winter, the lake surface freezes, sediment discharge into the lake progressively diminishes and finally ceases, and fine sediment in suspension in the lake settles on the bed as a thin, fine-grained horizon which grades down into the coarser unit beneath. In the following spring, the first meltwater floods suddenly bring in coarse-grained sediment and the cycle begins again. This annual cycle leads to the accumulation of annually-layered **varves** (Figures 14.1, 20.47). By counting varves, the life of the glacial lake and the variation of climate, which controls sediment input can be estimated.

Icebergs containing debris also calve into many glacial lakes, and any debris contained in them falls to the lake floor as the iceberg melts. Large stones, which fall onto the lake bed and become embedded in finer lake-bottom sediment are termed **dropstones**.

20.17 GLACIOMARINE ENVIRONMENTS

Most of the perimeter of the Antarctic ice-sheet and 10 per cent of the perimeter of the Greenland ice-sheet calve into the sea, i.e. their main ablation is by iceberg calving. Large parts of the perimeters of the European and North American ice-sheets also lay in the sea. The marine environment adjacent to the Greenland and Antarctic ice-sheets is dominated by **glaciomarine** processes, whilst a large proportion of Late Quaternary sediment on the north-eastern continental shelf of North Amer-

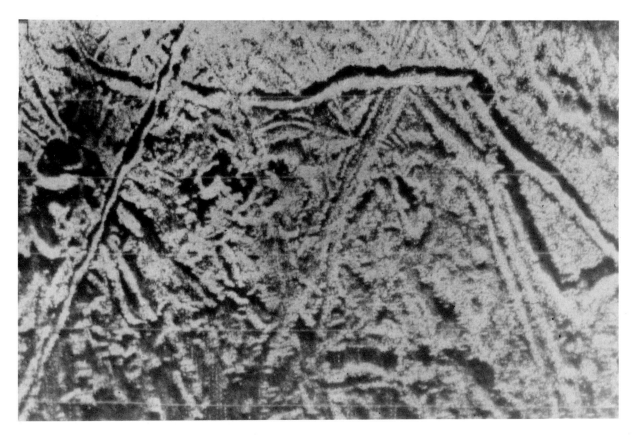

Figure 20.48 Sidescan Sonograph of iceberg tracks on the floor of Itirbilung Fiord, Baffin Island, Canada. They lie in 100 m of water and are produced when the keels of large icebergs moved by wind and tide, drag through fjord-bottom muds. The largest tracks shown are 20 m across and up to 5 m in depth. (*J. Syvitski, Bedford Institute of Oceanography, Canada*)

ica and the north-western continental shelf of Europe is of glaciomarine origin. Sedimentation in these environments is strongly influenced by enormous volumes of meltwater pumped into them from the marine glacier margins, and the flux of icebergs.

The marine glacier margin is a major locus of sedimentation at which outwash fans and push moraines form in deep water just as on land, and eskers are progressively exposed on the sea bed. However, the high velocity meltwater discharge from the ice front, after depositing its coarse sediment load in ice-marginal fans, does not flow across the sea bed, but rises buoyantly to the sea surface through the heavier salt-water to form a brackish water plume which is laden with fine clay- and silt-rich suspended sediment. As this plume spreads out, clay and silt fall to the sea bed over wide areas, forming a thick muddy blanket on the sea floor, which may extend many tens of kilometres from the ice margin. The other distinctive sedimentary components are dropstones which fall from icebergs calving from marine margins. Icebergs may travel hundreds of thou-

sands of kilometres from their source, and deposit dropstones as they travel.

Icebergs have a specific gravity of 0.9. A tabular iceberg rising 50 m above the sea surface will therefore extend as much as 450 m below the surface. As a result, many icebergs ground in relatively shallow water. Large forces can be imposed upon them by wind and tide, so that if they are grounded in the soft sea-bottom mud so characteristic of glaciomarine environments, they may plough through it, impelled by wind and tidal forces, leaving large **plough marks**, many metres deep and up to tens of metres wide (Figure 20.48). Ancient icebergs plough marks of this type are found extensively at the continental shelf edges of north-west Europe and north-western and north-eastern North America, as are extensive zones of glaciomarine sediments. Moreover, such sediments are also found on the deep ocean floor in relatively low latitudes in the Atlantic, dating from periods of extensive mid-latitude glaciation in the Late Quaternary. In the areas which were strongly glaciostatically depressed around these ice-sheets, glaciomarine sediments may be found high above modern sea-level as a consequence of land inundation by the sea around the ice-sheets.

20.18 FURTHER READING

Drewry, D.D. (1986) *Glacial Geologic Processes*, Arnold, London.

Sugden, D.E. and John, B.S. (1976) *Glaciers and Landscape*, Arnold, London.

Washburn, A.L. (1979) *Geocryology*, Arnold, London.

ICE AGES AND CLIMATIC CHANGE

For Hot, Cold, Moist, and Dry, four
champions fierce,
Strive here for mastery. . .

John Milton (1608–74)

21.1 THE TEMPO OF CHANGE IN THE LATE CENOZOIC ICE AGE

It is often implicitly assumed that the Earth's environment has been stable during the geologically-recent past, and that the drama of human history has been played out against a static backdrop. Neither of these assumptions is correct. Averaged over the last 500 000 years, the 'normal' environment of Edinburgh, where this is being written, is one with a mean annual temperature of about 0°C, compared with a modern mean temperature of 8°C, a treeless landscape, shrub tundra, the horizon to the north dominated by glaciers flowing south from the Highlands, and sea-level so low that it would be possible to walk across the south-central North Sea to Denmark. Only 10 000 years ago, when early Jericho was a flourishing city, the site of Stockholm was overlain by more than 2 km of glacier ice, whilst more recently the extensive Viking settlements of southern Greenland, established in the early Middle Ages, were cut off from European contact for over 300 years by the extension of polar sea ice, to be found abandoned when they were finally revisited in the early eighteenth century.

The recent past has known dramatic and fundamental changes of climate and environment which have affected the whole Earth, from the top of the highest mountains to the bottom of the deepest oceans. Moreover, many of these changes have occurred at surprising speeds. Although the Earth's environment may now be changing in response to human activities, even without them, rapid and dramatic changes in the environment would occur quite naturally.

The particular characteristics of recent environmental and climatic change arise because the Earth is in an **Ice Age**. This is a period when a large proportion of the surfaces of the continents are covered by ice-sheets. By this criterion the Earth is currently in an ice age in which $15.3 \times 10^6\,\text{km}^2$ of the continental area of $114.8 \times 10^6\,\text{km}^2$ is covered by glacier ice, with a volume of about $30 \times 10^6\,\text{km}^3$. This ice mass is predominantly made up of the ice-sheets of Antarctica $(27 \times 10^6\,\text{km}^3)$ and Greenland $(2.6 \times 10^6\,\text{km}^3)$. From time to time in the recent past however, the global ice mass has been very much larger, and has included ice-sheets in North America (about $30 \times 10^6\,\text{km}^3$) and north-west Europe (about $7 \times 10^6\,\text{km}^3$). For the last million years of this Ice Age at least, the Earth has oscillated between two predominant climate modes. One, an **interglacial** mode, in which the Earth has been much as it is today, with only two great ice-sheets (Figure 21.1a); and the other, a **glacial** mode, in which these have been joined by further ice-sheets in the middle latitudes of Europe and

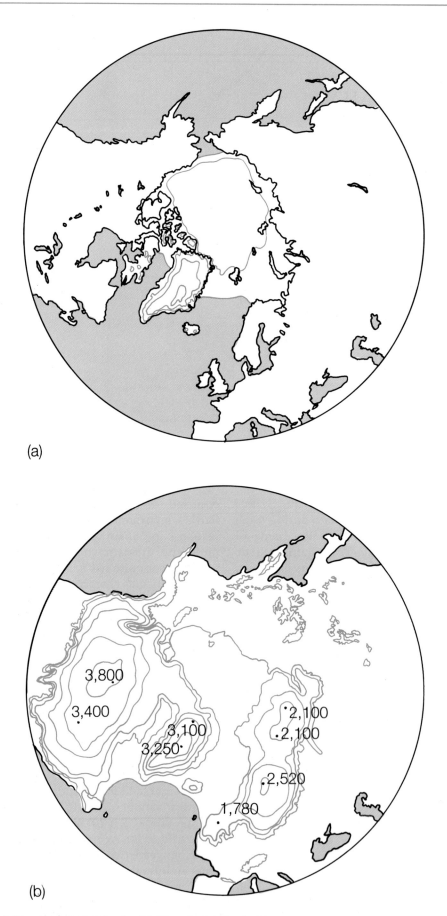

(a)

(b)

Figure 21.1 (a) The interglacial Earth. (b) The glacial Earth (showing elevations of ice-sheets). They show the northern hemisphere, including land and sea areas, the polar sea ice and ice-sheets.

North America (Figure 21.1b). The growth of these mid-latitude ice-sheets has been accompanied by a sea-level fall of a hundred metres or more because of transfer of water from the oceans to the growing ice-sheets, by expansion of cold polar waters into lower latitudes in the oceans, and invasion of middle latitudes on land by polar animals and plants.

The key to the history of long-term environmental change in the Earth's recent past is to be found in the oceans. Mud accumulates on the deep ocean floor at rates typically between 10 and 150 m per million years. Interred within this mud are the skeletons of small marine organisms which have either lived on the sea floor (**benthonic** organisms) or floated in the water column (**planktonic** organisms) during life. The shells of micro-organisms such as **foraminifers** (Figure 21.2) are particularly useful palaeoenvironmental indicators. They are largely made up of calcium carbonate, are abundant in the ocean and are readily found as fossil remains in ocean floor muds. They are also very sensitive to the ocean environment: ocean temperature, salinity, the availability of nutrients,

whether there is sea ice on the surface, etc. Their value as palaeoenvironmental indicators is illustrated in Figure 21.2. This shows the occurrence of two closely-related foraminifers in a core from the Norwegian Sea. *Neogloboquadrina pachyderma* is a species with two forms, one coils to the right and lives in ice-free surface waters in high latitudes, the other coils to the left and can live in surface waters where sea ice is common. As the core is sampled down from the surface, we first find evidence of the right-coiling form which is then replaced by the left-coiling form until the right-coiling form reappears. This reflects two periods during which the ocean was ice free in this region, separated by a period when it was covered by sea ice. This latter period was from about 75 000 years before present (BP) to about 13 000 years BP; the approximate duration of the coldest part of the last glacial period.

Micro-organisms such as foraminifers also show strong changes in shell chemistry during glacial–interglacial cycles. The oxygen (O) molecules in the calcium carbonate ($CaCO_3$) of their shells contain a heavy isotope with an atomic weight of

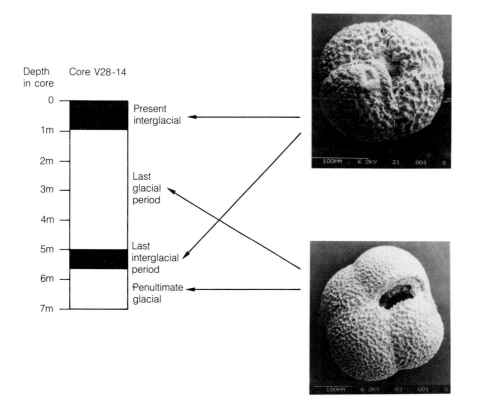

Figure 21.2 A core from the North Atlantic (V28–14) showing the interglacial periods (the present from 13 ka to 0 ka, and the last, 128–112 ka) marked in black, and the glacials in white. The right-coiling form of *Neogloboquadrina pachyderma* (dextral) dominates in interglacial periods, and the left coiling form (sinistral) dominates in glacial periods.

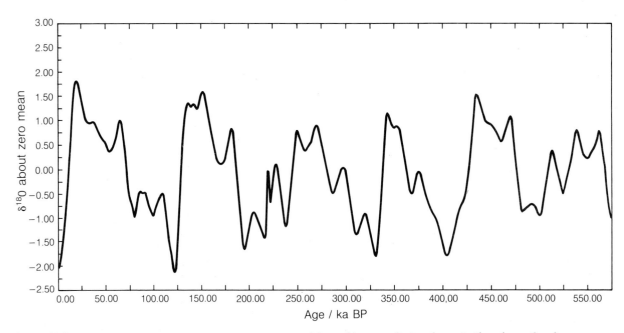

Figure 21.3 The variation in the oxygen-isotope composition of bottom-living foraminifers from the deep oceans during the last 575 000 years. Their composition reflects the isotopic composition of ocean water, and this largely reflects the mass of land ice on Earth. During glacial periods, the oceans are isotopically heavy (high values of $\delta^{18}O$) and light during interglacials. This record therefore shows six major glacial cycles with intervening temperate interglacials. Note that the glacial periods are much longer than the interglacials.

18 (^{18}O) and a light isotope of atomic weight 16 (^{16}O). During a glacial period, there is proportionately more ^{18}O than ^{16}O in the shells than during the preceding and succeeding interglacial periods (Figure 21.3). This appears to reflect the changing chemistry of the oceans during glacial and interglacial periods. The growth of large ice-sheets takes place by the transfer of water to them from the ocean surface. When this water evaporates from the ocean surface (Figure 21.4), the vapour is relatively enriched with the light isotope of oxygen (^{16}O). The ice-sheets which form from this water are therefore also relatively enriched in ^{16}O and the oceans relatively enriched in ^{18}O. The change in the oxygen isotope composition of the shells of oceanic foraminifers therefore largely reflects changes in the volumes of ice-sheets on Earth (Figure 21.3, 21.5), and by studying the oxygen isotope composition of suitable foraminifers down long cores from the oceans, we can reconstruct the fluctuations of ice-sheet volume. Figure 21.5 shows the variation in oxygen isotope composition in the shells of planktonic foraminifers in a (120 m) length of core from the equatorial Indian Ocean for the last 2 600 000 years. The isotopically heavy troughs in this curve represent cold, glacial periods when the global ice volume is large, and the isotopically light periods represent relatively warm periods, similar to the present day, with a smaller ice volume.

Records such as these reveal the fundamental tempo of long-term climatic change on Earth in the recent past, and also reflect the way in which the ice age has become progressively more intense.

Figure 21.6 shows an even longer record, which reflects strong cooling of the Earth between about 30 and 15 m.y. ago, probably associated with the formation of the Antarctic ice-sheet. At about 2.4 m.y. ago (Figure 21.5), the magnitude of climate change between cold and warm periods appears to have increased, and a steady rhythm of change with a wave length of about 40 000 years becomes established. At about this time, the Greenland ice-sheet may have formed for the first time. Between about 0.8 and 1.2 m.y. ago a 100 000 year rhythm begins to develop, which is dominant by about 800 000 years ago, whilst at the same time the magnitude of change between warm and cold periods becomes even greater. This may have been the time when large ice-sheets developed for the first time in the mid-latitudes of Europe and North America.

A great deal of information about past global climate has been gained from study of the modern ice-sheets of Antarctica and Greenland. These can be thought of as containing successive layers of frozen atmosphere (Figure 20.6b) which, in the case of Antarctica, contain a record going back for more than 200 000 years. Studies of cores which penetrated 2200 m of ice at Vostok Station in East Antarctica, where the ice-sheet is 3500 m in thick-

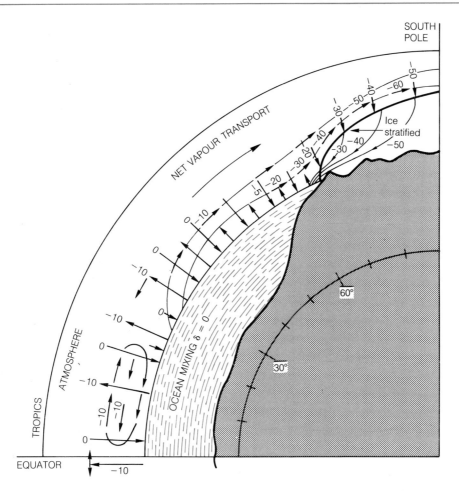

Figure 21.4 The oxygen-isotope fractionation process. When water is evaporated from the ocean surface, proportionately more ^{16}O is added to atmospheric moisture than is contained in ocean water. As this moisture is moved by storms over land and over glaciers, precipitation from it contains proportionately more ^{16}O than is present in the atmosphere. The atmosphere is therefore further enriched proportionately in the light isotope. The snow which falls on ice-sheets is thus isotopically very light. If ice-sheets grow, they preferentially take light isotopes from the ocean, which becomes isotopically heavy. The isotopic composition of the oceans, measured in the shells of organisms which lived in it, can therefore be used as an index of global ice volume change.

ness, permit us to reconstruct temperature on the ice-sheet surface, and the carbon dioxide (CO_2) and methane (CH_4) composition of the atmosphere for up to 185 000 years BP (Figure 21.7). Palaeotemperatures on the ice-sheet surface can be reconstructed from deuterium isotopes whilst the concentration in the atmosphere of CO_2 and CH_4 at the time of the original snowfall can be obtained by analyzing the gas trapped in bubbles in the ice at that time. As both gases are well-mixed in the Earth's atmosphere, this must be a record of average CO_2 and CH_4 concentration change in the global atmosphere.

21.2   THE CLIMATE SYSTEM AND ITS SOLAR DRIVE

These records of change in global ice volume, sea surface temperature and atmospheric composition can be thought of as indicators of the state of the Earth's **climate system**. We normally think of climate as simply related to the behaviour of the atmosphere. However, over long periods of time, changes in other parts of the Earth's surface fundamentally change climate. Thus, if the pattern of circulation of the ocean changes so as to warm some parts of the ocean surface and cool others, patterns of atmospheric moisture content, temperature, pressure and wind direction and strength will also change; if biological productivity changes so as to change the atmospheric content of carbon dioxide or methane, which are greenhouse gases (Table 21.3), the surface temperature of the Earth will change. If ice sheets grow, they will cool parts of the Earth by a latent heat effect, change patterns of wind circulation around them, increase the Earth's reflectivity, and so cool the Earth, lower the ocean level and change the ocean circulation pattern. We therefore think of the atmosphere, oceans, ice-sheets and the biosphere as parts of the climate system. This is a strongly

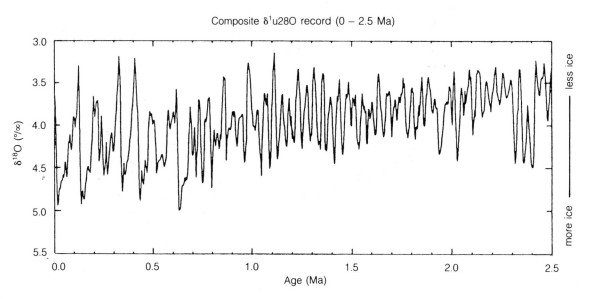

Figure 21.5 The variation of oxygen-isotope composition in surface-living foraminifers from the equatorial Indian Ocean over the last 2.5 m.y. It suggests that the global ice volume has increased during glacial periods, especially during the last 0.75 m.y., and that the average length of the cold periods has increased.

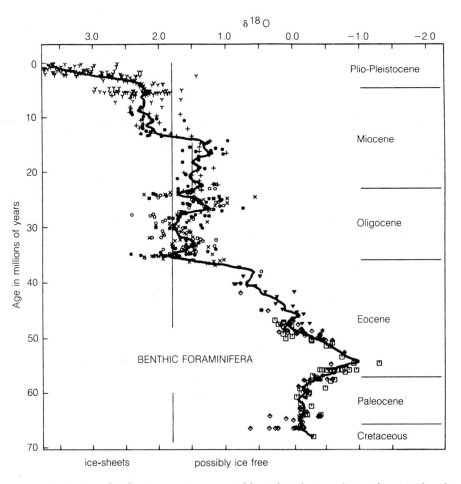

Figure 21.6 An oxygen-isotopic and palaeotemperature record based on bottom-living foraminifers from the Atlantic for the last 70 m.y. It reflects a gradual cooling from 55 m.y. ago. Major, high latitude ice-sheet growth may have begun between 35 and 15 m.y. ago. (The symbols mark the positions of various cores taken from the oceanic crust.)

coupled system, meaning that change in any one part will lead to changes in the other components.

Each of these components of the climate system operates on a different timescale. The global atmo-

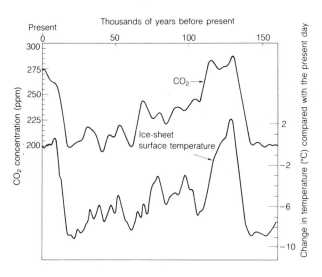

Figure 21.7 A stratigraphic record from the Antarctic ice-sheet at Vostok Station near the ice-sheet summit. It shows palaeotemperatures inferred from the deuterium isotope composition of ice, and carbon dioxide composition of the atmosphere. The CO_2 measurements are made on air contained in bubbles trapped in the ice. They show high carbon dioxide concentrations during the relatively warm conditions of the last interglacial at about 125 000 years ago, and very low concentrations during the succeeding glacial period.

sphere adjusts to change very quickly, over periods of weeks, the oceans and the biosphere adjust more slowly, over periods of hundreds to thousands of years, whilst ice-sheets are slowest of all, adjusting over periods of thousands to tens of thousands of years. This is why the climate system is so complex and why prediction of its future is so difficult. Figure 21.8 reflects this complexity by showing the different timescales on which we have evidence for climatic change on Earth during the last 180 million years.

Notwithstanding this complexity some fundamental and simple patterns of cause and effect have been established. James Croll, (a Perthshire born scientist who became Secretary of the Scottish Office of HM Geological Survey), suggested in 1875 that long-term changes in the amount of solar radiation reaching the Earth were controlled by rhythmic variations in the Earth's orbit around the sun and that the solar radiation variations produced the periodic climate changes which were being discovered by contemporary geologists.

In 1941, a Yugoslav astronomer, Milankovitch, calculated the magnitude and frequency of the changes of solar radiation which the Earth would receive as a consequence of orbital changes. He identified three orbital processes which would control these changes:

1. Axial **tilt** variation (Figure 21.9a).

The Earth's axis of rotation is not at right angles to the plane of the Earth's orbit around the sun, but inclined at an angle of about 23.5° to the orbital plane. This angle is not however constant, but varies between 24.5° and 21.5°. When the angle of tilt is greatest, there is the greatest difference in seasonal heating at any latitude, and least difference when the angle of tilt is least. The length of the tilt cycle is 40 000 years.

2. **Eccentricity** of the orbit (Figure 21.9b).
   The Earth's orbit is not circular, but elliptical. At some times it approaches close to a circle, and at others is more strongly elliptical. The length of the eccentricity cycle is about 100 000 years.

3. **Precession** of the equinoxes (Figure 21.9c).
   The Earth wobbles slightly on its axis due to the gravitational pull of the Sun and Moon on the Earth's equatorial bulge which changes the timing of the solstices (June 21, December 21) and the equinoxes (March 20, September 22) relative to the extreme positions the Earth occupies on its elliptical path round the Sun. Thus, 11 000 years ago the Earth's nearest point to the Sun occurred when the northern hemisphere was tilted towards the Sun (northern hemisphere summer), rather than during the northern hemisphere's winter, as is the case today. The length of the precession cycle is 23 000 years.

Figure 21.5 shows long-term variations in properties of the Earth which reflect the operation of the Earth's climate system. Careful statistical analysis shows that the pattern of change between 800 000 years BP and the present day is largely made up of three dominant rhythms: one with a period of 100 000 years, one of 40 000 years and one of 20 000 years. From this we conclude that variations in the Earth's orbit around the sun are the pacemakers which determine the long-term tempo of climatic change on Earth. This cannot however be the whole story, because of three fundamental problems which arise when we compare the Milankovitch theory with geological data:

1. The global temperature change which occurs between the coldest part of glacial periods and the warmest part of interglacials is about 4–5°C (although in some parts of the Earth the change is locally very much larger, and in others there is almost no change). However, the difference in the intensity of solar radiation reaching the Earth due to orbital variations between the periods of maximum and minimum radiant heating, would not in itself be enough to change the global average temperature by more than 0.4–0.5°C. There must

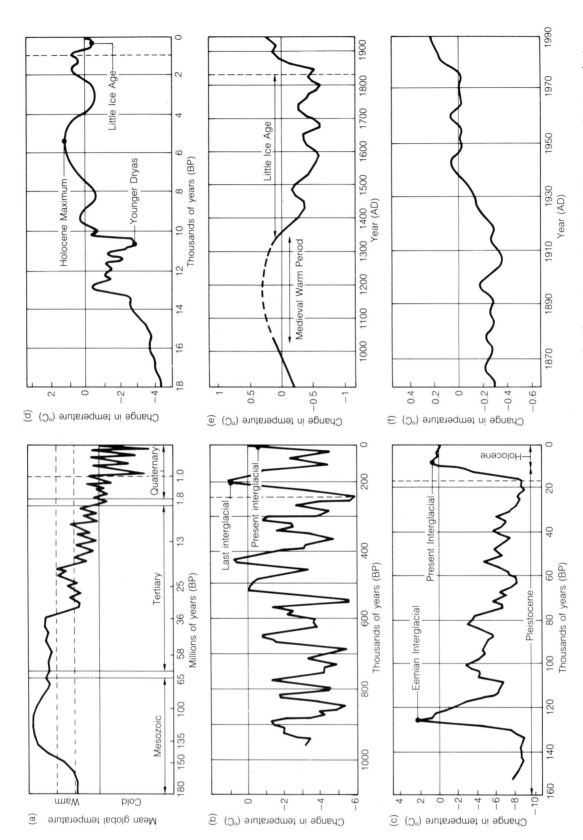

Figure 21.8 Patterns and scales of climate change during the last 180 m.y. (a) Climate change on a 10^7–10^8 year timescale, reflecting long-term climate evolution as a result of plate tectonic changes. (b) Change with a 10^5–10^6 year timescale of glacial/interglacial periods, showing the 100 000 year cycle. (c) Change on a 10^4–10^5 year timescale showing the structure of change within a glacial/interglacial cycle, dominated by 20 000 and 40 000 year Earth orbital changes. (d) Change over the last 10^4 years and the transition from a glacial to interglacial period. (e) Change over the last 10^3 years showing the Medieval Warm Period and the Little Ice Age. (f) Change during the last century, with a possible human influence on climate.

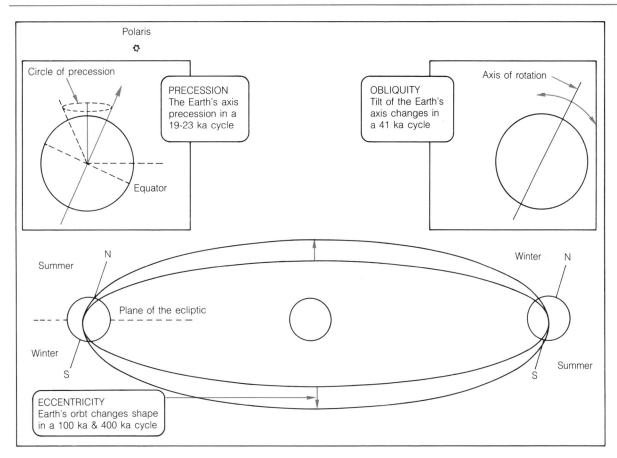

Figure 21.9 The three principal changes in the Earth's orbit around the sun which produce a periodic change in the intensity of radiation reaching the Earth. (a) Axial tilt variations, with a 40 000 year period. (b) Orbital eccentricity, with a 100 000 year period. (c) Precession changes, with a 20 000 year period.

therefore be some mechanism, internal to the Earth, which amplifies the solar signal to produce an effect ten times larger than the unaided effects of radiation changes.

2. Although there is no fundamental change in the rhythm of the orbital pacemaker, there is a switch in the dominant rhythm of climatic change through time. After at least 2 400 000 BP and prior to 800 000–1 000 000 years BP, the 40 000 year rhythm is dominant; after 800 000 years BP the 100 000 year rhythm dominates.

3. There is no fundamental change in the rhythm of the orbital pacemaker which might explain the progressive intensification of the ice age over the last 2–3 million years.

21.3   FEEDBACK MECHANISMS IN THE CLIMATE SYSTEM

The amplification of the solar radiation signal required to produce the observed climatic effect requires a positive feedback mechanism. This can be understood by the analogy of a radio receiver which receives a signal from a transmitter, but is capable of amplifying the signal. There are several such large-scale mechanisms within the climate system.

If an ice-sheet (which is highly reflective) begins to grow in area in response to Milankovich variations, the **albedo** or reflectivity of the Earth (Table 21.1) will tend to increase. This causes a larger proportion of solar radiation to be reflected back into space than previously, and thus the net cooling of the Earth will be larger than it would have been as a consequence of Milankovitch cooling alone, thereby producing greater glacier expansion and a still larger increase in albedo. A similar consequence can be produced by sea-ice albedo feedback. A Milankovitch cooling which produces an extension of sea ice will increase the Earth's albedo, leading to enhanced cooling, which encourages further extension of sea ice and so on.

There will however be natural limits to these processes. Extension of land ice or sea ice towards low latitudes will inevitably be halted by increasingly high radiation fluxes.

Another important feedback mechanism, though not yet entirely understood, involves the **greenhouse gases**. Greenhouse gases, such as CO_2 and NH_4, appear to have played a fundamental

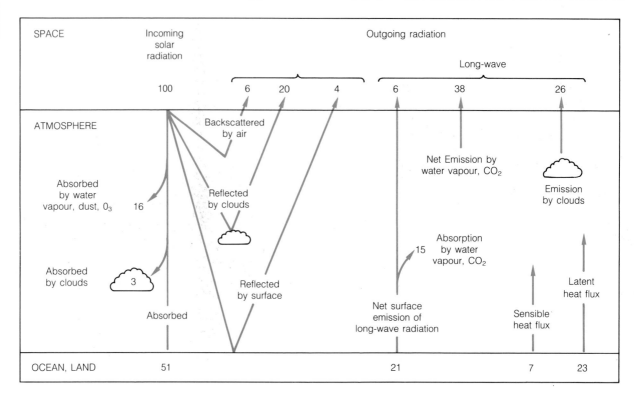

Figure 21.10 Radiation and heat balance of the atmosphere relative to 100 units of incoming solar radiation. Short-wave solar radiation is partly absorbed by vapour, dust and clouds, partly reflected back from clouds, air and the surface, and partly absorbed by the surface. The heated surface emits long-wave radiation, a large proportion of which is absorbed by atmospheric water vapour and CO_2, which re-emits long-wave radiation to warm up the surface further. This is the 'greenhouse effect'.

role in controlling the Earth's heat balance through its history (Table 21.3). Figure 21.10 shows the pattern of absorption and reflection of

Table 21.1 Albedo of various surfaces
(albedo = ratio of reflected to incident radiation)

Surface	%
Fresh snow	80 – 85
Old snow	50 – 60
Sand	20 – 30
Grass	20 – 25
Dry earth	15 – 25
Wet earth	10
Forest	5 – 10
Water (sun near horizon)	50 – 80
Water (sun near zenith)	3 – 5
Thick cloud	70 – 80
Thin cloud	25 – 50
Whole Earth and atmosphere	35

solar radiation by the Earth. Solar radiation passes readily through the Earth's atmosphere, some is absorbed by clouds, some reflected from the Earth's surface and some acts to warm up the surface. The warmed surface emits long-wave radiation which is absorbed by so-called greenhouse gases, which are able to re-emit this energy and thereby further warm up the Earth's surface and the lower atmosphere. This is the **'greenhouse effect'**.

If the concentration of CO_2 (or any other greenhouse gas) in the atmosphere increases, so does the absorption of long-wave radiation by the atmosphere, thereby heating up the atmosphere. A reduction of CO_2 causes the reverse effect. The concentration of CO_2 in the atmosphere largely depends upon biological activity within the oceans, ocean temperature, and ocean circulation so that it tends to be inhaled by the ocean in polar latitudes and exhaled in lower latitudes. CO_2 is more soluble in cold than in warm water. The ice-core record shown in Figure 21.7, shows dramatic global changes in CH_4 and CO_2 concentration in the atmosphere which parallel the changes in temperature and global ice volume through the last glacial cycle. The decrease in CH_4 and CO_2 at the beginning of the cycle and the increase at the end of the cycle must have had a substantial cooling and warming effect respectively. It suggests that the oceans inhale CO_2 during periods of cooling, contributing further to cooling, and exhale CO_2 during periods of warming, contributing further to warming. The oceans will naturally tend to increase their absorption of CO_2 from the atmosphere during periods of cooling, and release it to the atmosphere during periods of warming (cf. Figure 21.18).

There are also internal mechanisms within the Earth capable of producing negative feedback. If, for instance, an ice-sheet were to grow in response to Milankovitch cooling, enhanced by albedo feedback, the additional local mass on the Earth's surface would cause a local depression of the Earth's crust. This sinking of the crust beneath the ice-sheet load can be very rapid and exceed the capacity of the ice-sheet surface to compensate for it by accumulation of snow and ice on its surface. This lowering of the ice-sheet surface due to crustal sinking will increase the ablation rate and produce a negative budget (Figure 20.11) thus causing the ice-sheet to waste away.

Because of the higher angle of incidence of the Sun's rays on the Earth's surface in the equatorial regions compared with the polar region, the Earth receives over four times more solar energy per unit area between the equator and latitude 10°N than it does between latitudes 80° and 90°N. The temperature of the surface depends however on the relationship between solar radiation receipt, and energy loss by reflection and long-wave emission. In the equatorial zone there is a surplus of energy received over that lost by reflection and emission, whilst in the polar zone there is a greater loss than receipt (Figure 21.11). In the absence of other processes, the energy surplus in low latitudes would cause these latitudes to become progressively hotter, and the energy deficit at high latitudes would cause them to become progressively colder. Instead of this happening, the energy surplus in low latitudes is transported polewards to make up the high latitude energy deficit (Figure 21.11). This heat energy is transported through the circulation of the atmosphere and the oceans. The rate at which the atmosphere and the oceans need to transport energy depends on the geographical variation of the pattern of energy surplus and deficit. If both the equatorial

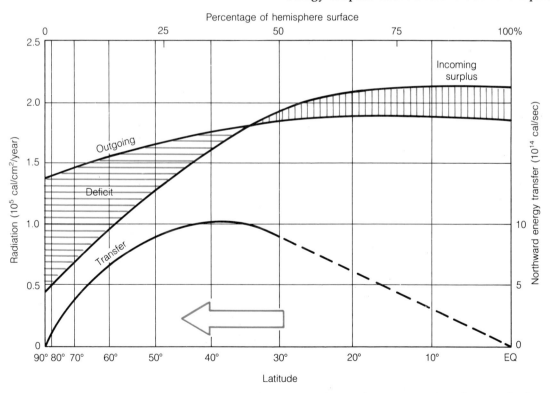

Figure 21.11 The Earth's latitudinal heat budget. In the equatorial zone there is a surplus of received solar energy over that emitted by long-wave radiation. In the polar zone there is a greater amount emitted than received. The energy surplus in low latitudes is transported north to make up the polar deficit. This transport takes place through the circulation of the atmosphere and the oceans. As the other process which drives the climate system circulation, the rotation of the Earth, is relatively constant through time, the variation of the heat budget through geological time produces a variation in the rate and pattern of circulation of the atmosphere and oceans.

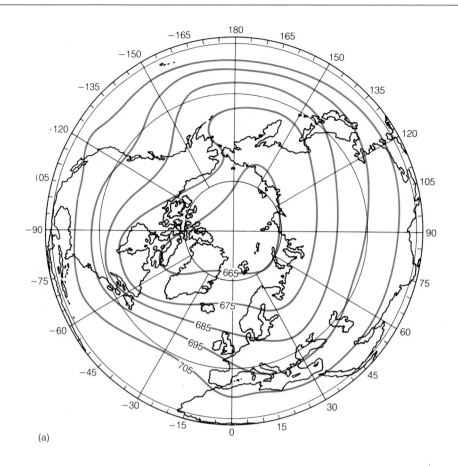

(a)

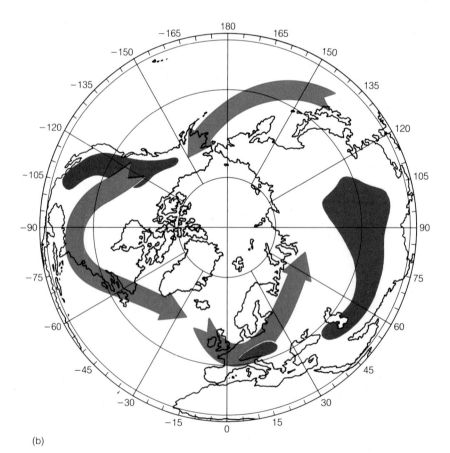

(b)

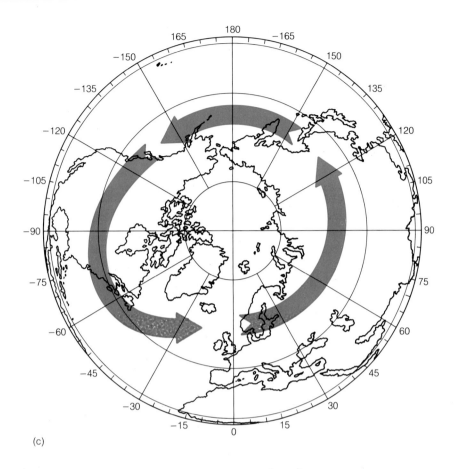

(c)

Figure 21.12 Standing waves in the northern hemisphere atmosphere (Rosby waves) and the effect of topography. (a) Shows the mean pressure at 3000 m in winter, reflecting the wave structure within the atmosphere. Large standing waves in the lee (eastern side) of the western North American mountains, of the mountainous western part of Europe and of the Tibetan plateau. (b) Shows a computer model of the storm tracks associated with this Rosby wave pattern. The darker shading shows the major mountain belts. (c) Shows the change in storm tracks if northern hemisphere mountains are removed. The standing waves and intervening troughs disappear, the circulation becomes smoother and the north–south component of circulation is reduced.

surplus and the polar deficit are smaller compared with the present day, the rate of transport by the ocean and atmosphere will be less than at present. The rate at which energy is transported by oceans and atmosphere depends on their rate of circulation, the average velocity of the global ocean currents and the atmospheric winds. They can be thought of as conveyor belts for heat energy, whose speed depends upon the relative magnitudes of the equatorial surplus and the polar deficit. The deficits in polar regions are greatest in areas of ice-sheets, sea ice and dense cloud cover, all of which have high albedo and reflect incoming solar energy. The surpluses are greatest in cloud-free areas of low albedo (Table 21.1). The strongest transport rates and the strongest circulation rates will be required in the transition zone between the areas of surplus and deficit. In the northern hemisphere, these are the zones of strong mid-latitude westerly winds. The ocean conveyor belt

is however the most powerful, because of the oceans greater mass compared with the atmosphere.

Changes in the global climate will occur if there are changes in the distribution of areas of surplus and deficit, or in the pattern and rates of circulation. Extension of ice-sheets and sea ice into middle latitudes extend the zone of deficit towards the equator and therefore pushes the zone of fastest circulation towards the equator (Figure 21.11).

21.5 GEOTECTONIC INFLUENCES ON THE CLIMATE SYSTEM

Because of the albedo contrast between continent and oceans (Table 21.1) and the fact that continents heat up and cool more quickly, the global patterns of energy surplus and deficit will

NORTH AMERICAN
ICE-SHEET

EUROPEAN
ICE-SHEET

0   500 km

Figure 21.13 The pattern of decay of the great ice-sheets which covered Europe and North America during the last glacial period from (1) their greatest extent between 18 and 15 000 years ago (the outermost concentric lines), to (2) their final decay 6–8000 years ago (the innermost concentric lines). The radial lines show patterns of flow.

be strongly affected by changing continental distribution. This will also affect the operation of the oceanic and the atmospheric 'conveyor belts', and the atmospheric circulation pattern will also be affected by the location of mountain masses.

Slow geotectonic changes which control continental distribution and mountain building are therefore the fundamental determinants of very long-term changes in the Earth's climatic regime, such as those represented by the periodic ice ages in the Earth's history, and the development of the Late Cenozoic ice age.

The two regions of the most extensive high-land terrain in the northern hemisphere are the Tibetan Plateau and Himalayan mountains in southern Asia and the western cordilleras of North America. They are broad plateau-like bulges on which narrower mountain ranges are superimposed. They lay at low elevations (below 1 km) 40 m.y. ago, but began to rise rapidly by 20–15 m.y. as a consequence of compression at convergent plate boundaries. They had reached half their present elevation by 10–5 m.y. ago and only attained their present elevation within the last million years.

The very large-scale circulation of the atmosphere is strongly influenced by these areas of plateau uplift. They produce large scale **standing waves** (**Rosby waves**) within the northern hemisphere atmosphere (Figure 21.12a,b). Rosby waves are equivalent to standing waves in the lee of large boulders in a stream. One occurs in the lee of the western edge of the western North American Cordilleras, one east of the mountainous leading edge of the European continent, and another in the lee of the Tibetan uplift. The consequences of geotectonic uplift can be clearly appreciated through computer models of the general circulation of the atmosphere in which the mountain areas are removed (Figure 21.12c). The large standing waves and the intervening troughs disappear, and the northern hemisphere circulation becomes much smoother. If Late Cenozoic uplift led to the intensification of atmospheric waves and troughs, these would produce stronger north–south components of circulation and a more varied climatic pattern than before. During a period of Milankovitch cooling, they would tend to draw down polar air more effectively into middle latitudes than if there were no waves, and thus intensify the cooling effect. If this effect became strong enough to cause mid-latitude ice-sheets to grow by about 1 m.y. ago, the albedo-feedback produced by these ice-sheets and by sea ice may have been

Table 21.2 Environmental changes

In the atmosphere	In the ocean	On land
atmospheric circulation	circulation	the extent of glaciers
temperature	temperature	the distribution of animals
precipitation	salinity	and plants and ecological
cloudiness	biochemical activity	zones (forest, steppe,
storminess	sea level	tundra etc.)
chemistry	distribution of	the extent of deserts
	organisms	the area of the land
	the extent of sea ice	nature of soils and rate of
		soil formation
		the size of lakes, the flow
		of rivers and the level of
		groundwater
		the extent of permafrost

(a)

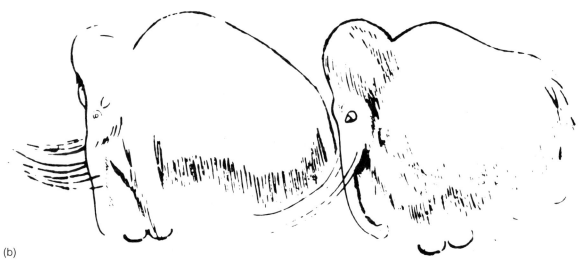

(b)

Figure 21.14 (a) A fossil mammoth dug out from peat near the Berezhovska river in Siberia. It is one of many mammoth fossils which have been found preserved in frozen ground in Siberia. (b) A sketch of mammoths made by a contemporary human.

enough to intensify the Milankovitch cooling to the extent shown by the ocean records in Figure 21.5. Thus, mountain uplift could have produced a transition from a temperate to an ice age Earth, and at about 1 m.y. ago caused some threshold to be passed which ushered in an even more intensive phase of the ice age.

21.6 ENVIRONMENTAL CHANGES DURING THE LATE CENOZOIC ICE AGE

The Late Cenozoic Ice Age can be characterized by three principal environmental modes:

Glacial Periods
Interglacial Periods
Periods of Transition

The principal environmental changes which occur

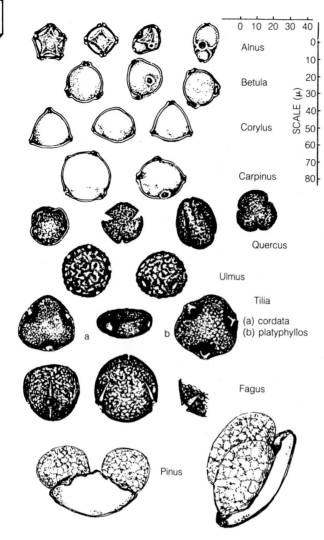

Figure 21.15 Pollen grains of temperate latitude trees. (*Alnus* – alder, *Betula* – birch, *Corylus* – hazel, *Carpinus* – hornbeam, *Quercus* – oak, *Ulmus* – elm, *Tilia* – lime, *Fagus* – beech, *Pinus* – pine).

between these modes are shown in Table 21.2.

Figure 21.1 shows diagrams of the Earth and the distribution of some of these features during the coldest part of the glacial period compared with the present day (an interglacial).

Transitions between interglacials and glacials are often complex. In general, global climatic curves for the last 0.8 Ma show a typically saw-toothed pattern (Figures 21.3, 21.5), with a slow though irregular climatic decline from the end of an interglacial to the coldest phase of a glacial period (generally at its end) over a period of 80 000 to 90 000 years, followed by a very rapid ameliora-tion to interglacial conditions over 10 000 to 15 000 years.

21.7 ENVIRONMENTAL CHANGES ON LAND

The most dramatic environmental changes on land

in the latest part of the ice age involve the growth and decay of large ice-sheets in the mid-latitudes of the northern hemisphere. These have included ice-sheets on:

> North America and the Canadian Rockies
> North-west Europe and Britain
> Barents Sea area
> Arctic Islands of Canada
> Possibly the Kara Sea and along the Siberian coastline

In addition, small ice caps similar to the ice cap of Vatnajökull in Iceland have formed in many mountain areas, valley glaciers have extended, and many mountain ranges which are unglaciated today developed glaciers.

During the last glacial period, the ice-sheets in North America and north-west Europe reached their maximum extents at about 18 000 BP. After an initial period of slow decay, they began to decay rapidly at about 14 000 BP, their decay stopped at between 11 000 and 10 000 BP, but then continued, and they had largely disappeared by 8500 BP in Europe and by 6500 BP in North America (Figure 21.13).

Associated with the growth and extension of large ice-sheets, the Earth's principal environmen-tal zones were also displaced equatorwards (Fig-ure 21.1). Immediately south of the great ice-sheets in the northern hemisphere lay a sparsely vegetated **tundra** zone, extending as far as north-ern France in western Europe and at least 150 km south of the ice-sheet margin in eastern North America. This zone was also characterized by the development of **permafrost** where the ground was permanently frozen for many tens or hundreds of metres below the surface. Surface water was unable to infiltrate into the ground which, cou-pled with the paucity of vegetation, led to imme-diate run-off of rainwater and highly variable stream and river discharges. The northern hemi-sphere tundra zone of Europe was inhabited by **Mammoth** (Figure 21.14) and **reindeer** and other currently polar animals. South of this, lay the **boreal** forests of pine and spruce extending as far as the Mediterranean in Europe and Florida in North America. Similar shifts in ecological zones occurred all over the globe.

The evidence for these changes largely comes from peat bogs and sediments which formed in lake basins. Trees, shrubs, flowering plants and grasses all have distinctive pollen grains (Figure 21.15). They are very small (generally less than 1/20 of a millimetre in diameter), very durable (and therefore preservable) and are produced in large numbers. They are dispersed by wind over large areas, so that accumulating sediments and peats preserve a representative sample of the pollen

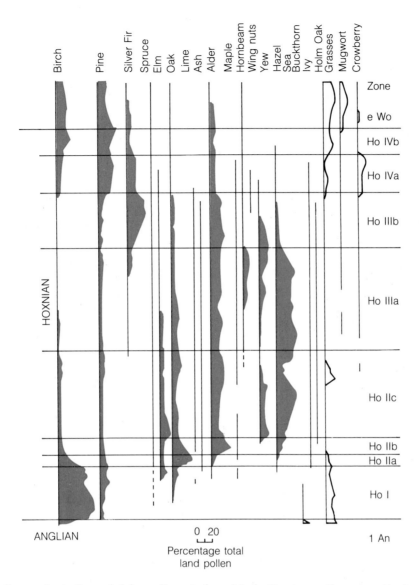

Figure 21.16 A pollen analysis through lake sediments from Marks Tey in south-eastern England. The sediments were probably deposited through the whole of the Hoxnian interglacial cycle, which is probably equivalent to the warm phase of the deep ocean record (Figure 21.3) between 280 000 and 330 000 years BP. It starts and ends with dominant vegetation currently typical of relatively cold sub-polar latitudes, and has a mid-interglacial phase dominated by broad-leaved trees such as oak, elm and beech. (*After C. Turner, 1970*)

from the regional vegetation. Using this means, it is possible to reconstruct vegetational changes over long time periods. Figure 21.16 shows the changing vegetational pattern through the Hoxnian interglacial cycle in part of south-eastern England, probably equivalent to the climatic cycle shown in the deep-ocean record between 280 000 and 330 000 BP (Figure 21.3). It shows a sequence from a cold, tundra phase at the end of the preceding glacial period, and successive transitions through birch and pine forest, into broad-leaved oak–elm forest, and then a reversion, through birch–pine forest and back into tundra. It does not merely reflect a warming and then cooling of climate, but also transition of the soils in which this vegetation grew, from raw mineral

soils, produced by grinding down of rocks by glacial erosion during the preceding glacial period, and churning by permafrost activity, to a humus-rich soil at the warmest part of the interglacial. The successive plant invasions depended on climate, but also on the progressive preparation of the rich soil needed by broad-leaved trees through the **humifying** effect of earlier, hardier species, such as birch and pine, capable of surviving on the raw, early-interglacial soils. The cycle in Figure 2.16 was a local representative of a global cycle in which the vegetable **biomass** progressively increased and then decreased. There is a seasonal cycle of vegetation growth and decay equivalent to a change of about 1 000 000 000 tons of carbon, which is associated with an annual

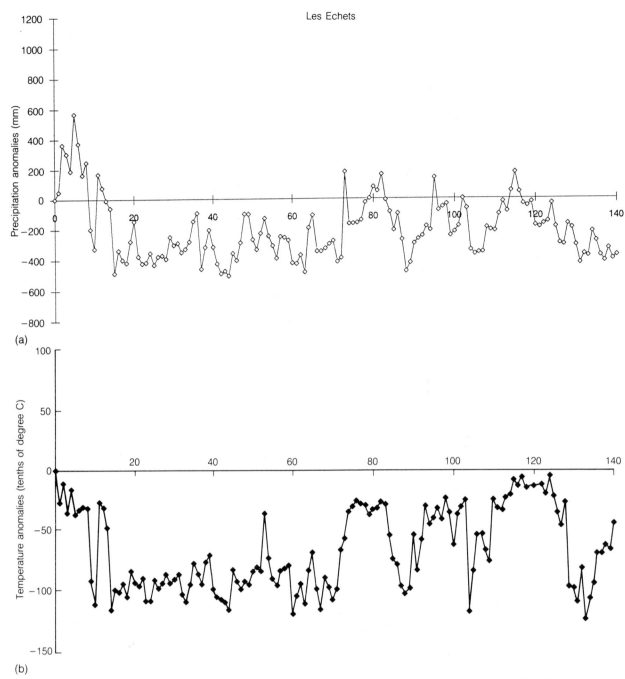

Figure 21.17 A computer reconstruction, based on a long pollen sequence from central France (Les Echets) of the local pattern of precipitation (a) and temperature (b) change in that area since the last interglacial, 120 000 years ago. The horizontal scale is in 1000s of years before present. The vertical scale is temperature/precipitation relative to the present day (Juiot)

change of about 5 parts per million by volume of CO_2 in the atmosphere (Figure 21.31). The global reduction in the area of forest during glacial periods, and its increase during interglacials, reflected locally by Figure 21.16, probably represents a change of about 250×10^{15} g of organic carbon on land, or half the total interglacial mass. The reduction of terrestrial carbon during glacial periods is associated with low CO_2 concentrations in the atmosphere (Figure 21.7), but an increased rate of carbon storage in the oceans (Figure 21.18).

Not only do pollen sequences tell us about past vegetation, but also about climate and the changing distribution of plants can be correlated with temperature and precipitation. Figure 21.17 shows a numerical reconstruction of temperature and precipitation in central France since the last interglacial, 120 000 years ago, based on a long pollen sequence. Not only does it show the magnitude of cooling during the last glaciation, but it also shows that precipitation diminished during the cold period, presumably because the cold atmosphere could carry much less water vapour from the colder ocean.

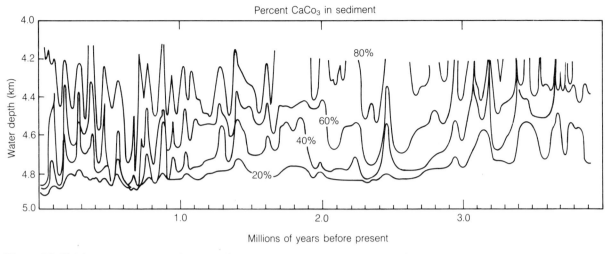

Figure 21.18 The variations in calcium carbonate concentration in ocean sediments in relation to their depth in the central equatorial Pacific during the last 4 million years. The data is derived from ocean cores. During glacial periods high carbonate concentrations reach greater depths than during interglacials, reflecting an increase of ocean storage of CO_2.

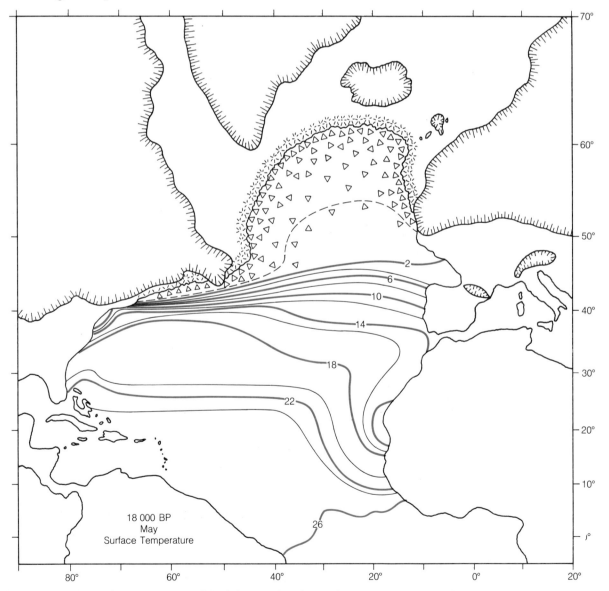

Figure 21.19 The surface temperature °C of the North Atlantic Ocean in May at the glacial maximum, 18 000 years ago, and the extent of sea ice.

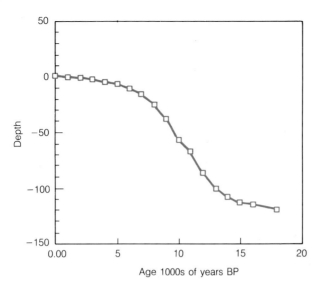

Figure 21.20 The rise of sea-level largely produced by melting of the mid-latitude ice-sheets of Europe and North America at the end of the last glacial period. It is derived from studies of coral reefs in Barbados.

These changes are also reflected in many other indices of global climate. In many parts of the world, but particularly impressively in China, a silt-rich wind-blown dust, called **loess**, accumulated during glacial periods. In north-central China, between Beijing and Xian, up to 100 m of loess accumulated in the last million years. Individual loess units approximately correspond to glacial periods. A 100 000 glacial phase is represented by about 10 m of loess, with individual loess beds separated by interglacial soils. The magnetic properties of the loess can be used as approximate indices of accumulation rate, and show a remarkable reflection of the Milankovitch frequencies of 40 000 and 20 000 years, suggesting that the wind regime over Asia is also ultimately driven by orbitally-determined radiation changes. There is also evidence from the other monsoon lands, of Africa, Arabia, India and Asia that the monsoon has been driven by changes in the global climate system on these same frequencies. Many lake basins in these areas show evidence of water-level fluctuations which reflect the behaviour of moisture-bearing monsoon winds.

Because of lower sea-levels during glacial periods, the Earth's land area was very much increased by the exposure above sea-level of many broad continental shelves. The exposure of these unvegetated sediment areas, and the intensification of winds in many low-latitude areas, produced a much greater amount of dust in the atmosphere, which is reflected in wind-blown dust in those parts of ice-sheet cores which represent glacial periods, and in increased terrestrial dust in deep-ocean cores in the tropics.

During the last million years, the oceans have exhibited fundamental changes in structure, composition and volume which parallel the glacial and interglacial cycles on land.

The ice core record from the Antarctic ice-sheet (Figure 21.7) shows how the concentration of CO_2 in the atmosphere increases from glacial to interglacial periods. This represents a massive net exhalation of CO_2 by the oceans into the atmosphere, and is also accompanied by a great increase in biomass on land. A great deal of calcium carbonate ($CaCO_3$) is found in ocean sediments, but, below a depth of 4–5 km, the concentration of $CaCO_3$ in sediments decreases very rapidly because of the dissolution of $CaCO_3$ in bottom waters under great pressure. However, in the Pacific Ocean for example, we find that the depth below which strong dissolution occurs changes between glacial and interglacial cycles. During glacial periods this critical depth increases, whilst in interglacial periods, it decreases (Figure 21.18). As a result, the total amount of CO_2 absorbed by the ocean increases during glacial cycles and reduces during interglacials. As the oceans are the Earth's principal store of CO_2, release of CO_2 from oceanic storage during interglacials increases CO_2 concentration in the atmosphere, enhancing the greenhouse effect and increasing plant growth, whilst loss to oceanic storage during glacial periods produces the reverse effects.

Fundamental changes of ocean circulation also occur between glacial and interglacial periods. During glacial periods, polar sea ice extends into lower latitudes, thus helping to cool the oceans. The polar and middle-latitude North Atlantic also receive large quantities of cold, ice-sheet meltwater. This extends the oceanic polar front, the region of thermal deficit, to the south, thereby pushing the region of strong thermal transport by ocean currents equatorwards. Figure 21.19 shows the surface temperature distribution in the North Atlantic at the last glacial maximum at 18 000 years BP. The zone of strong temperature gradient and strong ocean circulation lay some 2000 km further south compared with the present day.

A consequence of large-scale oscillations of global ice volume is to produce large-scale oscillations of global sea-level as ice-sheet growth is ultimately fed by ocean water. Thus, if the oceanic oxygen-isotope record (Figure 21.3) is primarily a record of global ice volume fluctuations it is also a record of sea-level change. Sea-levels are significantly lower than at present during glacial periods and similar to modern values during interglacial periods. Such global sea-level changes are termed

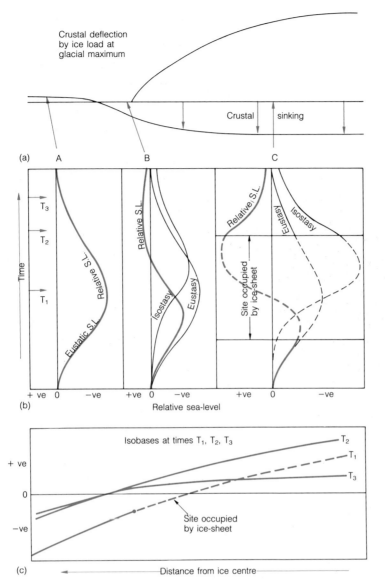

Figure 21.21 (a, b) Changes of relative sea-level produced near to ice-sheets are largely a consequence of the combined effect of global sea-level change due to changes in ice volume (glacial eustasy) and crustal depression by ice loading (glacial isostasy). At A, far from the ice-sheet, there is little isostatic effect and sea-level change through a cycle of global ice-sheet expansion and decay is largely eustatic. At B, near to the maximum extent of the ice-sheet, there is a significant isostatic depression of the crust during the glacial cycle so that relative sea-level does not fall as low as in case A, indeed, during deglaciation, relative sea-level is above the modern value. At C crustal depression is far greater than eustatic sea-level lowering, thus producing a high relative sea-level during deglaciation. (c) This shows isobases, being the current elevations of surfaces which lay at sea-level at times T_1, T_2 and T_3.

glacio-eustatic. An ideal way of measuring them more directly is by the study of coral reefs. Coral growth ceases as soon as they are exposed above sea-level. As the sea-level begins to fall, reefs are exposed above it, and coral growth ceases. When sea-level rises, younger reef corals grow above older corals, and a series of coral units are superimposed one above the other. One of the best records of eustatic sea-level rise (and therefore ice volume change) at the end of the last glacial period is derived from a series of superimposed reef corals in Barbados (Figure 21.20). This reflects an accelerating phase of ice-sheet melting between 14 000 and 12 000 years BP, a deceleration of melt-

ing between 12 000 and 10 000 BP, followed by a further acceleration, with the disappearance of the European ice-sheet just after 8500 BP, and the North American ice sheet by 6500 BP. The rates of sea-level rise during this deglaciation were as high as 2.5 m per 100 years at 9500 years ago and may be the source of Biblical flood legends at a time when urban centres were already established in coastal areas of the Middle East.

The general patterns of long-term glacio-eustatic sea-level change represented in Figures 21.3 and 21.20 are probably representative of a large part of the Earth's surface, but they are not applicable to the immediate vicinities of glaciated areas.

Figure 21.22 A 'staircase' of raised beaches in north-eastern Spitsbergen in the lower left of the photograph. Sea-level was very high immediately after deglaciation about 10 000 years ago, and has fallen by 90 m since then. The sea-level history is equivalent to that shown at site C, Figure 21.21. (see Fig. 20.3 for another view of raised beaches.)

When ice-sheets grow, they not only cause eustatic lowering of sea-level, but also cause **isostatic** depression of the lithosphere beneath them. It is believed that this occurs because of flow of material in the asthenosphere from beneath the area of ice-sheet loading to permit the stiffer, elastic lithosphere to sink beneath the ice-sheet. Because the lithosphere is relatively stiff, it will sink not only beneath the ice-sheet, but for some distance beyond it. Unlike the eustatic response to ice-sheet growth, which is instantaneous, the isostatic response is much delayed as it depends on relatively low rates of flow within the asthenosphere. Thus, if an ice-sheet begins to grow, global sea-level will fall, but after a while, in the area near to the ice-sheet, the lithosphere will begin to be depressed by the ice-sheet load. If the lithosphere sinks by the same amount that eustatic sea-level falls, there will appear to be no **relative** movement of sea-level with respect to fixed points on the Earth's surface. It is therefore important to define a **relative sea-level**, which is the sea-level that an observer standing on the Earth's surface would see. This is:

relative sea-level change = eustatic change –
isostatic change

Thus, if the eustatic change is –80 m, and the isostatic change is –100 m, the relative sea-level change will be:

relative sea-level change = –80 – –100 = +20 m

To an observer, sea-level will appear to rise by 20 m.

These effects are most dramatically seen during deglaciation. The lithosphere will be most strongly depressed in the area occupied by an ice-sheet. As this decays, eustatic sea-level will rise, and flood into the isostatically depressed area. Although the crust will tend to **rebound** isostatically after being depressed this will be a delayed response, and a large amount of **residual uplift** will remain to be achieved after complete removal of the ice-sheet and cessation of eustatic sea-level rise.

Figure 21.21 (a–b) shows some of those complicated relationships. At a point A, some distance from the ice-sheet, only the eustatic component of relative sea-level change will be felt. Sea-level will fall during ice-sheet growth and rise during ice-sheet decay. At point B, a small distance beyond the maximum extent of the ice-sheet, the early phase of ice-sheet growth, when the ice-sheet is

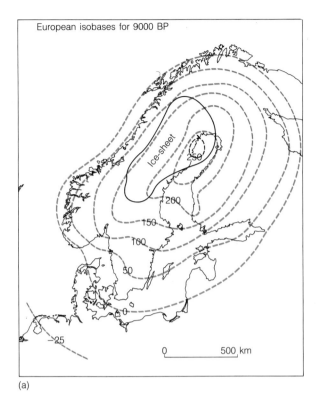

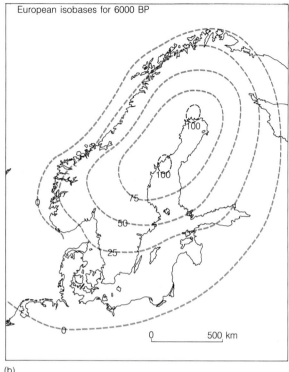

Figure 21.23 Isobases for 9000 and 6000 BP for the area occupied by the last European ice-sheet. They show the present elevation of surfaces which lay at sea-level at those times and reflect the pattern of isostatic uplift which has occurred due to glacial unloading of north-west Europe (cf Figure 21.13). The solid line over northern Sweden in (a) shows the location of the decaying ice-sheet.

far from the site, will be reflected by a eustatic sea-level fall. As the ice-sheet approaches the site, it will cause some local crustal depression, thereby reducing the magnitude of relative sea-level fall so that it is smaller than the eustatic component. At some time during ice-sheet decay, the curve of rising eustatic sea-level will cross the curve of crustal uplift, and relative sea-level will rise above the final value, before falling when ice-sheet decay and eustatic sea-level rise is complete, but the lithosphere is still rising. At point C, crustal depression beneath the ice-sheet load is very much greater than the eustatic lowering of sea-level. Relative sea-level then falls from a great altitude as the site is exposed beyond the marg-inof the retreating ice-sheet. In such areas a 'staircase' of **raised shorelines** can be found high above modern sea-level (Figures 20.3, 21.22), reflecting falling relative sea-level during the post-glacial period.

The geometry of the crustal effects of ice-sheet loading are reflected in **isobases**. These represent the present geometry of surfaces which formed at sea-level at specified times in the past. Thus, in Figure 21.21, the beach which formed at sea-level

at time T_2, will now be just below sea-level at site A, just above sea-level at site B, and well above sea-level at site C. The line connecting them is an isobase, and rises towards areas which were strongly isostatically depressed at the time the beach formed, and have subsequently rebounded. Figure 21.23 shows a map of isobases for 9000 and 6000 years BP in the area occupied by the last European ice-sheet. Evidence of the sea-levels of 9000 years ago occurs up to 230 m above modern sea-level in the Gulf of Bothnia, lifted to that elevation by isostatic uplift.

21.9 THE EARLY CENOZOIC AND MESOZOIC HOTHOUSE

Figures 21.5 and 21.6 reflect the progressive development and intensification of the Late Cenozoic ice age. If we work back in time, towards the beginning of the Cenozoic and into the Mesozoic, we find evidence for a warmer Earth, with much higher sea-levels than today, and a lack of evidence of large ice-sheets. Using planktonic foraminifers to reconstruct the surface temperatures of the Eocene ocean across a complete latitudinal transect from the south to north polar regions (Figure 21.24), suggests that although ocean surface temperatures in the equatorial zone were cooler than the present day, they were very much warmer in the polar zone. The heat transport in the ocean, the dominant part of the poleward energy conveyor belt, must have been more efficient than at the present day. The Earth was therefore more equable, with less severe temperature contrasts than at present. This is reflected in the fact that fossils of temperate tree species are found in early Cenozoic and late Mesozoic rocks in north Greenland and Spitsbergen (currently at 70°N), and in Antarctica. Continental reconstructions for the Cretaceous indicate that coal swamps occurred at latitudes in excess of 70°. There is no reason to believe that there were significant differences between the strength or pattern of solar radiation reaching the Earth from that of the present day, and so the reasons for the contrasts must be internal to the Earth rather than external. The most powerful internal influence on the operation of the energy conveyor belt is likely to have been the different locations of the continents and their influence on the circulation of the oceans and atmosphere. At the Cretaceous/Tertiary boundary (66.2 Ma), the North Atlantic was still largely closed, and there was no continuous oceanic passage around Antarctica. It was blocked by the Australian continent which was in contact with Antarctica. In contrast, a continuous oceanic passage existed in the sub-tropics of the northern hemisphere between the North American and Eurasian continents and the South American, African, Indian and Australian continents. The

ancient Pacific was not blocked to the north by the Bering Strait, but had a broad oceanic pathway to the north polar region.

The different geometry of the early Cenozoic/Mesozoic world from that of the present day would have produced a fundamentally different pattern of ocean circulation and therefore of heat transport. At the present day, the absence of a barrier across the circum-Antarctic ocean isolates the surface waters of the polar region from the surface water of the rest of the ocean and maintains a very cold water mass around Antarctica. Present-day deep waters around Antarctica do however flow strongly into the other oceans. This deep Antarctic water forms from the sinking of cold surface water, which then flows at depth into the other oceans. Because these cold surface waters dissolve CO_2 readily from the atmosphere, the Antarctic ocean plays a dominant role in the inhalation of CO_2 by the oceans from the atmosphere. The existence of a barrier to west–east circulation around Antarctica in the Cretaceous would have deflected latitudinal ocean circulation to produce stronger surface water and atmosphere interchange thereby warming the circum-Antarctic oceans and atmosphere, and reducing solution of CO_2 by the oceans. This would lead to an increase in atmospheric CO_2 concentration compared to the present day, an enhancement of the greenhouse effect, and relatively easier precipitation of $CaCO_3$ in the oceans. Figure 21.25 shows the result of a computer model of the near-surface temperatures of the mid-Cretaceous Earth compared with the present day. It shows how the different continental geometry would have led to warmer polar seas as a result of changed circulation patterns.

Evidence from sediment accumulations on continental margins shows that the Cretaceous was a period of high sea-level, and that there has been a general pattern of subsequent sea-level fall (Figure 21.26). The cubic capacity of the ocean basins is largely related to the size of the worldwide mid-ocean ridge system. Newly-formed oceanic crust is relatively hot, but as it moves away from the mid-ocean ridge spreading centre, it cools and shrinks, leading to sinking of the ocean floor away from the ridge. The rate of cooling, shrinkage and sinking is approximately constant, but when sea-floor spreading rates are high, the flanks of ridges move laterally relatively more quickly than they sink vertically in comparison with periods when spreading rates are low. As a consequence, the volume of the ridge system is large and the volume of the ocean basins relatively small during periods of fast spreading, whilst ridges have smaller volumes and the ocean a larger volume during periods of slow spreading. The fast spread-

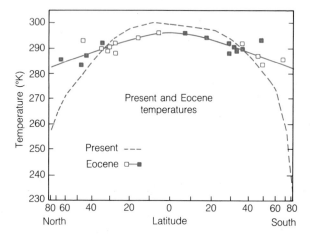

Figure 21.24 The latitudinal variation in the surface temperature of the Eocene Ocean (50 m.y. ago) compared with the modern ocean. The poleward transport of heat must have been more efficient than at the present day (cf Figure 21.11).

Present-day surface temperature (°K)

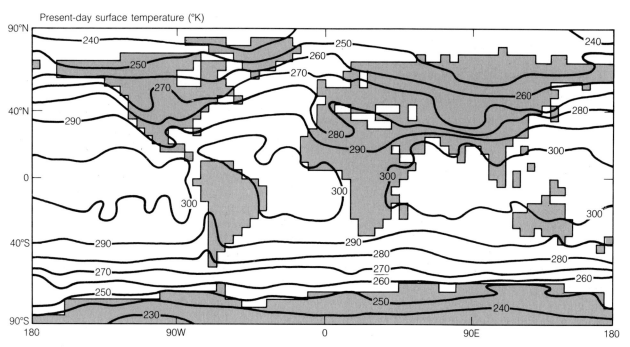

Mid-Cretaceous surface temperature

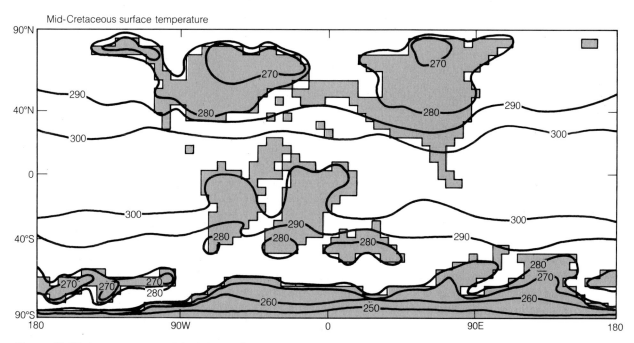

Figure 21.25 A computer model of the surface temperature of the Mid-Cretaceous Earth compared with one of the present day. Note the different continental geometries. The different geometries of the continents and the ocean basins compared with present day produced a pattern of circulation which more effectively warmed up the polar seas. There was much greater connectivity between the ocean basins compared with the present day.

ing rates of the Cretaceous (Figure 21.26) therefore led to high eustatic sea-levels, which subsequently fell as spreading rates diminished. Furthermore, during periods of strong continental compression leading to uplift, lateral squeezing of continents produced a high continental **freeboard**, for instance during Tertiary mountain building, so as to increase ocean basin volume and further decrease eustatic sea-level.

Widespread Cretaceous flooding of continental surfaces coupled with the relatively poor solubility of CO_2 in the Cretaceous ocean, led to a great amount of direct carbonate precipitation in warm, shallow Cretaceous seas producing the widespread chalk sediments.

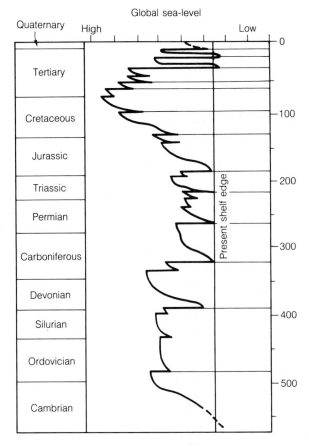

Figure 21.26 The pattern of global sea-level change on Earth since the Cambrian. The Late Cretaceous was a period of very high sea-level, which led to flooding of continental surfaces and deposition of much carbonate sediment in warm, shallow seas. Since then there has been an irregular decline of sea-level.

21.10 PRE-CENOZOIC ICE AGES

Within the Late Cenozoic Ice Age the Earth shows a relatively fast tempo of climatic change (Figure 21.5). Although the amplitude of climatic change on Earth during this Ice Age may be extreme, it is probable that much of the Earth's history has been characterized by climatic change of similar tempo, and that change in the climate system at orbital frequencies pervades the entire stratigraphic record.

However, there also appears to be a lower frequency, larger-scale tempo of change, of which the late Cenozoic Ice Age and the Late Mesozoic warm period are merely the latest expressions. This larger-scale global climate variation is reflected by the widespread evidence of pre-Cenozoic Ice Ages. This evidence is most dramatic at sites which are located for instance in Late Ordovician rocks in the Sahara (Figure 21.27) and at the base of Permian sequences in South Africa (Figure 20.27).

Ice ages characterized by widespread glaciation have now been identified (Figure 21.28) between the Late Devonian and mid-Permian (260–370 Ma), between the late Ordovician and early Silurian (410–440 Ma), several times during the late Proterozoic (1000–600 Ma), at about 2500 Ma, and possibly at some time during the Archaean, prior to 2600 Ma.

The Palaeozoic ice ages (260–370 and 410–440 Ma) primarily affected the Gondwanaland super-continent, which consisted of the fused continental plates of South America, Africa, Arabia, India, Antarctica and Australia (Figure 21.29). During the Ordovician/Silurian Ice Age, the evidence of glaciation is first apparent in North Africa, but occurs progressively later in southern Africa and South America. During the Devonian/Permian Ice Age, evidence for glaciation is first apparent in South America, and found progressively later in Africa, Antarctica and Australia Figure 20.27, Plate IXd). Figure 21.29 shows the apparent shift through time of the area of glaciation. The obvious explanation is that widespread glaciation occurred in those areas which were located in a polar position as the super-continent moved through the polar zone. The Silurian to Devonian interval during which there is no evidence for Gondwanaland glaciation, was one in which Gondwanaland had moved away from a polar position (Figure 21.29).

It therefore seems that the primary prerequisite for an Ice Age is the position of continents in polar positions. During the early Palaeozoic, the continent was Gondwanaland, during the Cenozoic, it has been Antarctica. However, the **intensity** of glaciation appears to be strongly affected by other factors, principally the oceanic circulation (partly controlled by continental locations which determine ocean basin geometry) and atmosphere circulation (partly controlled by continental location and the location of mountain chains). Thus, the timing of ice ages and their intensity is primarily determined by geotectonic processes, whilst the high frequency tempo of change is determined by solar and orbital processes.

The precise location of ice-sheets is also determined by geotectonically-steered patterns of ocean and atmosphere circulation. Such patterns are able to produce Cenozoic ice-sheets in the lowlands of North America and Europe which extended to latitudes as low as 40°N; but only in an Earth which had already been cooled by the existence of polar ice-sheets such as Antarctica and Greenland. The Late Proterozoic glaciation, is particularly puzzling in this regard, for evidence of glaciation is not only found in rocks which formed at high latitudes, but also in those which formed at low latitudes of less than 45°. This is an

Figure 21.27 A glaciated roche moutonnée (cf. Figure 20.28) from the Late Pre-Cambrian of the south-western Sahara.

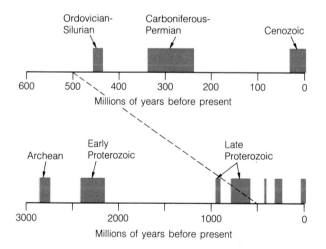

Figure 21.28 The known frequency of ice ages through the last 3000 m.y. of geological time. The upper diagram shows the timing of ice ages during the last 600 m.y., and the lower diagram their timing through the last 3000 million years.

unsolved enigma for which several explanations have been suggested: that the angle between the Earth's axis of rotation and the orbital plane was greater, thereby reducing the net incident radiation in the equatorial zone; a general refrigeration

of the Earth occurred, (which would have led to low sea-level, for which there is no evidence); a pattern of ocean currents existed which carried icebergs into low latitudes (most late-Proterozoic evidence is from iceberg-dropped boulders).

21.11 THE FUTURE OF CLIMATE AND POSSIBLE HUMAN INFLUENCES

In the late nineteenth century, James Hutton, the famous Edinburgh geologist, coined the axiom that **the present is the key to the past**. We might now add that the **past is the key to the future**, in that if we can establish a repetitive tempo of change in the past, it is reasonable to extrapolate that pattern so as to predict the future. If we asked how long the Late Cenozoic ice age might last, we might conclude that it would take as long as it takes the Antarctic continent to drift out of a polar position. To predict when this might be however requires more understanding of mantle convection than we currently have.

We may however have a better basis for prediction on a shorter timescale. The cyclical pattern of glacials and interglacials in the last million years reflect the Earth's response to changes in incom-

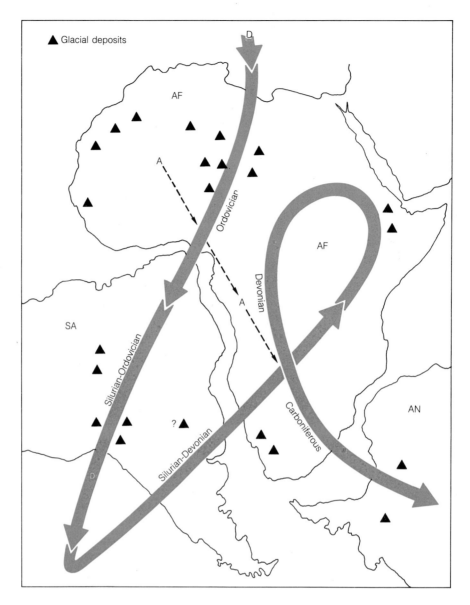

Figure 21.29 The apparent shift of the centre of glaciation between the Ordovician and Carboniferous over the super-continent of Gondwanaland, comprising the South American, African and Antarctic plates. This probably reflects the movement of the super-continent through the polar zone.

ing solar radiation produced by cyclical changes in the Earth's orbit around the sun. It has been suggested however that the precise nature of the response, such as the large amplitude 100 000 year cycles, of the last 0.75 m.y., may depend on the disposition of continents on Earth and the uplift of particular mountain masses such as Tibet and the western North American Cordilleras. The rate of continental drift is slow on million year time-scales, and thus continental positions can be assumed to remain constant in the near future. What of northern hemisphere mountains? The rate of northward drift of the Indian sub-continent has been rapid over the last 50 m.y., and the compression it has developed against the Eurasian plate has been the cause of Tibetan uplift. If these rates are maintained, we might expect uplift to hold its own against erosion and the Tibetan plateau at least to maintain its elevation, and thus for the area to maintain its recent role in regulating the Earth's response to insolation changes. As insolation changes can be predicted into the future, the correlation between reconstructions of past insolation and global climate can be used to predict future global climate. Figure 21.30 shows the result of one such exercise, which predicts the onset of the next glacial period, predicting a 400 km southerly shift of climate zones in western Europe within 5000 years; a subsequent strong warming followed by more intense cooling by 25 000 years and widespread glaciation of western Europe by 60 000 years into the future.

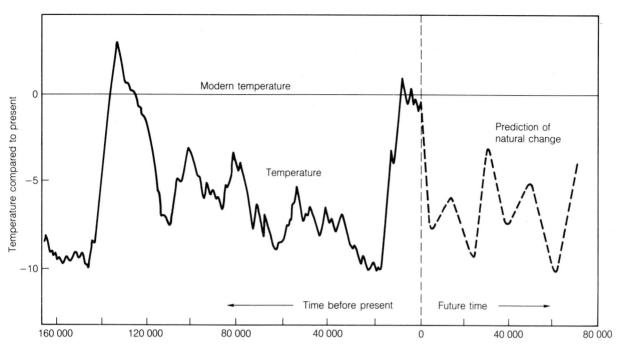

Figure 21.30 A prediction of future climate based on an extrapolation from the 'natural' pattern of variation in the Earth's recent past.

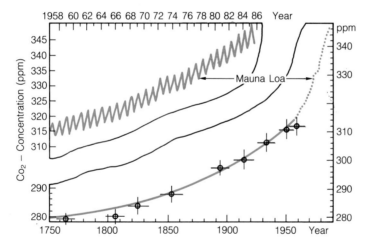

Figure 21.31 The increase of atmospheric CO_2 due largely to burning of fossil fuels since the 'Industrial Revolution'. The earlier part of the curve is derived from the study of ice cores in Antarctica, and the latest part from direct measurements on Mauna Loa, Hawaii.

However, such a prediction only takes into account solar driving of climate on long timescales. Figure 21.8 shows that climate also varies on **sub-Milkanovitch** timescales of hundreds of years, such as the so called **Little Ice Age** lasting from 1350 to 1870, when much of the North Atlantic region at least suffered from a significant period of cooling. It is probable that such events reflect internal interactions within the Earth's climate system. Similar events are likely to occur again and make prediction of the immediate future of the **natural** climate difficult.

A further difficulty in prediction arises from human activity. The greenhouse effect is of fundamental importance in maintaining the surface temperature of the Earth at levels needed by the types of animals and plants which live on its surface (Figure 21.10). The principal greenhouse gases are water vapour (H_2O), carbon dioxide (CO_2), ozone (O_3), nitrous oxide (N_2O) and methane (CH_4). Their concentration in the atmosphere and their present warming effect is shown in Table 21.3, which indicates very clearly how dependent the Earth's surface environment is on a

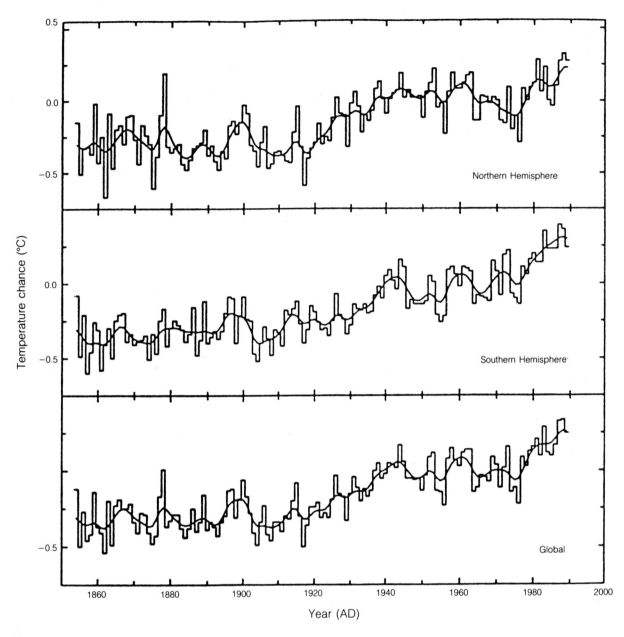

Figure 21.32 The pattern of global temperature change over the last 100 years.

greenhouse effect. If the greenhouse gases were lost from the atmosphere, the Earth's average temperature would fall by over 30°, ten times greater than the cooling from interglacial to glacial periods.

The Antarctic ice cores (Figure 21.7) indicate that global atmospheric CO_2 concentration increased from about 180 ppm at the time of the last glacial maximum, 18 000 years ago, to 280 ppm by the middle of the present interglacial; an increase which may have played an important role in the glacial/interglacial temperature increase. A value of 280 ppm was sustained until about 1750–1800, just before the industrial revolution, but since then has risen dramatically to 353 ppm by 1990 (Figure 21.31).

Extrapolation of these trends of atmospheric greenhouse gas concentration suggests a doubling of pre-industrial concentrations by about 2050. Computer models suggest that this will lead to an increase in global air temperature of between 1.5 to 4.5°C. One effect of such a temperature increase would be to warm the surface waters of the oceans, and thus cause sea-level to rise because of thermal expansion. The calculated temperature rise would produce a sea-level rise of between 30 and 110 cm by the year 2100. There would be many other effects, including shifting ocean currents, changing patterns of rainfall, glacier melting, changes in the distribution of animals and plants, and enforced changes in patterns of human agriculture, fisheries and habitation.

It would be wrong to suppose however that changes of similar magnitude and at greater rates have not occurred in the recent geological past. The principal difference is the presence of a human community which is highly adapted to the present climate and environment and therefore sensitive to even small changes.

These predictions however are only based on modelling. Is there evidence that an enhanced greenhouse warming is taking place? There appears to have been a global temperature increase of above 0.4°C since the beginning of the century (Figure 21.32) with the five warmest years of the last 100 occurring in the last decade, though it is not yet possible to assess whether this is a reflection of natural variability or due to a man-made greenhouse effect. For a geologist, a fundamental question arises: if a human-induced greenhouse effect is beginning, will the natural tendency of climate towards long-term cooling into another glacial period, eventually reassert itself, or has a new phase of human-controlled climate evolution just begun?

Table 21.3 'Greenhouse' gas concentrations

Greenhouse gas	Present atmospheric concentration (ppm)	Present warming effect (°C)
H_2O	2–3000	20.6
CO_2	345	7.2
O_3	0.03	2.4
N_2O	0.3	0.8
CH_4	1.7	0.8
Others		0.6
	Total	34.4

21.12 FURTHER READING

Bowen, D.Q. (1978) *Quaternary Geology*. Pergamon, Oxford.

Bradley, R.S. (1985) *Quaternary Palaeoclimatology*. Allen and Unwin, London.

Crowley, T.J. and North, G.R. (1991) *Palaeoclimatology*, Oxford University Press, Oxford

Houghton, J.T., Jenkins, G.J. and Ephraums, J.J. (eds) (1990) *Climate Change*. World Meteorological Organisation/United Nations Environment Programme, Intergovernmental Panel on Climate Change. Cambridge University Press, Cambridge.

Imbrie, J. and Imbrie, K.P. (1979) *Ice Ages: Solving the Mystery*. Macmillan, London.

Lowe, J.J. and Walker, M.J. C. (1984) *Reconstructing Quaternary Environments*. Longman, London.

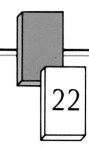

WIND ACTION
AND DESERT
LANDSCAPES

Where for many centuries only the camel
has been able to penetrate, the
helicopter . . . is now dropping geologists.

Georg Gerster, 1959

22.1 CIRCULATION OF THE ATMOSPHERE

The circulation responsible for the winds is primarily a response to the familiar fact that air is
cold over the polar regions and hot over the
equatorial belt. If the Earth did not rotate, heated
air would ascend at the equator and blow towards
the Poles where, having become chilled and
heavy, it would descend and return towards the
equator, becoming steadily warmer as it followed
the meridians into lower latitudes. This would be
a simple convection system like that illustrated in
Figure 3.16. But because the Earth is rotating, a
powerful deflecting force is at work, called the
Coriolis force, after the physicist who first recognized it in 1835. If anything (e.g. air, water,
projectile) moves relatively freely from south to
north in the northern hemisphere, it starts from a
place where it shares in the Earth's rotational
velocity to the east (over 1600 km/hour at the
equator) and passes through places where that
velocity is much lower, according to latitude (e.g.
about 1300 km/hour for New York; 1000 km for
London; 0 for the north pole). Consequently the
moving mass or object tends to travel eastward
faster than does the Earth immediately beneath it,
and the farther north it goes the more it turns
towards the east. Similarly, if the movement is

towards the Equator, where the rotational velocity
is higher, the tendency is to be left behind by the
Earth, i.e. to turn increasingly westward. Put more
generally, the deflection is always to the right in
the northern hemisphere, to the left in the southern one.

The Coriolis force is one of the minor factors
concerned in river erosion, the channel being
more effectively undercut on the side towards
which the force is directed. The persistent westward migration of the southward-flowing Kosi
river (Figure 17.35) can be partly ascribed to this
force. The less confined movements of projectiles
and aeroplanes, and of the winds themselves, are
far more powerfully affected. The high-altitude
winds that blow from the Equator to the poles are
deflected to the east. The return winds which
complete the convective circulation near the
ground would therefore be expected to be
deflected to the west, i.e. to be 'easterlies'. And so
they are across a broad belt on each side of the
Equator (the north-east and south-east trade
winds) and, less regularly, in the polar regions
(Figure 22.1). But we find that in each hemisphere
there is a belt of disorderly 'westerlies' separating
the polar easterlies from the tropical easterlies (the

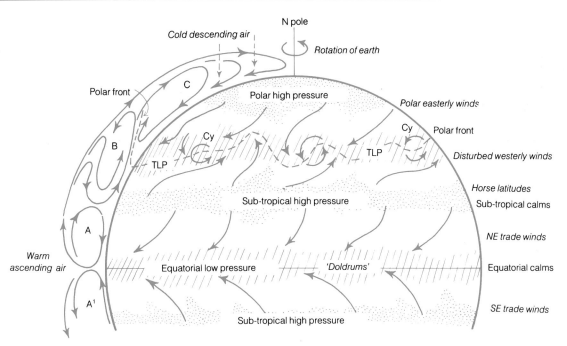

Figure 22.1 The general planetary circulation of the lower atmosphere in three main cells, A, B and C in the northern hemisphere (and similarly A1, B1 and C1 in the southern hemisphere). TLP, temperate low pressure belt; Cy, cyclones.

trade winds).

To account for this complication we must consider another global effect. Heated air ascending from the equatorial 'doldrums' turns poleward at a height of about 10–13 km, and passes into latitudes that are shorter than the Equator. Latitude 30°, for example, is 13 per cent shorter than the Equator. At about this position, but ranging between 25° and 35° (north and south according to the season), the crowding of the air is sufficient to raise the pressure so that air is obliged to descend towards the surface. Here, then, are the high-pressure subtropical calm belts that came to be known as the 'Horse Latitudes' in the old days of sailing ships. At the surface the descending air divides into (a) the trade winds that blow towards the equator (so completing the tropical convection cells A and A' of Figure 22.1), and (b) the disorderly westerlies that spiral **towards** the poles in cell B. But surface winds are already blowing **from** the poles, in cell C, and where the two opposing wind systems B and C meet, the weather becomes very disturbed and variable. The cold polar air tends to wedge itself southward, whilst the warm moist air from lower latitudes flows up the surface of this wedge and so becomes cloudy and a source of rain or snow, often accompanied by strong winds. The surface of the cold wedge is called the **polar front**.

In the northern hemisphere the polar front advances far to the south in winter and retreats to the north in the summer, its range over land being much wider than over the oceans. The latter have a stabilizing effect because of the relative slowness with which water gains or loses heat. A corresponding winter advance to the north and summer retreat takes place in the southern hemisphere. We now have the basic scheme for a threefold tandem arrangement of convection cells in each hemisphere (A, B and C in Figure 22.1). But there are many further complications. The high-altitude westerlies are found to be concentrated into jet-like streams between the tropics and the poles. Instead of maintaining a nearly uniform direction as they blow towards the poles, the air-streams follow sinuous courses, alternately surging far to the north and south of their mean paths. These surges are probably mainly responsible for the atmospheric eddies familiar as cyclones (with lowest pressure in the centre) and anticyclones (with highest pressure in the centre). Each cyclone is like an enormous vortex with winds spiralling round in an anticlockwise direction in the northern hemisphere, and clockwise in the southern. For anticyclones these directions are reversed, the general rule being a direct consequence of the Coriolis force. It is because of the continual recurrence of these broad eddies that the westerlies are described as 'disturbed' or 'disorderly'. In the British Isles the most settled type of weather is, paradoxically, 'unsettled'.

Over tropical oceanic areas that have been abnormally heated, rapidly ascending streams of air generate the devastating vortices of winds known as **hurricanes** in the Atlantic and its great Mexican, Carribean and Mediterranean embayments. The very term hurricane comes from a Carribean word meaning 'the spirit of evil'. In the west Pacific (e.g. between Australia and Japan) similarly violent whirlwinds are called **typhoons**, after the malevolent monster Typhon of Greek mythology. No special term refers to those originating in the Indian Ocean, where they are commonly referred to as **tropical cyclones**.

Whereas the comparatively mild cyclones of temperate regions are usually 1600 km or more across, the hurricane and typhoon systems of rotary winds may be only 300 or 400 km in diameter. But this means that from outside, where the pressure is highest, to the relatively calm but menacing 'eye' in the middle, where the pressure is lowest, the pressure gradient is correspondingly steeper. The resulting winds of hot moist air drawn into the spiralling updraught reach speeds of 120 to 200 or even 300 km/hour, and as the rising air expands and cools, its water vapour condenses and falls in an overwhelming deluge of rain. After passing their climax, hurricanes tend to broaden into ordinary cyclones as they travel away from the tropics. Some cross the Atlantic (Figure 22.2) and bring stormy weather to western Europe.

The **tornado** (Spanish tornar, to twist or turn), which often begins on land as a satellite to a severe hurricane, is a very much narrower column or funnel of swiftly spinning air rarely a kilometre and a half across and generally much less. The fiercely twisting winds create such havoc that any instruments that might have measured their speeds are inevitably destroyed. Estimates based on the fantastic effects of the winds sound incredible. The general destruction is due not only to the extreme violence of the winds, but also to the phenomenal reduction of pressure that takes place in the heart of a tornado. This results from the intensity of the surrounding rotation, just as happens in the vortex that develops in the water above the plug-hole when a bath is being quickly emptied. When a house is suddenly struck by a tornado it may be literally burst open by the excess of its internal air pressure, although the latter is likely to be something less than normal. A characteristic feature of a tornado is the long black funnel-shaped or snaky cloud that stretches towards the ground from the great thunder cloud riding above the storm. From this sinuous column the rainfall is catastrophic, and wherever the column itself touches the ground the ruination of trees and structures of all kinds is most severe. Over the sea the pendant column swings about like an elephant's trunk and finally joins up with leaping peaks of water and spray to form a **waterspout** (Figure 22.3). The sea makes its contribution where the pressure is so low that a kind of temporary isostasy comes into play. In response to the relief of pressure the water spurts upwards to complete its swirling union with the cloud from which the bulk of the waterspout is derived.

Small tornadoes, known as **dust devils**, are a fairly common event in flat desert and semi-desert

Figure 22.2 Mass of coral-rock thrown up a cliff about 12 m high, carried 30 m from the cliff and turned upside down by a hurricane; east coast of Barbados, Windward Islands (*C. T. Trechmann*).

Figure 22.3 Waterspout off the Island of Rhodes, Greece, October 1930. (*Syndication International Ltd*)

areas; each comprises a column of whirling dust and near the ground, sand, just a few metres across but several tens or even hundreds of metres high, that meanders across the barren desert surface. Dust devils usually collapse if their base encounters dense scrub vegetation or a sudden change in relief (hill or deep gulley), which breaks up the vortex.

In addition to the various planetary complications introduced into the basic scheme illustrated by Figure 22.1, there are further modifications dependent on the distribution of land and sea. Of these, only the Asiatic monsoon circulation need be mentioned here. This is caused mainly by the intense winter cold of Siberia and the high plateaus and ranges of central Asia, alternating with the summer heating-up of the continent, which leads to extremes of high temperature over vast areas. In winter the monsoon flowing out from central Asia strengthens the north-east trades and brings cool dry air over India. In summer the directions of the monsoon winds are reversed. Moisture-bearing winds then blow in from the oceans (e.g. from the south-west over India) and so control the rainy season.

Before proceeding to consider the direct action of winds on the land surfaces over which they blow, there are some general aspects that should not be overlooked. Because the wind distributes moisture over the face of the Earth it is one of the primary factors responsible for weather and the weathering of rocks, and for the maintenance of rivers and glaciers. The more northerly and southerly of the atmospheric surges mentioned above introduce streams of moist oceanic air far into

Greenland and Antarctica respectively, so providing sources of precipitation for the nourishment of ice-sheets. Moreover, in blowing over the oceans and other bodies of water, the wind transfers part of its energy to the surface waters and so becomes responsible for waves and their erosive work. Hurricanes and typhoons in particular generate unusually high waves and locally increase the height of high tides by driving water towards the land, so adding much coastal flooding to the catalogue of their capacity for destruction.

22.2 WEATHERING AND FLUVIAL ACTIVITY IN THE DESERT

The belt of deserts between the Atlantic and the Arabian Gulf lies mainly to the south of latitude 30°. Here, the high-pressure descending air is dry to begin with (Figure 22.1) and the resulting trade winds become increasingly desiccated so long as they are blowing over land. This belt continues north of latitude 30° from Iran to the Gobi of central Asia, though with several interruptions, rainfall being low because of distance from the oceans or the intervention of mountainous rain barriers. High mountains near the western coasts are mainly responsible for the deserts of North and South America. In all these arid regions the rainfall is rare and sporadic, both temperature and wind intensity are subject to violent fluctuations – daily and seasonal – and vegetation is extremely scanty or entirely lacking. Under these conditions mechanical weathering is dominant, involving the splitting, exfoliation and crumbling of rocks by alternating scorching heat and icy cold (Figure 15.4). Nevertheless, chemical weathering, though extremely slow and probably assisted by micro-organisms, plays a far from negligible part. By decomposition and solution, rocks that would otherwise successfully resist the stresses set up by temperature changes are gradually weakened until they can be shattered. By evaporation minute quantities of dissolved matter are brought to the surface. Loose salts are blown away, but the force of crystallization of salt within pores or minor fissures can crack the rocks. Oxides of iron and manganese, however, form a red, brown or black film which is firmly retained. The surfaces of long-exposed rocks and pebbles thus acquire a characteristic coat of 'desert varnish'.

Although most desert localities are not subjected to rainfall for years on end, no part can be regarded as permanently free from rain. Taken over a period of years, the rainfall averages only a few centimetres a year, approximately 25 cm being reached on the fringes of the desert. In the adjoining semi-arid regions the annual rainfall averages between 25 and 50 cm, but even here long periods of drought are usual.

The capacity for evaporation in deserts may exceed the amount of rainfall by as much as 20 or 30 times, so the growth of lakes is effectively prevented unless supplied with water from beyond the desert boundary (e.g. Lake Chad). Permanent streams cannot originate under such conditions, although well nourished rivers, like the Nile, with adequate sources in humid regions, may cross the desert without entirely dwindling away. Outflowing streams are otherwise short and intermittent, and even in coastal areas carry too little sediment during their brief period of flow for anything but temporary deltas to form. Over large areas the desert drainage is internal, and directed towards the lowest parts of the many depressions which, owing to earth movements and wind erosion, characterize the desert surface. The poetical generalization that 'the weariest river winds somewhere safe to sea' does not apply to desert regions.

The gorges and steep-sided wadis that dissect the desert uplands (Figures 22.4 and 22.5), some with extinct 'waterfalls'; the alluvial spreads that floor the depressions; the salt deposits and terraces of vanished lakes (Figure 22.6); the buried soils that only require turning over and irrigating to blossom afresh; the rock carvings and paintings of prehistoric artists, and other relics of ancient man from stone implements to actual habitations; all these point to pluvial climates when running water was far more actively concerned than it is today in developing the landforms of the desert which we actually see.

The Sahara has been described as 'the corpse of a once well-watered landscape'. But whereas man originally retreated before the spreading aridity, now he has returned for oil and gas and iron ores, and above all for the essential artesian water that alone can restore the long-vanished fertility and make the desert locally habitable. Between the Hoggar Mountains and Mauritania there is a vast expanse of desert-varnished **reg** or gravel desert known as the Tanezrouft and hardly heard of until

Figure 22.4 Gorge of the Wadi Barud, Egyptian Desert. (*O. H. Little*)

Figure 22.5 Gorge of the Wadi Gasab in the plateau of Ma'aza between the Nile and the southern end of the Gulf of Suez. (*W. F. Hume*)

Figure 22.6 Shore terrace of former pluvial lake, Colorado Desert, California. (*W. C. Mendenhall, United States Geological Survey*)

the French exploded their first atom bombs. Here, during the preliminary quest for water, French geologists discovered a thick fossil soil just below the surface. Pollen from the soil is of vegetation similar to the kinds that now flourish in a Mediterranean climate, and radiocarbon dating shows the soil to be about 7000 years old. Drilling has revealed deep-lying water-bearing sandstones fed by the rain that falls on the Atlas mountains to the north, where the intake area extends for hundreds of kilometres along the upturned edges of the strata. On the northern side of the Sahara this great aquifer already supplies artesian water to all the developing oilfields, making the settlement's market gardens very fertile (Figure 22.7). Throughout this region official policy – not everywhere strictly adhered to – is that the rate of withdrawal of water for irrigation, industrial and social purposes must not exceed the rate of replenishment along the intake.

Despite the multiplication of artificial oases, extreme desiccation of the desert is still its most striking natural feature; and it remains something of a puzzle that rare rainstorms and the spasmodic stream action should occur at all. When they do occur they are the result of cyclonic, funnel-shaped breakthroughs from the moist air of high-level westerlies (Figure 22.1). The chief characteristics of the rare desert rainfall are their erratic distribution and brief duration, and – apart

from the occasional light showers – their intense violence. Houses of dried salt mud are turned into a miry pulp and washed away when a sudden 'cloudburst' descends on an oasis. Travellers have been drowned in the floods that race down the dry wadi channels with little or no warning; indeed, probably more people drown in the desert because of flash floods than die of thirst. Such torrents, swiftly generated in a distant upland rainstorm, carry a heavy load of mixed debris, prepared for them by years of weathering and wind erosion. At the foot of the mountains or escarpments, where carrying-power is reduced by seepage or loss of gradient, the load is dropped to form alluvial fans (Figure 22.8) and 'deltaic' deposits: first the coarser and later the finer debris. Between the pediment or sediment-filled depression that commonly surrounds a desert mountain range or flanks an escarpment, there may be a more or less continuous slope of waste formed by the coalescence of alluvial fans; this feature is called a **bajada**, pronounced *bahada* (Figure 22.9).

Choked by their own deposits, the short-lived streams subdivide into innumerable channels that spread out laterally to form a braided pattern (Section 17.9) across the plain, which thus

Figure 22.7 El Oued, 'the city of domes' in south-east Algeria, where deep artesian wells now support a population of over 100 000, and basins dug through the surrounding dunes enable the roots of palms to tap a shallower source of flowing ground-water. (*Ritchie Calder*)

Figure 22.8 Alluvial fan, east wall of Death Valley, south of Badwater, California, passing into a salina of salt and mud in the foreground. Notice the miniature delta near the middle in front, formed by a temporary stream from the other side of the depression. (*J. S. Shelton and R. C. Frampton*)

Figure 22.9 The outer part of a bajada that originally continued across the foreground to the gullied escarpment on the Pacific side of the San Andreas fault zone, above which the photograph was taken, and against which the coalesced alluvial fans originally reposed. The valley in the foreground was formed at the expense of the bajada by the headward erosion of a stream draining into the Pacific. Now only a thread of water, the stream must have been much more voluminous during the pluvial phases of the Pleistocene. The view shows one of the beheaded channels leading down towards the playas and salinas of the Mohave Desert. (*J. S. Shelton and R. C. Frampton*)

becomes covered with a veneer of finer sediment. The waters of major flash floods may overflow the channel banks and cover very extensive areas. With large amounts of unconsolidated sediment and a diminishing supply of water, a point may be reached where the high water : sediment ratio, which is typical of stream-flow in permanent rivers, may change to the high sediment : water ratio of a mud-flow that is capable of carrying large boulders.

After surface flow has ceased, the water percolating downstream through the stream bed commonly causes cementation by dolomite of the near-surface sediments. The surface sands and silts dry out rapidly, however, and generally are not cemented; they can be readily carried away by the wind. Clay lags in stagnant pools may dry, curl and crack. Thin and fragile clay flakes are removed by the wind, but thicker and heavier flakes can be preserved in place, at least temporarily, by a covering of wind-blown sand.

If the rain falls on the gentle slopes of a depression, a shallow sheetflood may carry suspended fine material towards the middle as a mud flow. If the water reaches the lowest part before it is lost by further seepage and evaporation, a temporary lake is formed; the remnants of clay pans deposited from suspension within temporary interdune lakes of the Namib Desert during the floods of 1986 are illustrated in Plate XIIa. In temporary lakes, the dissolved material is concentrated by evaporation and finally deposited to form salt muds or glistening white sheets of rock salt and other evaporite minerals; gypsum crystals grow preferentially within the sediment. In west America the alluvial plains of arid or semi-arid enclosed basins are called **playas** (Figure 22.9), the more saline tracts being distinguished as **salinas** (Figure 22.10); in North Africa and Arabia these latter are known as **sabkhas**, or more particularly **inland sabkhas** to differentiate them from the marine-influenced **coastal sabkhas** (Section 6.8).

So long as more material is brought into desert basins by rare torrents and sheet floods than is removed by the wind, such depressions continue to be filled.

When sand, silt or dust is blown across a moist sabkha surface, the particles stick to the damp ground to form typically wavy **adhesion ripples**. In many inland sabkhas, crystallization forces the thin salt crust to adopt a hemispherical blister-like shape, which is hollow beneath. Wind-blown sand and dust adheres to the hygroscopic surface of the blisters; the blisters themselves may be breached and infilled by saltating sand grains, and their irregular surface can be covered by variable thicknesses of sand to form a much more complex type of adhesion ripple with a greater and very irregular amplitude.

22.3 THE GEOLOGICAL WORK OF THE WIND

As an agent of transport, and therefore of erosion and deposition, the work of the wind is familiar wherever loose surface materials are unprotected by a covering of vegetation. The raising of clouds of dust from ploughed fields after a spell of dry weather and the drift of windswept sand along a dry beach are known to everyone. In humid regions, except along the seashore, wind erosion is limited by the prevalent cover of grass and trees and by the binding action of moisture in the soil. But the trials of exploration, warfare and prospect-

Figure 22.10 Salt marsh (salina) in Death Valley, California. (*Dorien Leigh Ltd*)

Figure 22.11 Dust storm approaching Port Sudan, west coast of Red Sea. (*Paul Popper Ltd*)

ing in the desert have made it hardly necessary to stress the fact that in arid regions, the effects of the wind are unrestrained. The 'scorching sand-laden breath of the desert' wages its own war of nerves. Dust storms darken the sky, transform the air into a suffocating blast and carry enormous quantities of material over great distances (Figure 22.11). Vessels passing through the Red Sea often receive a baptism of silt from the desert winds of Arabia.

By itself, the wind can remove only dry incoherent deposits. This process of lowering the land surface is called **deflation** (*L. deflatus*, blowing away). Armed with the sand grains thus acquired, the wind near the ground becomes a powerful scouring or abrading agent. The resulting erosion is described as **wind abrasion**. By innumerable impacts the grains themselves are gradually worn down and rounded. The larger 'millet-seed' sand grains (1–2 mm diameter) are typically well rounded (Figure 22.12), but finer grains (0.06–0.2 mm) are commonly quite angular. The winnowing action of the wind effectively sorts out the transported particles according to their sizes.

Supplies of such materials are slowly liberated by weathering and by the occasional short-lived torrents that flush out the wadis. Particles of silt and dust are then picked up by the wind, whirled high in the air and transported far from their source to accumulate beyond the desert as deposits of loess (Section 22.6) or as ocean-floor silts and clays (e.g. west of the Sahara).

Sand grains are swept along near the surface,

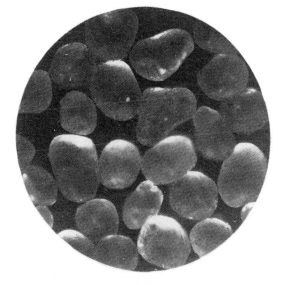

Figure 22.12 Frosted 'millet seed' sand grains, St Peter Sandstone (Ordovician), upper Mississippi States of USA. Photomicrograph × 28. (*P. G. H. Boswell*)

travelling by leaps and bounds (**saltation**), until the wind drops or some obstacle is encountered. The dunes and other accumulations of wind-blown sand thus come to be composed of relatively clean and uniform grains, the finer particles having been sifted out and the larger fragments left behind. It follows that pebbles and gravel are steadily concentrated on the wind-swept surfaces of the original mantle of rock waste.

It takes a stronger wind to force a sand grain into motion (**threshold velocity**) than to maintain that motion. In general, the size of grain that can be dislodged from a smooth surface increases in proportion to the wind velocity. However, because of the increased smoothness associated with very fine silt and clay (below a diameter of about 0.1 mm), it takes an increasingly strong wind to dislodge ever-finer particles unless they are disturbed by larger grains already in saltation. Thus fine clay and coarse-grained sand require about the same threshold velocities to set them in motion; once in the air, these fine particles are easily held in suspension by even a light wind. This explains why vehicles crossing any dry sand-free surface often leave behind a billowing cloud of dust.

As a result of continual attrition due to the friction of rolling and impact, the sand grains themselves are gradually worn down and rounded. The size of sand grain transported by the wind varies with its density and shape, and with the velocity of the wind. Volume for volume, a grain of quartz sand is only 2½ times heavier than water but is some 2000 times heavier than air. This, combined with the generally more rapid fluctuations in the velocity of the wind than of water currents, explains why aeolian sand can exhibit much sharper differences in the grain size of adjacent laminae than those of fluvial or marine origin. This ability to sort a range of grain sizes is well illustrated in the desert wherever gravel, sand and mud are worked over by the wind.

The prolonged action of the wind is far more effective in rounding sand grains than that of running water because of (a) the greater velocity of the wind; (b) the greater distances traversed by the grains as they bounce and roll and collide with each other across wide stretches of desert; and (c) the absence of a protective sheath of water. Some of the millet-seed sands of the desert are almost perfect spheres with a matt or frosted surface like that of ground glass. It is also noteworthy that visible flakes of mica, such as are commonly seen in water-deposited sands and sandstones, are very rare in desert sands and dunes except where extensive mica-rich rocks are exposed up-wind as, for example, in the Namib Desert. The easy cleavage of mica facilitates constant fraying during the wear and tear of wind action. Mica is thus reduced to an impalpable powder that is winnowed away from the heavier sand grains. These contrasts between water-laid and aeolian sands are of value in deciding whether ancient sandstones were accumulated subaerially or under water (see also Section 22.8). The Penrith Sandstone of the Eden valley, between the Pennines and the Lake District, is a well-known example of a Permian desert sand. Its rounded grains, the absence of mica and the cross-bedding of the formation all testify to the desert conditions of the time. Figure 22.12 illustrates an American example of millet-seed sand grains, rounded by Mid-Ordovician winds when Iowa, Illinois and the southern parts of the Canadian Shield were desert lands.

The most serious effects of wind deflation – from the human point of view – are experienced in semi-arid regions like the Great Plains of the United States where, during the 1930s, vast quantities of soil were blown and washed away from thousands of formerly productive wheat-growing farms. Originally an unbroken cover of grass stabilized the ground, but long-continued ploughing and over-exploitation finally destroyed the binding power of the soil and exposed it as a loose powder to the driving force of the wind. This national menace became critical during a period of severe droughts, culminating in 1934–5, when great dust-storms originating in the 'Dust-bowl' of Kansas and adjoining States swept eastward towards the Atlantic. Rain-wash and the creeping disease of badland erosion extended the devastation. Widespread measures of reclamation and protection have been undertaken to minimize the growing wastage and to conserve and improve the soil that remains. Wherever rain is deficient, deforestation, overgrazing or other misuse of land invariably leads to soil erosion. In the Mediterranean area, with its long dry summers, the green lands have been shrinking for centuries, especially in Spain and Algeria, while the population has been increasing; Libya used to be the granary of the Roman empire. Only relatively recently has a start been made to break this vicious circle by afforestation and irrigation, and the application, enforced if necessary, of a 'balanced' use of the land. The over-hasty development of agriculture in parts of the USSR quickly ran into similar difficulties, dramatically indicated by the great dust-storms just prior to the Second World War, blowing out of the semi-arid lands east of the Caspian and Aral seas. These troubles are now being successfully corrected, and the lessons are being utilized in the development of Africa and elsewhere.

A characteristic result of deflation, especially over regions where unconsolidated clays and friable shales are exposed, is the production of wide plains and basin-like depressions. The excavation of hollows is limited only by the fact that even in deserts underground water may be present. Once the desert floor has been lowered to the level of the groundwater, the wind cannot easily loosen and pick up the moistened particles, though it may drive pebbles across the surface when blowing hard and continuously. The base-level for wind

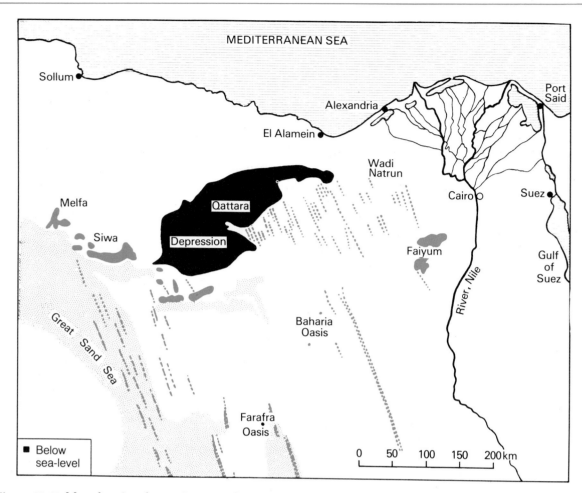

Figure 22.13 Map showing depressions, sand seas and lines (seifs) of dunes in the Egyptian Desert.

action is that of the water table, which may be far below sea-level. The 'pans' of South Africa and the Kalahari, and the depressions of the North African and Mongolian deserts, have all been excavated by ablation.

West of Cairo there is a remarkable series of basins with their floors well below sea-level (Figure 22.13), reaching −125 m in the salt marshes of the immense Qattara depression. Some of the smaller basins tap a copious supply of groundwater at depths of −15 to −30 m and have become fertile oases. To the north, the surface rises by abrupt escarpments to terraced tablelands formed on hard sandstones and limestones that formerly extended across the softer rocks of the depressions. To the south, following the direction of the prevailing wind, long stretches of sand dunes represent part of the removed materials. The other well-known oases of Egypt – Baharia, Farafra, Dakhla and Khargal – are above sea-level, but they have originated in the same way. All of them are margined by steep escarpments of resistant strata, underlain by shales in which the floors have been

excavated. These depressions are not crustal downwarps, like the shotts of Tunisia and Algeria, or Death Valley in California. Nor have they been hollowed out by water, for the sheetfloods caused by rare cloudbursts tend to fill them up with debris. The wind has been the sole excavator.

Deflation does not necessarily result in the creation of depressions, however. In Oman, a sequence of superimposed criss-crossing Plio-Pleistocene river channels has been exhumed by the wind and now stand up in inverted relief above the level of the modern wadi sediments. This excavation was possible because the flanking sediments were not so well cemented as those of the main channels. Deflation probably occurred in the later Pleistocene and, over an area of some 20 000 km^2, has resulted in the removal of most of a sequence of poorly consolidated fluvial sediments up to 50 m thick.

The effects of wind abrasion are unmistakably expressed in the forms and surfaces of the desert bedrocks. Just as an artificial sand-blast is used to clean and polish building stones and to etch glass,

Figure 22.14 An illustration of undercutting by wind erosion: Balanced rock, Garden of the Gods, Colorado. (*United States Geological Survey*)

so the natural sand-blast of the wind destructively attacks everything that lies across its path. Cars driven against wind-blown sand may have their windscreens frosted and their paint scoured off. The action on exposed rocks is highly selective. Like a delicate etching tool, the sand-blast picks out every detail of the structure. Hard pebbles, nodules and fossils are left protruding from their softer matrix until they fall out. Variably cemented rocks are fretted and honeycombed like fantastic carvings. Where there are thin alternations of hard and soft strata the soft bands are scoured away more rapidly than hard, which thus come to stand out in strong relief, like fluted shelves and cornices with deep grooves between. Where the wind blows steadily in one direction over strata of this kind, especially if the beds are tilted rather than flat, the softer materials are excavated into long passageways between deeply undercut overhanging ridges. Such fantastically carved 'cockscomb' ridges are common in parts of the Asiatic deserts, where they are called **yardangs**; some, such as in the Lut desert of south-east Iran or along the northern border between Niger and Chad, are large enough to be clearly visible on satellite images.

Undercutting is everywhere a marked feature of wind abrasion, the process being most effective within 30 to 60 cm of the surface where the saltating sand is most abundant (Figure 22.14). Telegraph poles in sandy stretches of desert have to be protected by piles of stones or even a metal sheath against the cutting action of the sand grains hurled against them. Along the base of an escarpment alcoves and small caverns may be hollowed out. As always, the effect of undercutting on slowly weathered formations is to maintain steep slopes. Joints are readily attacked and opened up, and these commonly determine the outlines of rock towers and pinnacles, left isolated like detached bastions in front of the receding wall of an escarpment (Figure 22.15). The volcanic plugs of the Hoggar (Ahaggar) and Tibesti in the Sahara exhibit wind-steepened walls and towers, rising from a sun-scorched pavement of basalt. In the 'cold deserts' of Antarctica similarly shaped pinnacles from igneous intrusions have been produced by wind erosion, and rise as much as 800 metres above the present ice-surface (e.g. Birger Bergensenfjellet, Monts Sør-Rondane).

Where the bedrock of the desert floor is exposed to blown sand it may be smoothed or pitted or furrowed, according to its structure. Compact

Figure 22.15 'Cathedral Spires', pillars, detached by erosion along steeply dipping joint planes, remain in front of a receding escarpment on the left. Garden of the Gods, Pike National Forest, Colorado. (*F. E. Colburn, Courtesy of United States Forest Service*)

limestones become polished, massive granites are smoothed or pitted, and gneisses are ribbed and fluted, especially where their foliation approximates to that of the dominant winds. Where pebbles have been concentrated by removal of finer material they become closely packed, and in time their upper surfaces are worn flat. In this way mosaic-like tracts of **desert-pavement** are developed. Isolated pebbles or rock fragments strewn

Figure 22.16 Pebbles from a coastal terrace near Wanganui, North Island, New Zealand, that were facetted by sand-blast with adjacent beach/dune sand (dreikanter or ventifacts). (*K. W. Glennie*)

on the desert surface are bevelled on the windward side until a smooth face is cut. If the direction of the wind changes seasonally, or if the pebble is undermined and turned over, two or more facets may be cut, each pair meeting in a sharp edge. Such wind-facetted pebbles, which often resemble Brazil nuts, except that their surfaces are polished, are known as **dreikanter** or **ventifacts** (Figure 22.16).

Three distinctive types of desert surface are produced by the effects of wind erosion, transport and deposition:

1. the rocky desert (the **hammada** or **hamada** of the Sahara), a desolate surface of bedrock with local patches of rubble and sand (Figure 22.17);
2. the stony desert, with a surface of rubble, gravel or pebbles (the **reg** of the Algerian Sahara; the **serir** of Libya and Egypt or the **gibber plain** of Australia); and
3. the sandy desert (the **erg** of the Sahara).

22.4 DESERT DUNES AND SAND SHEETS

Around one third of the land surface is desert, and on an average about a quarter of the desert areas are mantled with sand. A high proportion of the desert floor is an erosion surface of rock waste

Figure 22.17 The rocky wastes of Ahmar-Kreddou, viewed from the Col de Sfa, Algerian Sahara. (*Paul Popper Ltd*)

(Figure 22.17). Regions of shale and limestone provide little or no sand, but granites and other igneous rocks can be important sources of sand grains. The most effective sources of sand for direct incorporation into dunes, however, are either well-supplied coastal beaches or the deposits of ephemeral streams. Such accumulations of relatively fine sediment are readily reworked by the wind; the dust-size particles go into suspension and are carried beyond the limits of the desert to be deposited as **loess** (see later), but the unconsolidated sand can saltate downwind across the desert surface to the nearest sand dunes.

Complications caused by vegetation and moisture occur mostly around oases or in the transition zones where the desert merges into steppe or savannah country. But nevertheless, as the masterly study by R.A. Bagnold has made abundantly clear, the factors controlling the form of the sand accumulations are far from simple. They include the nature, extent and rate of erosion of the sediment source; the sizes of the sand grains and associated fragments; the varying strength, duration and direction of the wind; and the roughness or smoothness of the surface (e.g. the presence or absence of pebbles) across which the sand is transported and deposited. Of the resulting sand forms, three main types can be distinguished:

1. **Sand drifts**, which form in the wind-shadows of cliffs, protruding rocks or vegetation.

2. **Dunes**, which are of two basic types: those whose long axes are essentially either transverse (**transverse** and **barchan dunes**) or parallel (**longitudinal dunes**) to the dominant wind direction.

3. **Sand sheets** in interdune and extradune areas; they are commonly almost horizontal, and can be of fairly wide extent.

A dune is a mound or ridge of wind-blown sand that rises to a definite summit or crest. Dunes can form wherever a sand-laden wind deposits sand. This will happen during a fall in wind velocity, perhaps aided by a local increase in surface roughness, and seems to take place preferentially over an existing patch of sand.

One of the remarkable features of desert dunes is their apparent power to collect all the sand in their neighbourhood. The explanation appears to be that the wind exerts a shepherding effect on sand. Sand is transported over the relatively hard and smooth interdune areas more readily than over the dunes, where the soft unconsolidated sand exerts a greater drag on the wind. Thus saltating sand is blown obliquely towards existing dunes and the intervening surface is swept clean (Figure 22.18). It follows that this factor also plays a role in the type of dune formed.

Transverse and barchan dunes: transverse

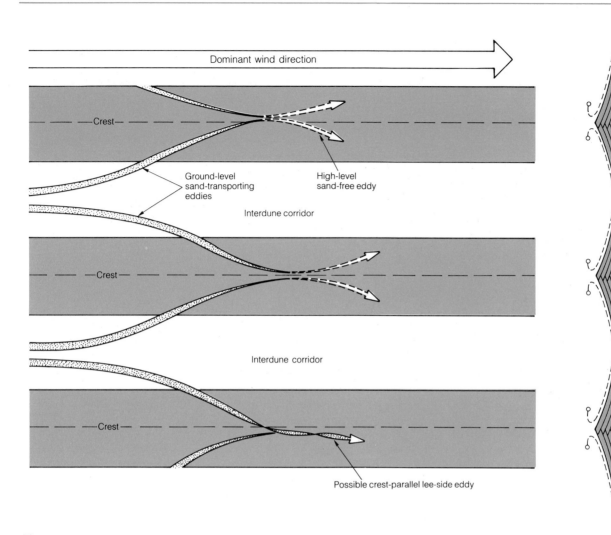

Dominant wind direction

Crest

Ground-level
sand-transporting
eddies

High-level
sand-free eddy

Interdune corridor

Crest

Interdune corridor

Crest

Possible crest-parallel lee-side eddy

Figure 22.18 Diagram to illustrate the shepherding effect of wind in sweeping sand from interdune areas towards adjacent longitudinal dunes which are built by flat-topped contra-rotating vortices. The wind is strongest between the dunes and is retarded by friction against them.

dunes have a long, low-angle windward slope rising to a crest, and a much steeper leeward slope (Figure 22.19). The latter is determined by the fact that sand blown over the crest falls into a wind shadow, and comes to rest at its natural angle of repose which, for dry sand, has a maximum slope of 34°. Any tendency for that angle to be increased by further deposition results in an avalanche of sand; the coarsest and roundest grains rise to the surface and travel the farthest, while the avalanche slope, or slip face, is reduced to about 30°. Deflation of sand grains from the windward edge of the dune and their redeposition on the leeward avalanche slope results in downwind migration of the dune. The smallest avalanche slope is some 30 cm high. Some large transverse dunes have avalanche slopes over 100 m high and have very slow rates of migration. The dunes of many transverse systems are equally spaced.

Two depositional features seem to be unique to aeolian sands. Over horizontal sand sheets and the

windward slopes of transverse dunes, a type of very low-angle climbing ripple is deposited; and cross-winds sometimes form large ripples across the surface of the slip face with their long axes aligned up and down the approximately 30° slip face.

Where the supply of sand is insufficient for broad transverse dunes to form, a series of smaller crescent-shaped dunes result (**barchans**; a Turkestan name that has been generally adopted). They also migrate by the transport of sand from the windward edge, over the crest to the leeward avalanche slope. We have already seen, however, that sand is transported more readily over a hard interdune surface than over the soft sand of the dune itself, so that not only is the limited supply of sand attracted to the dune, but the dune's flanks get drawn out in a down-wind direction to form the horns of the barchan. The width of a barchan is commonly about a dozen times the height, which ranges up to a maximum of 30 m or so. With winds of constant direction, colonies and elon-

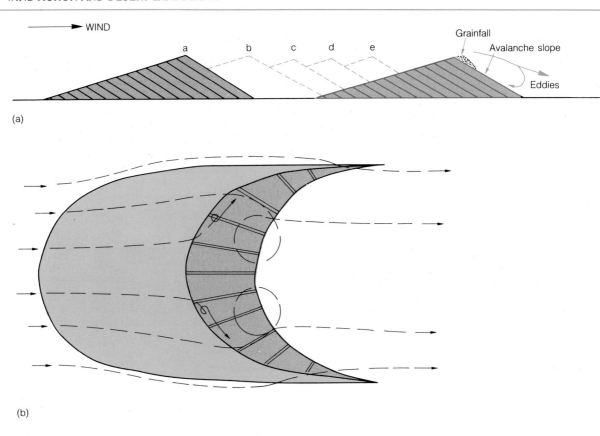

(a)

(b)

Figure 22.19 (a) A series of profiles illustrating the stages (a to e) of down-wind migration of a transverse dune by a combination of up-wind erosion and lee-slope deposition. (b) Idealized profile and plan of a barchan showing the opposing effects that the bulk of the dune and its surface roughness (cf. Figure 22.18) has on wind flow over and around it. Note that the orientation of the avalanche slope (double lines) varies through almost 180°, although the bulk of the sand dips in the direction of the prevailing wind. In ancient rocks such bedding distribution can be used to deduce palaeowind directions (Section 22.9).

gated swarms of barchans march slowly forward like a stream of vehicles in a one-way street (Figures 22.20 and 22.21). The rate of progress varies from about 2 or 3 mm a year for high dunes, to almost 50 m a year for very small ones.

Longitudinal dunes: transverse dunes are unstable at high wind velocities, when sand is deposited in rows of long linear dunes whose axes are fairly equally spaced and parallel to the regional wind direction (Figure 22.22). Although their origin is still controversial, dunes of this type are thought to form because of the differential rates of transport over dune and interdune described above; the wind is believed to create flat-topped contra-rotating helical vortices parallel to the dunes' axes (Figure 22.18), which transport sand obliquely up the flanks of opposing dune slopes, the dunes extending in a down-wind

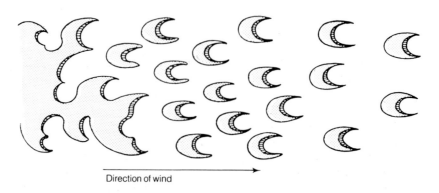

Direction of wind

Figure 22.20 Plan of a typical procession of barchans in the Libyan desert.

Figure 22.21 A colony of barchans in Mauritania, West Africa, advancing towards the south-west. (*I. F. N. A*)

direction. Much of the **Rub al Khali** (Empty Quarter of southern Arabia) is covered by parallel rows of longitudinal dunes in an area some 1000 km long by 500 km across. As we have already seen, south of the Qattara depression (Figure 22.13) there is a long tract of parallel longitudinal dunes, whose sand was deflated from the depression, and which are separated by corridors of bare desert floor.

Different dune types form where the wind patterns are more variable. Where seasonal winds blow from different quarters, the dune extends down the resultant wind direction with a slightly sinuous form; such dunes are known in the Sahara as **seifs** (Arabic for sword). With more complex wind patterns, the shepherding effects of desert winds can cause systems of small barchanoid dunes migrating across the interdune surface to be incorporated into almost stationary **star dunes**, which may grow to heights of 200 m or more above the desert floor (Plate XIIa). By analogy with ocean waves, extensive areas of desert that are covered by dunes of one or more types are known as **sand seas** or **ergs** (Figure 22.23).

In desert-margin areas and along temperate coastlines, dunes may become fixed by vegetation. Occasionally, strong gales may remove a swath of sand through these dunes, referred to as a **blow-out**. With the sand-covered flanks anchored by vegetation, the re-worked sand will form a new dune at the far end of the blow-out. The resulting form is known as a **parabolic dune**; its long horn-like stabilized flanks point up-wind, in the opposite sense to the barchan.

Sand sheets are one of the more enigmatic

Figure 22.22 Vertical aerial photo of the northern end of the Wahiba Sands, Oman, where they are truncated by Wadi Batha. These asymmetric dunes are up to 100 m high with a crestal spacing of 1 to 2 km. Fine diverging 'feather' dunes in the interdune areas, and small north-trending dunes over the wadi plain are mostly 1–5 m high; some are barchans with one long drawn-out horn. The main dune-forming winds of the South-west Monsoon blow roughly parallel to the SSW–NNE trending large dunes. The dark palm-garden oases are surrounded by the houses of three villages.

features of interdune and extradune (region beyond the limit of a sand sea) areas. These are sheets of almost horizontal aeolian sand that are devoid of dunes. The surface is normally covered

Figure 22.23 Sand dunes of Death Valley, California, which here form a small sand sea. Cottonwood Mountains in the background. (*United States Department of the Interior*)

with ripples, but pits dug into the sand generally display horizontal laminae of two distinct (**bimodal**) grain sizes. Although still not fully understood, it is possible that the bimodality results from coarser grains moving mostly by rolling over the surface while the finer grains saltate, deposition of the two grain sizes alternating with differential wind strengths. The horizontal, dune-free nature of sand sheets possibly also result from the effects of strong winds that alternate between deflation (leaving a **lag** of coarser grains) and deposition. Such strong winds, especially when

heavily sand-laden, seem to suppress any tendency to form even ripples except in the dying stage of a gust of wind. Widespread sand sheets are also characteristically developed on the borders of deserts, where a scanty vegetation diversifies the surface (Figure 22.24). Grass and scrub break the force of the wind and inblown sand is more or less evenly distributed.

The movement of sand is suppressed by increasing the surface roughness of the desert. A patch of pebbles increases the drag on the wind to such an extent that the velocity near the ground is

Figure 22.24 A pediplain of semi-arid 'desert scrub', with basalt-capped mesa behind; north-east of The Solitario, Texas. (*G. C. Gilbert, United States Geological Survey*)

less than it is over a patch of clean sand. Sand is not deflated from between the pebbles but, to the contrary, the resulting eddies tend to blow from the sand patch towards the pebbly surface until the sand is evenly distributed between the pebbles. Grass and scrub also break the force of the wind, and incoming sand is trapped between their blades and branches.

Among the curiosities of the desert, including dust devils and mirages, perhaps the most mysterious is the sudden booming noise that occasionally interrupts the silence, apparently when the surface layer of sun-baked sand on the avalanche slope of a dune becomes unstable and begins to slide down. This is perhaps most often heard when trucks are driven down the slip face of a dune, the truck sinking deeply into the avalanching sand, which produces a low sound like rolling thunder. Dr W. Ramsden suggested that the boom is the sound of electrostatic discharges occurring when very dry sand slips down the slope of a dune. Examination of sands by the scanning electron microscope reveals that the sand grains of booming sands are more highly polished than those of silent sand. Many known booming sands in the USA are composed essentially of quartz sand, though in Hawaii, some are composed principally of calcite sand. In these latter cases, how-

ever, the degree of aridity on the shores of a humid Pacific island may not have been high enough for electrostatic charges to build up within the sands. An alternative explanation in this case could be that the booming sound is caused by a low-frequency vibration of the avalanching sand grains.

22.5 COASTAL DUNES, SANDHILLS AND SABKHAS

Along low-lying stretches of sandy coasts and lake shores, where the prevailing winds are onshore (Plate XIIc), drifting sand is blown landwards and piled up into dunes, which form a natural bulwark of sandhills (Figure 22.25). In humid regions the conditions governing growth and removal are very complex, with the wind varying in strength and direction. Deposition may start when sand is trapped behind surface irregularities including grasses and trees. Vegetation and moisture tend to trap and fix the sand, but fixation is often incomplete. During severe gales old dunes may be breached and scooped out into deep 'blow-outs' with the creation of **parabolic** dunes. The resulting confused assemblage of hummocks and hollows gives coastal sandhills a characteristically chaotic

Figure 22.25 A stage in the landward drift (from upper right toward lower left) of the Culbin Sands, Moray Firth area, Scotland. Most of this area has now been stabilized by plantations. (*British Geological Survey*)

relief. Where the water table is reached the ground becomes marshy. As one belt of dunes migrates inland away from the beach, another arises in its place so that a series of huge sandy billows, as it were, is continually on the move towards the interior (Figure 22.25). Wide expanses of sandhills have spread inland in this way from low-lying coasts well supplied with sand, such as those of Holland and north Germany, and the Landes of Gascony adjoining the Bay of Biscay. Landward migration from the seashore is convincingly demonstrated by the existence of beaches and dunes that are partly composed of ground-up marine shells and other calcareous fragments. Excellent examples occur at St Ives and Perranporth along the north-west coast of Cornwall. The abundance of land snails in coastal dunes is a clear indication that the sand is at least partly composed of calcareous fragments.

The Culbin and Maviston sandhills, near the mouth of the Findhorn on the southern shore of the Moray Firth, furnish a classic example of the destruction of cultivated lands and habitations by advancing sand. Prior to 1694 the sandhills had already reached the fringe of the Culbin estate. In that year a great storm started a phase of accelerated encroachment which finally led to the complete obliteration of houses, farms and orchards,

and even to the burial of fir plantations (Figures 22.26 and 22.27).

Where the prevailing wind blows roughly parallel to the coast, as in south-east Oman and Namibia (Plate XIIb) the dunes are commonly of transverse type with their axes at right angles to or strongly oblique to the coast (Figure 22.28). This obliquity presumably represents the influence of different degrees of drag imposed by land and water.

Cemented carbonate dune sands, made up almost entirely of foraminifera and fragments of other marine shells, are common along some desert coastlines, especially in the Middle East. The calcareous material was derived mostly from the exposed floor of the adjacent continental shelf when the global sea-level was lowered during the last, and probably also an earlier, Pleistocene glaciation. Such **aeolianites** are cemented fairly rapidly under the influence of rare rainfall or when buried below the water table. They can be traced below today's sea-level around the coast of south-east Arabia; indeed, late Pleistocene dunes probably floor the axial parts of the Arabian Gulf, whose post-glacial marine flooding may have prompted the building of Noah's Ark!

Most of the dunes now covering the Rub al Khali or Empty Quarter of Arabia (Figure 22.29)

Figure 22.26 Advancing sand overwhelming a plantation of Scotch firs; Maviston Sands (south-west continuation of the Culbin Sands, Fig. 22.12. (*British Geological Survey*)

Figure 22.27 Remains of an exhumed plantation formerly buried by sand which has now migrated farther on; Maviston Sands. (*British Geological Survey*).

were sourced by sand transported southwards from northern Saudi Arabia and Jordan by the **Shamal** (north wind). Since the post-glacial flooding of the Persian Gulf, however, the supply of sand has ceased to several coastal areas. A flooded Gulf of Salwa has cut off the supply to Qatar, where dunes of quartz sand are now concentrated in the south-east part of the peninsula; the surface of Qatar is otherwise entirely of dolomite. Further to the south-east, not only has Sabkha Matti stopped receiving dune sand from Qatar, but the complete deflation of older systems of dunes has been prevented only because the water table rose in harmony with flooding of the Gulf. Evaporation of the groundwater has led to gypsum cementation and consequent preservation of the dune sands within capilliary range of the water table; the seasonally damp interdune areas are now sites of extensive deposition of adhesion ripples.

Along the coastline of the United Arab Emirates (Figure 22.29), the post-glacial Shamal induced the construction of longshore sand spits, which built up to sea-level and cut off extensive lagoons. Under the conditions of clear, tropical warm water, marine life flourished, resulting in the rapid manufacture of calcareous skeletal material (corals, bryozoa, molluscs) and, in addition, oolite banks were formed. On-shore winds deflated the beaches and helped to fill-in the lagoons, which acquired a binding mat of rubbery algae (Figure 22.30) as the sediment approached sea-level. In this way, the coastline has apparently prograded out into the Gulf by up to 30 km or more during the past 15 000 years. With extra high tides generated by storm winds, the former lagoonal surface is regularly flooded and, assisted by the algal mat which regenerates with each wetting, has built up to just above normal sea-level. In the prevailing arid climate, any water on the supratidal flats and shallow lagoons rapidly becomes hypersaline, gypsum crystals grow within the sediment and the whole area is covered with a film of halite. Such a salt-covered area is called a **coastal sabkha**. If the sabkha surface is deflated, the fine calcareous clays go into suspension but the gypsum crystals may break into sand-size grains that can be incorporated into gypsum dunes.

22.6 LOESS

We must next consider what happens to the vast quantities of dust which have been winnowed from the deserts and exported by the wind. Loess is an accumulation of wind-borne dust and silt, the particles being between 0.01–0.05 mm in diameter, but mostly of 0.02–0.03 mm.

From the deserts of Central Asia the wind

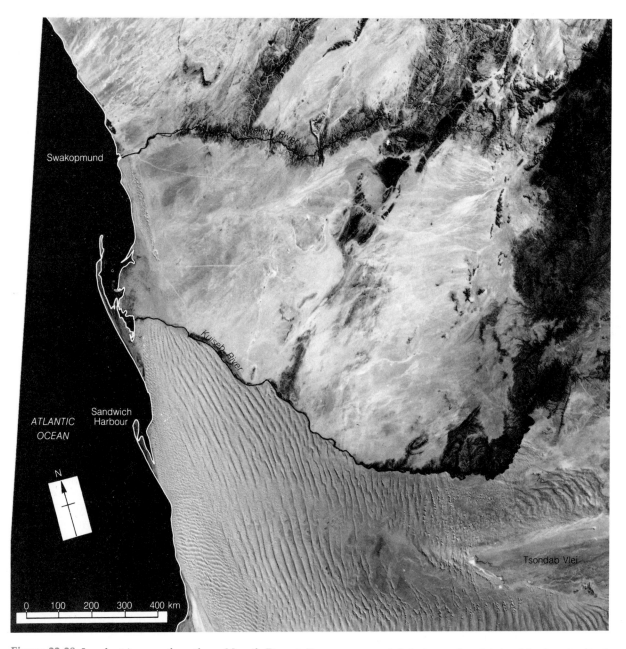

Figure 22.28 Landsat image of northern Namib Desert. Transverse coastal dunes replaced inland by longitudinal dunes that are underlain by the older Pleistocene sediments of Tsondab Vlei. The northward extending linear dunes have probably been trapped in the 1/2 km-wide gorge of the Kuiseb River for tens of thousands of years or more. Today, the Kuiseb River water rarely reaches the Atlantic Ocean. The sides of the image cover 185 km.

carries dust to the south and south-east where it is deposited over vast areas of the grassy regions of China as a thick blanket of loess (Figure 22.31). Further west, loess accumulates against the foothills of the Pamir plateau, where it provides a narrow belt of well-irrigated and fertile soil which has become one of the most densely populated agricultural regions of the world. From the deserts of North Africa much of the dust reaches the countries to the south and west, such as southern Sudan, northern Zaire, Nigeria and Mali, where it is retained by the soil. Some of the dust blows over the Atlantic where its distribution on the ocean floor has now been mapped; some is trapped by the Mediterranean while more occasionally turns Alpine snow a delicate shade of pink; and the Red Sea receives passing contributions from Arabia.

Other very important supplies of fine silt and dust were formerly provided by the rock flour of glacial and glaciofluvial deposits. During and after the retreat of each of the successive ice-sheets, the finer material was sifted out by the wind and deposited over the surrounding country (Figure

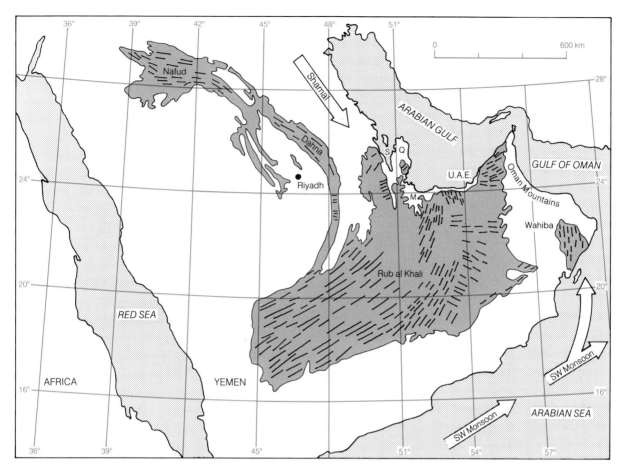

Figure 22.29 The trends of both longitudinal and transverse dunes in southern Arabia were controlled by the Shamal (north wind) and follow a clockwise rotation centred south-west of Riyadh. The Wahiba Sands were deposited by a north-blowing branch of the South-west Monsoon. M – Sabkha Matti; Q – Qatar; S – Gulf of Salwa; UAE – United Arab Emirates.

22.32). Thus it happens that a long belt of loess, mainly derived from glacial material in the west and from desert waste in the east, stretches from France to China. The term loess comes from a town of that name in Alsace. Beginning as local patches in France and Germany, the deposit becomes thicker and more extensive as it is traced across Russia and Turkestan, until in Shansi and the adjoining provinces of China it has its maximum development and locally reaches thicknesses exceeding 300 m.

Although some loess may be washed down from the air by rain, the bulk is deposited in areas where a natural reduction in wind velocity occurs, such as where mountainous terrain is replaced by broad plains. Each spring the grass grows a little higher on any material collected during the previous year, leaving behind a ramifying system of withered roots. Over immense areas many metres have accumulated, whole landscapes having been buried, except where the higher peaks project above the blanket of loess. The material itself is yellowish or light buff and devoid of stratification.

Figure 22.30 Lagoon and coastal sabkha formed behind the protection of a longshore sand spit at Rams, north of Ras Al Khaimah, United Arab Emirates; note the tidally-formed oolite delta growing into the Arabian Gulf. Mountains of the Musandam Peninsula in the background. (*K. W. Glennie*)

Figure 22.31 'Badland' erosion of loess in Kansu province, North China, brought about largely by destruction of timber and over cultivation in the past. Terrace conservation has checked this wastage in the valley, which is drained by a tributary of the upper Hwang Ho (see Figure 17.41).

Although it is friable and porous, the successive generations of grass roots, now represented by narrow tubes partly occupied by calcium carbonate, make it sufficiently coherent to stand up in vertical walls that do not crumble unless they are disturbed. The passage of traffic along country roads loosens the material, clouds of dust are removed by the wind, and the roads are worn down into steep-sided gullies and miniature canyons.

In the loess provinces of China, rain and small streams carve the surface into a maze of ravines and badland topography (Figure 22.31). The larger rivers flow in broad and fertile alluvial plains bordered by vertical bluffs. Here, and in the lowland and deltaic plains to the east, most of the alluvium is simply loess redistributed by water. In the loess uplands cultivation of the slopes is made possible by terracing. The steep-sided cliffs and walls, whether natural or artificial, are often rid-

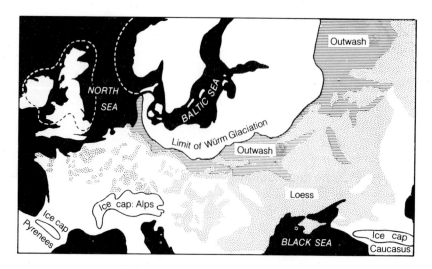

Figure 22.32 Map showing the European distribution of loess and loamy soil formed from it, and the marginal relation to the last major glaciation. (*After S. von Bubnoff*)

dled with entrances to cave-dwellings, many of which have their chimneys opening into the fields above. This mode of habitation has occasionally led to great disasters. In 1556, for example, widespread landslides and floods were started by a catastrophic earthquake and nearly a million peasants lost their lives.

In the Midwest of the United States there are deposits of loess, called **adobe**, that correspond in all essentials to those of Europe and Asia. In Kansas and Nebraska much of the wind-blown silt has come from the semi-arid lands of the west; but elsewhere it consists largely of rock flour, winnowed by the wind from the outwash and temporary lake deposits left by the shrinking ice-sheets of the Pleistocene, to find lodgement on grass-protected surfaces like the prairies. By subsequent erosion much of the original loess of the upper Mississippi Basin has contributed to the alluvium of the downstream flood plains. When these dry up in times of drought they form an important source for a second generation of loess.

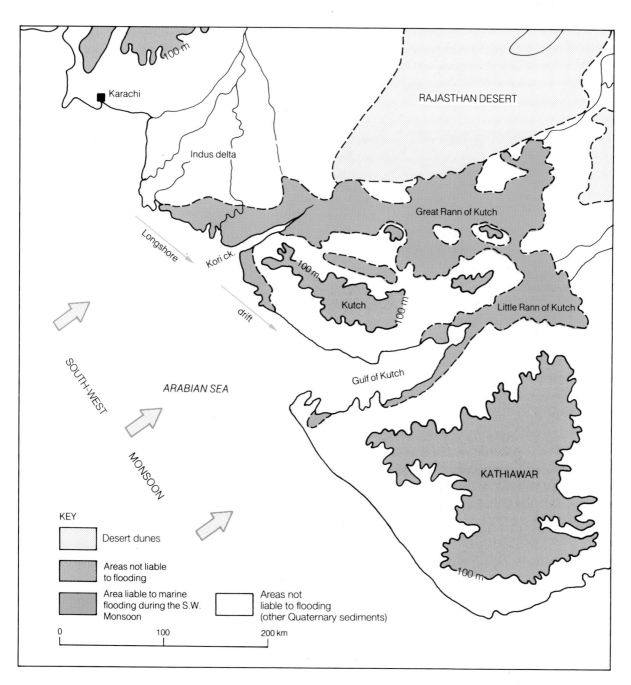

Figure 22.33 Outline of the Ranns of Kutch, India, which are flooded each year by seawater driven by the South-west Monsoon. Early winter evaporation leads to precipitation of a salt crust over the whole area, which is dissolved again during the next summer monsoon.

22.7 LONGER-TERM CLIMATIC CHANGES

We know that there are climatic differences from one year to the next, especially in areas like western Europe; and historically, Britain's climate, for instance, was positively warm some 6000 years ago, relatively mild during the Middle Ages, colder in the 17th and 18th centuries during the **Little Ice age** (Section 21.11), and currently is rather warmer. There is increasing evidence to indicate that during the past 10 000 to 20 000 years there have also been considerable climatic changes over the areas bordering today's deserts. Rock carvings in the northern Sahara show that not only could early man survive there, but he hunted giraffe and other animals more typical of a savannah than a desert climate; in Central Africa, linear dunes, now covered with vegetation, occur within less than 5° both north and south of the Equator; and along the southern margins of the Sahara, satellite imagery shows that the latest system of high dunes is superimposed upon the deflated remnants of an older system that had different axial orientations. It also seems likely that during the last glaciation, fairly strong and continuous winds built relatively simple systems of fairly large dunes (**draas**), whereas today's winds are capable of building only much smaller dunes that migrate over the surfaces of the draas like parasites. In North America, Florida was the site of a desert climate during the last Pleistocene glaciation whereas the existing North American deserts adjacent to the Rocky Mountains were then well-watered by summer meltwaters from nearby glaciers.

Figure 22.34 Isachsen salt dome (78°N Latitude), Ellef Ringnes Island, one of the Queen Elizabeth group of Arctic Canada. The core of the dome has a diameter of 6 km and consists of late Silurian evaporites which penetrate Cretaceous strata. Away from the Isachsen and other domes of the Island these beds lie nearly flat, but within 3 to 5 km of a dome they begin to be upturned, outcropping as concentric rings which become nearly vertical against the core. (*Royal Canadian Air Force*)

These features collectively indicate that during polar glaciations:

1. the world's climatic belts were compressed towards the Equator, possibly with an associated increase in the intensity and duration of aeolian activity, especially in deserts;
2. modern deserts seem to have had higher rainfall (**pluvials**) only at their higher-latitude boundaries;
3. because of a global fall in sea-level, many shallow-marine areas around the existing desert coasts were then the sites of dune deposition; thus some desert areas were much more extensive than now.

In the long term, of course, not only have we had a series of glacial and interglacial periods affecting global climate especially during the Pleistocene, but plate-tectonic re-orientation of our continents and oceans has, to a large extent, controlled continental climate over longer periods of geological time.

Neither plate tectonics nor the waxing and waning of polar ice caps can be blamed for all changes in local climate, however. Some of these effects are caused by other geological processes. For instance, there is historical evidence of a considerable climatic deterioration in north-west India and Pakistan during the past few thousand years. The river cultures of Mohenjodaro and Harappa declined some 4000 to 5000 years ago when the supply of water to the Indus became less certain after the River Jumna was captured by the Ganges. Even so, during the campaigns of Alexander the Great (356–323 BC) to the Punjab and back, his army of 110 000 men travelled through countries that were sufficiently well-watered and forested to provide for their needs, including shipbuilding. Today, these lands include the Rajasthan desert, an area 'so desolate that they cannot support passing caravans of even 100 men and animals' (D.N. Wadia, 1960). Alexander was able to sail his ships into the adjacent Ranns of Kutch (Figure 22.33), which were then shallow marine gulfs covering an area of some 30 000 km² east of the Indus delta. Since then, however, the summer winds of the Southwest Monsoon have driven seawater and entrained sediment along the coast from the Indus into the Ranns, which have now silted up. The present surface of the Ranns is just above normal sea-level; flooded during every monsoon by wind-driven high tides, the Ranns develop a salt crust every winter as they dry out. Deflated salt is blown inland over the Pleistocene dune sands of the Rajasthan desert, helping to destroy most of any post-glacial fertility the area may have had.

Most of the dunes of the Rajasthan Desert were probably formed by monsoon winds during the last polar glaciation, an interpretation that is in keeping with the occurrence of foraminifera-rich late Pleistocene cemented dune sands now over 100 m above sea-level on the hills of Kutch; these sands were derived from the adjacent shallow continental shelf, which was then exposed to deflation.

If, for any reason, the cover of vegetation is destroyed in areas of relatively low rainfall (fire started by lightning, cutting trees for firewood, overgrazing by sheep and goats), the soil can be deflated away to leave a barren rock surface on which nothing will grow. Indeed, rain can sometimes be seen to fall from thunder-clouds towards the hot, barren rock or sand surface of the desert, only to evaporate before reaching the ground, just one or two of the largest drops actually hitting the surface. Such a hostile environment can be prevented only if the ground can be protected by a cover of vegetation, and this is unlikely to happen naturally until such time as a much wetter climate is reintroduced to the area. Soil will not readily regenerate under arid conditions; thus a desert is more readily increased than reduced in area.

22.8   ANCIENT DESERTS

Aeolian dunes, but not necessarily desert conditions, have existed sporadically on the Earth's

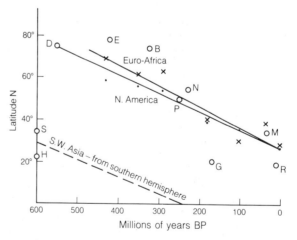

Figure 22.35 Diagram to illustrate the distribution of major salt deposits of the northern hemisphere; mean present latitudes plotted against ages (data from *F. Lotze, 1957, P.M.S. Blackett, 1961* and, open circles, other sources). The data indicates that Europe, North America and south-west Asia have been drifting northward at a rate averaging about 10 km per million years. B – Barents Sea, Carboniferous; D – Devon Island, Cambrian; E – Ellef Ringnes Island, late Silurian; G – Gulf of Mexico, mid Jurassic; H – Hormuz Salt, Cambro-Precambrian; M – Mediterranean, late Miocene; N – North Sea, late Permian; P – Peri-Caspian, late Permian; R – Red Sea, Miocene; S – Salt Range, Pakistan, Cambro-Precambrian.

surface for at least 2000 m.y. The distribution and proportions of land and sea relative to the Poles and the Equator, and the freedom of movement of water between these latitudinal extremes, may be among the factors that controlled the presence or absence of deserts in the past.

Prior to the development of land plants in the Late Silurian, the deflation of unconsolidated sediment was prevented only by moisture. In areas of seasonal rainfall, therefore, dunes could form during the dry season without the climate being very arid. We have seen that deserts occur today where the potential rate of evaporation greatly exceeds that of rainfall.

Much effort has gone into attempts to find criteria for differentiating between sandstones formed in aeolian as opposed to fluviatile environments, but a full discussion is outside the scope of this book. However, obvious aeolian indications include the presence of well-rounded coarse quartz grains and a lack of mica and drapes of clay. To these can be added continuous vertical thicknesses of sandstone – from a few centimetres to

several hundreds of metres – in which individual sets (e.g. sand deposited on the lee slope of a dune) can exceed 30 m and have distinctive types of laminae, ripple shapes and transport directions deduced from cross-bedding. The cumulative data will generally provide a clear distinction.

The deposition of halite does not occur today on the same scale as at some periods in the past. This is probably an effect of the very recent flooding of our coastlines following the melting, especially of the northern hemisphere ice caps, at the end of the Pleistocene. Extensive and thick deposits of halite and anhydrite must indicate arid conditions and either limited circulation of seawater or a terminal lake in a basin of inland drainage. Evidence of the existence of evaporite minerals some 3500 m.y. old has been found in Australia; although the original sulphate minerals have now been replaced diagenetically by others, it is possible to recognize their environments of deposition as both terrestrial and shallow marine.

Thick halite was deposited over large areas of Oman and Iran and in the Salt Ranges of northern

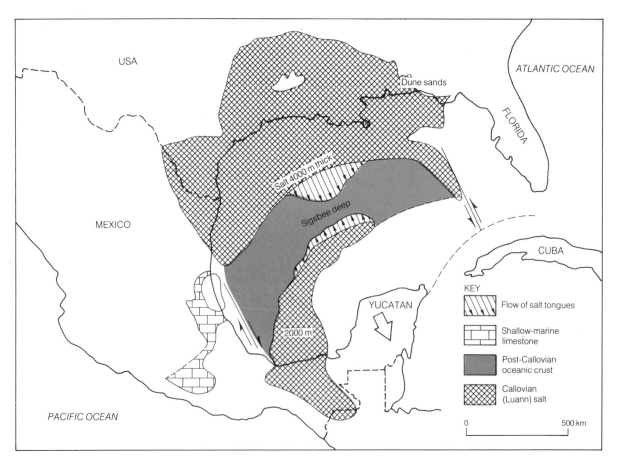

Figure 22.36 Simple map of the Gulf of Mexico in the Callovian (Middle Jurassic), where thick halite was precipitated in a basin floored by thin continental crust. Marine water is thought to have entered the basin from the west. A later phase of ocean-floor spreading seems to have split the halite into two parts; the limited development of salt tongues suggests that there was no void for the halite to flow into, but rather that the new oceanic crust was up-arched into a local high. Salt diapirs are now widespread within the area of the thick halite.

Pakistan during the late Precambrian and earliest Cambrian; dune sands overlie the salts in southern Oman (Amin Formation) and in Iran (Lalun Formation).

An extensive evaporite basin occupied much of the American Midwest, extending into Canada, during the Silurian. Perhaps more striking in terms of the latitudinal difference between their site of precipitation and present occurrence are the large salt-domes, one of which is 6 km across (Figure 22.34), of the Queen Elizabeth Islands of Arctic Canada. These thick evaporites range in age from Late Cambrian to Late Silurian. The very existence of such deposits across latitudes 75° to 78°N suggest that at the time of their accumulation, the Canadian Shield was not far from the Equator. Here, the northerly component of drift is of the order 40° in 450 m.y., giving an average rate of about 10 km per million years (Figure 22.35).

By far the most important salt deposits of Europe are of Permian age, although Devonian salts are known from the western USSR and Carboniferous salt from beneath the southern Barents Sea. Late Permian salt was also precipitated over large parts of the south-western United States. Of marine origin in the Delaware Basin of West Texas, salt is preserved as far north as the Williston Basin in North Dakota. There, however, it was laid down essentially in a non-marine environment but with a probable limited volume of marine brine supplied from the Phosphoria Sea to its west. Dune sands were deposited in areas marginal to the arid salt basins.

Another extensive salt sequence is of Mid-Jurassic age, and is found in a broad belt extending across the northern half of the Gulf of Mexico and onshore areas of Texas, Louisiana and Mississippi (Figure 22.36). A major basin formed in the Gulf following extensional subsidence of the continental crust as North and South America pulled apart during an opening phase of the central Atlantic Ocean. With only a limited connection

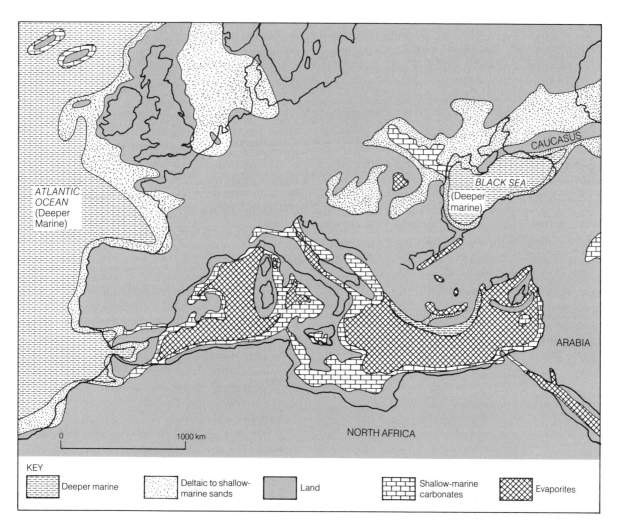

KEY
Deeper marine | Deltaic to shallow-marine sands | Land | Shallow-marine carbonates | Evaporites

Figure 22.37 Simplified map of the Mediterranean Sea some 6 m.y. ago. Almost cut off from the open ocean, the water evaporated to virtual dryness and halite was precipitated in the deep basins. Halite was likewise precipitated in the Red Sea, which was then closed at its southern end and had a narrow connection with the Mediterranean.

with the open ocean, probably in the west, evaporation resulted in the precipitation of thick halite, known in onshore areas of the United States as the Louann Salt. This halite was split into two parts when new oceanic crust later formed in the region of the **Sigsbee Deep** north of a line between the Yucatan peninsula of eastern Mexico and southern Florida.

Dune sands (Norphlet Formation) were deposited to the north of, and even over, the thin edge of the salt at the Alabama coast (Figure 22.36). Further to the west, the well-known **Entrada** dune sands extend over parts of Arizona, New Mexico, Utah and Colorado, reaching a thickness in excess of 300 m in southern Utah.

Palaeomagnetic data indicate that during the Permian and Early Jurassic, both the European and North American sequences of evaporites were deposited between about 10° and 30° north of the Permian equator.

Yet another 'Salinity Crisis' occurred about 6 m.y. ago when collision between southern Europe and the northward-migrating African plate brought about the isolation of the Mediterranean Basin. Evaporation of the Mediterranean waters led to hypersaline conditions and eventual complete dryness within the basin, the floor of which at a depth of some 6 km below sea-level, must have reached very high temperatures. Such a large 'hot spot' caused the climate of much of the adjacent western Europe to become semi-arid. Sabkha-type carbonates and anhydrite were deposited around the basin margin and thick halite, now forming diapirs, was precipitated over the floor of two deep sub-basins, which are separated by a saddle between Tunisia and Sicily (Figure 22.37). Narrow gorges were cut around the basin margins some hundreds of metres below present sea-level, and at the site of the Aswan dam in Egypt, 1200 km from the Mediterranean, the River Nile cut a channel over 200 m deep. Once ocean waters broke into the basin through the Strait of Gibraltar, it probably took something approaching 100 years to fill the basin again to oceanic levels.

22.9 ROTLIEGEND DESERT AND ZECHSTEIN EVAPORITIC SEA

In Europe, evidence from outcrops and from North Sea wells confirm that during the Late Permian two major depositional basins existed within this trade-wind desert belt. Ever since the Devonian Period at least, the united North America and north-west Europe had been migrating passively northward from the southern hemisphere, with an overall anticlockwise rotation cen-

tred over the Gulf of Mexico. Within the upper half of the Carboniferous sequence, widespread coals indicate deposition in equatorial rain forests. By the end of the Carboniferous, this area was moving into a geographical location similar to that of the modern Sahara, with the added aridity factor that it lay within the trade-wind rain shadow of the newly formed Variscan Mountains. Two major desert basins began to subside towards the end of the Early Permian, the Southern Permian Basin, which extended from the east coast of England to the Russo-Polish border, and the Northern Permian Basin lying between northern Denmark and east of the coast of Scotland (Figure 22.38).

The axial part of the Southern Permian Basin was occupied by a permanent desert lake whose waters were derived mostly from the Variscan Mountains in the south and from the east; numerous horizons of bedded halite represent periods of increased aridity. A marginal zone of anhydritic mudstones and sandstones represents the fluctuating sabkha coastline of the desert lake. Ephemeral streams and perhaps even more permanent seasonal rivers carried sediment and water from the mountains towards the lake. Clay-size particles were probably blown to the south-west out of the Permian desert, but unconsolidated fluvial sands were redeposited in a sand sea mostly as transverse dunes. Analysis of the dune bedding in many wells of the southern North Sea basin indicates that the prevailing wind blew from the north-east, an interpretation in keeping with a similar analysis of Permian dune sands in Britain made in 1937 by F. Shotton at, among other places, Mauchline in south-west Scotland (Figure 22.39). In the northern basin, bedding attitudes imply that there, the prevailing wind blew from the north-west.

Possibly because the winds were stronger at the basin margin in County Durham, the dunes there have their axes parallel to the prevailing Permian wind. Both salt precipitation in the desert lake and the strongest aeolian activity may have coincided with extensions of the Gondwana ice cap. The Permian desert sediments are known generally as the **Rotliegend** (from the German *Rotliegendes*, a miner's term for the red layers below the **Zechstein** carbonates).

Under the conditions of Permian aridity, the rate of sedimentation failed to keep up with that of basin subsidence, so that by late in the Permian, the surface of the Rotliegend desert lake was probably some 200–300 m below global sea-level. Towards the end of the Permian, a reduction of the Gondwana ice cap is believed to have caused a global rise in sea-level. Seawater probably broke across a low barrier possibly somewhere between

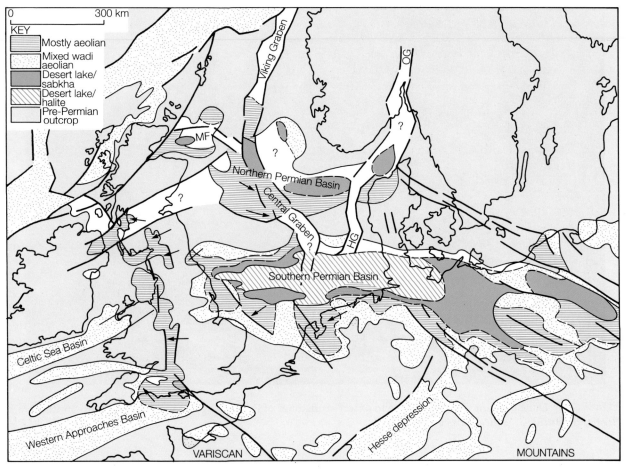

Figure 22.38 Map of north-west Europe showing the distribution of the main areas of Rotliegend desert sedimentation north of the Variscan Mountains (Southern and Northern Permian basins) with their axial desert lakes and marginal sands of sabkha, fluvial and aeolian origins. Arrows indicate wind directions deduced from outcrop and well data. MF – Moray Firth Basin; HG – Horn Graben; OG – Oslo Graben.

northern Scotland and Norway, and rapidly flooded the Rotliegend basins to create the Zechstein Sea. This land-girt sea was surrounded by desert highlands, the rate of evaporation was still high, and gypsum (anhydrite after burial) and halite were precipitated throughout the basins. Repeated waxing and waning of the Gondwana ice cap is thought to have controlled four major cycles of Zechstein flooding and evaporation.

The Rotliegend dune sands beneath the southern North Sea, The Netherlands and north Germany now form the reservoir for large volumes of methane gas derived from the underlying Carboniferous Coal Measures. Because salt has the ability to flow and so seal any fault-induced fractures, the cap of Zechstein halite (up to 1000 m thick) prevents the upward escape of gas now trapped in Rotliegend sandstones. Later diapiric movement of Zechstein salt has been an important factor in creating traps for oil and gas in younger sedimentary sequences, especially within the Northern Permian Basin.

At the beginning of the Triassic period, the Zechstein marine waters were temporarily replaced by relatively fresh water of terrestrial origin to create a large lake. With redeveloping aridity, the lacustrine (Bunter) shales were gradually covered by sands of fluvial and, locally, of aeolian origin (Bunter Sandstone). Three separate sequences of salt (Röt, Muschelkalk, Keuper halites) were then precipitated in sub-basins beneath the modern North Sea, with approximately time-equivalent dune sands developing within different sub-basins of the British Isles. Before a marine transgression brought terrestrial desert conditions to a close at the end of the Triassic period, arid conditions had prevailed over much of the north-west of Europe for some 85 m.y. The Late Triassic (Keuper) halites of the Cheshire Basin have provided much of Britain's domestic salt since the time of the Romans and, in the 18th century, were an important ingredient for establishing a flourishing chemical industry.

Figure 22.39 Dune bedding of barchan type in desert sandstones of Permian age, Mauchline Quarries, Strathclyde, Scotland. (*British Geological Survey*)

22.10 LATE PALAEOZOIC AND MESOZOIC DESERTS OF THE UNITED STATES

The western interior of the United States contains what must be the most widespread and best exposed sequence of desert sediments in the world; aeolian sandstones such as the Pennsylvanian Weber and Tensleep, the Permian Cedar Mesa, De Chelly and Coconino, and the Jurassic Wingate, Nuggett, Navajo, Glen Canyon and Entrada (already mentioned) are classical formations that are especially well displayed in Arizona, New Mexico, Colorado, Utah and farther north in Wyoming (Figure 22.40). The northward drift of North America was slower than that of north-west Europe, partly because it involved some 40° of counterclockwise rotation centred over the Gulf of Mexico since the western interior arrived within the northern hemisphere trade-wind desert belt during the Late Carboniferous (**Pennsylvanian**). For the next 160 m.y. until late in the Jurassic, a desert environment dominated sedimentation, punctuated by occasional marine transgressions, many of which were associated with the precipitation of evaporite minerals. The changing palaeogeography was controlled by differential basin subsidence and the creation of highland areas.

An analysis of dune bedding indicates that throughout most of this time span, the winds in the United States blew from roughly north to south (present north). After removing the effects of continental rotation, these palaeowinds conformed mostly to a trade-wind pattern modified to varying degrees by extensive incursions of shallow seas. By the end of the Jurassic, however, this vast area was already mostly north of the normal trade-wind belt, and any later deserts (as today) would have resulted from mid-continental distance from oceanic moisture or other rain-shadow effects.

22.11 CONCLUSIONS

Winds are a global phenomenon; they conform to a pattern generated by temperature differences between the Equator and the Poles, which is modified by differences in the Earth's surface velocity between those extremes. At their strongest, winds can cause catastrophic destruction on land as well as providing the driving force for ocean currents and for waves. Winds cause aridity over deserts and evaporation at the sea surface, and when the air is cooled, they are associated with the generation of clouds and rainfall. Thus winds play an important role in many of the geological processes at the Earth's surface.

Deserts are areas where the potential rate of

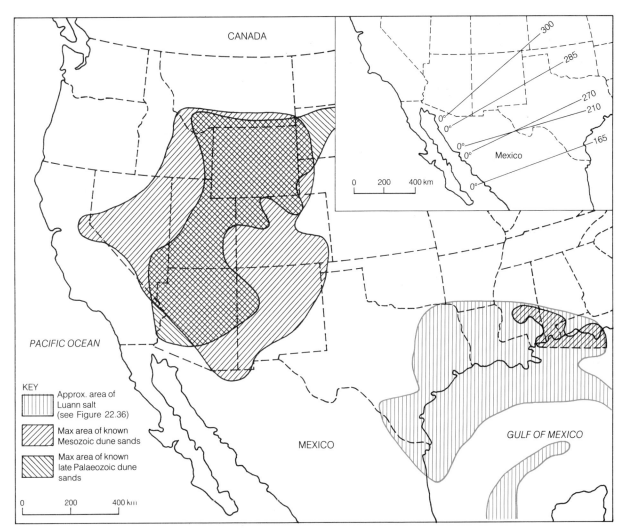

Figure 22.40 Outline map showing the areas of the Western Interior of the United States that were covered by dune sands during the Late Palaeozoic and Mesozoic. Most winds blew from the north. Flanking areas were generally hills and lowlands with, especially in the Late Palaeozoic, shallow seas with marginal sabkhas. The location of the Mid-Jurassic Luann salt (simplified from Figure 22.36 is included in south-east of map for reference). Inset: the changing location and orientation of the Equator from the Late Carboniferous (300 m.y.) to the Late Jurassic (165 m.y.).

evaporation greatly exceeds the rainfall; where the rainfall is too low or spasmodic to support vegetation. These usually occur in the trade-wind belts north and south of the Equator, in the rain shadow of mountains or far from the sea in the centres of large continents. They are not constant features of the world but have existed spasmodically for some 3500 m.y. or more, and currently occupy about 30 per cent of the existing land surface.

Deserts are characterized by dunes which, however, occupy only about a quarter of their surfaces. The remainder comprises barren rock, the deposits of ephemeral streams and lakes (inland sabkhas) and coastal sabkhas. Dunes are of two basic types: transverse dunes which have

their axes at right angles to the sand-transporting wind; and longitudinal dunes, whose axes are parallel to the sand-transporting wind. The latter seem to form under conditions where wind velocities are too high for transverse forms to be stable.

Desert conditions existed in north-west Europe throughout the Permo-Triassic, but in the western interior of the United States, such conditions lasted from the Late Carboniferous to the Late Jurassic.

Land-locked seas within the trade-wind belts are subjected to intense evaporation; such conditions have occurred several times during the Earth's history, and have resulted in the precipitation of considerable thicknesses of salts.

22.12 FURTHER READING

Bagnold, R.A. (1973) *The Physics of Blown Sand and Desert Dunes*, Chapman and Hall, London.

Blackett, P.M.S. (1961) Comparison of ancient climates with the ancient latitudes deduced from rock magnetic measurements, *Proc. Roy. Soc.*, London, A. vol. 263, 1–30.

Cooke, R.U. and Warren, A. (1973) *Geomorphology in Deserts*, Batsford, London.

Glennie, K.W. (1970) Desert Sedimentary Environments. *Developments in Sedimentology* **14**, Elsevier, Amsterdam.

Glennie, K.W. (1990) Lower Permian – Rotliegend, in *Introduction to the Petroleum Geology of the North Sea*, 3rd ed (ed K.W. Glennie), Blackwell, Oxford.

Lotze, F. (1957) *Steinsalz und Kalisalz*, Borntraeger, Berlin.

Pye, K. (1987) *Aeolian Dust and Dust Deposits*, Academic Press, London.

Pye, K. and Tsoar, H. (1990) *Aeolian Sand and Sand Dunes*, Unwin Hyman, London.

Thomas. D. S. G. (1989) *Arid Zone Geomorphology*, Bellhaven Press, London.

Wadia, D.N. (1960) *The post-glacial desiccation of Central Asia*, Monograph of the National Institute of Sciences of India, New Delhi.

COASTAL SCENERY AND THE WORK OF THE SEA

23

I with my hammer pounding evermore
The rocky coast, smite Andes into dust,
Strewing my bed and, in another age,
Rebuild a continent for better men.

Ralph Waldo Emerson (1803–82)

23.1 SHORELINES

Nearly all coastlines have been initiated by relative movements between land and sea. Rise of sea-level or subsidence of the land leads to the submergence of a landscape already moulded by sub-aerial agents. The drowning of a region of hills and valleys gives an indented coastline of bays, estuaries, gulfs, rias, fjords and straits, separated by headlands, peninsulas and off-lying islands. The Rio coast of Brazil (Figure 23.1) is one of the most striking and familiar examples of this type. In contrast very broad bays, like the Great Australian Bight, result from the submergence of plains. Coasts that have originated in these ways are called **coasts of submergence**. Because of the rise of sea-level that followed the last glacial period most coasts belong to this class. Indeed without that rise harbours would have been few and far between. The occurrence of 'raised beaches' along some of these coasts looks like a contradiction. But it only means that what we now see is the algebraic sum of a long history of ups and downs. The last major eustatic rise of sea-level has ensured that a majority of shorelines still retain the characteristics of submergence.

There are, however, formerly glaciated regions that have been rebounding more rapidly than the rise of sea-level, and have continued to emerge during the last few thousand years of nearly stable sea-level. The results, such as can be well seen at Stockholm and Helsinki and along the neighbouring coasts, can be described as submerged glacial topography that is now emerging (Figure 23.2). More typical **coasts of emergence** occur where tectonic uplift outstripped the rising sea (or has since done so), e.g. along the Pacific coast of South America.

Other varieties of coastline include those determined by volcanic activity (Figure 23.64), faulting (Figure 24.26), outwash plains (Figure 23.56) and the growth of coral reefs and atolls (Section 24.10). From a structural point of view Suess was the first to recognize two contrasted types which he distinguished as **Atlantic** and **Pacific**. Coasts of Atlantic or **transverse** type are determined by fractures and subsidences that characteristically cut across the strike or 'grain' of the folded rock formations (Figure 23.3); they characteristically border relatively young oceans that are widening as a result

Figure 23.1 A celebrated coast of submergence: Rio de Janeiro, Brazil. (*Brazilian Government Public Relations*)

Figure 23.2 An emerging coast due to post-glacial isostatic uplift, Örö Island and neighbouring parts of the Finnish Archipelago, Hitis, south-western Finland. (*Copyright: Ilmavoimat – Finnish Air Force. Reproduced by permission*)

of sea-floor spreading. Coasts of Pacific or **longitudinal** type border or lie within mountain chains, including island festoons like those of Asia, and follow the general 'grain' of the land. When par-

tially drowned such coasts are said to be of **Dalmatian** type (Figure 23.4).

The general outlines of a newly formed coast are soon modified by marine erosion and deposition,

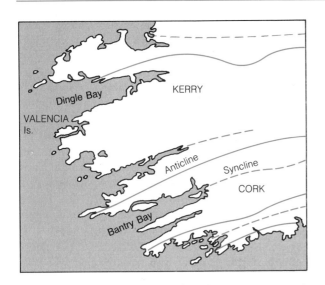

Figure 23.3 An example of submerged 'Atlantic' or transverse coast. Old Red Sandstone crops out along anticlines which have remained as uplands or broad ridges that jut out as promontories. Less resistant Carboniferous strata outcrop as synclines in the valleys, which pass seawards into long bays or *rias*, south-west Eire.

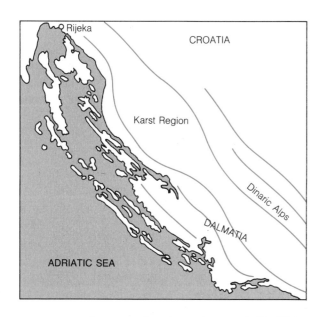

Figure 23.4 An example of a submerged 'Pacific' or longitudinal coast (Dalmatian type), Yugoslavia.

with development of a wide variety of shore features and coastal scenery. By the incessant pounding of waves, which break up the rocks and wear back the cliffs, the sea cuts its way into the land like a horizontal saw. The liberated rock fragments are rounded by innumerable impacts and continual grinding as the line of breakers is carried backwards and forwards over the foreshore by the ebb and flow of the tide. The worndown material is partly dealt with by the waves

themselves and partly by currents. Most of the finer sediment – including contributions from rivers and glaciers and the wind – is carried into deeper water coming to rest on the sea floor. The coarser sediment is swept towards or drifted along the shore to form shoals and beaches or to build out spits and bars (long embankments of sand and shingle), where the coastline abruptly changes its direction. Inlets and lagoons sheltered from the sea in this way develop into marshes, which in time are silted up by contributions from the landward side, or blanketed with sand dunes that advance from the seaward side. In these and other ways new land is added to the fringe of the old in compensation for the losses suffered elsewhere.

The waters of the seas and oceans readily respond by movement to the brushing of the wind over the surface; to variations of temperature and salinity; to the gravitational attraction of the moon and sun; and to the Coriolis force. The work of erosion, transport and deposition carried out by the sea depends on the varied and often highly complex interplay of the waves, currents and tides that result from these movements.

It may be noted in passing that lakes, especially the larger ones, behave in much the same way as enclosed seas. In consequence the shore features of lakes and seas have much in common. A lake formed by obstruction (e.g. the lava-dammed Lake Kivu, Rwanda/Zaire) drowns the surrounding land and so acquires a shoreline of the submergent type. A lake which has shrunk from its former extent in response to climatic or other changes (e.g. the Great Salt Lake of Utah) is margined by flats and terraces of sediment (Figure 22.6), and so acquires a shoreline of the emergent type. Tides are negligible in lakes, but seasonal variations of rainfall may cause the water to advance and recede over a tract of shore which is alternately covered and exposed, though much less frequently than the tide-swept foreshore of a sea coast. Waves and currents operate exactly as in land-locked seas of similar extent and depth, and are responsible for erosion and deposition on a corresponding scale. There are of course important biological contrasts. The swamps into which shallow lakes degenerate, with their luxuriant growth of aquatic vegetation and accumulations of peat, are very different from the mangrove swamps and tidal marshes that locally border the sea. On the other hand lake shores have nothing to be compared with the coral reefs of tropical seas.

23.2 TIDES AND CURRENTS

The tide is the periodic rise and fall of the sea which, on an average, occurs every 12 hours 26 minutes. Tides are essentially due to the passage around the Earth, as it rotates, of two antipodal bulges of water produced by the differential attraction of the moon and sun. The bulges are the crests of a gigantic wave, low in height but of enormous wave length. It is easy to understand that the water facing the moon should bulge up a little, but it is less obvious why there should be a similar up-bulge in the opposite direction on the other side of the Earth. The basis of the explanation is that the water centred at A (Figure 23.5) is attracted towards the moon more than the Earth, centred at E, while the Earth in turn is attracted more than the water centred at B. The water at the far side is thus left behind to almost the same extent as the water on the near side is pulled forward. From places such as C and D the water is drawn away and low tide results. As the Earth rotates, each meridian comes in turn beneath the positions of high and low tide nearly twice a day; not exactly twice, because allowance must be made for the forward movement of the moon. Nor are these positions exactly in line with the moon (as for simplicity they have been drawn in the diagram), because the tides are affected (a) by the Earth's rotation; (b) by the great continental obstructions met with during their circuit of the globe; and (c) by friction against the sea floor, especially in shallow seas.

The effect of the sun is similar to that of the moon but considerably less powerful. When the Earth, moon and sun fall along the same straight line, the tide-raising forces of sun and moon help each other, and tides of maximum range, known as **spring tides**, result. The moon is then either new or full. When the sun and moon are at right angles relative to the Earth, the moon produces high tides where the sun produces low. The tides are then less high and low than usual and are called **neap tides**.

In the open ocean the difference in level between high and low tide is only a metre or so. Enclosed basins have still weaker tides – only 30 cm or so in the Mediterranean and no more than 10 cm in the Black Sea. In shallow seas however, and especially where the tide is concentrated between converging shores, ranges of 6 to 9 m are common.

Where tidal range is small, coasts are dominated by wave activity. On macro-tidal coasts, strong tidal currents may be generated, especially at coastal constrictions where they may combine with the effects of Coriolis force to give whirlpools at certain states of the tide. The notorious Corryvreckan whirlpool in the narrow strait between the islands of Jura and Scarba in the Inner Hebrides of Scotland is one such example. Elsewhere, however, these tidal currents serve the useful purpose of maintaining by scour a deep channel at the narrow mouths of estuaries and bays. High tidal ranges are also associated with the development of mud flats and salt marshes (Figure 23.6).

The tides around the British Isles are of special interest. After passing up the western coasts the crest of the tidal wave swings round into the North Sea and proceeds southwards. The Coriolis force drives the water towards the right, that is, the British side, which in consequence has far higher tides than Norway and Denmark. In passing northwards up the Irish Sea the Coriolis force results in the tides being at least twice as high on the Welsh and English coasts as on the Irish side. Similarly the tides running up the English Channel are also forced to the right, giving the French coasts higher tides. It will be readily appreciated that conditions of exceptional complexity arise in the Straits of Dover and the southern part of the North Sea.

A current of about 2 knots accompanies the flood tide as it advances up the English Channel. In the Bristol Channel however, which is like a

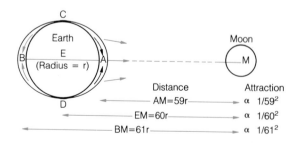

Figure 23.5 Diagram to illustrate the generation of tides.

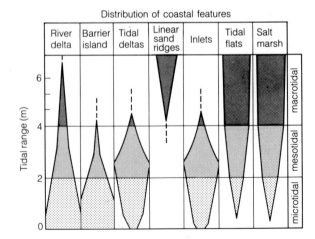

Figure 23.6 Tidal range and coastal type (after M.G. Hayes).

Figure 23.7 The 'bore' of the Severn advancing up-river at the peak of a spring tide. (*Keystone Press Agency Ltd*)

wide funnel leading into the Severn, the inflowing tide is forced into a rapidly narrowing passage. It therefore rises in height. Spring tides may reach as high as 12 or 13 m and the inward current may attain a speed of 10 knots. In these circumstances, and especially if the wind co-operates, the over-crowded tidal waters eventually travel bodily up the river with a wall-like front of roaring surf. This is the Severn **bore**, a vigorously advancing flood of powerful waves and breakers (Figure 23.7), which may ride up the river almost as far as Gloucester before subsiding completely.

Near the shore and between islands, tidal currents are often sufficiently powerful to transport sand and even shingle, and so to scour and erode the sea floor. In estuaries, where the outward flow of river water is added to the ebb current, transport is dominantly seaward. But since the fresh

river water, carrying a load of silt and mud, tends to slide out over the heavier salt water that has crept in along the bottom, it is the upper suspended load that is mainly swept out to sea, while the coarser debris is stranded and tends to accumulate as sand bars.

Powerful currents are generated by differences of salinity, which depend on (a) inflow of rivers, rainfall and the melting of ice, all of which freshen the water; and (b) evaporation, which increases the salinity.

Evaporation over the Mediterranean lowers its surface and increases the salinity and density. Surface currents therefore flow into the Mediterranean through the Dardanelles from the Sea of Marmara and the Black Sea (where the evaporation is more than balanced by the inflow of rivers), and through the Strait of Gibraltar from the Atlantic (Figure 23.8). In each case undercurrents of higher salinity flow outwards from the Mediterranean. Shore deposits are affected by the surface currents, while farther out the floor is scoured by the deeper current. A similar interchange of water takes place between the highly saline Red Sea and the Indian Ocean, and also between the comparatively fresh Baltic and the North Sea.

The main current systems of the oceans are primarily of a convectional nature, brought about by density differences due to heating in the tropics and cooling in the polar regions, and also to variations in salinity. They are greatly modified by the configuration of the continents and the dominant winds. Apart from the superficial movements charted in every atlas, cold water from the Arctic spreads south at depths between 1800 and 2400 metres, is warmed by the undercurrent from the Strait of Gibraltar, and continues south until it rises over the cold currents moving northwards from Antarctica. The latter can be traced into all the oceans, including the North Atlantic, while deep Arctic and Atlantic water also passes into the Pacific and Indian Oceans.

Oceanic currents have important climatic and biological as well as geological effects, and will be

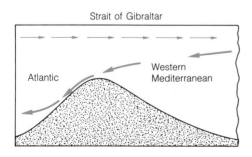
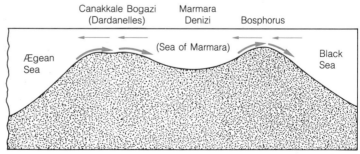

Figure 23.8 Salinity currents of the Mediterranean and Black Seas. Water of high salinity (about 37–38%, thick arrows) flows out of the Mediterranean, while water of lower salinity (thin arrows) flows in.

referred to where appropriate. More localized currents, due essentially to interactions between wind, tides and waves (e.g. surges, longshore drift and rip currents), or to slumping and churning up of bottom sediment on submarine slopes down which muddy water can flow (turbidity currents), are also dealt with on subsequent pages.

23.3 WAVES

The ordinary waves seen on the sea are almost entirely due to the sweeping of winds over the surface of the water. The exceptions are the far-travelled descendants of giant waves generated by catastrophic 'accidents' which suddenly displace immense volumes of ocean water. Such occurrences may result from earthquakes, volcanic eruptions, submarine landslides and avalanches of rock from high cliffs or steep mountain sides into deep water. Here we are concerned with waves that derive their motion and energy from the wind. To begin with, the surface is ruffled into undulations that move forwards and gradually increase in height and speed. The **height** of a wave is the vertical distance from trough to crest (Figure 23.9). The horizontal distance from crest to crest – or from trough to trough – is the **wave length**. This is measured at right angles to the wave front or crest line. Although the latter may be very long it should not be confused with the wave length, which is parallel to the direction of wave advance. The height ultimately attained by a wind-driven wave, where it is not restricted by shallowing water, depends on the strength, duration and **fetch** of the wind, the fetch being the length of the open stretch of water across which the wind is blowing. When the loss of energy involved in the propagation of the waves through the water is just balanced by the amount of energy supplied from the wind, the height reaches its maximum. The height cannot however be calculated from the speed and wave length of the waves. According to a very rough numerical rule adopted as a guide by mariners, the limiting height in feet is about half the speed of the wind in miles per hour.

Where waves are being generated by the gusty

winds of a storm new waves are continually forming. The resulting **sea**, as the assemblage of waves is called (e.g. a stormy or choppy sea), is likely to be a jumble of many different wave lengths. Here and there crests and troughs coincide and cancel out. Elsewhere crests may coincide and rise to form a high peak, and where two troughs coincide the sudden drop may be perilous to a passing ship. From the generating area the longer and higher waves grow at the expense of the shorter ones. Thus it happens that a series of more uniform waves is eventually sorted out of an initial chaotic 'sea'. Such waves may travel into regions of calm weather far beyond the fetch of the wind where they originated and started their progress towards the bordering shores. The height gradually declines when waves are no longer maintained by the wind; they are then described as a **swell** or **groundswell**.

A set of uniform waves (called a 'train') progressing through deep water can be described in terms of wave length λ, height h and period T. The period is the time taken for two successive crests to pass a given point. The speed v is then given by the simple formula $v \times \lambda/T$. An approximate relationship that is often useful is λ (in metres) $\times$ $1.56T^2$ (in seconds). This means that a deep-water wave with a period of 10 seconds has a wave length of about 156 m, the corresponding speed (v) being 56 km per hour. In the open ocean heights of 1.5 to 5 m are common, increasing to 12 or 15 m in high seas. The corresponding wave lengths range from 60 to 245 m, but as they flatten out into swell they may reach 300 m or more, about 760 m being the maximum recorded, with a speed of nearly 130 km per hour.

The sea waves we watch from the shore are generally a mixture of swells from distant storms and 'seas' from local winds. In other words they can be described as a **spectrum** of 'trains' of waves of different lengths, just as sunlight is a blend of light waves of different lengths which, when separated (as in the rainbow), appear as a spectrum of different colours. This is the origin of the traditional belief that 'every seventh wave is the biggest'. Actually the heights can vary irregularly within wide limits, according to the happenings in various parts of the globe.

It is important to realize that in the open sea – apart from wind drift – it is only the wave form that moves forward: the shape, not the water itself. The wave is a device for the transport of energy. Each particle of water moves round a circular orbit during the passage of each complete wave, the diameter being equal to the height of the wave (Figure 23.10). This is demonstrated by the behaviour of a floating cork under which a train of waves is passing. Each time the cork rises

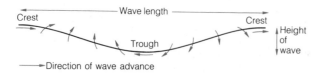

Figure 23.9 Profile of an ideal wave of oscillation from crest to crest, showing the direction of movement of water particles at various points.

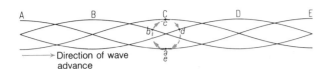

Figure 23.10 Diagram showing the orbit of a water particle during the passage of a wave of oscillation. A, B, C, D, E mark successive positions of the crest as the wave moves forward; *a, b, c, d, e* are the corresponding positions of the particle. AE = wave length; C*a* = height of wave.

and falls it also sways to and fro, without advancing appreciably from its mean position. Such waves are called **waves of oscillation**. If the wind is strong, however, each water particle advances a little farther than it recedes and the waves may become strongly asymmetrical (Figure 23.11). Similarly in shallow water, where friction against the bottom begins to be felt, each particle recedes a little less than it advances. In both cases the orbit, instead of being a closed circle, resembles an ellipse which is not quite closed, and a certain proportion of the water then slowly drifts forward in the direction of wave advance.

When gusts of wind brush over a field of corn the stalks repeatedly bend forward and recover, and waves visibly spread across the surface. Here it is obvious that the wave motion is not confined to the surface, since it is shared by the stalks right down to the ground. In the same way the energy contributed by the wind to a body of water is transmitted downwards as well as along the surface. Owing to friction the diameters of the orbits rapidly diminish in depth until at a depth of the same order as the wave length they become negligible. Indeed, at a depth of half the wave length the orbital diameter is only 4 per cent of that at the

surface. The greatest depth at which sediment on the sea floor can just be stirred by the oscillating water is called the **wave base**. Since wave length rarely exceeds 400 m and is generally much shorter than this, wave base never extends below a depth of 200 m. Ripple marks have been photographed at depths of hundreds of metres, but such effects are far beyond the influence of surface waves. There are, however, deeper waves generated at the boundaries of currents travelling in opposing directions. It has also been suggested that deep-sea ripple marks may be the work of gentle currents flowing over loose sediment while the particles are kept vibrating to and fro by the passage of earthquakes.

23.4 WAVES IN SHALLOW WATER

As waves approach the shore and pass into shallow water, several changes of great importance occur. 'Shallow' in this context means a depth of water less than half the wave length of the approaching waves. After the waves begin to 'feel bottom' they gradually slow down as the water shallows, each progressing more slowly than the one behind. The only characteristic of a train of waves that remains constant is the period T. Remembering the formula $v \times \lambda/T$, it follows that as the speed falls off the wave lengths become correspondingly shorter. Indeed the tendency of the wave fronts to crowd together as the shore is approached can often be clearly seen. Consequently when waves approach a shelving shore obliquely, as in Figure 23.12, the crest lines swing towards parallelism with the shore. This change of direction with change of speed is called **refraction**; it is essentially the same phenomenon as the

Figure 23.11 An asymmetrical wave driven by a high wind. Here the orbits are nearly elliptical, but with a net advance for each rotation, so that a little of the water moves forward in the direction of the wave-motion. (*T.M. Finlay*)

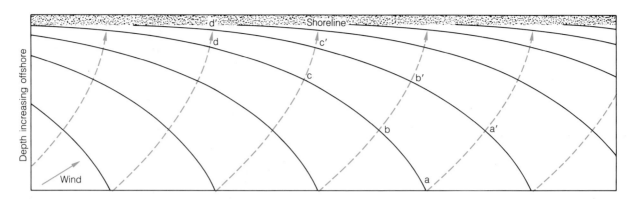

Figure 23.12 Wave refraction: diagram to illustrate the swing of oblique waves in shallowing water towards parallelism with the shore. While the crest at *a* advances to a^1, the crest at *b* advances a shorter distance to b^1... and so on. The crest lines *abcd* thus become curved, as shown.

sudden bending of a ray of light when it passes from air into, say glass, in which its speed is reduced by 30 per cent or more. Refraction of oblique waves around a headland and islet into a bay is well illustrated by Figure 23.13.

The effects of wave retardation off an indented coast are illustrated by Figure 23.14. The waves advance more rapidly through the deeper water opposite a bay than through the shallower water opposite a headland. Thus the practically straight crest of a wave at *a* moves to *a'*, while the crest at *b* moves only to *b'*. Each wave crest in turn begins to approximate to the curves of the shoreline. Consequently when the shore is reached by a wave such as *abcde*, all the energy from the long stretch *ac* converges on the headland AC (and that from *de*

Figure 23.13 Aerial photograph showing the refraction of waves around a headland and islet into a bay at San Clemente Island, California. The pattern of short-crested waves approaching from the upper left results from the interference of swells from two different directions. (*Official U.S. Navy photograph, by permission*)

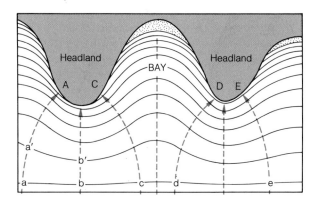

Figure 23.14 Diagram to show the effect of wave retardation of an indented coast.

on DE). In contrast the very much smaller amount of energy from the short stretch *cd* is dispersed around the shores of the bay from C to D. Thus while headlands are being vigorously attacked by powerful waves, deep embayments are not unduly disturbed and their waters provide safe anchorage for vessels sheltering from a storm. In the same way waves entering a harbour between or over piers spread out and rarely do more than ruffle the water inside.

In a train of waves advancing at right angles to a straight shoreline, each metre of the crest line of a wave represents about the same amount of energy. The wave energy from a deep column of water is concentrated into a shallow one. Since energy is proportional to mV^2, it follows that as V is reduced, as it is on a shelving sea floor, m must be increased (where m is the mass of water per metre of crest line). mV^2 turns into Mv^2. Each wave therefore increases in volume and becomes higher as it passes into shallower water. At the same time

Figure 23.16 Small breakers of the spilling type, coast of New Jersey, USA. (*Paul Popper Ltd*)

the wave front becomes correspondingly steeper. Finally a critical point is reached where the vol-

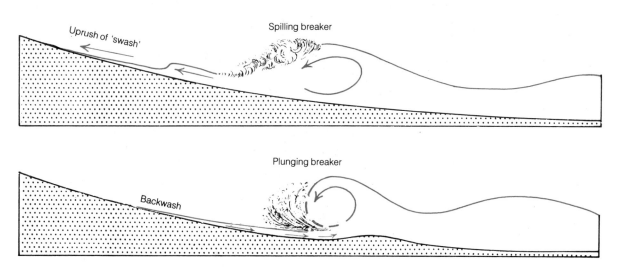

Figure 23.15 Diagrammatic indication of the difference between a spilling breaker with an effective uprush (constructive type) and a plunging breaker with an effective backwash (destructive type). (*Modified after W.V. Lewis*)

ume of water in front is insufficient to fill up the wave form as required by the orbital movement. The crest of the uncompleted wave is then left unsupported, and because the orbit is broken the wave itself breaks. The depth in which waves break (i.e. the mean depth as it would be in still water) ranges between 1.25 and 0.75 times the height of the unbroken wave following on behind.

Dependent on a variety of factors, chief of which is the steepness of the wave front, two contrasted types of breakers are recognized. Between the two extremes there are naturally all gradations, according to the nature of the shore and the waves, and the strength and direction of the wind. Advancing over a gently shelving floor a wave front steepens gradually, until the crest merely spills over into the trough in front. There is a **spilling breaker**, a foaming mass of water that surges forward as a

Figure 23.17 A plunging breaker viewed from above, approaching the coast of New Jersey. The marked break in the crest is an indication of a rip current (Fig. 23.40) flowing seawards. (*Paul Popper Ltd*)

Figure 23.18 Storm wave from the Atlantic breaking against the cliffs of St Ives, Cornwall, England. (*Fox Photos Ltd*)

cially if the waves are large, the rise of the crest is greatly augmented. The wave front steepens until an unstable-looking hollow appears in front. Over this the crest begins to curl, sometimes hovering a little while the wave continues its advance. But finally it plunges down, often with great violence. This is the **plunging breaker** (Figures 23.15 and 23.17).

What were **waves of oscillation** before they began to 'feel bottom' have now passed largely or wholly into **waves of translation**, in which the water advances bodily up the shore. This sheet of surf may re-form into smaller waves that break again higher up, so that there is a zone of breakers rather than a single break-point. The final translation of water up the beach is called the **uprush** or **swash**. The return of water down the slope is called the **backwash** (Figure 23.15).

Considerable volumes of air may be rapidly trapped by plunging breakers and compressed so strongly that the air reacts with explosive violence. It escapes by 'atomizing' the water into masses of foaming spray which reach astonishing heights in severe storms. Geyser-like displays may sometimes be seen when high waves are breaking on steep beaches, but uprushing jets of spray are more familiarly associated with the crashing pressure of breakers against rocky obstructions (Figure 23.18) and breakwaters (Figure 23.19).

23.5 MARINE EROSION

turbulent sheet of surf (Figures 23.15 and 23.16). But where the floor shallows rapidly, and espe-

The sea operates as an agent of erosion in four different ways:

Figure 23.19 A breaker crashing against the promenade at Hastings, Sussex, England, and bursting into jets of spray from the explosive expansion of entrapped air. (*Judges Ltd*)

1. by the **hydraulic action** of the water itself, involving the picking up of loose material by currents and waves, and the shattering of rocks as the waves crash, like giant water-hammers, against the cliffs (Figure 23.18);

2. by **corrasion**, when waves, armed with rock fragments, hurl them against the cliffs and, co-operating with currents, drag them to and fro across the rocks of the foreshore;

3. by **attrition**, as the fragments or 'tools' are themselves worn down by impact and friction; and

4. by **corrosion**, i.e. solvent and chemical action, which in the case of seawater is of limited importance, except on limestones and rocks with a calcareous cement.

In addition, alternate wetting and drying of rock surfaces encourages weathering processes, such as chemical attack by hydration and oxidation and physical disintegration in response to salt crystal growth. Frost action is important on polar coasts. Finally, softer and calcareous rock surfaces will be weathered by the chemical secretions of shore flora and fauna.

The destructive impact of breakers against obstructions is often far greater than is generally realized. The pressure exerted by Atlantic waves averages over $9700\,kg/m^2$ during the winter, while in great storms it may exceed even 30 000. Thus not only cliffs but also sea walls, breakwaters and exposed lighthouses are subjected to shocks of enormous intensity. Cracks and crevices are quickly opened up and extended. Water, often in the form of high-pressure spray, is forcibly driven into every opening, tightly compressing the air already confined within the rocks. As each wave recedes the compressed air suddenly expands with explosive force, and large blocks as well as small become loosened and are sooner or later blown out bodily by pressure from the back. The combined activity of bombardment and blasting is most effective as a quarrying process on rocks that are already divided into blocks by jointing and bedding, or otherwise fractured, e.g. along faults and crush zones.

The undercutting action of waves is illustrated by Figures 23.21 and 23.22. Cliffs originate and are maintained by similar undermining of the sea-

Figure 23.20 The Needles: stacks of Chalk in line with the cliffs at the western extremity of the Isle of Wight, England. The boldness of the cliffs reflects the unusual resistance of the Chalk, here due to folding and induration, and contrasts strongly with the low shoreline of Cretaceous sands and clays seen in the background. (*Aerofilms Ltd*)

Figure 23.21 Pedestal or 'mushroom' rock, of tough metamorphosed basalt, deeply undercut by marine erosion; near Maloy on the north side of Nordfjord, about 145 km north of Bergen, Norway. (*Mittet Foto, Oslo*)

Figure 23.22 Pedestal of calcareous sandstone (Jurassic) supported by an undercut pillar of shale on the foreshore at Sheepstones, Yorkshire. The sandstone is pitted with corrosion hollows caused by spray from breaking waves at high tide. (*British Geological Survey*)

excavated at the base of the cliffs, the latter gradually recede and present a steep face towards the advancing sea. But where the cliffs are protected for a time by fallen debris, and especially if they are composed of poorly consolidated rocks, the upper slopes may be worn back by weathering, rain-wash and slumping. At any given place the actual form of the cliff depends on the nature and structure of the rocks there exposed, and on the relative rates of marine erosion and sub-aerial denudation (Figure 23.20).

The most striking evidence of undermining is provided by caves. There are few stretches of coast along which the rocks are equally resistant to wave attack. Caves are excavated along belts of weakness of all kinds, and especially where the rocks are strongly jointed (e.g. Fingal's Cave, Figure 1.2). By subsequent falling-in of the roof and removal of the debris, long narrow inlets are developed. In Scotland and the Faroes a tidal inlet of this kind (Figure 23.23) is called a **geo** ('g' hard – Norse *gya*, a creek). The roof of a cave at the landward end of a geo – or indeed of any sea cave – may communicate with the surface by way of a vertical shaft which may be some distance from the edge of the cliff. A natural chimney of this kind (Figure 23.24) is known as a **blow-hole** or **gloup** (a throat). The opening is formed by the hydraulic action of wave-compressed air already

ward edge of the land. By falls from an over-steepened rock-face or by collapse of rocks overhanging the notch which may have been

Figure 23.23 Sea cave and development of inlets by roof collapse in cliffs of Old Red Sandstone. The Wife Geo, near Duncansby Head, extreme north-east coast of Scotland, looking seawards. (*British Geological Survey*)

Figure 23.24 Blow-hole or 'gloup', due to inland collapse of roof of sea cave. The part of the roof that still remains is known as The Devil's Bridge. Holborn Head, north-east Highland Region, Scotland, looking seawards. (*British Geological Survey*)

Figure 23.25 Arch cut through a headland of Dalradian quartzites; 'Great Arch', Doaghbeg, north of Portsalon, Co. Donegal. (*Irish Tourist Board*)

Figure 23.26 Stacks and cliffs of Old Red Sandstone, looking north. The Stacks of Duncansby, near John o'Groats, Scotland. (*British Geological Survey*)

Figure 23.27 The Old Man of Hoy, Orkney Isles, Scotland. A stack of Old Red Sandstone, 137 m high, rising from a platform of Devonian lava. (*British Geological Survey*)

described. The name blow-hole refers to the fact that during storms spray is forcibly blown into the air each time a breaker surges through the cave beneath.

When two caves on opposite sides of a headland unite, a natural arch results, and may persist for a time (Figure 23.25). Later the arch falls in, and the seaward portion of the headland then remains as an isolated stack. Well-known examples of stacks are the Chalk pinnacles at the western extremity of the Isle of Wight, England, known as the Needles (Figure 23.20), and the impressive towers of Old Red Sandstone near John o'Groats (Figure 23.26) and in the Orkneys (Figure 23.27).

As the cliffs are worn back a **shore platform** is left in front (Figure 23.28), the upper part of which is visible as the rocky foreshore is exposed at low tide (Figure 23.29). There may be patches of sand and pebbles in depressions, and beach-like fringes strewn with fallen debris along the foot of the cliffs, but all such material is continually being broken up by the waves and used by them for further erosive work, until finally it is ground down to sizes that can be carried away by currents. The platform itself is abraded by the sweeping of sand and shingle to and fro across its surface. Since the outer parts have been subjected to scouring longer than the inner, a gentle seaward slope is developed. In massive and resistant rocks this is an extremely slow process. Consequently, as the cliffs recede and the platform becomes very

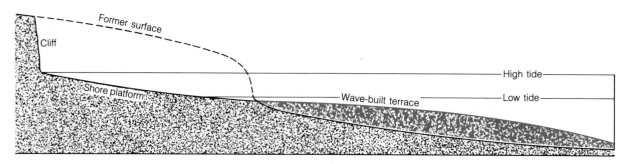

Figure 23.28 Idealized section to illustrate a temporary stage in the development of a sea cliff, shore platform, and wave-built terrace.

Figure 23.29 Shore platform cut across steeply dipping Silurian strata, near St Abb's Head, Borders, south-east Scotland. (*British Geological Survey*)

wide, the waves have to cross a broad expanse of shallow water, so that when they reach the cliffs most of their energy has already been used in transporting the abrading materials. Thus the rate of coast erosion is automatically reduced. In high latitudes, however, the cliffs may still continue to be worn back by frost and thaw, provided that the waves are able to remove what would otherwise become a protective apron of scree (cf. Figure 15.3). The shore platform off the rocky coast of west and north-west Norway – there known as the **strandflat** – has reached an exceptional width by this co-operation of processes, locally up to as much as 60 km (Figure 23.30). The most extensive level now stands 15 to 18 m above sea-level

Figure 23.30 Section across the *strandflat* north of Bergen, Norway. Length of section 52 km. (*After Fridtjof Nansen*)

because of recent isostatic uplift. Above this platform innumerable stacks and skerries, mostly flat topped, rise to heights of about 30 m, suggesting an older strandflat that has since been dissected. Similar strandflats have been described from parts of Spitzbergen and Greenland and other fjord coasts.

Where the sea is encroaching on a coast of poorly consolidated rocks, the platform in front is quickly abraded and normal coast erosion proceeds vigorously (Figure 23.31). In some localities the inroads of the sea reach alarming proportions. In Britain serious loss of land is suffered in parts of East Anglia and along the Humberside coast south of Flamborough Head, where the waves have the easy task of demolishing glacial deposits of sand, gravel and boulder clay. Since Roman times the Holderness coast has been worn back 4 or 5 km, and many villages and ancient landmarks have been swept away (Figure 23.32). During the last hundred years the average rate of cliff recession has been 1.5 or 2 m per year. The rate is not uniform, however, for severe storms and localized cliff falls are more destructive in a short time than the normal erosion of several average years.

As an extreme example of rapid coast erosion the great North Sea surge of 1953 may be cited. A **surge** is an abnormal rise of sea-level that occurs when spring tides, wind and storm waves all act together in the same direction in a more or less confined and shallow sea. Low-lying and easily erodable coasts are particularly vulnerable to these disastrous coincidences. On the last day of January 1953 very low atmospheric pressure over part of the North Sea caused a rise of sea-level which, though only 30 to 60 cm, meant the inflowing of immense volumes of additional water from the Atlantic, where the atmospheric pressure was higher. At the same time a violent gale from the north continuously drifted still greater quantities of surface water southwards. Whereas the usual flood tides hug the British coasts, this huge swell of water began to spread across the North Sea, particularly south of the latitude of the Firth of Forth. Only a limited amount of the excess could escape through the Straits of Dover. The peak of the growing surge had no alternative but to swing round towards The Netherlands. There the piling up of the storm waters reached its maximum. The sea poured in through countless breaches in the coastal defences and the most ruinous floods of salt water for hundreds of years devastated the low-lying areas previously reclaimed from the sea.

Along the lowland coasts of East Anglia and the Thames estuary the disaster, though less widespread, was locally just as destructive. Around the time of high tide the theoretical height (which is normally quite reliable) was exceeded by 2.4 m in the Thames estuary (Figure 23.33) and by 2.7–3.4 m in The Netherlands. Apart from the breaching of protective barriers by enormous

Figure 23.31 Coast erosion by the North Sea, south of Lowestoft, Suffolk; showing the disastrous effects of rapid recession of cliffs of glacial deposits due to destructive storm waves in 1936. (*British Geological Survey*)

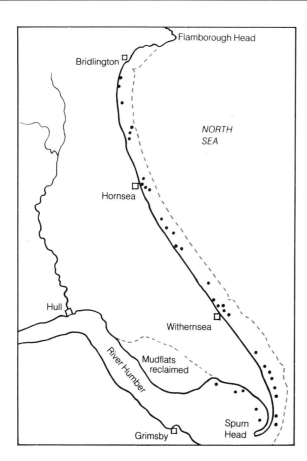

Figure 23.32 Sketch map showing the loss of land and villages by marine erosion along the Holderness coast, Humberside, England. The broken line indicates the approximate position of the coast in Roman times; former settlements are shown by black dots. (*After T. Sheppard*)

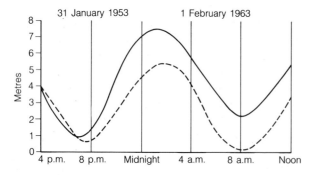

Figure 23.33 The North Sea surge of 1953. Tidal curves for Southend, near the mouth of the Thames: predicted, broken line; and actual, continuous line. (*After W.W. Williams*)

waves, erosion was most serious where low cliffs of glacial drift bordered the shore; 11 km south of Lowestoft, for example, a 7.6 km cliff was cut back nearly 11 m in two hours. Not far away, where the height of the cliff was less than 2 m, the frontal attack by undercutting was powerfully reinforced

at high tide by the shattering blows of 4-metre breakers crashing down from above, and 27 m of land was lost to the sea in a fraction of a single night. It was the threat of a repeated storm surge which led to the building of the Thames barrier to protect over a million Londoners who were at risk from surge flooding before its construction.

The great storm of February 1962 would have generated an even more serious surge had it not occurred during the moon's third quarter. As it was, the main atmospheric depression and the resulting winds directed the peak of the surge towards Hamburg and the adjoining coasts. The sea rose 6 m above the predicted height and the flooding of Hamburg was on an unprecedented scale.

23.6   SHORE PROFILES

As indicated in Figure 23.28, sediment in transit across a shore platform may accumulate in the deeper water beyond, to form a wave-built terrace in continuity with the platform, though generally with a less uniform surface than is suggested by the figure. The combined foreshore and offshore surface is a product of the joint action of erosion and deposition, each of which varies considerably from time to time and from place to place. The supply of sediment, for example, is irregular both in rate and distribution, since contributions are received from rivers and currents as well as from cliff wastage and platform abrasion – all widely varying sources of income. The processes concerned in the removal of sediment (also widely variable) are themselves largely controlled by the slope of the shore and its seaward continuation – that is to say, by the profile of the surface taken at right angles to the shore. A relatively steep slope favours destructive waves and removal of sediment from the landward side, so that the slope becomes less steep. Conversely a relatively gentle slope favours constructive waves and beach deposition on the landward side, so that the slope becomes steeper. The surface is therefore being continually modified, and in such a way that at each point it tends to acquire just the right slope to ensure that incoming supplies of sediment can be carried away just as fast as they are received. A profile so adjusted that this fluctuating state of balance is approximately achieved is called a **profile of equilibrium**. It is akin to the graded profile of a river, but along the shore the variable factors are still more numerous and difficult to evaluate. Theoretically speaking there must be a profile of equilibrium for any given set of conditions, but because of the alternating changes from tide to tide and from season to season, and the particularly violent changes from calm to storm, only a short-lived approximation to equilibrium can ever be achieved. The actual profile is continually being modified, especially along shores fringed with sand or shingle – loose materials which are easily moved.

The concept of the ideal profile of equilibrium has its uses however. At any given time the seaward slope at a given place may be either steeper or gentler than this ideal. Suppose AB in Figure 23.34 represents a relatively steep initial slope on a shore of submergence. In transforming this as nearly as possible into a profile of equilibrium CD, the waves cut a cliff-backed platform, while the resulting sediment is deposited as an offshore terrace, as already illustrated in Figure 23.28. Next, suppose *ab* in Figure 23.35 represents a relatively gentle initial slope, as when a broad valley floor becomes a bay by submergence. In transforming this into the profile of equilibrium *cd*, waves and currents build up a beach around the shores of the bay. Initial surfaces that may be almost flat are provided by the drowning of extensive alluvial plains, and on a still more widespread scale by emergence of the sea floor. In these cases, as described in Section 23.9, barrier beaches of sediment driven landwards are built up by the waves.

Seasonal and other short-term periodic changes lead to alterations of the effects portrayed in Figures 23.34 and 23.35. Shingle accumulations under the cliffs or upbuilding of beaches by the constructive waves of, say, the summer, result in a slope like *cd*. In winter however this is likely to be reduced to the gentler slope CD. Destructive storm waves bring this about by transporting material seawards, so that beaches become thin or even disappear for a time. These considerations cover various intermediate cases: e.g. parts of a shore platform may sometimes be covered with a temporary veneer of beach material, such as can be seen in Figures 23.22 and 23.29.

It must be clearly understood that Figures 23.34 and 23.35 are merely diagrams to illustrate the basic principles. The real profiles are commonly far from simple. Figure 23.36 attempts to depict a more realistic winter profile for a sandy shore.

Figure 23.35 The development of a profile of equilibrium cd from a more gently sloping initial surface, ab.

23.7 BEACHES: LANDWARD AND SEAWARD TRANSPORT

Leaving the longshore drift of sediment to be considered later, let us now review the migration of sediment towards the coast and away from it. From the outset it has to be realized that the actual processes are so complicated that satisfactory explanations in terms of physical principles are still lacking for most of the phenomena observed. A good deal can be learned from scaled-down experimental models, but the application of these is limited by the fact that sediment sizes and hydrodynamical processes cannot simultaneously be scaled down in the same proportion as, say, the dimension of a beach to be investigated or a harbour to be designed. The preparations for D-Day, during the Second World War, emphasized the critical importance of knowing what really happens on the Normandy beaches under varying conditions. Since then many other types of shores have been studied over long periods of changing winds and currents. Offshore, direct observations are made by frogmen, and radioactive tracers added to pebbles and sand grains show how sediment migrates under natural conditions.

In a classic paper of 1931 W. V. Lewis distinguished between constructive and destructive waves (Figure 23.15). **Constructive waves** are those in which the backwash is relatively weak, so that it does not seriously obstruct the uprush from the next breaker. Such waves include those that give spilling breakers with a fairly long period: say about 8 to 10 seconds, corresponding to a frequency of about 6 to 8 per minute. The elliptical shape of the orbit ensures a relatively strong horizontal component of motion, i.e. a strong uprush. The long period gives the backwash time to return to the trough that helps to feed the next oncoming wave; moreover some of the backwash may sink into the shingle or sand. When a mixed assemblage of sediment is swept up the beach the coarser material is left stranded at the top. Taking the ebb and flow of the tides into account, the result of a long spell of constructive waves is the building-up of a beach headed with a broad **berm** of coarse sand or shingle. The berm commonly has a more or less steep front leading down to the more gently graded profile of the foreshore. This

Figure 23.34 The development of a profile of equilibrium CD from a more steeply sloping initial surface, AB.

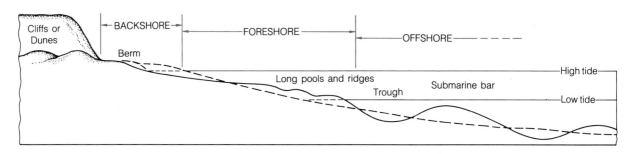

Figure 23.36 Diagrammatic indication of the subdivisions of a shore, with characteristic beach and offshore profiles in summer (broken line, showing berms built up by constructive waves) and winter (full line, showing berms cut back by destructive storm waves).

rapid change in slope tends to correspond with a break in the grain sizes of the sediment; e.g. if the berm is shingle the foreshore is likely to be sand, the latter tending to become progressively finer towards low-tide mark. The jump in grain size probably results from the accumulation of the coarser debris well above the reach of ordinary waves and tides, while the rest is dragged to and fro with the tides and subjected to long periods of attrition.

There are several definitions of the grain sizes of sedimentary fragments and particles, including beach materials. The British Standard recommendations are summarized in Figure 23.37.

Destructive waves are those in which the backwash is strong. Such waves include plungers with a somewhat short period: say about 4 to 5 seconds, corresponding to a frequency of about 12 to 15 per minute. These waves break on the backwash from the previous breaker, and because the orbit is nearly circular the main component of motion is downwards. The plunging breaker churns up the beach material, scouring out pools and troughs like those shown in Figure 23.36. Part of the sediment may be swirled back a little way, building up temporary structures like the ridges and submarine bar also shown in Figure 23.36.

Predominant grain sizes in millimetres

	Coarse	Medium	Fine
Shingle			
Boulders	200		
Pebbles	200 – 60		
Coarse gravel	60 – 20		
Gravel	60 – 20	20 – 6	6 – 2
Sand	2 – 0.6	0.6 – 0.2	0.2 – 0.06
Silt	0.06 – 0.02	0.02 – 0.006	0.006 – 0.002
Clay or mud			< 0.002

Figure 23.37 British Standard grain sizes.

Another part is carried forward in suspension by the surf of the uprush. The latter is relatively weak from the start and is further obstructed by having to advance over the returning backwash. Thus it is the backwash that is mainly in contact with the beach materials, and the dominant transport of sediment for the time being is seawards. Plunging breakers at high tide attack the berm, carrying part of it away and leaving the rest with an unstable and weakened front, which is readily susceptible to further 'combing down' during the next spell of high tides.

However there is one constructive activity which, although rarely coming into operation, is of major importance over a long period. Violent splashes of spray from the more powerful plungers – generally during spring tides and winter storms – may fling pebbles and boulders to the highest part of the berm or even beyond, so building up a strong coastal defence against all but highly exceptional attacks from the sea. Chesil Beach (Figure 23.38 and Section 23.9) is one of the finest examples of this kind of protective barrier.

Figures 23.16 and 23.38 both show that the shoreward edge of the uprush is conspicuously indented at roughly equal intervals, instead of being smoothly curved, as might have been expected. This pattern reveals the presence of shallow troughs and low ridges on the upper part of the beach. These are rudimentary **beach cusps**, which in appropriate circumstances may become deeply scalloped features with a relief of 3 or 6 m, especially along the seaward slopes of a berm built largely of shingle overlying less permeable materials. Tapering ridges or 'horns' of shingle point seawards and are separated by embayments floored by beach-sediments of finer grain. It has been generally considered that the height and width of the cusps increase with the height of the waves, but this is only true within certain limits. M. S. Longuet-Higgins and D. W. Parkin have made a careful study of the conditions favouring cusp building along parts of the south coast of England, including Chesil Beach. They

Figure 23.38 Chesil Beach, showing the highest part or 'berm' at Chesilton, Dorset, England, which is 13 m above high-water level. (*Aero Pictorial Ltd*)

found that the dimensions of the cusps are more closely related to the length of the swash or uprush than to the wave height. Moreover, particularly high and powerful waves may wash over the horns and disperse the shingle into the depressions, so that the cusps are soon destroyed. Oblique waves have a similar effect. Consequently cusps are formed and maintained only by waves advancing with their crestlines parallel to the shore. Under these conditions each advancing wave divides at the seaward extremities of the cusps, dropping the coarser shingle there, while water carrying the rest of its load floods each embayment from both sides, so that the backwash there is strong and results in deepening the embayment. Within the limiting conditions the horns are built up and the depressions are eroded. The writers mentioned above demonstrated this by conspicuously dyeing a number of pebbles occurring in cusp-depressions; it was found that the dyed pebbles were washed out and that those which returned were deposited on the horns. After the beach cusps have been smoothed out, the problem arises how they develop again. A possible explanation is that, starting with a smooth slope, the backwash acts like a sheet-flood and produces a series of rills or gullies. Of these the bigger ones grow at the expense of the smaller, until as they develop into cusp depressions they all come to have a similar

size. The size changes from time to time in accordance with the length of the swash and the height and direction of the waves.

The contrasted conditions of constructive and destructive waves referred to in previous paragraphs tend to be correlated with quiet and stormy periods respectively, and therefore in a general way with summer and winter. The operation of 'combing down' – thinning or removal of beaches by destructive waves – is sooner or later followed by restoration of the material by constructive waves. In some places the alternation amounts to little more than a seasonal exchange of material between the berm at the top of the beach and the offshore submarine bar or bars. But beach removal by a catastrophe like the North Sea surge of 1953 naturally takes much longer to restore, if indeed the damage should not be beyond hope of recovery. On that occasion, for example, the beaches of Lincolnshire were completely scoured away in one night, and it was not until 1959 that their former condition was restored.

Both thinning and thickening of beaches may be effectively assisted by winds in the appropriate directions. Gales blowing strongly offshore cause a surface drift of water away from the coast. To balance this an undercurrent is directed towards the shore and, though feeble, it co-operates with the waves of construction and so helps to reinforce the rather slow processes of beach growth and restoration. On the other hand, onshore gales not only whip up destructive waves, they also raise the hydraulic head of water along the shore and so strengthen the backwash and the outward currents that necessarily result from the piling-up of water against the coast. It should be noticed that the direction of sedimentary transport on the sea floor that results from onshore or offshore winds is opposite to that of the wind. The growth and maintenance of beaches are therefore favoured by offshore winds.

But if onshore winds blow some of the sand landwards to form sand dunes, the profile is altered so that the backwash is weakened. Constructive waves are then assisted in restoring the appropriate profile by washing in materials from the sea floor. Proof that this happens has already been presented in Section 22.5. Offshore winds then further assist in beach maintenance by returning some of the dune sand. Where this has become part of the fluctuating balance of nature, artificial removal of the dune sand (e.g. for building etc.) can only promote increased erosion along that part of the coast.

Similarly, artificial interference with natural profiles by offshore dredging introduces a factor that will have dangerous results if it causes one-way changes outside the normal limits of thinning and thickening. Look back at Figure 23.28. The wave-built terrace has been deposited from sediment transported seawards. **Erosion** on a highly abnormal scale is then suddenly introduced by dredging. This increases the slope from the shore and therefore promotes beach and cliff erosion, so providing material to make good the loss on the wave-built terrace. The actual changes in the processes at work are mainly (a) refractive concentration of wave energy from the dredged area towards the coast; (b) strengthening of breakers; and (c) strengthening of backwash and seaward currents. In 1897, for example, dredging was begun off the coast north of Start Point (Figure 23.39) to furnish shingle for use in the harbour works then in progress at Plymouth. Thinning of the beach at Hallsands then also began to be a one-way process. By 1902 the beach had been lowered nearly 4 m and cliff erosion had become a steadily increasing menace. The licence to dredge offshore shingle was cancelled. But it was too late. Storm waves were already attacking the little fishing village of Hallsands, built on a post-glacial raised beach, and by 1917 only the walls of a few ruined cottages still remained.

The seaward currents which have been mentioned require some discussion. When breakers and onshore winds pile up water against the coast,

Figure 23.39 Start Point, Devon, England, a promontory of Precambrian rocks. The remaining walls of the ruined cottages of the village of Hallsands can just be seen on the cliff edge in the background, on the left of the photograph, in front of the new white cottages. (*Fox Photos Ltd*)

the raising of sea-level must be counterbalanced by currents flowing away from the shore. These were formerly grouped together as the **undertow**, against the dangers of which bathers were warned, particularly in the deeper water over offshore depressions, where the seaward pull of the 'undertow' was recognized to be a source of peril, even to strong swimmers. It is now known that most of the water that would otherwise accumulate inside the zone of breakers finds an exit through occasional depressions or 'lows' in the offshore bars. These localized outflowing currents are much more than concentrations of undertow, since near the shore not only the bottom water flows outwards, but most of the water from sea floor to surface. Farther out these **rip currents** (Figure 23.40) weaken and die out in depth, but at the surface they may either continue for long distances or merge with other currents – longshore or shoreward – which they may encounter. Seen from the cliffs rip currents sometimes appear as long lanes of foamy or turbid water stretching far out to sea. From the shore they are most easily located by noticing places where the wave crests are lower than usual and the breakers interrupted and less active. A good example of the latter effect on a plunging breaker can be seen in Figure 23.17. Swimmers caught in a rip should not attempt to free themselves by swimming towards the shore against the current. By turning parallel to the shore and swimming at right angles to the rip they should soon escape the danger zone and pass into water where the currents are directed towards the shore.

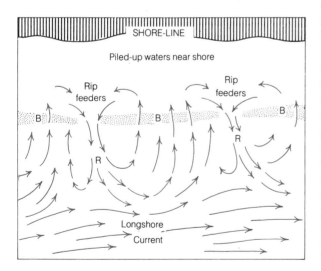

Figure 23.40 Schematic indication of the relation of rip currents, R, to depressions in the submarine offshore bars, B. The currents here shown are purely diagrammatic, since at best they represent only a momentary stage in continuously changing systems. (*Simplified from diagrams by F.P. Shepard*)

It will have been noticed that rip currents are the only ones that are easily capable of transporting sediment seawards. They are far more effective than the average ebb-tide currents and the largely discredited undertow. At best the latter may extend the backwash movement a little way in special circumstances. In the other direction there is a much stronger tendency to return sediment landwards. Once ordinary waves 'feel bottom' the sea-floor sediment is stirred to and fro, and as the 'to' is greater than the 'fro' there is a slight shoreward movement on balance, which increases as the zone of breakers is approached, and increases again as the breakers turn into waves of translation. The landward movement is reinforced by inflowing tides, locally on a powerful scale.

Thus, on the whole, any migration of material seawards at the expense of the wasting land is extremely slow and is confined to the finer grades of sediment. So far as the British Isles are concerned, gain of land – mainly mud-flats, shingle banks, sand dunes and salt marshes – slightly exceeds the loss. But the latter gets more publicity, being destructive of land and property, and sometimes of life.

However, all along the shore freshly broken fragments are being steadily abraded and reduced in size to boulders, pebbles and grains. Some of the fine products of attrition may be entrapped for a time in the interstices of the coarser materials, but much is carried seawards in suspension by the rip currents. This is eventually deposited on the sea floor beyond the range of normal wave and current action. Thereafter, only a lowering of sea-level or a rise of the sea floor, or a sudden incorporation into a turbidity current (Section 24.4) is likely to disturb its rest.

Given a stable sea-level, one would expect the sediments on the sea floor to become finer and finer as the depth increases, the full outward sequence being shingle, sand, silt and mud. Corresponding to this sequence, in strata of the same geological age traced laterally, we often find conglomerates against old shore-lines, passing outwards through sandstones into shale or mudstones. But in Recent geological time the sea-level has not been stable; sea-level 20 000 years ago was over 100 m lower than today's (Figure 21.21). Nearly every part of the continental shelf has in turn been the coastline of the past and berms and ridges of shingle, and the coarser sediments of glacier-fed rivers, have been left at various depths, to puzzle the oceanographer of former years who often found them far from the coast, where they were least expected. Obviously the present-day distribution of sediment on the continental shelves is highly abnormal, and must have been so throughout much of Pleistocene time. These 'abnormalities' however are consistent with what we have learnt already of the vicissitudes of Pleistocene history.

23.8 BEACHES: TRANSPORT ALONG THE SHORE

Longshore drift of sediment is brought about in two ways: by beach drifting, due mainly to oblique waves on the foreshore, and farther out by transport due to longshore currents. When waves are driven obliquely against the coast by strong winds, debris is carried up the beach in a forward sweeping curve. The backwash may have a slight forward movement at the start, owing to the swing of the water as it turns, but otherwise it tends to drag the material down the steepest slope, until it is caught by the next wave, which repeats the process (Figure 23.41). By the continual repetition of this zigzag process, sand and shingle are drifted along the shore.

The direction of drifting may vary from time to time, but along many shores there is a cumulative movement in one direction, controlled by the dominant or most effective winds. A subsidiary factor which aids or hinders beach drifting is the direction of the advancing flood tide. Farther out from the shore oblique winds and waves generate intermittent and fluctuating currents, since both have a component parallel to the general direction of the coast. These are the longshore currents, and again they are strengthened or weakened by the set of the tides. Longshore drift is by no means confined to the immediate shoreline.

In the English Channel the dominant winds and the inflowing tides both come from the southwest, and the cumulative direction of drift is up-Channel and through the Straits of Dover almost to the estuary of the Thames. Down the east coast of Britain the drift is mainly southwards, the dominant winds being from the northeast and the flood tide advancing from the north. There are a few exceptions: for example, the north coast of Norfolk, west of Cromer, stands athwart the main drift which, striking the coast obliquely, is deflected westward towards the Wash.

Wherever it is thought desirable to check the drift of sand and shingle, barriers known as **groynes** are erected across the beach. On the side from which the beach drift comes, sediment accumulates, and sometimes reaches a height a metre or so above the beach level on the other side, where sediment is being washed away, to be retained in turn by the next groyne (Figures 23.42 and 23.43). Groynes, like other structures designed to interrupt the natural flow of sediment, such as moles and piers constructed to protect harbours and river mouths, interfere with the replenishment of beaches farther along the coast by robbing them of their former sources of supply. Drifting does not stop along these starved beaches until they disappear. The coast behind and beyond them, which they had been protecting from erosion, is thereupon exposed to increasing attack by the waves. To the east of Brighton and Newhaven, England, for example, the wastage of the Chalk cliffs (Figure 23.60), has greatly increased since the erection of groynes.

In The Netherlands these problems are of vital importance. A supply of sand at the appropriate rate is found to give better coastal protection than groynes and sea-walls. To take a simple case: if a river mouth lies across the drift direction it was formerly protected by groynes on the up-drift side. This however weakened the coastal defences on the other side of the river mouth. Now, using suction pumps, sand on the up-drift side is delivered into pipelines through which it is conveyed beneath the river to the down-drift side. There, having left the river mouth clear, it provides the coast beyond with its natural income of sediment to balance the losses by beach drifting.

23.9 SAND AND SHINGLE SPITS AND BARS

Where shore drift is in progress along an indented coast, spits and bars are constructed as well as beaches. Where the coast turns in at the entrance to a bay or estuary the sediment transported by beach drift and longshore currents is carried more or less straight on, and some of the coarser material is dropped into the deeper water beyond. The shoal thus started is gradually raised into an

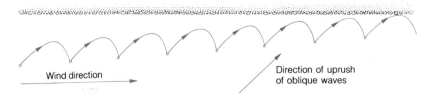

Wind direction

Direction of uprush
of oblique waves

Figure 23.41 Diagram to illustrate beach drifting, showing the path followed along a sloping beach by a pebble or sand grain under the influence of the uprush and backwash of successive oblique waves during an advancing tide.

Figure 23.42 Beach drift impeded by groynes at Eastbourne, NE of Beachy Head, Sussex, England. The direction of drifting is to the north-east i.e. up-Channel. (*Aerofilms Ltd*)

Figure 23.43 Beach drift impeded by groynes, St Margaret's Bay, north-east of Dover, England, showing that the direction of drifting continues up-Channel through and beyond the Straits of Dover. (*Aerofilms Ltd*)

embankment. This grows in height by additions from its landward attachment until a ridge or mound of sand or shingle is built above sea-level in continuity with the shore from which the additions are contributed. The ridge increases in length by successive additions to its end, like a railway embankment, until waves or currents from some other quarter limit its forward growth.

If the ridge terminates in open water it is called a **spit**. Storm waves roll and throw material over to the sheltered side, especially when they approach squarely. Spits thus tend to migrate landwards, often becoming curved in the process. Curvature is also brought about by the tendency of oblique waves to swing round the end (i.e. to be refracted) in places where the sea floor beyond slopes rapidly into deeper water. A spit may thus be developed into a **hook**, as indicated in Figure 23.44. Cross currents may assist or modify hook formation, and it is usual for spits to lengthen by the addition of a number of successive hooks. Hooked spits of simpler structure occur in large lakes where tides are absent and currents are negligible (Figure 23.45). This suggests that the dominant winds and waves are the agents essentially responsible for the curving of spits.

A good example of a curved spit is Spurn Head (Figure 23.46), which extends into the Humber in streamlined continuity with the Holderness coast. The latter is almost everywhere fringed with sand and shingle that drifts steadily from north to south,

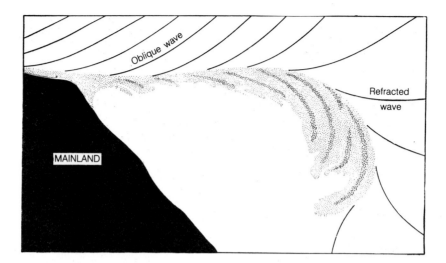

Figure 23.44 Diagram to illustrate the development of a hooked spit by the refraction of oblique waves.

Figure 23.45 Hooked spit, Duck Point, Grand Traverse Bay, Lake Michigan. (*I.C. Russell, U.S. Geological Survey*)

fresh supplies being continually furnished by the rapid erosion of the coast. Nearly all (97 per cent) of the transported material is carried beyond Spurn Head, cumbering the estuary with shoals on its way towards the Lincolnshire coast, where most of it is added to the seaward-growing coastal flats.

Southward drift is also very active along the east coast of Norfolk and Suffolk. Ten centuries ago the Yarmouth sands had already spread across the estuary of the Yare, forming an obstruction which deflected the river towards the south (Figure 23.47). The spit then continued to grow southwards, hugging the coast as closely as possible, with the river confined between itself and the mainland. By 1347 the end of the spit and the outlet of the river had reached Lowestoft. Since 1560, however, an artificial outlet has been maintained at Gorleston, where the spit now terminates. The truncated part has long ago drifted south. At Aldeburgh, half-way between Lowestoft and Harwich, the longest spit on the east coast has similarly diverted the outlet of the Alde (Figures 23.48 and 23.49).

A **bar** or **barrier beach** extends from one headland to another, or nearly so. When the bay inside is completely enclosed it becomes a marsh, or, if it receives streams from the mainland, a shoreline lake. More usually, however, outflowing drainage escapes through a deep and narrow channel kept open by vigorous tidal scour. Between Gdansk and Klaipéda, on the south-east Baltic coast, there are two very long bay-mouth bars, surmounted by sand dunes, with extensive tidal lagoons on the landward side (Figure 23.50).

A bar connecting an island to the mainland or to another island is called a **tombolo**. The name comes from an Italian example of a double tombolo, 130 km north-west of Rome, where a high rocky island is tied to the adjoining coast by a pair of shingle banks with a broad lagoon between.

By far the most impressive barrier beach of shingle in Britain is Chesil Beach (Figure 23.51). For 10 km south-east of Bridport the beach fringes the shore. Near Abbotsbury the shore recedes and for the next 13 km the beach continues in front of the tidal lagoon of the Fleet as a bar well over 6 m in height (Figure 23.52). Finally, becoming a tombolo, it crosses 3 km of sea to the Isle of Portland, which is thus tied to the mainland (Figure 23.38). Chesil Beach is a composite structure, its shingle having accumulated from local sources as well as by drift from each end. At the north-west end

Figure 23.46 Spurn Head, built by beach drifting into the estuary of the Humber, in continuation of the Holderness coast, Humberside, England; looking north-north-east (cf. Fig. 23.32). (*J.K. St Joseph*)

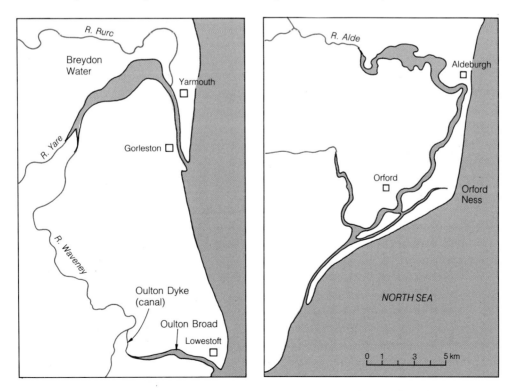

Figure 23.47 River Yare, Norfolk, England.   **Figure 23.48** River Alde, Suffolk, England.
Examples of river deflection in East Anglia by the southerly extension of sand and shingle spits. Both drawn to same scale.

Figure 23.49 Deflection of the River Alde south of Aldeburgh by the southerly growth of Orford spit, Suffolk. (*Aerofilms Limited*)

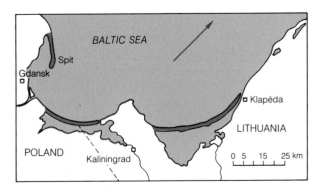

Figure 23.50 A spit north of Gdansk and two smoothly curved bay-mouth bars enclosing broad lagoons along the south-east Baltic coast of Poland and Lithuania.

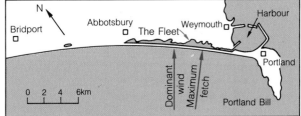

Figure 23.51 Map of Chesil Beach, Dorset.

there are pebbles or rocks from Cornwall and Devon; at the south-east end the shingle includes larger pebbles, supplied from the Portland promontory. Between the extremities the pebbles are mainly relics from an eroded land, now submerged, that formerly stretched in front of the Fleet when the sea-level was much lower. Although towards the Portland end the beach rises to the quite exceptional height of 13 m, the sea sometimes bursts over it during storms and pours through breaches into the low-lying area beyond. Since two villages were demolished in 1824, the worst disaster of this kind occurred late in 1942, when the railway between Portland and

Weymouth was partly washed away and the lower parts of Portland itself were seriously flooded. The beach is in fact very slowly migrating towards the Fleet, as a cumulative result of its alignment almost exactly at right angles both to the direction of the average dominant winds and to that of the maximum fetch (Figure 23.51). This alignment is one of great stability, since the beach ridges squarely face the more powerful storm waves and longshore drift is reduced to a minimum.

Because of the tendency of spits tied to the mainland to continue in the direction of the latter, or to curve inwards towards the coast if the water is shallowing in that direction, as it usually is, it is exceptional for wave-built structures to swing sharply away from the general trend of the coastline. When they do so the change of direction may be due to shallowing of water seawards, e.g. towards an island or shoal. But in some examples

Figure 23.52 Chesil Beach, Dorset, viewed from the West Cliff, Portland (see also Fig. 23.38). (*British Geological Survey*)

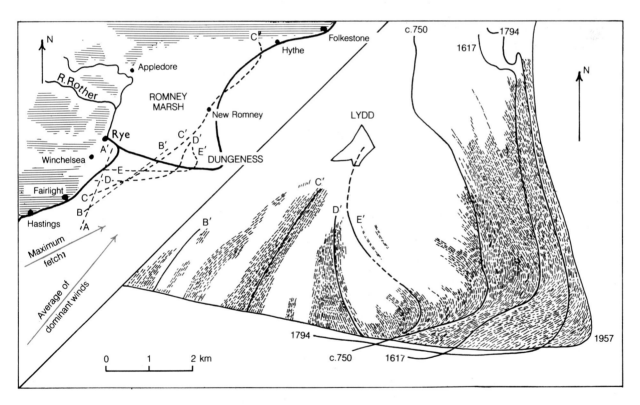

Figure 23.53 Map of Dungeness showing the pattern and sequence of the shingle ridges up to 1957. The outline of the shingle revealed by recent Ordnance Survey maps suggests that Dungeness is still extending eastward. The scale is about 6 times that of the regional map on the left. The latter shows Romney Marsh and (horizontal shading) the higher ground bordering it on the north and west. It also illustrates the Lewis (1932) hypothesis of the development of Dungeness as described in the text. (*After W.V. Lewis and W.G.V. Balchin*)

the change appears to be at least in part a reaction to the average direction from which the ridge-building storm waves approach. This in turn depends on the directions of the dominant winds and of the maximum fetch. But the particular mechanism involved often remains a puzzle, as in the case of Dungeness in Kent, England.

Dungeness is a triangular outgrowth of shingle ridges forming the seaward margin of Romney Marsh, which, now silted up and reclaimed, was once a broad bay between the end of the cliffs near Fairlight (east of Hastings) and Hythe (Figure

23.53). During the Danish invasion in the year AD 893 a fleet of 250 vessels sailed past Lydd to Appledore. An old map of about AD 750 shows that Lydd was then at least 1.6 km to the west of the sea, suggesting that the invaders took advantage of a gap in the shingle barrier, possibly the outlet of the river Rother, which at that time probably reached the sea near the New Romney of today. In 750 Dungeness was already clearly outlined, as indicated on Figure 23.53, and can be recognized as having had, then as now, the outward form of the peculiar type of construction known as a **cuspate foreland**.

Most cuspate forelands (e.g. Cape Hatteras, Figure 23.54) are the points of accumulation where two curved spits happen to meet. In accordance with this view Dungeness was originally supposed to represent the meeting-place of up-Channel beach drift with that coming down the east coast and through the Straits of Dover. But this idea is easily shown to be untenable. The two seaward sides of the triangle behave quite differently: while the southern side has been retreating

before the attack of the waves, the eastern side has been conspicuously built out. The foreland itself has advanced nearly 2.5 km during the last twelve centuries. Moreover, the piling of shingle against the groynes seen in Figure 23.43 shows that up-Channel drifting remains dominant beyond the Straits of Dover and continues along the east coast of Kent to accumulate between Deal and the Isle of Thanet – an 'isle' which in consequence has ceased to justify its ancient name.

The oldest storm ridges now to be seen are those on the west. W. V. Lewis supposed the first stage to be represented by a spit or bar that extended from the end of the cliffs as they were at that time, e.g. from A somewhere south of Fairlight to A' near Rye (see small-scale map of Figure 23.53). At later stages this direction AA' began to swing round, through BB', CC' and so on, towards the present-day direction, which is much more nearly at right angles to the dominant direction of advancing storm waves, and so is approaching stability. The oldest spit probably dates from Neolithic times, when the Flandrian rise of sea-level reached its maximum. Later there was slight emergence, and while cliffs were being eroded back the lengthening spit became a bar that may have extended from CC' through the shingle ridges near New Romney to those near Hythe marked C'', with a tidal outlet somewhere for the outflow of the Rother and other inland streams. But since no ridges are now exposed between Lydd and New Romney, it is equally possible that D' or E' were connected with C''. One of these stages, most probably near that marked E', may have been reached about the time when Julius Caesar landed near Hythe. From then a series of successive stages can be dated from old maps, leading up to this survey of 1957.

Unfortunately the pre-Roman history outlined above is purely hypothetical. The numerous explanations given for both earlier and later developments fail to carry conviction, particularly as there are innumerable bars that have not developed into structures like Dungeness. Guilcher indeed could only conclude that Dungeness is one of the 'true wonders of nature'. And so it will remain until the missing factors concerned in its evolution have been discovered.

23.10 OFFSHORE BARS AND BARRIER ISLANDS

The Atlantic and Gulf coasts of the United States are bordered by long stretches of barrier beaches which are separated from the mainland by lagoons or expanses of sea, except where they are locally tied to headlands (Figure 23.54). These are known as **offshore bars** or, if they should be discontinuous at

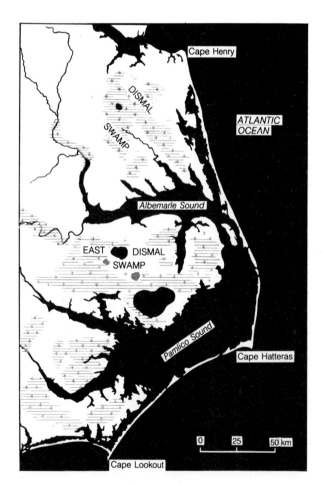

Figure 23.54 Offshore bars and barrier islands, with lagoons, sounds and swamps behind, along the coast of North Carolina, USA.

Figure 23.55 Miami, Florida, USA, with its offshore bar and narrow lagoon. (*Fairchild Aerial Surveys Inc.*)

both ends, as **barrier islands**. Their mode of origin was not easy to understand while they were thought to be diagnostic features of emergent shorelines. A nearly flat coastal plain passing into a wide offshore zone of shallow water was first envisaged. In accordance with the principle illustrated in Figure 23.35, loose sediment from the sea floor would then be driven landwards in order to restore an appropriate profile of equilibrium. To bring this about the waves would begin to drag the bottom several kilometres from the low-lying mainland, and as they would lose much energy in crossing the shallows, the coarser parts of the stirred-up sediment would be dropped before the shore was reached. The foundations of a mound or shoal could thus be laid offshore, but then the difficulty remains to explain how it could be built up into a bar with its crest above sea-level. As soon as any such submarine feature reaches a certain height, it is swept by breakers and material is transferred from the seaward to the landward side. The submarine bar therefore advances towards the shore. But it cannot become other than a submarine bar until it finally joins the shore or a headland, by which time it is a beach or a bay barrier, and is no longer an 'offshore' feature.

Figure 23.54 shows that the coast depicted is clearly one of submergence – not emergence. A borehole drilled at Cape Hatteras passed through about 3000 m of marine sediments (from Recent to Lower Cretaceous) before penetrating a pre-Cretaceous land surface of Precambrian rocks. This offshore belt has been subsiding almost con-

tinuously for well over 100 million years. But keeping in mind the Pleistocene changes of sea-level the difficulties can be overcome. 20 000 years ago the shoreline lay far out on the continental shelf, where it was probably margined by beaches with storm ridges already well above sea-level. As the latter rose these ancestral beaches continued to advance, maintaining contact with the coast where the hinterland was high enough, but losing contact wherever the surface on the landward side happened to be below the sea-level of the time – as in many places it is today (Figure 23.54).

The famous Florida beaches of Daytona, Palm Beach and Miami are offshore bars which are now nowhere far from the land (Figure 23.55) and in many places make contact with it. One of a long series of offshore bars and coastal barriers along the south-east coast of Iceland is shown in Figure 23.56. This coast is one of some complexity because it is being rapidly built out by glacifluvial outwash from the glaciers of Vatnajökull (see Figures 3.5 and 12.9). The offshore bars represent the return from the sea floor of some of the waste delivered from the land when sea-level was lower.

23.11 CLASSIFICATION OF COASTS

Valentin (1951) classified coasts, as **retreating** and **advancing** coasts; this has the practical advantage that it is primarily based on the criterion of whether land is being gained or lost. Loss can result from either submergence or erosion, and

Figure 23.56 Offshore barrier beach, with lagoons and outwash plains from Vatnajökull, behind Lón Bay, south-east Iceland. (*L. Hawkes*)

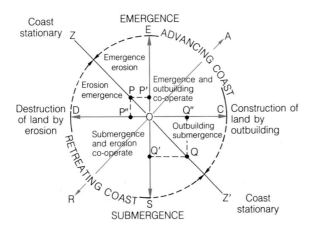

Figure 23.57 Graphical representation of H. Valentin's classification of coasts.

coastal retreat tends to reach its maximum when these two act in combination. Similarly gain of land can result from either emergence or outbuilding by constructive processes, such as deposition, and coastal advance tends to reach its maximum when these two co-operate. But there are many coasts in which the operative processes are in opposition; it is then the dominant one that determines whether on balance, i.e. over a suitably long period of years, the coast advances or retreats.

Valentin expresses his classification graphically by a diagram of which Figure 23.57 is a modification. The vertical axis through O represents emergence in the positive (upward) direction; submergence in the negative (downward) direction. The rate of coastal advance or retreat is proportional to distance from O. Similarly the horizontal axis represents outbuilding in the positive direction (to the right); erosion in the nega-

tive direction (to the left). Advance and retreat reach their maxima at points such as A and R on the diagonal AOR. The diagonal at right angles, ZOZ', represents a 'zero' line, along which points like P have equal components of gain by emergence, P', and loss by erosion, P''; or along which points like Q have equal components of loss by submergence, Q', and gain by outbuilding, Q''. ZOZ' represents balanced conditions in which the coast remains stationary on average over the period concerned.

The importance of taking time into consideration is illustrated in a remarkable way by the intermittent outward growth of the coastal plain near Cayenne in French Guiana. In the tropics mangroves spread vigorously over the tidal zone of low muddy shores – particularly of estuaries – forming an impenetrable thicket of interwoven roots which provide shelter for fearsome hordes of organisms. Practically all the sediment washed in by the rising tide is trapped and the mangrove swamp therefore continues to extend itself offshore across what had previously been shallow sea. In French Guiana this outward growth has added 10 to 16 km to parts of the coastal plain since the first surveys were made in 1751. But progress has not been uninterrupted. Over the course of the years a mysterious and still unexplained periodicity has been recognized. The north equatorial current flows from the south-east almost parallel to the coast, bringing from the mouth of the Amazon the silt and mud that is trapped by the mangroves. The curious feature here is that outward growth proceeds for eleven years, most rapidly during the middle of the period; but then the outermost mangroves begin to wither away, leaving the unprotected mud to be

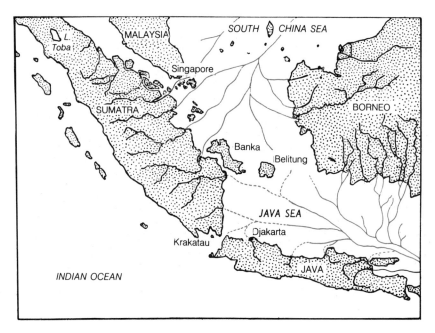

Figure 23.58 Map showing the drowned river system of south-east Asia, still identifiable across the shallower parts of the Java and South China Seas. (*After J.H.F. Umbgrove*)

removed by the backwash from the waves and carried farther along the coast. This phase also waxes and wanes over a period of eleven years. The periodicity has rarely varied by as much as a year. There was a phase of silting up and coastal advance from 1947 to 1958, but by 1962 the sea was gaining at the expense of the fringes of the mangrove swamps. Possibly the alternation may have something to do with the cyclic changes in the waters of the equatorial current. These might be variations of salinity, or of the abundance of trace elements acting like fertilizers; or periodic 'bloomings' of minor organisms injurious to the mangroves. But why the periodicity? So much remains to be discovered that it would be futile to guess.

23.12 RETREATING COASTS

The widespread occurrence of coasts of submergence at the present time has already been emphasized with numerous illustrations (e.g. Figures 23.1, 23.4, 23.58, 23.62). On a regional scale one of the most remarkable examples of a submerged land surface, with drowned valleys that can still be recognized on the sea floor, is that which formerly connected the Malay Peninsula with the islands of Sumatra, Java and Borneo, and innumerable smaller ones. The betrunked and dismembered modern valleys can be traced down to depths of about 100 m in the South China and Java Seas, and as shown in Figure 23.58 they join up into a small number of river systems. Such land connections

must have existed several times during the Pleistocene.

Another important land bridge connected Asia and North America during phases of lowered sea-level. Siberia and Alaska were joined across the Bering Straits when the sea was 46 metres or more below its present level: that is, up to about 10 000 years ago on the latest occasion. It has generally been supposed that man, as well as a great variety of plants and animals, first migrated into North America by this route. Now that it is known that the deepest parts of the straits and the Chukchi Sea to the north are no more than 55 m beneath the present surface, there is no doubt that the land bridge – of tundra type, but not glaciated – was well over 1600 km wide for long intervals of Pleistocene time.

The first effect of marine erosion on a newly formed coast of submergence is usually to intensify the initial irregularities of outline (see Figure 23.3). Where the rocks vary in resistance the waves pick out all the differences. Soft and fissured rocks are worn back into coves and bays, while the harder and more massive rocks stand out conspicuously. The Dorset coast north-east of Portland shows this process in active operation. Here there is a long coastal strip of soft Lower Cretaceous beds, backed on the landward side by an upland of Chalk, and formerly protected from the sea by a continuous rampart of hard upfolded Jurassic limestones. The sea has breached the latter in places and excavated the softer rocks behind, until slowed down by the more resistant Chalk. The Stair Hole (Figure 1.4) illustrates the breaching

Figure 23.59 Lulworth Cove, Dorset, England, a beautifully curved bay scooped out by the sea after the breaching of the resistant barrier of Portland limestones still forming the cliffs on both sides of the entrance to the Cove. (*Aerofilms Ltd*)

stage. Lulworth Cove (Figure 23.59) is a beautiful example of a bay scooped out by waves and rip currents, and backed by Chalk.

Eventually, however, erosion and deposition co-operate to smooth out the intricate outlines of a youthful shoreline. Bay-head beaches and deltas, fed by lateral currents from the headlands and by additions from streams, locally extend the shore in the seaward direction. Meanwhile the headlands recede before the concentrated attack of the waves, and stretches of cliff become longer and straighter (Figure 23.60). Spits and bars bridge across the inlets where oblique waves favour beach drifting, and these structures generally advance landwards to keep in line with the retreating cliffs. Embayments protected in this way are rapidly shoaled up by contributions from the land, aided by wind-blown sand and the growth of salt-marsh vegetation. Finally the bars and sand dunes encroach on, and coalesce with, the lagoon and marsh deposits, and a coast of smoothly flowing outline is evolved. Theoretically it might be supposed that if this cycle of marine erosion were to continue without interruption, the shoreline would slowly retreat as a whole. Lengthening lines of cliffs would join up

at the expense of the deposits representing the embayments that originally separated the headlands. But as we have seen, the latest rise of sea-level has been so recent that in most places only the earlier stages of the cycle have been completed. The later stage of slow retreat outlined above has been reached only locally and it is doubtful if it could continue far, given a stable sea-level. Retreat of cliffs implies a complementary broadening of the shore platform in front. This in turn reduces the energy of the waves that reach the cliffs until it becomes negligible. Coastal retreat by marine erosion is probably a limited process in the absence of continued submergence.

23.13 ADVANCING COASTS

Typical coasts of emergence are not common at present. Finland for example is steadily rising isostatically, and the south and south-west coasts are fringed with tens of thousands of islands as a result of the emergence of the higher parts of a hummocky ice-moulded surface (Figure 23.2). But although this archipelago owes its existence to

Figure 23.60 The Seven Sisters, south-south-east of Newhaven, Sussex; looking towards Beachy Head from the shore platform and shingle beach of Cuckmere Haven. Coastal erosion has produced a continuous line of cliffs with the valleys of former tributary streams left hanging at different heights according to their original gradients. (*British Geological Survey*)

Figure 23.61 The Motunau coast of north Canterbury, New Zealand. The coastal plain has an almost undissected cover of recent shelly deposits, indicating emergence by relatively rapid upheaval. Indented cliffs of older rocks are seen at the bottom right. (*V.C. Browne*)

Figure 23.62 A typical coast of recent submergence, Queen Charlotte Sound, Marlborough, New Zealand. (*R.H. Clark*)

emergence it is merely part of a drowned land surface that has become progressively less drowned than it was a few thousand years ago. Nevertheless the coasts are advancing because the emerging crystalline rocks are massive, and have been streamlined in form by previous glaciation, so that they resist erosion and are not easily cliffed. Similarly fjords, though fundamentally submergence features due to deep glacial erosion and invasion of the sea, are mostly along coasts that are still actively rising. The structurally weaker belts of rock having been already gouged out, marine erosion is extremely slow, and on the whole land is being gained.

A really typical shoreline of emergence is one in which the sea floor with its veneer of sediments has been uplifted to form a nearly flat coastal plain with a smoothly flowing shoreline margined by widespread stretches of shallow water. New Zealand furnishes ideal examples of uplifted sea floors (Figure 23.61), which are now fertile coastal plains that have lost little by cliffing, as well as others which illustrate submergence (Figure 23.62). Similar emergent plains occur along the south coast of Honshu (Japan), although to the west submergence has reduced a former landscape of hills to an offshore archipelago. At the present time, however, this and much of the Honshu coast is rising at rates ranging between 13 and 20 cm per century. Corresponding figures for Taiwan (Formosa) are 18 cm, and for Hong Kong 15 cm. These rates are slow compared with the post-glacial rise of sea-level, which averaged about 80 to 85 cm per century up to 6000 years ago. It follows that where

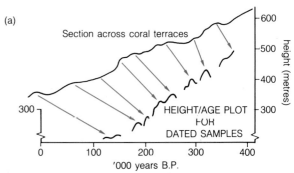

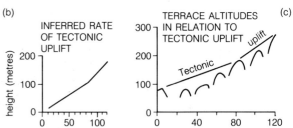

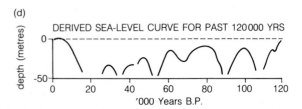

Figure 23.63 The separation of eustatic sea-level changes from tectonic uplift for raised coral terraces, Huon Peninsula, New Guinea. (*After Chappell*)

the sea floor has become a coastal plain by uplift

the rise must have been both considerable and relatively rapid.

Figure 15.3 shows an uplifted sea floor on the south-east coast of Iceland, backed with screes which for the time being are protecting the cliffs against renewed attack by the sea. But for the uplift the scree materials would have been abraded and removed soon after their fall, while at the same time being used by the sea as ammunition in undermining the cliffs.

One of the best documented examples of a coastline of rapid uplift is Chappell's study of the Huon Peninsula, New Guinea (Figure 23.63). Here a most remarkable staircase of abandoned coral terraces has been created by uplift of the peninsula by 180 m over the last 120 000 years, an astonishing average rate of 1500 B. Dating of the terrace fragments by ^{230}Th/^{234}U techniques allows the tectonic component of terrace formation to be isolated from the eustatic component and a global sea-level curve for stable coasts over the last 120 000 years is derived.

Examples of coastal advance by sedimentary outbuilding include the accumulation of beach-drifted materials (e.g. along the shore of Lincolnshire); the outward growth of storm ridges into forelands (e.g. Dungeness); offshore deposition of glacifluvial outwash (e.g. Iceland); and retention of washed-in mud by the intergrown roots of mangrove swamps (e.g. French Guiana and Surinam coast of South America). Deltas also contribute here and many of them exemplify the case where new land continues to be built out despite the opposing submergence due to the isostatic depression that results from the growing weight of sediment.

Outbuilding of coasts by lavas and other volcanic products that flow or fall into the sea is not uncommon when eruptions occur on the seaward side of volcanoes situated near the shore (Figure 23.64). Advancing volcanic coasts are naturally most characteristic of growing island volcanoes that have succeeded in establishing their craters above sea-level. The eruption on Tristan da Cunha in 1961, which led to the evacuation of the settlement, culminated in the seaward flow of lava from the central cone. Before the end of the year the lava field had extended the coastline by about 365 m,

Figure 23.64 Coastal advance by formation of a 'delta' of lava; Barcena (also known as El Boquerón) at the southern end of San Benedicto Island, Mexico. This new volcano originated on 1 August 1952, as a cinder cone which reached a height of 300 m in 12 days. In November lava flowed into the crater. In December a fissure opened at the base, which was already cliffed, and lava flowed out into the sea until February 1953. Since then only fumarole activity has been reported. (*Paul Popper Ltd*)

with a nearly vertical front in 15 m of water. Older volcanoes, like those of Hawaii, present many examples of lobe-shaped extensions of the coast where voluminous outpourings of lava have advanced over the sea floor.

On the other hand many more volcanic coasts are of the retreating type: (a) when the sea occupies a newly formed caldera; and (b) when an island volcano becomes extinct and no longer makes good the land it loses by marine erosion and isostatic subsidence. This item is referred to here, although it belongs to the class of retreating coasts, because it serves to introduce atolls and coral-reef coasts which, in tropical seas, are important members of the advancing class of coasts. The coral reefs of today have grown upwards and outwards during conditions of submergence. Since coral communities can grow upwards at a rate of about 300 cm per century, they can easily keep pace with all normal rates of either subsidence or rise of sea-level. Many atolls are built on foundations provided by ancient volcanoes that have subsided to considerable depths while the

corals and their associates were maintaining a living front at sea-level. Coral reefs are in fact involved in so great a variety of the problems of oceanic geology that – notwithstanding the fascination of their coastal scenery – they can be more appropriately dealt with in the chapter that follows.

23.14 FURTHER READING

Bird, E. C. F. (1984) *Coasts*. Blackwell, Oxford.

Davies, J. L. (1980) *Geographical Variation in Coastal Development*. Longman, London.

King, C. A. M. (1972) *Beaches and Coasts*. Arnold, London.

Pethick, J. (1984) *An Introduction to Coastal Geomorphology*. Arnold, London.

Shephard, F. P. (1972) *Submarine Geology*. Harper and Row, New York.

Steers, J. A. (1969) *The Coastline of England and Wales*. Cambridge University Press, Cambridge.

MARINE SEDIMENTS AND THE OCEAN FLOOR

24

Learn the secret of the sea? Only those who brave into dangers comprehend its mystery

H. W. *Longfellow* 1807–88

24.1 LIFE AS A ROCK BUILDER

In earlier chapters many aspects of the geological work accomplished by living organisms have been reviewed. These include the growth and protection of soils, the fixation of sand dunes; and the differentiation of water and carbon dioxide into the fuels we burn and the oxygen we breathe. And besides the vast accumulation of peat and coal there are immensely greater deposits which are largely composed of shells or other protective and supporting structures of once living organisms. Most of these hard parts consist essentially of calcium carbonate secreted from the sea by (a) animals such as the tiny single celled foraminifera (Figure 24.1) and (b) by plants of which coccolithophorids are the chief (Figure 24.2). These are made of the mineral **calcite** which allows them to exist in sediments situated several kilometres below the sea's surface. Although calcite has the propensity to dissolve in abyssal waters it is not nearly as soluble as another major form of calcium carbonate, **aragonite** from which pteropods (small molluscs often called sea butterflies) secrete their shell. As such, pteropod shell contents in ocean sediments tend to be confined to the less deep areas and are rarely found below 2 km of water depth. After death, these hard parts, if not dispersed through a superabundance of silt or mud,

or through dissolution in carbonate-rich waters, accumulate as shell deposits, which are commonly called deep-sea oozes, and the like. Other organisms such as corals, algae and molluscs constitute calcareous material to shallow water marine deposits and are often found as reef-like structures. The biochemical precipitation of calcium carbonate as aragonite by the product of bacterial activity, can also contribute significantly to the calcium carbonate content of shallow marine deposits, especially in subtropical and tropical latitudes.

Silica is extracted from seawater by minute single-celled plants (diatoms, see Figure 24.3) and animals (radiolarians, see Figure 24.4), both of which encase themselves within microscopic shells of opal. Siliceous deposits, in which one or the other of these predominate, constituted two important groups of deep-sea oozes. Today the siliceous remains on the ocean floor are dominated by diatoms, but during the Mesozoic radiolarians were dominant organisms producing siliceous oozes. Their occurrence in sediments in many ways highlights the overlying intensity of biological productivity in seawater, and as such their contents have often been used in sediment sequences to investigate the palaeoproductivity of ancient seas.

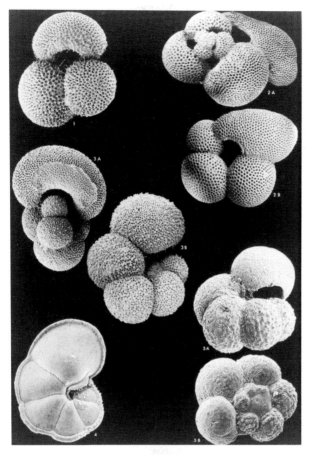

Figure 24.1 Deep-water shells of modern species of planktonic foraminifera. These shells, composed of calcite, can dominate many of the oceanic calcareous oozes. Magnification about ×80. (*From A.R. Fortuin, 1981*)

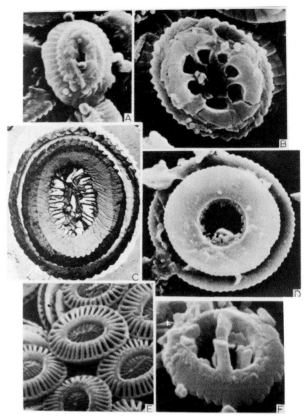

Figure 24.2 Examples of modern calcareous nannofossils (coccolithophorids). Scanning electron microscope of coccospheres made up of a number of shields or coccoliths. (*D. Kroon*)

Siliceous deposits formed by freshwater diatoms, which abound in the lakes of recently glaciated districts, are distinguished as **diatomaceous earths**. Into many of these lakes there is a considerable input of dissolved iron derived from soil reactions within the catchment area. In those lakes where there is a plentiful supply of dissolved oxygen within the water, the precipitation of iron often occurs in the form of limonite (or goethite) brought about by iron-fixing micro-organisms; varieties of bacteria and algae that coat their cells with ferruginous filaments. Since the iron-fixing and the silica-fixing organisms tend to flourish at different seasons and have phenomenal outbursts of fertility in certain years, it comes about that some of the deposits are made up of alternations of finely-banded iron ore and silica.

Similar alternations have been found on the sea floor. Some authorities have suggested that the observed biochemical deposition of limonite and silica may be the clue towards understanding some of the most enigmatic of the Precambrian rocks. In every continent these ancient rocks,

usually metamorphosed and crystalline, include what are variously described as haematite-quartzites, banded iron formations and banded jaspers. All of these have bands of iron ore alternating with bands of quartzite or jasper, the latter being a compact quartz-rock flecked with ferruginous inclusions which gives it a variety of bright colours, frequently brick-red or yellow and occasionally green. In some occurrences haematite has been changed to magnetite, and bands of black (magnetite) and white (quartz), or of black and red (jasper), or of all three colours – or more – may be seen.

Strongly supporting this idea is the discovery within the banded iron ores of Ontario and Minnesota of structures that can be identified as the fossilized spores and filaments of premature fungi and algae. These rocks have been dated at about 1900 Ma. Even more remote ancestors of reef-building algae have been detected, for instance in Zimbabwe. It is now well established that algae have flourished on the Earth from at least 3000 Ma.

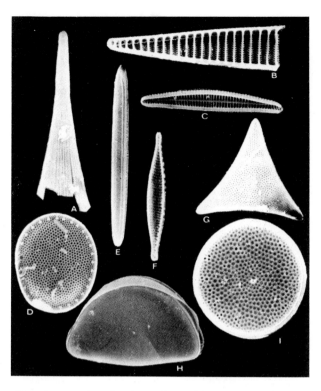

Figure 24.3 A selection of modern marine diatoms showing a range of morphologies. (*After L.H. Burkle, 1978*)

During the 1970s it was discovered that similar, but not identical, deposits are today forming on the ocean floor on and near the axes of active spreading ridges, particularly in the Pacific and Atlantic oceans. Here the discharge and precipitation of metal-laden waters, emanating upwards, from the reaction of seawater and hot basalt, form extensive deposits of iron oxides, but in addition oxides of manganese as well as unusual enrichments of barium, zinc, copper and other elements. Today it is known that virtually all active spreading ridges can discharge these hot waters (350°C) from the underlying hydrothermal activity, and this mechanism is a major source of metals to the oceans. Often the released iron is accompanied by silicon and these two elements often react in seawater to form the clay mineral, **nontronite**. Almost invariably within the region of these discharges occur exotic colonies of animals; their presence and existence is primarily due to the production of sulphide-consuming bacteria.

Phosphates are of greater economic value because of their vital importance as fertilizers and because workable deposits are far less common than we could wish. Most of them have resulted, directly or indirectly from organic activities. **Phosphorite** (calcium phosphate) deposits are almost exclusively found on continental shelves where

there is unusually high biological productivity in the surface seawaters, a result of high nutrient levels. Calcium phosphate is particularly concentrated in the bones, teeth and excrement of vertebrates, especially fishes. Vast numbers of fish are sometimes killed by the phenomenal reproductive outbursts of dinoflagellates, responsible for 'red tides' which leads to catastrophic mortality of nearly all forms of coastal life. Even so, most phosphorites have no easily identified biological form, and usually occur as pellets and small concretionary forms. There seems no doubt that their origin is related to the rain of organic matter to the sea bed and its degradation within surface and subsurface sediments, releasing phosphorus at or near the sediment – seawater surface so that it can precipitate with calcium. Due to occasional periods of winnowing, perhaps caused by either sea-level change and/or sea bed current activity variations, the phosphorite materials such as bones and concretions, can occur as a lag accumulate on the sea bed.

Another interesting store of phosphates has been provided on rocky coasts and islands by the droppings of countless generations of fish-eating birds. The inorganic matter of the fresh deposit is a mixture of phosphates, nitrates and other compounds of calcium and ammonium. Warm climates with intermittent rain favour the removal of the soluble constituents and reaction products, until eventually a residual crust of **guano** is formed. In some localities crusts up to 30 m thick originally blanketed the rocks. But most of the thicker and therefore more valuable deposits have already been fully exploited and little now remains. Geologically it is of interest to note that where the base of the guano has been uncovered, parts of the underlying rocks – generally volcanic – are found to have been metasomatically replaced by phosphate.

24.2 THE FLOORS OF THE SEAS AND OCEANS

It is important not to be misled by the smooth curves of Figure 24.5, which is merely a diagram to illustrate the nomenclature used in describing marine environments. The results of echo sounding, magnetic surveys, coring, submersible descents, and a variety of geophysical techniques have upset practically every hypothesis for understanding marine environments based on the scanty knowledge of a little more than a generation ago. A few submarine valleys, carved into the slopes of continental margins, have long been known, but echo sounding and submersible observations have revealed the presence of great

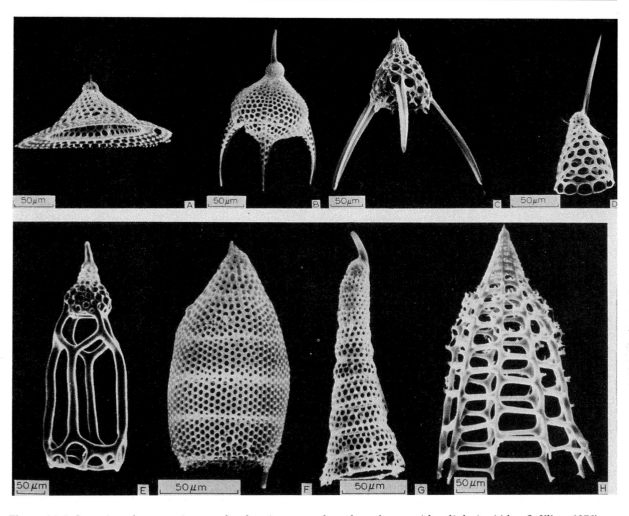

Figure 24.4 Scanning electron micrographs showing several modern theroperid radiolaria. (*After S. Kling, 1978*)

numbers of submarine canyons with tributary systems. Oceanic islands, mainly volcanic peaks and coral atolls rising from great depths, were once thought to be surrounded by almost feature-less plains of enormous area. These regions of the ocean floors are now known to be traversed by great submarine mountain ranges, built of basalt, which encircle the Earth and are known as mid-oceanic ridges. Great numbers of isolated peaks, volcanic cones and their relics, commonly occur-ring in rows or clusters, characterize the ocean floors (Figures 13.23, 26.10). Some of the highest peaks rise above sea-level to form islands, but the vast majority fall short of the surface. These are known as **seamounts** or, if they have nearly flat tops, like truncated cones, as **tablemounts** or **guyots** (Figure 24.36). Deep open trenches, like those margining the Pacific garland of Island Arcs (Figure 30.12) have been familiar features since the first HMS *Challenger* Expedition of 1872–6, but their significance remained a mystery until the 1960s when the theory of plate tectonics was substantiated. As a result of numerous geophysi-cal investigations, much has been learnt about

their topography and structure. Most are now associated with locii of subduction of one plate under another. They are sharply V-shaped in cross-section, sometimes free from sedimentary infilling, commonly more than 8000 m deep and sometimes over 3000 km long. Some trenches con-tain sediments of continental origin, and their floors are then flat and up to 8 km wide. Echo sounding has revealed that their infilling sedi-ments may be bedded and undisturbed.

Innumerable faults, characterized by earth-quakes, have been discovered intersecting and seemingly offsetting the mid-oceanic ridges (Fig-ure 13.13). The ocean floor is no longer thought to be a static feature of the Earth; newly formed at the mid-oceanic ridges, it moves away on either side, and eventually disappears into the mantle at the deep ocean trenches (Section 26.9, Figure 28.26).

The transition from abyssal plain to continental land – except where interrupted by the trenches mentioned above – is known as the **continental margin**. This is usually divided into three parts (Figure 24.5):

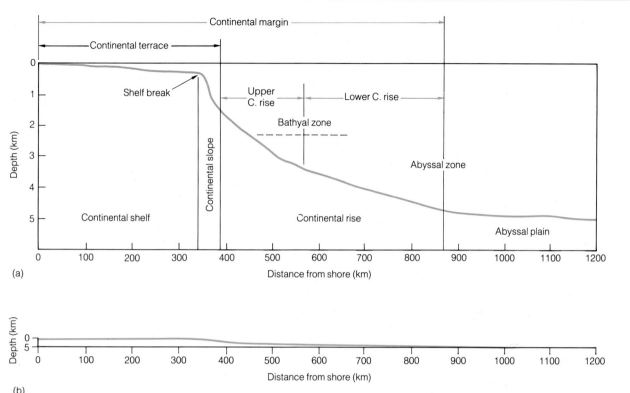

Figure 24.5 Principal features of the continental margin. (a) Vertical exaggeration of 1/50. (b) No vertical exaggeration. (*J. Kennett, 1982*)

1. the **continental rise**, which links the ocean basins to
2. the **continental slope**, which is locally quite steep (4°), shares some of the features of an escarpment and continues to the edge of
3. the **continental shelf**, which extends shorewards at a very much gentler slope (1°). The depth of the outer edge of the shelf is about 200 m.

Actually the margin varies considerably from place to place, like the width of the shelf. On average the steepening that marks the change from shelf to slope occurs at depths of about 150 m.

24.3 MARINE DEPOSITS

According to the location on the sea bed or ocean floor, the marine deposits forming today can be classified as follows (Table 24.1):

Littoral deposits, which are formed between the extreme levels of low and high tides; they include beaches and other bay deposits.

Shallow-water deposits, often called **neritic**, col-

lect on the continental shelf and at similar depths on the flanks of oceanic islands.

Deep-sea deposits. Beyond the edge of the continental shelf are the sands, silts, muds and oozes referred to as deep-sea deposits. The muds, etc., of much of the continental slope and continental rise, and of similar environments around oceanic islands, belong to the **bathyal zone**. Depending on currents, the depth to which the **bathyal zone** extends varies considerably from place to place, the average in round figures being about 2000 m. At greater depths lies the **abyssal zone** (2000–6000 m). Its characteristics are red clay and most of the deep-sea oozes. By the exploration of the deep oceanic trenches by photography and bathyscaph descents, previously unsuspected forms of living organisms have been discovered on their floors, and it is becoming necessary to distinguish a still deeper marine environment. For this the name **hadal zone** (from Hades) has been proposed. It is provisionally regarded as beginning at depths of 6000 m

An estimate of the volume of post-Palaeozoic sediments from these various environments is shown in Table 24.2. Although the continental slopes and rises contain about 60 per cent of the total sediment volume, they represent only about 10 per cent of the Earth's surface.

Table 24.1   Modern marine deposits

Dominant kinds of material / Zones of deposition	Terrigenous deposits	Chemical (authigenic) and biochemical precipitates	Biogenous deposits	
			Neritic (mainly *Benthos*)	Pelagic* (mainly *Plankton*)
Littoral zone and Shallow water or Neritic zone	Shingle Gravel Sand Mud	Oolite sands Calcareous muds Evaporites Cementing materials Phosphorites Glauconite Pyrite, etc. ....(8%)............	Shell gravels and Shell sands Coral reefs and Coral sands	(rare: usually overwhelmed by other materials)
Bathyal zone	Deep-sea muds and sands (With varied amounts of plankton remains) (15%)	coral muds		
	Green, black and blue muds Volcanic muds	Glauconite Pyrite, etc. Cementing materials		Pteropod ooze Deep-sea oozes ....(47%)........
Abyssal zone	Deposits from turbidity currents			Globigerina and Coccolithophorid ooze Diatom ooze Radiolarian ooze
...................Red clay (30%)...................				
	Mainly terrigenous	Manganese nodules	(rare)	(variable)

The figures in brackets represent the approximate areas covered by the various groups of deposits, expressed as percentages of the area of the ocean floor. (* 'middle' depths of oceans).

Classification of sediment constituents. According to the source of the dominant materials, three main groups of marine constituents are recognized:

1. **Terrigenous**: derived from the land by rivers, glaciers, wind and coast erosion. (Note that deposits on land are referred to as 'terrestrial'.)
2. **Biogenous**: derived wholly from ocean water with the co-operation of organisms. Consists mainly of two substances that are produced as the hard parts of organisms. These are calcite (calcium carbonate) and opal (amorphous silica) precipitated by living plankton and other organisms described above. Organic matter comprising mostly of carbon should also be included in this group, but usually sediments contain less than 1 per cent organic carbon, except on some continental shelves; for instance off Peru and Namibia where the rain rate of organic matter from plankton is exceptionally high.

Table 24.2 Sediment volumes of post-Palaeozoic sediments ascribed to different geological regions. (*After J Kennett 1982*)

	Volume ($\times 10^6$ km^3)
Continents	45
Shelves	75
Slopes	200
Marginal basins	35
Rises	150
Deeper basins	25
Total	530

3. **Authigenic**: although quantitatively small (about 1 per cent), this important group of sedimentary constituents are formed by a spontaneous crystallization or precipitation within the water column, or *in situ* within the sediments on the ocean floor. They record the physicochemical and biological reactions occurring during deposition and alteration of ocean floor sediments. The origin of the elements that precipitated in this way is diverse, including hydrothermal outpourings to the ocean, seawater itself and the resistant remains of biological organisms. Most authigenic constituents form from slow precipitation. The term **diagenetic** is often associated with authigenic sediments and recognition of authigenic, as opposed to diagenetic, materials is not easy. Strictly, diagenetic minerals form due to the chemical alterations of certain components in a sediment during burial and these are often bacterially mediated. Five groups of authigenic/diagenetic substances persist in deep-sea sediments; metal-rich sediments including iron and manganese oxides, manganese nodules (Section 24.8), phosphorites, zeolites (Table 4.3) and barite (barium sulphate). They are mostly recognized in deep-sea sediment such as in red clays where there is not excessive quantitative dilution by terrigenous and biogenous constituents.

Terrigenous Deposits. Terrigenous deposits are the most voluminous and are naturally associated with those areas bordering lands and especially off the mouths of great rivers. Sediment that is gently swept over the edge of the continental shelf, or carried more vigorously by turbidity currents, comes to rest on the slopes and plains beyond, or in the submarine canyons that diversify these regions. It must not be forgotten that during the Pleistocene episodes of low sea-level, silt, sand and gravel were deposited on the outer and deeper parts of the continental shelves where the shorelines were then situated. The discovery of coarse sediments far out on the shelf seemed to be anomalous until it was realized that when they

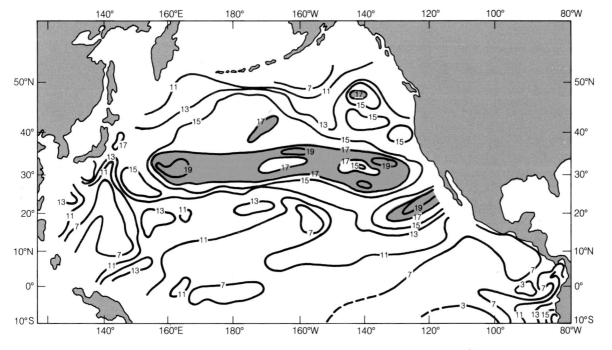

Figure 24.6 Distribution of quartz contents of Pacific Ocean surface sediments expressed on a calcium carbonate and biogenic silica-free basis. The narrow defined mid-latitude of high values represents an aeolian dust component which is identified with fallout from a high altitude jet stream. (*After T. C. Moore and G. R. Heath, 1978*)

were deposited the depth of water was appropriately shallow. Recognition that coarse sediments and heavier minerals might be found on the deeper parts of the shelves has led to discoveries of great economic value. Gravels containing concentrates of cassiterite ('tinstone', SnO_2) have long been dredged from the floor of the Java Sea between Borneo and Java and off Malaysia. Rich finds of diamonds have been made off the desert coast of Namibia. For 80 km north of the mouth of the Orange River gravels beneath coastal sand dunes constitute a most prolific source of diamonds of gem quality. Subsequently found fields on the sea floor are the beaches of several thousand years ago, when sea-level was 18 m or more lower than today. Sucking up diamond-bearing gravels with equipment something like a giant vacuum cleaner, occurs here and off the coasts of other diamond-producing countries (e.g. Sierra Leone).

The only really continuous supplies of land detritus that reach the abyssal ocean floor come from airborne dust (Figure 24.6) and, in high latitudes, from melting icebergs. The normal supplies from the former are so scanty that the rate of accumulation is extremely slow. But **discontinuous** supplies of typical terrigenous sediment are brought in by slumping and turbidity currents (see later) when plumes of suspended sediment carried from the shallower parts of the continental shelf to the ocean. It has been found in many of the core samples from the ocean floors that layers of shallow-water deposits alternate with pelagic oozes and red clay. In all cases these sedimentary units have been introduced by turbidity currents and can vary in thickness from less than 1 cm or so to greater than 1 m.

24.4 SUBMARINE CANYONS AND TURBIDITE FANS

A century ago it was discovered by soundings that the Hudson and Zaire (Congo) river valleys continue over the sea floor, cutting through the sup-

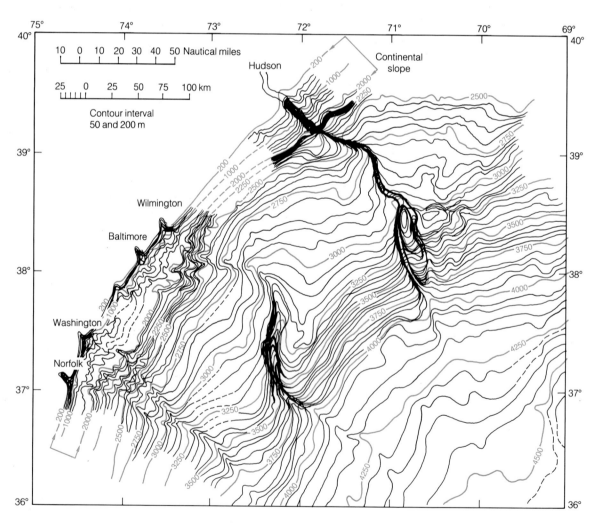

Figure 24.7 Bathymetry of the shelf slope and continental rise off the eastern United States showing the position of several canyon heads including the Hudson submarine canyon. (*After R.M. Pratt, 1967*)

posedly featureless floor of the continental shelf and become comparable in their dimensions with deep canyons where they traverse the continental slope and beyond. 190 km south-east of New York the Hudson submarine canyon is 10 km across from rim to rim, and over 1100 m deep (measured from the rim (Figure 24.7)); the Zaire example is on a vaster scale and is unique because it penetrates as a deep estuary into the continent at the river's mouth (Figure 24.8). Deep, trough-shaped or V-shaped valleys of average floor slope of 58 m/km (but considerably less in the longer canyons) occur off the deltas of the Ganges, Indus, Mississippi, Colorado and Rhône. Indeed there are few continental shelf/slope systems without attendant canyon systems though not all are obviously related to neighbouring river systems. The first described canyons were naturally a source of great perplexity and the mystery of their significance and mode of origin presented geologists with still another challenge of great interest.

It is now known that submarine canyons serve as major conduits of terrigenous sediment from the continents to the deep basins. Most canyons begin on the continental shelf, most commonly at or near the mouths of large rivers, and branch out normally to the coastline. Others may be deflected by structural boundaries. Most are cut to the base of the continental slope and extend further over the continental rise as channels to fan systems. Overall, canyons have a mean length of 50 km with the longest canyon (370 km) being found in the Bering Sea. The existence of such deeply-incised features implies an enormous amount of erosion at some time, although today only a few can be identified as dynamic features. These latter are generally associated with active margins (see Chapter 30), while the present relatively non-eroding ones are usually found on passive continental margins. Research submersibles have allowed direct observation of the sediments and many of the minor topographic features of the canyon walls. All canyons display features of vigorous relief on the steeper slopes of the continental margins, which thus turn out to be far more rugged than expected. A dentritic system of tributary valleys, often showing steep floors, as illustrated in Figure 24.9 is characteristic and the resulting submarine topography closely resembles that of a youthful land surface dissected by minor erosion. The trunk canyons range between broad valleys with V-shaped profiles, often stepped, and gorges with steep, even, vertical sides. Observations also show that the larger canyon floors are often flat, displaying transverse ripple marks, with scour marks around boulders. The sands within them are commonly graded, cross-laminated and indicate former powerful bottom currents. The presence of many large boulders on canyon floors demonstrates the occurrence of mass movement of bedrock down the steep canyon sides, while the main channels terminating on the continental shelf, act as chutes for sand and other sediments. In the case of the Scripps Canyon off southern California, which terminates shorewards within 50 m of low-tide level, the main valley is a conduit for transporting beach sediments.

Sediment cores brought up from the canyon mouths and beyond, include layers of shallow-water types of sediment and fossils which could have got there only by being transported through the canyons themselves. Closer inspection of these sediments suggests that their movement down canyons is probably by a combination of slow creep and turbidity currents.

From samples dredged from the walls it has

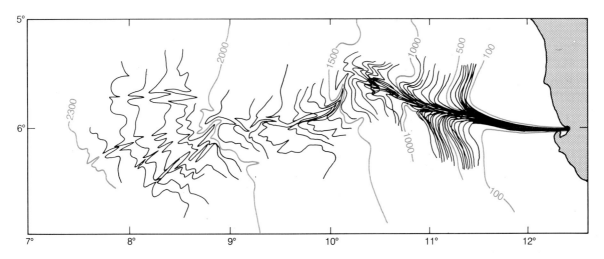

Figure 24.8 The Zaire (Congo). Submarine canyon showing the outer fan and channels. Note the canyon penetrates as an estuary at the river's mouth. Contour interval in fathoms. (*After B.C. Heezen et al, 1959*)

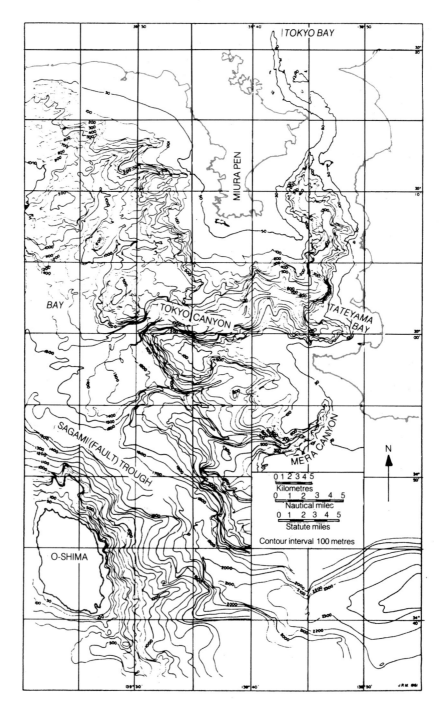

Figure 24.9 Canyons in the vicinity of Tokyo, based on a 1960 Scripps Institution of Oceanography, University of California and various Japanese surveys. Note the winding course of the Tokyo canyon and the innumerable tributaries off both walls. (*After F. T. Sheppard, 1963*)

been argued that some of these canyons must be geologically quite young. For example, the edges of Pliocene marine beds are exposed on the sides of the Monterey Canyon off the California coast, suggesting they must have been cut after the Pliocene, i.e. during the Pleistocene. But such evidence may only mean that the canyon has continued to develop since the Pliocene, e.g. by landslipping of the steep walls into the depths of a canyon that was already there. As we shall see

presently, conditions were often especially favourable for the formation of submarine canyons during the Pleistocene Ice Ages, but there is no reason to suppose that the Pleistocene had a monopoly of this erosion. Before the very great depths reached by some of the longer canyons were realized it was not unreasonable to suppose that they must have been formed by river erosion, like their counterparts on the continent. However, with the progress of submarine exploration it became evident that the relative changes of level required by the subaerial hypothesis (involving emergence and submergence of land and/or lowering and raising of sea-level) amounted to 3000 m or more, all over the continental margins. This was beyond belief and it is now known that sea-level change during the Pleistocene was barely more than 100 m. The remaining probability therefore is that most submarine canyons were formed beneath the sea, presumably by the erosive action of turbidity currents.

Evidence suggests that turbidity current frequency accelerated downcutting of the canyons, especially during periods of low sea-level stand, i.e. during glacial episodes of the Quaternary. In contrast, the post-glacial rise of sea-level would drown the heads of the canyons with the formation of an often wide shelf separating them further from sediment supply. Moreover during that time, as indeed today, most sediment introduced to the shelf is actively accumulating in depressions such as drowned lagoons and estuaries; little enters into the canyon system. As an example of this, rates of sedimentation at the foot of canyons since the low stand of sea-level during the last glacial event, have decreased from about 100 cm in a 1000 years to as low as 5 cm in a 1000 years.

Turbidity currents, also known as density or suspension currents, were mentioned in Section 17.11 and Figure 17.38 in connection with deltas.

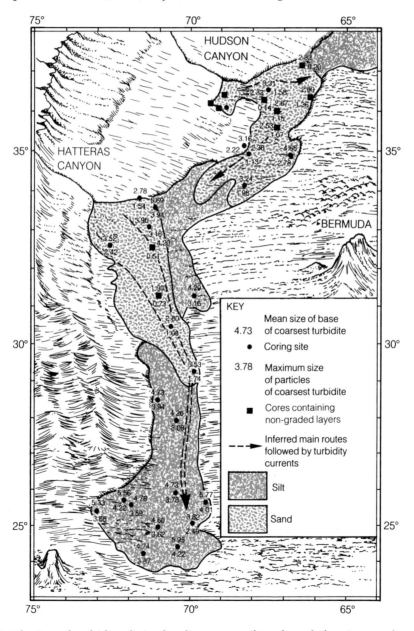

Figure 24.10 The distribution of turbidite-derived sediments as silt and sand, forming an abyssal plain off the eastern United States, and fed by discharge from the Hudson and Hatteras submarine canyons. Numbers represent mean and maximum grain sizes (in *phi* units: low numbers indicate coarse-grained sediments) of the coarsest turbidite in sediment cores. (*After D. R. Horn and others, 1971*)

There is no longer any doubt about their efficacy as agents of transport, deposition and erosion when they take the form of relatively heavy underflows moving rapidly along the floor of a lake or down the submarine slopes of the continental margins. Seawater containing sediment in turbulent suspension brings about a submarine topography that bears all the hallmarks of erosion by a flowing medium – somewhere between swiftly running water and avalanching mud flows.

Turbidity currents are usually short-lived, powerful, gravity-driven currents consisting of dilute mixtures of sediment and water with densities higher (1.03 to 1.39 m/cm^3) than surrounding seawater (1.0279 m/cm^3); the motion is maintained by internal turbulence. Most terrigenous sediments on the continental rise and beyond were introduced by such currents, and the vast amount of shallow-water derived sediment transported in this way has built up the relatively flat abyssal plains (Figure 24.10). The deposits of sediment from turbidity flows form **turbidites**. The mechanism of the flow of turbidity currents and their characteristics have been well studied, but their origin is still uncertain. Given a subaqueous slope, the essential requirements are a source of sediment and a trigger mechanism to start off the flow. A river in flood with a heavy bed load at the mouth, produces both sediment and flow. This combination would account for submarine canyons that are direct extensions of rivers. Such rivers, like the Zaire, have their deltas at the head of their canyons. Another example is the Magdalene River, which flows northward into the Caribbean Sea from the great highland valley of

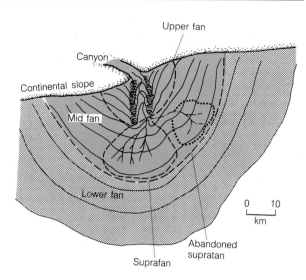

Figure 24.12 Idealized model for submarine fan growth, which involves the formation of a fan valley or upper fan with levees. The levees decrease in height and disappear down the fan. The suprafan is a small delta or fan-like deposit probably formed as the turbidity currents spread out upon leaving the confinement of the leveed valley. (*After W. R. Nordmark, 1970*)

Columbia between the East and West Cordillera of the Andes. The real trigger required by these rivers is whatever causes a sufficiently heavy rainfall to yield the discharge needed to ensure a heavy bed load; thus, a powerful hurricane might start a turbidity current. Turbidites and submarine slides and slumps are often associated and it is possible that the turbidity currents originate from slides and slumps when supply of sediment to the system is high. Alternatively, an obvious trigger is an earthquake that sets off submarine sediment slides or shakes loose unstable masses of more consolidated sediment which then slump downslope. Moving sediment and turbulent water readily mix together and rapidly develop into a turbidity current. Slumping may also be started by exceptional tides and waves over slopes of freshly deposited sediment that are just on the point of becoming too steep for stability. A major problem concerning the generation of turbidity currents is how the marine suspension becomes diluted to the low densities quoted above. An intermediate step between slumping and sliding and the turbidity current probably involves some form of debris flow.

Turbidites are by far the most important sediment type of terrigenous source that occur on the continental margins and fill up the abyssal plains. Because the sediment input to these regions is episodic, each turbidite layer may have a different thickness and areal extent, with units becoming thinner and finer-grained away from the source. Such units can vary from greater than a metre to a

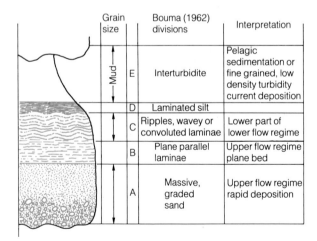

Figure 24.11 An idealized stratigraphic sequence of a turbidite bed, often called a Bouma sequence after Arnold Bouma. The interpretation of the flow regime is given on right; the thickness of the sequence, which is rarely complete, can range from a few centimetres to greater than 1 metre. (*After G. V. Middleton and M. A. Hampton, 1976*)

centimetre or so. Normally, the turbidites consist of sand and silts and are interbedded with inter-turbidite pelagic muds, which usually have a higher biogenous component. Occasionally a single coarse grain-sized bed extends over several hundred thousand km^2 in abyssal plains and represents one single turbidity event. Such sediment units tend to be absent on local submarine hills and other topographic highs.

More often than not the stratigraphic sequence of a turbidite unit is well defined and usually comprises most, if not all, of a characteristic succession of subunits which is often known as the Bouma sequence (Figure 24.11). The lowest sub-unit rests on the underlying sediment, usually a pelagic clay representing an interturbidite. This basal sediment often displays massive grain-size grading implying rapid deposition. Often, there are orientated erosion and fill marks at its base, known as sole markings. Above this the turbidite usually comprises a subunit of sediment showing plane parallel-laminae, above which the third sub-unit is more silty showing ripple marks and some convolute bedding. The uppermost turbidite sub-unit of the succession is a fine-grained, parallel-laminated silt. This succession is often complete in thicker turbidites, but those more distally placed, especially on the abyssal plains, often are of silts without graded bedding at the base.

When turbidity currents reach the base of the continental slope and their velocity decreases, many deposit their load as deep-sea turbidite fans. These are often seen at the mouths of canyons and are especially observed as thick successions on the margin of the north-west Pacific. Most fan systems (Figure 24.12) tend to show a very complicated morphology of submarine terraces, channels, levees and interchannel areas and slump features. In radial profile upper, middle and lower fan environments can sometimes be distinguished on the basis of grain-size changes and other features. **Abyssal cones** are similar features but are usually associated with major river deltas, e.g. off the Amazon, Zaire, St. Lawrence and Niger rivers, and also those off the Indian subcontinent. The Bengal Abyssal Cone received much of the products of denudation of the Himalayas during their late Cenozoic uplift. Indeed, the majority of modern fan systems tend to show the most rapid accretion during the Quaternary, especially during low sea-level stands associated with glacial events.

Before leaving this topic it should be mentioned that the thin layer of terrigenous sediment over much of the Pacific floor, would be much thicker but for the fact that the area is almost everywhere barred off from turbidity-currents supplied by the deeps around the margins of this great ocean. The

deep trenches associated with zones of plate sub-duction (Figure 30.2) are particularly obvious traps for sediment carried by turbidity currents, which may flow into them from the continental side either sideways or lengthwise.

What was probably the most gigantic turbidity current of modern times – and one that provided convincing evidence of its reality – was triggered off by the Grand Banks earthquake of 1929 (Figure 24.13). This powerful earthquake not only devastated the southern coast of Newfoundland, it also broke twelve submarine telegraph cables in at least twenty-three places. Nearly all the breakages

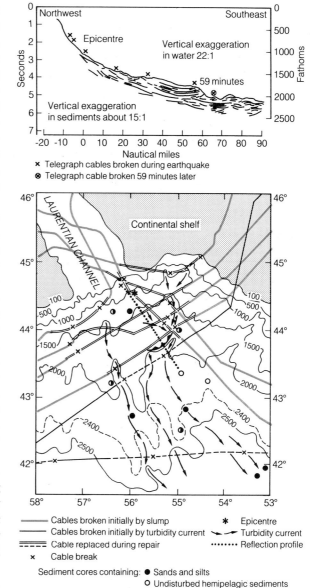

Figure 24.13 Map and seismic reflection profile of the Grand Banks area where cable breaks signalled the beginning of a slump and turbidity current in 1929. (*After C.B. Heezen and C. L. Drake, 1964*)

occurred in pairs, near the edges of a trough-like submarine canyon, i.e. where the cables were already in tension under their own sagging weight. The trough is in line with the Cabot Strait, or Laurentian Channel, which is itself a deep submarine valley cut into the continental shelf. The trough reaches a depth of about 2100 fathoms (1 fathom = 1.829 m) and is more than 200 fathoms below the abyssal plane on the western side and about 600 fathoms on the eastern side. Initially it was thought that the disaster to the telegraph services was caused directly by the earthquake. Twenty years later Heezen and Ewing wondered if turbidity currents might be strong enough to break the cables. If so, the breaks should have occurred in a definite time sequence, depending on the speed of the current concerned. They therefore looked up the records of the Grand Banks earthquake and to their great delight found the evidence to be just what they had anticipated.

The next step was to verify that the predicted deposit was actually there. Unfortunately, attempts to take cores from the sea floor south of the cable area had been frustrated by hurricanes in 1950 and again in 1951. Meanwhile, when Heezen and Ewing announced these results in 1952, Kuenen, who was chiefly responsible for setting up tank experiments to understand the mechanisms of turbidity flows, estimated that this turbidite layer would probably have an average thickness of 40–100 cm over perhaps 260 000 km^2 of the abyssal plain. Before the 1952 papers were published, Heezen had succeeded in bringing up six cores from the south of the southernmost cable. The first two were examined in 1953 and found to consist of foraminiferal clay of abyssal type, with a top layer of graded silt, 130 cm thick in one core and 70 cm in the other.

In 1954 an earthquake originating near El Asnam (Orleansville) about halfway between Oran and Algiers, was followed by the breaking, twisting and local burial of five cables which were spaced out on the continental slope of the Mediterranean floor. In this case the breaks occurred where the slope was steep, and far outside the area within which the ground was fissured and underground pipes ruptured. No cables were broken directly by the earthquake shock, but the passage of earthquake waves was sufficient to release slumps of unstable sediment. These quickly developed into turbidity currents. The times when telegraph messages were suddenly interrupted showed that the cables were broken one after another in the order of their distances from the coastline.

In recent years loose sediment from the lower reaches of the Magdalena River, Columbia, has several times contributed to turbidity currents which eventually come to rest far out on the abyssal plain of the Caribbean Sea. Jetties constructed at the river mouth have been shattered by the slumping of their foundations and considerable lengths of them have disappeared without trace. On the sea floor, to the north, a cable has been repeatedly broken at distances of 24 km to 56 km from the land. The sediments of the Caribbean abyssal plain vary in thickness between limits of about 1000 and 2000 m – several times thicker than is usual – and a high proportion is obviously turbidity-current sediment, easily recognized by its richness in the decayed remains of marsh grasses that grow abundantly in the swampy lowlands between the Cordillera and the mouth of the Magdalena.

24.5 MARINE ORGANISMS AND BIOGENOUS SEDIMENTS

The marine organisms that contribute most conspicuously to the sediments of the littoral and shallow-water zones belong to a group known collectively as the **benthos** (bottom dwellers) which includes mollusca, sea urchins, bivalves, corals and other forms that live on the sea floor. Many of these are firmly attached to the bottom or live within the sediment. Deposits of shells or of their wave-concentrated fragments accumulate in favourable situations, while elsewhere similar remains are dispersed as fossils through the terrigenous deposits. The North Sea is mainly floored with terrigenous material mostly of glacial origin, but between Kent and The Netherlands, there are patches of several square kilometres consisting almost entirely of large shells. More extensive accumulations occur off coasts where biogenous remains are not smothered by sand and mud. The reefs and atolls constructed by corals and their associates in tropical latitudes illustrate limestone-building on a particularly spectacular scale.

The biogenic oozes and red clay of the continental margins, particularly of the abyssal zone, are distinguished as **pelagic** deposits (Greek, *pelagos*, the open sea). These are largely composed of the remains of marine organisms belonging to a group called the **plankton** (the wanderers). This includes unicellular marine algae and plants (diatoms and coccolithophorids) and animals (foraminifera and radiolarians); certain floating molluscs known as 'sea butterflies' or pteropods; most of the eggs and larvae of the benthos including benthic foraminifera and all other forms which, unlike fishes, have no means of self-locomotion. The pteropods are blown along the surface by the wind or travel on currents, but the others, nearly all microscopically small, are mostly passively suspended in the water.

Diatoms, being silica-precipitating plants, cannot live below the depth of effective sunlight penetration, which in the open ocean reaches a maximum of about 100 m. Though individually quite invisible to the unaided eye, the diatoms are present in such prodigious numbers that they turn the sea in which they live into a kind of thin vegetable soup. This forms the main food supply for the predatory plankton, whose habit of feeding is therefore similarly confined to the sunlit or **photic** zone. Much of the silica devoured in this way goes into circulation again by being redissolved, especially in warmer waters.

The abundance pattern of biologically amorphous silica (opal), in deep-sea sediments (Figure 24.14), resulting mostly from diatom fallout, is closely governed by the pattern of biological productivity of diatoms and to a lesser extent radiolarians in the overlying seawater. The latter are confined to the more tropical latitudes. High opal contents in deep-sea sediments occur off Antarctica and the northernmost Pacific Ocean; here high nutrient upwelling especially of silicate, promotes a high productivity of silica-, rather than calcium-bearing organisms. Only when silicate has been depleted from surface waters will other plant types flourish. Diatom production can also be exceptionally high in coastal regions, especially those situated on the eastern margins of oceans. Even so, the sediments are rarely a reflection of this because surviving opal that has not dissolved in the seawater and at the sediment surface tends to be greatly diluted by terrigenous debris. Exceptionally, shelf sediments off Peru, Namibia, the Sea of Okhotsk (north of Japan) and the deeper sediments within the Gulf of California can contain 50 per cent or more diatom remains.

From the prolific surface waters penetrated by sunlight, the sea floor receives a slow but steady 'snowfall' of plankton shells, both siliceous and calcareous, that have escaped destruction. In the shallow-water zone, the tiny shells are generally hidden in an overwhelming abundance of terrigenous materials, and the remains of benthos. In the bathyal zone where the supply of terrigenous sediment is less overpowering they make a bigger show and can be readily found in the grey-green muds. In the abyssal zone, however, plankton shells (generally a mix of foraminifera, coccolithophorids and diatoms), accumulate over vast areas with but little contamination from other sources, to form the deep-sea oozes which, together with the red clay, constitute the pelagic deposits.

Fishes, whales and other marine animals which go actively after their food supply are grouped as **nekton** (swimmers). These contribute to all the marine deposits on a limited scale but concentrated remains, especially of fish bones and teeth, can be found in some oceanic as well as continental shelf sediments.

24.6 BIOGENOUS OCEAN SEDIMENTS

Until a century or so ago, nothing was known of the deep-sea deposits. Globigerina (planktonic foraminiferid) ooze, dredged up by one of the cable-laying steamers in 1852, was the first to be discovered. A systematic exploration of the ocean floor was carried out by the famous HMS *Challenger* expedition of the years 1872–76. The hundreds of samples then recovered are described in one of the fifty bulky volumes in which the scientific results of that great enterprise are recorded (1880–1895).

Table 24.3 summarizes the percentage of pelagic sediment coverage in the world oceans. Here the percentages of terrigenous sediment derived by turbidite flows and other mechanisms from the continental shelves is omitted.

The composition and distribution of the deep-sea oozes depend on the temperatures and nutrient content of the photic zone and on the depth and circulation of the oceanic waters. It is convenient to start with the main food supply – the diatoms (Figure 24.3). Although they flourish wherever there is sunlight and nutrient-rich water, diatoms make their greatest contribution to

Table 24.3 Percentage of pelagic sediment coverage in the world oceans. (*From Berger, 1976*)

	Atlantic	Pacific	Indian	Total
Area of deep-sea floor = 268.1 × 10⁶ km²				
Foraminiferal ooze	65	36	54	47
Pteropod ooze	2	0.1	—	0.5
Diatom ooze	7	10	20	12
Radiolarian ooze	—	5	0.5	3
Brown clay	26	49	25	38
Relative size of ocean (%)	23	53	24	100

the oozes where conditions are often unfavourable to other organisms (e.g. foraminifera) that feed on them, i.e. in the cold region around Antarctica and in the north Pacific (Figure 24.14). Coccolithophorids (Figure 24.2) as well are only an important component of subpolar and warmer waters and hence they will not dilute the diatom content of sediments in high latitudes. Minute box-like shells, and other of more intricate design, accumulate there as **diatom ooze** (greater than 30 per cent). The content of opal in the sediments off Antarctica can exceed 80 per cent by weight, while levels in the deep north Pacific and elsewhere are generally much lower, rarely exceeding 10 per cent by weight (Figure 24.14). Contamination by mineral matter is usual in these sediments, but whereas most of this consists of clay, often of aeolian origin and of other terrigenous sources as is found in the north Pacific between Japan and Alaska, it is mainly derived from floating ice around Antarctica.

Radiolarians flourish more, and their assemblage of species is more diverse in warm subtropical and tropical waters compared with their production in polar waters. Their remains are generally subordinate to those of diatoms and **radiolarian oozes** (greater than 30 per cent) are only typical in the equatorial area of divergence. However, in the past, and particularly in the early Cenozoic, radiolarians were much more abundant in sediments than their siliceous counterparts diatoms. Today radiolarian ooze is essentially a variety of red clay that is notably rich in other opaline remains; diatoms and sponge spicules. It occurs principally on an east–west strip between Central America and Christmas Island in the Pacific Ocean. Its abundance along this belt, some 12° latitude north of the Equator is due principally to the dissolution of biogenous carbonate rather than to an exceptional radiolarian production in the overlying water (Figure 24.14).

Of all the varieties of ooze, **foraminiferal**, composed of calcite, is by far the most widespread, because foraminifera (Figure 24.1) abound in the seas of both tropical and temperate regions, while a few distinctive species live only in very cold waters. Foraminiferids are especially characteristic of the Atlantic from the neighbourhood of Bear Island, north of mainland Norway, to South Georgia in the south Atlantic. It also makes extensive spreads in the western Indian Ocean and in parts of the south-west and south-east Pacific.

Biogenous carbonate production in surface seawater is geographically more uniform than that of opal as coccolithophorids and foraminifera require only nitrate and phosphate among the principal nutrient elements for their growth. This is one of the main reasons for the differences in the geographical distribution of foraminiferal (and coccolithophorid), $CaCO_3$, oozes (Figure 24.15)

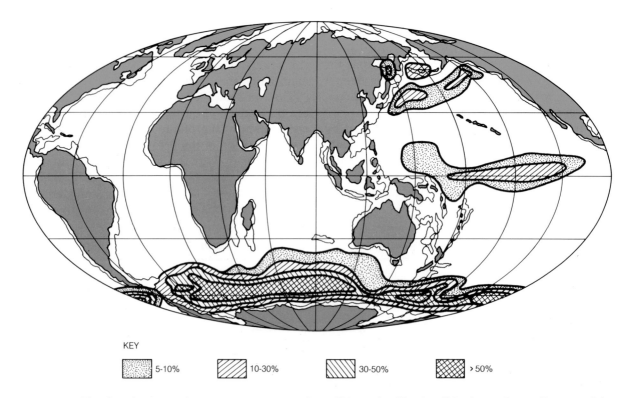

KEY

5-10%   10-30%   30-50%   > 50%

Figure 24.14 The distribution and percentage concentration of biogenic silica (opal) in the surface sediments of the world ocean, expressed on a $CaCO_3$ free basis. (*After S. E. Calvert, 1974*)

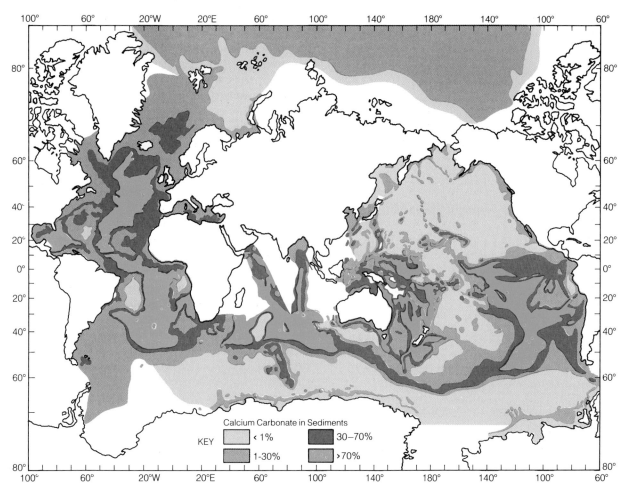

Figure 24.15 Calcium carbonate (calcite) contents of surface sediments of the world (expressed as percentage of dry sediment). (*After A. P. Lisitzin, 1971*)

and diatom oozes. Carbonate oozes (foraminiferal and coccolithophorid) are found on ridges and other topographic highs at all latitudes except in the high Arctic and Antarctic. **Pteropod** (sea-butterfly) oozes generally occur at even shallower depths than foraminiferids and their distribution is still further restricted to tropical and subtropical regions.

The distribution of calcite (mostly foraminifers and coccolithophorids) in deep-sea sediments (Figure 24.15) also reflects the extent of its solution in the sea. This property is, in places, more important than that of dilution from terrigenous sediment. Numerous experiments have been employed to measure the extent of solubility of the mineral with water depth. Small bags of carefully prepared foraminifers, machined calcite spheres or crystals of calcite in small submarine pump systems have been suspended at different depths in the ocean for extensive periods of time. By using the relative weight losses of the calcite it has been shown that not all parts of the ocean floor show water undersaturated with respect to calcite.

Theoretical calculations on calcite solubility in a medium of variable activity of dissolved carbonate, especially in respect to its position and depth in the oceans, tend to support the results of these experiments. The water depth at which the effects of solution of calcite are first seen is often termed the **calcite lysocline** and is below the boundary between supersaturation and undersaturation (Figure 24.16). Given that pteropod shells are made from aragonite and that this mineral is even more saturated than calcite the **aragonite lysocline** exists at even shallower depths. One of the real problems in oceanography is that the respective lysoclines of these two minerals are not of the same depth everywhere. In the northern Atlantic, calcite lysocline, which is situated at a depth of almost 5 km but shallows to 4 km in the south Atlantic and western Indian Ocean, becomes still shallower, less than 3.5 km, in some parts of the Pacific Ocean. Calcite shells, in the respective oceans above these depths, show little or no evidence of corrosion and the content of calcite in sediments here is controlled by its production in

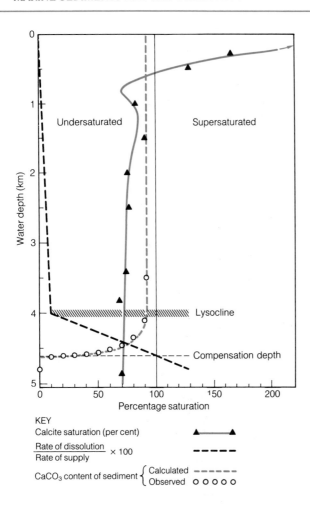

KEY

Calcite saturation (per cent)   ▲————▲

$\dfrac{\text{Rate of dissolution}}{\text{Rate of supply}} \times 100$ – – – –

CaCO₃ content of sediment $\begin{cases} \text{Calculated} \quad \text{– – – – –} \\ \text{Observed} \quad \circ\ \circ\ \circ\ \circ\ \circ \end{cases}$

Figure 24.16 The effect of water depth on the distribution of calcite. Values shown typify conditions in the central parts of the Pacific Ocean. Note that the lysocline occurs several hundred metres above the calcite compensation depth and is the depth which separates well-preserved from poorly-preserved assemblages of plankton remains. (*After Tj H. van Andel, G. R. Heath and T.C. Moore, 1975*)

surface seawater and its dilution primarily by terrigenous materials, especially clay minerals and quartz. Below the calcite lysocline, sediments display corroded shells of foraminifers and the extent of their solubility will be shown by a proportional increase in terrigenous substances. In the deepest parts of these oceans virtually all the calcite will have been solubilized leaving a residue of terrigenous and authigenic constituents as well as diatoms. These sediments are the red clays.

The dissolution of aragonite (pteropods) and calcite (foraminifera and coccolithophorids) is known to increase with pressure as well as a lowering of temperature, and the increased corrosion of shells with depth must be in part due to these factors. But the differences in the respective depths of the aragonite and calcite lysoclines between oceans cannot be caused by these proper-

ties alone, as both temperature and pressure at a given depth in the deeper regions of the oceans are either the same or at least very similar almost everywhere. The reason for the differences in depth of the onset of carbonate mineral dissolution is more related to the chemical properties of seawater and in particular to the concentration of the carbonate ion. Although the concentration of carbonate ion is only a minute proportion of the total amount of dissolved inorganic carbon in seawater, it is known to vary in a very systematic way; from relatively high values in the deep-water masses of the north Atlantic to values less than a half lower in the north Pacific. This change has a direct bearing on carbonate mineral dissolution which is facilitated at shallower depths in the Pacific, in particular near its margins.

Although it may seem logical to consider that much of the dissolution of aragonite and calcite occurs in seawater at depths below their respective lysoclines, in reality only a very small proportion of the rain of carbonate mineral fallout dissolves in this way. Most of the corrosion and dissolution actually occurs on the seabed. Even in the deepest parts of the ocean, calcite and aragonite occur in almost unmodified form at the sediment surface. To a very large extent this is due to the rather rapid descent of the minerals through the water column (80 m/day), especially if they have been ingested by larger zooplankton and packaged as faecal material. However, long exposure there, due to very slow sediment accumulation, to seawater undersaturated with respect to both aragonite and calcite means that these minerals come to their fate at or immediately below the sediment/seawater interface.

Some authors have referred to the level at which carbonate minerals disappear from surface sediments as the **calcite (aragonite) compensation depth** (CCD) or (ACD) which can be defined as that depth in the ocean where the rate of dissolution of carbonate balances the rate of its supply (Figure 24.16). In most instances its position is marked by a rapid transition, often over a few tens of metres of water depth, from carbonate ooze to red clays. Of course the CCD always underlies the calcite lysocline, sometimes by several hundred metres. From its definition, the CCD normally will be deeper in those regions where there is high biological productivity in surface waters producing a high rain-rate of carbonate shells falling to the sea bed, as for instance in the eastern equatorial Pacific Ocean.

24.7 AUTHIGENIC CONSTITUENTS

Red Clay. Roughly 100 million square kilometres of the ocean floor are known to underlie the CCD.

These areas are carpeted with red clay and are much more prominent in the Pacific Ocean than in the Atlantic and Indian Oceans. They include most of the North Pacific (except for the diatom belt in the extreme north and the diatom/radiolaria belt some 12°N of the Equator), much of the central parts of the South Pacific, the eastern part of the Indian Ocean and rather isolated deep basins in the Atlantic. The chief ingredients of this remarkable deposit, which was first mapped and described by the HMS *Challenger* scientists are:

1. clay minerals derived from the finest wind-borne dust and from alteration of the next two items;

2. wind-borne volcanic ash, and pumice fragments that may have floated far from their sources before sinking;

3. volcanic materials from submarine eruptions. Much more importantly the products of hydrothermal alteration of hot basalts by downward seeping seawater immediately under, or at least very close to, ridge crests. The discharge of metal-laden hydrothermal waters from vents to the ocean includes iron, manganese and several chemically associated metal ions;

4. rock-waste dropped from far travelled icebergs or floating trees;

5. insoluble organic relics such as fish, teeth and bones and the earbones of whales. Carbon from forest fires has also been recognized in these deposits;

6. dust from meteors and meteorites, and very rarely larger fragments that have fallen from the sky;

7. 'manganese nodules', composed mainly of manganese and iron oxides, with more or less contamination by clay.

The red clay is chocolate coloured rather than red, although red varieties occur locally. The colour is due to staining or filming of the ingredients by poorly crystalline iron oxides. In terms of content, the red clays consist mostly of wind-blown clay minerals of various kinds, quartz and feldspars the mix of which often relates to the different climatic zones of adjacent continents. Study of the change in abundance of quartz and the types of clay minerals in sediment cores of red clays has provided information on palaeowind patterns and their intensity and the climatic regime of the source region.

The enrichment of iron in these deposits is likely to have come from several of the sources listed above. Besides the coating on terrigenous clay-sized material much of the iron, and especially the manganese, is likely to be associated with aerial and submarine volcanic activity in the

form of fine dust, or even chemical precipitation from seawater. Over the last two decades observations on especially volcanically active ocean ridges, show that huge amounts of chemically charged waters emanate from them as a result of hydrothermal interactions of percolating seawater and hot basalts (Figure 24.17). The release and transport of iron- and manganese-charged waters and their overall instability as dissolved constituents implies precipitation, usually as oxides, far from their source. Dissolved iron, in contrast to manganese, tends in this respect to be much less stable and is mostly precipitated locally, whereas manganese can be transported thousands of kilometres before it becomes totally oxidized and precipitates.

'Cosmic dust' is produced from the fusion of minute meteorites and from the surface skin of larger ones as they plunge through the outer atmosphere, and overall is constantly settling down over the Earth's surface. It is only in those regions where the red clays are accumulating very slowly, generally less than 1 millimetre per 1000 years, that the remains of such dust forms a measurable part of the red clays. The larger 'cosmic dust' particles are tiny spherules, rarely more than 0.1 per cent of a millimetre across. These minute pellets are magnetic and a score or so were separated from a sample of red clay, and correctly identified by John Murray in 1876, shortly after his return from the *Challenger* expedition. Renewed interest in these spherules has been aroused by the discovery of their occurrence in deep-sea cores of red clay from depths representing ages of 10–15 Ma. A meteoritic origin is confidently ascribed to

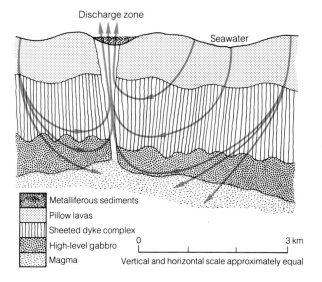

Figure 24.17 Pathway of cold seawater seepage, heating and hydrothermal interaction at active spreading ridges. Note the zone of recharge flow is ~6 km. (*After E. Bonatti 1981*)

these, and to those now accumulating, because they contain proportions of iron, nickel, cobalt and copper that match those found in iron meteorites. It has been also realized that as a result of man's activities 'industrial spherules' formed from electrical power stations, etc. can also contaminate the Earth's surface. Although their sizes are similar to those of cosmic dust, proportions of elements in their chemical composition are very different.

Chemical analyses of red clays show besides unusual enrichments of iron and manganese, values of nickel, copper, zinc, lead, molybdenum, etc. which like iron and manganese occur at levels much higher than are found in average crustal igneous rocks or fine-grained sediments such as those found on continental shelves. The origins of many of these heavy metals is rather unclear, and they probably represent the residuum of a number of sources. They could be due to precipitation out of seawater, especially of hydrothermally released waters, as is the case, for example, with lead. Alternatively they may also represent the insoluble residues of biological debris, the main elements of which have either dissolved or oxidized during their downward fall through seawater.

24.8 MANGANESE NODULES

Among the most intriguing of authigenic constituents on the sea floor are the manganese nodules. These were first discovered during the travels of HMS *Challenger* and later it was found that their occurrence is not confined to the calcareous and siliceous oozes and the red clays. They are also to be found in lakes and on continental shelves, especially in northern temperate waters where adjacent lands release plentiful amounts of manganese and iron through their river systems.

The oceanic manganese nodules (Figure 24.19) have created much interest. They are particularly associated with the Pacific rather than the Atlantic or Indian Oceans and overall the Pacific Ocean contains an average of ten per cent areal coverage of them on the seabed. In the southern central Pacific the level of coverage can double or more (Figure 24.18), with nodules occurring here and elsewhere ranging in diameter from < 1 mm to 2–4 cm (Figure 24.19). Exceptionally, slabs of man-

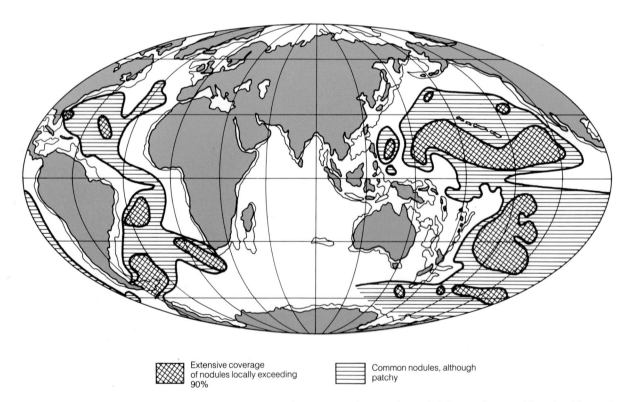

Extensive coverage of nodules locally exceeding 90%

Common nodules, although patchy

Figure 24.18 The general distribution of manganese nodules in the Pacific and Atlantic Oceans (their incidence in the Indian Ocean is not shown). Note their relatively higher abundance in the Pacific Ocean where the distribution of nodules are most abundant in east to west provinces, lying north and south of the biogenous oozes of the equatorial Pacific. (*After D. S. Cronan, 1977*)

10 millimeters

Figure 24.19 Cross-section cut of a manganese nodule showing two growth centres. Crude growth rings (3 mm/10⁶ yr) represent temporal changes in texture and composition. The nodule has a radius of 2 cm. (*Photograph by R. K. Sorem and A. R. Foster. From W. S. Broecker and T-H. Peng, 1982*)

more elements. As such they are very important potential ores, but the technical 'knowledge' to recover them in quantity has not been perfected as yet. Of interest are the changes in the ratio of manganese to iron and the relative values of the other elements in defined belts of ocean floor, suggesting that there must be some systematic control on the supply of elements and on their mode of accretion. Many suggestions have been put forward for their origin, including one associated with submarine volcanism. However, currently it seems generally agreed that much of their accretion is due to the authigenic/diagenetic modification of surface manganese and other elements due to the fallout of organic matter and other substances on the surface sediment. The organic decomposition there and within the immediate subsurface sediment causes locally reducing environments sufficient to mobilize manganese and associated sediments which consequently recycle and reprecipitate, where conditions are more oxidizing, on to nodule surfaces. The likelihood is that a nodule can grow from such recycling as well as from precipitation of the manganese in the overlying seawater. As a matter of interest the growth rate of the nodule bottoms, which are partly buried, often is much higher than nodule tops which select their elements mostly from the surrounding seawater. The chemical composition consequently can vary between the bottom and top of the nodules.

24.9 PALAEOCEANOGRAPHY

Cores from the Ocean Floors. At one time it was planned to drill through the ocean crust and through the Moho into the mantle; this great enterprise was known as Project Mohole. In 1961, using the drilling barge CUSS 1, a preliminary trial was made east of the island of Guadalupe, west of the peninsula of Baja California. The first hole ran into trouble after passing through 71 m of fine clay. A second boring then penetrated through 1790 m of sediment (mostly Upper Miocene) and reached the basaltic crust, but after penetrating 13 m of basalt the drilling bit had to be withdrawn. It soon became apparent that the cost of boring through the whole ocean crust would be too expensive and the project was abandoned. Such early investigations spurred planning by several American institutions to combine in a Joint Oceanographic Institutions' Deep Earth Sampling Programme (JOIDES) set up in 1964. The 1968 programme was extended as the Deep Sea Drilling Project (DSDP), using the ship *Glomar Challenger* with a drilling rig midships. Today, the equivalent Ocean Drilling Programme (ODP) is

ganese concretionary material and very large nodules up to 1 m in diameter have been found. Although some subsurface nodules occur, most are confined to the sediment surface, and presumably if they become buried, they are prone to destruction through dissolution. Ocean manganese nodules grow extremely slowly; estimates employing the decay of long-lived isotopes of uranium-series elements show an accretion rate of 3–4 mm per million years, although some ocean nodules, especially near the margins and in the northern equatorial Pacific, can grow an order of magnitude faster. One of the more perplexing features of such a slow growth is that it is up to three orders of magnitude slower than the accumulation of associated sediments. The fact that they are a surface sediment feature implies that there is some mechanism which is poorly understood that keeps the nodules 'afloat' at the surface during sediment accumulation.

Besides manganese, occurring as various oxide minerals, the nodules contain iron oxides and often up to per cent levels of nickel, copper, zinc, cobalt, molybdenum with minor amounts of many

truly international, and virtually all our knowledge of deeply buried sediments and crustal rocks come from these ventures. Using the technology developed by the petroleum industry, very long sediment sections (up to 1500 m) can be drilled in water depths of up to 5500 m. Since August 1968 some 138 legs, with work carried out on each leg for about ten weeks, have drilled some 800 sites throughout the oceans, except the Arctic. The vessels used were the first ever to be equipped for full dynamic positioning (fully computerized) using sonar beacons placed on the sea floor. Positioning is accomplished by the main screws and by thrusters at the bow and stern for lateral motion. Re-entering into the same hole is now a fairly routine procedure. Recovery of sediments during the early years of these ventures tended to be variable and often poor, but with the introduction of hydraulic piston coring, a complete sediment record is usually recovered to interpret the history of the present oceans.

Sediment dating. Information of fundamental geological significance is obtained by collecting and investigating these cores, as well as those collected by ordinary piston corers which can collect under ideal conditions some 30–40 m of the sediment record. For any interpretation of ocean history, as many as possible selected horizons of these recovered sediments must be accurately dated to obtain a base for estimating accumulation rate changes of the sediments.

A great variety of ingenious methods have been established to do this and all have special merits in establishing datum levels. Investigations centred on understanding the palaeoceanography of the Quaternary when glacial and interglacial conditions influenced sediment composition, may include radiocarbon or other radiometric-dating methods (Chapter 14) and the record of oxygen isotope change of calcium carbonate in the shells of foraminifera.

Growing foraminifera and other organisms extract the stable isotopes of oxygen from dissolved carbonates in seawater in proportions that are almost the same as that contained in seawater itself. Although there is a temperature effect, which can be used to understand palaeotemperatures, it is the changing ratio of the two principal non-radioactive isotopes of oxygen (i.e. ^{16}O and ^{18}O) in seawater that has been used to correlate sediments.

In the hydrological cycle of evaporation and condensation, the lighter ^{16}O isotope is preferentially retained in the atmosphere as water vapour and when cooled precipitates as rain and more importantly as snow. Precipitation at very high latitudes can show a $^{18}O/^{16}O$ ratio in snow/ice that is some 4

per cent lighter than ocean water. The Quaternary period is noted for its changing climate, oscillating between glacial and interglacial events. During glacial episodes a measurable proportion of the lighter isotope of the Earth's water is preferentially locked up in continental ice, and of course during interglacial warming the melting of much of this ice would release it back to the sea where it would be rapidly mixed throughout the oceans. Seeing that planktonic and benthonic calcareous organisms almost involuntarily extract the oxygen isotope composition of seawater during their growth, a study of oxygen isotope ratio changes of foraminifera remains in sediment cores will reflect global ice volume change with time. Relative to a standard, such $^{18}O/^{16}O$ ratios (i.e. $\delta^{18}O$) would become higher or more positive during glacial events and lower (more negative) during interglacials (Figure 24.20). This method of investigation, often termed **oxygen isotope stratigraphy**, has become a major tool for correlating marine sediments, particularly Quaternary sequences. Attention is now being focussed on the Tertiary as the technique is becoming potentially useful for that time period. Changes in the oxygen isotope record can be precisely matched and correlated between the Indian, Atlantic and Pacific oceans.

Marine sediment cores investigated for the oxygen isotope record can be subdivided into distinct isotopic events and over the last 1 m.y. have been assigned stage numbers based on the early work of C. Emiliani and more recently by N. J. Shackleton. These isotopic stages represent alternating interglacial (odd numbers) and glacial episodes (even numbers) starting from the present

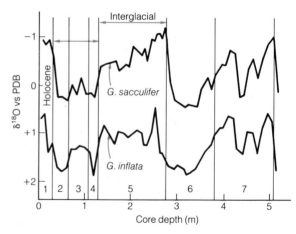

Figure 24.20 The oxygen isotope record of two planktonic foraminifera species through the late Quaternary (approximately 260 000 yr) from the south Indian Ocean. Negative oxygen isotope values of $\delta^{18}O$ (parts per thousand) generally represent interglacial conditions and these are represented by odd numbered stages. (*After J. C. Duplessy, 1978*)

Holocene (Figure 24.20). The most important aspects of oxygen isotope stratigraphy are that the events are isochronous, widely occurring, and are of different amplitude with time. They are so ideally preserved that they can be intercalibrated with other scales such as magnetostratigraphy and with biostratigraphic zonal schemes based on the evolutionary and abundance trends of foraminifera and associated organisms. Oxygen isotope stratigraphy is not a method of absolute dating; for this we have to intercalibrate it with, for example, volcanic ash horizons dated by radiometric methods (see also Chapter 21).

The deep-sea sediment record. Much of the above descriptions of the terrigenous, biogenous and authigenic constituents of ocean sediments relate to what we see in different parts of the oceans today. Early concepts of deep-sea sedimentation presumed that such features seen in surface sediments extended into the past geological periods, the assumption being that the deep sea was a tranquil unchanging environment with continuous sedimentation everywhere, and at any one place was compositionally uniform with time. Now it is known that there have been many physical and chemical aspects of change. These relate to alterations in the distribution of currents and temperatures of surface and subsurface waters that have been caused by climate changes over the Earth's surface. These are also likely to influence the type and measure of biological productivity in surface waters. Further, changes are known that alter the flow regime and chemical characteristics of bottom water masses and these can affect the redistribution of sediment types of the sea floor as well as altering the solubility of calcareous remains relative to those of biogenous silica. In the present oceans, sedimentation (from late Mesozoic times) has occurred on horizontally-moving tectonic plates which slowly subside as they cool. An attempt to evaluate change in oceanographic environment by inspection of the lithological changes in cores may need to reconstruct the palaeogeographic positions and the palaeodepths of water to which the sediments of different ages have been subjected. Such interpretation is termed **plate stratigraphy**. As an example, the older sediment of a core could have accumulated on the upper flanks of a mid-oceanic ridge (2.5–3.0 km depths) which would be above the calcite compensation depth (CCD), but subsequent sedimentation is likely to have occurred when the lithosphere was further down its flanks at depths below the CCD. The resulting stratigraphy of such a core (Figure 24.21) could show basalt, overlain by a thin layer of basal sediment of hydrothermally precipitated iron and manganese,

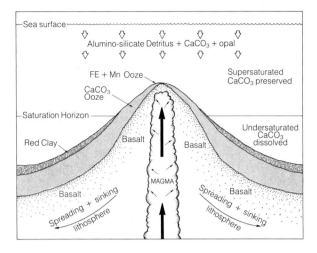

Figure 24.21 Sediment facies related to plate stratigraphy showing the sequence of sediment types accumulated by the lithospheric plates as they move away from the crest of mid-oceanic ridges. (*After W. S. Broecker and T.-H. Peng, 1982*)

followed by accumulation of carbonate sedimented on the upper flank and overlain by a thinner sequence of red clay or siliceous sediment or both.

Changes in global ice volume and causal palaeoclimates that have occurred in the latter part of the Cenozoic, and especially during the Quaternary, have had an indirect influence on input rates of wind-borne dust to the oceans. The oscillations between glacial and interglacial conditions have also profoundly influenced the production of calcareous and silica secreting organisms in surface seawater. In some parts of the oceans, for instance in the north-west Indian Ocean, biological productivity is thought to be higher during interglacial periods while in the eastern Pacific the reverse seems to be the case. Equally so the history of ocean sedimentation, particularly of terrigenous constituents, was also affected by sea-level changes. Sea-level as a long period feature can be influenced tectonically, especially with respect to changes in the rate of lithospheric plate formation. However, during the Quaternary it is almost wholly influenced by glacial/interglacial events (Section 21.1). For instance, since the time of late Quaternary maximum glaciation 18 000 years ago and 10 000 to 6000 years ago when extensive ice melting caused the **Holocene transgression**, sea-level has risen an incredible 100 to 130 m. Sea-level changes, especially since the last glacial event have created the broad, relatively flat coastal plain or continental shelf as we know it today. Today, at a time of exceptionally high sea-level stand most ocean basins are starved of water-borne terrigenous input; in contrast during periods of low sea-level, shelf sediments are subject to resuspension and are flushed to the deep sea increasing sedimentation rates.

Palaeoceanographic changes have been examined using changing rates of total sediment accumulation obtained from ocean drillings in the major ocean basin. Average accumulation rates have been calculated for a number of Cenozoic time intervals (Figure 24.22). In general, the results from different oceans suggest that the accumulation patterns are globally synchronous indicating some alternation between periods of high (e.g. late Oligocene) and low (e.g. late Eocene/middle Oligocene) accumulation. These correlate with global sea-level fluctuations; higher accumulation rates occurring at times of low sea-level. The Pacific basin, in particular, is subjected to a very limited continental drainage and is mostly surrounded by trenches which act as sediment traps or dams for debris being transported off its continental shelves. Under these circumstances, if the cause of accumulation rate changes was solely due to terrigenous supply, amplitude changes in accumulation rate for the Pacific should be less than for the other oceans. As this is not the case (Figure 24.22), Davies and Worsley assumed that the correlation between sea-level and sediment accumulation is due to biogenous constituent accumulation. Such

a model was considered by Berger, who advocated that high sea-level allows higher rates of biogenous precipitation to occur on continental shelves, starving the ocean basins of carbonate. He suggested that at times of low sea-level dissolved river loads containing calcium and nutrients would more easily reach the deep sea.

Changes in accumulation rates of biogenous carbonate may not simply reflect biological productivity, but may be related to its preservation in seawater and sediments. Figure 24.23 shows that the level of the calcite compensation depth (CCD) has changed dramatically throughout the late Mesozoic and Cenozoic, and estimates on its variability suggest a maximum change of about 2000 m. Such change may also involve the balance of distribution of carbonates on shelves and the deep ocean. Figure 24.23 indicates that variations in sea-level with time have a configuration that is similar to the CCD. Moreover, the CCD pattern is similar for the three major oceans and thus requires a mechanism that is global in its influence. Depression of the CCD is synchronous with periods of low sea-level. During the Eocene the CCD of the Pacific remained shallow (3200 m) but at the Eocene–Oligocene boundary (38 m.y.) it drops abruptly (to 4500 m) and is more spectacular in the Pacific than in other oceans where only a general fall occurs. This difference between oceans may be related to bottom water circulation. In the middle Miocene a spectacular drop in the CCD is timed to a fall in sea-level. Berger has interpreted these results as a partitioning of carbonate between shallow and deep seas. During global transgressions, the development of broad shallow-shelf seas allowed for extensive carbonate precipi-

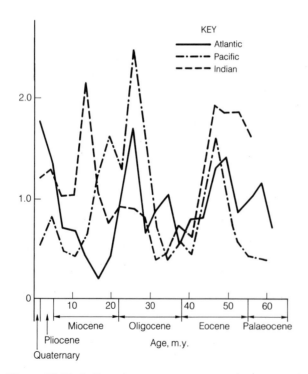

Figure 24.22 Sedimentation rates expressed in units of grams of total sediment/cm²/10³ yr in the major oceans for the last 65 m.y. based on the stratigraphic distribution of unconformities and the recovery rate differences in DSDP sites. (*After T. A. Davies and T. R. Worsley, 1981*)

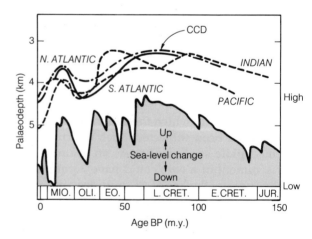

Figure 24.23 Correlation of the changes in the calcite compensation depth (CCD) for the Indian, Pacific and Atlantic Oceans over the last 150 m.y. (*after Tj. H. van Andel 1975*) and global eustatic sea-level change. (*After A. R. Vail et al, 1977 and Tj. H. van Andel, 1979*)

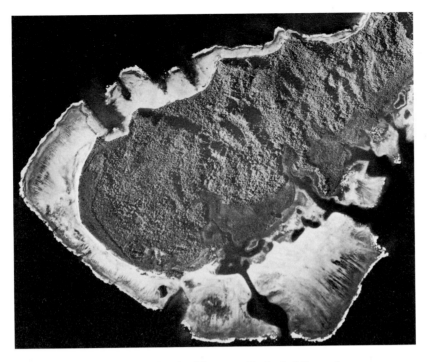

Figure 24.24 Fringing reef, British Solomon Islands. (*Overseas Geological Survey*)

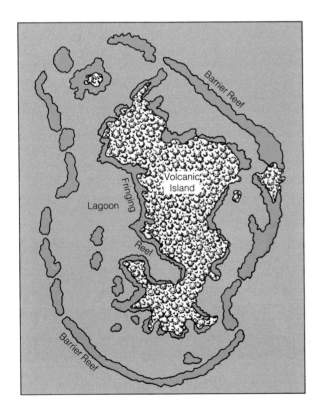

Figure 24.25 Fringing and barrier coral reefs of Mayotte, Comoros Islands, at the northern end of the Mozambique Channel. The embayed outlines of the island indicate cumulative subsidence.

tation, especially as coral and other reefs. This extracted substantial amounts of the dissolved carbonate in seawater and tended to impoverish or lower the flux of calcite to the ocean floor. The CCD, as defined, will be shallower at such times.

24.10 CORAL REEFS AND ATOLLS

In favourable situations in tropical seas, corals, together with all the organisms to which they give shelter and attachment, grow in such profusion that they build up reefs and islands of very considerable size. Clothed in vivid green, crowned by the coconut palm, and fringed with the white foam of the ceaseless surf, the 'low islands' of the Elizabethan mariners have a reputation for dazzling but treacherous beauty. Dangerous to navigation and difficult to explore, they have been equally tantalizing to geologists who sought to account for their existence. Darwin was the first to face the problems in a scientific spirit and by him coral reefs were divided into three main classes:

1. **Fringing reefs** consist of a veneer or platform of coral which at low tide is seen to be in continuity with the shore, or nearly so. The width may be 800 m or more, and where the reef is facing the open sea (Figure 24.24) the windward side may have a protective algal ridge before sloping steeply down to the surrounding sea floor, which is commonly deep. Many fringing reefs, however, have grown on the inner side of relatively shallow lagoons enclosed by barrier reefs (Figure 24.25).

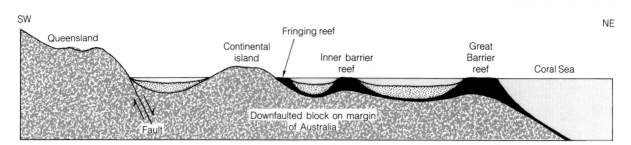

Figure 24.26 Schematic section to show the tectonic relationship of the Great Barrier reef and its associated islands to the mainland of Queensland. Reef rock, black; lagoon and channel sediments, dotted. (*After J. A. Steers*)

Figure 24.27 Hao Atoll, Tuamotu Archipelago (half-way between Fiji and Peru). (*J. P. Caplin, American Museum of Natural History, New York*)

Figure 24.28 Ifaluk Atoll looking west, western Caroline Island, western Pacific. (*U. S. Navy*)

2. **Barrier reefs** are situated up to 300 km offshore, with an intervening lagoon. The 2000 km complex of reefs known as the Great Barrier Reef, which forms a gigantic natural breakwater off the north-east coast of Australia, is by far the greatest coral structure in the world (Plate XIId, Figure 24.26). Most barrier reefs, however, of which there are countless examples, are island-encircling structures forming irregular rings of variable width, more or less interrupted by open passages on the leeward side (Figure 24.25).

3. **Atolls** resemble barrier reefs, but are without the central island (Figures 24.27, 24.28 and

Figure 24.29 Coral growths exposed at very low tide on a fringing reef off the coast of Queensland, near Port Denison. (*American Museum of Natural History, New York*)

24.30). They are essentially low-lying ring-shaped reefs enclosing a lagoon which again is generally connected with the open sea by passages on the leeward side. Within the lagoon may be isolated patches of coral forming annular or pinnacle shaped structures.

Reef-building corals live in colonies of thousands of tiny individuals (polyps), each occupying a cup-shaped depression in a calcareous framework which is common to the whole colony. As the successive generations of corals grow upwards and outwards through the restless waters in their competition for food, the stony frameworks also branch upwards and outwards and simulate the forms of certain plants, some being like shrubs and others like cushioned rock-plants (Figure 24.29). The interspaces between the dead structures are cemented and bound together by coralline algae. These precipitate calcium carbonate within themselves, and still more as incrustations which coat their surfaces and cover the coral growths to which they may be attached. Where waves are strong the outer ridge-like margin of a reef is likely to be largely constructed by purplish red coralline algae, because these organisms can withstand exposure better than corals, so long as they are kept moistened by spray. Other calcareous contributions are made by shelled molluscs, foraminifera, bryozoa, worms and bacteria,

and inorganic cementation by aragonite and magnesium-rich calcite. The whole assemblage forms a white porous limestone which gradually becomes more coherent as it is buried and subjected to prolonged saturation by seawater.

The development and maintenance of coral reefs depend upon the conditions that favour a vigorous growth of the living colonies. A thriving reef has to contend not only with the waves, but also with grazing organisms, such as fish, molluscs and echinoids, that feed on individual coral polyps and algae, and boring organisms such as bivalves, sponges, worms, shrimps, algae and fungi which penetrate the hard skeleton, weakening the structure and creating calcareous debris (Plates XIIe, f). The reef represents the margin of success in a never-ceasing struggle against death and destruction. Not only have the corals and calcareous algae to supply material to maintain a flourishing living face; they have also to provide the broken masses of coral rock and other debris that accumulate to form the low land surface of the reef and the voluminous seaward foundations required for continued outward growth. The seaward face of the living reef passes downwards into a talus slope that may descend steeply to very great depths. On the lagoon or landward side of the growing face is the **reef flat**. This may be a platform with isolated clumps or patches of coral that remain submerged except at low tide, or it may be an irregular built-up surface consisting of material thrown up by breakers to a height of about 3 to 5 m (Figure 22.2). A certain amount of debris is also washed into the lagoon when the reef is swept by heavy seas.

Reef-building corals flourish best where the mean temperature of the surface water is about 23° to 25°C, with 18°C as the lower limit of tolerance – and that not for long. Reefs and atolls are therefore restricted to a zone lying between latitudes 30°N and S, except locally where warm currents carry higher temperatures beyond these limits. The reefs of the Bermudas, for example, are dependent upon the warmth of the Gulf Stream. Along the torrid belts of the oceans the equatorial currents drift towards the west, becoming warmer on the way, and consequently reefs flourish far more successfully in the western parts of the oceans than on their colder eastern shores.

The water must be clear and salty. Opposite the mouths of rivers, where the diluted seawater carries suspended silt and mud, corals cannot live and few reefs appear. Conversely, corals grow best on the seaward side of the reef, where splashing waves, rising tides, and warm currents bring them constantly renewed supplies of oxygen and food, where the fore-reef slope carries away the smothering calcareous debris, and where, on the windward side, they are protected from violent breakers by the algal ridge. Corals cannot long survive exposure above water, nor can coralline algae if above the level continually reached by spray. Consequently living reefs can never grow much above low-tide level. Dead reefs are found above sea-level, but they are exposed as a result of a drop in the position of sea-level relative to land which can result from both uplifting earth movements as well as eustatic sea-level falls. Reef flat levels are therefore commonly used as indicators of former sea-levels.

On the other hand, as most reef-building corals live in symbiotic relationship with unicellular algae (zooxanthellae) which require abundant sunlight they do not grow in great profusion at

Figure 24.30 Map of the Suva Diva Atoll (about 80 by 60 km), Indian Ocean, showing the depth of the lagoon floor in fathoms. (*After R. A. Daly*)

depths greater than about 30 m. Coralline algae are similarly restricted to shallow water, because they, too, are dependent on an adequate supply of radiant energy. Light intensity diminishes rapidly with depth and also with distance from the tropics, becoming negligible during the long nights of the polar regions. A few corals – many still living – have been dredged up from the cold sea floor of oceans at high latitudes, for example at 260 m on the margins of Rockall Bank in the North Atlantic. These deeper-water corals do not have zooxanthellae and they live in an environment incapable of supporting the vast communities of reef-builders. Another clear indication that the latter can flourish only in warm, shallow, sunlit seas lies

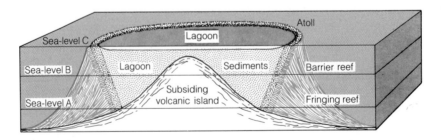

Figure 24.31 Diagram to illustrate Darwin's theory of the successive development of fringing reef, barrier reef and atoll around a subsiding island. No account is taken of the interruptions caused by the Pleistocene fluctuations of the sea level.

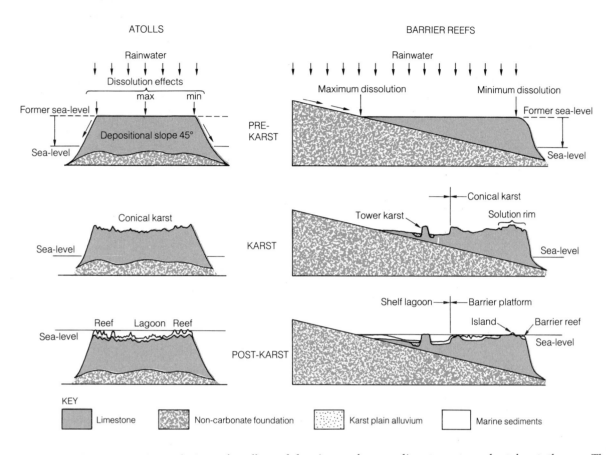

ATOLLS

BARRIER REEFS

Figure 24.32 Diagrammatic evolution of atolls and barrier reefs according to antecedent karst theory. The sequences begin with subaerial exposure and terminate with depositional consequences of a drowned karst topography. (*After E.G. Purdy, 1974*)

in the fact that reefs and atolls are easily 'drowned' if their upward growth cannot keep pace with any submergence (eustatic or tectonic) that may be in progress. Many reefs and atolls killed off in this way have been found on the sea floor. The rate of linear growth of corals varies with the species, the range for massive forms being from 4 mm to 30 mm per year, and for branching forms from 4 mm to 200 mm per year. Allowing for the necessity to provide material for the reef-flats and for the talus slopes on the seaward flanks, the average rate of reef growth obtained by ^{14}C dating in situ corals from Holocene reef cores, is about 5 mm a year.

24.11 THE ORIGIN OF REEFS AND ATOLLS

It follows from the above considerations that an essential condition for the initiation of a coral reef is the pre-existence of a submarine platform not far below sea-level. The origin of fringing reefs is easy to understand. The minute larvae of corals and the spores of coralline algae are drifted along

by ocean currents, and those that reach suitable shores find attachment and start new reefs that gradually grow upwards and outwards towards the surface and the sunlight.

Barrier reefs and atolls, however, are remarkable in that they generally rise from depths where no reef-building corals or coralline algae could live. There are two possibilities: either (a) the reefs could have grown upwards from submerged banks within 30 or perhaps 60 m of the surface; or (b) they could have grown upwards and outwards from fringing reefs during the submergence of the land or island to which they were originally attached. Another feature that calls for explanation is that the lagoons have nearly flat floors (Figure 24.30), and depths that vary with the widths, the range being from about 90 m for the larger examples (80–160 km across) to 50 m for the smaller ones.

The first general explanation was offered by Darwin in 1837 as a result of the observations he made during his celebrated voyage in the *Beagle*. He visualized all reefs and atolls as different stages in a single process (Figure 24.31). Growth begins

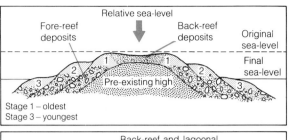

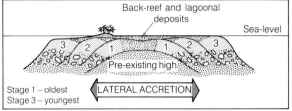

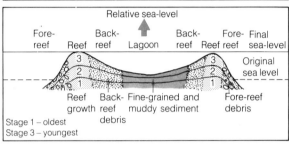

Figure 24.33 Responses of reef growth to regression, stable sea-level and transgression. (*After M. W. Longman, 1981*)

with the building of a fringing reef around, let us say, a mid-ocean volcanic island. Subsidence of the island (at rates up to 0.2 mm per year, by lithospheric cooling) combined with continuous growth converts the reef into a barrier reef. Since reefs can grow upwards at a rate of about 5 mm in a year it will rarely happen that they are unable to keep pace with the movement. The submerged area between the island and the rim of coral rock forms the lagoon. By further subsidence the summit of the central island sinks out of sight, and the barrier reef becomes an atoll.

Darwin's simple theory has not passed unchallenged, but it satisfactorily accounts for most of the features associated with reefs, including several that he had not considered. The reality of subsidence – or at least submergence – is proved by the embayed shorelines and drowned valleys of land areas within the lagoons of barrier reefs. Moreover, the shorelines are not cliffed, as they would have been but for the protection from breakers afforded by the barrier reefs. Yet cliffs there must have been when the drowned valleys were being eroded, because then the sea would have been slightly turbid and there could have been no reefs to break the force of the breakers. Any original cliffs, like the valleys, must have been drowned. The theory does not, however,

make it clear how the lagoons of the present day have come to be so remarkably uniform in depth. Figure 24.31 shows the enormous quantity of lagoon sediment necessary to fill in the 'moat' around a subsiding volcanic island. Alternatively, the flat lagoon floors of atolls and island-encircling barrier reefs suggest that the corals grew upwards from the seaward faces of platforms worn down by marine erosion and since submerged.

In 1910 Daly showed that these features are an inevitable result of the Quaternary oscillations of climate and sea-level. He had already noticed the narrowness – and therefore the youthfulness – of the reefs fringing the Hawaiian Islands. Connecting this observation with the discovery that a former glacier had left its traces on the flanks of Mauna Kea, he inferred that corals could not have flourished along those shores during the glacial stages and that the existing reefs must have grown there while the sea gradually rose as a result of the melting of ice stored up during the last glaciation. Daly thought that during the glacial stages most of the pre-existing reef-builders would be killed off, leaving only a few living reefs in sheltered spots from which the reefs of the interglacial stages and those of the present day could be colonized as the seas grew warmer. However, it is now realized that only the marginal belts of the coral seas were colder. On the whole, the Pleistocene reefs suffered much more from the oscillations of sea-level than from lowered temperatures.

The most important contribution made by Daly in his 'Glacial Control' theory was the recognition of the effects of these oscillations. Pre-Pleistocene volcanic islands and successive generations of coral reefs must have been vigorously attacked by the waves through the wide range of levels illustrated in Section 21.1 by Figure 21.5. Whenever sea-level was falling, the tops of the then existing reefs would be exposed and so become a ready prey to marine erosion and subaerial dissolution by rainwater. As the sea rose the reefs would build themselves up again. Daly was mainly concerned with the last of these upbuildings. This began some 20 000 years ago, at which time there must have been innumerable platforms and peripheral terraces of marine erosion, planed down to levels tens of metres lower than the ocean level of today.

The tops of limestone platforms would be lowered and dissected by fresh water dissolution into karst landforms, the general form of which would be dish-shaped. As sea-level rose, during the melting of the polar ice as the interglacial period started, corals would preferentially colonize the prominences (Figure 24.32). The Holocene transgression took place at an average rate of 10 mm per year up to 8000 years BP, at 5 mm per year from 8000 to 5000 years BP, at 2 mm per year from 5000

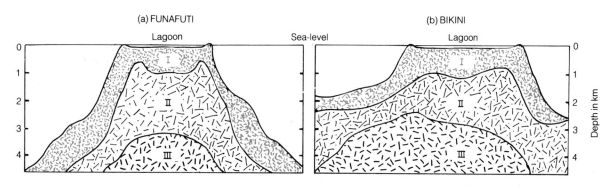

Figure 24.34 Cross sections of Funafuti and Bikini Atolls showing the submarine structures inferred from the velocities of seismic waves (Chapter 25) through the foundation rocks at various depths.
I represents reef and lagoon calcareous deposits.
II might be in part firm carbonate rock, but is more probably the equivalent of Layer 2 and mainly composed of volcanic rocks.
III corresponds to the main refracting layer of oceanic crust, composed of massive basaltic rock; its upper surface descends laterally to the normal depth of about 7 km. To the south-west of Bikini the base of the crust (the Moho) was found at a depth of 13 km. Under both atolls III is abnormally thick, as would be expected in a volcanic foundation.
   Note: the horizontal scales of (a) and (b) are not the same. Funafuti lagoon is 13 km across. Bikini lagoon is nearly 45 km across. (*Simplified from R. W. Raitt et al.*)

to 2000 years BP and at 1 mm per year since then. Sea-level has been at its present position for about 2000–3000 years (approximately 0.1 per cent of the duration of the Pleistocene). During the last marine advance many of the deeper corals would not be able to keep pace with the periods of most rapid rise of sea-level (up to 15 mm per year) and those parts of the reef would be drowned and eventually buried by deep-sea sediments.

Gradually as sea-level stabilized, shallow-water coral reefs started anew on features of positive topographic relief. The results of drilling through modern reefs reveal that the Holocene increment has, in the main, been merely a veneer of only a few metres above the pre-Holocene surface. Thicknesses vary depending upon which portion of a reef is drilled, but maximum values of about 30 m for Holocene growth are frequently reported. Where Holocene growth has been essentially vertical then the reef configuration closely duplicates that of the pre-Holocene foundations (which commonly have a karst landform) but where the pre-Holocene surface is close to sea-level then recent growth has been mainly lateral thus masking the form of the foundations. In an area of tectonic uplift, reefs rise relative to sea-level and renewed growth develops flanking terraces (Figure 24.33).

24.12   THE FOUNDATIONS OF ATOLLS

Darwin's subsidence theory refers essentially to submergence by earth movements, but it left the origin of platforms unexplained. Daly's glacial control theory makes good this deficiency and refers to submergence by a rising sea-level. Together, with the karst antecedent foundation theory, they account for all the major characteristics of coral reefs, but the results of recent borings and geophysical explorations have indicated that glacial control is only of the nature of an amendment to the more fundamental fact of subsidence, which has now been abundantly confirmed. In 1881 Darwin wrote to his friend Agassiz: 'I wish that some doubly rich millionaire would take it into his head to have borings made in some of the Pacific and Indian atolls.' Acting on this suggestion a few years later, the Royal Society of London sponsored the first of a series of drilling expeditions to Funafuti, one of the atolls of the Ellice Islands, north of Fiji. Coral rock and reef talus were penetrated to about 400 m. Some of the coral rock was thought to be in the position of growth and so to support Darwin's case, but much of the core material came from the talus slope, thus raising doubts that left the problem unsettled. More recently seismic exploration has revealed the presence of a foundation with the seismic properties of basalt underlying the whole of the lagoon at a minimum depth of about 900 m (Figure 24.34 (a)).

In 1936 a hole was drilled through the dry lagoon floor (Figure 24.35) of an uplifted atoll known as Kita Daito Jima or North Borodino. This is a small island south of Japan and west of the Ryukyu arc. A depth of nearly 560 m below sea-level was reached in reef-rock and reef sediments, the lowest strata being of Upper Oligocene or Lower Miocene age. Similar results were obtained by drilling through a lagoon north-east of Borneo

Figure 24.35 The raised atoll of Kita Daito Jima (or North Borodino) showing part of the old lagoon floor and the surrounding rim of what was once the living reef. (*R. A. Saplis, U.S. Geological Survey*)

(429 m), again without reaching a foundation for the reef.

Bikini, an atoll at the northern end of the Marshall group, became notorious when it was decided to test nuclear weapons there. Since 1946 it has been studied in great detail. Several holes were drilled, showing that corals and other organisms of shallow-water types occur in all the cores and chips brought up, the maximum depth reached being 779 m, in strata low down in the Miocene. This convincing evidence of subsidence

was followed by seismic exploration (Figure 24.34 b). The results indicated that the sediments were supported by a basement of hard rock, probably basaltic, at depths of between 2400 and 3000 m beneath the lagoon. It will be noticed that the peak of the foundation is not under the middle of the atoll. This lack of symmetry is probably due to coral growth having been more vigorous on the windward side, i.e. towards the south-east. The suspected nature of the foundation was confirmed in 1950, when basaltic rocks, some of them pyro-

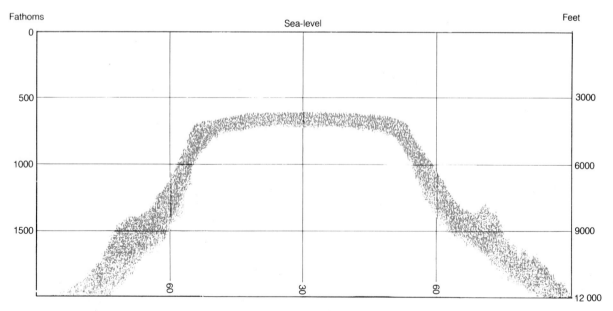

Figure 24.36 Echo-sounding record across a typical guyot south of Eniwetok Atoll. The truncated platform at a depth of 620 fathoms is nearly 15 km across, with a gently sloping rim down to 700 fathoms. The slopes then become abruptly steeper until the guyot is 35 km across at 2000 fathoms, after which they flatten out to the surrounding ocean floor at 2000 fathoms. (*After H. H. Hess, 1946*)

Figure 24.37 Deep-sea photograph of calcareous sands with pronounced ripple marks (averaging 20 cm from crest to crest), taken at a depth of 1710 fathoms (3127 m) over the platform of a guyot rising from the Atlantic floor west of Portugal; lat. 41° 12′N, long. 15° 14′W. Area photographed about 1.5 by 2.5 m. (*The National Institute of Oceanography*)

clasts, were dredged from the slopes of Bikini at various depths between 1800 and 3600 m.

In 1952 two borings were put down in Eniwetok, an atoll to the west of Bikini. One on the northern part of the rim passed in turn through reef-rock and coralline sediments (rich in foraminifera and algae, as well as corals) of Recent, Miocene and Eocene ages. Hard basement rock was encountered at 1405 m, but no samples of this were recovered. The absence of Pliocene sediments probably means that they were planed off by marine erosion during the exceptionally low sea-levels of the Pleistocene. Sediments of Oligocene age have not been recognized with certainty. The second boring, on the south-east part of the rim, passed through a similar sequence, but reached a basaltic foundation at 1283 m. This time a total of 4 m of olivine-basalt core was recovered. Thus it was definitely established that the basement supporting this atoll is an old volcanic edifice, the summit of which is still more than 3300 m above the surrounding ocean floor. To this level it has subsided during the last 50 m.y. or so, while a rim of coral reef was successfully maintained at about sea-level by upward growth. The reef also supplied sediments for the lagoon floor and the outer talus slopes. After a long period of doubt and controversy Darwin's explanation of coral

reefs can now be hailed as one of the very few Victorian 'theories' that has not become a casualty as a result of modern discoveries.

24.13 SEAMOUNTS AND GUYOTS

The evidence that the summits or flanks of worn-down volcanic cones served as the foundations of early Tertiary coral reefs, and afterwards subsided a thousand or more metres while the reefs grew upwards, at once suggests that there should be similarly drowned islands outside the coral seas but uncrowned by coral reefs. And even within the coral seas examples might be expected, provided that the islands subsided so rapidly through the critical zone that the coral community could not keep pace. These inferences would have been triumphant predictions had they not already been fulfilled. While on wartime service in the Pacific, H. H. Hess saw from the echo-sounder records that at least twenty flat-topped seamounts had been crossed during his voyages, the depths of their summits ranging from 900 to 1800 m below sea-level. In a celebrated paper published in 1946 Hess interpreted these remarkable features as volcanic islands which had been truncated by wave erosion and deeply submerged (Figure 24.36). For the flat-topped seamounts he proposed the name **guyot** (after Arnold Guyot, a Swiss geographer of the eighteenth century) to distinguish them from other seamounts, such as volcanic cones that had failed to reach the surface. This work was followed up by several expeditions organized by various governments and private institutions, culminating in the International Geophysical Year. Among many surprises, the unexpectedly high speeds of currents at various depths was not the least. The reality of these currents is impressively demonstrated by the ripple marks seen on many of the photographs that have been taken by cameras lowered to the appropriate depths. Figure 24.37, illustrating conditions at a depth of well over 3000 m, is an excellent example from the Atlantic, the exploring vessel being a later HMS *Challenger*. Other pictures taken over the same guyot show part of the bedrock margined by a thin cover of ripple-marked sediment. The half-scoured-out boulders seen in Figure 24.37 are probably wave-worn fragments of the same volcanic bedrock.

In addition to some 400 islands charted in the Pacific, well over 1000 deeply submerged seamounts have been recorded and at least 100 of these in sufficient detail to be recognized as guyots. Half of the guyots have been dredged, as well as some of the sharp-peaked seamounts, and in every case the materials brought up have been mainly volcanic.

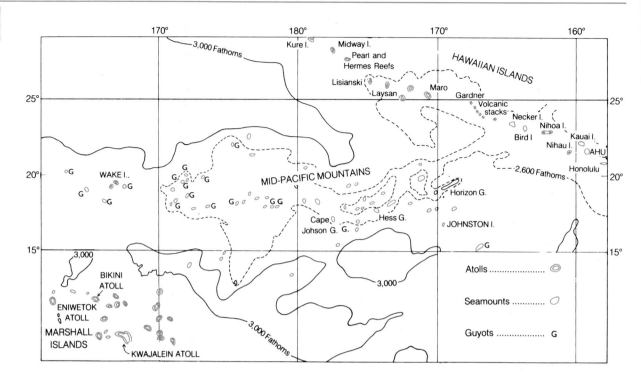

Figure 24.38 Simplified chart to show the sunken seamounts of the Mid-Pacific Mountains. The greater part of the Hawaiian Chain appears in the north-east and the northern group of the Marshall Islands in the south-west. (*After E. L. Hamilton, 1956*)

In 1950 the Scripps Institution and the US Navy joined in a Mid-Pacific expedition to determine the structure of Bikini Atoll and to investigate some of the guyots west of Hawaii. The guyots were found to be truncated peaks on a gigantic submarine range extending for 2400 km from Necker Island in the Hawaiian Chain to near Wake Island and now known as the Mid-Pacific Mountains (Figure 24.38). A typical example is Hess Guyot with a flat top, 19 × 13 km, standing about 1700 m below sea-level, but 3000 m above the surrounding ocean floor. Dredging from just below its top, and also from the edge of Cape Johnston Guyot, brought up reef-corals dating from the Middle Cretaceous, about 100 m.y. ago. Hollows and fissures in the flat tops of four of the guyots yielded globigerina ooze of Palaeocene and Eocene ages. But in all cases by far the commonest material of the hauls was olivine-basalt debris, ranging from rounded boulders and pebbles to erosional products of smaller sizes. The evidence is clear that the guyots formed a chain of basaltic islands that were eroded down to sea-level in Cretaceous times. Corals found lodgment here and there and many small reefs seem to have started, but the rate of submergence was such that they never developed into atolls.

E. L. Hamilton, who has graphically described the mid-Pacific range (Figures 24.38 and 24.39) compares it with the younger Hawaiian Chain,

each being a complex linear series of volcanic ridges and peaks rising from a broad swell on the ocean floor. Another great range of seamounts begins just beyond the west-northwest end of the Hawaiian Chain, but is aligned northwards towards the junction of the Aleutian and Kamchatka–Kurile trenches. This is known as the Emperor Range because its guyots and seamounts were first discovered by the Japanese, who named them after their former emperors. Like the higher mid-Pacific peaks, those of the Emperor Range probably passed through their subaerial history long before the Hawaiian volcanoes began their activities.

Radiometric dating of basalts can give an indication of the extent of lateral movement of volcanic islands riding on ocean plates, e.g. Hawaii 15 cm per year, Austral Island 9 cm per year (see also Section 28.2), Marquesas Island 10 cm per year. Dating and magnetic studies of successive basalts on volcanic islands and atolls show that Midway Atoll has migrated through 13° of latitude over 18 m.y. (a movement of 1400 km) and is moving out of the northern limit of coral seas whereas Pitcairn Island at 24°S is presently reefless and moving along a north-west trajectory into reef seas at a rate of 11 cm per year. Measurements of the rate of coral growth and percentage areal coral cover has recently shown a marked reduction at higher latitudes. The latitudinal limit beyond which coral

Figure 24.39 An artist's impression of the Mid-Pacific Mountains. 'If the Pacific Ocean were drained away, the mile-deep sunken islands would emerge as truncated volcanic cones. The original oil painting is by the distinguished scientific illustrator Chesley Bonestell and is based on part of the bathymetric chart of the Mid-Pacific range.' (*Reproduced by permission of E. L. Hamilton*)

growth cannot keep pace with mid-ocean plate subsidence and thus the threshold for atoll formation has been termed the Darwin Point. The Darwin Point exists at the northern end of the Hawaiian Archipelago at 28°N latitude; atolls and coral islands transported north-west by tectonic movement on the Pacific plate appear to have been drowned near the Darwin Point during the last 20 m.y. (Figure 24.40).

24.14 FOSSIL REEF-CORALS AND ANCIENT CLIMATES

As indicators of past climates the distribution of fossil reef-building corals has a clear bearing on continental drift. Several geologists have pointed out that in Greenland and the Arctic Islands of Canada such corals flourished well within the Arctic Circle during Lower Palaeozoic times, and that since then the broad belt outlining their distribution has moved southwards towards the tropical zone of coral reefs of the present time. If reef-building corals were restricted in the past to the conditions which favour them today this means that the drift of Europe and North America had a strong northerly component. Alternative suggestions have been made: e.g. that Palaeozoic reef-building corals may have preferred cooler water, like some of the solitary species of today; or that the Arctic seas were then sufficiently warm, but that the tropical seas were too hot for coral growth. However, such *ad hoc* hypotheses are easily ruled out. Whether corals could survive in cold waters or not, they could only flourish with a continual and plentiful supply of sunshine. This condition could not be fulfilled during the long winter nights of the polar regions – unless,

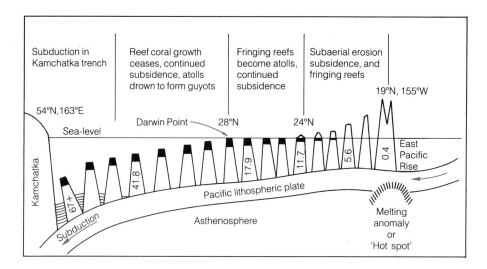

Figure 24.40 Schematic illustration of the Hawaiian Emperor chain migrating on the Pacific plate towards the Kamchatka trench. At a latitude of 28°N (the Darwin Point) reef growth slows to a rate that cannot keep up with subsidence, and guyots result. (*After Grigg, 1982*)

Figure 24.41 X-radiograph of *Montastrea annularis* coral slab, revealing seasonal density banding. Annual increment of linear growth is about 15 mm and the coral was about 10 years old when collected.

indeed, its equivalent was supplied by marine organisms radiating so powerfully that they could belong only to the realms of science fiction. When corals were living in the Lower Palaeozoic seas that covered the regions now known as the Queen Elizabeth Islands and the north of Greenland, those regions were surely situated far to the south of the Arctic Circle. Independent evidence of this

from the occurrence of salt domes has already been presented (Section 22.7). It seems fair to infer, as P. M. S. Blackett has done, that the mean latitude of the coral reefs of a given geological era or period may be taken as giving a first approximation to the average position of the equatorial belt during that interval. Blackett discussed the data very thoroughly and summarized it as shown in Table 24.4.

Russian and Chinese geologists have arrived at similar conclusions by measuring the convexities of the cushion-like forms of banded algal reefs of known ages. Here the principle involved is the well-known response of plants to sunlight, the turning towards the light technically called **heliotropism**. In the case of living algal reefs the axes of the convexities point towards the direction of maximum receipt of radiation from the sun. Algal limestones carry the climatic record far back into Precambrian time. A most interesting example comes from a region east-north-east of Peking, where algal reefs occur at intervals in unfolded Sinian strata. The time period represented is from 1200 to 600 Ma ago. The convex upper surfaces of the bands are of the same kind as those formed by algal communities now living. But detailed measurements of the directions of the axes of the convexities reveal a progressive change in the apparent angle of the sun's maximum radiation. This can be interpreted either in terms of continental drift to mean that during the last 600 Ma of Precambrian time the region was displaced southwards through 40° of latitude; or in terms of polar wandering to mean that the north pole migrated northwards by the same amount. The real explanation is likely to be a complex mixture of both.

Table 24.4 Coral reefs giving position of equatorial belt

Geological periods	Range of age in m.y.	Mean age in m.y.	Mean latitude of reef-coral distribution	
			America	Europe and Africa
Cretaceous ⎫ Jurassic ⎬ Triassic ⎭	70–225	150	16°N	35°N
Permian ⎫ Carboniferous ⎭	225–350	290	38°N	61°N
Devonian ⎫ Silurian ⎭	350–440	395	50°N	47°N
Ordovician	440–500	470	53°N	61°N

The external and internal structure and the skeletal chemistry of corals respond to changes in the physical and chemical conditions of growth and are thus sensitive indicators of past marine environmental conditions. The growth forms of coral reflects the setting of growth. When a massive hemispherical form reaches the air/water interface a distinctive micro-atoll form, with flat dead top and living sides, is created. Minute changes in the relief of the surface then accurately reflect the history of changing water level. Seasonal changes in climate are reflected in dense and less dense banding within the coral structure revealed by X-radiography (Figure 24.41). Dense bands are normally created when conditions of growth are not optimum, for example during a period of greater cloud cover in the wet season. As each couplet of dense and less dense layers represents a year, density banding provides an accurate chronology, and since some corals can grow to be several metres in diameter they may have recorded in their structure hundreds of years of the prevailing conditions of growth. We can thus use corals to indicate short-lived periods of extreme conditions (for example severely low or high temperatures) which can be preserved within the coral structures as unusually dense bands.

Corals are known to incorporate into their skeletons elements and compounds present in the seawater. Terrestrial fulvic acids which are washed from the land out to sea during rainy seasons are incorporated into near-shore corals of the Great Barrier Reef of Australia. These compounds fluoresce when subjected to ultraviolet light and the intensity of fluorescence is directly related to the runoff. Consequently long-lived near-shore corals can reveal centuries of coastal rainfall record, greatly aiding our understanding of past climates. Similarly the quantities of metals, such as lead, incorporated into coral skeletons have been used to indicate fluctuations in the history of the use of industrial lead by man.

24.15 FURTHER READING

Darwin, Charles (1842) *The Structure and Distribution of Coral Reefs*, Smith Elder, London.

Gross, M. G. (1986) *Oceanography*, 3rd edn, Prentice Hall, Englewood Cliffs, New Jersey.

Hopley, D. (1982) *The Geomorphology of the Great Barrier Reef*, John Wiley, New York.

Isdale, P.J. (1984) Fluorescent bands in massive corals record centuries of coastal rainfall, *Nature*, **310**, 578–9.

Kennett, J. P. (1982) *Marine Geology*, Prentice Hall, Englewood Cliffs, New Jersey.

Kuenen, Ph. H. (1958) No geology without marine geology, *Geologische Rundschau*, vol. 47, 1–10 (in English).

Pettersson, Hans (1960) Cosmic spherules and meteoric dust, *Scientific American*, **88** (2), 123–32.

Scoffin, T.P. and Stoddart, D.R. (1978) The nature and significance of microatolls, *Philosophical Transactions of the Royal Society of London*, **B284**, 99–122.

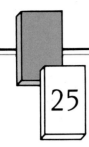

EARTHQUAKES AND THE EARTH'S INTERIOR

Man is learning to harness for his enquiring use the very wrath of the earth; the tremblings of our vibrant globe are used to 'X-ray' the deep interior

Reginald A. Daly, 1928

25.1 THE NATURE OF EARTHQUAKES

When a stone is thrown into a pool, waves spread through the water in all directions. Similarly, waves spread out through the earth when there is a sudden localized disturbance of the rock. An earthquake is such a disturbance. Near the source of an earthquake the shaking of the ground can be felt and the effects may be catastrophic, but further away the **seismic waves** become smaller until they can be detected only by delicate instruments called seismographs (Gr. *seio*, to shake; *seismos*, an earthquake).

Seismic waves may be generated by any sudden displacement. A quarry blast, or the passage of a train, or even a hammer blow is sufficient to create seismic waves which then propagate through the earth to distant points; but only underground nuclear explosions are comparable in size with large earthquakes. We now believe that the type of sudden displacement at the source of most earthquakes, including all larger ones, is slippage along a fault. A few earthquakes associated with volcanic activity may have a different, more explosion-like mechanism. Earthquakes along faults are distinguished as **tectonic** earthquakes. The term

tectonic (Gr. *tekton*, a builder) refers to any structural change in rocks brought about by their deformation or displacement.

Figure 25.1 shows a spectacular example of slippage along a fault plane associated with a tectonic earthquake. Here the slippage has reached the surface and the type of displacement is clear. Such examples are rare however, as most earthquakes occur along buried faults. Even when large displacements are observed at the surface these are often an indirect manifestation of the fault movement, because near-surface sedimentary layers often exhibit local effects not characteristic of the fault beneath.

Rocks under increasing stress eventually rupture. The moment at which this occurs, and the exact place, are impossible to predict, but once rupture has initiated the increased stress concentration at the edges of the ruptured area ensures that the rupture itself propagates along the fault; the sudden slippage occurs at each point as the rupture front passes, and stress is relieved over an extended region. The total area over which rupture occurs is related to the size of the earthquake,

Figure 25.1 One of two parallel scarps, 2–3 km apart, representing the surface expression of the movement and the resulting major earthquake that originated at a depth of 35 km beneath Quiches in the Andes of Peru on 10 November 1946. The scarp illustrated has a vertical throw of 4 m and an unbroken length of 5 km through country at an elevation of 3600–3900 m. The zone between the two scarps subsided about 3 m on average. (*Arnold Heim*)

and determines its duration; the longest ruptures may extend for several hundred kilometres, taking tens of seconds to reach the extremity. The amount of slippage which takes place at each point is another factor which is related to the size of the earthquake, and may be up to several metres for the largest. The direction of this slippage, and the orientation of the fault plane itself, both depend upon the type of stress that has built up. However, this relationship is complicated by the fact that earthquakes tend to happen along pre-existing faults, which may have been formed at a time when the stress orientation was different.

Types of fault movement have already been described in Section 8.2(10) and may be vertical, horizontal or oblique. After the great Alaskan earthquake of 1899, it was possible from the presence of barnacles clinging to the uplifted rocks of Disenchantment Bay to measure the uplift, which in this case reached an exceptional maximum of 14 m. Crustal blocks often move obliquely, both vertical and sideways movements being observed. Surveys carried out after the Kanto earthquake of 1923, when Tokyo and Yokohama were wrecked, showed that the floor of the Bay and the surrounding mainland had been twisted round in a clockwise direction, the measured shift of the volcanic island Oshima being over 3.5 m (Figure 25.2).

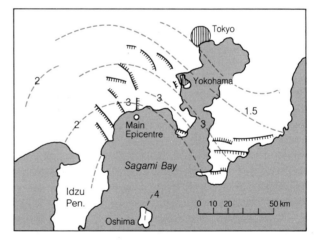

Figure 25.2 Sketch map of the Sagami Bay area, Japan, to indicate the surface movements associated with the disastrous Kanto earthquake of 1 September 1923. Vertical displacements ranging from several centimetres to a metre or two are shown by shading on the downthrow side of the fault lines. Horizontal rotation (clockwise) is indicated by dashed lines, with numbers representing the approximate displacements at various localities in metres.

Some earthquakes are preceded by smaller **foreshocks**, and larger earthquakes are always followed by **aftershocks**. These are usually interpreted as the readjustment of stresses in a volume of disturbed rock close to the fault plane, but aftershocks may also be associated with the main fault. The accurate location of aftershocks gives us clues about their relationship with the main earthquake. However, the almost complete absence of aftershocks following some smaller earthquakes testifies to the variety of conditions which are possible. For example, two earthquakes of similar size occurred in the western United Kingdom on 17th July 1984 and 2nd April 1990. The first one of these was in the Llŷn Peninsula of North Wales and had many aftershocks, which were still being recorded six years later. The second, near Bishop's Castle on the Welsh border, produced no significant aftershocks. There is evidence that previous earthquakes in these two parts of the UK had a similar contrast in their number of aftershocks; this suggests a dependence upon locality which is so far unexplained.

Earthquakes are newsworthy when there is great loss of life. This gives a very distorted picture of their number and size because most occur in places far from habitation, and because even a quite large earthquake may cause little damage in a town built to withstand earthquakes. The Loma Prieta earthquake in California on 17th October 1989 claimed fewer than 100 deaths, while that in Armenia on 7th December 1988 claimed about 25 000, although both were of similar size and close to major cities. Throughout the world there are hundreds of earthquakes each day large enough to be felt, and typically five capable of major damage. The largest earthquake for more than a decade occurred on 23rd May 1989. It was much larger than all those which hit the headlines

during the 1980s, but it went largely unnoticed except by seismologists, as it was located offshore beneath the Macquarie Ridge, midway between New Zealand and the Antarctic.

25.2 EFFECTS OF EARTHQUAKES

One of the most alarming features of a great earthquake experienced on land near the place of origin is the passage of ground waves which throw the surface into ever-changing undulations. These may appear to be only 30 cm or so high and 6 to 9 m from crest to crest, but the effect is so universally terrifying that accounts of the writhing of the ground are usually wildly exaggerated. Indeed, it is now believed that these accounts are the result of an optical illusion. During the main shock of the earthquake swarm that ravaged Chile in May 1960, the motion of the ground was described by scientific observers as 'slow and rolling like that of the sea during a heavy swell. Parked cars in Concepción rolled to and fro through a distance of half a metre and bobbed up and down'. The period of the waves was 10 seconds or more and they continued to pass for 3 or 4 minutes before ceasing to be visible.

Apart from the dislocation of structures built across a fault when stress is suddenly released in an earthquake, the most spectacular and perilous effects in great earthquakes are due to the passage of surface waves. In extreme cases fissures may open, although most such effects relate to the compaction of sediments as a result of the disturbance (Figure 25.3). Water pipes and gas pipes may be cracked open as well as roads (Figure 25.4).

Figure 25.3 Deformation of the quayside, Yokohama, left after the passage of surface waves; Kanto earthquake of 1923.

Figure 25.4 Gaping fissure left in a country road near Yokohama; Kanto earthquake of 1923.

Railways may be buckled and twisted (Figure 25.5); bridges may collapse and buildings crash to the ground. Frequent shaking through a few millimetres suffices to wreck most buildings not specially constructed to withstand earthquakes. A movement of as little as 0.01 mm can be distinctly felt if one is sitting or standing still, although it is primarily the acceleration of the ground, proportional to the force on any object or building, which determines the severity of the effect of a seismic wave. In a strong earthquake the effect of the up-and-down vibrations on the feet has been described as like 'the powerful upward blows of a monstrous hammer'.

In the region of destruction, landslides may be set moving on valley sides, and avalanches may be started in snowy mountains. Glaciers may be shattered and where they terminate and break off in the sea icebergs become unusually abundant. Vast masses of loose sediments may be so disturbed by submarine shocks that they slump for many kilometres down the continental slope. In Sagami Bay in 1923 landward parts of the floor were thus uncovered by 300 to 450 m, seaward parts being correspondingly built up where this enormous bulk of sediment came to rest. The passage of seismic waves through water-filled sands, especially in alluvial districts, brings about the compaction of the deposit. The resulting decrease of volume obliges the water to escape, which it often does with sufficient violence to ascend through fissures as powerful sandy jets. These may issue at the surface like isolated fountains around which sand-craters develop. It is common for soft sediments near the surface to behave as a liquid during a large earthquake, with disastrous effects on the stability of whole buildings.

Groundwater and its circulation may be greatly disturbed in other ways by earthquakes. Old lakes may be drained off through open cracks, and new ones formed in depressions. In 1935 Lake Solar was engulfed by a great fissure that opened across its floor in the Kenya rift valley. Not all fissures close up again, as Figure 25.6 clearly demonstrates.

The appalling losses of human life that accompany great earthquakes in highly populated areas are mainly due to secondary causes such as the collapse of buildings, fires, landslides and the giant waves called **tsunami**[1] from the Japanese (*tsu*, harbour; *nami*, waves). Gas mains may be

Figure 25.5 Buckling of railway lines near Tokyo; Kanto earthquake of 1923.

[1] It should be noted that this word is plural, but is also used in the singular, like our English word *sheep*.

broke out in all directions and completed the toll of death and destruction. At least 250 000 lives were lost and twice as many houses destroyed. In the loess country of Kansu in China, 200 000 people were killed in 1920 and another 100 000 in 1927 by catastrophic landslips of loess which overwhelmed cave dwellings, buried villages and towns and blocked river courses, so causing calamitous floods.

25.3 THE LISBON EARTHQUAKE OF 1755

The catastrophic earthquake that ruined Lisbon during the morning of All Saints' Day in 1755, is one of the largest known in western Europe. The destruction of Lisbon by this earthquake and the accompanying fires and tsunami shocked Western civilization more deeply than any other natural calamity before or since (Figure 25.7). In Portugal itself it seemed incomprehensible that such cruel blows should have occurred on a holy day, at the very time when the city churches were crowded with worshippers, many thousands of whom were crushed to death by the collapsing masonry. Abroad, the sense of horrified awe was intensified by a panic-born rumour, destined to become a widespread belief that still persists, to the effect that the broad marble quay alongside the Tagus, together with all the people who had fled to it in the vain hope of finding safety, had vanished for ever into a fissure that opened and closed again during the passage of the second and greatest of the main shocks. Engineers afterwards showed this story to be false. Although the marble blocks were loosened and dislodged they were still there. But the loss of life was no less real, for the quay and indeed all the lower parts of the city were swept by gigantic tsunami. Waves up to 9 m high rushed in and completed the ruin and destruction. The people on the quay were washed away and their disappearance added plausibility to the more dramatic story of their engulfment. Moreover, there was a similar report from a town in Morocco where many thousands of the population were said to have been 'swallowed up by the earth'. This is undoubtedly a wild exaggeration; there is to this day no confirmed report of anyone being 'swallowed' by a fissure during an earthquake. Nevertheless, the possibility of engulfment has always been the most terrifying of the many perils associated with the great earthquakes. In Christendom the horror was magnified by fear of a sudden descent into the fiery torments of hell.

The Lisbon survivors only described what they had been taught to expect to happen, as witnessed by the appalling account of an earthquake recorded in the sixteenth chapter of the Book of

Figure 25.6 One of the many permanent fissures opened during the Orleansville earthquake, Algeria, of 1954. (*Paris Match*)

torn open and, once started, fires can rapidly spread beyond control, since the water mains may also be wrenched apart. In San Francisco in 1906 far more damage was done by fire than by the earthquake itself. The Kanto earthquake of 1923 occurred just as the housewives of Tokyo and Yokohama were cooking the midday meal. Fires

Figure 25.7 Old wood engraving telescoping the story of collapsing buildings, fires, tsunami, and 'engulfment' of the marble quay that accompanied the Lisbon earthquake of 1 November 1755. (*Scribner, Walford and Co.*)

Numbers. There we read that Korah and his associates and 250 followers had the temerity to accuse Moses and Aaron of taking too much upon themselves. Moses 'was very wroth'. Not content with rebuking them, he threatened them with the dire consequences of provoking the Lord:

'And it came to pass, as he had made an end of speaking all these words, that the ground clave asunder that was under them:

And the earth opened her mouth and swallowed them up, and their houses, and all the men that appertained unto Korah, and all their goods.

They, and all that appertained to them, went down alive into the pit, and the earth closed upon them: and they perished from among the congregation.

And all Israel that were round about them fled at the cry of them: for they said, Lest the earth swallow us up also.'

This last verse well expresses the general consternation felt throughout Europe after the Lisbon catastrophe. An unprecedented outburst of public discussion followed. The earthquake was variously attributed to the Devil and his legions of evil spirits; God's anger against the scandalous behaviour of his worshippers in the Lisbon churches; the need for frightening sinners into repentance; the need to punish Portugal for the undue severity of its Inquisition; and the need to remind human-

ity of the flames of hell-fire within the earth. John Wesley was particularly eloquent on the last of these themes. He boldly asserted that earthquakes were punishments for sin, and that any attempt to explain them away as natural consequences of the earth's internal heat was itself a sin. In a word, trying to account for earthquakes was doomed to produce more earthquakes.

The philosopher Kant, who was only thirty in 1755, took a very different line. He firmly condemned all attempts to interpret God's purpose as acts of outrageous impertinence. He regarded the Lisbon earthquake as a purely natural occurrence and made the sensible suggestion that we should find out where earthquakes were likely to happen and then take care not to build cities in such places. (This still applies!) Rousseau also regarded earthquakes as a necessary part of nature and pointed out that we cannot expect to prevent an earthquake at a particular place merely by building a large number of churches there.

In the then narrow world of science the most notable contribution stimulated by the Lisbon earthquake came from John Mitchell (1724–93). Mitchell had already invented a torsion balance for determining the mean density of the globe and had done pioneering work in terrestrial magnetism and structural geology. Having collected what information he could about the effects of the great earthquakes and their distribution, he published his tentative conjectures in 1760, but not

Table 25.1 The 'modified Mercalli' scale of seismic intensities (adapted and abbreviated)

Intensity	Description of characteristic effects
I	**Instrumental:** detected only by seismographs
II	**Feeble:** noticed only by sensitive people
III	**Slight:** like the vibrations due to a passing lorry; felt by people at rest, especially on upper floors
IV	**Moderate:** felt by people while walking; rocking of loose objects, including standing vehicles
V	**Rather strong:** felt generally; most sleepers are awakened and bells ring
VI	**Strong:** trees sway and all suspended objects swing; damage by overturning and falling of loose objects
VII	**Very strong:** general alarm; walls crack; plaster falls
VIII	**Destructive:** car drivers seriously disturbed; masonry fissured; chimneys fall; poorly constructed buildings damaged
IX	**Ruinous:** some houses collapse where ground begins to crack, and pipes break open
X	**Disastrous:** ground cracks badly; many buildings destroyed and railway lines bent; landslides on steep slopes
XI	**Very disastrous:** few buildings remain standing; bridges destroyed; all services (railways, pipes and cables) out of action; great landslides and floods
XII	**Catastrophic:** total destruction; objects thrown into air; ground rises and falls in waves

without a salutary warning that he considered them to be still insufficiently supported by facts. What impressed him most was the enormous area over which the earthquake was felt and the evidence that it was something that travelled outwards from the neighbourhood of Lisbon, gradually dying out in all directions. He knew, for example, that distant lakes were set swaying, like water in a tilted bath, not only in Switzerland but as far away as Loch Ness, where the surface continued to rise and fall through a range of 60 to 90 cm for about an hour. This long-distance tilting he contrasted with the limited effects of volcanic earthquakes, which shake the surrounding country 'for only 10 or 20 miles'. Ascribing volcanic eruptions to the escape of vapour generated when large quantities of water come into contact with 'Subterranean Fire', he asks himself, *What is to be expected when the vapours are confined?* It was Mitchell who originated the idea that volcanoes act as safety valves in so far as they allow the upward escape of vapours. But, he thought, if the vapours originate too deeply to find a passageway to the surface, they must give rise to an explosive shock followed by a wave-like migration through the rocks in all directions. Mitchell was thus the first to recognize that from the region of the initial shock an earthquake spreads out in waves to distances so great that they can no longer be perceived. This was a remarkable achievement for two centuries ago. Mitchell may justly be hailed as the one who discovered seismic waves, and indeed as the founder of the vast subject that has since developed into geophysics.

25.4 EARTHQUAKE INTENSITIES AND ISOSEISMAL LINES

Mitchell's recognition that the destructiveness of an earthquake – or what we now call its **intensity** – decreases outwards from the source of the disturbance stimulated attempts to define the degree of intensity in terms of information that could be supplied by people living in the area where the earthquake could be felt. From the human point of view damage depends on population density, building standards and the nature of the ground, weak alluvium being more susceptible than strong bedrock. Intensity has for many years been expressed by reference to an arbitrary scale from 1 to 12, which is summarized in Table 25.1. The present scale is a modification of a similar one

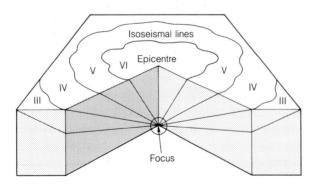

Figure 25.8 Block diagram showing isoseismal lines and their relation to the epicentre and to the wave paths radiating from the focus of an earthquake. The Roman numerals indicate intensities.

devised by the Italian seismologist, Mercalli.

In physical terms intensity is determined partly by the duration of the perceived disturbance, and also by the maximum rate of change of these movements of the ground, i.e. by its maximum acceleration. This can only be measured accurately using **accelerometers** designed for the recording of strong ground motion.

It is essential to recognize that intensity measures the effect of the seismic waves reaching the observer. This effect increases with their strength and also depends upon the ground conditions; it has already been stated that unconsolidated material and soft sediments have a greater effect than hard rock. Since the seismic waves become weaker as they spread out from the earthquake, the intensity decreases with distance. Thus the intensity measured at one point has no bearing upon the earthquake size, except perhaps to impose a lower limit, and only when intensity measurements are collected from the whole **felt area**, can an estimate of the size be made. Such surveys of **macroseismic** observations are commonly conducted after significant earthquakes using standard questionnaires.

Before seismographs, intensities provided the only way to locate earthquakes; the location of maximum intensity provides an estimate of the point immediately above the earthquake, while the intensity falls off more slowly with distance if the earthquake is deeper. So we can learn a good deal from the simple concept of intensity, as did the pioneer seismologists. A line drawn through all the places with the same intensity is an **isoseismal line** (Figure 25.8). Each one generally encloses a roughly circular or elliptical area. The place of origin is called the **focus**, and the point or line on the surface vertically above is the **epicentre**. From the intensities at the epicentre E and at a point G on an isoseismal line at a known distance from E, R. D. Oldham showed, at least in principle, how the depth of focus could be estimated (Figure 25.9).

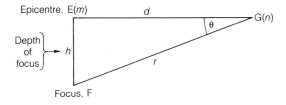

Figure 25.9 To illustrate Oldham's method for estimating to a first approximation the focal depth (h) of an earthquake

Intensity at E = m (known)

Intensity at G = n (known)

Distance between E and G = d (known)

Problem: to find h, the depth of the focus.

From the focus the intensity (expressed in terms of acceleration) theoretically decreases outwards inversely as the square of the distance. Thus to a first approximation we have $n/m=h^2/r^2=(\sin\theta)^2$. The angle θ being thus determined, $d\tan\theta=h=$the depth of the focus.

From the records of 5605 shocks in Italy, Oldham found that 90 per cent of the earthquakes originated at depths of less than 8 km; nearly 8 per cent at depths between 8 and 30 km; and the rest at greater depths. Tectonic earthquakes are now classified as:

Shallow, when the depth of origin is less than 60 km (or 70 km, the depth adopted by some seismologists).

Intermediate, when the depth of origin is between 60 (or 70) and 300 km.

Deep, when the depth of origin is more than 300 km, the maximum depth so far recorded being about 720 km. Most deep earthquakes originate between 500 and 700 km.

A few earthquakes are related to volcanic activity (Figure 25.10). These earthquakes are always shallow, and are small compared with the largest tectonic earthquakes.

25.5 EARTHQUAKE MAGNITUDE AND SEISMIC MOMENT

From a measurement of the amplitude of seismic waves made from a seismic recording or **seismogram** it is possible to estimate directly the size of an earthquake provided its distance from the seismograph is known. This was recognized by C. F. Richter, who in 1935 devised a scale for comparing the magnitudes of Californian earthquakes. His scale defined the **magnitude** of an earthquake and related this to the logarithm (to the base 10) of the maximum amplitude in micrometres (units of 10^{-6} m) measured on a certain seismograph with specified characteristics, at a distance of 100 km. A table of empirical corrections was compiled for observations made at different distances. A number of magnitude scales have been developed since, suitable for different types of seismograph and for either local, regional or global magnitude determinations, and relationships have been established to enable values to be converted between the different scales. However, the **Richter magnitude**, whose definition is closely related to the original one, has persisted as the international measure of earthquake size, and is universally recognized by the general public. From Richter's definition it is clear that the numerical value of the magnitude does not have to lie between 0 and 10 as is commonly supposed. Extremely small events have a negative magnitude, and only our experi-

Figure 25.10 Fissure opened during a swarm of earthquakes preceding a volcanic eruption, south-west rift zone of Kilauea, Hawaii. Such fissures, although spectacular, are always secondary features, perhaps resulting from landslides or slumping of near-surface material. (*Alex. Cockburn*)

ence of the largest earthquakes sets any upper limit. Also, because the magnitude is a logarithmic function of the amplitude, the difference in size corresponding to successive magnitude units increases rapidly as we move to larger magnitudes. Since 1904, when seismograms first provided information from which magnitudes could be calculated, only a few shocks to the end of 1991 have had magnitudes of about 8.5 or larger:

1906 Andes of Colombia/Ecuador
1906 Valparaiso, Chile
1911 Tien Shan, Sinkiang, China
1920 Kansu, China
1933 Japanese Trench
1950 North Assam, India
1960 Chile (three major shocks)
1964 Alaska
1968 Offshore Japan

The San Francisco earthquake of 1906 had a magnitude of about 8.25; that of the Kanto earthquake of 1923 (Section 25.2) was about 8.2. There are difficulties in determining Richter magnitudes for very large earthquakes, and any values quoted are always approximate.

The magnitude is closely related to the total amount of elastic energy propagated away from the focus as seismic waves, and the following (again empirical) relationship was found between magnitude and this energy by B. Gutenberg and C. F. Richter,

$$\log E = 4.8 + 1.5M$$

where the energy, E, is expressed in joules (J). For a magnitude of 8.6, which is very rarely exceeded, E amounts to 10^{17} J. The average annual release of energy from all earthquakes ranges from about 10^{18} to 20^{20} J and most of this, 80 per cent or more, is generally accounted for by a few really major shocks. Not all the energy released in an earthquake is propagated away in seismic waves as elastic energy; typically only one per cent of it is. The rest may be converted into potential energy associated with uplift of large areas of ground, or into kinetic energy which dissipates as heat.

In recent years seismologists have come to realize the shortcomings in all the intensity, magnitude and energy scales. The problem is that they are based upon observed relationships with no theoretical foundation. None of them measures directly a physical property of the earthquake itself. To address this problem new quantities have been introduced, which attempt to relate seismogram measurements to properties of the earthquake source through well-defined mathematical formulae. The most important of these is the **seismic moment**, which is a more rigorous measure of the earthquake size. Rupture and slippage along a fault plane can be thought of as a rotational motion: material on opposing sides of the fault moves parallel to the fault in opposite directions. This corresponds to a net rotational force about a point exactly on the fault. The seismic moment, M_0, is the summation of these rotational effects due to each element of rock, and is given by

$$M_0 = \mu A d$$

where μ is a constant depending upon the material, A is the area of the ruptured fault and d is the average slip. M_0 can be calculated from the seismogram in various ways. For example, if the amplitude of the observed seismic wave is plotted against wavelength, then the resulting curve tends towards an amplitude at long wavelengths which is proportional to M_0.

25.6 TSUNAMI

The giant sea waves often associated with earthquakes of high magnitude are caused by displacements of water due to sudden large-scale changes of level of the sea floor, e.g. by the fault movements responsible for the earthquake or by submarine slumping set off by the earthquake. Related to the latter is the landsliding into deep water of immense volumes of rock from the oversteepened, glaciated sides of certain fjords. One of the most impressive giant waves generated in this way occurred in an inlet of the Gulf of Alaska, south-east of the Malaspina Glacier, immediately after an earthquake in 1958. Heavy rainfall following frequent repetitions of frost and thaw had already weakened the highly fractured belt of rocks at the head of the inlet, and the earthquake shook loose a rock-avalanche so vast in volume that the displaced water started a wave with a steep front that rose to a height of 30 m or more and reached a velocity of 200 km an hour. The forest was destroyed for several kilometres along the shores and in places the momentum of the surging water carried it up to at least 525 m, as proved later by the height to which trees had been stripped of their bark and the bedrock of its covering of soil.

Tsunami are usually the result of topographic changes on the sea floor brought about by powerful earth movements occurring along deep-water coasts, such as those on the sides of the great Pacific trenches. Much less commonly the displacement of the sea floor results from the formation of a submarine caldera (e.g. the Krakatau tsunami of 1883), but in all cases the same general principles apply. If the displacement causes a large depression of the ocean surface, water is drawn in from all sides and throughout the whole depth, which may be very great. One manifestation of this inward flow is the menacing withdrawal of the sea from neighbouring coasts that commonly portends the onset of a dangerous tsunami. If the sea-floor displacement is upwards and causes a widespread upheaval of the ocean surface, water flows outwards in all directions and at all depths. However, whether the displacement is up or down, the momentum of the vast volume of moving water carries it far beyond its position of rest or equilibrium. The resulting ebb and flow takes the form of a series of gradually subsiding, oscillating waves. In deep water these waves have wavelengths of hundreds of kilometres and travel at speeds of hundreds of kilometres an hour, but they are typically no more than a metre high, so that they pass unnoticed by ships in the open ocean. Nevertheless the energy being transmitted is immense, since the whole depth of water is involved. Consequently, when these waves reach shallowing water and narrowing bays and inlets they rise to heights that swiftly become terrifyingly destructive near their source, and may again become dangerous after travelling and dispersing through thousands of kilometres (Figures 25.11 and 25.12).

Theoretically the velocity v in deep water is given by $v^2 = gd$, where g is the acceleration of gravity (about 9.6, or say 10 m/s^2) and d is the depth of water. If d is 4 m, then v^2 is 40000, and $v = 200$ m/s $= 720$ km/h. This was first verified in practice by the study of the Krakatau tsunami, which were picked up by tide gauges in all the chief harbours of the world. Velocities across the Indian Ocean varied between 565 and 720 km an hour, according to the differing depths of water along the paths traversed between Krakatau and the tidal stations, e.g. Aden and Cape Town.

Tsunami originating with a major earthquake on the western slopes of the Japanese Trench in 1933 (magnitude 8.5) raised mountainous waves, with surges up to 27 m high, along the shores of some of the Pacific bays and inlets of Japan, where thousands of people were drowned. The waves were recorded at San Francisco about ten hours later, having traversed the Pacific at 756 km an hour. The waves took nearly twenty hours to reach Iquique in the north of Chile. Iquique and other parts of the Pacific coast of S America have their own long history of destructive earthquakes and tsunami, some of the latter having been strong enough to be seen in Hawaii and recorded in Japan. Outside the Caribbean the Atlantic shores have not suffered seriously from tsunami since the outstanding Lisbon example of 1755, which was sufficiently powerful to have an effect in the North Sea after passing up the Channel.

Figure 25.11 Sandy Beach, near Hilo on the north-east coast of Hawaii, a few minutes before a tsunami, resulting from a powerful earthquake that originated beneath the Aleutian Trench, tore across the beach and the highway, with the results seen in Fig. 25.12. (*Honolulu Advertiser: Photo Y. Ishii*)

Following powerful earthquakes that originated beneath the deep trenches of the western Pacific, about 150 disastrous tsunami have afflicted the Japanese coasts since records of earthquakes first began to be kept there. In recent years it has been noticed that submarine earthquakes with magnitudes over 8 have invariably been followed by tsunami. Despite the enormous number of earthquakes of lesser magnitude very few of these have been associated with tsunami. It is therefore suspected that those few, and probably some of the others, were started by large-scale slumping and turbidity currents unloosed by the sudden movements. It should also be noticed that movements along strike-slip faults would be unlikely to cause tsunami by themselves, since the vertical crustal displacements would be negligible. But avalanching of thick layers of sediment down the steeper sides of the offshore trenches might cause sea-floor changes of level amounting to a hundred metres or so.

Surprise is often expressed that the inhabitants of the 'low islands' of the Pacific and other mid-oceanic atolls should escape being swept away by the passage of tsunami. Their safeguard is two-fold. These island homes rise steeply from the sea floor, like pinnacles with diameters that are very short compared with the wavelengths of the tsunami in deep water. It is the energy that is transmitted, not the water. Only where the ocean floor gradually shallows over a broad area, as for example across the great swell from which the Hawaiian Islands rise, is the low wave-front sufficiently retarded to steepen appreciably. Then, where the shore topography is appropriate, it may quickly become a wave of translation, like a bore (cf. Section 23.2). Figures 25.11 and 25.12 illustrate the swift growth of such a wave near Hilo, from a tsunami that originated over 3600 km away in the Aleutian Trench.

25.7 DISTRIBUTION OF EPICENTRES: EARTHQUAKE ZONES

Accurate location of earthquakes over a wide range of magnitudes and for all points on the globe really began in the 1960s with the installation of the worldwide seismograph network described in the next section. The cumulative record of well-located earthquakes is increasing continuously but it may take thousands of years to obtain a complete picture for the least active areas. Although the main features of the earthquake distribution on land were known before seismic networks were installed, the distribution beneath the oceans was only poorly known. Plate XIII shows the worldwide distribution of earthquakes recorded since 1964. Earthquakes may occur anywhere, but it is immediately clear that most of

Figure 25.12 Vehicles passing along the coastal highway were swept up the slopes by the sudden onset of the tsunami. A man fleeing for safety can be seen on the left. The position of the highway can be recognized by the road sign, right centre. (*Honolulu Advertiser: Photo Y. Ishii*)

them occur along well-defined narrow belts which divide the world into a number of 'plates'. The map is designed to highlight the continuous boundary of each plate. One seismic belt closely follows the deep ocean trenches and associated island arcs, the quakes being located on the continental sides of the troughs. From Alaska, down the western side of the Pacific, it lies parallel to the Kurile, Japan, Marianas and Philippine trenches. Then one branch crosses to the Kermadec and Tonga trenches to the north-west of New Zealand, whilst another swings round to the west parallel to the Java or Indonesian trench.

On the eastern side of the Pacific, the earthquake zone follows the west coast of North America, being particularly characteristic of California, although there is no ocean trench associated with it there. It continues southwards parallel to the Middle American trench, and on down the Andes where it is associated with the Chilean trench.

The colouring of the earthquakes shows that the Circum-Pacific belt includes shallow, intermediate and deep earthquakes. Indeed most deep earthquakes occur in the Circum-Pacific area, the foci deepening from shallow, through intermediate to deep in a landward direction from the margins of the trenches. This correlation between the deep ocean trenches and the earthquake belt obviously must have some high significance. As we shall see

in the next chapter, the evidence suggests that the earthquake zone depicts the routes along which slabs of oceanic lithosphere move downward into the mantle along the trenches.

Another earthquake belt, that of the Mediterranean and Trans-Asiatic zone, extends along the Alpine mountain arcs of Europe, and North Africa, through Asia Minor and the Caucasus, Iran and Pakistan, to the Pamirs, Himalayas, Tibet and China. This zone is characterized by most of the larger quakes of shallow origin, and others of intermediate origin. For a long time this earthquake zone was thought to be free from deep earthquakes, but in 1954 a powerful earthquake originated at a depth of 630 km under the southern slopes of the Sierra Nevada in Spain. The Mediterranean and Trans-Asiatic earthquake zone loosely follows belts of Tertiary and Recent mountain-building. There are no associated deep-ocean trenches that are visible; indeed this earthquake zone lies inland, but attempts are being made to discover whether now-vanished trenches once existed.

Most of the remaining earthquakes are shallow, and are located along the mid-oceanic ridges and transform faults that intersect them. Along the transform faults they are confined to those parts that lie between and connect seemingly off-set portions of the ridges. The **en echelon** pattern of ridge crest and transform faults can clearly be seen in the earthquake distribution along the mid-Atlantic ridge in Plate XIII. An equal angle projection is used so that abrupt changes of direction in the seismicity, as at these transforms, are correctly reproduced.

It is obvious from the Plate that the global distribution of earthquakes delineates a mosaic of about ten relatively aseismic masses, each of which may include continent, ocean or both. These are the plates of **plate tectonics** (see Chapter 28). The slow but inexorable motion of these plates is a fundamental cause of earthquakes.

25.8   SEISMOGRAPHS

From the focus of an earthquake, waves are propagated through the Earth in all directions, and when they arrive at a seismological station they are recorded on seismographs, provided they are not too vigorous to put the instrument out of action. Instruments for detecting seismic waves have been devised in great variety (Figure 25.13). Between the two world wars most of those in

Figure 25.13 The first recorded 'seismoscope', devised in AD 132 by the Chinese astronomer Chang Heng. A pillar, suspended in the dome-covered cylinder, swayed in the direction of the initial motion when the instrument was disturbed by an earthquake, and operated a mechanism which knocked a ball from the nearest dragon's head into the mouth of the frog beneath. (*The Science Museum, London*)

common use were delicately poised pendulums designed to record either the horizontal or the vertical components of the vibrations of the ground. A good seismological station would have two horizontal seismographs mounted at right angles – usually one to respond to N–S movements and the other to E–W movements – and one vertical instrument; the three giving a complete record of the movements in three dimensions.

The essential requirement for recording the passage of seismic waves is to have a point that remains stationary, or nearly so, while the body of the seismograph moves with the Earth as the wave passes. The horizontal pendulum illustrated in Figure 25.14 consists of a weighted boom pivoted against a massive support which is firmly attached to the Earth so that it follows the Earth's vibrations. Because of its inertia the weight tends to remain stationary. It is the relative movement between the end of the boom and the rest of the instrumentation that is recorded, after being suitably magnified.

A small mirror attached to the end of the boom reflects a beam of light on to a sheet of photographic paper wrapped around a cylindrical drum which rotates on a long screw; while rotating, the drum carries the paper along at right angles to the beam of light. The length of the beam determines the magnification. The vibrations are thus recorded continuously, except for short breaks which provide a record of the time. The clock mechanism that drives the drum also operates a shutter that cuts off the light at regular intervals, so making the breaks. Accurate timekeeping is essential and this is now generally assured by recording radio time signals directly on the seismogram. Each sheet of the latter gives a 12- or 24-hour record, convenient for regular replacement.

Seismic detectors have advanced considerably since the age of the Milne–Shaw seismograph, but the basic principle remains the same. Modern seismic detectors are separate from their recording

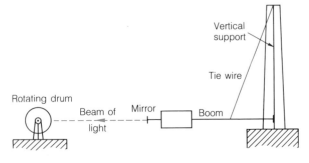

Figure 25.14 To illustrate the essential parts of a horizontal seismograph of the Milne–Shaw type.

systems, so it is usual to refer to the detector itself as the **seismometer**, reserving the term **seismograph** for the whole apparatus if it includes a visual recording system. Most seismometers nowadays use an electromagnetic method of converting the motion into a recorded signal. The elements of an electromagnetic seismometer are a cylindrical permanent magnet, usually called the **mass**, suspended by means of springs inside a coil of wire which is fixed to the casing. Motion of the coil with respect to the magnetized mass induces a current in the coil, which is fed to an electronic amplifier and recorded or displayed. Miniature versions of these detectors, called **geophones** and sensitive to higher frequency seismic waves, are used in large numbers for seismic prospecting.

A great advance in the study of earthquakes came in the early 1960s with the installation of the Worldwide Standard Seismograph Network (WWSSN) – a global network of over 100 stations of a standard design, recording ground motion on special paper mounted on a slowly-rotating drum. The drum rotates once per hour, and the pen slowly moves along the axis, so that a continuous record is drawn as if along the thread of a screw. Each day the paper is replaced. A similar principle was used for the older seismographs previously described.

A seismic station usually has more than one seismograph. We have already seen that three seismometers are needed to detect the vertical, N–S and E–W components of ground motion. Vertical component seismographs are the most common, but WWSSN stations include all three. Until recently it has been usual to distinguish seismographs of the **short period** and **long period** types, recording seismic waves close to 1 second and 20 seconds period respectively, so a WWSSN station has a total of six seismographs.

Another development has been the installation of a number of seismometers some distance apart in a pattern, all linked to a central recording system. An installation at Eskdalemuir, near Dumfries in the south of Scotland, which was opened in 1962, consists of 20 seismometers 1 km apart arranged in the form of a cross. Such an installation is called a **seismic array**, and has two advantages over a single site. First, by adding the outputs with suitable time delays it is possible to improve the quality of the signal by reducing the proportion of seismic noise. This noise has many origins, including storms, local man-made activity and signals from other earthquakes, and it is always present to some degree on seismic records. Secondly, the vertical and horizontal angles of approach of the seismic waves can be determined. Various designs of seismic array have been tried, the largest being the Large Aperture Seismic Array (LASA) in Montana, USA, which operated for 15 years from 1965 and comprised 21 subarrays with a total of over 200 seismometers covering an area 200 km across. One problem with a large array is that differences in the geology beneath different parts of it can become sufficient to disturb the uniformity of the seismic signal observed, thereby nullifying its advantages.

One serious limitation until recently was that a seismometer needed a mass-spring system with a natural period of oscillation similar to the periods of the waves to be detected. This made the design of seismometers to record the longer period waves especially difficult, as they required a natural period of oscillation of 20 seconds or more. A recent advance has been the development of miniature seismometers, small enough to be placed in a borehole, and which can detect a whole range of frequencies of seismic waves, including the very long wavelengths. The fundamental need for a large mass or complicated suspension system has been removed in an ingenious way. The output signal is converted electronically into a force which is applied to the mass itself in order to inhibit its motion relative to the seismometer casing. The size of the force required to do this

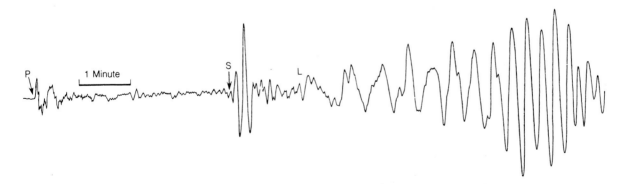

Figure 25.15 Seismogram recorded at Pulkovo Observatory, Russia, of an earthquake originating in Asia Minor on 9 February 1909. The time interval S–P is 3 minutes 43 seconds, corresponding to a distance of 2253 km from the epicentre. (*After B. Galitzin*)

varies with time as the wave passes. Because the mass motion is modified by this **feedback force**, the characteristics of the seismometer do not depend critically upon its size, and the range of frequencies recorded by the seismometer is governed primarily by the design of the feedback. It is now much more common to see seismometers which can record a wide range of seismic wavelengths; these are called **broad-band** seismometers.

25.9 SEISMIC WAVES

Seismograms recorded from distant earthquakes vary greatly depending upon the distance. The most simple seismograms are usually recorded at great distances, between 3500 km and 10 000 km. Such a seismogram is illustrated in Figure 25.15. Three conspicuous pulses can be picked out (labelled P, S and L). The first arrival is the *p*rimary or P wave, and is followed some minutes later by the second major arrival which is the *s*econdary or S wave. P and S are body waves which have travelled through the Earth along paths similar to that shown in Figure 25.16. The curvature of the path towards the surface results from an increase in the wave velocity with depth.

P or 'primary' waves are **compression–dilatation** (or **compression–rarefaction**) waves like those of sound, in which each particle vibrates to and fro in the direction of propagation (Figure 25.17). We are all familiar with the transmission of sound through air and water, and this exemplifies the more general fact that P waves pass through gases, liquids and solids alike. S or 'secondary' waves are **transverse** or **shear** waves, in which the motion of each particle is at right angles to the direction of propagation, like the loops in a taut rope shaken at one end. Shear waves pass only through solids: not through liquids. P and S waves travel at different velocities through rocks of the same kind, because they depend on different properties. The velocity of P depends on density and compressibility (resistance to compression), while that of S depends on density and rigidity (resis-

tance to distortion or shearing). As a rule P waves travel at about 1.7 times the velocity of S waves.

Besides transmitting body waves P and S, two other kinds of wave are possible associated with boundaries between materials of different velocities and densities. These are responsible for the third train of signals, the onset of which is marked L in Figure 25.15. They were first known collectively as L (*long*) waves because generally they have longer periods, but now the term **surface waves** is invariably used. Their prominence on seismograms recorded beyond a certain distance from the source is due to the fact that they spread out and disperse their energy over a surface nearly according to $1/d$ (d being distance from the source), whereas the body waves, P and S, disperse their energies throughout a volume according to $1/d^2$, which rapidly becomes much smaller than $1/d$ for increasing values of d.

Surface waves travel with velocities rather similar to those of the S wave in the shallow layers, but since they travel in these layers throughout their path they arrive after the S wave except at short distances. One of the two kinds of surface waves was predicted by the third Lord Rayleigh in 1887 20 years before it was recognized on seismograms. In **Rayleigh waves** the motion of particles is in elliptical orbits in the vertical plane of propagation. This is similar to the motion of a float on the sea as waves pass beneath it, but at the surface the orbit of a Rayleigh wave rotates in the opposite sense; they are retrograde rather than prograde. In the second kind of surface wave the motion is horizontal and at right angles to the direction of propagation – they are like horizontally polarized S waves. These waves were recognized on seismograms before they had been accounted for. It was an Oxford mathematician, A.E.H. Love, who explained them by an extension of Rayleigh's theory, and since then they have been known as **Love waves**.

Examination of Figure 25.15 shows that the surface waves have a very different character from the body waves. The length of a body wave is related to the duration of the rupture; it does not spread out as it travels, although it may be immediately followed by other signals resulting from reverberations of the wave in shallow surface layering either at the source or receiver. The surface waves however have an extended oscillatory appearance which is fundamental to their wave type. They travel at a velocity which is a sort of average of the velocity of S waves in the shallow layers they pass through, causing motion down to a depth of about one wavelength from the surface. The longer wavelength parts of the surface wave sample deepen as they travel. Since the velocity generally increases with depth, it follows that the

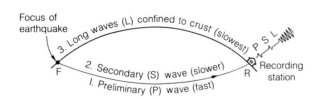

Figure 25.16 Section through a segment of the Earth, showing the paths followed by the P, S and L (surface) waves from an earthquake originating at the focus F and recorded at a station R.

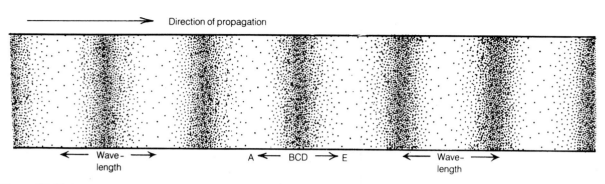

Direction of propagation

← Wave-
length →
A ← BCD → E
← Wave-
length →

Figure 25.17 Diagrammatic representation of compression–rarefaction waves (P waves). The initial shock sets up a zone of compression or rarefaction which rapidly moves outwards (velocity in granite about 5. 6 km/s), although the movements of the individual particles are very slight.
A zone of rarefaction behind C; C zone of maximum compression; E zone of rarefaction in front of C;
B compressed particles moving back towards A; D   compressed particles moving forward towards E.

longer period parts travel faster and arrive earlier, with the result that the surface wave spreads out in time as it travels, and the oscillations move towards shorter wavelengths further along the seismogram. This effect is called **dispersion**, and its analysis provides seismologists with one way to determine the variation of wave velocity with depth within the Earth (cf. Figure 26.5). For example, by this method it is possible to distinguish between continental and oceanic profiles.

When the **times** taken by P and S waves to travel to seismological stations in various parts of the world are plotted against the respective **distances** from the epicentres, they fall on smooth curves which are nearly the same for all distant earthquakes originating at any particular depth, irrespective of the location of the paths between source and station. This shows that to a good approximation the wave velocities deep inside the Earth vary with depth only. Figure 25.18 shows at a glance that the time interval between the arrival of P and S steadily increases with distance, and this is a simple consequence of the ratio of the P and S wave speeds being similar at all depths. H. Jeffreys and K.E. Bullen measured arrival times for many earthquakes at

many stations worldwide. They published their first set of travel time tables in 1935. The Jeffreys–Bullen travel time tables have been used routinely for the determination of earthquake locations until the present day, and although refinements can be made on the basis of modern knowledge, they are sufficiently accurate for most applications, and can be used as a standard to which more refined determinations can be referred.

When an earthquake is recorded at a distant station, measurement of the interval between the arrival times, S–P, serves to determine to a first approximation both the distance from the epicentre and the time of origin of the shock. At least three such distance determinations at well-spaced stations are necessary for a first indication of the position of the epicentre (Figure 25.19). Greater accuracy is attainable when several determinations are available, and especially when the depth of the focus has also been calculated from the records, since S–P really refers to the distance of the recording station from the focus. When the focus is deep the difference between focal distance and epicentral distance may be considerable (cf. Figure 26.12).

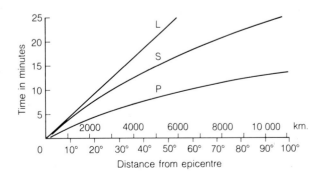

Figure 25.18 Time–distance curves for P, S and L (surface) waves (the curve for surface waves refers to those that have traversed the continental type of crust, and is approximate since for surface waves the velocity depends upon wavelength).

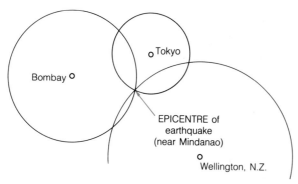

Figure 25.19 To illustrate how the epicentre of an earthquake can be fixed when its distances from three suitably placed stations are known. A circle is drawn on a globe around each of the three stations (e.g. Bombay, Tokyo, and Wellington), with a radius corresponding in each case to the respective distance of the epicentre. The epicentre lies at the point of intersection of the three circles.

There is close international cooperation for the effective exchange of information used in the determination of earthquake locations and origin times. The National Earthquake Information Center of the US Geological Survey calculate earthquake locations and magnitudes from worldwide arrival time and amplitude observations for a large proportion of recorded earthquakes. They publish a monthly list called the Preliminary Determina-

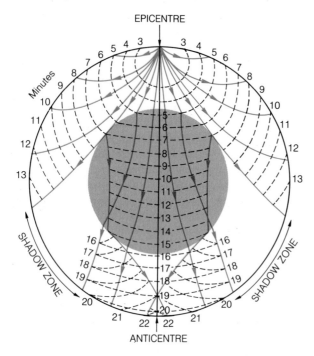

Figure 25.20 Section through the centre of the Earth showing the wave paths (firm lines with arrows), wave fronts (dashed lines), and arrival times (in minutes reckoned from the zero time of the shock). Since there is a shadow zone free from P and S waves for each such earthquake, it is inferred that the Earth has a core which refracts the deeper waves as shown in the diagram.

Figure 25.21 The P wave shadow zone cast by the Earth's core in the case of an earthquake originating in Japan.

tion of Epicenters. It takes many months to assemble the measurements from all seismic stations worldwide, yet as many observations as possible are needed for the preparation of a comprehensive and accurate list of located earthquakes. This role is performed by the International Seismological Centre (ISC), housed near Newbury, England. The ISC monthly bulletins are published two years later, and contain the location, magnitude and other details of every earthquake which the Centre can locate, together with all the measurements received. Through the work of the ISC we are able to present maps of earthquakes such as Plate XIII. Nowadays these enterprises are greatly assisted by computers and modern worldwide telecommunications.

25.10 THE INTERNAL ZONES OF THE EARTH

The idea of 'X-raying' the globe with waves generated by its own earthquakes was introduced in Section 2.3. Seismograms of distant earthquakes provide information of some of the properties of the interior at different depths, just as rays of light bring to our eyes the colours of, say, the internal parts of an intricately designed glass paperweight. We see the internal structure of the latter, and similarly, by a comparative study of seismograms of the same earthquake recorded at stations well distributed over the Earth, we can 'see' at least the dominant features of the internal structure of the globe. Whether an earthquake is regarded as 'near' or 'distant' depends on the distance of its

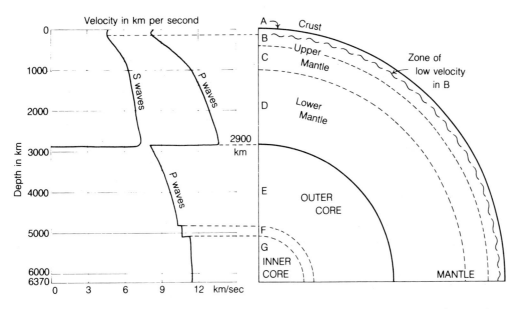

Figure 25.22 Diagrams showing the variations in velocity of P and S waves, from the surface to the centre of the Earth, and the inferred structure of a quadrant of the Earth, lettered in accordance with K. E. Bullen's classification of the successive zones.

epicentre from the station where it is recorded. If the station is within a few hundred kilometres of the epicentre and the earthquake originated within the crust, then the P and S waves that are recorded there include those that have travelled only through crustal rocks. The results are of great importance in working out the thickness of the crust and its variation from place to place. Moreover, seismic waves can also be generated by means of artificial explosions and picked up by a line of small instruments at suitable distances. Some of the details of crustal structure, continental and oceanic, are reviewed later. Here we shall first consider the P and S waves of distant earthquakes, which descend through the Moho into the mantle before coming up to the surface again (Figure 25.16).

The travel times of the P waves indicate that their velocities increase with depth from about 7.8–8.1 km/s near the top of the mantle (where there are some significant variations) to about 13.6 at the bottom, i.e. at a depth of nearly 2900 km. For S waves the corresponding increase is from 4.35 to 7.25 km/s. The waves that just attain this depth emerge at the surface at places about 11 500 km (105°) from the epicentre of the earthquake concerned. At stations beyond this distance there is a striking and critical change in the records. No P or S waves appear – although the surface waves come along at the appropriate times – and this type of record is all that is received by stations within the next 4500 km. At distances greater than 16 000 km (143°) from the epicentre the P wave reappears, and continues to do so right up to the antipodes of

the epicentre. Corresponding to each earthquake of sufficient magnitude to be recorded so far from its source there is a ring-like shadow, free from P and S waves, as illustrated in Figures 25.20 and 25.21. This P wave shadow zone was first noticed by Oldham in 1906. He at once inferred that the Earth has a core which acts like a spherical lens by refracting the deeper waves inwards, so concentrating them in the antipodal region at the expense of a surrounding zone of shadow.

In accordance with this strong refraction the velocity of the P waves is greatly reduced when they pass from the mantle into the core (Figure 25.22). But the S waves are not transmitted at all, so indicating that the core is fluid. It was at first expected that the core would be fluid throughout. However, in 1936, Lehmann, a Danish seismologist, who celebrated her 100th birthday in 1988, detected a still deeper discontinuity enclosing an inner core which she suspected to be solid. Seismic records of underground nuclear explosions (in which the time and place of origin are exactly known) confirmed her tentative conclusions. Until 1962 the inward passage from fluid to solid was thought to be a gradual change spread over a transitional zone (F in Figure 25.22) about 160 to 320 km across. B.A. Bolt then showed that agreement between observation and theory is greatly improved by regarding F as an independent zone with fairly sharp boundaries against the outer (E) and inner (G) parts of the core. More recent research has shown that this 'transition zone' in the core is no longer required to explain the observations. The change from mantle to core is

Table 25.2 Internal zones of the Earth (*from data computed by Harold Jeffreys, K.E. Bullen, E.C. Bullard and B.A. Bolt and updated from A.M. Dziewonski and D.L. Anderson*)

Zones		Depth (and radius) of boundaries in km		Velocity of P waves km/s	Density gm/cm^3	Pressure in units of 10^6 bar*
CRUST (continental)†	A	Sea-level	(6371)		2.7 2.8 2.9	
Mohorovičić Discontinuity		33	(6338)			9000
UPPER MANTLE	B	50 } 250 } Low velocity zone		7.9–8.1 7.8 8.1	3.32	
		400	(5971)	8.90 9.13	3.54 3.72	140 000 270 000
	C	(720: deepest earthquakes)				
		670	(5701)	10.27 10.75	3.99 4.38	382 000
LOWER MANTLE	D					
Core/Mantle Boundary		2891	(3480)	13.71 8.06	5.57 9.90	1 368 000
OUTER CORE	E/F					
		5149	(1222)	10.36 11.03	12.2 12.8	c.3 300 000
INNER CORE	G					
		6371	(Centre)		13.1	c.3 600 000

* Standard pressure of 1 atmosphere = 1 031 250 dyne/cm^2
1 bar = 1 000 000 dyne/cm^2

† In oceanic areas Zone A extends from the ocean floor at a mean depth of 4.8 km to the Moho at a depth of about 10–11 km below sea-level.

known to be sharply defined i.e. no transition zone can be detected, as it should be if a zone were more than 2 or 3 km thick. However, there does appear to be a distinct transition between the upper and lower mantle; this is discussed in the next chapter.

When the velocities of P and S waves at various depths within the Earth are known, it is possible by making certain reasonable assumptions to estimate the distribution of density with depth. The results are acceptable only if they provide the Earth with the correct moment of inertia as well as with the correct mean density, both of which are accurately known. The calculated density distributions that survive these severe limitations are all very similar and they all involve a heavy core. There is still some uncertainty about the nature of density changes within the mantle. K.E. Bullen found evidence of slight departures from a regular downward increase at depths of 413 and 984 km, suggesting a division of the mantle into three

parts. These are labelled B, C and D in his lettered classification of the Earth's internal zones, A being the crust. More recent evidence favours discontinuities at 400 and 670 km. Table 25.2 summarizes the data corresponding to our present state of knowledge. Densities and pressures are thought to be correct within 2 or 3 per cent.

25.11 FURTHER READING

Bolt, B.A. (1980) *Earthquakes*, Freeman, New York.

Bullen, K.E. and Bolt, B.A. (1985) *An Introduction to the Theory of Seismology*, Cambridge University Press, Cambridge.

Gere, J.M. and Shah, H.C. (1984) *Terra non Firma*, Freeman, New York.

Gubbins, D. (1990) *Seismology and Plate Tectonics*, Cambridge University Press, Cambridge.

Mogi, K. (1985) *Earthquake Prediction*, Academic Press, London.

THE EARTH'S CRUST AND MANTLE: COMPOSITION AND DYNAMICS

The Earth has a spirit of growth
Leonardo da Vinci (1452–1519)

26.1 THE EARTH'S LAYERING

We have seen from studying the propagation of earthquake waves that the Earth has a layered structure on a broad scale, being split compositionally into the crust, mantle and core. Major **compositional** changes such as the crust–mantle and core–mantle boundaries involve large, abrupt changes in seismic velocity, leading to distinct reflected and refracted waves. However, the Earth is also layered **mechanically**, giving rise to more gradual changes in seismic velocity where an increase in pressure or temperature causes a change in the mechanical properties of the rock without changing its chemical composition. For example in the previous chapter we have seen that in the upper mantle a relatively broad seismic low-velocity zone has been discovered from analysis of seismic waves. This low-velocity zone is consistent with mantle material at or near its melting point (about 1300°C at pressures corresponding to the depth of the low-velocity zone), and therefore marks a major change in the mechanical behaviour of mantle material. This kind of mechanical change is analogous to the change in the properties of road tar on a hot day. Above this boundary the mantle is relatively cool

and rigid, and can move as a single unit comprising the crust and part of the upper mantle, together known as the **lithosphere**. The thickness of the lithosphere varies between about 80–125 km, compared with an average crustal thickness of 35 km for the continents and 7 km for the oceans (overlain by a water column of about 5 km depth), so most of the lithosphere is actually comprised of mantle material. Below this boundary the temperatures and pressures are high enough to allow the sticky, or **viscous** flow of mantle material over long time periods, either completely in the solid state or with a small percentage of molten material in a form known as a **partial melt**. The mantle layer which can flow in this latter way is known as the **asthenosphere** and is associated with the seismic low velocity zone. Thus the Earth above the core is split compositionally into the crust and mantle, and mechanically into the lithosphere and asthenosphere.

Since the boundary between the base of the lithosphere and the asthenosphere is essentially a thermal one, it follows that an upwelling of heat from the asthenosphere or lower down in the mantle can raise the boundary towards the Earth's surface, effectively thinning the lithosphere. It is very important not to confuse the compositional unit of the crust with the mechanical unit of the lithosphere.

Other gradual changes in seismic velocity in the upper mantle are related to the dramatic changes in atomic structure, known as **phase changes**, which occur when a material is compressed too much for its previous atomic structure to bear. This results in a rearrangement of atoms within the minerals to accommodate the higher pressure, but again does not involve a change of chemical composition. The most important phase changes in the mantle are thought to be the transformation of olivine (believed to be the major component of the mantle) first to its **spinel** phase at a depth of about 400 km, and then to its **perovskite** phase at about 670 km. Both of these depths are marked by slight increases in seismic velocity, more gradual than would be expected for a major compositional boundary.

The nature of the seismic discontinuity at 670 km depth is a subject for great debate at the present time. It marks the onset of the spinel–perovskite phase change and is also approximately at the level of the deepest earthquakes (720 km), and so in this sense is also a mechanical boundary. Much research is also directed at establishing the extent to which the boundary may also reflect a subtle change in mantle chemistry, which would establish whether or not the mantle is in fact compositionally layered. The absence of a sharp seismic discontinuity rules out the possibility of a major compositional change there, but even a small change in density (5 per cent or so) due to a compositional change would be enough

to stop the transfer of large amounts of mantle material across this boundary. This would imply an upper mantle which has been isolated from the lower mantle over geological time periods.

Table 26.1 is a generalized summary of the major compositional, mechanical and phase boundaries of the Earth above the core. All of these discontinuities, whether sharp or gradual, have been discovered by seismic investigation, but their nature has largely been established by reference to laboratory studies of rocks and minerals at appropriate temperatures and pressures.

26.2 SEISMIC EXPLORATION OF THE CRUST

Records of P and S waves from near earthquakes and artificial explosions have been extensively used for estimating the nature and thicknesses of the dominant layers that make up the Earth's crust. The waves travel with velocities that depend on the densities and elastic constants of the rocks through which they pass. For P waves these range from very low velocities, such as 2 km/s for loose or poorly consolidated sediments; through a wide range of higher values for indurated sediments; to 5.2–6.2 km/s for sialic crystalline rocks, including granite and granodiorite; and to 6.2–7.2 km/s for basic rocks such as gabbro and its common metamorphic equivalents. Eclogite, however, goes up to 7.9–8.1 km/s. The velocity for a particular kind of rock is found experimentally to vary not only with its composition and structure, but generally to increase with rising pressure (i.e. with depth), while decreasing only slightly with rising temperature.

Table 26.1 Layering of the earth above the core–mantle boundary

Depth km	Compositional layering	Mechanical layering	Phase changes of olivine
	CRUST		
35		LITHOSPHERE	
100	UPPER MANTLE	ASTHENOSPHERE	OLIVINE
400	Transition zone	MESOSPHERE	
		Deepest earthquakes	SPINEL
670	----?----?----?----		
	LOWER MANTLE	(No earthquakes below 720 km)	PEROVSKITE

The principles involved in seismic exploration of the crust can be most easily illustrated by considering the simplest type of case that arises in seismic prospecting. Here the vibrations are provided by artificial explosions, which have the great advantage over natural shocks that they are under complete control and can be used wherever and whenever it happens to be convenient. On land a charge of dynamite is lowered into a shot hole previously drilled to a depth of a few metres. The charge is detonated electrically at an exactly recorded time. The time and place of origin is exactly known and the depth of focus is negligible. The problem is to determine the thickness h of an upper layer of rock (through which the P waves travel at velocity v_1) resting on a lower layer in which the velocity v_2 is considerably greater than v_1 (Figure 26.1).

The waves from the explosion spread out in all directions from the shot point. Some (a) reach the surface directly through the upper layer. Of those that reach the boundary between the two layers, some (b) are reflected back to the surface; some (c) are refracted along the boundary, where they travel with the higher velocity v_2 and are bit by bit returned to the surface from each point along the boundary; and some (d) are refracted into the lower layer until they in turn are reflected or refracted at the next boundary. At the surface the time of arrival of the P wave can be recorded wherever a suitable instrument is placed for its reception. (S waves following similar paths at lower speeds can also be recorded). In practice geophones G_1, G_2 ... up to, say, G_{20} are fixed in the ground in line with the shot point at known distances, and all are linked with the recording van. Alternatively the geophones may be placed nearer the shot and spaced more closely to pick up the reflected waves (b) to best advantage, as in echo-sounding at sea. This method, known as 'reflection shooting', is in common use in the exploration for oil-bearing structures. A different spacing of the geophone is adopted for 'refraction shooting', designed to pick up waves (c), and also, waves (d) that penetrate more deeply before returning to the surface.

Figure 26.1 refers to the refraction method in its simplest form. The waves (a) are the first to reach the geophones (G_1–G_3) situated near the shot point. Beyond a certain distance, l, the refracted waves from the top of the lower layer arrive first because of their greater velocity while travelling along the lower side of the boundary. l is the distance at which both direct and refracted waves arrive at the same time t. As indicated in the lower part of Figure 26.1, l is determined by the intersection of the graphs obtained by plotting the times of the first arrivals against the respective distances. In each case the inverse of the slope of the graph gives the corresponding velocity, e.g. $v_1 = l/t$. Knowing the two velocities, the thickness h of the upper layer is given by the relationship

$$h = \frac{l}{2} \sqrt{\left(\frac{v_2 - v_1}{v_2 + v_1} \right)}$$

26.3   CONTINENTAL CRUST

The large-scale structure of the crust, briefly out-

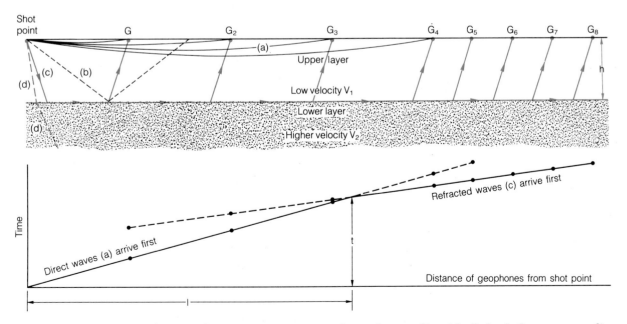

Figure 26.1 Wave paths from a shot point to a series of geophones, G, with (below) the corresponding time–distance graphs for the first arrivals of direct and refracted waves. For details see text.

lined in Section 25.10, was originally investigated by studying the records of near earthquakes; records made, not by a series of geophones, but by seismographs permanently established in observatories within a few hundred kilometres of the epicentre. The first major boundary to be recognized was the discontinuity between mantle and crust, now familiarly known as the Moho. In the course of a study of an earthquake that occurred in north-west Yugoslavia in 1909, A. Mohorovičić found well-defined pulses of P and S waves that had travelled directly through the crust in addition to those of P and S waves that had been refracted into the mantle before coming up to the surface again. The next advance was made by V. Conrad when studying the records of an earthquake that occurred in the Austrian Alps in 1923. Besides recognizing the four pulses just mentioned, Conrad found a fifth – a P wave intermediate in velocity between the other two. This he referred to a crustal discontinuity lying between the upper part of the crust and the mantle. Jeffreys identified the corresponding S pulses in the records of shocks that originated in Jersey and near Hereford, both in 1926. The boundary between the upper and lower crustal layers (then thought to be respectively of sialic and basic composition) was called the Conrad discontinuity. But this has never won the celebrity attached to the Moho, because there are many areas in which it cannot be found. In fact the striking outcome of many refraction and reflection surveys of the continental crust is how different it appears from place to place, and not just in terms of a Conrad discontinuity.

Much of the work of crustal and upper-mantle exploration is now carried out by utilizing the effects of artificial explosions just below the surface – as in ordinary geophysical prospecting, but on a larger scale. The focal depth is then zero and the time and place of origin are exactly known. This leads to considerable simplification, but it also leads to the unexpected result that the wave velocities are often found to be higher than those based on records of natural earthquakes. One explanation of this discrepancy is lack of exact knowledge as to the time and focus of an earthquake; another is that it is not everywhere true that the wave velocities increase with depth. One such exception is the occurrence of metamorphic rocks, enriched in basic minerals such as biotite and hornblende, overlying granitic rocks through which seismic waves travel more slowly. As may be gathered from Figure 26.2, waves originating from a focus within such a 'low-velocity channel' take longer to reach a given station, so that their apparent velocity is less than it would have been if the focus had been at the surface, or if the velocity

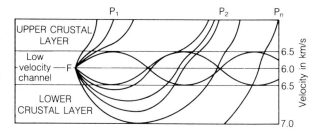

Figure 26.2 Diagrammatic illustration to show schematically how a 'low-velocity channel' tends to guide and confine some of the waves from an earthquake focus at F. Other waves take longer to reach the surface than where there is no low-velocity channel.

had continued to increase downwards without interruption. Such 'low-velocity channels' are often found in the mid-crust in recent orogenic belts.

Another cause for velocity variations, whether the waves are generated naturally or artificially, lies in the structure of the rocks through which the waves are propagated. Transmission through schistose metamorphic rocks, for example, is faster across the schistosity than parallel to it. This anistropy is well illustrated by an earthquake of 1946 that had its focus in the Haut-Valais, between the Mt. Blanc and Aar massifs (Figure 31.15(a)). P waves recorded at Coire, at the north-eastern end of the Aar massif, travelled parallel to the structural directions, the average velocity being 5.97 km/s. Those recorded at Neuchatel travelled at right angles, at an average rate of 6.3 km/s. Nowadays much more subtle changes in seismic anisotropy can be detected using S waves.

Whereas in many regions there is seismic evidence for a compositional division into upper and lower crust, there are others where such evidence is lacking. Moreover, where upper and lower crustal layers have been determined they are not always sharply divided, and their relative thickness varies and appears to be dependent on the geological structure. There is, however, a general increase in the velocity of seismic waves with depth through the continental crust. Generally speaking, the values for the upper part suggest rocks of a bulk composition comparable with that of granodiorite whereas those of the lower crust are more basic.

An important by-product of the exploration for hydrocarbons offshore in the North Sea has been a wealth of deep-sounding seismic reflection data which, after much computer processing, shows a distinct layering of the crust into a near-surface reflective sedimentary layer, a relatively 'clean' non-reflective upper crust and a reflective lower crust (Figure 26.3). Thus the upper and lower crust may be different in structure as well as in composition or

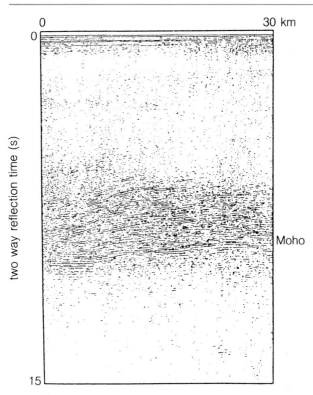

0   30 km

Moho

two way reflection time (s)

Figure 26.3 Stacked seismic reflection section over an area of continental crust showing well-developed lower crustal layering. In this area individual reflectors can be continuous for 10–20 km. The Moho is located at the base of the layering. (*After M. Warner, 1990*)

metamorphic grade. There are mechanical differences too. Shallow earthquakes commonly nucleate at the base of the upper crust at a depth of about 10 km and then propagate rapidly upwards to break the surface. In contrast, slip in the lower crust occurs more gradually and stably by plastic deformation at higher temperatures. For example, the minerals quartz and feldspar are known to break suddenly in a **brittle** fashion at temperatures and pressures similar to those in the upper crust, but to strain plastically under lower crustal conditions. An example of such **plastic** strain would be the development of schistosity in shear zones. (This has even been suggested as a possible cause of the reflective lower crust, but other explanations, including intrusive interlayering, are plausible alternatives.) This difference in mechanical properties is another important distinction between the upper and lower crust.

Another seismic technique which has been applied to the layering of the Earth's crust is that of **surface-wave dispersion**. This occurs because low-frequency surface waves have longer wavelengths, and so penetrate deeper into the Earth than high-frequency surface waves. Because of the general increase in seismic velocity with depth the lower-frequency waves in general travel at a greater velocity and arrive first on a seismogram. **Dispersion** is the name given to the characteristic

stretching out of the surface wave train that results. For example, Figure 25.16 shows the gradual increase in frequency of the surface waves after the onset of the surface wave train (marked L on the diagram). The technique is equally applicable when the deeper low-frequency waves encounter a low-velocity zone, when the opposite type of dispersion takes place and in this case the high-frequency waves arrive first. When there is no change in velocity with depth all the different frequencies arrive at the same time as an impulsive wave known as an **Airy phase**.

Dispersion is best seen over very long wave paths where the different frequencies are well spread out on a seismogram and are easy to measure. One of the longest continental paths over which surface waves have been recorded is between the Aleutian trough, where a great earthquake (epicentre marked A on Figure 26.4) occurred on 20 March 1958, and the seismic station of Lwiro in Zaïre, where both Rayleigh and Love waves were recorded and carefully analysed. The direct path was mainly continental, 13 240 km out of a total great circle distance of 14 240 km, with an average elevation of about 600 m. The path taken in the opposite direction, through the antipodes, was mainly oceanic, 21 790 km out of a total of 25 790 km, with a mean depth of water of 4.5 km. From the two sets of results it is possible to correct each for its 'contamination' by the respective oceanic and continental interruptions.

Figure 26.5, together with a great deal of other data, indicates an average crustal thickness of nearly 40 km along the path across Asia and Africa, as shown in Figure 26.4; and an average crystalline oceanic crustal thickness of 5.3 km plus 0.8 km of sediment. Including the water column this gives 10.6 km for the average depth of the Moho in the eastern Pacific and south Atlantic. This is very near the world oceanic average of 11 km for the depth of the oceanic Moho. The continental crustal estimate of nearly 40 km (Figure 26.5) may be on the high side. The crustal thickness of the Baltic Shield averages 37.5 km by the method based on Rayleigh waves, whereas P waves from artificial explosions gives 34 km, which is probably more accurate. For Africa as a whole, from Algeria to Cape Town, the Rayleigh method gives 35 km. It is indeed remarkable that all the old shield areas have nearly the same crustal thickness: 36 km for the Canadian shield; 35 km for the Australian; and 35 km for that of East Antarctica (i.e. without its carapace of ice).

Although generally speaking, the continental crust has an average thickness of about 35 km, it varies between 20 km and 70 km. The maximum of about 70 km occurs beneath mountain ranges; in other words, the mountains have roots.

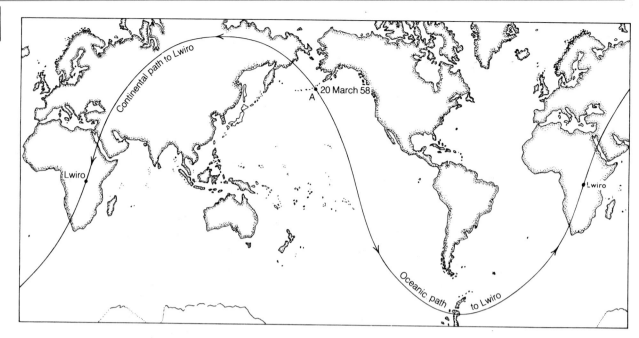

Figure 26.4 Diagram showing the continental and oceanic paths followed by Rayleigh waves recorded at Lwiro in Zaïre from an earthquake with its epicentre at A in the Aleutian arc.

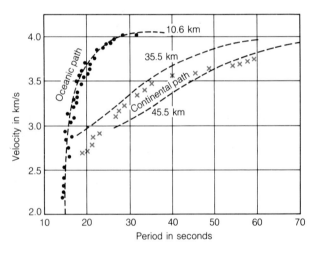

Figure 26.5 Velocities of Rayleigh waves plotted against their periods along the continental and oceanic paths of Fig. 26.4. The period is the reciprocal of the frequency. The broken lines indicate the results theoretically to be expected for the depths of the Moho below sea-level shown by the attached figures. The actual results are plotted as dots (oceanic) and crosses (continental). The thickness of the continental crust is obviously variable, but averages about 40 km. (*Both after R. L. Kovach, 1959*)

26.4 OCEANIC CRUST

In principle the method of seismic profiling at sea is the same as that illustrated in Figure 26.1, except for the presence of seawater, the depth of which must be determined. Explosives as a sound source are rapidly giving way to more advanced systems, such as arrays of 'air guns', and hydrophones (rather than geophones) are arranged in lengths of tubing maintained at a depth of a few metres for considerable distances behind the ship. The travel times are continuously recorded and appear as a profile of the ocean floor complete with the interfaces between the various layers of rock (Section 9.10).

Where the ocean is 4 to 5 km deep, that is away from swells and ridges on the one hand, and deeps and trenches on the others, the Moho is found at a depth of 10 to 11 km. It is overlain by the three following crustal layers:

Layer 1, the uppermost layer, is sedimentary with a thickness ranging from zero up to several kilometres. The thickness varies according to distance from continental margins and the degree of transport by bottom currents, and in the case of pelagic sediments according to food supply and ocean depth.

Over mid-oceanic ridges sediment is sparse or absent or occurs in thin isolated pockets where the relief is high (Figure 26.6), a fact confirmed by dredging, boring and by observations from highly mobile submersible craft that can dive to 3000 m and more. The sediment gradually thickens on the sides of the ridges.

The thickest sediments occur close to the foot of continental slopes, and where large rivers have built sedimentary fans and cones that extend hundreds of kilometres out to sea. A relatively thick sedimentary layer also characterizes the equatorial belt which is a rich breeding ground.

Layer 2 is about 1 to 1.5 km thick with P velocities that would fit many kinds of sedimentary as

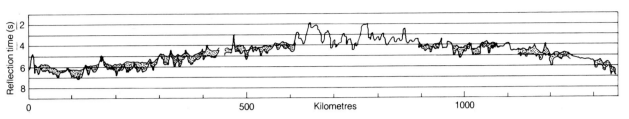

Figure 26.6 Drawing of a seismic profile at about 40°N across the mid-Atlantic ridge, illustrating the absence of sediment from the median part of the ridge, and its collection within pockets, where the topography is strongly diversified, on the flanks of the ridge. (*After J. and M. Ewing, 1967*)

well as basaltic rocks. However, wherever samples of this layer have been brought up by dredging or drilling, or are exposed on land in an **ophiolite** sequence (Section 31.5), it has been found to be basaltic. The layer is also highly fractured, allowing the penetration of seawater deep into the oceanic crust. Seismic profiles show layer 2 to have a rough upper surface, and from submersibles working within the North Atlantic rift it has been seen to consist of pillow lava, which forms when molten lava cools suddenly as it extrudes in 'blobs' directly into contact with seawater. The pillow lavas grade into a complex of vertical sheeted dykes of dolerite at the base of layer 2. These dykes are each approximately 0.5 m wide, and are in direct contact, side by side.

Layer 3, the lowest and main oceanic crustal layer has an average thickness of about 5 km, with P velocities indicative of a basic composition. The upper part is comprised of relatively coarse-grained gabbro, with a sequence of interlayered olivine gabbro and plagioclase peridotite forming at the base of layer 3. The base is much more

similar in composition to the mantle than to continental crust, so the oceanic Moho is more gradational rather than a sharp discontinuity.

Thus the thickness, composition and layering of the oceanic crust is completely different to the continental crust. Figure 26.7 shows a model for the formation of oceanic crust by a magma chamber fed from the underlying mantle at ocean ridges. The model assumes the sea floor is spreading away from the ridge axis to make way for new crust, and explains why oceanic crust appears to be so uniform compared with the diversity of the continental crust.

26.5 THE MANTLE

It has long been thought probable that meteorites, stony and iron, might be direct clues to the nature of the Earth's mantle and core. Stony meteorites are like our terrestrial peridotites in many ways and, moreover, a few varieties have a composition not unlike some of our basalts. For these and other

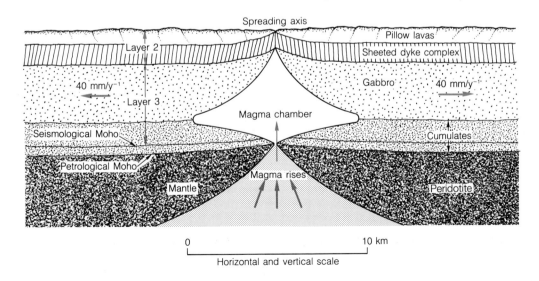

Figure 26.7 Diagrammatical cross-section of the structure of the crust and topmost mantle beneath an ocean ridge crest, drawn without vertical exaggeration. Layer 2 structure shows the tendency of the pillow lavas to dip inwards and the dykes to dip steeply outwards resulting from rotation near the spreading axis. (*After Bott, M. H. P., 1982*)

reasons a similar range of composition for the mantle, with ultrabasic materials predominating, has seemed to be a plausible guess.

Olivine-rich rocks (peridotites), at the appropriate temperatures and pressures, give seismic velocities and densities that are about right for most parts of the upper mantle. Eclogite would do equally well, and various associations of these rocks would also match the seismic and density requirements. That eclogite does occur in the mantle is proved by the fact that it occurs in diamond pipes as inclusions, some of which themselves contain diamond. There cannot be much eclogite in comparison with peridotite, however, because eclogite contains too high a proportion of the heat-generating radio-elements to be an abundant constituent of the mantle. Further evidence that the mantle is composed of various types of peridotite is provided by some peridotite nodules that occur as inclusions within lava and kimberlites (p61). The basalt has evidently carried these xenoliths up from the mantle of which it is probably itself a differentiate, perhaps a first-formed melt. However, the most striking evidence comes from sections of oceanic lithosphere which have been sheared off and are exposed at the surface. Such ophiolite sequences confirm the layered structure of the oceanic crust shown in Figure 26.7 and that oceanic crust is underlain by ultrabasic material in the form of peridotite. confirm the layered structure of the oceanic crust shown in Figure 26.7 and that oceanic crust is underlain by ultrabasic material in the form of peridotite.

26.6   THE LOW-VELOCITY ZONE OF THE UPPER MANTLE

As long ago as 1926, the celebrated Californian geologist, Beno Gutenberg, suspected that there was a low-velocity zone in the upper mantle. As may be gathered from Figure 26.2, waves originating from a focus within a low-velocity zone take longer to reach a given station, so that their apparent velocity is less than it would have been at the surface, or if the velocity had continued to increase downwards without interruption. Thirty years after Gutenberg's suggestion, the reality of a low-velocity zone within the upper mantle was established by worldwide records of the blasts from underground nuclear explosions. Remembering that there is considerable variation from place to place, the velocity decrease begins on average at a depth of about 100 km and, after passing the minimum and resuming the normal downward increase, the velocity, as it was just below the Moho, is only regained at an average depth of about 150 km or more. It should be noted that the actual depths are lower for oceanic regions

and higher for continental, indicating that the continental lithosphere is thicker on average than oceanic lithosphere.

The existence of a low-velocity zone within the upper part of the mantle is thought to depend on the temperature of the material in relation to the melting point or range, the velocity being least where the actual temperature makes its nearest approach to the melting temperature. We know that the low-velocity zone is not due to a total melting of the mantle at these depths because the low-velocity zone generally transmits S waves, which cannot pass through a liquid. Both temperature and pressure increase with depth and therefore affect the mechanical properties of the mantle. In the upper part of the low-velocity channel the temperature effect is dominant; that is, as the depth increases the temperature rises more rapidly than the melting point of the material that happens to be there. Evidence of at least partial melting in the low-velocity zone in some areas of the world has been recorded by Gorshkov, an eminent Russian volcanologist. In 1957 he found that S waves from earthquakes originating in the Japanese or Aleutian arcs failed to arrive at seismic stations so situated that the P waves, which did arrive, had passed through the mantle at depths of 55 to 60 km beneath the volcanic belt of Kamchatka and the Kurile Islands (see Figure 26.12). Again, since S waves cannot pass through liquids, whereas P waves can, Gorshkov concluded that the volcanic belt was underlain at these levels by pockets or layers of molten rock material. Judging from the lavas now being erupted the molten material would be of basaltic composition.

The low-velocity zone in the upper mantle is thus a kind of 'lubricated' zone, making relative movement between the more rigid overlying layers and the interior possible. As a result, it has become important to refer to the crust and the upper part of the mantle above the low-velocity zone as the **lithosphere**, and the low-velocity zone as the **asthenosphere**. This dynamic aspect of the Earth's mechanical behaviour is of fundamental importance, and the evidence for lithosphere creation, movement and destruction must be carefully evaluated before a general plate tectonic model is presented in Chapter 28. Here we shall concentrate on three of the major features of this dynamism: ocean ridges (where new lithosphere is formed); chains of oceanic volcanoes (which imply the lateral movement of lithosphere) and subduction zones, where old lithosphere disappears from the surface.

26.7   OCEANIC RIDGES AND RISES

An interconnecting network of submarine moun-

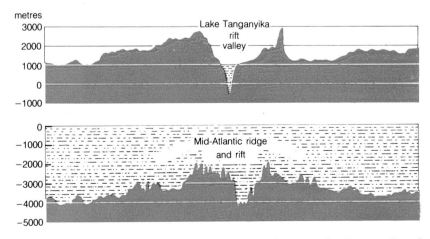

Figure 26.8 Topographic profile across the Atlantic floor to show the mid-Atlantic ridge and its median rift. Length of section about 480 km. (*After B. C. Heezen, 1959*)

tain ranges extends through the oceans, generally as mid-oceanic ridges. This extensive mountain system differs from continental mountains, not only in its greater length of about 64 000 km, but also because it is composed of basaltic lavas, and is, in fact, an outcrop of layer 2 of the oceanic crust (see Figures 26.6 and 26.7).

The Atlantic mid-oceanic ridge (Figure 26.8) has been known since 1855, being recognized by the *Challenger* Expedition. In 1953, *Discovery II*, working along the ridge north of the Azores, observed that it had a double crest with a deep and steep-sided trough between. There is indeed, as shown graphically in Figure 26.9, a close topographic resemblance between this high-level submarine rift valley and those of the high-level plateaux of Africa. Soon afterwards, the Lamont geologists, in their research vessel *Vema*, discovered that the mid-Atlantic rift continues along the crest south of the Azores. The rift has been detected at many other places along the crest, but there are long stretches where, instead of a conspicuous central rift, a number of parallel ridges and troughs occur.

In Iceland, which is a culmination of the mid-Atlantic ridge, the rift is well represented by the central graben (Figure 12.9), which bisects the island. Topographically, the graben is shallower than the submarine rift, because it is largely filled with the volcanic products of Pleistocene and recent eruptions. Indeed the volcanic activity displayed along the ridge by Iceland and Jan Mayen in the North Atlantic, and by Tristan da Cunha in

the South Atlantic, are vigorous reminders of the active growth of oceanic crust at the ridge axis. In 1970 Project FAMOUS (French–American Mid-Ocean Undersea Study) was founded. The intention was to use submersibles, specially constructed for manoeuvrability and ability to dive to depths of the order of 3000 m or more, and so descend into mid-ocean rifts and make detailed observations with the aid of specialized instruments and techniques. Starting in 1973, the mid-Atlantic rift was examined about 600 km south-west of the Azores. It was found in detail to be asymmetrical: topographically, thermally and magnetically. The western side, 300 m high, is bounded by a steep cliff that rises by narrow steps backed by nearly vertical fault scarps, whereas the eastern wall rises gently in broad steps backed by short steep slopes. Fresh pillow lavas, thought to be no more than 100 years old, were viewed at close quarters, and collected from fissures located in the rift floor. Subsequent laboratory examination showed the mineralogy and chemical composition of the basaltic lavas to vary very slightly outwards from the middle of the rift and to do so more gradually eastward than westward.

In the east Pacific a broad submarine bulge, 2000 to 4000 km wide, known as the East Pacific Rise, lies close to North America. Compared with the mid-Atlantic ridge, its topography is relatively smooth, with only occasional ridges and troughs parallel to its crest. As a submarine feature the crest of the East Pacific Rise passes into the Gulf of

Figure 26.9 Diagram to illustrate the similarity of form and scale between the Tanganyika rift of Africa and the submarine rift of the mid-Atlantic ridge. Each section is 725 km long. (*After B. C. Heezen. 1959*)

California and at the landward end is intersected by the San Andreas fault, only becoming submarine again to the south-west of the Island of Vancouver.

The mid-oceanic ridges are characterized by shallow earthquakes. Moreover, they are intersected at high angles by strike-slip faults of a variety known as transform faults (Section 28.3). The ridges are off-set against the faults, and those parts of the faults connecting the off-sets are also characterized by shallow earthquakes. In fact the idea that the ridges were connected was first proposed after noticing that shallow marine earthquakes formed continuous lines associated with ocean ridges and their transform faults.

26.8 CHAINS OF OCEANIC VOLCANOES

Before there was formal evidence (Chapter 28) that the oceanic lithosphere is moving away from the mid-oceanic ridges, where new basaltic crust is formed, and disappearing into the mantle at the trenches, Dietz (1961) and Hess (1962) had suggested that this was so. The idea was based on the thinness of oceanic sediments and the absence of Palaeozoic sediments from the ocean floors. The oceanic lithosphere was envisaged as moving as if

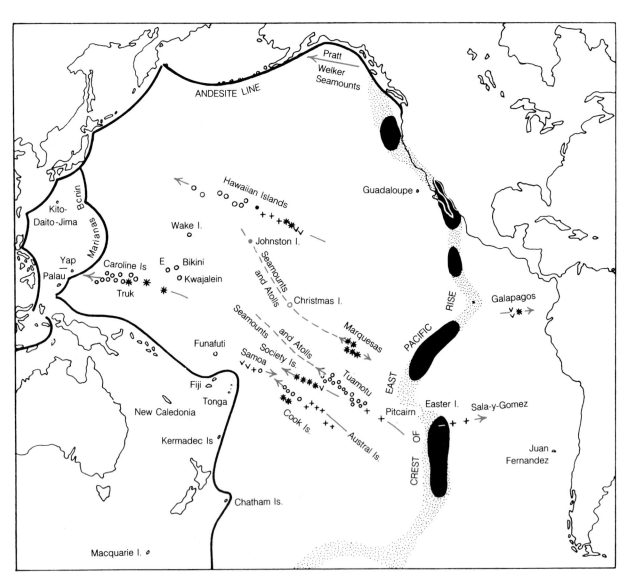

Figure 26.10 Map of the Pacific Ocean to show (i) the Andesite Line; (ii) the crest of the East Pacific Rise (black areas have heat-flow values > 3, ranging from 3.06 to 8.09); and (iii) linear chains of islands and seamounts, illustrating progressive increase of age and submergence in the directions indicated by arrowheads (v = active volcano; star = extinct but little eroded volcano; cross = deeply dissected and embayed volcanic island; solid circle = basaltic relics; open circle = atoll or seamount). The localities of atolls that have been drilled are named (Funafuti, Fiji, Bikini, E = Eniwetok, and Kito-Daito-Jima).

carried along on a conveyor belt, now equated with the viscous asthenosphere. To test the hypothesis of sea-floor spreading, as it was then called, Tuzo Wilson (1963) ascertained, for the volcanic islands of the Atlantic, (a) the oldest age reported for each island (by interbedded fossiliferous beds or by radiometric dating) and (b) the distance between each island and the ridge. The results showed that distance definitely tends to increase with age, apart from those volcanic islands that have remained on the ridge. The correlation is not perfect, as Wilson remarked, but in view of the roughness of the data the figures are sufficiently striking to support the interpretation that the volcanic islands originated on or near the ridge and have since been carried away from it.

Turning to the Pacific Ocean, it was long ago recognized by Dana (1890), in relation to the origin of coral reefs and atolls, that the Hawaiian chain shows a sequence of forms, starting from the active volcanoes of the east-south-east end, through volcanic domes showing various stages of erosion and increasing indications of submergence (like Pearl Harbor), followed by volcanic islets and stacks (some with fringing reefs), to atolls and seamounts at the west-north-west end. Since this first example was recognized, other chains and belts of islands and seamounts have been described from the south-west Pacific by L. J. Chubb (1927–57) and Tuzo Wilson (1963), all of which illustrate a similar sequence of volcanic activity followed by extinction and increasing signs of submergence (e.g. drowned valleys, barrier reefs and atolls, and guyots and other seamounts). The chains all trend approximately east-south-east to west-north-west, are nearly straight and, with the exception of the Samoan chain, they all have their active or more recently extinct volcanoes at the east-south-east end and their oldest and most deeply submerged members at the west-north-west end, as illustrated in Figure 26.10. In the case of the Hawaiian chain, Miocene fossils have been discovered on Oahu; lava from the same island has been dated at not less than 20 m.y., and Cretaceous fossils have been brought up from one of the guyots where the range of the Emperor seamounts begins (Section 24.13). These discoveries indicate that the evolution of the chain

has taken not less than 100 m.y. to reach its present state. Relative movement between the volcanoes and the magmatic source in depth has produced a chain about 3000 km long in 100 m.y. That is, there has been an average rate of movement of about 3 cm/year. There are two possible explanations:

1. the sub-crustal magma source slowly migrated linearly from the end of the chain with atolls, where it began millions of years ago, to the end with volcanoes, where it is still operating;
2. the magma source remained in its present position while the volcanic islets and underlying oceanic crust have migrated across this mantle 'hot spot' towards the west-north-west.

In this connection Figure 26.11 is of special interest. The floor of the Gulf of Alaska slopes gently towards the Aleutian trench and is variegated by seamounts and guyots that rise abruptly to heights of 1060 m to nearly 4000 m. The most significant feature of this chain of submarine mountains is the exceptional position and depth of guyot GA-1. All the other guyots have their platforms at depths of about 760 m. GA-1 is unique because it lies within the Aleutian trench and has its platform at a depth of over 2500 m, over 700 m deeper than the others. Clearly this guyot was carried down to its abnormal depth during the formation of the trench favouring explanation 2 above. Figure 26.11 gives the impression of the oceanic crust approaching the gigantic obstruction of the Aleutian Arc and bending down or subsiding in front of it.

26.9 OCEAN TRENCHES: SUBDUCTION ZONES

Deep-focus earthquakes occur beneath the continental sides of the deep ocean trenches; commonly on the continental side of island arcs which, like those of the Pacific, have their convex sides directed towards the ocean. Like the island arcs, the trenches that border them are also arcuate. A trench about 4800 km in length, borders the west coast of South America (Plate XVI), where there are no arc-like festoons of islands, and

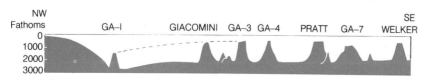

Figure 26.11 Profile of the Gulf of Alaska floor from the continental shelf south-east of Kodiak Island across the Aleutian trench (with guyot GA-1) and along the Pratt-Welker chain of seamounts and guyots. Length of section is 628 km.

related deep-focus earthquakes occur beneath the Andes.

The Kuriles are an excellent example of a **single** arc, with a submarine trench or trough on its outer convex side (Figure 26.12). Beneath the trench, and mainly to the landward side, there is a broad zone of shallow earthquakes with foci down to 60 or 70 km. This also roughly coincides with a belt of negative gravity anomalies, which is what would be expected from the bending down of oceanic lithosphere as it approaches the trench axis. Evidence for this bending is shown in the topography of the Aleutian Arc in Figure 26.11. The Kurile arc itself is a line of active or recently extinct volcanoes, usually rising from an erosion surface of Tertiary or older sedimentary rocks which have been folded, metamorphosed and granitized. This orogenic belt is well exposed around the volcanoes of Kamchatka to the north, and those of Japan to the south-west. Beneath the line of volcanoes the earthquake foci (apart from shallow volcanic shocks) fall mainly within the range of the intermediate class, i.e. at depths of 70 to 300 km. Since on average the depth of foci increases with distance from the open ocean, the epicentres of deep earthquakes occur much farther inland, or beneath marginal seas such as the Sea of Okhotsk between the Kurile arc and the Asiatic mainland. The deepest earthquakes occur down to a maximum of 720 km.

Hugo Benioff (1954), an American seismologist, assembled the data shown in Table 26.2 for a number of earthquake zones which dip beneath the continental margins.

Such earthquake zones, (illustrated in profile to the right of Figure 26.12), associated with ocean trenches, have come to be known as **Benioff zones**. Note in Figure 26.12 the tendency of dip to increase with depth and of magnitude to decrease with depth.

Certain island festoons, of which the Indonesian arc is an excellent example, differ from single arcs like those of the Kurile and Aleutian Islands in having an additional feature consisting of a sedimentary outer arc which has developed from within the submarine trench. The single Aleutian arc shows a gradual transition towards the **double** type where it links up with the Alaskan peninsula. The volcanic line of the islands is continued by the volcanoes of Alaska, but a second chain of islands appears between the volcanoes and the trench, almost in the middle of the negative gravity anomaly near the trench axis. To distinguish the two the line of volcanoes is said to be in the 'back arc', and the mainly sedimentary outer arc is known as the 'fore arc' (Table 30.1 and Figure 30.3). The sedimentary fore arc occurs just to the landward side of the trench, and the volcanic back arc about 150 km behind the trench axis (Figure 26.13). Note that the volcanic chain overlies that part of the Benioff zone at a depth of 150–200 km in Figure 26.14. Globally the position of the volcanic chain overlies this depth of penetration, and therefore depends fundamentally on the dip of the Benioff zone.

In the Indonesian arc, the fore arc is a submarine feature for long distances, e.g. south of Java (Figure 26.13), but in places it has risen above sea-level forming rows of islands. The Mentawai Islands, off the coast of Sumatra, are a good example (Figures 26.13 and 26.14). There the strata can be seen to have been strongly folded early in the Tertiary and again during the Miocene. The islands have recently been rising, as the terraces of

Table 26.2 Variations in the dips of Benioff zones

Marginal region	Intermediate foci		Deep foci	
	Depth km	Dip of Benioff zone	Depth km	Dip of Benioff zone
Aleutian Arc	70–175	28°		
Kurile–Kamchatke Arc	70–300	34°	300–700	58°
Japanese Arc (Bonin–Honshu)	70–400	38°	400–550	75°
Indonesian (Sunda) Arc	70–300	35°	300–700	61°
New Hebrides	70–300	42°		
Chile	70–290	23°	550–650	58°
Peru–Ecuador	70–250	22°	600–650	47°
Central America (Panama–Acapulco)	70–220	39°		
Average dips		33°		60°

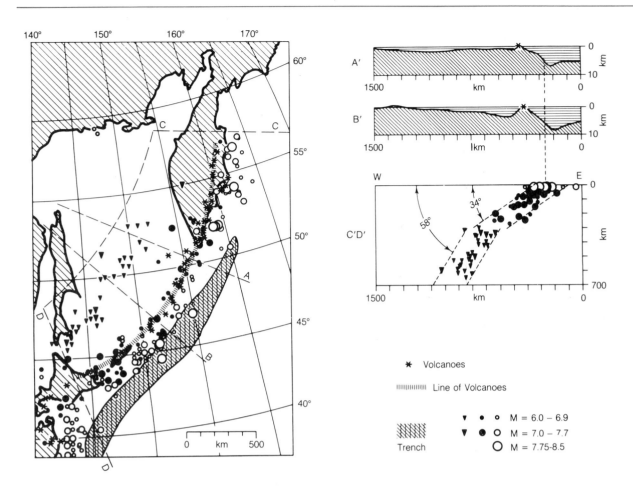

Figure 26.12 Map of the Kamchatka–Kurile volcanic arc with topographic profiles and a composite seismic profile. Epicentres (map) and foci (seismic profile) of shallow, intermediate and deep earthquakes are represented by circles, circular dots and triangular dots respectively. These symbols are of different sizes according to the magnitudes (M), as indicated below the profiles. A' and B' are vertically exaggerated profiles along the lines A and B on the map. C'D' is a composite seismic profile of the whole arc from CC across Kamchatka to DD across northern Japan. (*Hugo Benioff, 1954*)

uplifted coral reefs clearly prove; that they are still doing so is indicated by the effects of shallow-focus earthquakes. Farther north the ridge can be traced by way of the Nicobar and Andaman Islands into Burma, where it becomes a continuous land feature and passes into the high mountain range of the Arakan Yoma, west of the Irrawaddy. Far to the east of Java the ridge emerges conspicuously above sea-level as the Island of Timor, which is celebrated for the great heights to which its successive coral reefs have been upheaved, some of them now standing over 1200 m above sea-level.

The unity of this long arc of mountains, rows of islands and submarine ridges is indicated not only by its topographic continuity and association with a Benioff zone (Figures 26.13 and 26.14), but also by its coincidence with a belt of negative anomalies of gravity. These are particularly strong where the arc is still under water. The discovery of the major part of this negative belt (Figure 26.13), the

first of its kind to be made, was the great achievement of F. A. Vening Meinesz. In the course of a series of confined and tedious expeditions (1923, 1926, 1929–30) he carried out his pioneer work on the measurement of gravity at sea, using submarines lent by the Royal Netherlands Navy. These could be submerged to depths where the effects of waves on swinging pendulums became negligible. The 'Diving Dutchman', as Vening Meinesz came to be called, was eventually able to chart a well-defined negative belt from the ridge northwest of the Mentawai Islands to the Mindanao Trench off the Philippines, with only a short break near Sumba (west of Timor), 4000 km in all.

From the belt of negative gravity anomalies that characterize the trench of the Indonesian arc, Vening Meinesz inferred a down-buckling and consequent thickening of the oceanic crust to form a root. Since then, however, seismic as well as gravity surveys both here and in other regions, notably the Puerto Rico and Peru–Chile trenches,

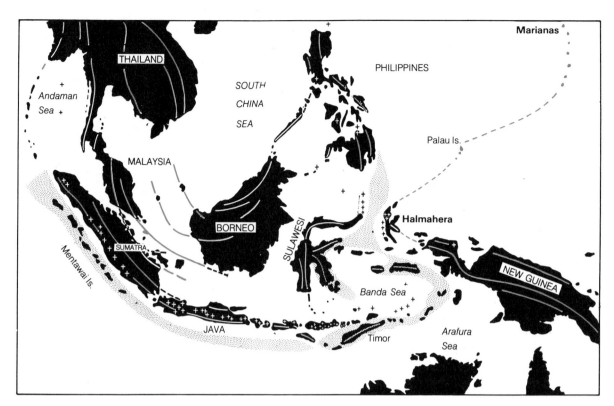

Figure 26.13 Tectonic map of the Indonesian arc and adjoining territories. The belt of negative anomalies of gravity discovered by Vening Meinesz is indicated by a shading of fine dots; this can now be continued through the Andaman Islands into Burma. The line of active volcanoes is shown by black or white crosses. The deepest part of the trench, south of Java, is known as the Java Trench.

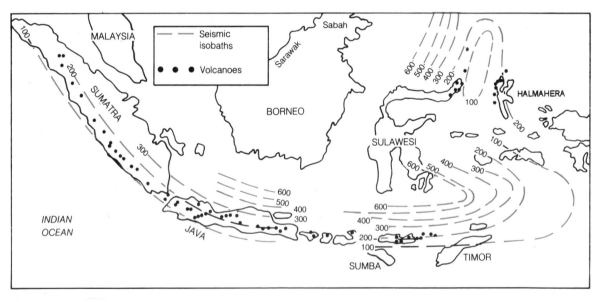

Figure 26.14 Map showing the generalized distribution of intermediate and deep earthquakes associated with the trench bounding the southern side of the Indonesian archipelago. The seismic isobaths depict the Benioff zone dipping towards the continental area of south-east Asia. The line of active volcanoes lies above the foci of earthquakes of intermediate depth (150–200 km). (*Adapted from T. Hatherton and W. R. Dickinson, 1969*)

have shown that although all of them are characterized by negative gravity anomalies, and are in fact out of isostatic equilibrium, the oceanic crust beneath them is of the same order of thickness as elsewhere in the transitional region between continental and oceanic margins. Moreover, the sedi-

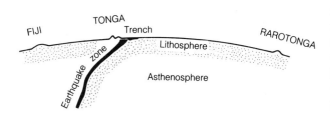

Figure 26.15 Diagrammatic section across the Tonga trench illustrating the subduction of oceanic lithosphere based on seismic correlation of the oceanic lithosphere with an anomalous seismic belt, within the mantle, characterized by low attenuation and high velocity of seismic waves relative to those of the adjacent asthenosphere. (*After J. Oliver and B. Isacks, 1967*)

mentary deposits within trenches are characteristically thin except for that part of the Peru–Chile trench south of Valparaiso where they are at least 2 km thick and form a flat floor to the trench. The negative gravity anomalies characteristic of trenches are now thought to be due to dynamic displacement of mantle resulting from the bending down of oceanic lithosphere at the trenches, the only substitution for the volume of high density mantle being low density seawater filling the trench.

In studying data from several hundred earthquakes recorded from the Fiji region, and associated with the Tonga–Kermadec Island arc and trench to the north-east of New Zealand. Oliver and Isacks (1967) found more conclusive evidence of oceanic lithosphere turning down into the mantle. They discovered an anomalous seismic zone in the mantle, the upper surface of which is roughly coincident with the upper surface of the Benioff zone that dips beneath the island arc down to a depth of about 700 km. Seismic body waves propagated in the anomalous zone are less attenuated than are waves of the same type propagated from similar depths in the mantle elsewhere. Oliver and Isacks suggested that if low attenuation can be equated with the mechanical property of strength then the anomalous zone can be equated with the lithosphere, and thus depicts oceanic lithosphere travelling down, i.e. being **subducted**, into the mantle (Figure 26.15). In other words, the anomalous zone related to the Tonga trench is a **subduction zone** revealing the disappearance of the moving oceanic lithosphere down into the mantle. The zone is at most 100 km thick and probably only 50 km thick.

26.10 HIGH-PRESSURE TRANSFORMATIONS

Examples such as the metamorphism of basic rocks into eclogite and the artificial production of hard translucent diamond from soft, black graph-

ite illustrate the effect of high pressure in reorganizing the lattice structures of minerals so that the atoms are more closely packed than before. Phase transitions of the graphite/diamond type are now known to be common throughout the mineral world. Silica, for example, passes through a number of phases ranging from the familiar quartz (density 2.65 g/cm^{-3}) to the much denser and harder coesite (3.01 g/cm^{-3}), which forms in the temperature range 500–800°C at pressures of 35 000 MPa or more, corresponding to conditions near the top of the mantle in undisturbed continental regions. Coesite is evidently not a normal crustal mineral, though it might be expected to occur in the roots of high mountains, if there were any free silica there to be so transformed. Natural occurrences of coesite are rare, but are sometimes found in the walls and exploded fragments of certain large craters where pressures and temperatures have been extremely high. Two kinds of such craters have been recognized: (a) those due to the searing and explosive impact of giant meteorites; and (b) those due to the explosive eruption of hot volcanic gases which had to overcome not only the weight of the overlying rocks, but their strength (e.g. the late Miocene Ries Basin in western Germany, east-north-east of the Black Forest). At pressures corresponding to the weight of overlying rocks at a depth of 300 km coesite is transformed into a still denser phase which was first made artificially and given the name **stishovite**. This man-made mineral is nearly twice as dense as quartz. It has the hardness of diamond, but is immensely more costly to make. In 1962 it, too, was detected in the glassy material formed from the shattered and fused granitic rocks from the floor of the Ries Basin.

At what may be called the ultra-high pressures produced in the laboratory either by diamond anvil presses or by shock waves, still more remarkable changes occur. At a pressure corresponding to a depth of 1300 km, diamond ceases to be stable and turns into a metallic form of carbon not unlike iron. As the pressure of the shock waves dies down, the metallic carbon quickly reverts to graphite. The metallic state is one in which the ions are much more tightly packed than in the crystals of rock-forming minerals; so tightly, in fact, that some of the electrons are forced out of their orbits and become detached, so that they are free to move through the crystal lattice. It is for this reason that metals are good conductors of electricity. An electric current along a wire is the swift rush of detached electrons through the interstices between the ions of the metal. When non-metallic elements are subjected to a pressure sufficiently high to strip off some of the outer electrons, they too are turned into metals. With further increase of pressure the process of 'atom squashing' or **pressure-ionization**, to give it its technical name, continues until more and more electrons are forced out of their orbits: first from the outer shells and then from the inner ones in turn. Finally, when the intensity of pressure-ionization reaches its limit – which it does in the stars known as white dwarfs – all the atoms are crushed like hollow eggshells and matter becomes completely **degenerate**. It then consists of tightly crowded atomic nuclei immersed in a homogeneous medium of electrons.

The above brief introduction to a difficult branch of physical chemistry will suffice to show that the properties of matter under the high pressures and temperatures of the Earth's deep interior may differ fundamentally from everything we are familiar with in everyday life.

Figure 26.16 shows the location of the major phase changes which affect mantle minerals compared with the seismic P wave velocity profile. The bulk composition of the upper mantle is also shown on this diagram, indicating that 57 per cent of the upper mantle is comprised of the mineral olivine, with smaller proportions of orthopyroxene (17 per cent), clinopyroxene (12 per cent) and garnet (14 per cent). Since olivine is the dominant mantle mineral, the major phase changes which affect the process of subduction are those of olivine/spinel and spinel/perovskite, which are thought to occur in the mantle transition zone

near the seismic discontinuities at 400 km and 670 km. The olivine/spinel phase change actually occurs in two stages, via an intermediate step known as a Beta phase. An increase in pressure results in a sudden increase in density as the lattice readjusts to the pressure with the release of energy as heat. A reaction such as this, which produces heat, is said to be **exothermic**. The increase in density does not occur at exactly the same depth everywhere within the descending lithosphere, but is spread out over a depth range of several kilometres because some areas within the lithosphere will be more susceptible than others. The increase in density speeds up the rate of sinking of the lithosphere into the mantle, and results in intermediate-depth earthquakes which show evidence from focal mechanism studies of tension down the dip of the slab. In contrast the deepest earthquakes show down-dip compression above the 670 km discontinuity, indicating that the slab of descending lithosphere is at least being slowed down, and possibly completely arrested, at levels in the mantle where we would expect the spinel/perovskite phase transition to occur. A plausible reason for this is that the spinel/perovskite phase transition absorbs the kinetic energy of the descending slab in the form of heat (in an **endothermic** reaction), and so resists motion. Another explanation for the resistance to slab descent at the level of the deepest earthquakes is a dramatic increase in the viscosity in

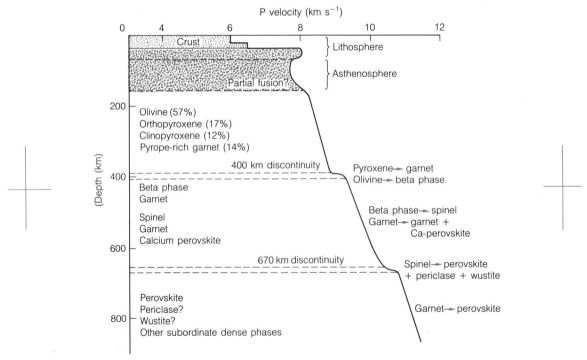

Figure 26.16 Interpretation of the *P* velocity-depth distribution of Burdick and Helmberger (1978) for the upper mantle and transition zone in terms of partial fusion (or melting) and phase transitions of the constituent minerals of peridotite. (*After Bott, 1982*)

the lower mantle inferred from the study of very long-wavelength gravity anomalies.

Increasing pressure also results in pyroxene going into solid solution in garnet, changing the composition of the garnet. This transition eventually leads to the conversion of eclogite, a garnet-pyroxene rock, into **garnetite** – a rock consisting entirely of garnet (of a composition unlike that of crustal garnet). This transition is particularly important in subduction zones where, with increasing pressure, the original basaltic material of the descending oceanic crustal slab is converted first to eclogite and then to garnetite.

If the 670 km discontinuity also represents a change in composition between the upper and lower mantle, with an associated density contrast of more than 5 per cent, then it is unlikely that the descending lithosphere can cross the boundary, and the upper and lower mantle can be regarded on a broad scale as isolated units. Despite a great deal of work on this question, we still do not know whether or not the mantle can be regarded as essentially one or two chemical units. The jump in viscosity and the absence of earthquakes in the lower mantle indicate that the mantle is at least mechanically different on either side of the 670 km discontinuity.

26.11 MANTLE HETEROGENEITY

Recently the quantity and quality of seismic data, and their availability in the form of digital records, has allowed the **tomographic** imaging of the Earth in three dimensions, similar to medical imaging techniques used in hospitals. (An example of a tomographic slice through the Earth is shown in Plate Ib. The property which is imaged is the deviation of seismic velocity from a reference layered model.) The seismic tomographer suffers from not having complete control over where to place his sources and receivers, but otherwise the principle is the same. The fuzzy images of the Earth's interior which have emerged during the last decade show that the mantle is in fact laterally heterogeneous as well as layered. For example, subduction zones are characterized by material in the uppermost mantle with relatively high seismic velocities; ocean ridges by material in the uppermost mantle with low seismic velocities. This implies that ocean ridges are underlain by hot asthenosphere material, and subduction zones are characterized by cool descending lithosphere. However, only the large subduction zones, notably the Circum-Pacific belt, have velocity anomalies which appear continuous throughout the mantle. Continental lithosphere is also heterogeneous, with the older cratons or shields having mantle lithosphere with faster seismic velocities than younger continental lithosphere. This implies that the lithosphere is cooler and thicker under the continental cratons. Chemical isotope studies on mantle xenoliths from cratonic areas are suggestive of a chemically distinct lithosphere underlying the Earth's oldest shield areas.

In summary, the mantle is laterally heterogeneous as well as vertically layered, with the physical and chemical characters of the uppermost mantle showing a good correlation with surface features, while the transition zone and lower mantle are dominated by large velocity anomalies at the site of the major subduction zones. The subtle, but significant, changes in seismic velocity revealed by seismic tomography have already changed our image of the mantle. As time goes on the quantity and quality of the data will improve this image, but a complete understanding of their meanings will continue to require the assessment of independent evidence, such as the global gravity field, geochemical analysis of mantle xenoliths and high-pressure laboratory simulation of mantle conditions.

26.12 FURTHER READING

Bott, M. H. P. (1982) *The Interior of the Earth*, 2nd edn, Edward Arnold, London.

Heirtzler, J. R., and Bryan, W. B. (1975) The floor of the Mid-Atlantic rift, *Scientific American*, vol. 233, No. 2.

Toksöz, M. N. (1975) The subduction of the lithosphere, *Scientific American*, vol. 233, No. 5, 88–98.

Wilson, J. Tuzo (1963) Evidence from islands on the spreading of ocean floors, *Nature*, vol. 197, 536–8.

Woodhouse, J. H., and Dziewonski, A. M. (1989) Seismic modelling of the Earth's three-dimensional structure, in *Seismic Tomography and Mantle Circulation* (eds R. K. O'Nions and B. Parsons), Royal Society, London.

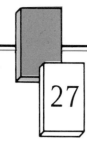

27

PALAEOMAGNETISM AND CONTINENTAL DRIFT

The latitude's rather uncertain,
And the longitude also is vague.

William Jeffrey Prowse (1836–70)

27.1   THE EARTH'S MAGNETIC FIELD

Although Thales, earliest of the Greek philosophers, was familiar with the lodestone (magnetite) and its apparently magic properties, and the compass was used as a direction-finder by Chinese navigators well over 2000 years ago, it was not until the year 1600 that William Gilbert, Physician to Queen Elizabeth I, published his celebrated book *De Magnete*, in which for the first time it was shown that 'the terrestrial globe behaves like a giant magnet'. Gilbert experimented with spheres of magnetite and found that iron filings placed on the surface aligned themselves in the same directions as freely suspended magnetized needles situated in corresponding positions on the Earth's surface. He concluded that the Earth's magnetism, being like that of a uniformly magnetized sphere, is therefore of internal origin. This was a remarkably accurate first approximation. In 1839 Gauss rigorously proved that by far the greater part of the magnetic field originates below the Earth's surface, but that there is a small and variable part that originates outside. The latter – the **transient** disturbances – include the magnetic storms which are often accompanied by widespread auroral displays. They originate high up in the atmosphere, where electric currents are generated by cosmic rays and by radiation and charged particles emit-

ted from sunspots, solar flares and the like.

While the main part of the magnetic field is like that of Gilbert's magnetized sphere, it is equally like that of a powerful bar magnet (a 'dipole') placed near the middle of the Earth along the axis joining the north and south geomagnetic poles (Figure 27.1). Gauss determined the dipole field that most closely matched the Earth's actual main field, and found this left over certain irregular deviations, even when local anomalies due to the magnetic effects of iron ores in the crust have been allowed for. The whole field can thus be subdivided as follows:

EARTH'S MAGNETIC FIELD

- **Transient field**, originating in the upper atmosphere

- **Magnetic anomalies**, due to concentrations or deficiencies of magnetic minerals and rocks in the crust

- Main Field of internal origin
  - **Regular** major field equivalent to that of a strong centrally placed dipole (Figure 27.1)
  - **Irregular** ('non-dipole') residual fields

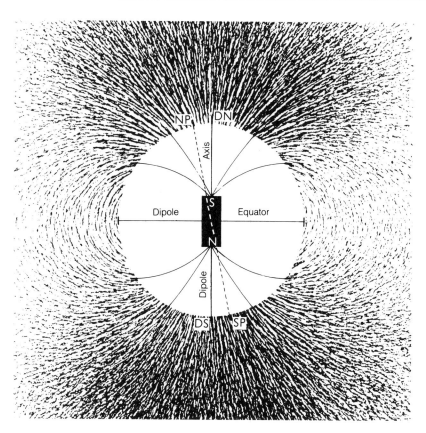

Figure 27.1 Iron filings sprinkled on a sheet of glass over a strong bar magnet represent the dipole part of the Earth's magnetic field. DN and DS are the north and south geomagnetic poles; NP and SP are the geographical North and South poles.

The main field is continually changing at easily measurable rates. The **secular magnetic variations**, as they are called, are recorded at various observatories where the details necessary to determine the total magnetic field and its parts are continuously measured, like the elements that make up records of the weather. Three magnetic elements are essential and those in common use are the following:

1. The **intensity**, which is the total magnetic force, F; the horizontal and vertical components are referred to as H and Z respectively.
2. The **magnetic dip** or **inclination**, which is the angle I, measured between the horizontal and the direction taken by a freely-suspended magnetized needle; i.e. the angle between the directions OH and OF in Figure 27.2. At the geomagnetic N and S poles the dip is vertical.
3. The **magnetic declination** is the angle D between the direction of H and the geographical north.

The rapidity of the changes is well illustrated by the declination. In 1546 a compass in London pointed 8°E of true north. Since then the direction has varied year by year, becoming true north in 1665 and reaching 24°W in 1823. The direction of

change then reversed, as shown on Figure 27.3, which also records how the inclination varied over the same period. The variation in the intensity, F, is both more rapid and more irregular. During the last hundred years the strength of the field in the London district diminished by 5 per cent, but recently it has been increasing again. About a century ago the intensity at Cape Town rapidly increased to about a third more than its present value. These remarkable changes are phenomenally fast compared with the sluggish processes occurring in the mantle. This highly significant fact leaves no alternative to the inference that the main field has its origin in the outer core, where there is the possibility of fluid motions such as might result from thermal convection, and where the material is metallic and capable of carrying electric currents.

Taken by itself, it is found that the pattern of the irregular part of the main field is continually changing and that it bears no relationship of any kind to the distribution of land and sea or to geological structures or rocks. This is another fact pointing to the core as its place of origin. Moreover, it drifts slowly westward at about 0.18° per annum, which implies that the magnetic field rotates more slowly than the Earth as a whole. In

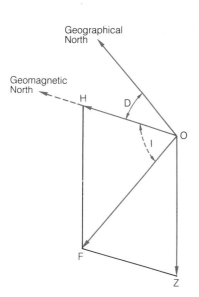

Figure 27.2 For description see text.

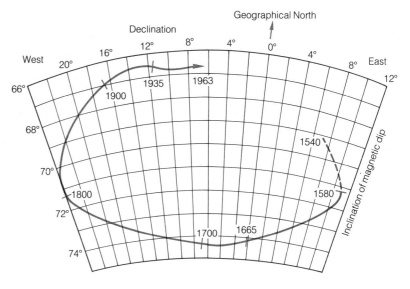

Figure 27.3 The change in the direction of the mean magentic force, F, in the London district since 1540, shown by records of magnetic declination and magnetic dip.

other words, the source of the field, i.e. the core, rotates more slowly than the mantle. The resulting shearing along the discontinuity would thus favour the generation of eddies. The latter, together with vortices in other parts of the outer core, might well be responsible for the irregular part of the field.

27.2 THE ORIGIN OF THE EARTH'S MAGNETIC FIELD

The magnetic field cannot be a permanent property of the material of the core. Above a certain temperature known as the Curie point all magnetic materials lose their magnetism. For iron at atmospheric pressure the Curie point is 770°C; for nickel, 330°C; and for magnetite, 580°C. With increasing pressure the Curie point drops. Thus the high temperature and the liquid-like state of the outer core both render permanent magnetization impossible. The magnetic field must therefore be continuously produced and maintained, and this again suggests the generation of electric currents in material of high electrical conductivity and capable of internal movements. Only the outer core, being what we can best describe as a metallic liquid, fulfils these conditions. The suggestion that the outer core acts like a dynamo has been investigated in great detail by W. Elasser in the United States and Edward Bullard in England.

The principle on which the dynamo hypothesis is based is that of the dynamo itself: namely, that when an electric conductor is moved through a magnetic field an electric current is generated in the conductor. Current is generated in an ordinary dynamo by the rapid rotation of wire coils in a strong magnetic field. In the outer core of the Earth the rotating coils are represented by thermal convection cells. Given a magnetic field to start the system working (a very weak field would do) convection provides the motion necessary to generate electric currents in the convecting material. The electric currents increase the intensity of the magnetic field, which in turn increases the strength of the electric currents . . . and so on until convective movement, electric current and magnetic field reach an approximate state of equilibrium. For this reason the system is described as a 'self-exciting dynamo'.

Two problems remain to be solved before the hypothesis can claim the status of a reasonably complete theory. One is the source of the original magnetic field. The other is the source of the energy that keeps the convection and the associated vortices going without a breakdown. The need for an initial magnetic field sounds like assuming what has to be explained. But a weak field could have started in a variety of ways at the time when the Earth was being formed as a planet. Since we do not know how the Earth originated, the matter is of no more than theoretical interest at present.

Three sources of energy for maintaining the field have been considered. The one usually appealed to is radioactivity. But it is difficult to see how there could be sufficient radioactivity in the core to supply the heat necessary to maintain convection currents; if there were, the mantle would also be endowed with so much radioactivity that the crust could not have cooled down to its present temperatures. Another hypothetical source of energy to which appeal has been made is that liberated (like latent heat) when phase changes occur as a result of decrease of pressure. This is a hypothesis based on the work of Dirac and Ramsey, and developed by Egyed.

The third source of energy, already referred to, is downward migration of iron from the mantle to the core, a process which would release gravitational energy and generate heat. This hypothesis bristles with difficulties. If the process happens at all, it should have come to an end long ago, or have now become too slow to supply any significant amount of heat.

A more likely possibility for the release of gravitational potential energy is the downward migration and freezing out of iron from the liquid outer core to form the solid inner core. If the core was completely molten when it formed and the inner core has gradually grown since, then a total cooling of only 80°C would be sufficient to provide 10^{12} Watts, over the lifetime of the Earth. This gravitational energy coupled with compositional convection driven by less dense fluids generated at the surface of the growing inner core would provide more than adequate energy to drive the geomagnetic dynamo.

27.3 PALAEOMAGNETISM: ROCKS AS FOSSIL COMPASSES

In describing the Earth's magnetic field in section 27.1 it was stated that the main part of the field is nearly equivalent to that of a powerful bar magnet (a dipole) placed near the Earth's centre and orientated along the axis joining the N and S magnetic poles (Figure 27.1). This **geocentric axial dipole field** accounts on average for about 95 per cent of the whole. Most of the balance is made up of feeble and irregular fields equivalent to a number of weak dipoles not situated centrally. At the present time the magnetic axis departs considerably from the axis of rotation. However, the departure is continually changing, and even during the few hundred years in which records have been kept (see Figure 27.3) the average divergence between the magnetic and geographical north poles has been very much smaller than the value at any given time. But to make it a reasonably sound assumption that the mean magnetic field is that of a geocentric dipole orientated along the axis of rotation, the period over which the average is taken should be not less than several thousand years. As noted below, this consideration has an

important bearing on the collection of rock specimens from formations likely to give information about the geomagnetic fields of earlier times.

In rocks such as basaltic lava flows and basic sills and dykes crystals of ferromagnetic material – mainly oxides of iron such as magnetite – acquire a stable magnetism which is 'frozen in' as it cools through its Curie temperature. The latter is about 575°C for magnetite, but may range between 700°C and 200°C, according to the mineral concerned. The newly acquired or **remanent** magnetism has the same direction (the declination, D) and the same dip (the inclination, I) as the local geomagnetic field at the time of consolidation. This can be demonstrated experimentally. It has also been proved for natural flows erupted at known dates from volcanoes situated where magnetic records for the same dates are available (notably Etna, Hekla, and some of the volcanoes of Japan).

Measuring the remanent magnetism is carried out with special types of highly sensitive galvanometers. The samples on which the measurements are made are usually slices cut from cylindrical cores drilled from the hand specimens (or *in situ* from the rock itself); non-magnetic tools being used. The specimens must be unweathered and must be accurately orientated at the place of collection: i.e. marked with the horizontal plane and dip (if any) as well as with the local direction of the magnetic north or south. To ensure that a statistically reliable mean value of the remanent magnetism can be calculated it is rarely enough to sample a single flow. Specimens from a series of flows, estimated to cover at least a few thousand years, are required to provide the data for calculating the approximately equivalent axial dipole field. A series of specimens taken across a sill or dyke from one margin to the other will provide an adequate time span if the intrusion is thick enough. Samples from the margins, which cooled quickly, will have passed through the Curie points thousands of years before those from the middle, where cooling was slow. Other samples of interest are those from the country rocks near their contacts with sills and dykes. Shortly after the time of intrusion such a rock will have been heated up well above its Curie points. It will then have lost its original magnetism and gained a new one on cooling, the latter corresponding closely to that of the margin of the intrusion.

Similar considerations apply to the sampling and orientation of sedimentary rocks collected for geomagnetic data. Detrital grains of magnetite, for example, inherit a remanent magnetism from their parent rock. Thus, when they finally come to rest in a semi-consolidated sediment, they tend to align themselves like compass needles along the direction of the geomagnetic field in which they then find themselves. Probably more important is the fact that ferruginous alteration products, e.g. haematite, which form coatings to the detrital minerals and act as cementing materials to newly deposited sediments, also become magnetized in line with the local geomagnetic field. Such 'chemical magnetization' as it is called, makes red sandstones and shales, like those of many well-known Permian and Triassic formations, particularly rewarding sources of palaeomagnetic data. Sampling of a given formation should be through a thickness of 100 m or more, and over as wide an area as possible, in order to give an ample time range for effective averaging. Measurements of dip and strike must be carefully made, so that correction for the effects of folding or tilting can be properly applied. The importance of freedom from weathering is obvious, since ferruginous weathering products become 'chemically magnetized' at the time of their formation, which for all practical purposes is the present time.

Apart from some earlier work in France, which was not systematically followed up – probably because of the World Wars – the study of palaeomagnetism was begun about 1950 by P.M.S. Blackett, who before long attracted an enthusiastic team of co-workers, some of whom, notably S.K. Runcorn, soon started teams of their own. Because of the obvious bearing of the results on the geological problems of continental drift and the physical problems of the origin of the Earth's magnetic field, the subject rapidly spread throughout the scientific world as one of the most fertile and actively developing branches of geophysics.

27.4 PALAEOMAGNETIC RESULTS: GEOCENTRIC AXIAL DIPOLE

The fundamental hypothesis on which palaeomagnetic investigations are based is that the geomagnetic field, averaged over an appropriate time, has always been very nearly that of a geocentric axial dipole. In other words, the geomagnetic poles have always, on average, coincided approximately with the geographical poles. If this coincidence were exact, then for rocks of a particular age from a given locality the averaged value of I (the palaeomagnetic inclination or dip) would give the latitude λ of the locality at that time from the relation

$$\tan I = 2 \tan \lambda$$

shown graphically on Figure 27.4. The latitude fixes the **distance** of the pole. The **direction** of the pole is given by the averaged value of D (the palaeomagnetic declination). Knowing both the distance and direction of the geographical pole

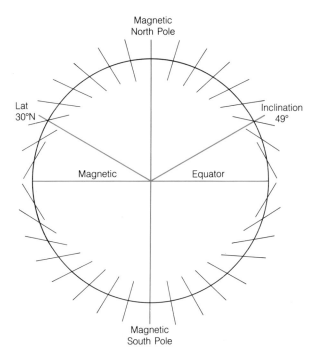

Figure 27.4 Idealized diagram to illustrate the relationship between magnetic inclination and the corresponding latitude for a geocentric axial dipole field. An inclination of 49°, for example, corresponds to a latitude of 30°, tan 49° = 1.15 = 2 tan 30°.

from a given locality fixes the position of the pole on the globe at the time when the measured rocks received their magnetization.

In practice such pole determinations can never be accepted as exact: because (a) the minor components of the Earth's field (Section 27.1) have been neglected; (b) the rocks tested may have suffered magnetic changes that have not been recognized and allowed for; and (c) slight orientational and experimental errors inevitably creep in. Individual pole determinations may be correct within about 15°, and possibly less if based on a dozen or more reliable samples. It should also be noticed that considerably greater variations may be found in the pole positions relative to a given continent and recorded for a given geological Period (e.g. the Permian, as illustrated in Figure 27.10). These differences are not necessarily errors. Most of them may represent genuine changes in the position of the continent during the very long interval of time involved, or in the positions of certain regions that have been displaced by great strike-slip faults, or have gone through the severe disturbances of orogenic movements. The effects of some of these difficulties have been reduced as increasingly detailed work has been accomplished year by year, and especially as it has become more generally possible to combine radiometric dating with magnetic measurements.

The dipole hypothesis is particularly well supported for the last 2 m.y. or so. Statistically, the mean position of the Quaternary magnetic pole, as determined for a large number of specimens from various parts of the world, does not depart significantly from the geographical pole of today (Figure 27.5).

27.5 REVERSED MAGNETISM: THE PALAEOMAGNETIC TIMESCALE

In a long sequence of rocks that have been adequately tested for remanent magnetization, two well-defined groups commonly turn up, within one of which the stable remanent magnetization is in the opposite direction to that of the other set. About half the rocks are N-seeking the remainder being S-seeking. Nearly all Permian samples, for example, are S-seeking. Initially it was thought that this phenomenon of **reversed magnetization** could be due either to alternating reversals of the Earth's polarity (**field reversal**) or to certain processes of **self-reversal** within the rock itself, such as might be brought about by physical or chemical changes in the magnetic minerals during or after the rock's formation.

Self-reversal is evidenced by the fact that it can

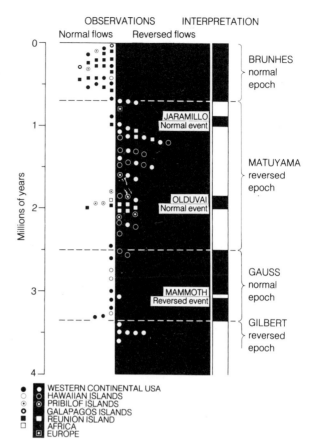

Figure 27.6 A timescale for reversals of the Earth's magnetic field, established from palaeomagnetic data with matching radiometric ages derived from nearly 100 volcanic formations located in both hemispheres. The data group themselves into 4 major polarity epochs, in which shorter polarity events are superimposed. (*A. Cox, G. B. Dalrymple, and R. R. Doell, 1967. Copyright by Scientific American, Inc. All rights reserved*)

be brought about by laboratory experiments – but only on certain specimens constituting a small percentage of the number tested. Moreover, a few cases are known (e.g. in the Tertiary lavas of Mull) where both normal and reversed magnetization are found within different specimens from the same flow, so indicating that self-reversal can be a natural phenomenon. In the Mull example the chief magnetized mineral in a **normal** specimen is titaniferous magnetite in well-formed crystals; whereas in a **reversed** specimen there is a considerable variety of ferro-magnesian minerals, representing the products of subsequent alteration of a kind involving oxidation.

As long ago as 1929 Matuyama found that rocks with reversed magnetization in Japan were all early Pleistocene in age, whilst younger rocks showed normal magnetization. Complete evidence that magnetic reversals are not fortuitous, since they occur at the same time all over the Earth, was not established, however, until 1964. By then there were sufficient determinations of

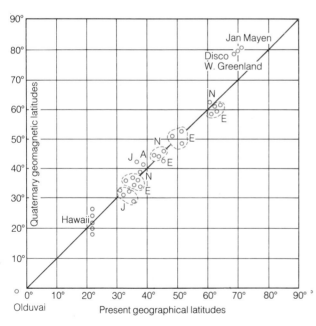

Figure 27.5 Geomagnetic latitudes, determined from Pleistocene and Recent lavas, plotted against the present latitudes of the latter. A, Australia; E, Europe; J, Japan; N, North America. The points nowhere depart far from the diagonal, which implies that within the limits of error the mean positions of the Quaternary magnetic poles were not significantly different from the geographical poles of today.

remanent magnetization correlatable with radio-metric age determinations, by the potassium–argon method, to establish a timescale of normal and reversed magnetism extending backwards in time for 3.6 m.y. The relevant data was obtained from close on 100 magnetized volcanic formations, occurring within both hemispheres. The investi-gations were mainly made by Cox, Dalrymple, and Doell, in the US Geological Survey Laboratory in California, and by McDougall, Tarling and Cha-malann at the Australian National University. By the late 1960s the time-scale had been extended back to about 4.5 m.y. The potassium–argon method, in this context, has insufficient precision

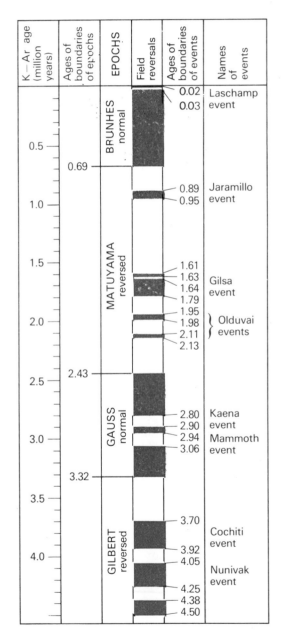

Figure 27.7 Timescale for geomagnetic reversals in which the time-span of events is based on data from sediments. (*Adapted from A. Cox, 1969*)

to carry the timescale farther back in time.

As can be seen from Figure 27.6, during the last 3.6 m.y. there have been two periods with normal polarity, as at the present day, and two with reversed polarity. These main periods are known as **geomagnetic polarity epochs**, and they are named after investigators, of various nationalities, who have made important contributions to knowledge of the Earth's magnetism. Within these epochs are shorter **polarity events**, to which place names have been given, commemorating the localities where they were first recognized. When the geomagnetic timescale was first established, three polarity events were known. By 1969, twenty more determinations having been made, five polarity reversals were known (Figure 27.7).

The magnetic reversal timescale has an impor-tant application in determining the ages of deep-sea sediments. In 1966, Opdyke, Glass, Hayes and Foster established what might be called the mag-netic stratigraphy within seven cores from the Antarctic, and they correlated this stratigraphy of normal and reversed epochs and events with fau-nal zones as defined by radiolaria. Here then was a method of correlating deep-sea sediments from widely separated regions, and also of determining the rate of deposition (Figure 27.8).

As described in Chapter 28 the interpretation of sea-floor spreading anomalies has greatly increased the time span over which polarity changes can be established. It has led to the determination of a geomagnetic polarity timescale back to 165 m.y. ago. Using this approach 336 polarity changes have been documented making the polarity timescale of Figure 27.9 a particularly useful stratigraphic tool.

27.6 PALAEOMAGNETIC RESULTS: DRIFTING CONTINENTS

By analysing older and older rock samples it is found that going backward in time palaeomag-netic poles increasingly diverge from the geo-graphic pole. For example Pliocene poles, from European rocks, cluster around the geographic pole (Figure 27.10) whereas for the Permian the differences amount to 40° to 50° of latitude (Figure 27.10). For the Lower Palaeozoic the divergence is still greater, and it reaches 90° or more in the late Precambrian. Similar results are obtained from the rocks of other continents, except that the 'polar-wandering curves' are found to be conspicuously different for the different continents (Figure 27.11).

Until about 1956 most geophysicists seem to have favoured **polar wandering** (without conti-nental drift) as a sufficient explanation for the

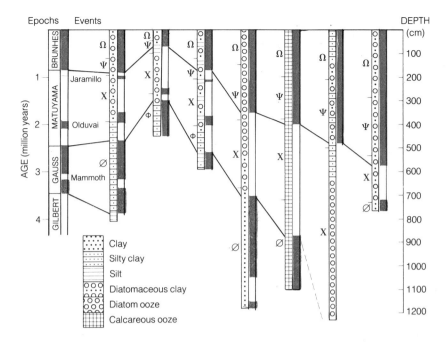

Figure 27.8 'Magnetic stratigraphy' established within deep-sea cores from the Antarctic, and correlated with radiolarian faunal zones (Greek letters). (*After N. D. Opdyke, B. Glass, J. D. Hayes and J. Foster, 1966*)

changes of latitude disclosed by the palaeomagnetic data then available. The polar-wandering hypothesis is generally understood to mean that an outer shell of the Earth, involving the crust and probably part of the mantle down to the low velocity zone, has shifted **as a whole** relative to the axis of rotation, which remains almost fixed, relative to the stars. If this were an adequate explanation by itself, then the geographical positions of the poles at any given time would be the same for all the continents. Assuming the continents to have remained fixed relative to one another, there would be a single 'polar-wandering curve'. But later work soon revealed that this is far from being the case. Every continent has its own 'polar-wandering curve' and these are so widely different that continental drift on a grand scale is obviously implied.

To fix the position of a continental area (regarded for the moment as a rigid slab) on the globe requires three items of information, of which the palaeomagnetic data yield approximate estimates of only two. The three requirements are:

1. the orientation relative to the contemporary poles (determined by the declination of the remanent magnetism);
2. the latitude of each locality tested (determined by the inclination or dip of the remanent magnetism); and
3. the longitude of each locality tested (not provided by palaeomagnetic measurements).

The third item, the longitudinal position, can be judged within certain limits since, at any given period of the past the continents of the present day must not overlap, and they must occupy positions consistent with geological evidence of the kind appealed to before the data provided by palaeomagnetism were available. The Tibetan plateau is regarded as the result of an overlap between Asia and a northern extension of India, but that northern extension is a hypothetical inference, not a 'continent of the present day'.

For convenience the longitudinal position is often judged relatively to some particular continent. Africa is commonly adopted for this purpose, following a convention started by all the pioneers, including Snider as well as Wegener. The meridian through Greenwich is similarly adopted as the conventional standard relative to which present-day longitudes are reckoned. There is no natural standard of longitude to be discovered; but the positions of the equator and poles and the intervening circles of latitudes can be discovered, and this applies equally to ancient times, with the aid of palaeomagnetism. The polar-wandering curve for Europe drawn in Figure 27.11 implies that the British Isles lay within 20° or so of the equator in Carboniferous and Permian times. But a more direct and clearer picture of the palaeomagnetic results is given by plotting the equator for a given pole and constructing **magnetic isoclines** (lines of equal magnetic inclination) between the two. The isoclines are small circles parallel to the equator of the time, which is the isocline for which $I = 0°$. As shown on

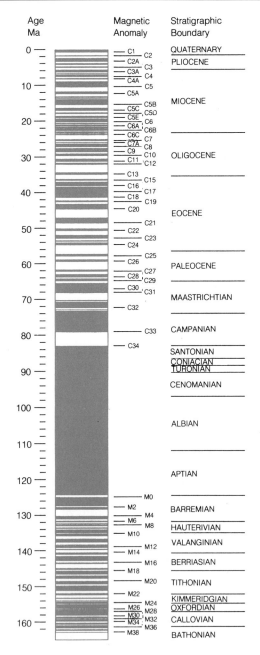

Figure 27.9 Geomagnetic polarity timescales. The magnetic field was normal, conforming with the present disposition of the North and South magnetic poles, during periods marked as black, while it was reversed during periods marked as white.

Figure 27.4, the inclination increases towards the poles, where it is 90°. This very convenient method of representing the palaeomagnetic results has been extensively used by D. van Hilten. Figure 27.12 is an example of one of his isoclinal maps. It shows at a glance that, during the Permian, Europe and the British Isles lay thousands of miles south of their present positions and also that the subsequent drift towards the north was accompanied by a clockwise rotation.

On the basis of the axial dipole hypothesis it can be inferred by use of palaeomagnetic determinations that in past geological ages the continents have occupied positions relative to the poles and to each other, very different from those of today's familiar geography.

27.7 CHANGING VIEWS OF CONTINENTAL AND OCEANIC RELATIONSHIPS

The major structural features of the globe fall into the following pattern (cf. Figure 27.13):

1. (a) The continents of the **Laurasia** group in the north, with
   (b) the intervening basins of the North Atlantic and Arctic oceans.
2. (a) The continents of the **Gondwana** group in the south with
   (b) the intervening basins of the South Atlantic and Indian oceans.
3. The **Pacific** Basin, lying everywhere outside 1 and 2.

Until comparatively recently the 'common sense' view was that these primary units had remained essentially fixed throughout geological time in the global positions they now occupy. But for nearly a century there was a vigorous debate as to whether continent and ocean could interchange as a result of vertical or radial movements. In the earlier days the widespread occurrence of marine sediments over the lands suggested that the continents could sink to oceanic depths and the ocean floors rise to become dry land. It was gradually recognized, however, that with only a few minor exceptions these deposits proved no more than temporary flooding of the lands by shallow seas. For this reason, amongst others, Dana expressed the view in 1846 that continents and oceans have never changed places, and that the general framework of the Earth has remained essentially stable. Nevertheless, Edward Forbes, tackling the subject from the biological side in the same year, found it impossible to explain how certain animals and plants had migrated from one continent to another unless some parts of the oceans had formerly been land. Thus began the long controversy regarding the permanence of the continents and ocean basins.

During the last century it was tacitly assumed that if interchanges between continent and ocean had to be postulated then the movements involved could hardly be other than vertical, except for the slight lateral movements in the crust of a supposedly contracting Earth.

The suggestion that there might have been lateral displacements of the continental masses on

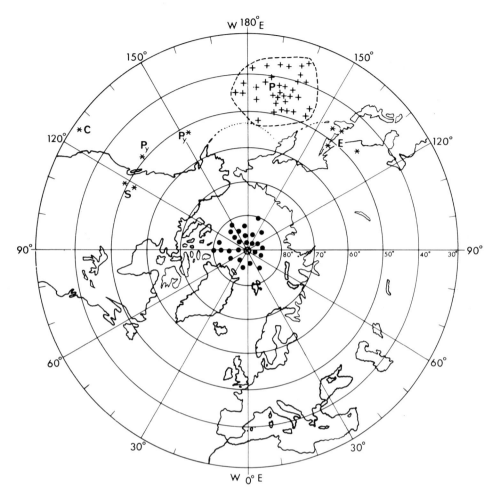

Figure 27.10 Stereographic projection of the northern hemisphere (from the North Pole to latitude 30°). Black dots clustered around the North Pole indicate the positions of Pleistocene and Pliocene geomagnetic poles. Crosses clustered around P indicate the positions of Permian geomagnetic poles determined from European sites undisturbed by Alpine orogenies (British areas include Ayrshire, Durham and Devon). The mean position of these poles, P, is lat. 43°N and long. 170°E. Stars, dispersed on both sides of the cluster around P, indicate the positions of Permian geomagnetic poles from European sites within the Alpine orogenic belts; C, Corsica; E, Esterel (south-east France); Py, Pyrenees; S, Southern Alps. (*Data mainly from R. R. Doell and A. Cox, 1961; and D. van Hilten, 1964*)

a gigantic scale is generally ascribed to F.B. Taylor in America (1908) and to Alfred Wegener in Germany (1910). For several years these pioneers developed their unorthodox hypotheses quite independently. Actually, however, the germ of the idea can be traced back to 1620, when Francis Bacon was sufficiently impressed by the parallelism of the opposing shores of the Atlantic to speculate as to its meaning. But he did not go so far as a Frenchman, one P. Placet, who published a fantastic memoir in 1668, entitled *La corruption du grand et du petit monde, où il est montré que devant le déluge, l'Amérique n'était point séparée des autres parties du monde*. Nearly two centuries later Antonio Snider united the continents in much the same way as Wegener did. In a book with the optimistic title *La Création et ses Mystères dévoilés* (Paris, 1858) he published the two maps here reproduced as Figure 27.14. Snider's reconstruction of Carbonif-

erous geography was intended to explain the fact that most of the fossil plants preserved in the Coal Measures of Europe are identical with those of the North American Coal Measures. Although the two diagrams reappeared in J.H. Pepper's popular and highly entertaining *Playbook of Metals* (London, 1861), the idea they embodied was evidently regarded as too outrageous to be worthy of serious scientific attention. For all practical purposes it was soon forgotten.

27.8 TAYLOR'S CONCEPT OF CONTINENTAL DRIFT

It was not until Wegener published his famous book on the subject in 1915 that the possibility of continental drift began to be widely discussed. But Taylor must be given credit for making an

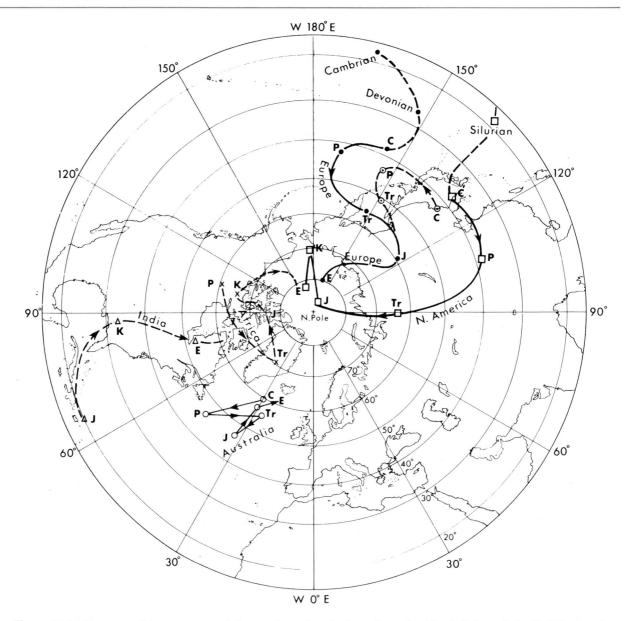

Figure 27.11 Stereographic projection of the northern hemisphere from the North Pole to latitude 10°, showing palaeomagnetic polar wandering curves for Europe (solid black circles) with an offshoot for Siberia (open circles with a central dot); North America (open squares); Africa (crosses); India (open triangles); and Australia (open circles). Geological ages of samples from which the data were obtained are indicated by letters: E, Eocene; K, Cretaceous; J, Jurassic; Tr, Triassic; P, Permian; and C, Carboniferous. It should be noticed that to avoid congestion these curves are drawn through the mean positions of clusters covering considerable areas (cf. the P cluster in Figure 27.10). To represent the data more accurately the curves should be broad bands instead of lines.

independent and slightly earlier start in this precarious field. In a pamphlet privately printed in 1908 he described the continents as 'huge landslides from the polar regions towards the equator', but he did not publish his ideas until 1910. His immediate object was to account for the distribution of mountain ranges. He pictured the original Laurasia as being a continuous sheet of sial and supposed it to have spread outwards towards the equator, more or less radially from the polar regions, much as a continental ice-sheet would do. Wherever the resistance was least, the crust flowed out in lobes, raising up mountainous loops and arcs in front. Such movements, of course, would be impossible without complementary stretching and splitting in the rear. And, indeed, there is ample evidence of down-faulting and disruption in the coastal lands and islands of the Arctic and North Atlantic, and especially in the highly-fractured region between Greenland and Canada, the map of which looks like a jigsaw puzzle with the separate bits dragged apart. In the southern hemisphere the originally continuous Gondwanaland similarly spread out, breaking up

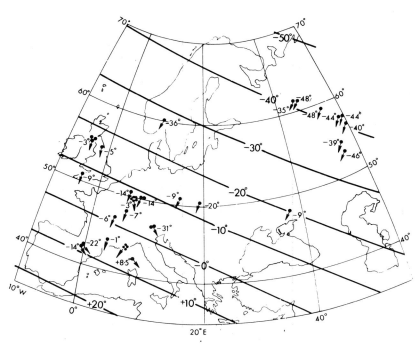

Figure 27.12 Isoclines in heavy black lines drawn about the mean Permian pole position (43°N; 170°E) for Europe north of the Alpine front. The source-localities of measured samples are indicated by dots, with inclinations in degrees and orientations shown by arrows. Negative signs north of the Permian equator (and positive to the south) imply reversal of the normal magnetic field. (*D. van Hilten, 1964*)

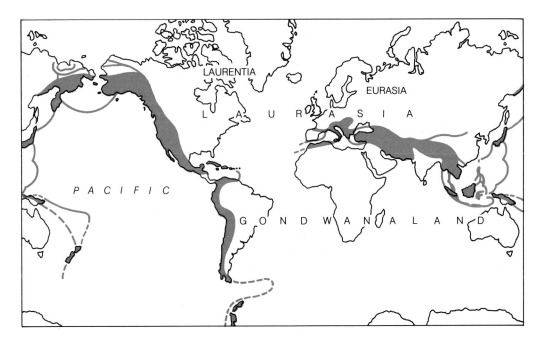

Figure 27.13 Map showing the distribution of the Mediterranean and Circum-Pacific Alpine orogenic belts, and continents of the Laurasian and Gondwana groups.

into immense rafts which also migrated towards the equator and raised up mountains in front. The basins of the South Atlantic and Indian Oceans were interpreted as the stretched and broken regions left behind or between these drifting continents.

For two reasons Taylor's hypothesis received scant attention. A considerable amount of lateral continental movement was thought to be implied by the structures of orogenic belts, but it seemed to be unnecessarily extravagant to invoke thousands of kilometres of horizontal displacement when from thirty to seventy according to the contraction hypothesis would have sufficed. Sec-

Figure 27.14 Maps published by A. Snider in 1858 to illustrate his conception of continental drift. The lefthand map represents his reassembly of the continents for late Carboniferous times.

ondly, Taylor's attempt to explain the alleged movements was quite unacceptable. He postulated that the moon first became the Earth's satellite during the Cretaceous, and that at the time of its capture it was very much nearer to the Earth than it is today. The resulting tidal forces were supposed to be sufficiently powerful not only to alter the rate of the Earth's rotation, but also to drag the continents away from the poles.

Apart from the improbability that the Earth was without a moon before the Cretaceous, there are two fatal objections to this conjecture:

(a) If the late Cretaceous and Tertiary mountain building is to be correlated with the supposed close approach and capture of the moon, then we are obviously left with no explanation for all the earlier orogenic cycles.

(b) If the tidal force applied to the Earth by the newly captured moon had been sufficient to displace continents and raise mountains on the scale required, then, as Jeffreys convincingly pointed out, the friction involved would have acted like a gigantic brake, bringing the Earth's rotation practically to a standstill within a year.

Taylor's 'explanation' is untenable, but from the criticisms one very important conclusion may be drawn. The fact that the Earth continues to rotate shows that neither tidal friction nor any other force applied from outside the Earth can be responsible for mountain building or for continental drift. Whatever the primary cause for the latter may be, it must be looked for **within** the Earth.

27.9 WEGENER'S CONCEPT OF CONTINENTAL DRIFT

Wegener's highly ingenious concept of the evolution of the continents and their distribution is graphically illustrated by his own maps, strange and fantastic on first aquaintance, but now widely familiar (Figure 27.15). His picture of the world in Carboniferous times is somewhat similar to Snider's, except that India and Antarctica are tucked in between Africa and Australia, with the horn of South America wrapping around West Antarctica. For this combination of Laurasia and Gondwanaland, making up the whole land area of the globe, Wegener proposed the name **Pangaea** (Gr., all earth). Snider had urged that the forests of the Carboniferous Coal Measures were tropical, and that in consequence Europe and North America were then near the equator, thus implying that South Africa lay near the south pole. Similarly, but because of the distribution of the Permo-Carboniferous glaciations, Wegener inferred that the Carboniferous south pole occupied a position just off the present coast of South Africa (Figure 27.16).

The present distribution of the continents was regarded as a result of fragmentation by rifting, followed by a drifting apart of the individual masses. The southern continents began to unfold during the Mesozoic era by being dragged away from wherever the south pole happened to be at any given time during the progress of the outward movements. Somewhat later North America began to break loose and to drift away to the west, Greenland being the last to go. The Atlantic is the immense gap left astern, filled up to the appropriate level by the inflow of sima from the mantle. By the time the continents had reached their present positions Antarctica found itself stranded over the south pole; Africa lay athwart the equator; India had been tightly wedged into Asia; and Australia and New Guinea had advanced far into the Pacific, by-passing the Banda arc and eventually separating it from the Pacific arcs.

The drift of the continents away from the poles

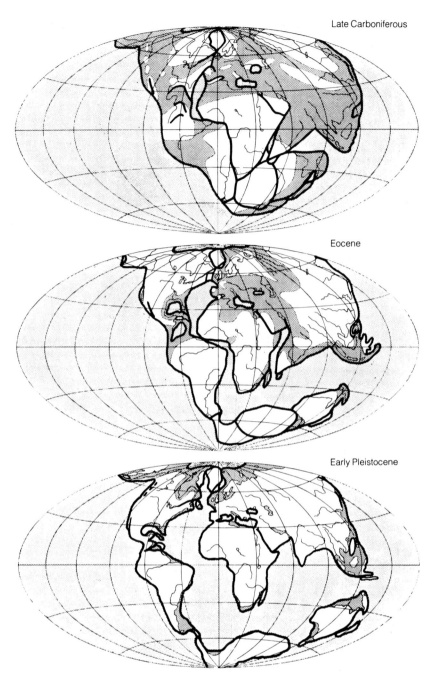

Late Carboniferous

Eocene

Early Pleistocene

Figure 27.15 Wegener's reconstruction of the distribution of the continents during the periods indicated. Africa is placed in its present-day position to serve as a standard of reference. The more heavily shaded areas (mainly on the continents) represent shallow seas. (*From A. Wegener, Die Entstehung der Kontinente und Ozeane, 1915*)

was dramatically described by Wegener as the **Polflucht** – the flight from the poles. He ascribed the force involved to the gravitational attraction exerted by the Earth's equatorial bulge. The force is a real one, but it is many millions of times too feeble to drag the continents from their moorings.

Wegener also postulated a general drift towards the west. As the Americas moved westwards against the resistance of the Pacific floor, their prows, he supposed, would be crumpled up into great mountain ranges. Between the two immense rafts a trail of fragments lagged behind and

formed the islands of the West Indies. The stretched-out isthmus connecting South America and Antarctica similarly lagged behind, forming the horns of the two continents and shedding bits of sial that now remain as the island loop of the Southern Antilles. The supposed effects of the westerly drift of Asia are less happily conceived. The great oceanic deeps were regarded as gaping fissures torn in the Pacific floor and not yet fully healed, while the island festoons were interpreted as strips of sial that remained attached to the mainland only at their ends.

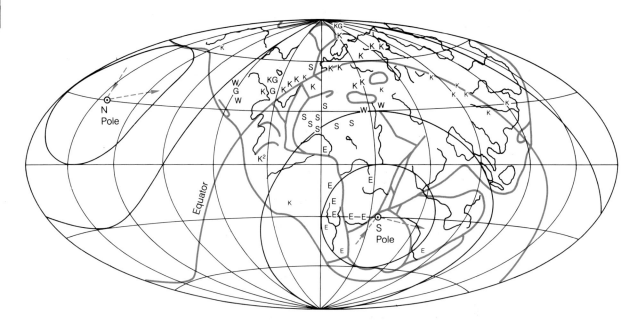

Figure 27.16 Reassembly of the continents for the late Carboniferous, based on the palaeoclimatic evidence available in 1924. Glaciated regions are indicated by E (*Eis*, ice), K shows the distribution of coal, mainly near the equator of the time (heavy line). The dotted areas on both sides of the Carboniferous equator were arid regions; with W, desert sandstone; S, salt; and G, gypsum. The South Pole is placed near Durban and the North Pole in the Pacific, north-east of Hawaii. It should be noticed that [E] on the equator near Boston, Mass., USA, (the Squantum tillite), of doubtful age in 1924, is now ascribed to the Lower Palaeozoic. (*After W. Kappen and A. Wegener, Die Klimate der geologischen Vorzelt, 1924*)

The westward movements were ascribed to the differential attractions of the moon and the sun on the continents. Tidal friction acts like a brake on the rotating Earth, and as the effect on protuberances is greater than that on lower levels of the crust the continents tend to lag behind. If they did lag behind, they would appear to drift to the west. But here again the force invoked is hopelessly inadequate to overcome the enormous resistance that opposes actual movement. The tidal force barely affects the Earth's rotation, and is about ten thousand million times too small to move continents and raise up mountains.

In support of the case for continental drift Wegener marshalled an imposing collection of facts and opinions. Some of his evidence was undeniably cogent, notably the distribution of the Permo-Carboniferous glaciations, already discussed in Sections 9.5, 21.10. But so much of his advocacy was based on inadequate data and speculation that a storm of adverse criticism was provoked. It was all too easy to demolish some of Wegener's particular views. Moreover, following the lead of many influential geophysicists, most geologists were reluctant to admit the possibility of continental drift, because no recognized natural process seemed to have the remotest chance of bringing it about. The 'flight from the poles' and

Figure 27.17 Comparison of Africa and South America by placing the 200-metre isobaths (depths below sea-level) in juxtaposition. (*S. Warren Carey, 1958*)

the westerly tidal drift have both been discarded as operative factors. Polar wandering originally seemed to be impossible on a scale of geological importance, but it has since become dynamically respectable and henceforth it can be taken into account as fully as may prove necessary. Continental drift, now embracing crustal separation and ocean-floor spreading and renewal, is known to be in operation at the present time. The controversial storm over continental drift has consequently abated and Wegener's name become an honoured one.

27.10 GEOLOGICAL CRITERIA FOR CONTINENTAL DRIFT

The chief geological criteria for continental drift are based on the following considerations:

(a) If the continents formerly occupied widely different positions on the Earth's surface, then the

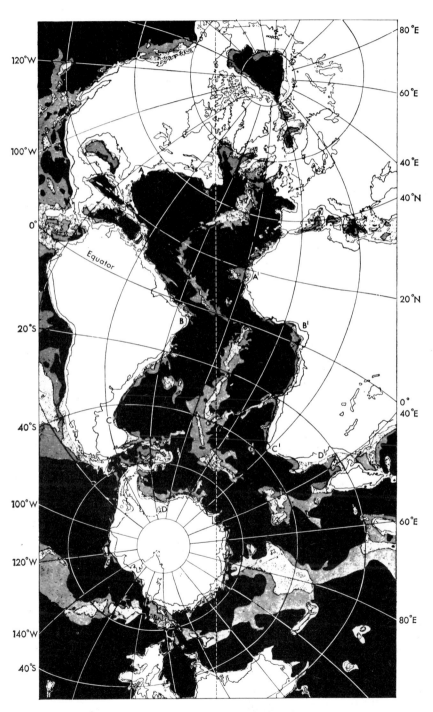

Figure 27.18 Map illustrating the morphological fit of the opposing lands of the Atlantic. (*S. W. Carey, 1958*) Carey envisaged North and South America, Antarctica and even Australia as having been swung as a whole, towards Eurasia and Africa, about the hinge (bend) in the Alaskan orogenic belt, suggesting that such motion through 20° or 30° would close the Arctic, Atlantic and Southern oceans. He proposed the term orocline to describe such an orogenic belt that appears to have been bent or flexed.

distribution of climatic zones, as inferred from geological evidence, should have correspondingly changed. Ample evidence of such changes has already been summarized in earlier chapters: e.g. dealing with the Permo-Carboniferous glaciations (Section 21.10); desert sands and wind directions and the distribution of salt deposits (Sections 22.8, 22.9); and of ancient coral reefs (Section 24.14).

(b) If two continents, now far apart, were formerly united, it should be possible to detect the fact by the recognition of certain geological features that were originally shared in common; e.g. orogenic belts of which the truncated ends can be naturally joined up; specific and unusual details and sequences of geological history as recorded in the sedimentary and other rocks of the separated lands; and the identity of the fossil remains of animals and plants (especially those of land and freshwater species) which could migrate freely across united continents but not across an intervening ocean.

(c) Indications from strike-slip (transform) faults, like the San Andreas fault, with cumulative lateral displacements sometimes amounting to hundreds of kilometres, that movements implying continental drift have long been in progress.

27.11 THE OPPOSING LANDS OF THE ATLANTIC

Ever since Wegener thought the Atlantic Ocean to be an enormously widening rift, with sides still matching 'as closely as the lines of a torn drawing would correspond if the pieces were placed in juxtaposition', the parallelism of its opposing shores has been discussed. Figure 27.17 shows the result of a direct and accurate comparison made in 1958 by Carey between South America and Africa, not along the actual shore-lines which have been modified by marine erosion and deposition, but offshore at a depth of 200 m, i.e. towards the top of the continental slope. Carey made this comparison by sliding transparent caps over the surface of the globe until he attained the best fit. A similar comparison at a depth of 2000 m, i.e. about half way down the continental slope, is, if anything, slightly better. Subsequently, E.C. Bullard in 1964, feeding all the relevant data into a computer, which was asked to find the level at which the 'fit' is most nearly perfect, received an answer between these two depths. Figure 27.18 was prepared by S.W. Carey in 1958 for his Symposium on Continental Drift, that did so much to reawaken interest in the subject. Of this 'fit' Carey wrote:

> Thus the east coast of North America is the mould of the great western bulge of Africa, the Gulf of Guinea is the negative of the bulge of

Brazil, and the embayment of the South American shelf to the Falkland Islands is the mould of the Cape of Good Hope and most of South Africa.

In 1965, E.C. Bullard, J.E. Everett and A.C. Smith published a best 'fit' map of the continents margining the Atlantic Ocean (Figure 27.19), constructed by using purely objective arithmetic methods. They moved the continents over the surface of the Earth with the aid of Euler's theorem, according to which a spherical surface displaced over a sphere can be moved to any other position by a single rotation about an axis passing through the centre of the sphere.

With the aid of a computer, Bullard, Everett and Smith found the best 'fit' to be at 500 fathoms, rather less than 1000 m below sea-level. The closeness of the 'fits' between South America and Africa, on the one hand, and between North America, Greenland and Europe on the other, not only exceeded their hopes but fully confirmed the results of Carey's earlier 'fits'.

In fitting together the continents margining the Atlantic, Bullard, Everett and Smith found, as Carey had done before them, that when Africa

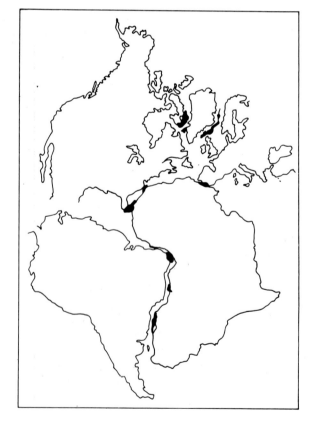

Figure 27.19 The best 'fit' of continents surrounding the Atlantic at the 500-fathom contour. The black areas indicate overlaps. (*After E. Bullard, J. E. Everett and A. G. Smith, 1965*)

was fitted to North America it overlapped Spain. In order to achieve a 'fit' without overlap, Spain had to be rotated so as to close the Bay of Biscay and bring its northern shore against the 500 fathom contour to the west of France.

27.12 GEOLOGICAL SIMILARITIES BETWEEN THE TWO SIDES OF THE ATLANTIC

For many years South Africa's eminent geologist, Alexander du Toit, was indefatigable in assembling the evidence bearing on continental drift. In his well-known book *Our Wandering Continents*, published in 1937, he showed that a striking series of correspondences can be recognized in the sediments, fossils, climates, earth movements, and igneous intrusions of the two sides of the Atlantic. Both regions had essentially the same geological history during the Palaeozoic and early Mesozoic times, and the combined evidence points very persuasively to the high probability that they were then very much closer together than now. Du Toit considered it possible that the original distance between the present opposing shores may have been as little as 400 km. But this was a minimum estimate. In 1961 H. Martin reviewed the subject afresh, with special reference to Namibia and Brazil. Discussing whether the more detailed knowledge now available has tended to increase or decrease the similarities between the two sides of the South Atlantic, he writes:

> There is not the slightest doubt that from the Silurian to the Cretaceous every correction of the statigraphy and lithologic columns has increased the similarities. The stratigraphic and lithologic columns for this period of some 200 m.y. have become almost identical. I do not think that, for a comparable length of time, a similar likeness between parts of any two continents can be found.

Just as striking evidence is the fact, recognized by radiometric dating, that a long and complex orogenic cycle reached its closing stages during the late Precambrian and Cambrian. The trend lines are sub-parallel to the coasts and appear equally on both sides of the South Atlantic.

While the Permo-Carboniferous glaciations of Gondwanaland still remain the most convincing line of geological evidence for continental drift (see Section 21.10, Figure 20.27, Plate VId), there can now be added the discovery of the Iapo tillite west-south-west of São Paulo in Brazil, in the same stratigraphical position as the Table Mountain tillite, near Cape Town. Both are late Silurian or early Devonian. Martin recorded that the direction of ice movement near Cape Town, and the

directions of transport of Devonian sediments in the Paraná Basin both indicate the existence of elevated areas in regions occupied today by the South Atlantic. He also drew attention to the striking contrast between Table Mountain and Brazilian glacial deposits and the strata of early Devonian age in Australia, which contain a warmwater fauna with corals. This climatic contrast had disappeared in the late Carboniferous, when Australia was glaciated like the other southern continents.

It should also be added that the widespread indications of late Precambrian glaciations (averaging about 650 ± 50 m.y.) in southern Africa (Katanga, Lower Zaïre, Namibia) have been matched by the discovery of similar deposits in Lavras (Brazil), the age of which lies between limits of about 600 and 800 million years. It is probable, though not rigidly proved, that all of those correspond approximately with the Sturtian tillites of South and Central Australia, and others that are correlated with them in Tasmania and Northern Australia.

Amongst a host of similarities too numerous to mention, attention should again be drawn to the great floods of basalt that inundated the Triassic deserts of both continents about the beginning of Jurassic time. In the Paraná Basin basalts cover an area of a million km^2 or more and reach a thickness of 1.5 km. In southern Africa a smaller area of basalts now remains, but originally, judging by the criss-crossing swarms of basic dykes and sills in the Karroo Basin and by the enormous thickness of flows still preserved on the eastern side from the Drakensberg to the Lebombo, the volume of lava erupted may have been still greater than in South America. Such spectacular signs of unusual activity in the depths would seem to be a significant preparation for the major break up of Gondwanaland and the appearance of seaways between the separating lands.

Palaeontological evidence has never proved decisive one way or the other in discussions on continental drift. In 1943 G.G. Simpson directed attention to the relevant fact that, of the known Triassic reptiles of South America, only 43 per cent of the families and 8 per cent of the genera are known in Africa, with no species in common. 'These figures', he concluded, 'are decidedly inconsistent with any direct union of corresponding parts of South America and Africa.' Simpson stressed the impossibility of inferring from what is actually known – the distribution of a given group of organisms – both the land connections of the time, and the probabilities of dispersal for the organisms concerned. But negative evidence may be destroyed at any moment by fresh discoveries, whereas positive evidence can never be explained away. And positive evidence is by no means lacking. Near the top of the Carboniferous in both South Africa and South America there is a thin band of lacustrine deltaic clay containing the bones of a small fresh-to-brackish-water reptile called *Mesosaurus*. Since remains of this little animal, about 45 cm long, have not been found anywhere else in the world, they suggest that South America and Africa were united towards the end of the Palaeozoic.

One of the early indications that there was no North Atlantic in Palaeozoic times was the recognition that similar fossils, including the Cambrian trilobite *Olenellus*, occur both in Scotland and Newfoundland. Moreover, the mid-Cambrian trilobites of Sweden are the same as those of Utah. When North America and Europe are placed together (Figure 27.20), the Caledonian fronts of Greenland and the North-West Highlands, and the corresponding front of the Appalachians are automatically joined up. Moreover, the western convergence across Europe of the outer Caledonian and Hercynian fronts, until they almost meet in the south-west of Ireland, is continued in North America, where the fronts eventually cross. Radiometric dating has shown that much of the northern part of the Appalachians was affected by

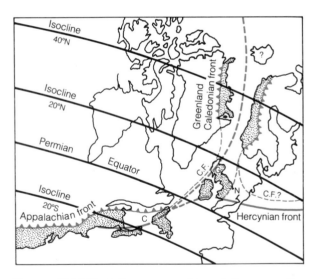

Figure 27.20 Tentative reassembly of the opposing lands of the North Atlantic, based primarily on moving Europe and North America so that the mean positions of their Permian poles coincide and the isoclines come into line as shown. The Caledonian and Appalachian orogenic belts are shaded, C.F., the Caledonian Front of the North West Highlands of Scotland; G, the Great Glen Fault of Scotland; C, the Cabot Fault of Newfoundland and Nova Scotia and its probable continuation along the New England coast; N, a possible median area (including part of the North Sea, East Anglia and The Netherlands, if the Scandinavian Caledonian Front should continue through Brabant and Poland.

orogenesis that would be called Hercynian or Variscan in Europe. Tuzo Wilson has suggested that the Great Glen Fault has its natural continuation in the Cabot Fault and this correlation is again found to be consistent with the reassembly of Figure 27.20. Both in Scotland and Newfoundland there are similar volcanic centres associated with dyke-swarms and granodiorite complexes of Lower Devonian age, like Ben Nevis; and the Old Red Sandstone type of Devonian, so characteristic of Scotland and the rest of Caledonian Britain and Ireland, is also found in Norway, Greenland, and Spitzbergen, as well as in parts of eastern North America.

27.13 ATTEMPTS TO REASSEMBLE GONDWANALAND

Du Toit, like Taylor, thought it more logical to suppose that a double landmass existed during the Palaeozoic, rather than the single landmass, **Pangaea**, envisaged by Wegener. However this may be, prior to the Tertiary Alpine–Himalayan orogeny, Laurasia and Gondwanaland were separated from one another by a long seaway that was called the Tethys by Suess.

Reconstruction of Laurasia is mainly a matter of closing the Atlantic Ocean, as discussed in the previous pages. Reconstruction of Gondwanaland is in many ways more difficult. Many attempts have been made, using all the geological evidence, and most of the results have a remarkable family resemblance. The greatest disagreement relates to the positions of India and Madagascar.

Figure 27.21 presents some examples of Gondwanaland based on geological evidence. In these reconstructions the difficulty has commonly arisen that some item of seemingly favourable evidence has had to be sacrificed in order to achieve a reasonable 'fit' of all the parts.

Figure 27.21(a) was an attempt by du Toit (1937) based on his idea that the Cape folds of South Africa continued into South America at one end, and along the eastern side of Australia at the other. He called this belt the Samfrau Geosyncline. It is of interest to note that the more recent investigations of Antarctica have justified du Toit's faith that one day this continent would provide evidence of Permo-Carboniferous glaciation.

Du Toit made two inferences that are of interest today:

(a) He saw in the rift valleys of Africa, the Red Sea and the Rhine graben, early stages in the separation of land masses. He also compared the escarpment-like edges of the tilted blocks, that

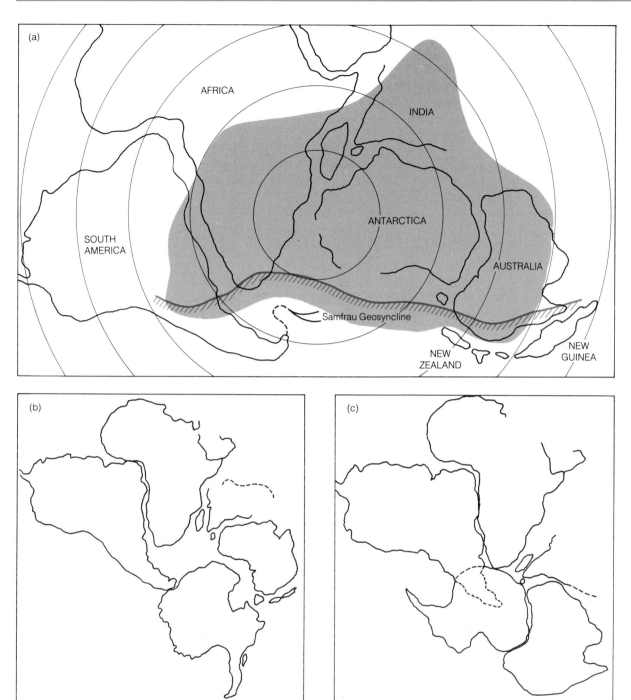

Figure 27.21 Sketch map illustrating various attempts to reconstruct Gondwanaland from geological evidence as it may have been in Permo-Carboniferous times: (a) A. L. du Toit (1937) (b) S. W. Carey (1958) (c) J. T. Wilson (1963).

margin these rift valleys, with the fault-like scarps that form the shorelines of continents like Africa and South America which have been separated by rifting. Furthermore, these escarpment-like coastlines, tilted gently inland, provided du Toit with an explanation of rivers which rise close to such high shorelines and flow inland, sometimes right across a continent before reaching the sea. Amongst others he cited as examples in South Africa, the Orange and Limpopo rivers, and farther north the Congo, now the Zaïre. Similarly the Paraná, Uraguay and São Francisco rivers in South America rise close to the shorelines and flow for long distances parallel to the coast before reaching the sea.

(b) Du Toit left spaces between the present-day continents in his reconstruction of Gondwanaland. In doing so he was guided by the sedimentary facies. For example, in comparing the Cape ranges of South Africa with the Sierra de la Ventana, Argentina, he concluded that the lateral variation of the Palaeozoic sediments and their structures required that they should be separated from one another by a distance of from 500 to 600 km. He verified this distance by the intersections of other lines of evidence, and he thought the gap to have been occupied by land at least during the Jurassic. Eventually, however, by the entry of the sea, du Toit envisaged the gap between South America and Africa to have become the proto-South Atlantic.

Figure 27.21b is a reconstruction of Gondwanaland by Warren Carey (1958). Here Ahmad's Indo-Australian link is adopted, but Antarctica gets badly in the way, and again Graham Land is unnaturally displaced.

Figure 27.21c is a reconstruction by Tuzo Wilson (1963). He also adopts the Indo-Australian link and uses the mid-oceanic ridges as a guide. But this reconstruction leaves no room for Antarctica; in consequence the rule forbidding overlapping is broken. India also seems to have been placed too far south and the whole assemblage is overcrowded.

Figure 27.25b is a reconstruction by Smith and Hallam (1970). With the aid of a computer they found the best 'fit' for Gondwanaland at 500 fathoms, using only those 'fits' that they judged to be geologically sound. They found Ahmad's 'fit' for India against north-west Australia to be geo-

metrically sound, but they discarded it because they found that India-Australia-Antarctica could not be fitted against Africa without leaving large gaps between them. From the stratigraphic evidence they considered that such gaps did not exist.

27.14 TESTING THE REASSEMBLY OF GONDWANA BY REFERENCE TO THE PALAEOMAGNETIC POLE POSITIONS

Creer (1965) calculated the positions of the palaeomagnetic poles for various geological periods from reliable palaeomagnetic data of the time, and showed that it is possible to discover whether or not there has been movement between two continents by comparing their 'polar wandering' curves. If the curves have the same shape, so that they can be superimposed one upon the other, then there has been no relative movement.

From the relevant palaeomagnetic data for South America, and by choosing a sequence of related longitudes, Creer found evidence for the movement of this continent across the south pole during the Palaeozoic and early Mesozoic eras (Figure 27.22a). This result can also be expressed as a 'polar wandering' curve (Figure 27.22b). Creer similarly constructed a 'polar wandering' curve for Africa, and found that the Palaeozoic parts of the 'polar wandering' curves for South America and Africa can be superimposed, the two continents at the same time being fitted together, indicating

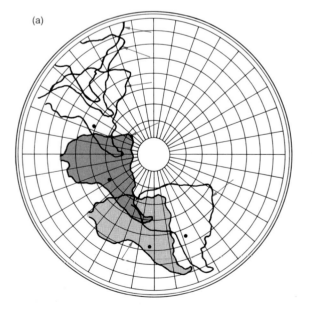

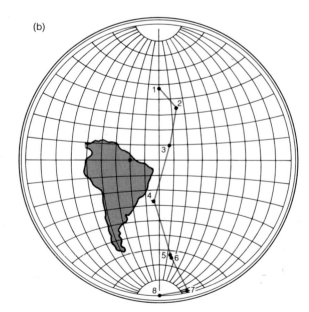

Figure 27.22 (a) The palaeolatitudes and orientations of South America for periods ranging from early Palaeozoic to early Mesozoic, constructed from palaeomagnetic data. (b) Palaeomagnetic data expressed as a polar wandering curve for the South Pole, with South America fixed, for the Palaeozoic era. (*K. M. Creer, 1965*)

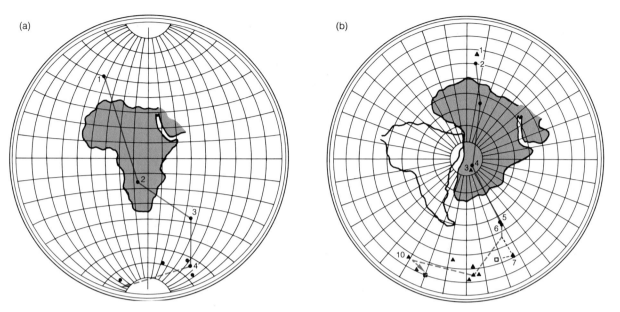

Figure 27.23 (a) The polar wandering curve for the South Pole, relative to a fixed Africa, constructed from palaeomagnetic data derived from African rocks. (b) The fact that the Palaeozoic parts of the polar wandering curves for South America and Africa have the same shape and can be superimposed indicates that there was no relative movement between the two continents during that time. (*K. M. Creer, 1965*)

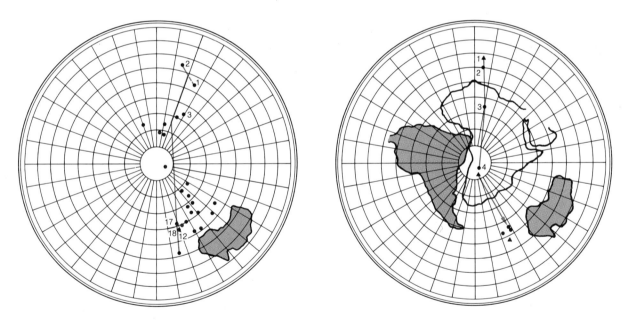

Figure 27.24 The relative positions of South America, Africa and Australia during the Palaeozoic era, determined by superimposing their polar wandering curves for that time. (*K. M. Creer, 1965*)

that there was no relative movement between them during this era (Figure 27.23b). Furthermore, for the Palaeozoic era, the 'polar wandering' curve for Australia can be superimposed on those for South America and Africa, indicating that there was no relative movement between these three continents during this era, and that their relative positions were approximately as indicated in Figure 27.24.

More recently the development of plate tectonic ideas (Chapter 28) has allowed continental reconstructions of the dispersal of Pangaea to be made by reference to seafloor-spreading anomalies. Impressively detailed and chronologically extremely precise reconstructions can be built for periods when marine magnetic anomalies have been preserved. The plate tectonics method has turned out to be so successful that the marine anomaly approach has completely revolutionized reconstruction methods and replaced all other techniques for determining continental drift over the last 170 m.y. For earlier periods, prior to the breakup of Pangaea, marine magnetic anomalies cannot be used and geological and palaeomagnetic reconstruction methods must still be employed.

As described in Chapter 28 the plate tectonic history of all the major ocean basins has been worked out in immense detail allowing the drift of continental blocks such as India to be traced back through time by plate-rotation calculations. Such plate tectonic histories place important constraints on the earlier positions of the continents as part of Gondwanaland. The reconstruction of Du Toit, with India adjacent to Antarctica, has in this way been vindicated through plate tectonic studies. Du Toit's more northerly location of Madagascar, within the jigsaw fit of Gondwanaland (Figure 27.21), also tends to be strongly favoured. Debate

about the geography of Gondwanaland now centres on such details of the Du Toit fit.

Figure 27.25 shows the range of reconstructions that can still be accommodated by the plate tectonic history of the Indian ocean. All are very similar to Du Toit's 1937 reconstruction. Palaeomagnetic results can be used to assess these various reconstructions directly. The scatter of palaeomagnetic poles about Gondwanaland apparent polar wandering paths, for each of the three new reconstructions of Figure 27.25, has been calculated by Thompson and Clark as follows:

Reconstruction	Average deviation
Smith and Hallam	19.8°
Tarling	21.6°
Powell *et al*	22.0°

The reconstruction that produced the lowest scatter and is consequently judged to be the best,

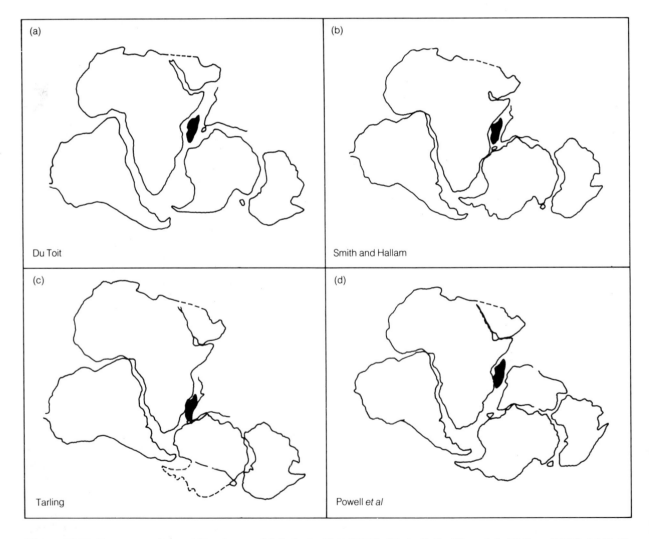

Figure 27.25 Reconstructions of Gondwana. (a) A. L. du Toit (1937). (b) A. G. Smith and A. Hallam (1970). (c) D. H. Tarling (1972). (d) C. McA. Powell, B. D. Johnson and J. J. Veevers (1980). (See also Figure 28.13.)

from a palaeomagnetic point of view, is that of Smith and Hallam. A best-fitting polar wandering path for the Smith and Hallam fit is drawn in Figure 27.26. Confidence circles are plotted at 10 m.y. intervals along the path, for between 550 and 160 m.y. ago, in order to illustrate the degree of accuracy with which each part of the polar wandering path is known.

27.15 PALAEOMAGNETISM: COLLISION TECTONICS

In addition to determining the relative displacements of continental cratons, palaeomagnetic methods are being increasingly applied to local tectonic problems in mountain belts. Palaeomagnetic studies have pointed to many large scale displacements and rotations of rock bodies that had previously escaped detection. Regions in which palaeomagnetic methods have proved to be particularly valuable are the Alpine system, the Caribbean, New Zealand, Tibet and the mountains of north-west America. Figure 27.10 includes examples of Permian pole positions from Corsica, the Pyrenees and the southern Alps which demonstrate anticlockwise rotations of between 20° and 40°. Before these masses were caught up in the Alpine orogeny they would have had identical pole positions to the undisturbed European sites. Subsequently as the rock masses have rotated anticlockwise their magnetizations have also been

swung around with the rocks and now serve as testimony to the ancient rotations. By measuring magnetizations of different ages palaeomagnetists are able to decipher and date quite complex rotation histories.

About two thirds of palaeomagnetic poles from the Cordilleran margin of North America deviate from the main North American cratonic polar wandering path. Here, as shown in Figure 27.27, the great majority of poles fall to the right of the main path. This right-handedness points to predominantly clockwise rotations. In addition many of the deviatory poles lie further away from the collection sites than is the case for poles obtained for stable parts of North America. This 'far sidedness' of the Cordilleran poles is caused by the magnetizations being shallower than in the cratonic areas. The shallow Cordillerian inclinations are interpreted in terms of the source rocks having been transported thousands of kilometres northwards since the time when they acquired their magnetic remanences in geomagnetic fields of low latitude and low inclination.

Deviatory poles from western America were

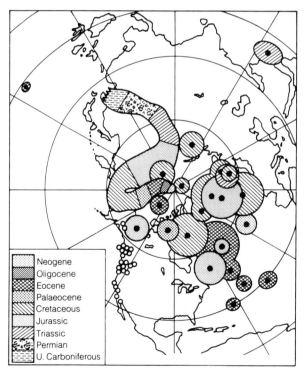

Figure 27.27 Palaeomagnetic pole positions from the western margin of North America, compared with the apparent polar wandering path derived for the stable continental interior. Small open circles mark the palaeomagnetic sites in the Cordilleran margin. The shading indicates the stratigraphic age of the palaeomagnetic path and of the deviatory poles. (*After R. Van der Voo and J. E. T. Channel, 1980*)

Figure 27.26 95 per cent confidence limits drawn at 10 m.y. intervals for the Gondwana apparent polar wandering path, based on the reconstruction of Smith and Hallam (1970), calculated using the method of R. Thompson and R. M. Clark (1982).

first noted as early as 1957 but these early results were explained in terms of non-dipolar geomagnetic behaviour or by local faulting and tectonics. The large proportion of 'righthanded' directions since discovered means that the geomagnetic explanation is no longer tenable and the consistency of results within tectonostratigraphically distinct blocks supports palaeomagnetic explanations involving large scale geotectonics rather than local structural deformations. Plate tectonic reconstructions of the interaction between plates in the Pacific ocean and the western edge of North America (Figure 28.27) show how low-latitude 'Pacific' lithosphere has been transported northwards and eastwards to collide with North America. This independently derived tectonic history ties in remarkably well with the palaeomagnetic expectations of large-scale northward displacements and clockwise rotations.

Palaeomagnetic evidence ever increasingly indicates that many fault-bounded rock bodies, or **terranes**, in orogenic belts were formed in widely-separated regions and that they were subsequently bodily moved into their new associations. It is now recognized that the transportation and assembly of terranes in orogenic fold belts is a fundamental part of mountain building and continental growth.

27.16 FURTHER READING

Bullard, E.C., Everett, J.E. and Smith, A.C. (1965) The fit of the continents around the Atlantic. *Philosophical Transactions of the Royal Society*, **258**, pp. 41–51.

Creer, K.M. (1965) A Symposium on continental drift: III, palaeomagnetic data from the Gondwanic continents. *Philosophical Transactions of the Royal Society*, **258**, pp. 27–40.

Du Toit, A.L. (1937) *Our Wandering Continents*, Oliver and Boyd, Edinburgh.

Powell, C. McA., Johnson, B.D. and Veevers, J.J. (1980) A revised fit of East and West Gondwanaland. *Tectonophysics*, **63**, pp. 13–29.

Smith, A.G. and Hallam, A. (1970) The fit of the southern continents. *Nature*, **225**, pp. 139–44.

Thompson, R. and Clark, R.M. (1982) A robust least-squares Gondwanan apparent polar wander path and the question of palaeomagnetic assessment of Gondwanan reconstructions. *Earth and Planetary Science Letters*, **57**, pp. 152–8.

Van der Voo, R. and Channel, J.E.T. (1980) Paleomagnetism in orogenic belts. *Reviews of Geophysics and Space Physics*, **18**, pp. 455–81.

Wegener, A. (1966) *The Origin of Continents and Oceans*, translated from the fourth (revised) German edition of 1929, by J. Birman, Methuen, London.

Wilson, J.T. (1963) Continental drift. *Scientific American*, **208**, pp. 86–100.

SEA-FLOOR SPREADING AND PLATE TECTONICS

*It is my opinion that the Earth is very noble
and admirable . . . and if it had continued
an immense globe of crystal, wherein
nothing had ever changed, I should have
esteemed it a wretched lump of no benefit
to the Universe.*

Galileo (1564–1642)

28.1 INTRODUCTION

The greatest obstruction to acceptance of continental drift earlier in the century was the difficulty of understanding how continents composed of sial could possibly sail on and displace sima just as ships sail on and displace water. At that time it was thought that the oceanic crust was a thick continuation of a continental basaltic layer. In a lecture to the Geological Society of Glasgow, in 1928, Holmes suggested a possible way out of the difficulty. By comparison with the planetary system of winds he envisaged thermal convection currents within the Earth's substratum (now the upper part of the mantle), with subsidiary cyclonic and anticyclonic systems induced by regions of greater and lesser radioactivity. In the subsidiary systems hot currents rose beneath the continents, and spreading out laterally dragged the continental blocks apart (Figure 28.1). Where continental currents met oceanic currents they both turned downwards, and here Holmes envisaged the high-pressure facies eclogite as being formed from basalt or amphibolite as the currents descended.

The resultant increase in density he thought to result in the subsidence of eclogite, and the speeding up of the descending currents, thus making room for the continents to advance. The ascending currents, where continents were torn apart, healed the gaps by forming new ocean floor (Figure 28.1). This concept, which made little if any impact on geological thought at the time when it was conceived, now appears as a prelude to future geophysical discoveries. Indeed as Keith Runcorn has stated Holmes' concept 'in all the essentials is the sea floor spreading theory'.

Since 1964 there has been a complete change of viewpoint relating to continental drift, arising out of the wealth of new magnetic data relating to the ocean floor, and its imaginative interpretation and confirmation, with the resulting establishment of the reality of ocean-floor spreading. Drifting continents are now regarded as parts of rigid plates of lithosphere, about 100 km thick, composed of crustal rocks and upper mantle down to the low velocity zone (Section 26.6). Individual plates may

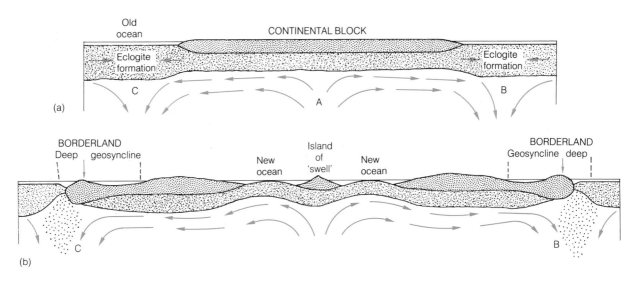

Figure 28.1 Diagrams to illustrate a convective-current mechanism for 'engineering' continental drift and the development of new ocean basins, proposed by A. Holmes in 1928, when it was thought that the oceanic crust was a thick continuation of the continental basaltic layer (lower stippled layer). (a) A current ascending at A spreads out laterally, extends the continental block and drags the two main parts aside, provided that the obstruction of the old ocean floor can be overcome. This is accomplished by the formation of eclogite at B and C, where sub-continental currents meet sub-oceanic currents and turn downwards. Being heavy, the eclogite is carried down, so making room for the continents to advance. (b) The foundering masses of eclogite at B and C share in the main convective circulation and, melting at depth to form basaltic magma, the material rises in ascending currents: e.g. at A, healing the gaps in the disrupted continent and forming new ocean floors (locally with a swell of old sial left behind, such as Iceland). Other smaller current systems, set going by the buoyancy of basaltic magma, ascend beneath the continents and feed great floods of plateau basalts, or beneath the 'old' (Pacific) ocean floor to feed the outpourings responsible for the volcanic islands and seamounts. (*Arthur Holmes, Transactions of the Geological Society of Glasgow, 1928-1929, vol. 18*)

include both continental and oceanic crust, so that the expression 'continental drift' is no longer strictly appropriate (Figure 28.2). The plates are the inert aseismic regions of the Earth, bounded by narrow mobile belts, characterized by earthquakes and volcanic activity.

28.2 MAGNETIC 'STRIPES': SEA-FLOOR SPREADING

In 1952, Mason, of the Scripps Institute of Oceanography, towed a magnetometer, tied to the stern of a research ship, halfway across the Pacific Ocean. The results proved to be of so much interest that he and his colleagues subsequently surveyed a broad strip of the Pacific floor offshore from Mexico to British Columbia. When the magnetic anomalies were plotted on a chart, like contour lines, it was found that they revealed a roughly north–south pattern of stripes.

At the time when they were discovered, the cause of the magnetic anomalies on the Pacific floor could be no more than a matter of inference by analogy. As Raff records 'a single glance at the map was enough to show that we had something

quite new in geophysics'. Since the anomalies are sharply and steeply bounded, showing that whatever is responsible for them comes up to the ocean floor, or nearly so, it appeared at the time that they might be caused by structural changes as, for example, by the presence of lava flows, or dyke-like basic intrusions, or by variations in the topography of the main crustal layer.

In 1963 Vine and Matthews, after a detailed magnetic and bathymetric survey of the middle part of the Carlsberg ridge in the Indian Ocean, as part of the International Indian Ocean Expedition of 1963, computed the magnetic profile across the Carlsberg ridge, from the bathymetric profile, assuming normal magnetization. They found it differed completely from the observed anomaly profile. On the other hand, when they assumed that the basaltic crustal layer was divided into blocks 20 km wide, with alternately normal and reversed magnetization, the computed profile closely resembled the observed profile (Figure 28.3).

Assuming spreading of the oceanic floors, according to the hypothesis of Dietz and Hess, and periodic reversals of the Earth's magnetic field (Figure 28.4), Vine and Matthews suggested that if

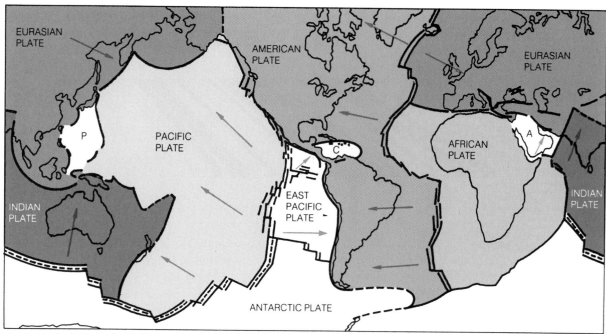

A = Arabian P = Philippine C = Caribbean

Figure 28.2 The earth's surface divided into 6 major plates, according to Le Pichon (1968), and a few smaller ones, all of which move in the directions of the arrows.

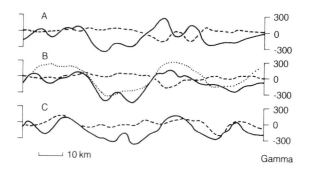

Figure 28.3 Profiles perpendicular to the trend of the Carlsberg Ridge illustrating (a) the lack of resemblance between the observed (solid lines) and computed (dashed lines) profiles, assuming normal magnetism, and (b) the resemblance between the observed profile (solid line) and the dotted profiles in B, which was computed for a model in which the basaltic layer is divided into parallel blocks with alternating normal and reversed magnetization (*After F. J. Vine and D. H. Matthews, 1963*).

the basaltic crustal layer was formed by rising convection currents along the middle of the oceanic ridges, and the resultant extrusions and intrusions of basalt each assumed the direction of magnetization of the Earth's magnetic field at the time when it consolidated, then the striped pattern of magnetic anomalies, like those of the North Pacific ocean, could be explained. The idea is that each new intrusion and extrusion of basalt, along the middle of the oceanic ridges, splits the previ-

ous intrusion into two, and the two halves move away on either side. The striped pattern should therefore be bilaterally symmetrical on either side of the median rift. In other words, according to the Vine–Matthews hypothesis, the magnetic anomalies might provide a magnetic record of ocean-floor spreading that could be read from each ocean floor outwards from the mid-oceanic rift.

At the time when Vine and Matthews proposed their hypothesis there was much resistance to its acceptance, because it was not then generally agreed that magnetic field reversals were a reality. Many earth scientists still regarded it as possible that reversals might be self reversals. The following year (1964), however, A. Cox, R.R. Doell and G.B. Dalrymple established the reality of reversals of the Earth's magnetic field. Since the reversals took place at the same time in various parts of the Earth it would be too much of a coincidence to consider them to be other than field reversals, and the Vine–Matthews hypothesis gained favour. It gained complete acceptance after Opdyke *et al* (1966), as recorded in Figure 27.8, found a similar magnetic timescale for the last 4 m.y, fossilized within the stratigraphic columns of sedimentary cores collected from the Antarctic. The reliability of palaeomagnetic epochs and events, whether determined from basaltic or sedimentary rocks, was thus accepted as a proven means of dating events during the last 4.5 m.y.

In 1966 the results of an aeromagnetic survey by Heirtzler *et al*, over an area of 350 km^2 over the

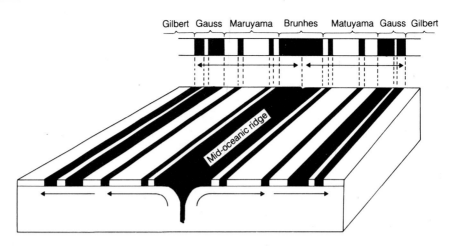

Figure 28.4 Block diagram illustrating how 'magnetic stripes' on the ocean floors can be explained by spreading of the oceanic lithosphere away from oceanic ridges on either side, according to the Vine–Matthews hypothesis. (*Adapted from A. Cox et al, 1967*)

Reykjanes ridge, part of the mid-oceanic ridge, were published. This investigation not only showed a sequence, from the ridge outwards on either side, of normally and reversed magnetized stripes conforming to the newly established sequence, but it also showed that the anomalies are symmetrical about the crest of the ridge, as the hypothesis of ocean-floor spreading requires (Figure 28.5).

Since teams from Lamont–Doherty Observatory manning research ships had collected magnetic data as a matter of routine on their many cruises since the early 1960s, a wealth of magnetic data already existed, but it had to be deciphered. Records across mid-oceanic ridges were not directly available, because the research ships had changed course and speed, according to the prime object of their investigations. Cox in 1973 recorded that Heirtzler undertook the task of organizing and correlating the existing data and, with the aid of a computer, constructed magnetic profiles across the Atlantic. By 1968, profiles across the mid-oceanic ridges had been published for the South Atlantic, the South Indian Ocean, and the North Pacific. It was a proven fact that the striped patterns of magnetic anomalies characterized extensive areas of these three oceans. For the middle parts of the oceanic ridges the ages of the magnetic anomalies had by now been estimated by comparison with the known sequence of polarity events for the past 4.5 m.y. (Figure 27.7). By assuming constant spreading rates for individual profiles, and using the relative widths of magnetic 'stripes' as a guide to the rate of spreading, it was found that the calculated ages of the marine magnetic 'stripes' matched those that were ascribed by analogy with the ages of polarity epochs and events determined from lava-flows on land. Con-

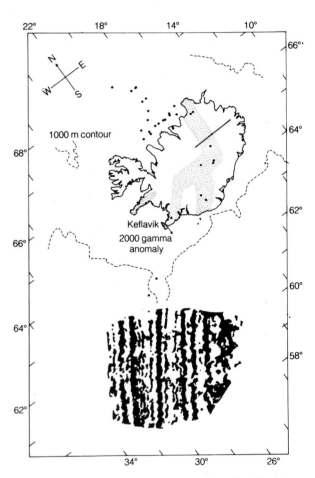

Figure 28.5 Magnetic anomalies over the Reykjanes Ridge, south-west of Iceland, showing 'magnetic stripes' (positive anomalies black) and their bilateral symmetry. (*After J. R. Heirtzler et al, 1966*)

versely, dating the marine 'stripes' by analogy, provided a means of estimating the rates of spreading in the various oceans. The widths of the magnetic 'stripes' vary according to the spreading rate, being wider where the spreading rate is rapid, as in the equatorial Pacific.

There is no direct way of dating marine stripes older than 4.5 m.y. Heirtzler *et al* (1968), however, by assuming constant spreading rates for the individual oceans, constructed a timescale for the magnetic 'stripes' back for the past 75 m.y, that is back into the upper Cretaceous. It was then possible to construct an isochron map of the ocean floors, by mapping known magnetic anomaly 'stripes' at time intervals of 10 m.y. Isochron lines marked 10 lay on the mid-oceanic rift 10 m.y ago; isochron lines marked 20 lay on the mid-oceanic rift 20 m.y. ago and so on. Spreading rates are calculated from the ages of the isochron lines and their distances apart. The slowest rates of 1 or 1.5 cm a year occur in the Atlantic and Indian Oceans, and the most rapid, 6 cm a year, on the east Pacific rise (see also Section 24.13). If the rate of separation of oceanic crust at the mid-oceanic ridges is under consideration, the spreading rates have to be doubled. For example new oceanic crust with a width of 2 or 3 cm is emplaced in the Atlantic Ocean each year.

In 1970, the result of coring across the South Atlantic Ocean confirmed the calculated magnetic timescale of Heirtzler *et al.* The cores penetrated down to the basaltic layer (Layer 2), Upper Cretaceous being the oldest stratigraphic level penetrated. The age of the oldest sediment immediately overlying the basaltic layer was determined from microfossils within each core, and this age was found to increase with increased distance from the mid-oceanic rift on either side (Figure 28.6). The data reveal a constant spreading rate of 2 cm a year extending backwards in time for about 80 million years.

28.3 TRANSFORM FAULTS

Associated with and commonly striking approximately at right angles to the mid-oceanic ridges are fracture zones that appear to offset the ridges (Figure 28.7a). Amongst the earliest to be studied in detail were those off the western coast of North America, which have sliced the rise into long crustal slabs, striking roughly east–west. These fractures were first recognized as faults from the vertical displacements which have produced great submarine scarps rising hundreds of metres above the ocean floor on the deeper side. Because they are practically straight they were at once suspected to be strike-slip faults. The Mendocino, Murray, Clarion and Clipperton faults were first described by Menard in 1955, and since then many more have been discovered.

The 1600 km long Mendocino fault, for example,

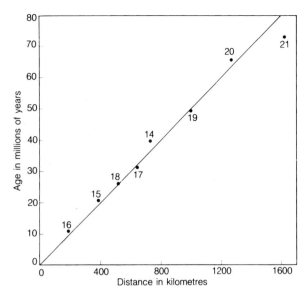

Figure 28.6 Graphic construction revealing a constant spreading rate for the oceanic lithosphere of the South Atlantic. The ages, in millions of years, of sediments (from numbered sites) immediately overlying the basaltic layer are plotted against their distance from the South Atlantic ridge. The close agreement of the plotted points with a straight line representing a spreading rate of 2 cm a year is strong evidence of a constant spreading rate. (*After A. E. Maxwell et al, 1970*)

has a southward facing undersea cliff rising in height from 1500 to as much as 3000 m. The block to the north has been only slightly tilted as a whole, but in detail it is considerably fractured. The relative horizontal displacement along the Mendocino fault, and similar faults to the south, were determined by reference to the magnetic anomaly 'stripes', which are cut through by the faults, and in each case are quite different on the two sides of the faults (Figure 28.7b). The amount of lateral displacement along the Murray fault was the first to be measured. It was found that the east–west profile of the magnetic anomalies on the northern side (plotted as in Figure 28.7c) could be matched with the corresponding profile on the southern side when the latter was shifted 154 km to the east, so that an anomaly E (Figure 28.7c) was found to have its continuation at D.

At that stage no matching could be recognized along the two sides of the other great faults. The survey had not, in fact, been carried far enough west. In 1958 Vacquier and Raff continued the magnetic survey westward until they eventually found a convincing match along the Pioneer fault, which indicated a displacement of 265 km. But along the Mendocino fault no match was detected until the Scripps men had patiently extended their survey farther west throughout two seasons of

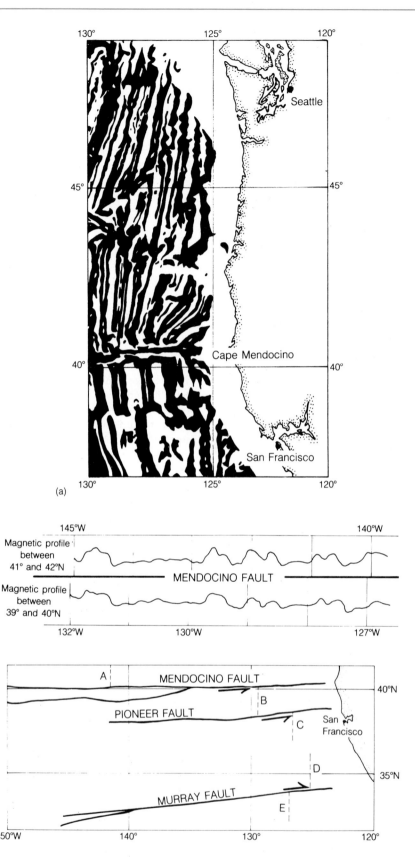

Figure 28.7 (a) Pattern of positive magnetic anomalies discovered in the Pacific floor off the coast between Seattle and San Francisco. (b) Matching of profiles of magnetic anomalies from north and south of the Mendocino fault, indicating a lateral displacement of 13° of longitude. (c) The magnetic anomalies on the north side of the Mendocino fault can be matched with those on the south side only when two anomalies such as A and B, 1160 km apart, are brought together. Similarly, the displacement along the Pioneer fault is the distance between B and C, 265 km; and that along the Murray fault the distance between D and E, 154 km. (*After A. D. Raff and R. G. Mason, 1961*)

work. In 1961 they succeeded in locating a pattern on the north side that matched the pattern on the south side – but 1160 km to the west (Figure 28.7b and c).

Similar faults offsetting the mid-oceanic ridges have been found to be a worldwide phenomenon. Prior to 1965, faults characterized by horizontal displacements like those just described, and the San Andreas fault in North America and the Great Glen fault in Scotland, were thought to be transcurrent faults, and inferred to be younger than the phenomena, such as the mid-oceanic ridges, which they offset. The great difficulty about this interpretation was that it seemed impossible for great strike-slip faults like the San Andreas fault to terminate. Yet there was no evidence that they continued as small circles right round the Earth. For many years geologists had recognized that earth movements, expressed by earthquakes, ranges of young fold mountains and major faults, like the San Andreas with great horizontal displacements, are restricted to narrow belts, sometimes described as mobile belts. In 1965 Tuzo Wilson recognized that these mobile belts form a continuous network over the Earth, dividing its surface into several large rigid plates like a mosaic. Faults with major horizontal displacements can be traced into movement belts of other types, such as fold mountains and island-arcs, or mid-oceanic ridges. Wilson suggested that the junctions between movements of different kinds should be called **transforms**, and that the faults concerned be called **transform faults**.

Transform faults differ from transcurrent faults in that they are intimately related either to the creation of new crust, as at mid-oceanic ridges, or to the annihilation of crust where it disappears into the depths at ocean trenches. The faults that strike across mid-ocean ridges connect portions of the ridge crests that are offset in relation to one another. With great intuition, Wilson recognized that the apparent offsets of the ridges might be original features formed when continental crust split apart and new oceans were born (Figure 28.8). The mid-Atlantic ridge, including its offsets, for example, is essentially parallel to the shores of the bordering continents. The direction of horizontal movements along transform faults, formed in this way, would be in the opposite direction to that required if the faults were transcurrent faults responsible for offsetting the ridges after they were formed (Figure 28.9). The directions of movements along transform faults intersecting mid-oceanic ridges have now been amply established by seismic studies, so that there is no doubt that the directions of movement along them conform with Wilson's concept of transform faults. Moreover, the associated earthquakes, which are all shallow, are confined to the ridges themselves and to the portions of transform faults between the offsets. Where the faults extend away from the ridges, beyond the offsets, no earthquakes occur; the faults are here dead.

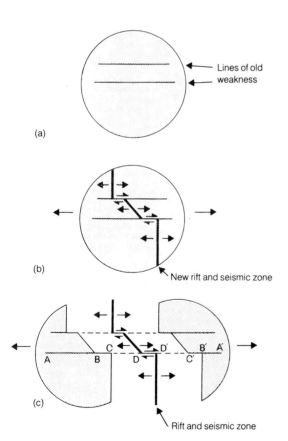

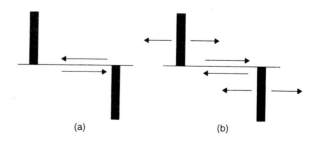

Figure 28.8 Diagrammatic illustration of three stages in the rifting of a continent into two parts, and their separation resulting in the evolution of new ocean floor. Exemplified by the separation of South America and Africa with the opening of the South Atlantic ocean. (*After T. Wilson, 1965*)

Figure 28.9 Diagrammatic illustration of (a) the directions of movement along a transcurrent fault responsible for the off-setting of the black band and (b) the directions of movement along a transform fault connecting off-set portions of a mid-oceanic ridge (black).

28.4 PLATES OF LITHOSPHERE

In 1967 McKenzie and Parker outlined 'a paving stone' hypothesis, in which ocean crust was newly formed at mid-oceanic ridges and destroyed at the trenches. By reference to Euler's theorem they determined the pole of rotation for the north-west Pacific from slip directions along marginal fault-planes, and located it at 50°N 85°W. If their paving stone hypothesis was correct then the North Pacific should have moved as a rigid slab, and the slip directions of all earthquakes within this slab should lie on small circles around the pole of rotation. Examination of the published data revealed that this was so, a conclusion that was confirmed by Isacks and Sykes the following year.

In 1968 Morgan outlined the hypothesis of plate tectonics. He established that transform faults, like those intersecting mid-oceanic ridges, which form parts of the boundary between two plates that are moving relative to one another, lie on small circles that are concentric about a pole of rotation (Figure 28.10). The movement of a plate is defined by the position of its pole of rotation and its angle of rotation about the rotation axis; its rate of movement varying with distance from the pole of rotation, being nil at the pole and reaching a maximum on the equator relative to the pole of rotation. Morgan divided the Earth's surface into twenty plates, and determined their poles of rotation by drawing great circles at right angles to relevant groups of transform faults. If the members of a group of transform faults have a common axis of rotation then such great circles should intersect at the pole of rotation, e.g. the great circles intersection at A in Figure 28.10. In actuality the intersections have a small spread; an inevitable result of small errors that arise in mapping the ocean floors.

Le Pichon (1968) simplified the concept of plate tectonics by dividing the Earth's surface into six major plates, and a few small ones. He determined the poles of rotation, about which the plates may be supposed to have rotated, by two different methods. In one he utilized the spreading rates calculated from the ocean-floor magnetic anomalies, and in the other he made use of the strike of transform faults intersecting mid-ocean ridges, as described above. The directions of movement of these plates are illustrated in Figure 28.2.

28.5 VARIETIES OF PLATE MARGINS

The margins of plates can be grouped into three varieties, as follows:

1. **Transform margins** where plates glide past one another along transform faults;
2. **Convergent margins** where plates disappear. If the earth remains the same size then the amount of crust consumed must equal the amount of new crust that is formed. Examples of consumption are found at ocean trenches where oceanic crust disappears down into the mantle along Benioff zones.
3. **Divergent margins** where new ocean crust is formed, and plates grow. These occur at mid-oceanic ridges, and where continental crust splits apart and new oceans are born, e.g. the Red Sea.

28.6 TRIPLE JUNCTIONS

Places on the Earth's surface where three plates meet are called **triple junctions**. They can be succinctly classified in terms of the types of plate boundaries that join each other at the junction. McKenzie and Morgan have shown that there are sixteen theoretical types of triple junction. In practice several of these types are unstable or have geometries which quickly evolve into new configurations and so they are not found on Earth. Five types of triple junction are found in the present-day plate tectonic geometry. One common triple junction type is where three active ridges meet as found for example at Galapagos and in the central Indian Ocean. Other stable types are the trench-trench-fault and fault-fault-ridge geometries. An example of each of these two types are the intersection of the Peru–Chile trench with the Chile

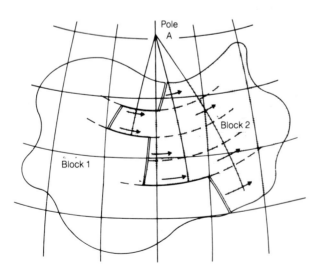

Figure 28.10 Transform faults drawn on a sphere lie on small circles concentric about a pole of rotation (A), the position which can be found at the intersection of great circles drawn at right angles to the relevant transform faults. The movement of Block 2, relative to Block 1 must result from rotation about A. (*After W. J. Morgan, 1968*)

Ridge and Bouvet Island in the South Atlantic. The final two stable present-day types of triple junction of fault-fault-trench and ridge-trench-fault are to be found at the northern end of the San Andreas fault and at the mouth of the Gulf of California respectively.

28.7 ANCIENT PLATE MOTIONS: EULER ROTATIONS

The basis for reconstructing past plate movements is the detailed mapping of magnetic stripes and transform faults in the ocean basins. Extremely detailed histories of plate movement have been deciphered from ocean-floor studies by combining the positions of magnetic anomalies with the trends of ancient transform boundaries.

The mathematics, or spherical trigonometry, needed for this task were first worked out by Leonhard Euler, the leading mathematician and theoretical physicist of the eighteenth century. Euler's pre-eminent studies were in pure mathematics. Indeed he was probably the greatest algorist in history and his total output was immense. One of Euler's many mathematical achievements was to demonstrate formally that the outcome of any series of translations on a sphere could be represented by just one rotation about an axis through the centre of the sphere. Any ancient plate tectonic movement can thus be succinctly described in terms of a single rotation about a Euler axis or pole.

In plate tectonic studies Euler rotations have traditionally been found by using a computer-search technique to 'home in' on the rotation that best fits the ocean-floor data. By fitting the transform faults under study by small circles the best Euler pole position is selected. The angle of the Euler rotation is similarly found by searching for the angle which best accounts for the distances between magnetic stripes, of the same age, on opposite sides of ridge crests. More recently a statistical spherical regression approach has been applied to magnetic anomaly and transform fault data by R. Thompson and M. J. Prentice which obviates the need for a trial and error search method and is able to calculate Euler rotations directly, along with an error estimate. Figure 28.11 plots the motion of the Euler pole for the North American and African plates for the Mesozoic anomalies designated M25 to M0 along with the elliptical-shaped confidence limits, as calculated from the spherical regression method.

Comprehensive studies have been made of all known sets of magnetic anomalies and transform faults in order to build up series of Euler rotation poles which concisely describe the history of sea-

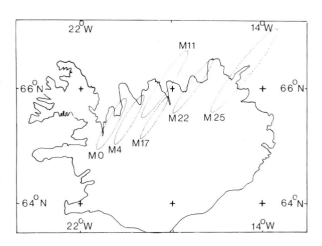

Figure 28.11 90 per cent confidence regions around the Euler finite rotation poles for the North America–Africa plate pair at anomalies M25, M22, M17, M11, M4 and M0. Euler poles calculated using the spherical regression method of R. Thompson and M. J. Prentice (1987).

floor spreading from the earliest known anomalies and transforms formed 180 m.y. ago right through to the present day. The accuracy with which reconstruction can be made for this period of time is very high, allowing the complete history of the breakup of Pangaea to be described quite precisely.

In plate tectonics everything is in motion, nothing remains fixed. This continuous relative motion includes not only the plates, but also all three types of plate boundary, even the triple junctions and the Euler rotation poles. Plate tectonic reconstructions thus show only the relative position of the plates and the continents. Motion with respect to a fixed frame of reference, such as the Earth's spin axis, must be found from alternative methods. Palaeomagnetism provides one technique whereby palaeolatitudes may be estimated. Backtracking of volcanic chains to relatively fixed hot spot sources provides another approach to providing approximate palaeolatitudes and palaeolongitudes.

28.8 THE BIRTH, GROWTH AND DECLINE OF OCEAN BASINS

Major oceans are formed by the rifting and breakup of supercontinents. The oceans grow through the process of sea-floor spreading. Inevitably during the life-time of an ocean subduction will begin and come to dominate the tectonic processes. As a consequence the ocean will begin to shrink as more sea floor is lost through subduction than is created by sea-floor spreading. Eventually the ocean will close in a major period of continental collision and mountain building. This complete sequence in the life of an ocean is known as the **Wilsonian Cycle** and takes several hundred million years to run its full course.

The oceans of the world today are all at different stages of the Wilsonian Cycle. The Atlantic, at some 160 m.y. old, is at the early growing stage, the Indian Ocean, with a more complex history, is more mature but still growing. The Pacific Ocean is further into the cycle, with subduction dominating over spreading processes. It is in a state of decline, but it still has some 200 m.y. to run before closing completely.

A description of the tectonic history of our present oceans follows along with a discussion of approaches to integrate their histories into an all-encompassing global picture.

28.9 THE ATLANTIC: AN OPENING OCEAN FLOOR

Of the world's major oceans the Atlantic has had the simplest history. The key to understanding the tectonic history of oceans lies in the identification of the sea-floor spreading magnetic anomalies. In the Atlantic the magnetic anomalies are well-preserved and generally straightforward to interpret.

At the present time sea-floor spreading in the Atlantic involves the movement of the four major plates of Eurasia, North and South America and Africa. New sea floor is being generated along the whole length of the mid-Atlantic Ridge from the Nansen Ridge in the Arctic Ocean to the Bouvet triple junction at 55° south in the South Atlantic, a distance of over 17 000 km. A short zone of subduction is found associated with the small Caribbean plate. Here sea floor is being subducted westwards beneath the Lesser Antilles at a rate of around 1 cm/year.

The oldest Atlantic magnetic anomalies have been found off the east coast of North America and the north-west coast of Africa (Plate XIV). They are interpreted as anomaly M25 (see Figure 27.9). 160 m.y. ago they would have been located on either side of a new ridge crest which formed when spreading started as Pangaea first began to breakup with the rotation of Laurasia away from western Gondwana. This first spreading in the newly-formed Atlantic involved areas of the central Atlantic, the Caribbean and the Gulf of Mexico (Figure 28.12a). The central Atlantic has continued to widen steadily as the North American plate has rotated away from the African plate. New sea floor has been continuously welded on to the African and the North American plates at the mid-Atlantic ridge as the two plates have spread apart (Figure 28.12a–f).

The oldest magnetic anomaly in the South Atlantic is M12. This points to opening beginning around 140 Ma, slightly later than in the central Atlantic. Spreading here marked a further split in Pangaea with South America and Africa pivoting apart. Like the central Atlantic the whole of the southern Atlantic has progressively widened with only relatively minor changes to the direction or rate of spreading. The tectonic history of the South Atlantic has thus been beautifully simple, only involving the rotation apart of the two plates of Africa and South America.

For the North Atlantic the history is once again one of a gradually opening ocean basin, with the slow drift of the North American and Eurasian plates away from each other. However, in the case of the North Atlantic, ocean-floor spreading progressively worked its way northwards. The North Atlantic began to form 110 m.y. ago when Europe first began to separate from North America. By 95 m.y. ago spreading had extended to the Labrador sea and by 65 m.y. ago the Norwegian–Greenland sea was also starting to form.

This progressive northward propagation of spreading in the North Atlantic has been very accurately reconstructed from the magnetic anomalies. The oldest anomaly found between Spain and Newfoundland is M0 (Plate XIV), whereas in the southern Labrador sea the oldest anomaly is C32, in the northern Labrador sea C28, and in the Norwegian Sea C24. This northward-spreading progression was achieved by small plates spalling off from the main European and North American plates. These micro-plates have all been later reincorporated into the main plate pair so that at the present day spreading in the North Atlantic involves just the European and North American plates. The first micro-plate to break away was Iberia which rotated 34° anticlockwise between 150 and 80 m.y. ago forming the Bay of Biscay. This was followed by Rockall Bank which moved as a separate plate between 110 and 85 m. y. ago. Separate movements of the Hatton Bank, Greenland, Jan Mayen and Svalbard all followed at later stages as spreading moved northwards and the active ridge crest switched from the Labrador Sea branch to the present configuration which passes through the Norwegian Sea.

In summary the opening of the Atlantic Ocean has primarily involved a four-plate system of the North and South Atlantic plates rotating and spreading away from the Eurasian and African plates. All the sea floor created at the Atlantic constructive plate boundaries remains today, with the exception of a small amount lost in the Caribbean. The Atlantic is thus an example of a young ocean whose history is one of progressive opening and growth.

Figure 28.12 Reconstructions of the positions of the continents bordering the Atlantic at six periods between 200 million years ago and the present day. The relative dispositions of the continents are known from the Atlantic Ocean floor magnetic lineations, while the palaeolatitudes are determined from palaeomagnetic evidence. (*After J. D. Phillips and D. Forsyth, 1972*)

28.10 THE INDIAN OCEAN: A MATURING OCEAN FLOOR

The Indian Ocean has had a much more complicated tectonic history than the Atlantic Ocean, making interpretation of its magnetic anomalies more difficult. In addition, thick sediment sequences such as in the Bay of Bengal and off East Africa have tended to obscure many of the old magnetic anomaly patterns. As a consequence the identification of magnetic lineations has undergone a number of revisions, as more and more data have been accumulated, and the precise details of the early-spreading history of the Indian Ocean are still much debated. Nevertheless magnetic reversal lineations are preserved in many parts of the ocean allowing much of the Indian Ocean tectonic history to be accurately deciphered.

Today a ridge-ridge-ridge triple junction lies in the centre of the Indian Ocean separating the African, Antarctic and Australian plates. Active sea-floor spreading is taking place on the boundaries between the three plates. The fastest spreading is found on the Australian–Antarctic boundary on the south-east India Ridge where rates reach a maximum of 7.5 cm/year at a point 90° away from the Euler rotation pole. In the north of the Indian Ocean Arabia moves as part of a separate plate with boundaries running from the Carlsberg Ridge into the Gulf of Aden and the Red Sea and along the Owen fracture zone. India also moves separately, spreading away from Africa along the Carlsberg Ridge but converging slowly at around 1 cm/year with the Australian plate along a diffuse zone of discordance in the region between the Ninety-east Ridge and the Chagos Bank.

As in the Atlantic Ocean the old magnetic anomalies are to be found adjacent to the continental edges, demonstrating that the tectonic history has largely been one of spreading. The oldest anomalies are found on the north-western Australian continental margin (Plate XIV). Anomalies of M26 age stretch to the north across the Argo Abyssal Plain through to anomaly M16 at the Java Trench. The next oldest anomaly sequences begin with anomaly M22 and are found at two localities, namely the Mozambique Basin and off Dronning Maud Land. Anomalies beginning at M10 age are located off much of western Australia, while south of India and Sri Lanka, to the east of Madagascar and between Australia and Antarctica the oldest anomalies are of C34 age. These four anomaly sets, beginning with M26, M22, M10 and C34, mark distinct phases in the fragmentation of Gondwana and are the key observations that allow the early history of the Indian ocean to be understood.

The evolution of the Indian Ocean can be

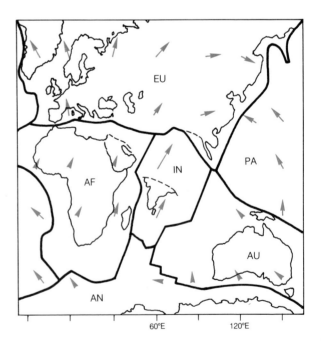

Figure 28.13 Possible break between East and West Gondwana 150 million years ago. Ridge segments shown by thick lines and transform faults by thin connecting lines. At this time South America and Africa were part of one plate while Antarctica, Australia, India and Madagascar formed another plate. (*After J. J. Veevers et al, 1980*). The continued addition of new magnetic data in combination with the use of isochrons (see Plate XIV) means that such reconstructions are regularly modified.

Figure 28.14 Reconstruction of plate motions 60 million years ago to illustrate the tectonic evolution of the Indian Ocean. At this time India was drifting rapidly northwards, Australia had separated from Antarctica and Madagascar was part of the African plate. (*After R. G. Gordon and D. M. Jurdy, 1986*)

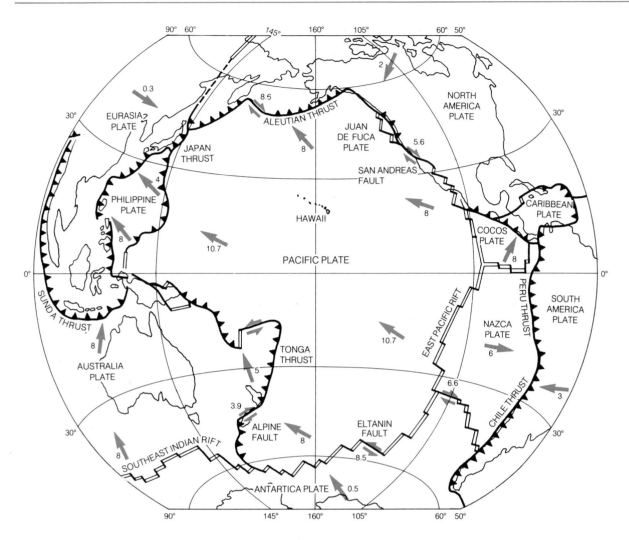

Figure 28.15 Present-day plates and plate movements of the Pacific and the Circum-Pacific area. Active ridge crests shown by double lines, transform faults as single connecting lines and destructive boundaries as barbed lines with barbs pointing down the subducting plates. Rates of plate movements shown by full arrows. Relative movement between plate pairs shown by half arrows. Note how tectonic movements from as far apart as the Japan trench to the East Pacific Rise and from the San Andreas Fault to the Antarctic plate boundary, a region encompassing almost one quarter of the Earth's surface, can all be related to the north-westwards rotation of the Pacific plate parallel to the trend of the Hawaii Island chain.

divided into nine main phases. Prior to the rifting and dispersal of Gondwana active spreading was taking place along the linked Tethyan Sea and Pacific Ridge crests. However, by anomaly M26 time the Tethyan spreading centre had migrated south and begun to tear fragments from the northern margin of Gondwana. In this first recorded phase of Indian Ocean spreading a continental block was transported north away from Australia, as shown by the Argo Plain anomalies mentioned above. The second tectonic phase in the history of the Indian Ocean marks a major step in the breakup of Gondwana. This was the time when East and West Gondwana began to separate. This important event separated the western continents of Africa plus South America away from the eastern Gondwana

continents in a two plate system (Figure 28.13). The spreading rearrangements that caused this split heralded the eventual demise of the Tethys Sea. In phase three a further major rifting episode began at anomaly M10 time as India plus Madagascar separated from the Antarctic/Australian plate. A fourth spreading phase must have taken place in the Cretaceous magnetic quiet zone. As the geomagnetic field was not reversing at this time no marine magnetic anomalies mark the event, but a change in spreading is needed in order to produce continuity between the Mesozoic and Cenozoic spreading anomalies. This invoked fourth phase would have taken place between 105 and 95 m.y. ago as Africa, India and Madagascar once again moved together as part of the same rigid plate. At the end of this fourth drift phase a further big reorganization in the spreading pattern of the Indian Ocean took place. India broke away and began its spectacular northward journey. Spreading rates of 26 cm/year and giant north–south transform faults caused India to drift rapidly northwards (Figure 28.14). At the same time much slower spreading got under way between Antarctica and Australia (Figure 28.14). Around 55 m.y. ago India began to impinge into Asia causing further spreading reorganizations, and in an eighth spreading phase beginning 32 m.y. ago the Ninety-east transform locked as India slowed after its northward flight. The final, ninth phase of spreading has seen the formation of the Gulf of Aden and the Red Sea. All these later phases of spreading are well recorded by the Cenozoic anomalies preserved in the centre of the Indian Ocean (Plate XIV).

The history of India through these events is worth reflecting on. Following the breakup of the supercontinent of Gondwana India first moved southwards along with Antarctica and Australia. It then moved eastwards with Madagascar and later drifted northwards, as part of the African plate, before rapidly moving northwards by itself to collide with Asia. Throughout most of this history subduction was taking place to the north of India. A likely future phase in the evolution of the Indian Ocean would be for subduction to recommence, but this time to the south of India, where old lithosphere is being heavily loaded by sediment fans and for the Indian Ocean sea floor to resume a northward drift.

28.11 THE PACIFIC: A WANING OCEAN FLOOR

The Pacific Ocean has had a longer and more complex history than that of the Atlantic or Indian Oceans. Large tracts of sea floor generated at Pacific spreading centres have been destroyed at subduction zones around the edge of the Pacific. Nevertheless many magnetic anomalies are preserved which accurately record many of the details of Pacific sea-floor spreading history. Fortunately magnetic anomalies now found on only one Pacific Basin plate can even be used to infer the presence of plates which have long since disappeared since all magnetic lineation anomalies are produced in congruent pairs.

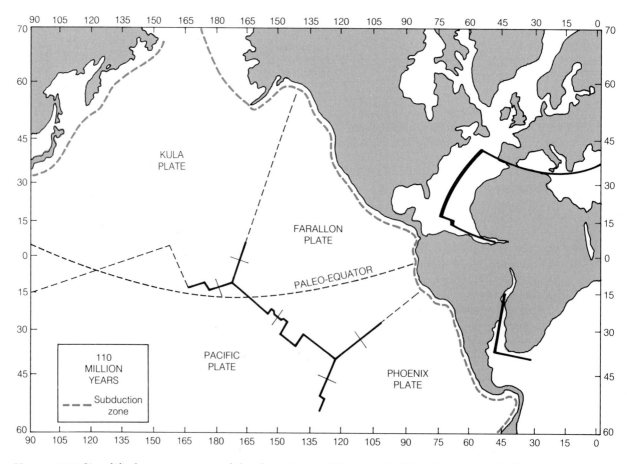

Figure 28.16 Simplified reconstruction of the disposition of the paired ridge-ridge-ridge Pacific triple junctions about the palaeo-equator 110 million years ago. (*After R. L. Larson and W. C. Pitman, 1972*)

At the present time the Pacific plate is the largest in the world – it encompasses almost one quarter of the Earth's surface. In the south-east and eastern parts of the ocean we have the Nazca and Cocos plates (Figure 28.15) and to the south the Antarctic plate. A few minor plates such as the Philippine, Juan de Fuca, Explorer and Easter Island micro-plates complete the make-up of the Pacific basin.

On the southern and eastern borders of the Pacific plate, spreading is occurring along the East Pacific Rise. Spreading rates of between 6 and 10 cm/year occur on the Cocos-Pacific and Antarctic-Pacific boundaries while extremely fast opening at 16 cm/year is taking place at the Nazca–Pacific Ridge crest. In the centre of the Pacific the Hawaiian Island chain leads away north-westwards from the active volcano of Mauna Loa towards Midway Island (Figure 28.15). On its northern and western borders the Pacific plate is colliding with the Australian and Eurasian plates and here subduction at rates of up to 9 cm/year is producing deep seismic zones sloping down from the Kermadoc, Tonga, Marianas, Japan and Kuril Trenches. Motion between the Pacific and North American plates is roughly parallel to the west coast of North America and follows the San Andreas transform which is slipping at 5 cm/year. However, along the Alaskan coast collision is taking place between the Pacific and North American plates with subduction rates of about 5 cm/year (Figure 28.15). All these various movements are neatly explained by a single north-westward rotation of the Pacific (Figure 28.15) in one of the best demonstrations of the great elegance and truly global nature of the plate tectonic theory.

180 m.y. ago the Pacific plate was a small micro-plate situated to the north-east of New Guinea. Its subsequent growth to its mammoth present-day size is encapsulated by the evolution of a pair of ridge-ridge-ridge triple junctions. Originally the two triple junctions would have been separated by an active spreading ridge a few hundred kilometres long and magnetic lineations from M38 age onwards record this early plate geometry. The distinctive angular pattern of these anomalies in the north-west Pacific Basin and near the Magellan Rise record the way the triple junctions migrated away from each other with time. The southern angular anomaly pattern can be followed past the Phoenix Islands across the southern part of the Pacific plate to where it merges into the present Pacific–Nazca–Antarctic triple junction (Plate XIV). The northern angular anomaly pattern can similarly be followed north-eastwards right across the Pacific to the Gulf of Alaska (Plate XIV). These two sets of angular magnetic anomalies demonstrate that there must have been a long history of a four-plate system in the Pacific. There would have been five active spreading ridges separating the four plates as illustrated in Figure 28.16.

During the Mesozoic the Pacific plate continuously had new material added to it along its northern, eastern and southern boundaries. New sea floor was also being continuously added to the other three plates of the four-plate system. To the east was the Farallon plate (Figure 28.16). Small

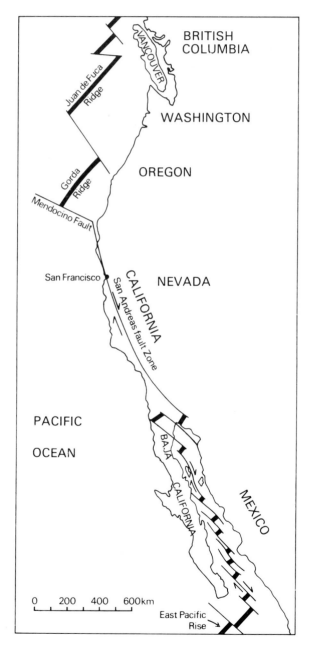

Figure 28.17 Map illustrating the continuation of the east Pacific Rise along the Gulf of California, where it is offset by many transform faults. At the head of the Gulf the Rise is replaced by the San Andreas fault-zone, and appears again as an oceanic feature in the Gorda and Juan de Fuca Ridges.

remnants of this once giant plate now make up parts of the Nazca, Cocos and Juan de Fuca plates. To the north was the Izanagi and later the Kula plate (Figure 28.16). Small remnants of these northerly plates have been caught up in the collision tectonics of the North American Cordillera and preserved as terranes (Section 27.15), but otherwise these once large plates have been totally consumed by northward subduction beneath Eurasia and Alaska. While, completing the Pacific picture, to the south spreading took place between the Pacific and Phoenix plates (Figure 28.16). This ridge crest has evolved into the present Pacific–Antarctic plate boundary.

A large tract of the central Pacific is without magnetic anomalies having formed in the Creta-ceous quiet interval. However, long palaeo-transforms such as the Mendocino, Pioneer and Murray (Figure 28.7) run westwards from the area of Mesozoic anomalies across to the Cenozoic anomalies in the eastern Pacific. They point to continued evolution of the paired triple junction arrangement through this quiet zone period. In the eastern Pacific anomalies are again well preserved with a clear record from A34 right through to the present day. They show that further east–west spreading continued between the Pacific and Farallon plates.

Throughout this whole Pacific–Farallon evolution of 180 m.y. of east–west spreading the distance between the spreading ridge and the subduction zones on the western edge of North

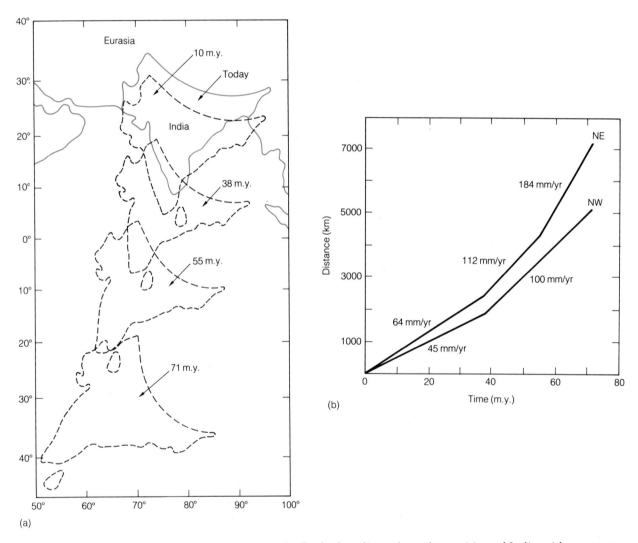

(a)

(b)

Figure 28.18 The collision of India with Eurasia. (a) Dashed outlines show the position of India with respect to Eurasia at four times since the Late Cretaceous. The relative positions of the two continents have been determined by adding Euler finite rotations around the India–Africa–North America–Eurasia plate circuit. Collision of the two landmasses took place about 55 million years ago. (b) Distance of the north-east and north-west tips of India from their present positions with time. Before collision the rate of movement was more than 10 cm/year, whereas afterwards it decreased to half its earlier value. (*After P. Molnar et al, 1981*)

and South America has been getting smaller. Around 30 m.y. ago the two boundaries first came together. The San Andreas fault has developed in response in order to accommodate the relative movement of the Pacific and North American plates (Figure 28.17). In the southern Pacific the ridge axis and subduction zones are still gradually approaching each other, as the Nazca and Cocos plates continue to spread eastwards away from the Pacific plate and to subduct beneath South America. Eventually these small remnants will also disappear completing the long history of Pacific–Farallon spreading.

In contrast to the Atlantic and Indian Oceans, Pacific tectonics has seen subduction dominating over spreading. More sea floor has been destroyed than generated as the Pacific Basin has progressively decreased in area and the ridge system migrated eastwards across the Ocean.

28.12   THE SOUTHERN OCEAN

Magnetic reversal lineations are found in three additional areas in the Southern Ocean. Anomalies are preserved on the Antarctic plate between Bouvet Island and Queen Maud Land, and demonstrate that the plate geometry of the South American, African and Antarctic plates has remained little altered as the three continents have drifted away from each other. The triple junction at the centre of this three-plate system has, however, behaved in a more complex manner switching back and forth between a ridge-ridge-ridge and ridge-fault-fault geometry as it has slowly migrated northwards along the mid-Atlantic ridge.

In the Tasman Basin, marine magnetic anomalies take the form of a conjugate set, exposing a former period of spreading between North Island, New Zealand and south-east Australia which began at anomaly C34 time and ceased after the formation of C24 age anomalies. Since this time the Tasman Basin has moved as part of the Australian plate.

The third and most revealing set of anomalies in the Southern Ocean are the peninsula series which stretch out across the Antarctic plate from the Antarctic peninsula towards the central Pacific. Intriguingly these anomalies young towards the presently passive Antarctic margin, ranging in age from C6 to C32. On account of their younging direction the peninsula-series lineations must have originated as part of an earlier plate which has since been incorporated by the Antarctic plate. With the exception of this small section of captured sea floor the rest of the plate has been totally subducted and all other evidence

for it has disappeared. It is the southern hemisphere equivalent of the Kula plate and has waggishly been christened the Aluk plate.

28.13   THE GLOBAL LINKAGE PROBLEM

The sea-floor spreading movements described above as deduced from the magnetic lineation patterns in the various ocean basins can be linked together to buildup a global picture of plate movements. The relative movements of plates in the Indian Ocean can be linked directly and precisely to the Atlantic relative movements as both sets of plate rotations involve the African plate. The Pacific relative plate movements can similarly be linked to the Indian Ocean movements through the joint involvement of the Antarctic plate. In this way a complete global pattern of plate movements can be built up and relative movements can be estimated at all plate boundaries, even across ancient destructive boundaries. As an example the collision history of India with Asia (Figure 28.18) is found by summation of Euler rotations around the India–Africa–North America–Eurasia plate circuit. Likewise subduction on the south-east Pacific rim has been calculated through the Nazca–Pacific–Antarctic–Africa–South America circuit.

The plate circuit method normally works well,

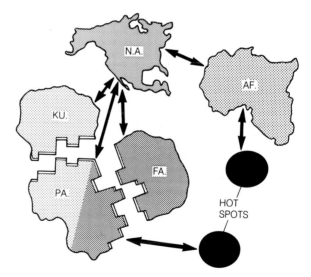

Figure 28.19 Schematic illustration of the combined plate circuit/hot spot method of linking plate movements in the Pacific Basin to the movements of continents around the Atlantic and Indian Oceans. PA: Pacific plate. KU: Kula plate. NA: North American plate. AF: African plate. HOT SPOTS: Hot spots in the Pacific and Atlantic Basins are assumed to have remained at a constant distance from each other. (*After D. C. Engebretson et al, 1985*)

but the peninsular series anomalies highlight a problem. Just one break in a chain of plate rotation additions is sufficient to spoil the calculations. In order to account for the young peninsular anomalies immediately adjacent to the Antarctic coast, a former split is needed through Antarctica and this inevitably must break the linkage between the Pacific and Indian Oceans. This linkage problem can only be overcome by using a hot spot reference frame approach. With this second method the Pacific-Farallon-Kula-East Antarctic plate complex is tied to the hot spot traces of the Emperor and Line Islands chains, while the North and South America–Eurasia–Africa–India–Australia–West Antarctica plate movements are tied to hot spot traces in the Atlantic and Indian Oceans. The two systems are then joined to form globally based reconstructions on the assumption that the hot spots forming the ocean island chains have remained fixed (Figure 28.19).

The linked global picture shows how spreading has been taking place around a relatively stationary Antarctica causing the other continents to move northwards in addition to their drift away from one another. Also many of the Pacific plate movements have had a northerly component. This combination of globally encircling northward movements has led to a long history of collision tectonics in the northern hemisphere. In the last 100 m.y. 7000 km of lithosphere has disappeared down the Japan trench and 6500 km down the Aleutian trench. The paired ridge-ridge-ridge triple junctions of the Pacific are found to have moved progressively eastwards while the Americas have drifted westwards causing 6000 km of oceanic lithosphere to be subducted beneath

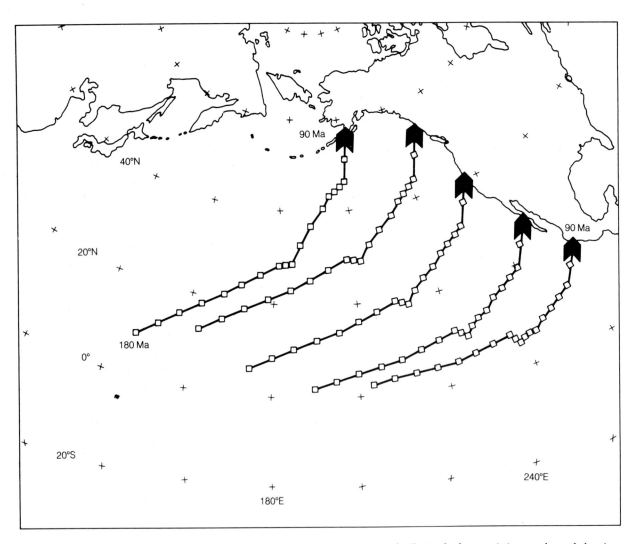

Figure 28.20 Trajectories of Farallon plate lithosphere crossing the Pacific Basin before arriving at the subduction zones along the west coast of North America at 90 million years ago. The open squares show the movement at 5 m.y. intervals. Relative movements derived from the combined plate circuit/hot spot method of Figure 28.19. (*After D. C. Engebretson et al, 1987*)

North America and 5000 km beneath South America.

Cox and his colleagues at Stanford have used the second, hot spot reference frame, approach to trace out the movements of Pacific plates before their convergence with North America and Eurasia. Displacements capable of transporting terranes (Section 27.15) large distances across the Pacific were found. Figure 28.20 shows their calculations of Farallon plate trajectories for terranes docking 90 m.y. ago at five localities along the western edge of North America. Later displacements involving the Kula plate are found to display more pronounced south–north movements, all in good agreement with the palaeomagnetic work described in Chapter 27.

28.14 THE THERMAL EVOLUTION OF PLATES: THERMAL MODEL

New ocean floor is first formed by the intrusion of hot magma at mid-ocean ridges above the rising limbs of convection cells. When the magma cools it solidifies and anneals so adding new material to the plates. As this hot material moves away from the ridge crests it loses heat and cools further. On account of the low thermal conductivity of igneous rock and the great thickness of the lithospheric plates the heat losses take place gradually over millions of years while the plates spread slowly but inexorably away from the ridge crests.

Exceedingly simple and elegant thermal models have been advanced that account mathematically for this cooling. The thermal models have been found to fit very well with observations of oceanic heat flow, sea-bed bathymetry and lithosphere thickness.

In the simplest model, due to McKenzie, plates are assumed to have flat lower boundaries at a constant temperature, close to the melting point for basalt, of around 1200°C (Figure 28.21). As the plates pull apart from the ridge crest hot material wells up, filling the gap between them, and producing the same temperature (1200°C) along this vertical edge (Figure 28.21). The top surface of the plate is in contact with the oceans and so can be taken to have a constant temperature of around 0°C (Figure 28.21). A key aspect of the model is that the plates are assumed to cool only by conduction. On this basis isotherms of equal temperature can be calculated in the plates by using the Law of the Conservation of Energy. The shapes of the isotherms predicted for this flat, moving plate model are shown in Figure 28.21b.

In moving-plate models, heat is transported horizontally by the movement of the plate and vertically by conduction. As a result many of the

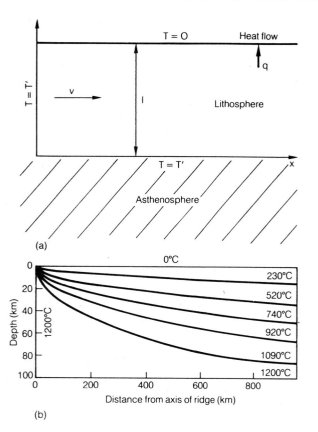

(a)

(b)

Figure 28.21 Cooling of ocean plates. (a) As a plate spreads away from a ridge crest with velocity, v, material at temperature, T′, accretes onto the plate. The plate cools by conduction, losing heat through its upper surface. (b) Isotherms within a 100 km thick plate moving at a half rate of 2 cm/year calculated using the thermal spreading model of (a). The isotherms can be used to calculate the heat flow, q, and ocean floor bathymetry. Both heat flow and bathymetry are found to depend on the square root of the sea-floor age.

characteristics of the oceanic lithosphere would be anticipated to be dependent on the age of the sea floor.

28.15 COOLING OF OCEAN PLATES

As Hess recognized in his seminal studies on the history of ocean basins in 1962, temperature isotherms must dip away from the ocean ridges and as a consequence there will be an accompanying decrease in heat flow and an associated subsidence of the ocean floor as it moves away from the ridges. These effects are neatly quantified by the thermal moving-plate model. For example the heat flow through the upper surface of the moving plates can be derived from the vertical separation of the isotherms in Figure 28.21b by use of Fourier's Law of Conduction. The moving plate model predicts that, for sea floor younger than about 100 m.y, the heat flow should decrease as the square root of the age of the sea floor. A remarkable agreement is found between measurements of oceanic heat flow and this predicted pattern of heat flow (Figure 28.22).

Similarly, sea bed topography can be calculated from the temperature distribution in the moving plate. On the assumption of isostatic equilibrium the topography of the sea floor depends only on the density distribution within the plate. As the ocean floor cools its density will increase and it must therefore sink if it is to maintain isostatic balance. Density changes can be found from the isotherms of Figure 28.21b by use of a suitable thermal expansion coefficient, and hence ocean floor subsidence is also obtainable directly from the thermal cooling model. Once again a relationship with the square root of the age of the sea bed is to be expected. This relationship is plotted in Figure 28.23. Excellent agreement is found between predicted and observed sea-bed depths.

28.16 SUBDUCTION OF OCEAN PLATES

Ocean floor is destroyed at the deep trenches by being pulled by convection forces into the upper mantle in a process known as subduction. Inclined planes of seismic activity slope down from the trenches many hundreds of kilometres into the interior of the Earth along the direction of motion of the descending sea floor. Large numbers of earthquakes are caused by the frictional movement of the downgoing slab as it underthrusts the overriding plate. These zones of intense seismic activity are called Benioff zones following the work of Hugo Benioff who, in the late 1940s, established the geometrical form of these inclined

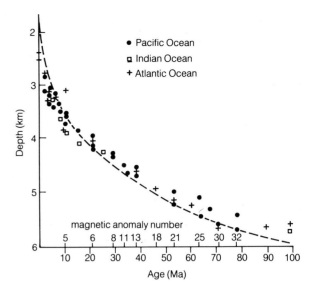

Figure 28.23 Variation of the depth to the top of the oceanic lithosphere with age. The dashed line is the theoretical elevation curve calculated from the simple conductive plate model of Figure 28.21. The age of the oceanic lithosphere is found from the magnetic lineation anomalies. The depths are averaged values. (*After J. G. Sclater and D. P. McKenzie, 1973*)

planes and their association with deep trenches around the edge of the Pacific basin.

As an ocean plate moves down into the mantle it begins slowly to warm up. At a depth of around 100 km it will start to melt partially and release hot hydrous fluids and siliceous melts. These ascend and give rise to the surface volcanism of island arcs and active continental margins that lie adjacent to the deep ocean trenches. The bulk of the downgoing ocean plate can retain its thermal and mechanical properties for around 10 m.y, by which time it can have been transported deep into the mantle. Earthquakes continue to be found along the path of the descending plate while it remains cold and brittle. Eventually the plate must warm to the temperature of the surrounding mantle, when its mechanical behaviour will be that of a plastic and not a brittle solid, and earthquake activity will consequently cease. In some regions such as the Sunda Arc, the Tonga Trench and the Kurile–Kamchatka area the Benioff zones reach down to depths of almost 700 km.

McKenzie has shown that exactly the same thermal model as is used to explore the cooling of plates as they spread away from ridge crests, can be used to determine how the plates will warm as they are driven down into the hot mantle. Once again it is assumed that conduction is the mechanism by which the plate changes temperature. In this situation of subduction both the upper and lower plate surfaces are taken to be hot. The

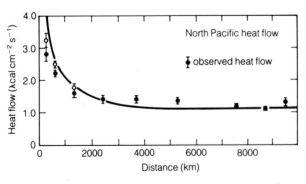

Figure 28.22 Variation of heat flow with distance from the ridge crest in the North Pacific. The line shows the prediction from the simple conductive cooling model of Figure 28.21. The model does not give a good prediction near the ridge, where non-conductive heat loss is associated with active hydrothermal circulation. (*After J. G. Sclater and J. Francheteau, 1970*)

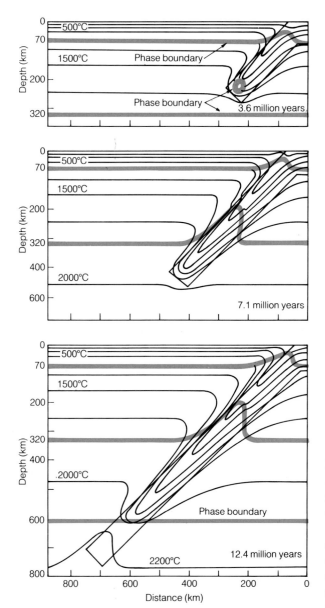

Figure 28.24 Evolution of a descending lithospheric slab. The series of diagrams plot out the temperature distrubution in a slab which begins to subduct at an angle of 45° at a rate of 8 cm/year. The subducting slab carries cold material down into the mantle. Phase changes in the slab occur at shallower depths than in the surrounding mantle because of the lower temperatures. When the slab has penetrated to 700 km it has warmed up to mantle temperatures and loses its original identity. (*After M. N. Toksoz et al, 1973*)

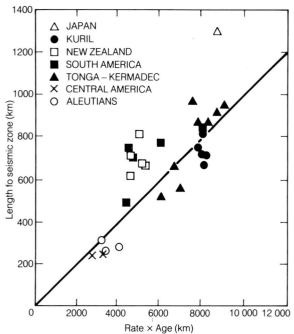

Figure 28.25 Downdip length of seismic Benioff zones plotted as a function of convergence rate and age of subducted lithosphere.

starting distribution of the isotherms as the plate begins to subduct are based on the distribution calculated in the cooling model for a plate that has spread a long distance from the ridge crest. In addition, as the pressure will increase in the sinking plate, two additional effects have to be taken into account. These are (i) adiabatic heating i.e. 'the heat of compression' and (ii) the latent heat associated with pressure-induced phase transitions. The isotherms of Figure 28.24 show the results of plate-warming calculations for a subducting slab.

Using the moving-plate thermal model, the depths to which cold ocean lithosphere would be expected to be transported down into the mantle can be calculated. Fast moving plates would be expected to carry cool isotherms to great depths and so have long Benioff zones. Unusually deep earthquakes would in this way be predicted for particularly old, cold ocean floor being subducted at high velocity.

P. Molnar and his group have calculated that the length of Benioff zones should be proportional to the rate of collision multiplied by the age of the sea floor being subducted. Figure 28.25 compares the results of their mathematical calculations with seismological observations on the lengths of eight Benioff zones. A very good agreement is seen between the thermal model calculations and the seismological observations, as young, slow moving ocean floor produces only moderately shallow earthquakes in short Benioff zones, while old, fast-moving oceanic lithosphere produces the longest zones. That the depths and positions of deep earthquakes can be estimated so well from simple thermal model calculations is another remarkable verification of the ideas of sea-floor spreading and plate tectonics.

28.17 DRIVING MECHANISMS

It is now widely accepted that convection in the mantle is the primary physical process driving the evolution of the Earth's surface. As a mode of propagating heat, convection was discovered in 1797 by Count Rumford. Subcrustal convection within the Earth was suggested by William Hopkins in 1839 and specific applications to geological problems were discussed by Osmond Fisher in his *Physics of the Earth's Crust* of 1881. He pointed out that the frictional drag on the underside of the crust would be expected to promote mountain building along the margins of the continents, where two systems of currents approached one another and turned downwards. He also recognized that 'The existence of convection currents beneath the cooled crust of the earth at once

furnishes a means of obtaining those local increments of temperature which in some form or another appear to be needful in order to explain volcanic phenomena'.

Unfortunately, instead of serving as a constructive stimulant, Fisher's ideas were regarded as wild speculations and ignored. One very notable consequence of this is worth recording. Because Fisher's ideas had been forgotten, and because Wegener himself saw no reason for exploring the possibility that continents might be carried along like ice floes drifting with the current, his first lecture on continental drift provoked a storm of indignation. It is tempting to suppose that the seed destined to grow into the great concept of continental drift may have been implanted in Wegener's mind during his first expedition to Greenland in 1906–08. Off the north-east coast he

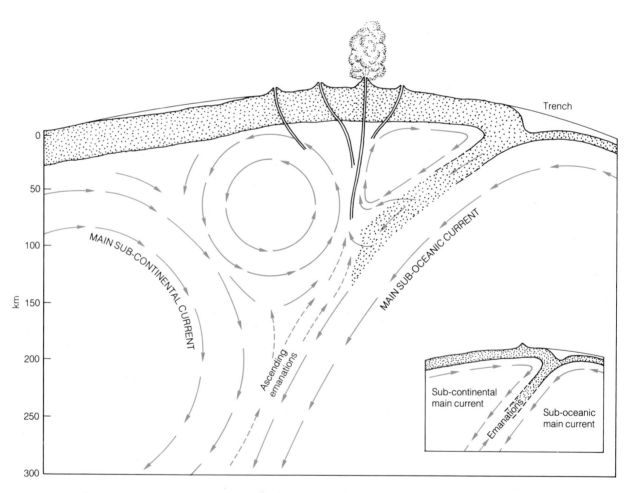

Figure 28.26 Diagrammatic section across a typical island arc (e.g. Japan or Kamchatka) to illustrate a speculative arrangement of convection currents which might account for the origin of andesitic lavas and their associates, and for the concentration of volcanoes over the region of intermediate earthquakes. The small inset diagram indicates the downturn of the major currents along the belt of intermediate and deep earthquakes. This dips beneath the continental crust and, being subject to continual fracturing, provides passageways through which hot emanations from great depths can ascend towards the surface. At the same time various crustal rocks, including sedimentary rocks charged with brine, are being dragged down by the main currents.

could not have failed to observe both the splitting up of the polar ice pack into gigantic floes, as undulating ocean swells surged beneath, and the formation of lanes of open sea as wind or currents drew some of the adjacent floes apart. It appears, however, that the analogous idea of the drifting apart of Africa and South America was not consciously inspired until the Christmas of 1910, when he had an opportunity of examining at leisure the beautiful maps of the André *Handatlas*. The near-parallelism of the opposing coasts of the South Atlantic at once fired his imagination. Despite warnings of the wrath to come, he shortly afterwards announced his hypothesis of 'moving continents' to a gathering of German geologists. Their hostile reaction foreshadowed the fierce international controversy that followed the publication of Wegener's first papers on the subject in 1912. More than forty years had to elapse before the possibility of continental drift could claim more than a handful of supporters.

In 1912 and for years afterwards, while Wegener was looking for natural processes that might be equal to the task of displacing continents he overlooked the most effective one – subcrustal currents. Only in the 1928 and last edition of his famous book is there any reference to the possibility that convection currents in the mantle might provide the transporting agent which, above all else, he required in order to answer his critics. However, it is only fair to remember that at that time convection in the mantle was widely supposed to have ceased at an early stage in the Earth's history. Moreover, it was argued by geophysicists that there was no more possibility of continents overcoming the resistance of the ocean floor than there was of icebergs ploughing through layers of boulder clay thicker and stronger than themselves. Convection in the 'solid' Earth seemed to be as impossible as continental drift. And even if subcrustal convection were to be admitted as a physical possibility and not merely regarded as an *ad hoc* hypothesis invented to bolster up the continental drift hypothesis, it could not explain displacements of the continents unless the ocean-floor obstruction could be overcome, e.g. by the transformation of basaltic rock into eclogite, which was heavy enough to sink out of the way by joining the down-turning currents. This was adding yet another hypothesis. Wegener's lack of enthusiasm for convective transport of

his continental fragments can be readily understood.

Today the position is different as plate tectonics demonstrates that the surface of our planet is very mobile. This surface mobility is taken to imply that the underlying mantle is also mobile, and that some form of convection must be responsible for driving the plates. Although the precise geometry of mantle convection remains the subject of much debate, previous objections to convection have been swept away by the realization of the importance of creep processes and theoretical calculations now indicate that the mantle should be in a state of vigorous, fully developed convection. Radioactive decay of the long lived isotopes ^{232}Th, ^{238}U, ^{40}K and ^{235}U is recognized as being able to account for the great majority of the global heat loss of 4×10^{13} watts and so provide the energy that powers convective processes in the mantle.

Holmes was the first scientist to appreciate the importance of convection both as a means of transporting heat from the interior of the Earth and as a method of powering large horizontal crustal movements. Figure 28.1, taken from the First Edition and Figure 28.26 taken from the 2nd Edition of Holmes' *Principles of Physical Geology* show his ideas on the form the convection systems might take. They remain in remarkable concert with present thinking.

28.18 FURTHER READING

Engebretson, D.C., Cox, A. and Debiche, M. (1987) Reconstructions, plate interactions, and trajectories of oceanic and continental plates in the Pacific Basin, in *Circum-Pacific Orogenic Belts and Evolution of the Pacific Ocean Basin* (eds J.W.H. Monger and J. Francheteau), American Geophysical Union, pp. 19–30.

Runcorn, S.K. (1981) Wegener's Theory: The Role of Geophysics in its Eclipse and Triumph. *Geol. Rundschau*, **70** (2), pp. 784–93.

Scotese, C. R., Gahagan, L. M. and Larson, R. L. (1988) Plate reconstruction of the Cretaceous and Cenozoic ocean basins. *Tectonophysics*, **155**, pp. 27-48.

Sullivan, W. (1977) *Continents in Motion, the New Earth Debate*, Macmillan, London.

Thompson, R. and Prentice, M.J. (1987) An alternative method of calculating finite plate rotations. *Physics of the Earth and Planetary Interiors*, **48**, pp. 79–83.

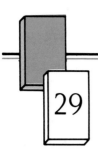

29

PLATEAUS, RIFT VALLEYS AND CONTINENTAL BASINS

Ex Africa semper aliquid novi
(There is always something new out of
Africa)

Pliny

29.1 SURFACE EXPRESSIONS OF EPEIROGENIC MOVEMENTS

This chapter considers the crustal structures and the resulting modifications of surface relief brought about by movements of uplift and subsidence. Ample reference has already been made in earlier chapters to the emergence or submergence of land areas due to eustatic changes of sea-level and to isostatic readjustments. Here we shall be concerned with the effects of up-and-down movements of blocks and strips of the crust: with what has come to be broadly described as **vertical tectonics**. The general idea of epeirogenesis, as opposed to orogenesis, was introduced in Section 7.7. If taken too literally these terms suffer from some overlapping of meaning, since mountain building involves vertical uplift, and the foundations of the continents are constructed of woven orogenic belts. However, in current practice **orogenesis** is used to imply that the internal structure of the belt concerned has been characteristically transformed (e.g. by folding, metamorphism), whereas in **epeirogenesis** there is no such

internal transformation, the movements being essentially vertical, with warping on various scales but no intensive folding. Areas which have not been subjected to orogenic movements for hundreds of millions of years are known as **stable shelves** or **platforms**. These have a folded and generally crystalline foundation or **basement**, parts of which are hidden by an uncomformable cover of nearly flat-lying sedimentary strata or volcanic rocks. Where the basement itself is exposed at the surface over extensive areas the region is more specifically described as a **shield**. Thus one speaks of the Russian platform and the Baltic shield (Figure 7.13). The corresponding features in North America are the Great Plains and Interior Lowlands, and the Canadian shield. All these are areas that since the Precambrian have had an epeirogenic type of geological history.

The crust of platforms and shields behaves somewhat like a badly cracked pavement on a poorly laid and shifting foundation. Widespread swells and sags, e.g. plateaus and basins, are produced by differential warping on a regional scale and are usually accompanied by marginal and interior faults. In turn ridges and troughs, e.g. block mountains and rift valleys, are produced by differential movements of the fault-margined blocks and strips into which the crust has been shattered. Whilst it is true that the floors of some basins and rift valleys have subsided far below sea-level (Figure 29.17), the cumulative effect of epeirogenic and related movements has been to elevate the greater part of the continental surfaces well above sea-level.

Plateaus are broad uplands of considerable elevation. Tibet, the Colorado plateau and the plateaus of Africa are outstanding examples. **Basins** are relatively depressed regions of roughly rounded or oval outlines. The term is used very widely and is applied to all broad sags of the crust, whatever the surface levels may be; from sea basins, like the Black Sea or the North Sea, to mountain-rimmed plateaus which are often, like the Tarim Basin of Asia, characterized by internal drainage. Ideally, the drainage from a plateau would be outwards and that of a basin inwards; but as many plateaus have locally dimpled or down-broken surfaces, and many basins such as the Congo and Sudan Basins have drainage exits through marginal depressions in the rims, this simple criterion is far from being of general application. The term 'basin' is also given to ancient crustal sags which have been filled with sediments and in some cases, as in Africa, subsequently uplifted into plateaus.

In the course of time the surface relief which results from epeirogenic movements becomes greatly modified by denudation and deposition, and locally by volcanic activity. Nevertheless, as illustrated by the examples already mentioned, the face of the Earth of today is diversified by many new or boldly preserved topographic features which are primarily due to differential vertical movements of the crust.

29.2 PLATEAUS

The mean heights and areas of a selection of plateaus are given in Table 29.1. Unlike the linear orogenic belts, they are often equidimensional (often being of equal breadth and length), and were not subject to folding during their uplift. Plateaus can of course be floored by folded rocks, but the folding took place many millions of years before the elevation of the plateaus.

A glance at the world map shows that plateaus such as the Tibetan Plateau, the Altiplano of Bolivia and the Colorado Plateau are rimmed by active fold belts and lie adjacent to destructive plate margins; it is difficult to escape the conclusion that their origin must be related to plate collision. Conversely, other plateaus, such as those of southern and eastern Africa, are distant from plate margins and unconnected with collision tectonics. Hence, it is convenient to have a subdivision into **orogenic plateaus** and **epeirogenic plateaus**.

29.3 OROGENIC PLATEAUS

Plateaus are geomorphological features that have come into being in the very recent past. In the following descriptions of the Tibetan and Colorado plateaus, some wider-ranging discussion is needed to place the plateaus in their geological context.

The Tibetan Plateau is the most extensive and highest plateau on Earth, with much of it at a height of more than 5000 m. To the north of the plateau are a series of major fault-bounded basins and mountain ranges (Figure 29.1). These trend east–west, as do most of the faults that are major wrench-faults. The rocks flooring the basins and exposed within these ranges are the eroded remains of a northerly-aging succession of orogenic belts wrapping round the southern margin of the Siberian shield. The belt closest to the shield

Table 29.1 Some major plateaus

	Mean elevation m	Area 1000 km²
German Alpine Foreland	500	35
French Massif Central	700	70
Anatolia Plateau	1000	500
Southern African Plateau	1500	2100
Tarim Basin	1100	600
Gobi Plateau	1100	1650
East African Plateau	1200	1000
Colorado Plateau	1800	500
Bolivian Altiplano	3800	350
Tibetan Plateau	4500	2000

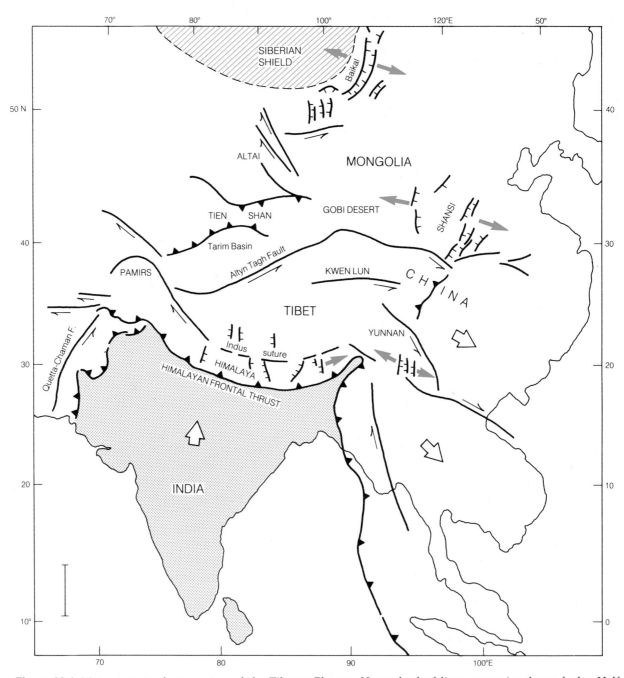

Figure 29.1 Major tectonic features around the Tibetan Plateau. Heavy barbed lines are major thrust faults. Half arrows indicate the relative movement along transverse faults. Large open arrows indicate movements of major crustal blocks relative to the Siberian Shield during the Tertiary. Brown arrows indicate direction of crustal extension; note the areas of crustal extension in the Himalayas and Tibet, normal to the line of the Himalayas, which are of Recent age, post-dating the main compression. (*Modified after Tapponier, 1986*)

is of mid-Palaeozoic age, whilst the Hercynian orogeny is well represented in the Altai and Tien Shan mountains; the folding of the Kwen Lun and Tsingling Shan continued into the Jurassic but these ranges, like the Tibetan Plateau, only achieved their present heights as a result of major uplifts which began in the Pliocene. Asia thus appears to have grown by successive additions to the Siberian nucleus. This appearance of outward

growth refers only to the times of major folding, because the terrains now represented within the intermontane basins and ranges are often much older than the age of folding. Also the warped and faulted surfaces of the depressions have received a veneer of desert or lake sediments, or clastics derived from the erosion of their mountainous rims. It must be emphasized that from Siberia to the Himalayas the present relief is primarily the result of vertical movements, the latest phase of which began in the late Tertiary and is still in progress. To the south of Tibet is the modern mountain range of the Himalayas; although it is the most impressive mountain range in the world, it is merely a fringe along the edge of the plateau.

The plateau, together with the Himalayas, owes its origin to isostatic uplift resulting from the collision between the Asian and Indian lithospheric plates. Peninsular India is a block of Precambrian lithosphere that, after the breakup of the ancient southern-hemisphere continent of Gondwanaland, migrated north-eastwards till it impacted on the southern margins of the Asian continent (Figure 28.18). The margins of the impacting Indian Plate are marked by the north–south Quetta–Chaman wrench fault in Baluchistan to the west and a similar fault in Burma to the east. Although some of the force of the impact between India and Asia was absorbed by crustal shortening, much of it was also taken up by major east–west shear faults, such as the Altyn Tagh Fault, running approximately at right angles to the main impact; as a result, large blocks of crust, bounded by these faults in China and South-East Asia, were displaced eastwards and south-eastwards (Figure 29.1). As a result of some of these easterly movements, certain areas, such as parts of the Himalayas, Yunnan and the Shansi depression, underwent east–west crustal tension that resulted in the formation of rift valleys. These rifts formed by compression-induced crustal tension, unlike the epeirogenic rift valleys that we shall discuss later in this chapter.

One major result of the India–Asia collision is that the crust beneath Tibet is very thick – of the order of 65–80 km. At one time it was believed that this was due to a doubling effect as a result of the Indian crust underthrusting the Asian crust during the collision. This older model has been revised in the light of more modern data, particularly as the result of geological traverses and geophysical surveys across the Plateau by Chinese, French and British scientists in the 1980s. Whilst it is still acknowledged that Indian crust exists beneath the Himalayas, the thick Tibetan crust is now believed to be due to shortening and doubling up of the Asian crust (Figure 29.2). Furthermore, the geophysical evidence is that there is plunging of cold lithosphere into the mantle beneath southern Tibet as the Indian plate slides beneath the Himalayas. Secondly there is isostatic buoyancy caused by upwelling of hot mantle beneath north-central Tibet (Figure 29.2c), resulting in modern extrusions of basalt. The crust under north-central Tibet must be thinner than elsewhere to provide the overall uniform isostatic uplift indicated by the uniform plateau height.

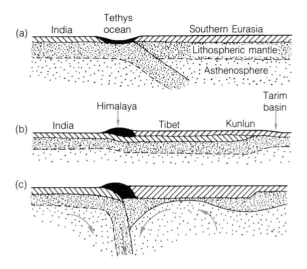

Figure 29.2 Diagrammatic sections across the Tibetan Plateau. (a) The relative positions of the Indian and Eurasian Plates, separated by the Tethys Ocean prior to collision. (b) A section to illustrate early ideas of the origin of the Himalayas and Tibet, with the Indian crust underthrusting Eurasia and causing crustal doubling. (c) A modern hypothesis showing thickening of the crust under Tibet by shortening of the Eurasian crust; the Indian crust penetrates only below the Himalayas. The mantle section of the lithosphere beneath Tibet is thinned by upwelling hot asthenosphere. (*After Molnar, 1988*)

The geophysical evidence is that the Moho is not flat and continuous beneath the Tibetan Plateau. The relatively uniform height of the Tibetan plateau thus masks this very important feature – the existence of relatively thin crust overlying relatively hot mantle beneath north-central Tibet.

29.4 THE COLORADO PLATEAU AND THE GREAT BASIN

The Colorado Plateau and the Great Basin lie within the great block of western North America known as the Cordillera (Figures 29.3 and 29.4). The Cordillera includes all the ranges and plateaus west of the Great Plains, and is a broad assemblage of mountain ranges belonging to orogenic belts of various ages, together with their associated plateaus and intermontane basins. The Cordillera consists of the **Coast Ranges** on its western side, the **Rocky Mountains** on the east, and in between are the Colorado Plateau and the **Great Basin**, the latter characterized by upstanding blocks, known as the **Basin Ranges**, rising to a height of over 3000 m; the Basin Ranges mainly strike north–south, though in Oregon they swing to strike east–west.

The Colorado Plateau has a magnificent variety

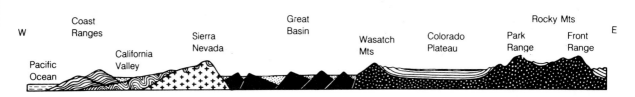

Figure 29.3 Schematic section across the Cordillera of the western United States. Length of section about 3000 km.

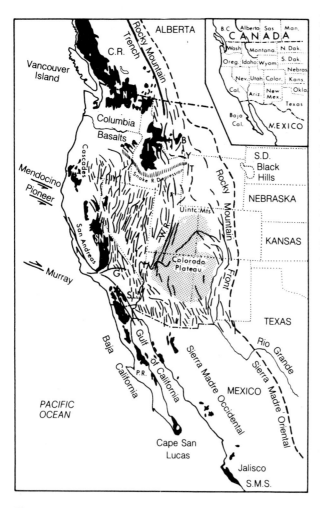

Figure 29.4 Sketch map to indicate some of the major tectonic features of the North American Cordillera. The main outcrops of granitic batholiths (black) from north to south are C.R., Coast Range, extending north-west to Alaska, where it swings round as the core of the Alaska Range (including Mt McKinley, Figs 31.1, 31.44a); N, Nelson; B, Boulder; I, Idaho; S.N., Sierra Nevada; P.R., Peninsular Ranges; and in part S.M.S., Sierra Madre del Sur. The chief faults of the Basin and Range Province (between the Colorado Plateau and the Sierra Nevada) are drawn as portrayed by J. Gilluly (1963) for the USA; similar faults continue into Mexico, but have not been mapped in detail. Other localities indicated by letters are: G, Garlock fault; S.L., Salton Lake; T, Teton Range; W.R., Wasatch Range; Y, Yellowstone Park.

of features of unrivalled interest, from the Grand Canyon, the natural bridges, and the Petrified Forest to the Henry Mountains and other famous clusters of laccoliths, the many volcanic fields, large and small, and to Meteor Crater, a witness to the impact of a genuine visitor from space. To these may be added salt-domes and uranium deposits and an endless list of other geological attractions. But the most intriguing problems are those that seem to be the simplest at first sight: the reason for the great height of the Plateau and the essentially unfolded condition of its sediments. Familiar illustrations have perhaps given the impression that the uplifted strata are everywhere flat-lying. There are, however, local interruptions caused by faulting or by **monoclines** (i.e. flexures in which beds are bent from one level to another) which take the place of faults. These structures can be thought of as small-scale relatives of the strongly developed 'basins and swells' of the encircling Laramide belts.

At the close of the Cretaceous the region was nearly at sea-level. Since then the average uplift in the south has been about 2500 m, and in the north about 1800 m, but as the Plateau is almost as extensive as the British Isles, the regional tilt is very gentle. The uplift occurred mainly in two pulses: one accompanying the great outbursts of volcanic activity around the borders and the intrusion of laccoliths in the interior (late Miocene to early Pliocene); the other, considerably greater, making possible the deep incision of the Grand Canyon (late Pliocene to present) and also accompanying scattered volcanic manifestations of crustal perforation similar to that of the Swabian tuffisite pipes and the volcanic fields around Ruwenzori (e.g. south of the Henry Mountains in the Najavo district, where the volcanic products are potash-rich; and in the Hopi Buttes, south of the Grand Canyon, where the products are soda-rich).

The Great Basin is, for the most part, an area of internal drainage made up of more than a hundred undrained troughs and basins lying between long block mountains which trend approximately north and south. In the extreme south-west of the area the drainage escapes into the Colorado River at a point which is now the site of the great Boulder Dam. The upstanding blocks, known as the Basin Ranges, rise to heights of 2000 to over 3000 m (Figure 29.5). Most of the ranges are tilted blocks,

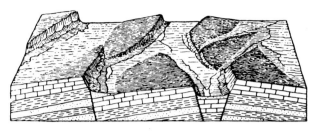

Figure 29.5 Diagram to illustrate the fault-block structure of the ranges of the Great Basin, Utah. (*After W.M. Davis*)

generally 18–24 miles across, with steep fault scarps on one side and gentler back slopes on the other. The bordering ranges of the Sierra Nevada and the Wasatch Range are gigantic tilted blocks of the same type, with their high fault scarps facing inwards in each case. The older scarps have been considerably eroded into ravines and spurs, and the lower slopes are commonly aproned with screes and fans of rock-waste. Both these and the spurs are truncated by triangular facets along the lower slopes of some of the ranges, showing that the latest fault movements have been quite recent. The severe earthquake of Owens Valley in 1872 was caused by a displacement of several feet along a normal fault at the eastern foot of the Sierra Nevada. This, and a few later dislocations, notably in 1954 and 1957, show that movements have not yet ceased.

The basins and troughs are mostly levelled up by sediments from the ranges, and in places by volcanic contributions, to heights that are usually between 1000 and 2000 m. Some are barren desert wastes, others support scanty vegetation, and a few contain lakes, mostly temporary and saline and surrounded by **salinas** or **playas**. Great Salt Lake, west-north-west of the Uinta Mountains, is the salt-saturated relic of a much larger freshwater lake, known to geologists as Lake Bonneville. The terraced shorelines of this ancestral lake make conspicuous horizontal features along the slopes of the many ranges that formed its margins or stood out as islands above its surface. The Precambrian floors of these depressions are generally far below sea-level, and even the present surface falls to exceptionally low levels in a few places along the western margin. Death Valley, for example, (Figures 22.8 and 22.10) is 85 m below sea-level, although the range only a few miles to the west attains a height of over 3000 m.

The Great Basin is only part of the 'Basin and Range' province. To the north this structural province extends beyond the Snake River depression, which subsided across the general trend of the Cordillera to make room for 15 000 m of lavas, mainly basalts, which were erupted during

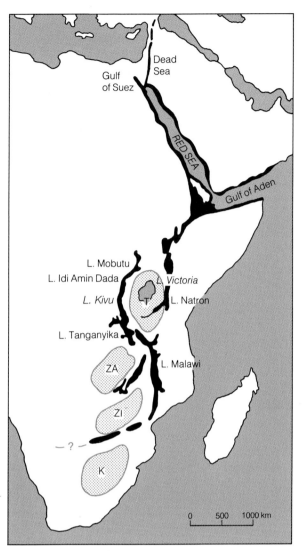

Figure 29.6 Sketch map of the Afro-Arabian rift valleys. Stippled areas are cratonic nuclei: T Tanzania craton; ZA Zambia Block; ZI Zimbabwe (Rhodesia) craton; K Kaapvaal craton.

Pliocene to recent times. Both this vast volcanic field and also that of the older Columbia basalts to the north-west have been similarly broken by faults into tilted blocks and graben-like depressions. This is another indication that the development of Basin and Range structures has continued through a long period of time. Stratigraphical evidence shows that in the type area of the Great Basin some of the faults began in late Oligocene times. The main uplifts, however, followed the voluminous Pliocene eruptions, which in the south and west of the Great Basin took the form of enormous flows of ignimbrites, the volume of which amounts to at least 35 000 cubic miles (140 000 km³).

Nearly all the faults are normal and require extension of the crust, mainly in an east–west direction, but far from exclusively so, as in places the faults trend at a variety of angles to the

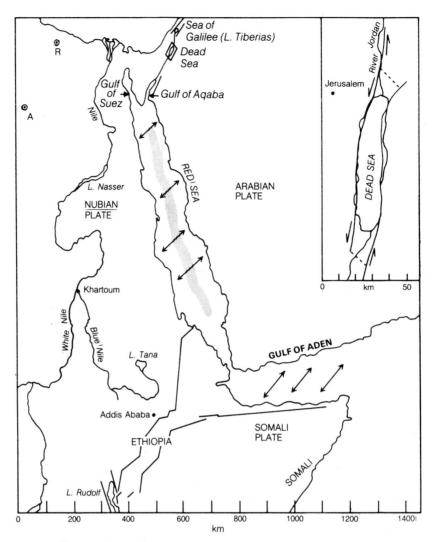

Figure 29.7 Diagrammatic illustration of the crustal separation of Africa and Arabia resulting from the opening of the Red Sea and the Gulf of Aden in the direction of the arrows. R and A are the poles of rotation active respectively during the opening of the Red Sea and the Gulf of Aden. The stippled ornament along the Red Sea axis represents an area of oceanic crust. (*After R.W. Girdler and B.M. Darracott, 1972*) The Red Sea rift continues southward into Ethiopia and northward between Israel and Jordan; the inset map illustrates the movements associated with the formation of the Dead Sea rift. (*After A. M. Quennell, 1958*)

dominant direction (Figure 29.4). A few of the faults are transcurrent, notably on each side of the Idaho batholith, which appears to have drifted northwards relative to the great Snake River depression. The latter is consequently interpreted as a tensional effect in which the crust was fissured to depths where basaltic magma was tapped and so given access to the surface, which consequently subsided with the underlying crust. As we have already seen, the structure of the Uinta Mountains also points to superficial extension in a north–south direction. For the Great Basin between the Wasatch Range and the Sierra Nevada, Eardley estimates that the faults imply an extension of 50 km during the last 15 m.y.

Some recent interpretations of the tectonics of western North America suggest that the extension seen in the Great Basin may be due to incipient or aborted back-arc spreading. Of major importance to other tectonic interpretations of the area is the fact that the North American plate has migrated westward and over-ridden the Mid-Pacific Rise and the former ocean basin and deep oceanic trench that lay between the Rise and the western margin of the North American continent (Section 31.12). The uplift of both the Colorado Plateau and the Great Basin are tied in with the regional uplift of the American West. The whole area is one of high heat flow and active magmatism, and the uplift could be due to thermal expansion resulting from either a deep mantle plume, or due to heat from the overridden mid-ocean ridge; in a regional context the latter is plausible. Another explanation is that there has been volume expan-

Figure 29.8 Map showing in a generalized way the tectonic basins of Africa and the intervening swells, plateaus and rift valleys.

is at the highest topographic level, and subsequent younger ones occur at successively lower levels. The overall disposition of the surfaces creates the strong impression that the whole subcontinent has been subject to long quiescent periods during which erosion has taken place interspersed with sudden episodes of vertical uplift. Although the erosion surfaces have been flexured to a limited extent, they have not been thrown into the same major upwarps that are cut across by the major, continental rift valley systems. The reasons for this continent-wide even uplift, that has affected this immense block of continental crust sporadically for at least the past 150 million years, are still unknown.

sion as a result of hydration of subducted lithosphere. This hypothesis is supported by the presence of xenoliths of serpentinized peridotite in the diatremes of the Colorado Plateau; other xenoliths of glaucophane schist indicate the presence of hydrated subducted oceanic crust beneath the Plateau. Two major unresolved problems are: why is the crust beneath the Colorado Plateau 45 km thick, compared with thicknesses of approximately 30 km for the surrounding terrain, including the Great Basin; and why has the Plateau survived as a structural entity whilst the surrounding country has been modified so extensively by active faulting and erosion?

29.5 EPEIROGENIC PLATEAUS AND RIFT VALLEYS

Present-day high plateaus that lie well away from plate collision margins must have arisen by processes unconnected with thickening of the crust by doubling up as a result of underthrusting or crustal shortening. Parts of the African continent provide classical examples, together with parts of Europe, eastern Siberia, and the Great Plains of the mid-west USA.

The South African Inland Plateau. Much of southern Africa is over 2000 m high although the Kalahari Basin drops below this level. A feature of this part of the continent is that it is cut across by a series of almost-horizontal erosion surfaces. These have been described already in Section 18.14, and Figures 18.43 and 18.44 are particularly relevant. The oldest erosion surface, the Gondwana surface,

Many continental areas of major uplift are cut across by elongate depressions known as rift valleys which form some of the larger intra-plate geomorphological features. The Afro-Arabian Rift system, for example, extends from northern Syria, via the Red Sea, East Africa and Malawi, into southern Africa, a distance of some 7000 km (Figure 29.6). Although the system does contain the Red Sea segment that is floored by oceanic crust, there is no doubt that it is part of the same overall fracture system. The term **rift valley** was introduced by J.W. Gregory for the 'Great Rift Valley' of East Africa, which he recognized as a tectonic feature due to faulting. Gregory defined the term to mean a long strip of country let down between normal faults – or a parallel series of step faults – as if a fractured arch had been pulled apart by tension so that the keystone dropped *en bloc* or in strips. The floors of some exceptionally deep rift valleys have obviously subsided (Figure 29.17) but in far more examples it is clear that they have merely lagged behind the surface of the adjoining plateaus during the course of a general uplift. The keystone analogy is a misleading one in that a keystone has an empty space into which it can fall until its inwards-sloping sides again find lateral support; a rift-valley wedge could not sink unless its weight displaced mobile material at depth, in which case volcanic activity might be expected on a far greater scale than is actually observed. In fact some of the deepest troughs, such as Lakes Tanganyika and Malawi, show no signs of volcanism.

Modern rift valleys are intimately associated with topographically-high crustal uplift areas that may be major domes or elongate, undulating upwarps. Characteristics of the rifts are:

1. A relatively narrow width of some 30 to 50 km,

irrespective of the length of the rift valley; the width is very close to the thickness of the rifted continental crust;

2. The rifts are generally bounded by major, opposed normal faults that are not necessarily continuous but often *en echelon*; the rift floor may contain numerous minor normal faults that parallel the main structure;

3. They are the loci of numerous shallow earthquakes;

4. Seismic, gravity and isostatic data indicate that they are underlain by anomalously light mantle that exists at relatively-shallow depths;

5. They are zones of relatively-high heat flow and, where cutting across the higher parts of undulating upwarps, are characterized by volcanic activity and hot springs;

6. The magma types are chemically distinctively different from those of collision zones and the oceanic spreading ridges; we shall return to this particular point later.

Most of these characteristics are consistent with crustal extension, thinning and rupture, and it is impossible to disassociate the major features of rifting and crustal upwarping. In many of the features listed above, the continental rift valleys are identical to the mid-ocean ridges, and a particularly relevant juxtaposition of oceanic and continental rift systems is at the southern end of the Red Sea. Here the Gulf of Aden and Red Sea spreading zones are in contact with the northern end of the Ethiopian sector of the East African Rift (Figure 29.7). The Arabian Plate, until very recently part of the African Plate, has moved north-eastwards away from Africa leaving the Gulf of Aden (the zone of maximum extension) occupied by the oceanic crust of the western extension of the Carlsberg Ridge; this relatively-new oceanic crust has typical magnetic stripes and the ridge is offset by transform faults. Extension across the Red Sea is not so great and there is only a thin zone of oceanic crust along the axial trough.

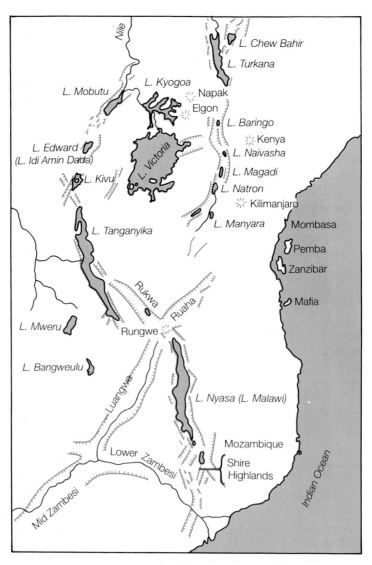

Figure 29.9 East Africa, showing localities mentioned in the text.

29.7 THE AFRICAN RIFT VALLEYS

For hundreds of millions of years, the movements of the African shield have been persistently epeirogenic, giving a structural pattern of broad basins separated by irregular swells which rise towards the east to form a coalescing series of plateaus, the latter being traversed by a spectacular system of rift valleys (Figures 29.8, 29.9). It is often considered that the rift valleys represent an incipient stage of the separation of continental crust, in this respect being as the Red Sea or the Gulf of Aden might have looked several million years ago. The rifts are characterized by a broad zone of negative gravity anomalies up to 1000 km wide on which is superposed a narrow zone of positive gravity anomalies, of the order of 40 to 80 km wide. It is important to realize, however, that no oceanic crust has been found within the rift valleys, and geological evidence shows that they are floored by continental rocks, largely of Precambrian age. The positive gravity anomalies can be accounted for by intrusions of basic and ultrabasic rocks.

The African rift valleys do not constitute one long continuous trough with a curving branch to the west. Some individual faults can be traced for long distances, but others are shorter and arranged *en echelon*. But looking at the East African rift faults as a whole, it is impossible not to recognize that they are all closely related parts of a single system of major tectonic features that extends from the Limpopo to the Red Sea, over a distance of about 3500 km.

The component rift valleys are usually grouped as follows:

1. The Lake Rudolf and Ethiopian section
2. The Eastern or Gregory Rift, east of Lake Victoria
3. The Western Rift from Lake Tanganyika, through Lake Kivu, Lake Idi Amin Dada (L. Edward) and Lake Mobutu (L. Albert) with the uplifted massif of Ruwenzori between the last two sections (Figures 29.15 and 29.16)
4. The Southern Rift, the Malawi (Nyasa) section and its bifurcations into the southern end of the Central Plateau.

There is a remarkable uniformity in the widths of most of the rift valleys, as indicated by the following measurements:

Lake Mobutu (Albert)	45 km
Lake Tanganyika (north)	50
Lake Tanganyika (south)	40
Rukwa	40
Lake Rudolf	55
Lake Natron	30–50

Figure 29.10 A 6 km stretch along the south-eastern fault scarp of the Albert (Mobutu) Rift, viewed from the air over the winding road that leads from the plateau to the lake port of Butiaba, Uganda. (*Overseas Geological Surveys*)

The Ethiopian Rift, although a very obvious zone of extension, has undergone relatively little spreading. These three spreading zones reflect differing responses to the north-easterly migration of the Arabian Plate, which itself reflects the geometry of an upwelling cell in the upper mantle. They are also a composite present-day picture of the different stages of early continental breakup, such as that believed to have happened during the breakup of Gondwanaland (see Section 27.13). In this context, the Ethiopian Rift Valley might be regarded as a future Red Sea or Gulf of Aden. Conversely, it may never develop into a major spreading zone and may end up as a 'failed' or 'aborted' rift, such as the Benue Trough of Nigeria or the North Sea.

Figure 29.11 The western escarpment of the Rift Valley between Lake Manyara and Lake Natron, Tanzania. The maximum height of the escarpment, which cuts across the lower, eastern flanks of the Ngorongoro volcano (partly covered in cloud), is 1500 metres. (*Photo J.B. Dawson*)

Ruaha (north)	40
Lake Malawi (Nyasa)	40–60

Rift valleys in other continents also have similar widths, e.g. the Rhine graben, 30–45 km, and Lake Baikal south 55, north 70 km (Figures 29.21 and 29.24). These widths are all of the same order as the thickness of the continental crust, a significant relationship also brought out in the scaled-down model experiments carried out by Cloos (Figure 29.32).

Some of the oldest scarps of the African rifts, such as those of the Ruaha Rift of southern Tanzania, or the Luangwa graben of Zambia, date from Karroo (Carboniferous) times. By comparison with the modern rift scarps that are sharp and steeply defined (Figures 29.10, 29.11), these older scarps are deeply dissected by erosion. This illustrates that the modern rift features, most of which are not older than 20 million years, are the result of a recent movement of more ancient fracture systems that traverse the African continent – some recent rift faults are clearly reactivated older fractures. A very obvious feature is that many of the recent rift fractures have been controlled by the 'grain' of older fold belts that themselves wrap around the ancient Archaean cratons (Figure 29.12). For example, the rift fractures in Ethiopia and Kenya follow the north–south trend of the Late-Precambrian/Cambrian Mozambique Fold Belt. Likewise, the western branch of the rift (the Ugandan, L. Tanganyika and L. Rukwa section) follows the Precambrian fold belts wrapping round the western side of the Tanzania Shield (Figure 29.12). Further south, the Luangwa Rift, and the Limpopo Rift lying between the Zimbabwe and Kaapvaal cratons, both follow the trend of Precambrian circum-cratonic fold belts. An interesting exception to this general rule is the Carboniferous-age rift fractures of southern Tanzania that cut across the grain of the Mozambique Belt. Their criss-cross pattern is reminiscent of the reticulate Mesozoic fracture patterns of Western Europe (Figure 29.36); these are also similar in that they are both the relatively-minor continental expressions of much larger, contemporaneous fracturing and spreading events, the one being the breakup of Gondwanaland, the other being the opening up of the North Atlantic.

Unlike the South African Inland Plateau referred to earlier, the plateaus of East Africa and Ethopia in particular are high upwarped areas. Since both areas are volcanically very active, it is relevant to question whether the present-day topographic elevation is genuinely due to upwarping and not just due to large piles of extruded lavas? The evidence comes from distortion of a formerly-horizontal, major geological datum plane. In East Africa, there is a widespread, deeply-lateritized, sub-Miocene erosion bevel. Contours on this ero-

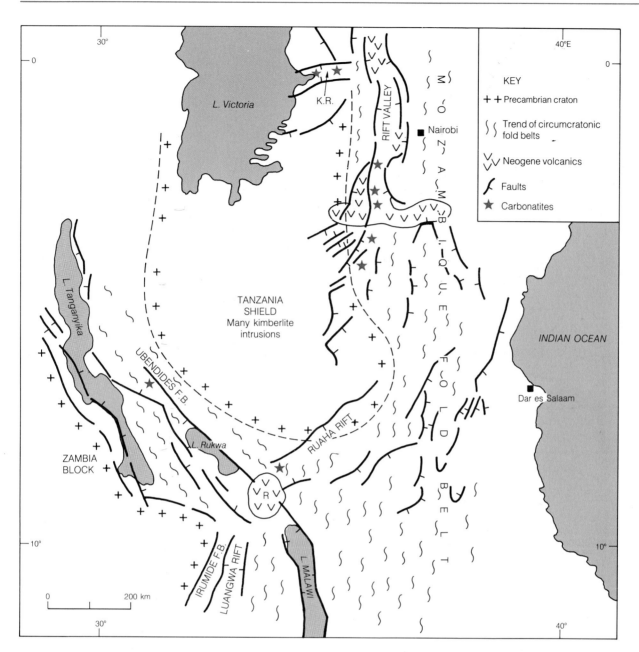

Figure 29.12 Relationship of the rift structures in Tanzania and Zambia to the Tanzania Shield, and to the Mozambique, Irumide and Ubendides fold belts. The Rungwe Volcanics (R) lie at the intersection of the three rift systems. Note the location of carbonatites (stars) within the rift valleys (see also Fig. 29.25). Abbreviations: K.R. – Kavirondo Rift; R – Rungwe Volcanics; F.B. – fold belt.

sion surface show that it rises from the coastal belt, where it is buried by late-Tertiary and Holocene sediments, towards the Kenya Highlands where it attains a height of nearly 2000 m on the shoulders of the Rift (Figure 29.13). Many major volcanoes, including Mount Kenya, have been extruded upon this upwarped surface. The contours depict an inequidimensional dome, elongated along a north–south axis, that decreases in height towards Tanzania in the south and towards Lake Turkana in the north; a similar dome rises to the north of Lake Turkana to form

the base of the volcanic highlands of Ethiopia. Similarly, a linear, undulating major upwarp delineates the Baikal Rift zone of Siberia (Figure 29.24).

Turning next to the geometry of the faulting, there was considerable debate during the earlier part of this century as to whether the faulting was normal, due to extensional forces, or reversed due to compression. The evidence in favour of reversed faulting has been shown to be invalid, and there is now ample evidence that the dominant type of faulting associated with the East

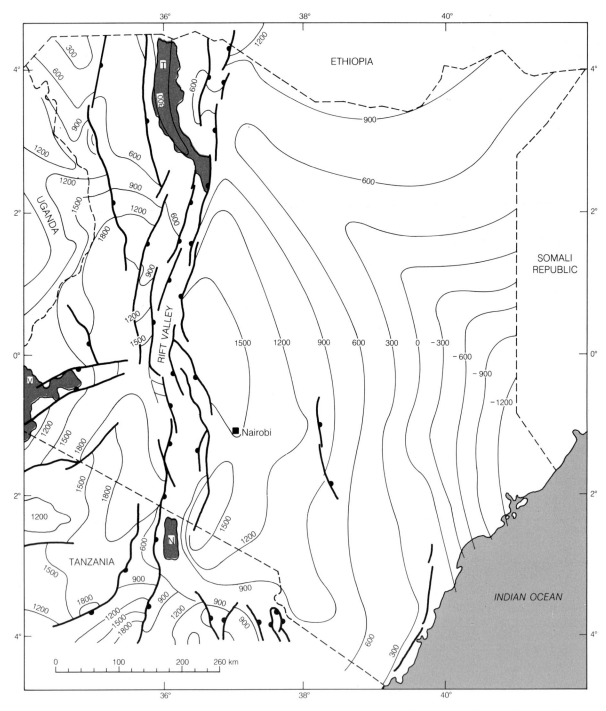

Figure 29.13 Contours (in metres) on the present-day elevation of the sub-Miocene erosion surface in Kenya and northern Tanzania. Abbreviations: T – Lake Turkana; N – Lake Natron; V – Lake Victoria. (*From Baker et al, 1972*)

African rifts is normal. The north-west side of the Mobutu Rift, for example, is a normal fault (Figure 29.14). On the south-eastern side of the Mobutu Rift deep boreholes have been sunk through gently folded sediments at various distances from the foot of the scarp. As the distance increases so does the depth at which the fault and crystalline basement rocks are encountered, the range being from 130 m to 1233 m. The south-eastern fault is thus proved to be a normal fault dipping inwards towards the lake at 65°. Further south, where the depressions around Ruwenzori begin to converge to form the Mobutu Rift, the 'nose' of Ruwenzori (Figure 29.15) has also been shown to be bordered by normal faults both by visible geological evidence and by the results of borings.

If confirmation is required of the important part played by uplift associated with rift-formation,

Figure 29.14 Normal fault between Miocene beds (on the right) and Precambrian basement rocks; exposed on the northern bank of a tributary flowing into the Semliki River near its outlet from the southern end of Lake Mobutu. The tributary has cut through one of the main boundary faults on the Zaire side of the Rift. (*J. Lepersonne*)

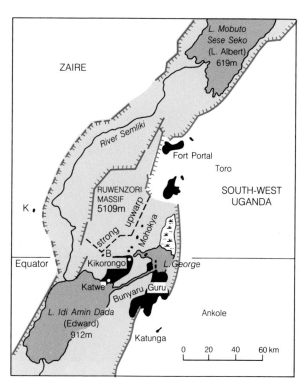

Figure 29.15 Map of part of the Western Rift Valley, showing the massif of Ruwenzori and the adjoining volcanic areas (black). Two small areas are marked B (Bukangara) and K (Karibumba).

there is the astonishing feature of the Western Rift – unique in its immensity – the towering horst of Ruwenzori, which is by far the highest non-volcanic mountain in Africa. It rises from within a bifurcation of the rift valley (Figure 29.15) to snow-clad peaks (the 'Mountains of the Moon') up to 5119 m high, i.e. to nearly 4 km above the general level of the plateau (Figure 29.16). Towards Lake Mobutu (L. Albert) the great massif narrows to a long 'nose' flanked by fault scarps. Here recent uplift is proved by the occurrence of raised terraces of alluvium which were originally part of the valley floor of the Semliki River.

Another effect of rifting has been to form major elongate drainage basins, some immensely deep (Figure 29.17). Some, such as L. Turkana, L. Tanganyika and L. Malawi, have rivers flowing out of them and hence remain fresh. Others, however, have become internal drainage basins and developed into major evaporating basins in which there are major deposits of sodic evaporites; examples are L. Magadi in southern Kenya, and Lakes Natron, Manyara, Balangida and Eyasi in northern Tanzania (Figure 29.9).

29.8   THE RED SEA AND THE GULF OF ADEN

The **Red Sea**, as illustrated by Figure 29.18, has an axial trough about 50 or 60 km across at its widest part, margined by shallow shelves. Gravity anomalies over the axial trough are strongly positive, and in 1958 were interpreted by Girdler as indicating the presence of a mass of basic rock beneath the sea floor. The presence of such heavy rock, approximately 2 km below the sea floor was confirmed by seismic surveys, and found to be overlain by consolidated sediment and/or lavas, surfaced with loose sediments (Drake and Girdler, 1964). Magnetic surveys, whereby magnetic anomalies were located over the central trough, further confirmed the presence of basic rock. The linear pattern of the magnetic anomalies, resembling that which characterizes the ocean floors, led to the interpretation of the basic rock beneath the Red Sea as new oceanic crust, emplaced during the separation of the continental shores of the Sea (Figure 29.7).

In 1966 Vine calculated the spreading rate across the Red Sea, from data recorded by Drake and Girdler (1964), and Allan *et al.* (1964). He found it to be 1 cm a year, away from the median line, during the last 3 or 4 million years. Subsequently a

Figure 29.16 An exceptionally clear view of the uplifted massif of Ruwenzori, looking across Lake George 913 m from Mweya Lodge towards the summit of the range 5119 m. From the southern end to the 'nose' on the right the distance is about 80 km. (*Paul H. Temple*)

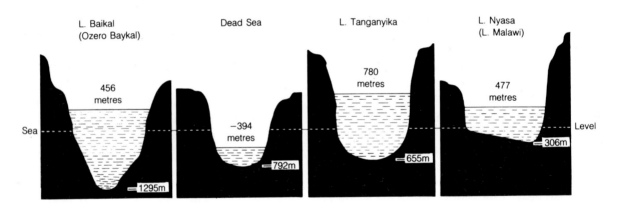

Figure 29.17 Sections through rift-valley lakes to illustrate local depression of the floor below sea-level. Vertical scale greatly exaggerated. Each section is 80 km across.

more detailed assessment by Allen and Morelli (1969) indicated a spreading rate of 1.1 cm a year away from the median line. That is, the shores of the Red Sea have separated at a rate of 2.2 cm a year.

A further seismic refraction survey by Davies and Tramonti (1970) suggests that the basaltic ocean floor may be wider than the median trough, but as yet the exact boundary between continental and oceanic crust has not been defined. No transform faults have so far been mapped, although apparent offsets of the magnetic anomalies suggest that they exist.

The **Gulf of Aden** has been found by seismic refraction investigations to be underlain by oceanic crust; a branch from the Carlsberg mid-oceanic ridge extending along its median zone. Accordingly the median zone, like the mid-Atlantic ridge, has a rough topographic surface. The median zone is intersected at high angles by scarps that are not continued through the bordering continental crust (Laughton, 1966). Epicentres are located along the rough median zone, and extend into the axial trough of the Red Sea. Seismic solutions for two earthquakes whose epicentres were located on scarps in the Gulf of Aden, show the directions of movement to be consistent with those to be expected along transform faults.

The rough median zone is characterized by 'striped' magnetic anomalies, and the spreading rate they depict, outward from the median line of the Gulf, varies from 0.9 to 1.1 cm a year. That is the Gulf of Aden is opening at the rate of 1.8 to 2.2 cm a year.

29.9 THE DEAD SEA

Unlike the Gulf of Aden and the Red Sea in which the continental crust has pulled apart, the **Dead Sea**–Jordan Valley sector of the rift system is mainly characterized by strike-slip motion, with the edge of the Arabian Plate sliding northwards relative to the crust on the western side of the depression. The Dead Sea occupies part of the depression, with mountains rising about 1000 m on either side. The structure was recognized in 1958 by A.M. Quennell on the basis of the relative positions of distinctive rock types, structural features and erosion surfaces on opposite sides of the rift. One clear piece of evidence for the sideways movement is the absence of deltas at the mouths of the Wadi Zarga and the Wadi Mujib, but the presence of a delta with no sediment source in the form of the El Lisan peninsula (Figure 29.19b). Assuming that the El Lisan deposits resulted from sediment transported down the two wadis in the Pleistocene, when the climate was less dessicated than at present, Quennell inferred late-Pleistocene horizontal northerly displacement of the wadis by about 45 km. The overall displacement along the rift is about 107 km, resulting from two major movements – one of 62 km in the early Miocene to early Pliocene, followed by the 45 km late-Pleistocene movement. Another important observation by Quennell was that the major fractures of the Dead Sea–Jordan rift were sinuous or sharply deflected, so that the transcurrent movements resulted in increasing separation of the crustal blocks to form a deep crustal chasm (Figure 29.19a). This type of structure is known as a **rhombochasm**. The Dead Sea rhombochasm has

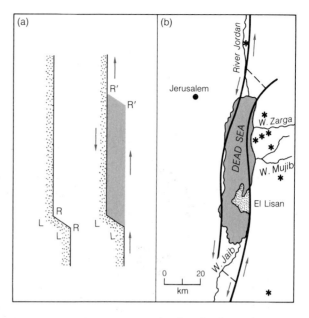

Figure 29.19 (a) Formation of a rhombochasm by transverse faulting. Points R′ R′ have moved from RR, leaving the rhomb-shaped gap. (b) The Dead Sea basin; compare with (a). Stars – volcanic centres.

not been filled by sea-floor material or ascending basalt, but was filled at depth by upward plastic flow of dense material from the upper mantle or lower crust. Much sediment has been washed into the depression (Figure 29.20), in which there are also thick evaporite deposits since the basin is one of internal drainage and high evaporation. There is only a small amount of basaltic activity, mainly in the Syrian sector of the rift, though there are a few volcanic centres on the plateau to the east of the Dead Sea (Figure 29.19b).

29.10 THE RHINE GRABEN

The Rhine graben is part of a fracture system that mainly affected the continental edge of north-western Europe during the spreading associated with the onset of the formation of the North Atlantic in the Triassic (Figure 29.21). The structure is Y-shaped, with the Lower Rhine graben running north-west to link into the North Sea fracture system (Figure 29.36). The Hesse depression and the Upper Rhine graben are controlled by basement structures that follow a coherent zone of basement weakness, originating in the Hercynian, that traverses the basement of western Europe from southern Norway to southern France (the Mediterranean–Mosja Zone). Of the three branches of the structure, the Upper Rhine graben has attracted most attention. To the west and east, the crystalline massifs of the Vosges and the Black Forest have been upwarped above the floor of the

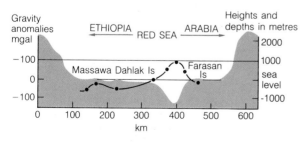

Figure 29.18 Section across the Red Sea to show the relation between the bottom topography and the gravity anomalies (broken line with actual determinations marked by dots). (*After R. W. Girdler, 1958*)

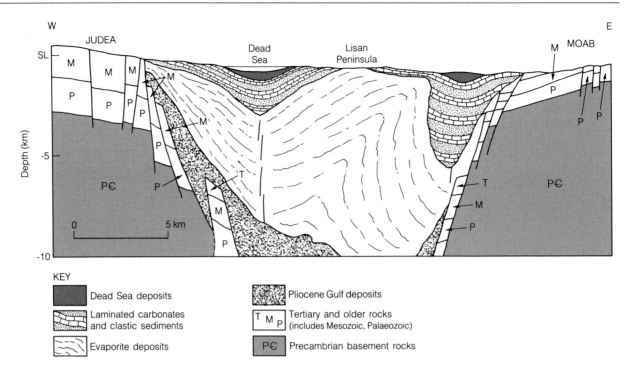

Figure 29.20 Geologic section across the Dead Sea, showing thick sediment fill since the end of the Tertiary. (*After Zak and Freund, 1980*)

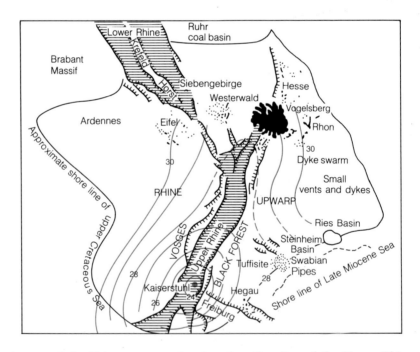

Figure 29.21 Sketch map of the Rhine upwarp, showing the rift valley of the Upper Rhine, and its northerly bifurcation into the rifts of the Lower Rhine and Hesse. Volcanic areas are in black. The contours are depths to the Moho. Horizontal ornament shows the distribution of the Triassic and Jurassic marine sediments.

rift, the regional stratigraphy showing that an updomed area had already emerged above sealevel in the late-Cretaceous. This has been attributed to the upwelling of a mantle plume that has come closest to the surface in the south of the graben below the Miocene Kaiserstuhl volcano (Figure 29.21).

The first graben subsidence took place during the Eocene in the south on the top of the dome, and the rifting and centre of sediment deposition migrated northwards during the upper Eocene and Oligocene. The result is a classical graben structure (Figure 29.22). A further phase of fracturing in the Pliocene was controlled by sinistral

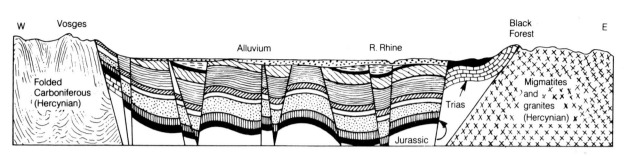

Figure 29.22 Section across the Rhine graben, north of Mülhausen, showing the folded and faulted structure of the strata at the top of the rift-valley block as determined by numerous borings. Miocene beds appear locally beneath the alluvium, but most of the strata shown here are of Oligocene age, resting on Jurassic. Length of section about 48 km. Width of the rift valley at the surface 32 km.

shear motion along the Hercynian basement fracture zone. This coincided with the last major movement in the Alpine orogenic belt just to the south: in this respect the basement shear movement may be analogous to the extensional shears of the Tibetan Plateau and western China discussed earlier. Hence the Rhine graben results from a combination of crustal stresses due both to mantle upwelling, and to reactivation of old basement structures by orogenic compression.

29.11   THE BAIKAL RIFT VALLEY, SIBERIA

The Baikal Rift (Figure 29.23) has many features in common with the East African Rift system. To the north of Baikal is the Precambrian nucleus of the Siberian or Anabar shield which is covered with almost horizontal lower Palaeozoic sediments. Wrapping around the shield on its southern side are the Yablonovoi Ranges, the eroded remnants of a fold belt of mid-Palaeozoic age that has a complex contact with the undisturbed sediments

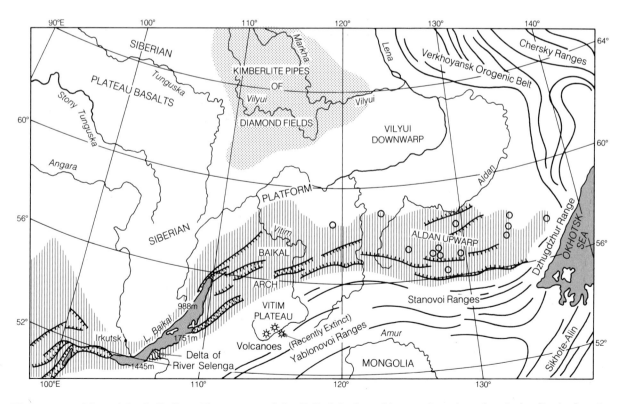

Figure 29.23 Map of the Baikalian rift system and the Baikal Arch and its continuations (vertical ruling), showing their relationship to the Caledonian folds to the south and to the Siberian Platform to the north with its diamond fields (dotted) and vast areas of plateau basalts. The figures attached to Lake Baikal refer to the maximum depths, in metres, of water in the three depressions occupied by the lake.

overlying the shield. The Sayan, Baikal and Aldan upwarps largely follow the trend of the contact. These upwarps, that have been delineated on the basis of an upwarped Tertiary erosion surface (as in East Africa) (Figure 29.24), have been rifted and one of the rifts contains Lake Baikal. The lake is distinguished for being the world's deepest lake (1750 m – see Figure 29.17) and also the largest body of fresh water. The section of the rift submerged by Lake Baikal occupies a topographic low between the Sayan and Baikal upwarps, in this respect resembling Lake Turkana that lies in the lowlands between the Kenya and Ethiopian Domes. Volcanic activity has taken place very recently on the Baikal and Sayan upwarps, and older rift-related activity occurs on the Aldan Upwarp.

29.12 RIFT VALLEY MAGMATISM

Although some quite extensive sections of rift valleys, such as the Lake Tanganyika and Lake Malawi sectors of the East African Rift and the immediate Lake Baikal area of the Baikal rift, show little or no magmatic activity, rift valleys in general are the sites of often very extensive igneous activity. This is due to the crustal extension that is required to give freedom of movement to the rift blocks and the associated fracturing through the whole crust. Indeed the uplift, extension, fracturing of the crust, and ascent of volatiles and magma are related to the development of magma sources in the underlying mantle. Magmatism is not confined to the rifts themselves, being also found on the shoulders of the structures; to be more correct, the overall magmatism is connected with the upwarp upon which the rifting is imposed. For example, the Eifel and Swabian volcanics of Germany (Figure 29.21), Mts Elgon and Kenya in East Africa (Figure 29.9), and the Vitim Plateau volcanoes of the Baikal Arch (Figure 29.23), all lie on the shoulders of their respective rifts.

Regarding the igneous rock types found within the rift valleys, the reader may find it helpful to refer to page 697 where the more unusual rock types are classified. A characteristic of the rift valley igneous rocks is that they contain unusually high concentrations of the alkalies Na and K, and also volatiles such as CO_2, F and Cl. The widespread olivine basalts are rich in alkalies compared with those found on the mid-oceanic ridges or in fold mountain belts, and fractionate to hawaiites, peralkaline trachytes and phonolites rather than dacites and rhyolites. Rhyolites do occur in the rift

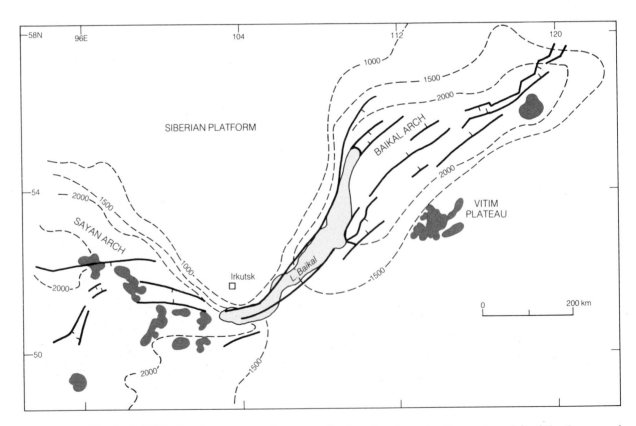

Figure 29.24 The Baikal Rift showing contours in metres (broken lines) on the Cenozoic uplift of the Sayan and Baikal Arches. Recent volcanic areas are shown in black. For clarity not all faults on the arches are shown. (*Modified after Logatchev, 1978*)

valleys but they are the unusual varieties known as **comendites** and **pantellerites** that are rich in alkalies, F, Cl and boron; moreover, on isotopic evidence, the ones in the Naivasha area of Kenya are believed to contain some component derived from fusion of lower-crustal rocks, rather than being the simple fractionation products of a parental alkali basalt. At the other end of the silica-saturation scale are olivine melilitites and olivine nephelinites – rocks so poor in SiO_2 that they contain nepheline ($NaAlSiO_4$) and melilite ($Ca_2Al_2SiO_7$) rather than plagioclase feldspar. Another rock type that is commonly found upon major upwarped areas and also within the rifts, is the magmatic carbonate rock **carbonatite**; this rock type, which is found in association with undersaturated silicate rocks, represents high degrees of volatile concentration. A good example of a carbonatite complex is the dissected volcano Napak (Figure 29.26) that lies in eastern Uganda on the upwarp between Lake Victoria and the Gregory Rift Valley; a calcitic carbonatite central plug, together with surrounding ijolite (nepheline-clinopyroxene rock) and nepheline syenite, forms an annular sub-volcanic complex beneath a partially-eroded pyroclastic volcano of nephelinitic and phonolitic composition; two points are worthy of note – the nephelinite and the phonolite are the extrusive equivalents of the plutonic ijolite and syenite (the nephelinite is olivine-free), and the pyroclastic extrusives testify to highly-explosive eruptions of volatile-rich magma. Carbonatite lavas and tuffs have been found at a small number of eruptive centres. Most, such as those of the Kaiserstuhl (Figure 29.21) and the Fort Portal volcanic field in the foothills of Ruwenzori (Figure 29.15) are calcitic. Until very recently these extrusions of calcium carbonate were an enigma, because it was believed that, at the temperatures required to form molten melts at atmospheric pressures, calcium carbonates would dissociate. However, recent high-temperature experiments, drawing on the observation that fluorite and fluorapatite are common minerals in carbonatites, have shown that, by the addition of fluorine to the experimental system, calcite is stabilized. The modern highly-alkali carbonate lavas from the active volcano Oldoinyo Lengai in the rift valley of northern Tanzania (Figure 29.25 and Plate IXf) differ from calcite carbonatites in having a high

Figure 29.25 The carbonatite volcano Oldoinyo Lengai (3000 m) towers above, and close to, the western boundary escarpment of the Gregory Rift Valley, 20 km south of Lake Natron. Its shape and radial erosion are typical of a highly-explosive, pyroclastic volcano. The volcanic carapace of Napak (Fig. 29.26) would, before erosion, have looked like Oldoinyo Lengai (Plate IXf). (*Photo J. B. Dawson*)

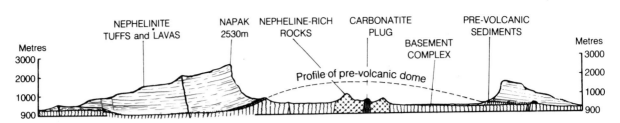

Figure 29.26 Generalized section across the deeply eviscerated Napak volcano, southern Karamoja, Uganda, showing the core of carbonatite (black). Length of section 30 km. (*After B. C. King, 1949*)

alkali content; they contain 33% Na_2O, 14% CaO, 8% K_2O, 35% CO_2, 2% SO_3, 3% F and 3% Cl, and represent the ultimate in alkali and volatile concentration. Their unusual chemistry is reflected in the very low eruption temperature of 585°C and the lowest viscosities yet measured for terrestrial lavas. A major debate revolves around whether lavas of this composition are unique to Oldoinyo Lengai or whether they previously existed at other extrusive centres but, being largely water-soluble, have been leached away.

It is clear that, within a given swell structure or rift system, magmatic activity can vary in type both in time and location. Examples from the Rhine and African rifts will suffice. The onset of the Rhine uplift was accompanied by eruptions of volatile-rich olivine-melilitite that form the famous tuff pipes of the Swabia area on the south-east side of the swell. Within the rift igneous activity comprises the late-Tertiary Kaiserstuhl carbonatite–nephelinite volcano in the south and the alkali basalt volcanoes of the Hesse depression in the north. The most recent Holocene activity is in the rift in the Eifel district where the rocks are basalts, nephelinites and phonolites.

The activity in the East African Rift is complex (Baker *et al*, 1972). The oldest Tertiary activity was the extrusion of thick series of fissure basalts in Ethiopia and Somalia during the Eocene and Oligocene, followed by extrusions of more basalts from shield volcanoes in the Miocene. Basaltic and silicic activity within the Ethiopian rift, which formed in the mid-Miocene, have continued to the present day. In Kenya, there was no activity before the early Miocene when large nephelinitic volcanoes erupted along the Kenya–Uganda border, at the same time as huge extrusions of basalt in the pre-rift Turkana depression. This was followed by extrusion of flood phonolites from the crest of the Kenya Dome. Following the first faulting of the rift in the early-Pliocene, volcanic activity was largely confined to the rift floor and its margins, though Pliocene fissure basalts trachytes and phonolites, erupting within the rift, locally spilled over onto the eastern plateau. Numerous central volcanoes (some with big calderas) erupted in the rift during the Pliocene; most were basaltic or trachytic, but

nephelinite–carbonatite volcanoes erupted in the Kavirondo rift valley (Figure 29.12), and the phonolitic shield volcano of Mt Kenya built up to 75 km east of the rift. The latest Pleistocene activity within the rift was of fissure trachyte, while basaltic activity continued away from the rift. In north Tanzania, the earliest (Pliocene) eruptions were from major basalt shield volcanoes; some have major collapse calderas, of which Ngorongoro, the world-famous wildlife reserve, is the largest. Kilimanjaro, the highest mountain in Africa is, in fact, the coalescence of three basaltic shields (Shira, Mawenzi and Kibo) topped by the trachyte dome of Kibo.

Following a major phase of Holocene (1.2 m.y.) rifting, which has given rise to the present-day asymmetric form of the Tanzania section of the Gregory Rift, there was another phase of activity dominated by nephelinitic/carbonatitic volcanism, and a chain of carbonatite volcanoes, including the active volcano Oldoinyo Lengai, extends southwards from the Kenya border (Figure 29.12). This phase of activity, in its limited volume, highly explosive nature and magma type, contrasts with the earlier major quiet effusions of basalt, and with the contemporaneous trachytic activity in Kenya to the north.

In the Western Rift Valley of East Africa, Pleistocene volcanic fields occur in the southern and eastern foothills of Ruwenzori (Figure 29.15). The activity here was very explosive, the whole area is deeply cratered, and lava flows are few in number but highly unusual. Calcium carbonatite lavas occur in the Fort Portal area, and further south in the Bunyuraguru and Katunga areas are flows of highly potassic ultrabasic composition. They contain the minerals leucite ($KAlSi_2O_6$) and kalsilite ($KAlSiO_4$) that are the silica-poor analogues of potassium feldspar; kalsilite is also the potassic equivalent of nepheline. Basically these lavas, known as **ugandite, mafurite** and **katungite** are the potassic equivalents of olivine nephelinites and melilitites. The major point to be noted is that there is a complete absence of any **basaltic** activity. Further south, to the north of Lake Kivu, is the Birunga volcanic field (Figure 29.27) in which there are eight major volcanoes (six extinct) and

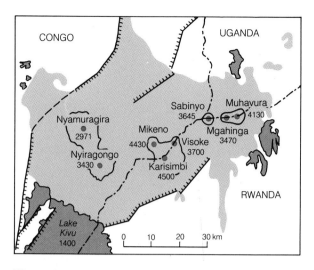

Figure 29.27 Sketch map of the eight major volcanoes (heights in metres) of the Birunga volcanic field. The Uganda part of the field is known as Bufumbira.

containing leucite and/or potash feldspars. The two active volcanoes illustrate this difference quite clearly. The 'leucite basalt' is typical of Nyamuragira and its satellites, whereas the lavas of the other active volcano Nyiragongo, only 16 km distant, are astonishingly different, consisting almost entirely of leucite, nepheline, kalsilite and melilite. To the south of Lake Kivu is another volcanic field in which the lavas consist largely of olivine basalts, together with minor flows of trachyte, and a few rhyolitic pyroclastic units. Thus, the various volcanic fields of the Western Rift show remarkable variety over a relatively small geographic area. They also highlight another major regional variation – that of the dominance of potassic magmatism in the Western Rift as opposed to the dominantly sodic magma types of the Gregory Rift.

Figure 29.28 Aerial view of Lake Magadi, southern Kenya, an inland drainage basin on the floor of the Rift Valley. Except for a few dark-coloured brackish lagoons, the surface of the hypersaline lake is covered with an evaporite crust of white trona (hydrated sodium carbonate). The elongate shape of the lake, its peninsulas and island are controlled by tightly-spaced minor horsts and grabens on the floor of the Rift Valley. The length of the lake is approximately 22 km. (*Photo J. B. Dawson*)

numerous smaller vents and cones. Some of the volcanic rocks are similar to those around Ruwenzori but there are also many flows of plagioclase-bearing lava which differ from olivine basalt in

Hot springs. The high heat flow along the rifts ensures that circulating groundwater is heated up before it comes to the surface, and most rifts have numerous hot springs along their length. Some

Figure 29.29 Trona dredge, Lake Magadi, Kenya. The floating dredge mines the floating crust of trona which is being continuously renewed by evaporation of the hypersaline waters. The trona is converted into soda ash, which is used as a flux in the glass industry. (*Photo J. B. Dawson*)

springs contain high amounts of dissolved gases; those rich in CO_2 give rise to the sparkling mineral waters for which the spas of the Rhine valley have long been famous, while hot springs in the Tanzania rift contain large amounts of the rare gas helium. Magmatic heat is a very valuable source of energy and geothermal power stations have been functioning for some time in volcanic areas in Italy, New Zealand and the western USA. More recently, in the Naivasha area of the Kenya Rift, the heated groundwater has been tapped as a source of geothermal energy, and electricity generated at Naivasha is carried by power lines to the nearby city of Nairobi. Further south in the rift, hot springs discharging into the rift valley lakes of Natron and Magadi carry within them large quantities of sodium, carbon dioxide, chlorine and fluorine leached from the mainly-volcanic rocks through which the heated groundwater has been circulating. The evaporating waters of the lakes then precipitate commercial deposits of **trona** (hydrated sodium carbonate) and halite (Figures 29.28 and 29.29). Like the salts of the analogous Dead Sea, they are perhaps unique amongst the world's exploited mineral deposits in that they are being constantly replenished.

Finally, we must mention **kimberlites**. Although not characteristic of rift valleys, they are,

nonetheless, associated with epochs of epeirogenic uplift of the ancient continental cratonic nuclei around which the rifts run; examples of areas of kimberlite intrusion are the Kaapvaal craton of southern Africa, the Tanzania Shield (Figure 29.12), and the Siberian Platform (Figure 29.23).

29.13 HOW AND WHY DO RIFTS OCCUR?

We have seen earlier that rifts have major opposed normal faults with numerous smaller faults between the main fractures; this is shown well by a section, based on geological observation, geophysical surveys and drilling, across the Rhine graben (Figure 29.22).

The earliest attempts to replicate rift structures were made in 1927 by S. Taber who floated wooden blocks in water (Figure 29.30); the central wedge settled deeper into the water with subsequent back-tilting of the marginal blocks. H. Cloos later devised some ingenious experiments in which updoming was simulated by the slow upheaving of the swelling surface of an inflating rubber hot-water bottle coated with moist clay. The pattern of miniature faults produced in the clay by the uplift and extension (Figures 29.31 and

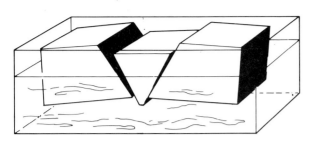

Figure 29.30 Wooden blocks (cut through as shown) floating in water; to illustrate the formation of a rift valley bounded by inward-sloping normal faults, and the consequent tilting of the marginal blocks. (*After S. Taber, 1927*)

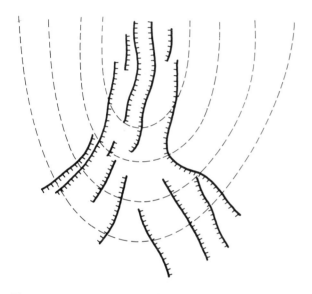

Figure 29.31 The bifurcating pattern of rift faults developed towards the end of an elongated dome or upwarp, here produced by the swelling of an oval-shaped hot-water bottle coated with moist clay. (*After Hans Cloos*)

29.32) faithfully reproduce the splaying-out pattern that is developed at each end of the Rhine graben (Figure 29.21). Similar patterns are seen at the northern end of the Red Sea (Figure 29.7), in the North Sea (Figure 29.36), at the northern end of Lake Malawi (Figure 29.9) and in other sections of the African rift valleys. Many models for rift systems have major opposed normal boundary faults, with numerous minor normal faults on the graben floor running roughly parallel to the main structure. Although it is acknowledged that the boundary faults may be discontinuous, the elongate, continuous-width pattern of many rift valley lakes has perhaps contributed to the concept of the simple graben structure. An entirely new facet of rift structure has resulted from detailed bathymetric and seismological studies on Lake Tanganyika during the 1980s by B.R. Rosendahl and his co-workers. Although it had been known for a long time that the depth of the lake is not consistent along its length, the depression has now proved to be, not a simple graben, but a series of alternate opposite-facing asymmetric basins, separated by high ridges of basement rocks (Figure 29.33). They are, in essence, a chain of linked half-grabens. A simplified model for the individual basins is that they are bounded on one side by a major fault whilst the opposite side is made of a downwarp, minor faults or a combination of the two (Figure 29.34); the area of maximum sediment deposition in each basin – the 'depocentre' – is close under the major border fault. The interbasin ridges, across which the sinuous major boundary faults change their throw (Figure 29.35), are termed 'accommodation zones' and are highly

Figure 29.32 Experimental production of a rift valley by slow upheaval of layers of moist clay. Note that the surface width of the rift valley is of the same order as the total thickness of the clay 'crust'. (*Hans Cloos, 1939*)

faulted. Similar structures have recently been found in Lake Malawi and Lake Turkana, and also in the Gulf of Suez, and it is very possible that Lake Baikal, with its three basins, may have a similar structure.

The experiments of Cloos showed that a substantial swell can produce the required fracture pattern of many rift valley structures. Although it is now generally recognized that upwelling mantle cells cause active rifting and crustal separation, what causes the broad swell structures is not so clear, though they must be mantle-controlled. J.H. Illies suggested that a broad 'cushion' of low-density mantle exists beneath the Rhine swell. Gravity and seismic evidence point to the presence of 'anomalous' mantle beneath the Kenya Dome. Whether due to the presence of low-density mantle, dyke injection into normal mantle, or the occurrence of small amounts of intergranular melt, is not geophysically resolvable. Whereas geophysical methods are intrinsically indirect, xenoliths of assumed upper mantle material are brought to the surface in certain types of volcanic rock. These latter tend to be the relatively rare alkali basalts, nephelinites, melilitites and kim-

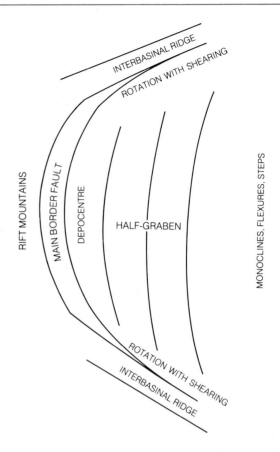

Figure 29.34 Main characteristics of the individual basins of Lake Tanganyika. (*From Rosendahl et al, 1986*)

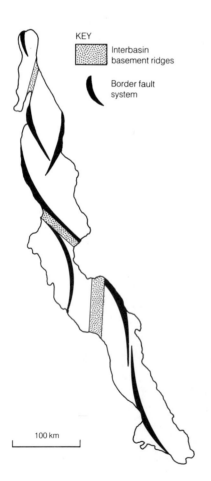

KEY

Interbasin basement ridges

Border fault system

100 km

Figure 29.33 Simplified tectonic map of Lake Tanganyika. (*From Rosendahl et al, 1986*)

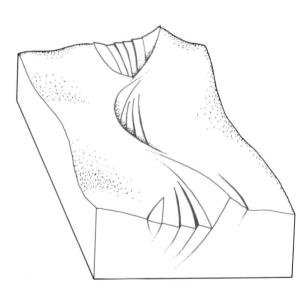

Figure 29.35 Block diagram showing alternation of half-grabens along a sinusoidal interconnection of border faults and intrabasin ridges. (*After Rosendahl et al, 1986*)

berlites associated with intra-plate swells and rifts; xenoliths are absent in the relatively large-volume basalts of the oceanic ridges or the tholeiitic flood basalts. Many xenoliths are of garnet- or

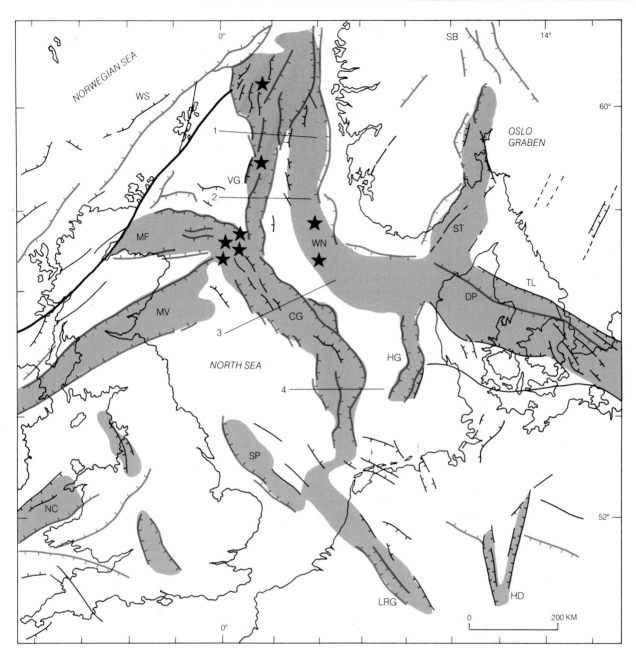

Figure 29.36 Mesozoic fractures of the North Sea areas. CG – Central Graben; DP –Danish Polish Depression; HD – Hessen Depression; HG – Horn Graben; LRG – Lower Rhine Graben; MF –Moray Firth Trough; MV –Midland Valley; NC – North Celtic Trough; SB – Sparagmite Basin; SC – South Celtic Sea Trough; SP –Sole Pit Basin; ST – Skagerrak Trough; TL –Tornquist Line; VG – Viking Graben; WN –West Norway Trough; WS – West Shetland Basin. Stars are Mid-Jurassic volcanic centres, lines 1–4 are cross sections shown in Fig. 29.37. (*Compiled from P.A. Ziegler, 1978*)

spinel-lherzolite (lherzolite is an olivine, enstatite, diopside rock) that are taken to be representative of the upper mantle at deeper or shallow levels respectively. However, in some xenoliths the original minerals have been partly replaced by less-dense mica and amphibole as a result of metasomatism. During metasomatism, elements such as potassium, rubidium and volatiles, derived from magmas migrating upwards from the asthenosphere (sometimes visible as frozen

veins in the xenoliths), react with lithospheric lherzolites. Metasomatized xenoliths have been found in the volcanic rock in the rifts in south-west Uganda, northern Tanzania, the margins of the Red Sea, the Eifel and Hesse depressions of the Rhine graben, and on the Hoggar swell of the Sahara. In the Tanzanian examples the veining and metasomatism have reduced the density from $3.4 \, \text{gm cm}^{-3}$ for unaltered mantle to between 3.14 and $3.35 \, \text{gm cm}^{-3}$, depending on the degree of veining and alteration; these low values bracket the density of $3.29 \, \text{gm cm}^{-3}$ required by geophysical gravity models for the mantle beneath the East African swell. Hence metasomatized mantle consistent with Illies' low-density 'cushion' does exist beneath some uplifted and rifted areas. Nonetheless, a note of caution is necessary. We still do not know whether metasomatism can be of the required areal extent to match the swell areas or, as has proved thus far, is more confined to the rifts themselves.

29.14 RECOGNITION OF OLD RIFT VALLEYS

It is a reasonable assumption that the mantle upwelling processes, leading to crustal swells, extension, rifting and magmatic activity, have existed in the past, and we have already mentioned the older Carboniferous rifts of East Africa. But how can we recognize older rifts that now have no geomorphological expression? If the rifts are blanketed beneath later sedimentary formations, the evidence must be geophysical. By seismic and gravity surveys it is possible to pick up linear anomalies that reflect unusually thick narrow belts of sedimentary or volcanic rocks. A good example is the North Sea (Figures 29.36 and 29.37) where seismic surveys from oil exploration have identified older buried rocks where major extensional faulting had given rise to several sedimentary basins and three major rifts – the Viking and Central Grabens and the Moray Firth Trough. These structures, buried beneath thick sequences of Cretaceous and Tertiary sedimentary rocks, have been confirmed and delineated by extensive drilling, which has also revealed several Jurassic volcanic centres of the alkali-basalt type associated with rifting; these are located at the junction of the three rifts. Incidentally, the fanglomerates and other coarse sediments that washed down from the old escarpments have proved to be excellent oil reservoirs.

Another geophysical anomaly that has been interpreted as an ancient rift occurs in central North America. A magnetic and gravity anomaly begins in Kansas, runs for hundreds of kilometres northeastwards into and along Lake Superior, before turning southeastwards beneath the Michigan Basin (Figure 29.38). In the Lake Superior Basin, the anomaly coincides with thick outcrops of late-Precambrian basalts, and it is believed that

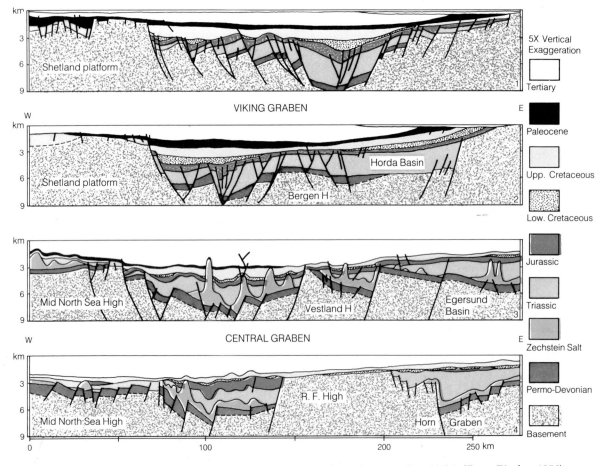

Figure 29.37 Sections across the North Sea Basin. Numbers refer to lines on Fig. 29.36. (*From Ziegler, 1978*)

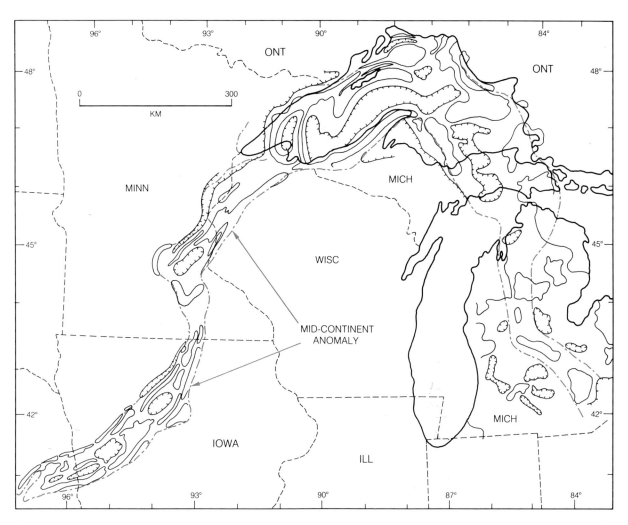

Figure 29.38 Magnetic anomaly map of the Central North America rift system. The dot-dash lines outline the margins of the Proterozoic Keeweenawan volcanic belt; note its north-south trend to the east of Lake Michigan. This anomaly is the same as a regional gravity anomaly. (*After Halls, 1978*)

linear extensions of the basalts in an old rift are the cause of the anomaly beneath the thick sedimentary sequences of the mid-West and the Michigan Basin.

If it can be assumed that alkali gabbro, syenite, nepheline syenite and peralkaline granite intrusions represent the plutonic centres beneath the former, now-eroded volcanoes of alkali basalts, trachyte, phonolite and comendite/pantellerite typical of the rift valleys, chains of these plutonic complexes may outline the position of old rifts, or at least old swell structures; the case may be strengthened if plutonic carbonatite complexes are present, or there is supporting stratigraphic evidence. Confidence in this means of identifying old rifts is provided by, for example, the Chilwa Igneous Province, at the southern end of Lake Malawi, where Jurassic alkali plutonic complexes, including the famous Chilwa Island carbonatite, lie within a recognizable, possibly-reactivated,

sector of the East African Rift, and on another continent by the Oslo Rift where the alkali complexes intrude a trough of Lower Palaeozoic sediments lying between Precambrian outcrops. Similarly a distinctive line of intrusions, including two carbonatites, that has been interpreted as a palaeorift, lies along the St. Lawrence valley of eastern Canada, while other chains, that may be rift-related, occur in South Africa (the Franspoort Line), Angola and Namibia. An exceptionally good example is found in the well-exposed fjord coast of South Greenland, where Precambrian sediments and lavas, which accumulated in fault-bounded troughs, are cut by Precambrian nepheline syenites, alkali gabbros, peralkaline granites and swarms of 'giant' dykes (up to 800 m wide, see Chapter 12) of alkali dolerite, trachyte and peralkaline microgranite. These intrusions were emplaced in three main episodes over a period of some 200 million years, during which

Figure 29.39 Profile across the shallow sag of Lake Victoria and the deep troughs of the Western and Eastern Rifts. Length of section 800 km.

the crustal extension was approximately 1.5 km (Upton and Blundell, 1978). The area is a particularly fine example of how crustal extension and associated alkali magmatism can repeatedly affect the same piece of continental crust reactivated over a protracted period of time.

29.15 SEDIMENTARY BASINS OF THE CRATONS

Sedimentary basins may form in a variety of tectonic settings (Miall, 1984), all requiring depression of the crust to accumulate water in which sedimentation can occur. Here we are concerned with the particular type of basin that forms as the result of epeirogenic flexing of the ancient shield areas; they are the downwarped equivalents of the epeirogenic upwarps discussed earlier in this Chapter. Two main characteristics of these basins are that they are very large and ovoid in shape, and they have depositional centres ('depocentres') rather than depositional axes.

We will confine ourselves to two major sets of basins – those of the African and North American cratons. Other good examples of cratonic basins are the Moscow Basin and the Arabian Basin – the world's major petroleum reservoir. A cratonic basin developing at the present day is the Eyre Basin of central Australia.

29.16 THE BASINS OF AFRICA

Africa is a vast continental shield, margined with mountainous folded belts only in the extreme north and south. Since the early Jurassic there have been many limited marine invasions across the coastal plains, with wide extensions during the Cretaceous, when Ethiopia and Somalia were submerged, and the sea spread across the Sahara from the Mediterranean to Nigeria. Otherwise, Africa has been a land area since the Ordovician, and much of it since the Precambrian. For hundreds of millions of years the movements of the shield have been persistently epeirogenic, giving a structural pattern of broad basins separated by irregular swells (Figure 29.8). The plateaus and

swells have been intermittently uplifted and denuded, with the result that they now consist largely of old rocks which were formerly very deep-seated. The basins have been the receptacle of thick deposits of continental sediments representing the material eroded from the uplifted tracts.

An impressive feature is that much of the margin of Africa is upwarped and the basins are separated from the coast by ridges of rock older than those in the basins (Figure 29.8). The major rivers originating in the basins, such as the Niger, the Nile, the Congo, the Zambesi and the Limpopo, cut through the ridges in a series of cataracts to reach the sea. Historically, it was these impediments to navigation that delayed exploration of a large area of Africa to a much later date than North and South America where major navigable rivers, such as the Mississippi and the Amazon, permitted easy access to the continental interior. Some of the basins are arid regions of internal drainage. The **Chad** Basin is one of the most remarkable. Lake Chad, fed by the Shari River from the swell to the south, is a shallow expanse of swamps and open water with no visible outlet. Yet the water is not stagnant, and although evaporation from its surface is high it does not become brackish. Despite appearances, Chad is not a terminal lake, for it drains underground and feeds the oases of the lowlands south of Tibesti, 450 miles to the north-east. **El Juf**, a vast desert depression north of Timbuktu, is one of the most awesome parts of the Sahara and until recent years the least accessible and least known.

In striking contrast is the equatorial **Congo** Basin, copiously watered by the great river system after which it is named. The whole region is underlain by thick continental sediments of Karroo age (late Carboniferous to Jurassic). Everywhere these dip gently inwards from the surrounding swells, as a result of warping which occurred in Tertiary times. The basin as a whole is slightly tilted towards the south-east. Where the Congo River meets the swell on the coastal side it escapes across the obstruction to the Atlantic by way of a series of cataracts. The basin of the **Sudan** is less easily traversed by the Nile. In the southern

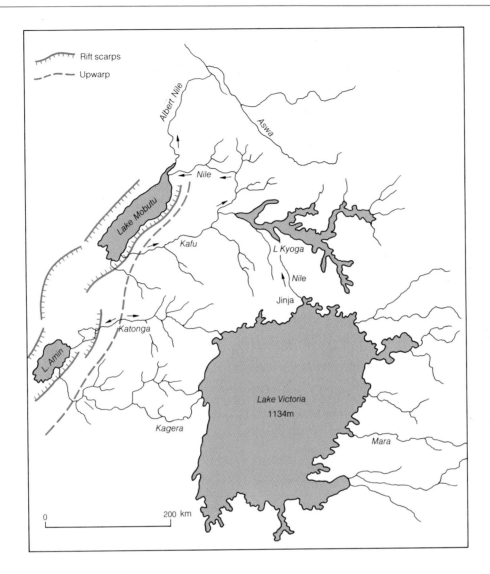

Figure 29.40 Drainage near Lake Victoria.

part of the basin the White Nile follows a sluggish and tortuous course through a wide expanse of papyrus swamps and lagoons. Thick floating accumulations of vegetable remains, known as the **Sudd**, obstruct the interlacing channels. These hot reedy swamps are the remains of a former lake, the extent of which is marked by vast spreads of sediment. The level rose until the waters overflowed through a notch in the northern rim, to become the vigorous Nile of the six cataracts between Khartoum and Aswan.

Unlike these other basins, the Lake Victoria Basin is at a high level, occupying a shallow crustal sag on the East African plateau between the eastern and western rift valleys (Figure 29.39). The lake, very recent in age, is no more than 80 m deep, despite its huge area of 70 000 km². Its formation is contemporaneous with that of the western rift, as a result of which the upwarped eastern shoulder of the rift impeded river systems for-

merly flowing westward (Figure 29.40). The drainage lines of the Katonga and the Kagera are continuous from Lake Victoria to Lake Idi Amin (L. Edward), and the Mara River, now flowing into Lake Victoria, was probably continuous with the Kagera; further north the Kafu flowed from Lake Kyoga into Lake Mobutu. The upwarping of the shoulder of the rift caused reversal of flow in the Kafu, Katonga and Kagera, and the middle course of the Mara-Kagera river was drowned when the downwarping caused the Lake Victoria Basin. Lake Victoria overflowed at the lowest point on its watershed near Jinja to form the stretch of the Nile between Victoria and Kyoga. Lake Kyoga itself was formed by back-tilting of the Kafu River; it subsequently flowed north up a former tributary, before overflowing westwards into the rift.

On the other side of the Equator lies the **Kalahari** Basin, partly grassy steppes and partly desert,

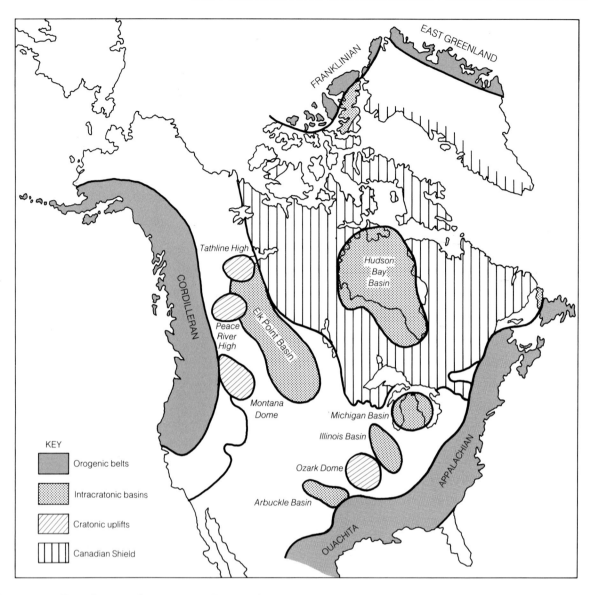

Figure 29.41 Distribution of intracratonic basins of North America.

with internal drainage in the north towards the brackish swamps of 'lake' Ngami. The name Kalahari is sometimes applied in a broader sense to include a northern extension, here distinguished as the **Cubango** Basin, after the swamps in which its internal drainage terminates. Though superficially continuous over vast areas, the two basins are separated structurally by a series of minor southwest–northeast arches along the crests of which Precambrian rocks outcrop at intervals. Belts of late Karroo volcanic rocks follow the same directions in both basins. To the south is the **Karroo** Basin, the type area of the Karroo system. This now stands high and rises eastwards to the Lesotho Highlands, culminating in the basalt-capped crest of the Drakensberg escarpment, already described and illustrated (Figure 18.42).

The African habit of 'basin and swell' is evidently of very old standing. The oldest of the basins so far mentioned owe the depression of their floors to movements that began early in the Karroo, i.e. in late Carboniferous times. But north of the Karroo Basin of South Africa is situated one of the oldest and best known of the world's structural basins: that of the Rand goldfields, which dates from nearly 3000 m.y. ago. Johannesburg stands on its northern rim, where the gold-bearing strata of the Witwatersrand System dip southwards and, like the inward-dipping beds on the eastern and western rims, are accessible to mining until depths are reached where the rocks are too hot to be worked without prohibitively expensive air-conditioning. North of Pretoria a younger basin, but also very old, since it dates

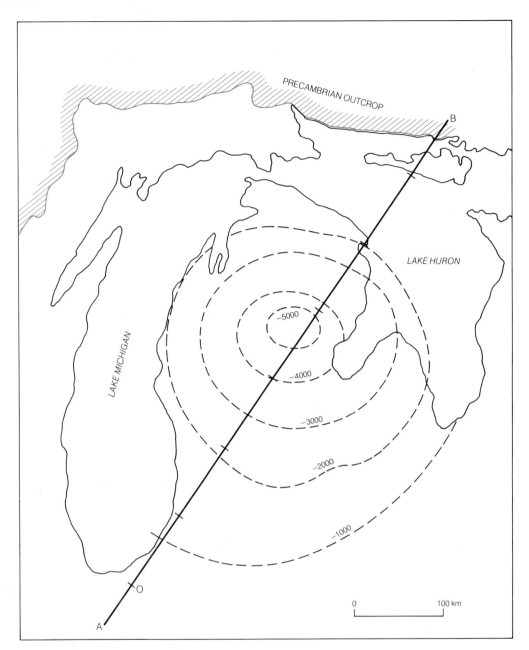

Figure 29.42 Structure contour map, in metres, to the Precambrian beneath the Michigan Basin. (*After J. H. Fisher et al, 1988*)

from 2000 m.y. ago, is occupied by the Transvaal System and the great Bushveld Complex, described in Chapter 12.

Continental sedimentary basins that concern us here are of two basic types: the graben type – associated with uplift and rifting discussed earlier, and the epeirogenic type – areas of broad regional subsidence unaccompanied by conspicuous faulting. These two basin types follow each other in time, the termination of rifting being followed by broadly-subsiding intracratonic basins centred over the earlier rifts but extending more widely into the surrounding areas (see, for

example, the sections across the North Sea Basin, Figure 29.37). The double-basin formation is currently interpreted as being due to the rise from the mantle of a thermal plume that initially causes crustal thinning and rifting with the formation of the elongate graben basins; at the close of the thermal event, there is broad crustal downsagging with the formation of the wide, intracratonic basins. It will be noted that in this model the spreading and crustal stretching is limited and does not develop into a new area of oceanic crust.

In most crustal areas, the formation of a major intracratonic basin is a unique event. In the case of

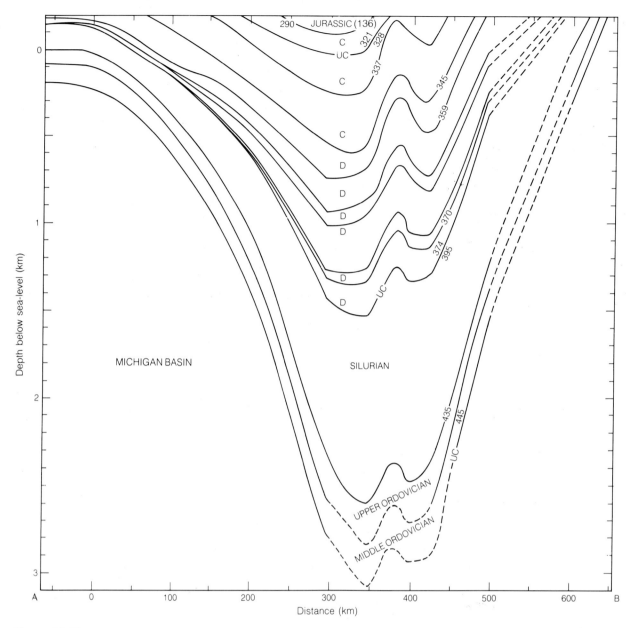

Figure 29.43 Post-lower Ordovician cross-section of the Michigan Basin, on line marked A–B on Fig. 29.42. UC marks major unconformities. C and D are Carboniferous and Devonian formations respectively. The anticline at 380 km is due to late Palaeozoic tectonics. (*From Sleep and Sloss, 1988*).

29.17 PALAEOZOIC CRATONIC BASINS OF NORTH AMERICA

southern Africa, as noted above, intracratonic basins have formed repeatedly during geological times. M.R. Cooper provided good evidence that basin formation in southern Africa is cyclic, with a periodicity of 320 million years. If the formation of these basins is due to the thermal expansion/ contraction model outlined above, a major outstanding problem is why has this portion of the crust been the site of such repeated thermal activity? And why has this thermal activity always been insufficient to cause major spreading?

The Michigan, Illinois, Elk Point and Hudson Bay Basins (Figure 29.41) are large ovate basins on the North American craton. All are surrounded by areas of low relief. The sediments are dominantly carbonates and evaporites with subordinate fine sands and shales. There is a marked absence of thick arkoses.

The almost-circular Michigan Basin provides a good example that has been well studied geophysically and extensively drilled during oil exploration. The basin deepens towards the centre

(Figure 29.42) which is some 5 km below sea-level and contains sediments ranging in age from Upper Cambrian/Lower Ordovician to Jurassic. There are sandstone, shale, reef limestone, salt and anhydrite units that thicken towards the centre of the basin, and sedimentation and subsidence were apparently in approximate balance, though unconformities indicate some interruption in the sedimentation (Figure 29.43). Throughout the history of the basin there were minor shifts in the position of the depocentre. The basin is floored by highly-fractured Precambrian rocks and it can be assumed that the pattern of north–south fractures seen in the Precambrian rocks north of Lake Superior exists below the basin. These faults moved repeatedly during the lower Palaeozoic, penetrating upwards into the basin strata, and a major movement took place at the end of the Carboniferous. Most of the faults in the basin sediments are situated in the centre of the basin, as are minor folds. However, an absence of folds along the margin of the basin suggests that the mid-basin folds cannot be due to compression and that, although some of the deeper folds may be due to compaction over Precambrian highs, the folding is probably due to vertical tectonics.

Surficial loading is inadequate to explain the magnitude of the subsidence. If thermal contraction of the lithosphere is invoked, at least two heating events (Cambrian and pre-Middle Ordovician) would be required; however deep drilling has found no evidence of a post-Cambrian thermal event. It is known that the basin overlies a linear gravity anomaly that is interpreted as a volcanic-filled Precambrian rift (see Section 29.14) but whether the effect of this rifting could have persisted throughout the Palaeozoic, and given rise to a focussed rather than linear depression, appears unlikely. As in the case of other cratonic basins, the cause of subsidence is not yet fully understood. A recent theory views the basin development as an on-craton response to events in the Appalachian fold belt on the eastern margin of the North American continent; it is proposed that the Appalachian orogenic activity caused the episodic subsidence of the basin, possibly through weakening of the lower crust and reactivation of a pre-existing upper-crustal isostatic imbalance (Powell and van der Pluijm, 1990). This theory implies the ability of the continental crust to transmit horizontal stress for hundreds of kilometres from the continental margin into the interior of the craton.

Definitions of the more unusual rift-valley igneous rocks.
Alkali basalt – basic volcanic rock that, for a given SiO_2 content, contains more $Na_2O + K_2O$ than normal basalts.

Hawaiite – an intermediate volcanic rock containing, like andesite, olivine, pyroxene and andesine, but also containing interstitial alkali feldspar.

Trachyte – an acidic volcanic rock consisting mainly of alkali feldspar, with subordinate pyroxene and amphibole. It may also contain small amounts of either quartz or nepheline in rocks transitional to rhyolite and phonolite respectively.

Phonolite – a volcanic rock consisting mainly of alkali felspar and >10% by volume of nepheline; the subordinate dark minerals are alkali pyroxene and amphibole.

Syenite and **nepheline syenite** are the plutonic equivalents of trachyte and phonolite respectively. The term "**peralkaline**" refers to igneous rocks in which the amount of $Na_2O + K_2O$ is in excess of what can be accommodated in feldspar and mica; the excess is incorporated into alkali-rich pyroxene (aegerine) and amphibole (riebeckite).

29.18 FURTHER READING

Bell, K. (1989) *Carbonatites. Genesis and Evolution*, Unwin Hyman, London.

Chang Chengfu, Shackleton, R.M., Dewey, J.F. and Yin Jixiang (1988). *The Geological Evolution of Tibet*, Royal Society, London.

Coward, M.P. and Ries, A.C. (eds) (1986). Collision Tectonics. *Geol. Soc. London Spec. Publn* **19**.

Frostick, L.R., Renaut, R.W., Reid, I. and Tiercelin, J. (eds) (1986). Sedimentation in the African Rifts. *Geol. Soc. London Spec. Publn* **25**.

Miall, A.D. (1984) *Principles of Sedimentary Basin analysis*, Springer, New York.

Powell, P.D. and Pluijm, B. van der (1990) Early history of the Michigan Basin: subsidence and Appalachian tectonics. *Geology* **18**, 1195–8.

Ramberg, I.B. and Neumann, E.-R. (eds) (1978) *Tectonics and Geophysics of Continental Rifts*, 2 vols, D. Reidel, Dordrecht.

Shackleton, R.M., Dewey, J.F. and Windley, B.F. (eds) (1988) *Tectonic Evolution of the Himalayas and Tibet*, Royal Society, London.

Sloss, L. L. (ed) (1988) Sedimentary cover – North American craton, in *Geology of North America*, vol D2, Geological Society of America, Denver.

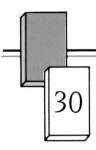

30

CONTINENTAL MARGINS AND BASIN EVOLUTION

So high as heaved the tumid hills, so low
Down sank a hollow bottom broad and deep,
Capacious bed of waters

John Milton (1608–1674)

Rift valleys are important sites of sedimentation within continents, but it is around the edges of the continental land masses that we find the greatest accumulations of sediments. Coastal plains on land, the shallow water which lies offshore, and the adjacent deeper parts of the oceans, are all important regions of sedimentation. Looking at present day sedimentary basins we can see features which we can use to interpret the depositional environments of sedimentary rocks in the stratigraphic record. The size and shape of a basin, the type of sedimentary fill, the distribution of material across the basin and the changes in sedimentation through time can all be used to characterize a basin. They can be used to interpret the mode of formation and evolution of ancient sedimentary basins. This insight into the nature of the surface of the Earth through time is of more than academic interest: sedimentary basins are sites of the formation of coal, oil and gas, while many minerals of economic importance occur concentrated in sedimentary rocks.

30.1 GEOSYNCLINAL THEORY OF BASIN FORMATION AND DEFORMATION

In the middle of the nineteenth century geologists

began to recognize that sediments accumulated in large-scale downwarps in the Earth's crust. These regional troughs or basins were called '**geosynclines**' (Figure 30.1 and Section 31.3) and many geologists considered that the sediments were derived from neighbouring large-scale crustal upwarps or 'geoanticlines'. The main tenet of the geosynclinal theory was that the troughs gradually subsided and filled up with sediment over many millions of years; outpourings of lava and volcanic ash and intrusions of igneous rock were associated with the downwarping. Folding and metamorphism took place once a thick pile of sediment had accumulated, and this deformation of the sedimentary rocks eventually led to the formation of a mountain chain. The mechanism that produced the deformation was never adequately understood, but the loading of the sedimentary rocks in the geosyncline was thought to be important.

Over the next hundred years a classification of geosynclines was developed on the basis of their position relative to the edge of the continent, on their deformation and the type of magmatism. Some of these terms are still occasionally used today (miogeosyncline, eugeosyncline, etc), but with the development of the plate

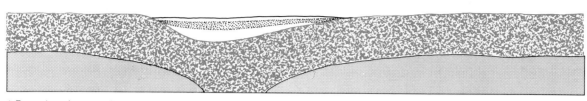

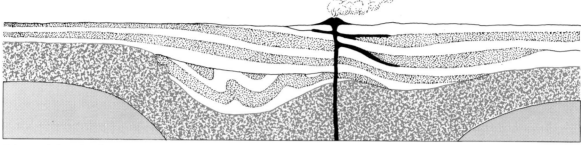

1 Formation of geosyncline

2 Accumulation of sedimentary and volcanic rocks

Figure 30.1 The geosynclinal theory of sedimentary basin formation. The mechanism for producing a crustal downwarp and the subsequent deformation of the sediments can now be explained in terms of plate tectonics.

tectonics theory in the mid-1960s a completely new way of looking at sedimentary basins evolved.

30.2 PLATE TECTONICS AND SEDIMENTARY BASINS

When it became accepted that most of the large-scale features of the Earth's crust were the result of the horizontal movement of large pieces of lithosphere, the geosynclinal theory of sedimentary basin formation and deformation became redundant. The formation of depressions on the Earth's surface could now be explained in terms of the thinning of the lithosphere by stretching, warping of crust by loading and the flexure of crust during collision.

As a first step in understanding sedimentary basin formation and deformation in terms of plate tectonics, J. Tuzo Wilson in 1966 recognized a cycle consisting of:

1. an opening phase due to sea-floor spreading
2. a period of closure by subduction of oceanic lithosphere, and
3. continental collision resulting in deformation.

The Wilson Cycle, as it became known, could be used to explain the formation and deformation of many sedimentary basins (Figure 30.2). This model assumes that extension and collision are orthogonal (that is, the plates move perpendicular to the plate boundaries). In recent years the importance of strike-slip motion between plates has received more attention. Oblique motion results in areas of local extension (**transtension**)

and local compression (**transpression**) which are also important in the formation and deformation of many smaller sedimentary basins. The strike-slip cycle described by H. Reading in 1980 is analogous to the Wilson cycle and the two can be regarded as end members of a continuum between purely orthogonal plate motion and purely transform plate motion (Section 28.3).

In a review of the tectonics of sedimentary basins Ingersoll (1988) drew up a classification of basins in terms of their tectonic setting. The three basic types of plate boundary – divergent, convergent and transform (strike-slip) – are taken as the starting point in the classification and in the full scheme 23 types of basin are recognized: we can simplify this to nine main tectonic settings for sedimentary basins (see Table 30.1 and Figure 30.3).

30.3 PASSIVE CONTINENTAL MARGINS

Rift and intracratonic sag basins have been considered in the previous chapter on plateaus and rifts: the other important sites of sedimentation at divergent margins are considered here.

The formation of a rift valley is the first stage in the separation of a piece of continental lithosphere into two tectonic plates. As the two pieces of continental crust move further apart, new oceanic crust forms in the area between and the process of sea-floor spreading begins. The Red Sea is presently at this stage in the evolution of a new region of oceanic crust as Arabia and Africa move apart. The tectonic plates of North America and Europe have been moving apart for a longer period of

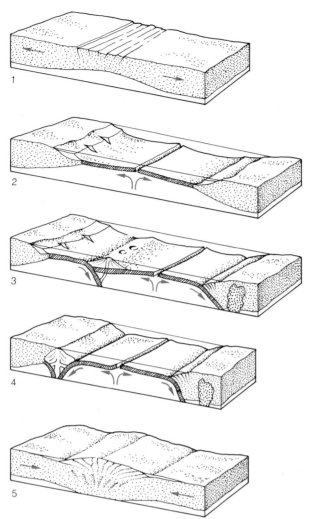

Figure 30.2 Sedimentary basin formation by plate tectonics: (1) rifting of continental crust; (2) formation of new oceanic crust by sea-floor spreading; (3) and (4) the ocean closes by subduction of oceanic crust; (5) following the subduction of all the oceanic crust the two continental plates collide to produce an orogenic belt. (*Based on Warren Hamilton, 1970, The Uralides and the motion of the Russian and Siberian Platform*)

time to form the Atlantic Ocean.

During the process of rifting and separation, the crust at the edges of the two new continents becomes thinned as extensional faults break up the crust as it stretches. Rifts and spreading centres are areas of high heat flow, so the continental crust at this margin is relatively warm and buoyant. As new oceanic crust is formed, the two continental margins become further away from the hot spreading centre and they cool down. Colder crust is more dense, so the edge of the continental crust gradually subsides.

The trailing edges of continents resulting from the formation of a new ocean are sites of little seismic activity as there is no relative movement

Table 30.1 Sedimentary basin classification (*Modified from Ingersoll 1988, Bull. Geol. Soc. Amer.*, **100**, 1704–19)

Tectonic Setting of Sedimentary Basins

A. Divergent settings

1. Terrestrial rift valleys and aulacogens within continental crust

2. Intracratonic basins: broad sag basins formed within continental crust

3. Passive continental margins in intraplate settings at continental–oceanic interfaces

4. Oceanic basins: basins floored by oceanic crust unrelated to arc-trench systems

B. Convergent settings

1. Foreland basins: basins formed at continental margins during crustal collision, including foredeeps and thrust-sheet-top (piggyback) basins

2. Trenches: deep troughs formed by subduction of oceanic lithosphere

3. Fore-arc basins: basins developed between subduction complexes and magmatic arcs

4. Back-arc basins: basins behind magmatic arcs, including retro-arc, inter-arc and marginal back-arc basins

C. Transform settings

1. Basins formed along strike-slip fault systems

occurring between the continental and oceanic crust. They are referred to as **passive continental margins** and they lie within a single tectonic plate. For example, North America and the western part of the Atlantic Ocean are moving westwards together as a single tectonic plate: the boundary between the oceanic and continental crust off the east coast of North America is hence a passive continental margin (Figure 30.4).

The edge of the continental crust, being thinner, sinks below sea-level as it cools to form a **pericontinental sea** on the continental shelf. Shallow marine conditions may spread far over the continent to form an **epicontinental sea**, such as the North Sea in north-west Europe. Most continental shelves extend as broad, very gently sloping areas from the coast to the steeper slope down to the deeper level of the oceanic floor (Figure 24.5). Continental shelves vary from a few kilometres wide (for example, west of Corsica) to over

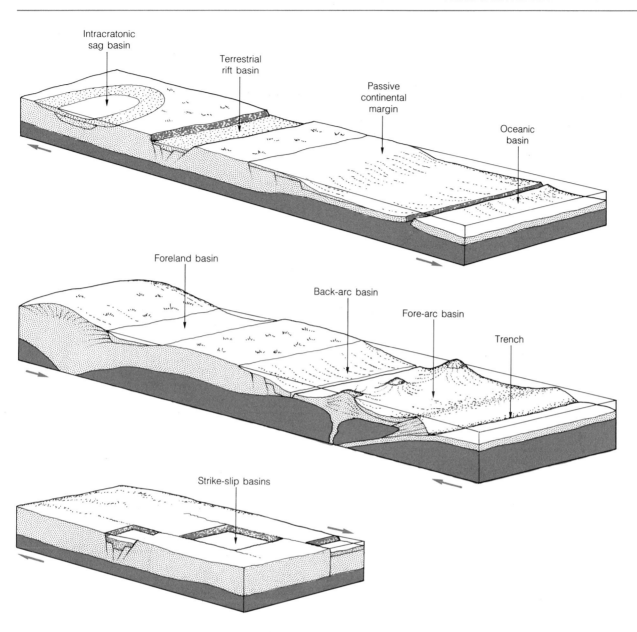

Figure 30.3 The nine principal classes of sedimentary basin.

1000 km in north-western Europe (Figure 30.4). They comprise areas of relatively shallow water, up to around 250 m in depth and are sites of sediment accumulation. As the continental margin subsides, thick sequences of sediment are built up over tens of millions of years.

Most of the sediment eroded from the continental landmass is carried by rivers to the continental shelf. Some of the world's biggest rivers (e.g. the Amazon and the Congo) feed onto passive continental margins and some (e.g. the Niger) may form large deltas which build out onto the continental shelf (Section 17.11). Deltas which build out onto the shelf as large bodies of sediment are locally very important parts of the passive margin, but much of the sediment load is distributed to other parts of the margin. Where wave activity is strong along coasts, sand accumulates in bars to form barrier islands. Barrier islands are important features along the coast of the south-eastern United States (the passive margin between the North American continent and the North Atlantic). In other parts sediment is carried further out onto the continental shelf by tidal and storm currents. Tidal currents can move large quantities of sand, building up large offshore bars and ridges, such as the Dogger Bank in the North Sea. Storms affect the deeper parts of the shelves, distributing sand in broad sheets.

The shallow-shelf seas are very favourable places for a wide variety of organisms to live and calcium carbonate produced by plants and ani-

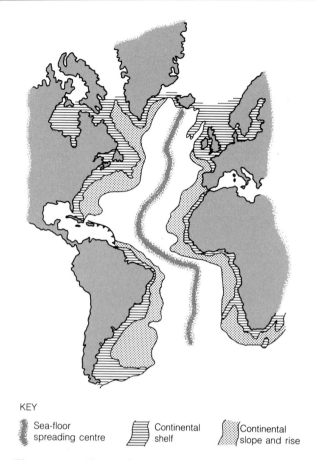

KEY

Sea-floor spreading centre

Continental shelf

Continental slope and rise

Figure 30.4 The Atlantic Ocean, with continental shelves bordering the continents on both sides of the ocean; the British Isles lie on the broad north-west European shelf area.

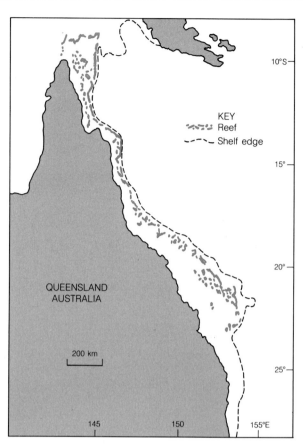

KEY
Reef
Shelf edge

10°S
15°
20°
25°

QUEENSLAND
AUSTRALIA

200 km

145   150   155°E

Figure 30.5 The continental shelf of north-east Australia; in this tropical area carbonate sedimentation is important and the shelf edge is lined with reef build-ups. (Great Barrier Reef, see Section 24.10.)

mals makes up an important part of the sediment on some shelves (Chapters 6 and 24). Warm seas in temperate and tropical regions are the most favoured sites for these organisms and hence the calcium carbonate deposits are most abundant in these areas. Under the most favourable conditions, in clear waters away from the sand and mud brought in by rivers, colonies of organisms such as corals can build up reefs. Reefs act as natural breakwaters, and the area between the reef and the coast becomes a shallow, quiet area where organisms can flourish. The Great Barrier Reef off the coast of Queensland in eastern Australia is over 2000 km long and lies up to 200 km offshore (Figure 30.5). The shelf between the reef and the coast is a broad area dominated by carbonate sedimentation. On the seaward side of the reef shelly debris and broken parts of the reef form an apron of material.

The parts of a passive margin are shown on Figure 30.6. At the edge of the continental shelf lies the **continental slope**, which extends down from the shelf to between 2 and 4 km water depth with a slope of around 3 to 5 degrees. Continental slopes are frequently cut by canyons which act as preferential pathways for the transport of material down from the shelf to the **continental rise**. The material is commonly carried in dense flows which are turbid mixtures of sediment and water known as turbidity currents (Chapter 24). These currents are initiated by earthquakes or sediment instability on the shelf edge, and in a matter of a few hours can carry sediment hundreds of kilometres out into deep water. Turbidites may form a

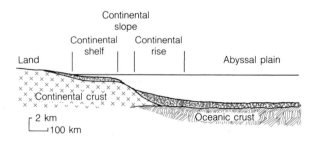

Continental
slope
Continental   Continental
shelf   rise
Land   Abyssal plain

Continental crust
2 km
100 km   Oceanic crust

Figure 30.6 A cross-section across a continental margin.

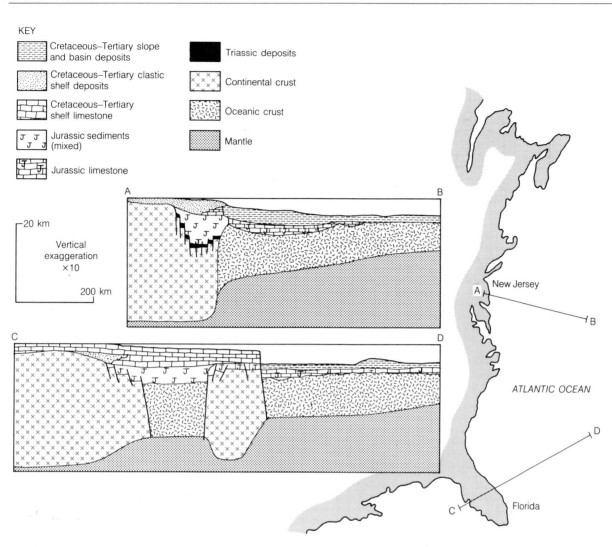

Figure 30.7 The continental margin of the east coast of North America. The cross-section near New Jersey shows the Triassic deposits of rift basins; sea-floor spreading in the Jurassic formed a broader shelf area of clastic and carbonate deposition. During the Cretaceous and Tertiary the New Jersey shelf was an area of clastic sedimentation with finer deposits and turbidites on the continental slope and rise. In the warmer waters off Florida carbonate sedimentation has dominated on the passive margin.

cone of detritus, a 'submarine fan', which radiates out from the bottom of a submarine canyon. The largest submarine fans occur close to deltas or major rivers; for example, the Amazon and Mississippi rivers both supply large submarine fans on passive margins.

Beyond the continental rise lies the **abyssal plain**, the deepest part of the ocean basin. Abyssal plains are sites of pelagic sedimentation, although some distal turbidites may supply clastic detritus from the continental shelf. Claystones, siliceous and calcareous oozes make up the pelagic sediments of the deep ocean basins. The rates of sediment accumulation in these basins are very slow, as little as 2 mm/1000 years.

Passive margins evolve from the edges of rift basins, and the sedimentary sequence found in

passive margin deposits usually reflect this history. The basement at the margin will be continental crust, but there will often be evidence that this crust has been thinned by extensional faulting. The oldest sedimentary rocks will be clastic deposits formed as the rifting occurred: these will be the deposits of alluvial fans, rivers and lakes – sometimes with volcanic rocks if volcanism was associated with the rifting (Chapter 29). The rift and graben topography will sometimes be preserved as these clastic sediments fill up the rift basins. As the pulling part of the continental blocks continues the area will become invaded by seawater and marine deposits will overlay the continental sediments. Initially subsidence at the continental margin will be relatively rapid as the margin cools, but the rate of subsidence will decrease through time as the spreading centre becomes further away. Sediment will also contribute to the subsidence and the passive margin becomes a site of steady subsidence and accumulation of sedimentary rocks. If the margin remains passive these rocks will show little sign of deformation.

Many of the features of passive margin evolution can be illustrated with reference to the Atlantic seaboard of North America (Figure 30.7). The oldest sediments in the passive margin sequence are Triassic continental clastic deposits localized in grabens developed in the earliest stages of rifting as outlined in the previous chapter. During the Jurassic further extension took place and this margin subsided to become an area of continental and shallow marine sedimentation. Close to the continental margin, clastic sediments and evaporites were the main Jurassic deposits but further east carbonate sedimentation took place directly on newly-formed oceanic crust. Rifting within the continental margin off Florida detached a piece of continental crust which formed a nucleus for reef development throughout the Mesozoic and Tertiary. In the Cretaceous extension and subsidence

at the margin continued: reefs developed at the shelf edge off New Jersey whilst at the southern part of the margin shallow marine limestone sedimentation covered the whole shelf area. Carbonate sedimentation persisted in the warmer climate of the shelf off Florida throughout the later stages of the opening of the Atlantic, in contrast to the cooler waters off New Jersey where clastic sedimentation was more important. The main phase of opening in the North Atlantic took place in the late Cretaceous and beyond the shelf break turbidites and pelagic sediments supplied the continental rise and abyssal plain of the Atlantic Ocean.

On the other side of the Atlantic, extension within the continental crust of Europe gave rise to a very broad shelf area covered by an epicontinental sea. The North Sea lies on the north-west continental shelf of Europe and was a broad area of extension and subsidence related to the Atlantic opening during Mesozoic and Tertiary times: the thick sequences of clastic and carbonate shelf deposits accumulated during this period are the sites of the North Sea oilfields (Chapter 9 and Section 29.14).

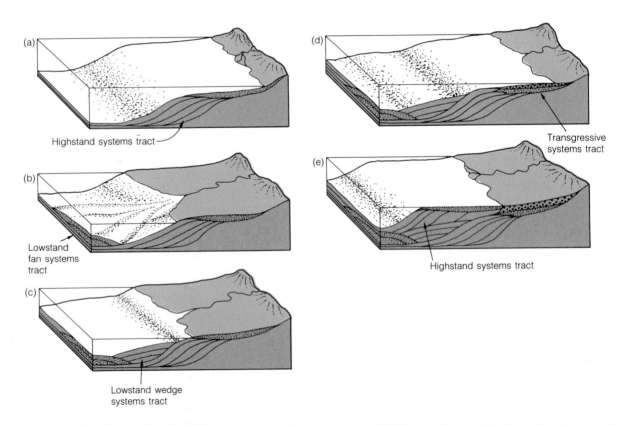

Figure 30.8 The effects of sea-level changes on a continental margin: (a) high sea-level with a broad shelf area and the deposition of a 'highstand systems tract'; (b) falling sea-level and shelf sediments are swept out to form a submarine fan – a 'lowstand fan systems tract'; (c) the shelf is largely exposed at low sea-level and a 'lowstand wedge systems tract' forms; (d) as sea-level rises sediments are once again deposited on the shelf as a 'transgressive systems tract'; (e) a new 'highstand systems tract' is formed at high sea-level. Note that the vertical scale is strongly exaggerated on these idealized diagrams. (*Based on H. W. Posamentier and P. R. Vail, 1988*)

All the deposits on continental shelves are sensitive to changes in sea-level. If the sea-level drops, the beach will be left stranded and a new beach will form further out on the shelf. This type of change is called a **regression**, as the sea has moved relatively away from the land. Conversely, a rise in sea-level results in a new beach forming on the old land surface: such a change is referred to as a **transgression** of the sea over the land. In other parts of the shelf the type of deposit found at a given point will change as a result of rises and falls in the level of the sea. By careful comparison of the types of deposits a pattern of rises and falls in sea-level can be identified in sequences of sedimentary rocks. On a larger scale, changes in sea-level affect the whole pattern of sedimentation on the continental shelf, slope, rise and even the abyssal plain.

On Figure 30.8, the sediments of a passive margin are grouped into packages or 'systems tracts' which can be related to high and low sea-level stages. At times of low sea-level ('low stands'), the shallower parts of shelf areas are exposed to subaerial erosion, and the deeper parts come within the influence of currents in the upper part of the water column. During these periods shelf sediments are transported to the edge of the continental shelf and supply turbidity currents which flow down the continental rise to form submarine fans on the continental rise and abyssal plain. Hence, in deep-water sediments, increased amounts of coarse sediment signify periods of low sea-level.

When the sea-level is higher ('high stands') the sediments pile up closer to the coast. Areas which were formerly regions of river and coastal plain sedimentation become inundated with seawater as a marine transgression takes place. During high stands the shelf is broad, and most of the sediment supplied from the land accumulates in shallow waters. At these times little sediment reaches the edge of the shelf and turbidity flows down to the abyssal plain are less frequent.

Vail *et al* (1977) showed that the sedimentary deposits of passive margins could be divided into sequences which represent deposition during periods of rising or falling sea-level. The boundaries between sequences are frequently unconformities and these can be recognized on seismic reflection profiles (Section 9.10) across passive-margin sedimentary rocks. Comparison of the deposits formed on passive margins in different parts of the world showed that similar patterns of rises and falls in sea-level can be seen in deposits of the same age. From this information Vail and colleagues drew up a sea-level curve through geological time (Figure 30.9) which they consider reflects global fluctuations in sea-level,

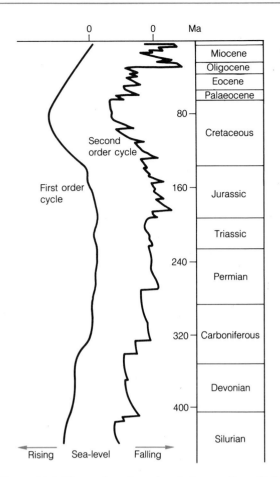

Figure 30.9 The cycles of sea-level change (*from Vail et al 1977*). The first order cycle is thought to be due to changes in the rate of spreading in the world's oceans. Second order cycles are related to the volume of ice in the polar regions.

i.e. eustatic sea-level variations.

Two factors are seen as important controls on global sea-levels. Climate is the most important short-term control, producing cycles of a few million years of sea-level rise and fall. When the world climate is in a cold phase (for instance, during an ice age) the volume of ice in the polar regions increases, and the volume of seawater in the world's oceans decreases. This results in a global 'eustatic' lowering of the sea-level. Calculations have revealed that the worldwide sea-level has changed by around 150 m between cold and warm phases during the Pleistocene. A second factor, acting over tens to hundreds of millions of years, is related to spreading at the mid-ocean ridges. During periods when the rate of spreading in the major oceans is high, a lot of young buoyant crust is generated in the ocean basins. This buoyant crust displaces water out of the ocean basins and onto the continental shelves, causing a relative rise in sea-level.

30.4 OCEAN–OCEAN AND OCEAN–CONTINENT COLLISION

Where lithospheric plates are in collision, or moving past each other, stresses in the crust result in areas of uplift and subsidence. The areas of uplift become regions of erosion which supply the sedimentary basins formed by crustal subsidence. The pattern of uplift and subsidence is determined by the nature of the lithospheric plates which are in relative motion at the active margin. Where collision is occurring this may be between two pieces of continental lithosphere, two oceanic plates or between an oceanic and a continental plate.

When a piece of oceanic lithosphere collides with a continental plate, the denser oceanic crust is usually forced down into the mantle beneath the continental margin to form a subduction zone. In the case of two oceanic plates colliding, one is subducted below the other. In both cases, the subducted slab of lithosphere begins to melt and the rising magma gives rise to volcanoes in the overriding slab. A chain of volcanoes – a volcanic arc – forms parallel to the destructive plate boundary. Four distinct morphological units can be recognized: the **trench**, the **fore-arc** region, the **volcanic arc** and the **back-arc** region (Figure 30.10). Together these components comprise **arc-trench systems**.

30.5 ARC-TRENCH SYSTEMS

Arc-trench systems are regions of convergence between two pieces of lithosphere; however, the overriding slab is more often in a state of extension than in compression. Old oceanic crust which is cold and dense readily subducts, and may do so at a rate which exceeds that of the convergence. When this occurs there is an oceanward migration of the hinge point where the oceanic crust subducts (Figure 30.10) and the trench moves away from the overriding plate in relative terms. This process is known as **roll-back** (Dewey 1980) and leads to extension in the fore-arc and back-arc regions. The Mariana Arc-Trench system in the west Pacific is an example of such an extensional arc. On the other side of the Pacific, younger and less dense oceanic crust is subducting in the Peru–Chile trench. This hotter, more buoyant crust does not subduct so readily and so trench roll-back does not occur. There is some evidence to suggest that this places the fore-arc and back-arc into compression. In some cases a balance is achieved between trench roll-back and the oceanward migration of the overriding plate resulting in a neutral arc which is neither under extension or compression: the Aleutian arc south of Alaska is an example of a neutral arc.

The **trench** formed where the subducting plate starts to go under the overriding plate is an important site of sedimentation. Trench

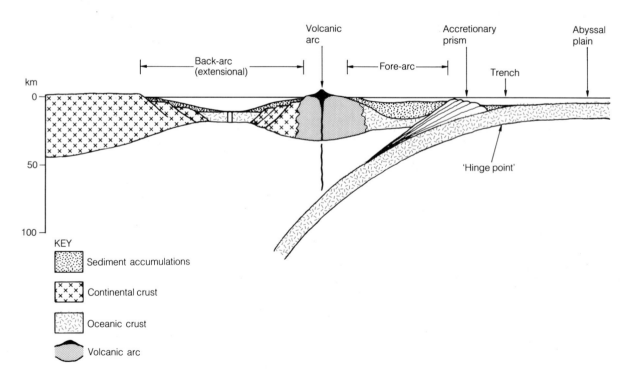

Figure 30.10 Arc-trench systems and the sites of sediment accumulation. The hinge point occurs where the downgoing plate starts to bend into the subduction zone. (*After W. R. Dickinson and D. R. Seely, 1979*)

sediments are frequently detached from the downgoing plate and form an **accretionary prism** (see below). Sedimentary basins also form in the fore-arc between the accretionary prism and the arc (**fore-arc basins**) and behind the arc (**back-arc basins**). The size, shape and development of these basins will depend on whether the region is in overall compression, extension or in equilibrium; the type of crust forming the overriding plate (oceanic or continental) is also important. The five basic types of arc-trench system and the sedimentary basins associated with them are shown in Figure 30.11.

30.6 TRENCHES

Where the downgoing slab bends to go down the subduction zone a deep trench is formed. These trenches are hundreds or even thousands of kilometres long and only a few tens of kilometres wide. They are typically around 2 km deeper than the level of the ocean floor at around 6 km below

sea-level. The trenches form the deepest parts of the world's oceans: the Mariana Trench is over 10 km below sea-level, formed where oceanic crust of the Pacific Plate is going down under oceanic crust of the Philippine Sea Plate (Figure 30.12). The slope down into the trench is relatively gentle, but steeper on the arc side, where the trench wall is the front of the overriding plate: they are therefore asymmetric in cross-section.

Oceanic trenches act as important sediment traps. They receive pelagic sediment (muds and biogenic oozes) from the water column above, but this is normally swamped by the deposits of mass-flows and turbidites. This detritus may be derived directly from the adjacent volcanic arc, in which case it will be dominated by volcanogenic material. However, trenches act as funnels for turbidity currents which can transport sediment hundreds or even thousands of kilometres from the source. The Sunda Trench (Figure 30.13) receives material from Sumatra to the east and from the Asian continent to the north. The detrital

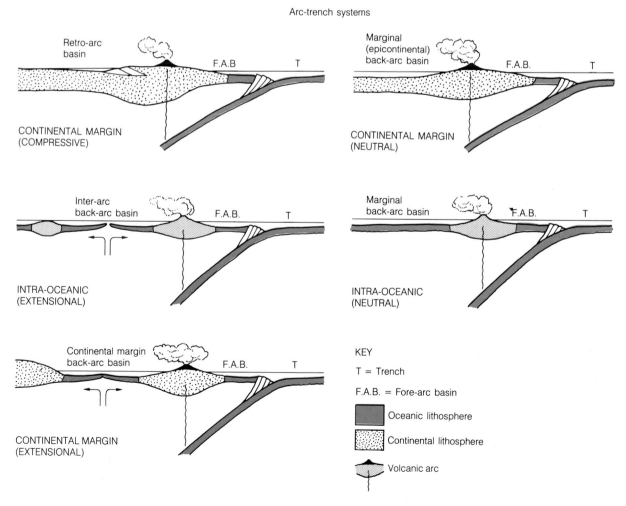

Arc-trench systems

Figure 30.11 The five basic types of arc-trench system. (*After W. R. Dickinson and D. R. Seely, 1979*)

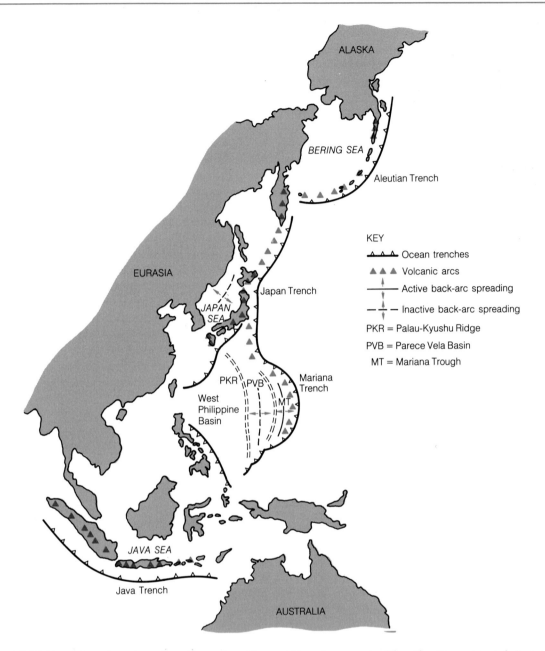

Figure 30.12 Examples of arc-trench systems along the east Eurasian margin. The Aleutian arc-trench is a neutral intra-oceanic system with old oceanic crust forming the floor of the Bering Sea. Back-arc extension within the Eurasian continental margin produced the Japan Sea. Behind the Mariana Trench, a series of intra-oceanic back-arc basins have formed during the Cenozoic. The Java Sea is floored by continental crust and is part of the neutral Java arc-trench system.

material from Asia, ultimately eroded from the Himalayas, can be distinguished because of its more quartz-rich 'continental' composition compared to the detritus from the volcanic arc. The volume of sediment in the trench is largely determined by the proximity of a landmass to supply detritus. Trenches which lie some distance from the nearest continental landmass or volcanic arc receive only pelagic sediment and very distal turbidites. 'Starved' trenches of this type are common in the Western Pacific area, e.g. the Mariana

Trench (Figure 30.12). Other trenches show a substantial variation in the volume of sediment depending on the local supply. At its southern end, the Peru–Chile Trench is completely filled by sediment from the adjacent highlands of Chile. Further north, little sediment is supplied from the arid coast of Peru and the trench is underfilled.

30.7 ACCRETIONARY PRISMS

As the downgoing slab moves under the overrid-

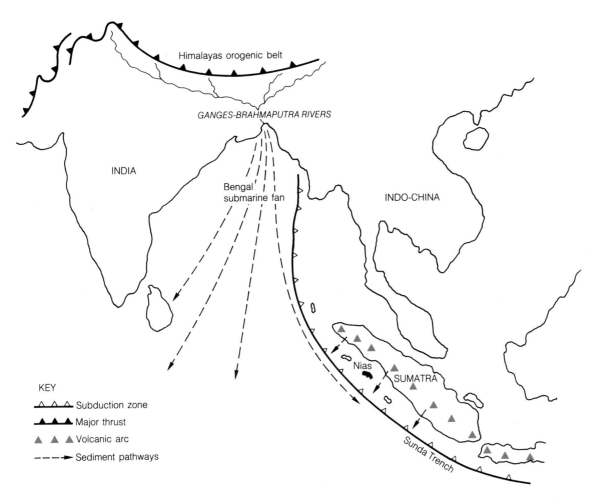

Figure 30.13 The Bengal submarine fan is supplied with material eroded from the Himalayan orogenic belt carried by the Ganges and Brahmaputra rivers. This continentally-derived material is quartz-rich and distinctly different from the volcanogenic detritus from the Sumatra arc.

ing plate the sediment in the trench may be taken down into the subduction zone or it may be scraped off by the overriding plate. The trench deposits which are scraped off are accreted onto the front of the overriding plate in a series of slices. Repeated addition of slices causes a build-up of material known as an **accretionary prism**. An accretionary prism is characterized by being composed of packets of strata deposited in a deep-water trench setting, each packet faulted against the next. Each individual slice will show a normal stratigraphy of older rocks at the base and younger at the top. However, as a new slice is accreted it is added below the previously emplaced unit. The accretionary prism as a whole will contain progressively older slices of stratigraphy further away from the trench (Figure 30.14). If the sedimentation in the trench is relatively rapid and most of this material is scraped off onto the front of the accretionary prism, the wedge of material can build close to or even above sea-level

to form an outer arc. The outer arc can be a source of detritus for the trench and the fore-arc area on the other side.

The predicted structure has been recognized in the area between the Sunda Trench and Sumatra (Figure 30.13). Seismic reflection profiles show the existence of low-angle faults which dip towards the arc in the lower part of the accretionary prism. The accretionary prism is a very well-developed feature offshore Sumatra and is emergent as a series of islands which lie parallel to the trench. On the island of Nias, Miocene rocks are exposed and are interpreted as trench deposits which have been uplifted by repeated accretion to their present position.

In the Southern Uplands of Scotland Lower Palaeozoic rocks are composed of sequences of black shales and sandy turbidites which are interpreted as deep marine deposits (Figure 30.14). The sequences are cut into slices by a number of reverse faults which are generally parallel to the

(a)

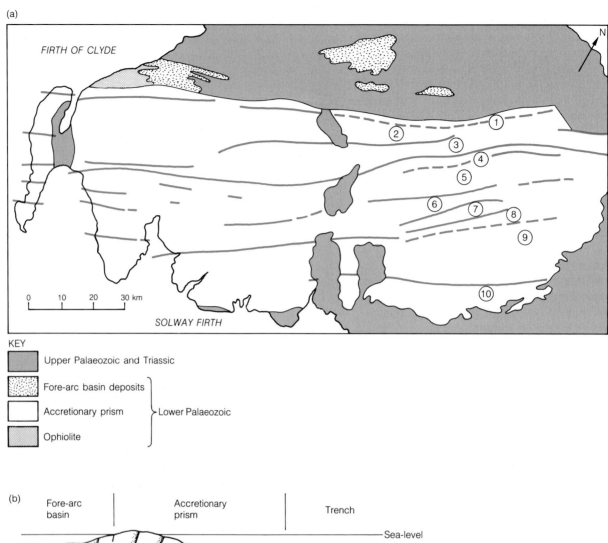

KEY

Upper Palaeozoic and Triassic

Fore-arc basin deposits

Accretionary prism   Lower Palaeozoic

Ophiolite

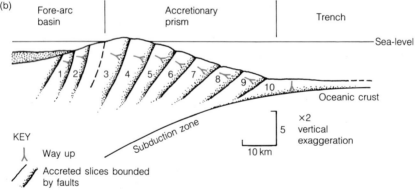

Figure 30.14 (a) The Lower Palaeozoic of the Southern Uplands of Scotland. Beds of deep-water mudstones and turbiditic sandstones are divided into ten slices by parallel faults: within each slice the beds young to the north, but the youngest slice is in the south and the oldest in the north. (b) A model for the formation of the Southern Uplands by accretion of trench sediments into an accretionary prism. (J. R. Leggett, 1982)

bedding. Within each slice, the stratigraphy youngs towards the north, but the oldest slices are in the north and the youngest in the south. This relationship is consistent with an interpretation of the Southern Uplands as an accretionary prism which formed by detachment of slices of material from the 'Iapetus Ocean' which was subducted northwards beneath southern Scotland.

The structure of an accretionary prism may be complicated by slumping down the inner wall of the trench. This occurs because the relatively steep front of the accretionary prism is unstable and readily fails: large quantities of material can fall back into the trench when slumping occurs. A form of diapirism (Section 10.2) occurs in accretionary prisms: water carried down into the subduction zone is heated and moves back up into the accreted material, mobilizing mud which then forms a diapir. The structural complexity of accretionary prisms results in depressions on the prism and on the inner slope of the trench. These trench-slope basins are filled with hemipelagic sediment and with turbidites derived from the higher parts of the accretionary prism. The structure and stratigraphy of accretionary prisms is hence very complex and may differ from the idealized form shown in Figure 30.14.

30.8 FORE-ARC BASINS

The distance between the trench and the volcanic arc is determined by the angle at which the subducting slab enters the mantle. The descending oceanic slab does not start to generate a melt until it has reached a depth where the surrounding mantle is hot enough. This occurs at between 90 and 150 km depth. The magma generated by the subducting slab results in a volcanic arc parallel to the trench. The distance between the trench and the volcanic arc, the **arc-trench gap**, will be determined by the angle of descent of the subducting slab. Steep subduction zones ($> 45°$) result in a narrow arc-trench gap and conversely a shallow subduction ($< 45°$) results in a broad arc-trench gap. The fore-arc basin lies between the accretionary prism and the volcanic arc (Figure 30.10) and, depending on the arc-trench gap and the size of the accretionary prism, it may vary between 50 and 100 km in width. The crust in the fore-arc region will be oceanic in all intra-oceanic arc-trench systems. In the formation of a continental margin arc, the rupture in the crust can either occur at the edge of the continental crust or some distance away from the margin within oceanic crust. Evidence from present-day fore-arc regions indicates that the rupture does not often occur at the continental margin. This is probably because continental margins are irregular in thickness and jagged in plan view and hence do not present favourable sites for a suture to occur. Trenches therefore develop as simple straight or arcuate structures within the oceanic crust. This leaves a piece of old, cold ocean floor between the trench and the arc to form the basement to the fore-arc basin (Figure 30.16). Locally, continental crust may also lie within the fore-arc region. If, due to steep subduction, the arc-trench gap is small the fore-arc region will simply take the form of a slope down to the trench (Figure 30.16): in these situations no fore-arc basin is formed.

The floor of a fore-arc basin will be formed by sediments deposited on the trapped crust before the arc-trench system developed. In the case of intra-

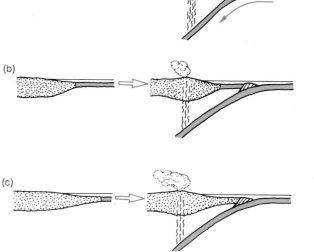

Figure 30.15 Fore-arc basins formed (a) within an intra-oceanic arc-trench system; (b) at a continental margin, but with the rupture occurring within the oceanic plate to leave oceanic crust in the fore-arc; (c) at a continental margin with continental crust lying beneath the fore-arc basin. (*After Dickinson and Seely, 1979*)

oceanic arcs, these will be abyssal plain sediments, principally pelagic oozes and clays. Fore-arc basins in continental margin arcs will be underlain by the deposits of the outer parts of a passive margin – principally turbidites of the continental slope. Once volcanism in the arc commences volcanic detritus is supplied to the fore-arc basin. In the early stages of fore-arc basin evolution, the main sediment source will be the evolving volcanic arc. The volcaniclastic material is as deposited primary pyroclastic sediments and as turbidites and debris flows of reworked detritus. Muds and oozes of pelagic origin area will also occur.

The thickness of deposits accumulated will depend on the depth of the fore-arc basin and this is largely controlled by the height above the ocean floor of the accretionary prism (Figure 30.16). Intra-oceanic arcs tend to have relatively small accretionary prisms because of the small volumes of material deposited in trenches in this setting (e.g. the Aleutian arc-trench system, Figure 30.12). Accretionary prisms may build up to sea-level where there is a sufficient volume of material added by accretion from the subducting slab. This is more often the case for arc-trench systems which lie close to an abundant supply of detritus (e.g. the Sunda Trench and Sumatran fore-arc, Figures 30.13 and 30.17).

Fore-arc basins which form in a continental margin arc-trench system are supplied with detritus from the continental landmass in addition to the volcanic and pelagic material. Turbidite deposits within the basin will therefore show both volcanic and continental provenance. Emergent accretionary prisms form an outer arc which may be eroded and transported into the fore-arc basin, and hence the accretionary prism represents a further source of detritus to fill the basin.

Through time, these deposits will gradually build up in the basin and hence the depth of water in which the sedimentation is occurring will become shallower through time. Complete sequences in fore-arc basins which have filled with sediment show a progression from deep-water turbidite facies up into shallow-marine shelf deposits in the younger parts of the basin fill. Deltas build out from the basin margins and some fore-arc basins fill completely to become sites of continental sedimentation. Isostatic subsidence due to sedimentary loading will occur as the basin fills.

30.9 BACK-ARC BASINS

Three types of basin can be recognized in back-arc regions. **Inter-arc basins** are extensional basins which form in intra-oceanic arc-trench systems where the 'roll back' of the trench results in rifting in the overriding plate. This leads eventually to the generation of new crust by sea-floor spreading behind the arc. **Marginal basins** lie between an island arc system and a continental margin. They may also form by rifting and spreading, but some occur in neutral arcs systems where the back-arc is neither in extension or compression. **Retro-arc basins** occur in continental margin arc-trench systems which are under compression. It is thought that movement on thrusts shift material away from the arc and stack it up on the continental crust: this thrust sheet stacking generates an excess load on the plate and causes subsidence. The mechanism of basin formation is hence the

(a)

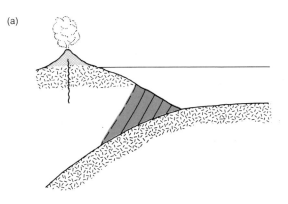

(b)

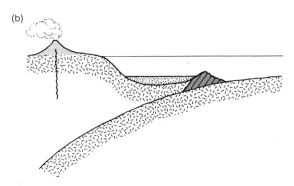

(c)

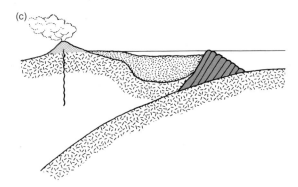

Figure 30.16 The configuration of fore-arc regions: (a) a simple sloped fore-arc with no fore-arc basin development; (b) a small accretionary prism forms the edge of a deep-water fore-arc basin; (c) a large, emergent accretionary prism allows the formation of a larger fore-arc basin which may fill up with sediment. (*After Dickinson and Seely, 1979*)

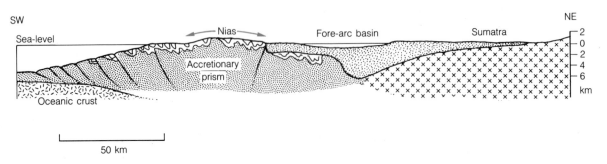

Figure 30.17 The Sumatran fore-arc (see Figure 30.13). The accretionary prism has built up to sea-level and formed outer-arc islands such as Nias; a broad fore-arc basin lies between the accretionary prism and Sumatra: the youngest deposits in the basin are fluvial and shallow marine close to the Sumatran arc. (*After D. E. Karig et al, 1980*)

same as for a foreland basin in areas of continent–continent collision. Retro-arc basins show similar characteristics to the foreland basins which form in intra-continental collision zones (see below).

The history of **back-arc inter-arc basins** in the western Pacific (Figure 30.12) shows that basin formation is intermittent. Since late Eocene times the Pacific Plate has been subducting westwards. Initially the Pulau-Kyushu Ridge was the active volcanic arc, but in the mid-Oligocene the arc rifted and Parece Vela Basin started to form by sea-floor spreading as a back-arc basin: the West Mariana Ridge became an active volcanic arc. At about 17 million years ago spreading ceased in the Parece Vela Basin and the volcanic arc rifted to form a new back-arc spreading centre further east. The active volcanic arc is now the Mariana arc, behind which the Mariana Trough is opening as a new back-arc basin. The Mariana Trench, east of the arc is the site of subduction of the Pacific Plate.

Inter-arc back-arc basins like the Parece Vela Basin and Mariana Trough are bounded by an active volcanic arc and a remnant arc left when the new back-arc spreading occurred. The principal source of detritus in inter-arc back-arc basins is the adjacent volcanic arc (Figure 30.18). Pyroclastic material is supplied as flows and falls of volcanic ash during eruptive periods, but there is also reworked volcaniclastic material eroded from the volcanoes and deposited by mass flows and turbidity currents in the adjacent parts of the back-arc basin. Primary and reworked volcaniclastic detritus build up an apron of material on the margin of the basin which may extend tens of kilometres out towards the basin centre, depending on the amount of material supplied. Another source of coarse clastic material is the remnant arc which lies on the opposite side of the basin to the active volcanic arc. The remnant arc is not a site of volcanism or uplift, so erosion reduces the relief and the supply of material from this source diminishes with time.

Pelagic sedimentation may be significant in the deeper parts of back-arc basins, away from the clastic supply at the margins. The nature of the pelagic sediment will depend on the depth of water (see Chapter 24). The oceanic basement in back-arc basins is typically about 1 km deeper below sea-level than crust of the same age formed by mid-ocean spreading centres. At these depths the pelagic sediments are predominantly red clays and siliceous oozes which lithify to form red shales and cherts. At the spreading centre, hydrothermal activity related to the rising magma results in localized metalliferous deposits (Sections 13.8, 24.7).

Back-arc basins cease to become regions of active spreading after a period of time – perhaps 10 or 20 million years. Generation of new crust ceases when rifting occurs in the arc and a new cycle of back-arc basin formation starts (Figure 30.18). The old inter-arc basin may then begin to fill, but this will be very slow because there is no longer a supply of detritus from the active volcanic chain. The only sources of non-pelagic material will be the remnant arcs; these soon become eroded to below sea-level and supply little detritus. Basins such as the Parece Vela Basin in the west Pacific are still areas of deep water, although basin formation stopped 17 m.y. ago.

Sedimentation patterns and sequences in **back-**

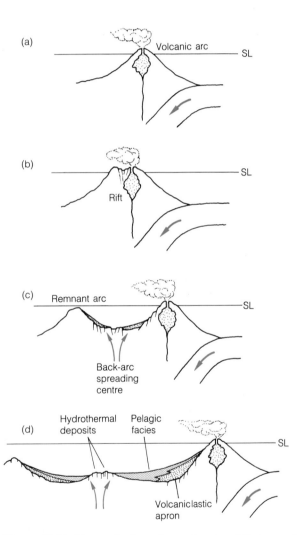

Figure 30.18 The evolution of an extensional (inter-arc) back-arc basin. The volcanic arc (a) undergoes extension and begins to rift apart; (b) as rifting continues a spreading centre develops separating a remnant piece of arc from the active arc; (c) volcanogenic detritus is supplied to the back-arc basin from the active volcanoes and eroded from the remnant arc. In a mature back-arc basin (d) an apron of volcaniclastic detritus from the arc interfingers with pelagic deposits in the basin centre; hydrothermal deposits may be associated with the spreading ridge. (*After Carey and Sigurdsson, 1984*)

arc marginal basins are more varied. Sediment derived from the continental margin is an important detrital component in addition to volcanogenic material. Drainage patterns and the climate of the adjacent continent will govern the volume and type of sediment supplied to the back-arc basin.

The Japan Sea is a back-arc marginal basin formed by spreading between Japan and the Eurasian margin (Figure 30.12). The initial suture occurred close to the continental margin and the volcanic arc of Japan developed on continental basement. Rifting of the margin followed by back-arc spreading has formed the Japan Sea Basin. Pelagic sedimentation dominates in the Japan Sea because clastic supply into this basin has been relatively sparse, despite the proximity of the landmass to the west: this is a consequence of the absence of large rivers feeding into the Japan Sea from the continental landmass.

Neutral arc-trench systems form a marginal basin by trapping a piece of ocean crust behind the volcanic arc. The Bering Sea between Alaska and Siberia is partly floored by old oceanic crust trapped behind the Aleutian arc which lies to the south (Figure 30.12). Sedimentation in marginal basins of this type shows the characteristics of passive margins but with some input of volcaniclastic detritus from the active arc.

Some of the best-studied ancient back-arc basins are in the Lower Palaeozoic of Wales. They formed during the closure of the Iapetus Ocean and volcanic arcs and arc-related basins developed as a result of the oceanic crust of Iapetus subducting beneath Britain. Much of the Ordovician geology of the Snowdon region can be interpreted in terms of the development of volcanic arcs and related back-arc basins (Kokelaar 1988).

30.10 CONTROLS ON SEDIMENTATION IN ARC-TRENCH SYSTEMS

The effects of eustatic sea-level changes may be difficult to recognize in arc-trench related basins because of the strong influence of tectonic and volcanic processes. Subduction at the trench gives rise to a high concentration of seismic activity which may act as a trigger for turbidites and debris flows in arc and trench basins. The subducting slab of lithosphere is commonly irregular in thickness and morphology due to seamounts, oceanic ridges and plateaus. When these thicker areas on the downgoing plate enter the trench they collide with the accretionary prism and cause a massive slump failure on the inner wall of the trench. Furthermore, as they enter the subduction zone they appear to cause uplift in the region of

the accretionary prism and fore-arc basin. Subsidence in inter-arc basins in the back-arc region is controlled by both the volumes of sediment entering the basin and the rate of crust generation at the spreading centre.

Volcanicity in arcs is not uniform along the arc, nor does it remain constant through time. Both the volume of volcanogenic material and the composition of the lavas and ashes are variable. Local variations in volcanic output are reflected in the supply of detritus to the adjacent basins. Inactive parts of the arc become denuded by erosion and the upper volcanic parts are removed to expose the plutonic rocks of the arc massif. This unroofing of the arc is reflected in the petrology of the sediments derived from it; volcanic lithic clasts become less common and are replaced by larger mineral grains of feldspar and quartz.

As mentioned above, the volume and distribution of non-volcanogenic detrital components is largely controlled by the proximity of continental landmasses, the climate and drainage patterns. Carbonate sedimentation occurs in arc-related basins where the supply of volcanogenic and terrigenous clastic material is sparse. Biogenic carbonate sedimentation is important in the back-arc Japan Sea and thick carbonate banks and reefs are known from parts of the Sumatra fore-arc.

30.11 FORELAND BASINS

Subduction of crust at a trench can eventually consume an ocean. As the ocean closes it brings

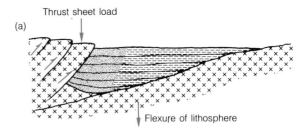

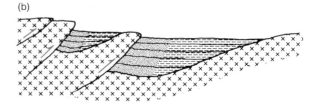

Figure 30.19 Foreland basins form as a result of lithospheric flexure following loading of the crust by thrust sheets. Two types of foreland basin can be recognized: (a) a simple foreland basin consisting of a foredeep forward of the thrust front; (b) a thrust-sheet-top (or piggy-back) basin formed on top of a thrust sheet in addition to a foredeep. (*After Ricci Lucchi, 1986*)

two continents together to the point where they collide. The Himalayas are the most spectacular example of the result of collision between two major continental plates, in this instance India and Asia (Section 31.1). The mountain chain which has formed is over 8 km high and is clearly an area of uplift and erosion. However, continent–continent collision also results in subsidence at the margins of mountain belts. As the plates collide, lithospheric material is forced both up into the air and down into the mantle below. The material forced up is unstable and spreads out sideways, moving as large thrusts and nappes. These thrust sheets add material onto the edges of the plate and this excess load causes the plate to develop a flexure (Figure 30.19). This lithospheric flexure or downwarp of the crust results in subsidence and the formation of a peripheral basin known as a **foreland basin**. The foreland basin is a site of deposition of material eroded from the adjacent mountain belt. The Indo-Gangetic plain is a large foreland basin which is the site of deposition of detritus eroded from the Himalayan mountain belt to the north.

The tectonic evolution of a foreland basin is very closely linked to the development of the orogenic belt (Figure 30.20). As collision leads to more uplift, the volume of material transferred away from the collision zone on thrusts increases. As the thrust sheets apply more load to the foreland, the crust undergoes more flexure causing the basin to deepen and widen. Subsidence in foreland basins is asymmetric because the load on the crust is on the side next to the orogenic belt. Foreland basins hence have much greater accumulations of sediment against the mountain front, and taper away from the orogenic belt. As well as becoming thinner, the sediments become finer towards the foreland as most of the sediment is supplied from the orogenic belt.

The simple geometry of a foreland basin becomes more complex as the orogenic belt evolves. Thrust faults propagate out towards the foreland and some of the earlier deposited sediments become involved in the deformation. These thrusts may completely detach the basin from the basement. A distinction can be made between foreland basins which lie beyond the thrust front, referred to as **foredeeps**, and those which occur within the fold-thrust belt, called **piggyback** or **thrust-sheet-top** basins (Figure 30.19). Thrust-sheet-top basins have a lower preservation

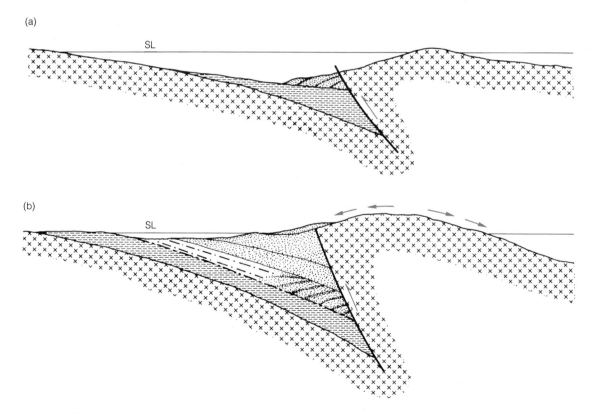

Figure 30.20 The evolution of a foreland basin: (a) in the early stages of collision between two continental blocks, thrust-sheet loading causes crustal flexure and a foreland basin is formed; however, uplift in the orogenic belt is relatively small and little erosion takes place and so the basin is underfilled with sediment; (b) with further orogenic uplift a mountain belt is formed and large quantities of detritus are supplied to the foreland basin; sedimentation exceeds subsidence and the basin fills up to or above sea-level.

potential because further thrust movement and uplift within the fold-thrust belt will result in erosion of the basin deposits. The eroded material is reworked into younger foredeep deposits. Through time, the centre of deposition migrates away from the orogenic belt as the thrust sheets carry the tectonic load further out onto the foreland. This load causes further flexure of the crust and the outer edge of the basin migrates onto the foreland.

In most foreland basins sedimentation follows a similar pattern through time. The basin forms in the early stages of collision (the early orogenic phase) due to loading causing flexure of the plate. However, until the uplift brings the mountain belt well above sea-level, little erosion occurs. During this phase, the subsidence in the foreland basin exceeds the rate of supply of sediment to the basin. The basin in the early orogenic phase is hence largely filled with water, and sedimentation occurs at the bottom of a deep basin. The sediments are largely deposited as deep-water turbidites and other mass flows of detritus from the mountain belt. A Tertiary foreland basin stretches from France into Germany lying parallel to the Alpine belt (Figure 30.21): the deep water, early

orogenic sediments in this region are referred to as 'flysch' and this term is often used for similar sediments in other foreland basins.

As uplift continues in the mountain belt, the rate of erosion increases and the sediment supply to the foreland basin increases, causing the basin to fill up with detritus. As the basin fills, the water depth decreases and the sedimentation shows the characteristics of shallower-water deposition. The late orogenic phase of sedimentation therefore occurs in a shallow-marine environment or in rivers and lakes on land. These deposits are referred to as '**molasse**' in the Alpine foreland basin, which is known as the 'Swiss Molasse Basin' (Figure 30.21). Hundreds of metres of shallow-marine and freshwater 'molasse' were deposited in this basin. The clastic shallow-marine sediments were deposited on deltas, beaches and in tidally-influenced sand bars. During the freshwater molasse phases, large alluvial fans of material formed against the mountain front, fringing a broad plain of river and lake deposits in the centre of the basin.

Subsidence in the foreland basin is not steady and continuous because the load which causes the flexure is not continuously applied. The load is the result of the emplacement of thrusts and nappes which move out from the core of the orogenic belt. With each episode of thrust movement and nappe emplacement, more load is added and the basin subsides. This causes a lowering of the basin floor, which means that there is a relative rise in sea-level. These changes in water depth can be recognized in the sediments deposited. For instance, in the Swiss Molasse Basin two main freshwater and two marine molasse stages of deposition are recognized; the marine phases occurred following emplacement thrust nappes in the Alps which caused loading and subsidence in the basin. This strong tectonic control on water depth in foreland basins makes it difficult to recognize the effects of global eustatic sea-level fluctuations.

There is also a strong structural control on the distribution of sediments in foreland basins. Between the core of the orogenic belt, where most of the detritus is ultimately derived, and the sedimentary basin lies the fold and thrust belt (for example, in the southern Pyrenees, Figure 30.22). The thrust belt is made up of a series of allochthonous (detached) thrust sheets which are complexly folded and faulted. These structures generate an irregular topography within the thrust belt which controls the pattern of drainage. Rivers flowing through the thrust belt tend to be funnelled to widely separated points along the basin margin from where the sediment-laden water spreads out into the foreland basin. As a consequence, the coarsest, most proximal sediments are not found

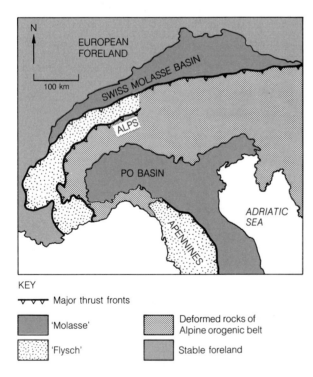

KEY

▽▽▽ Major thrust fronts

◼ 'Molasse'

▒ Deformed rocks of Alpine orogenic belt

⣿ 'Flysch'

▨ Stable foreland

Figure 30.21 Sedimentary basins associated with the Alps and the Apennines. North of the Alps the early foreland basin deposits (deep water 'flysch') have been deformed by later thrusting; at a later stage the Swiss Molasse Basin developed and was a site of shallow marine and continental deposition. (*After Ricci Lucchi, 1986*)

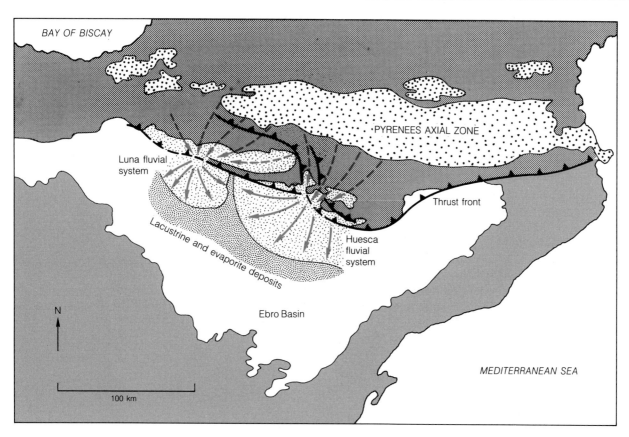

Figure 30.22 Oligo-Miocene alluvial distribution patterns controlled by the structure of the thrust front along the northern margin of the Ebro Basin, Spain. Rivers draining from the Pyrenean axial zone entered the Ebro foreland basin at points where there were structural lows in the thrust front; the rivers distributed sediment as broad fans of fluvial deposits. (*From Hirst and Nichols, 1986*)

as a uniform fringe along the margin of foreland basin but are concentrated in discrete areas.

The pattern of sediment dispersal within the foreland basin will principally be determined by whether the basin is 'open' or 'closed'. Open foreland basins have a connection to an ocean whereas closed basins are totally landlocked with no connection to the open sea. The Swiss Molasse Basin appears to have been connected to the Tethys Ocean through most of its history. There is evidence that water from the Alps distributed sediment along the axis of the basin towards the south-west. In contrast the Ebro Basin, the foreland basin on the southern side of the Pyrenees in Spain, was enclosed during the Oligocene and Miocene. The rivers flowing from the Pyrenees distributed sediment on river flood plains before reaching lakes in the basin centre. The climate during this period was relatively warm and these lakes dried out leaving thick deposits of evaporite minerals (Figure 30.22).

Very thick accumulations of sediment can occur in foreland basins. There are over 5 km of middle Miocene to Pleistocene sediments in the southern Himalayan foreland basin and the Pliocene to present Adriatic Basin (the foreland basin to the Apennines of Italy, Figure 30.21) contains up to 7 km of sediment. Sedimentation rates are high, up to 75 cm/1000 years.

30.12 TRANSFORM PLATE BOUNDARIES

Plate margins where one plate is moving past the other are referred to as transform or strike-slip plate boundaries. Crust is neither created nor destroyed at these margins. However, the movement between the two plates is not smooth, and if the plate edges are irregular, zones of local compression (**transpression**) and extension (**transtension**) are created as the plates move past each other (Section 31.18). Basins formed in these settings are characterized by steep faults at the edges and a rhombic shape. One of the most notable characteristics of basins in strike-slip belts is their great depth compared to the size of the basin. Major strike-slip faults cut very deep into the crust, and basins are a result of the void which occurs when the crust splits in this way. Subsidence is rapid compared to basins formed by other mechanisms.

Three stages of strike-slip basin development were recognized by Reading (1980), analogous to the J. T. Wilson Cycle of basin formation and deformation (Section 30.2). In the first phase of the cycle, transtension in the crust leads to basin formation: coarse alluvial sediments fringe the basin, passing laterally into fine-grained lacustrine or marine sediments. With continued extension crustal thinning takes place and allows basic magmas to rise, ultimately forming oceanic crust (for example, in the Gulf of California). The transtensional phase is followed by a period of basin filling as extension wanes. In this second phase sedimentation, turbidites and debris flows fill the basin and succeeding shallow-water deposits may spread out beyond the faulted basin margins. The final phase is one of transpression: reverse faults and thrusts at the basin margins deform and uplift the basin sediments leading eventually to erosion of the basin fill. Strike-slip belts alternate between transtension and transpression because of irregularities in the margins as the two plates move past each other.

The shape of sedimentary basins formed in strike-slip settings depends on the pattern of faulting (Figure 30.23). Movement along a single, major strike-slip fault can result in the formation of a basin: this occurs where a bend in the fault (a releasing bend) causes extension (Figure 30.23a). Sedimentary basins formed at releasing bends may be supplied with material by uplift at other parts of the fault where a bend causes compression (a restraining bend). Extensional stresses in the region of a fault termination (Figure 30.23b) also result in basin formation. Strike-slip plate boundaries are rarely marked by a single fault. Commonly a number of separate faults can be recognized, and it is the shape of these faults and the relative movements between them which govern the patterns of uplift and subsidence occurring in the region. Straight faults with sharp offsets (Figure 30.23c) form **pull-apart basins**: these basins are steep on all sides and may be very deep. Curved and branching faults form basins which are elliptical or wedge-shaped (Figure 30.23d): basins of this type may have either steep or gentle margins and they may form a complex pattern of small basins and highs.

Strike-slip basins formed within continental crust are normally filled with non-marine sediments. Typically a lake will form and act as a sediment sink. The amount and type of sediment

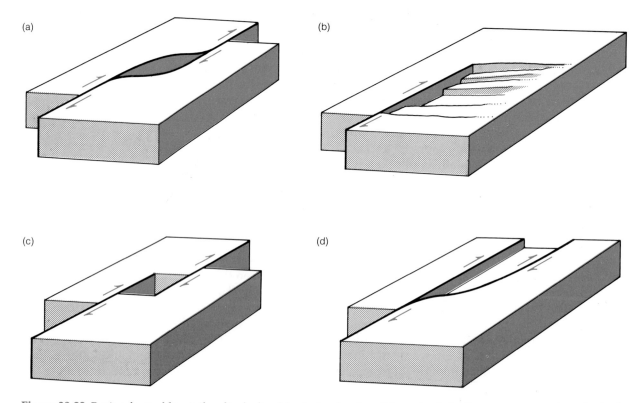

Figure 30.23 Basins formed by strike-slip faults: (a) a 'releasing bend' in a single fault creates a gap when the fault moves; (b) a region of subsidence may form at a fault termination; (c) an offset in a fault produces a 'pull-apart' basin; (d) a branch in a strike-slip fault produces a region of extension between the two strands. (*After H. G. Reading, 1980*)

supplied to the lake will depend on the nature of the surrounding rocks and the climate. The Dead Sea (Figure 30.24) has formed in an offset on a major strike-slip fault between the Palestine Plate in the west and the Arabian Plate to the east. A restraining bend on the fault further north in Lebanon has formed an area of high ground, but the supply of detritus is not great because of the arid climate of the region. The high temperatures and low rainfall are also responsible for the high salinity and low level of the Dead Sea: evaporation exceeds the input from rivers and the salts in the water have become concentrated by the evaporation. Apart from evaporite deposits, the principal sedimentation in the Dead Sea is along the margins where alluvial fans of debris have formed by erosion of material from the high land on either side.

The San Andreas fault system in western North America is an example of a strike-slip plate boundary (Figure 30.25); the east Pacific, with part of the coast of California attached, is moving northwards past the North American continent. The Pliocene Ridge Basin (Figure 30.26) in south-ern California is one of a number of basins related to the San Andreas/San Gabriel fault system. In this instance the climate was humid and the basin was filled with fluvial and lacustrine deposits. Alluvial fans formed along the margins of the basins whilst clastic material supplied by a river in the north was deposited in the lake which covered most of the basin. Over 10 km of sediment was deposited in the basin in approximately 5 million years.

The largest strike-slip basins form where the pull-apart results in sufficient extension to allow the formation of spreading centres within the basin. The Gulf of California lies at a strike-slip plate boundary and is a basin 1300 km long and up to 250 km wide (Figure 30.25). It is composed of a number of coalesced pull-apart sub-basins which have spreading centres in their deepest parts. In the northern part of the gulf the basin is filled with fluvial and deltaic deposits supplied by the Colo-rado River. Further south, the clastic supply from the adjacent arid lands is very low and the domi-nant sediments are pelagic muds.

It is not easy to recognize strike-slip basins in

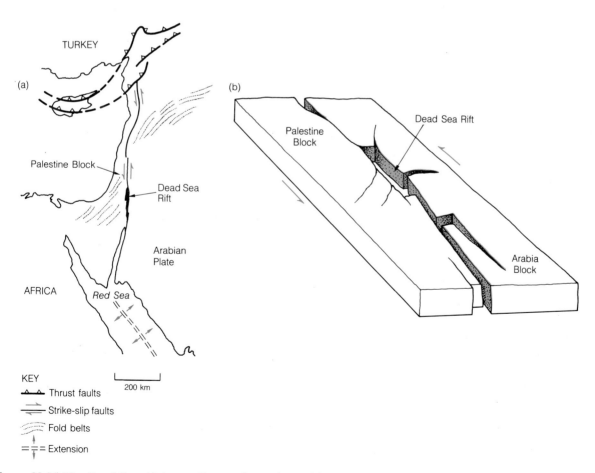

Figure 30.24 The Dead Sea rift is a pull-apart basin formed by an offset in a major strike-slip fault between the Palestine Block and the Arabian Plate. (*After A. M. Quennell, 1958*)

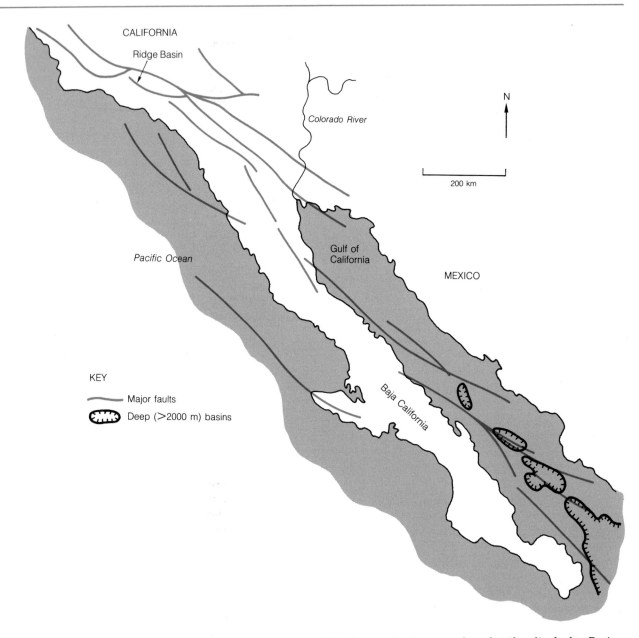

Figure 30.25 The boundary between the Pacific Plate and North America is a complex of strike-slip faults. Basins have formed inland along parts of the San Andreas fault system (e.g. the Ridge Basin) and there are deep strike-slip basins floored by oceanic crust in the Gulf of California. (*After J. C. Crowell, 1974*)

the stratigraphic record because the steep faults at the margin of the basin are often interpreted as normal faults developed in an extensional setting. However, there is one important characteristic of strike-slip basins which can be used to identify them. Alluvial fans frequently form at the edges of the basin supplied by erosion of the adjacent rocks: as strike-slip faulting continues, the fan will be moved away from this source to another point on the basin margin. The fan may then be adjacent to a different bedrock type. This mismatch between the fan sediments and the adjacent bedrock type is often taken as a characteristic of

strike-slip faulting and hence of basins formed in this setting. A strike-slip origin of the Midland Valley in southern Scotland, in the Devonian, has been inferred on the basis of this and other evidence. Alluvial fans of coarse material formed on both the southern and northern sides of the basin. A strike-slip motion on the faults can be shown by comparing clasts in the conglomerates with bedrock lithologies on the basin margin.

30.13 THE EVOLUTION OF SEDIMENTARY BASINS

Our classification of sedimentary basins is based

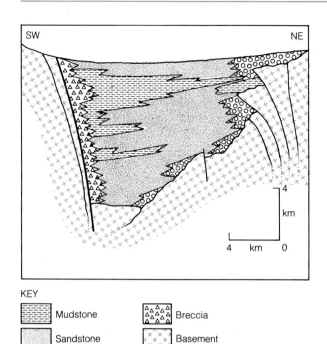

KEY

Mudstone   Breccia

Sandstone   Basement

Conglomerate

Figure 30.26 A cross-section through the Ridge Basin in California. This basin formed by strike-slip movement between strands of the San Andreas and San Gabriel faults; the earlier deposits are marine, but most of the sedimentary fill is alluvial along the margins and lacustrine in the centre. (*After H. G. Reading*)

on the plate tectonic setting: in determining the type of sedimentary basin in a given instance we try to determine the nature of the underlying crust, the proximity of a plate boundary and the type of plate boundary (Ingersoll 1988). However, plate tectonics is fundamentally a dynamic process, and through geological time, some or all of these controls on the sedimentary basin may change. An oceanic basin with pelagic sedimentation may become part of a fore-arc basin when an oceanic trench forms; an intracontinental rift basin will evolve into the passive margin of an ocean if plate separation continues; the continental margin will become the site of a foreland basin if continental collision occurs.

In referring to a sedimentary basin, we are therefore not considering a constant feature, but one which changes in character through time. In examining a sequence of sedimentary rocks, we may expect to find evidence of a number of different basin types as a result of changes in the tectonic setting. As an example, we may consider the history of sedimentation in the region of the southern Pyrenees of Spain since Permian times in terms of changes in the plate motions in the region. The stages in the evolution of the area shown in Figure 30.27 and discussed below are a

modified version of the views of Puigdefabregas and Souquet (1986).

Following the Hercynian orogeny, Permian sedimentation in the Pyrenees occurred in a series of intracontinental rift basins: the deposits are principally those of alluvial fans and braided rivers which formed conglomerates and sandstones known as the 'Bunter' pebble beds. During the Triassic and Jurassic, the first extension related to the opening of the Atlantic occurred. Fluvial and lacustrine conditions were followed by shallow-marine carbonate platform deposition as the whole region underwent lithospheric stretching. During this stage sedimentation was following the pattern of a passive continental margin.

In the early Cretaceous, some local plate movement commenced which profoundly influenced the subsequent history of the area. The Bay of Biscay began to open, and the new microcontinent of Iberia started to move south-west away from Europe. This resulted in the formation of a number of lozenge-shaped, transtensional basins in the Pyrenean region. These were sites of non-marine and shallow-marine sedimentation. As the opening of the Bay of Biscay continued, these basins deepened to become sites of deeper-marine carbonate sedimentation. A change in relative plate motions occurred at about 110 million years ago as Iberia began to rotate and move southeastwards relative to Europe. Deep transtensional basins of turbidite deposition with narrow carbonate platforms along the margins formed between Iberia and France.

A second change in plate movement occurred in the late Cretaceous, probably as a consequence of the northward movement of Africa relative to Europe. Iberia started to move towards France and the first signs of compressional tectonics occurred at about 75 million years ago. The transtensional basins went through a transition to foreland basins as uplift related to crustal shortening occurred in the Pyrenees. During the early Tertiary, continental and shallow-marine sedimentation gave way to relatively deep-water turbidite troughs as subsidence exceeded the rate of sediment supply from the evolving organic belt. The supply of detritus appears to have increased during the Eocene and turbidite deposition gave way to shallow-marine and continental sedimentation. By the beginning of the Oligocene the southern Pyrenean foreland basin was land-locked and a region of entirely continental sedimentation. Rivers spread material which had been eroded from the orogenic belt over the northern part of the basin and lakes occupied the basin centre. This pattern of continental sedimentation continued through the Miocene. Deformation related to the

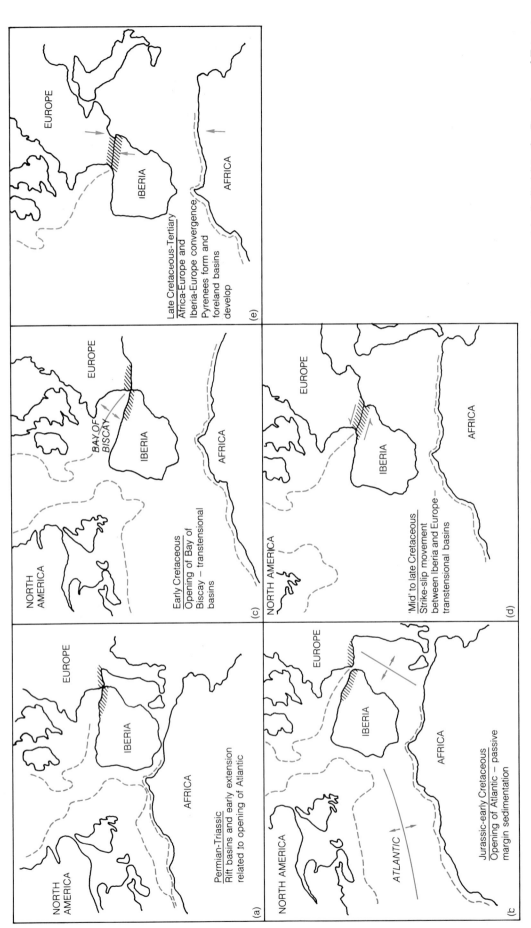

Figure 30.27 The changes in the plate motions and boundaries around Iberia and their effect on the evolution of the Pyrenees (shaded): (a) in the Permian and Triassic early extension related to opening in the Atlantic formed rift basins; (b) as opening in the Atlantic continued through the Jurassic and early Cretaceous the Pyrenees were an area of shelf sedimentation; (c) opening of the Bay of Biscay in the early Cretaceous led to a rotation of Iberia and the formation of transtensional basins between Iberia and Europe; (d) a change in the direction of movement of Iberia in the 'mid' to late Cretaceous formed new transtensional basins; (e) at the end of the Cretaceous Africa and Iberia started to move towards Europe, the Pyrenean orogenic belt was formed and foreland basins developed in northern Spain and southern France. (*After C. Puigdefabregas and P. Souquet*)

convergence between Iberia and Europe ceased in the Miocene.

30.14 FURTHER READING

Allen, P.A. and Allen, J.R. (1990) *Basin analysis, principles and applications*, Blackwell Scientific Publications, Oxford.

Dewey, J.F. (1980) Episodicity, sequence and style at convergent plate boundaries, in *The continental crust and its mineral deposits* (ed D.W. Strangeway), Special Paper, Geological Association of Canada, **20**, pp. 553–73.

Hamilton, W. (1970) The Uralides and the motion of the Russian and Siberian Platforms. *Geol. Soc. Amer. Bull.*, **81**, 2553–76

Mitchell, A.H.G. and Reading, H.G. (1986) Sedimentation and tectonics, in *Sedimentary environments and facies* (ed H.G. Reading), Blackwell Scientific Publications, Oxford, pp. 471–519.

Vail, P.R., Mitchum, R.M. Jr and Thompson, S. (1977) in *Seismic stratigraphy – applications to hydrocarbon exploration* (ed C.E. Payton), Mem. Amer. Assoc. Petrol. Geol. **26**.

Wilson, J.T. (1966) Did the Atlantic close and then re-open? *Nature*, **211**, pp. 676–81.

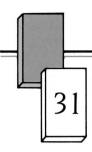

31

OROGENIC

BELTS

It is useful to be assured that the heavings of the earth are not the work of angry deities. These phenomena have causes of their own.

Seneca (4BC – AD65)

31.1   THE NATURE OF OROGENIC BELTS

Most orogenic belts are elongated zones of the Earth's crust that began as basins in which the accumulated sediments were abnormally thick compared with those of the same age on the adjoining platform or shield. The strata and any accompanying igneous rocks were eventually folded, intruded by granitic plutons, faulted and overthrust, and more or less intensely metamorphosed. This group of processes is collectively known as **orogenesis**. Associated uplift made it possible for long sections of the orogenic belt to become mountain ranges like the Alps, the Himalayas, the Andes and the North American Cordillera, with a varied relief of high peaks and deep valleys carved by the agents of denudation. Most of the great mountain ranges of today are parts of orogenic belts that came into existence at various times of major plate interaction from the Jurassic to the present, and particularly during the Cenozoic. The former basins within the orogenic belts are widely regarded as the deformed equivalents of various types of oceanic and intracratonic basins of the present-day Earth as described in Chapters 29 and 30.

From the foothills to the axes of the main ranges structural complexities and metamorphism tend to become increasingly intense. In many belts great masses of granitic rocks are emplaced as batho-liths and stocks. One of the most astonishing facts is that, while the original cover is being removed by erosion or tectonic denudation (e.g. detachment faults), the granitic heart or core of a mountain range may continue to rise until it is exposed in the flanking valleys, and eventually at the summits themselves. Nanga Parbat and some of its Himalayan neighbours (Figure 2.6) and Mt McKinley in the Alaska Range (Figure 31.1), and the high peaks of the Canadian Cordillera and the Andes of Tierra del Fuego, are celebrated examples of this culminating feature of mountain building.

Evidently orogenesis involves subsidence followed by uplift, i.e. by a reversal of the direction of **vertical** movements (a reversal now known as **inversion**). Most geologists think of an orogenic belt as a long, relatively narrow crustal zone that is endowed in some way with an effective source of mobility. The basins and bordering uplifts are the first manifestations of this mobility, and in some cases these uplifts, as for example in the Transantarctic Mountains (Section 31.11), may themselves constitute major mountain belts. The tectonic processes of basin development, deformation, metamorphism, granitic emplacement and uplift, may continue for very long periods before the belt loses its mobility and settles down to a state of relative stability. That

Figure 31.1 Part of the c.950 km long Alaska Range, looking south-west, showing the exceptional height reached by Mt McKinley (6096 m) which, like Mt Blanc in the Alps, occurs in the middle of the great bend of the Range, where the latter is 96 km wide. The mountains in the foreground and to the right beyond the river range from 1500 m to 2500 m and consist of Triassic and Cretaceous lavas, shales and greywackes. Tertiary sands and gravels, tightly folded, are truncated by the 600 m plain seen in the distance. (*Bradford Washburn*)

is to say, until the plate interaction responsible for the orogenic activity ceases due to some plate reorganization. By then the belt may have become welded to the platform or shield in which, or alongside which, the initial basins appeared. Thus orogenesis contributes to the growth of continents. The erosion and deposition that follow bring its surface-level and its crustal thickness into conformity with those of the surrounding regions. It then shares in the epeirogenic movements that are characteristic of platforms and shields, e.g. broad and gentle upwarping into swells and downwarping into basins; local faulting, monoclinal folding, and rifting; and occasional invasion from below by basic magmas from the upper mantle that may be tapped by through-going fissures during lithospheric extension. But although such regions may be 'stable' now, they have not always been so, since they are themselves assemblages of interlacing orogenic belts, each of which was once a mobile zone. Nor is there any certainty that they will everywhere continue to be stable, since new orogenic belts are destined to develop in the future as surely as the

old ones did in former ages. For example, as the indentation of Asia by India ceases, Himalayan uplift will diminish and a new plate boundary and incipient orogenic belt will develop elsewhere. There is already a hint of this in the Indo-Australian Plate to the south and east of India.

Some of the structures characteristically developed within the sedimentary and volcanic infillings of the basins of orogenic belts have already been illustrated. These include alternations of more or less open anticlines and synclines, tightly packed isoclinal folds, and recumbent folds (see Chapter 8); and thrusts and nappes (Figures 31.4 and 31.16a). Other examples follow in this chapter. The axes of the major structures of a range of fold mountains (e.g those that can be shown on small-scale maps) are generally roughly parallel to the trend of the range, but detailed mapping of small-scale structures on large-scale maps commonly reveals wide departures from the dominant trend. This complexity results from the fact that the structures of an orogenic belt are not produced by a single deformation, but represent the accumulated effects of what may have been a long sequence of orogenic events. These may be due to a variety of different processes acting on different materials and at different depths. Thus it is not surprising that the local 'grain' may depart considerably from the regional trend of the belt.

Despite the complexity of folded folds and folded thrusts it is usual to find that, taking the belt as a whole, the large-scale recumbent folds and thrusts are directed outwards from the interior of the orogen towards one or other of the relatively stable platforms or shields which bordered it or else towards the adjacent oceanic lithosphere in the case of continental margin or 'cordilleran' orogens. A continental crustal block towards or over which the structures splay out is called a **foreland**. Some orogens, particularly those resulting from continental collisions have two forelands and the resulting mountain system then has two bordering fold-thrust belts, in each of which the dominant structures splay out towards the neighbouring foreland (Figure 31.2).

This large-scale unity of structure is particularly well displayed by the Alpine system of ranges that resulted from continental collisions (Figure 31.3). The Alpine–Himalayan system as a whole extends from Gibraltar in the west, to the Himalayas in the east. It originated in a long, wide and composite system of basins which developed between the northern foreland of Europe and Asia (Eurasia) and the southern foreland of Africa (including Arabia) and India. To distinguish this immense seaway from the present Mediterranean it was called the *Tethys* by Suess. We now know it widened eastwards towards the Pacific between Gondwana to the south and Eurasia to the north (Figure 31.2). As a result of several orogenic phases resulting from the interaction of the northern and southern forelands as they moved independently, two main sets of ranges originated: a northerly set, including the Alps, the Carpathians, and the Caucasus, Elburz and Hindu Kush ranges; and a southerly set, including the Atlas Mts, the Apennines, the Dinaric Alps, the Tauric and Zagros ranges, and the Himalayas. Where these bordering ranges lie far apart there may be a broad

intervening region of sea, plain, or plateau. Examples of these are the Western Mediterranean, the Hungarian plain, the Iranian plateau, and the exceptionally high plateau of Tibet. Such intermontane basins or plateaus have been called **median** areas. Where there is no median area and the bordering ranges are back to back, the resulting bilateral structure is generally conspicuously asymmetrical. The Western Alps, for example, are so highly asymmetrical as to be almost unilateral, all the thrusts and recumbent folds, except in the extreme south, being directed towards the European foreland (Figure 31.16). Orogenic belts bordering ocean basins usually appear asymmetric, but this is rather deceptive. In the Andes (Section 31.9), for example, the rocks on the continent are thrown eastward towards the South American craton and the range appears asymmetric. Along the continental margin, however, thrusting in the submerged subduction zone is toward the Pacific Ocean. The oceanic side of a continental margin orogen is sometimes referred to as the hinterland.

The direction of overfolding and thrusting is naturally described as it is seen at the surface; that is to say, with reference to the underlying rocks, which are implicitly regarded as having remained stationary. Thus the overthrusts of the North-West Highlands of Scotland (Figure 31.4) or the nappes of the Western Alps (Figure 31.16a) are said to be directed towards the north-west. These movements have commonly been regarded as a result of transport from the south-east. But such a description is purely relative. The rocks have basically been sheared along major zones of lithospheric weakness during plate convergence.

Although plate tectonics now provides an explanation for relative movements of forelands and the transmission of stresses over large distances within the lithosphere, there is an alternative explanation for outward movements, including movements in opposite directions, which obviates the impossibility of transmitting forces for long distances through relatively thin slabs of rock by applying pressure from behind. If the rocks should have an opportunity of gliding down

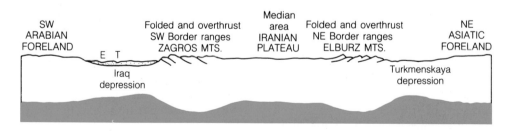

Figure 31.2 Diagrammatic section across the symmetrical composite orogenic belt of Iran and Iraq, with the bordering depressions of Iraq and Turkmenskaya [CIS] E, River Euphrates; T, River Tigris. Length of section, 2160 km.

sloping surfaces of low frictional resistance as a result of their own weight (as in landslides), then transmission of force is no longer a necessity. Gravitation, a body force, is not transmitted like a surface force, but acts throughout the layers of rock concerned as it does in a glacier. Gravitational structures in orogenic belts are, however, generally regarded as being secondary to plate interaction as a deforming agent.

31.2 OROGENIC BELTS OF EUROPE

The general outlines of the tectonic framework of Europe are shown on Figure 31.3. The oldest parts of the continent form the Baltic Shield, and in the extreme north-west a fragment of the Laurentian Shield. Here they will serve to emphasize the fact that by no means all orogenic belts are now mountainous tracts. Most of the Precambrian orogenic belts, and many stretches of the later

ones have lost their original mountainous relief by long continued erosion. Consequently the rocks now exposed to view (e.g. along the shores of Finland, Figure 31.13) include those which were formed from still older rocks by metamorphism operating at depths far below the level of the surface as it was during the time when such crustal belts were mobile and undergoing active deformation. It is obviously important to distinguish clearly between the geographical concept of a mountain range and the geological concept of an orogenic belt. The one refers to the height and relief of the land at a certain time in Earth history such as the present day; the other to the structures of the rocks, whether the belt or any part of it is mountainous or lowland or submerged beneath the sea. Old orogenic belts, while displaying structures indicating former compression and uplift, are now commonly eroded and peneplained.

The stable triangle of the Baltic Shield and

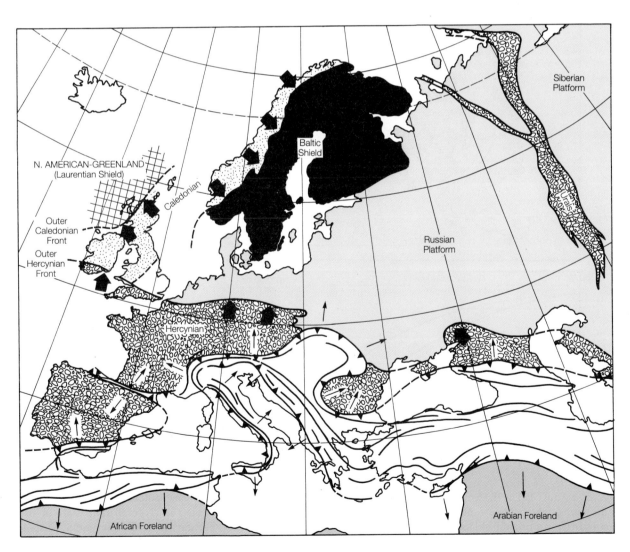

Figure 31.3 Tectonic map of Europe. The composite Alpine orogenic belt is outlined in thick brown lines with long arrows indicating the outward direction of thrusting and overfolding. The brown 'teeth' are on the upper plates of the thrusts. Outward thrusting of the Older Hercynian and Caledonian belts is shown by short, thick arrows.

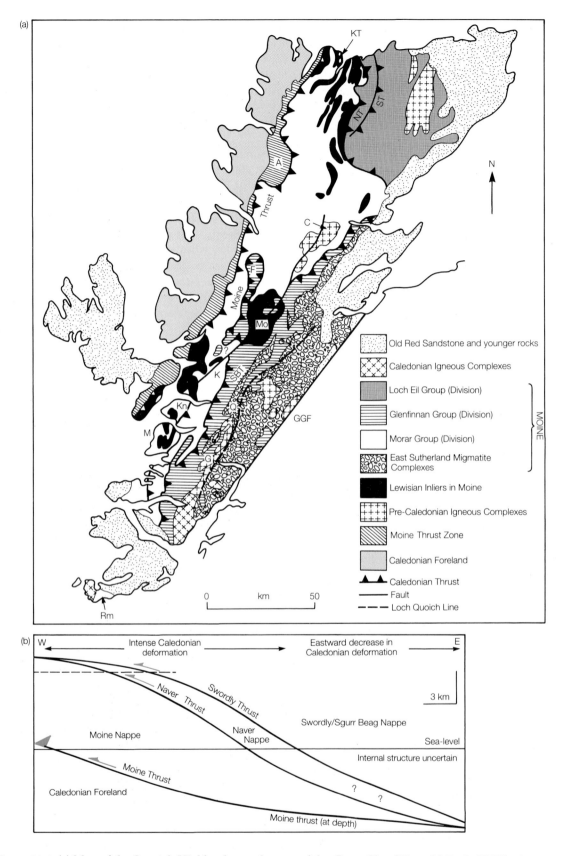

Figure 31.4 (a) Map of the Scottish Highlands, north-west of the Great Glen (Wrench) Fault (GGF), showing thrust faults along which Moine metamorphic rocks have over-ridden basement rocks of Neo Britain (part of the Precambrian continental shield of Laurentia). A, Assynt; C, Carn Chuinneag; F, Fannich; G, Glenfinnan; K, Kintail; Kn, Knoydart; KT, Kyle of Tongue; M, Morar; Mo, Monar; Q, Quoich; Rm, Ross of Mull; NT, Navar Thrust; SBT, Sgurr Bheag Thrust; SH, Strath Halladale Granite; ST, Swardly Thrust. (b) Section of main structural units in north of area. (*Adapted from G.Y. Craig, 1991*)

Russian Platform is bordered on its three sides by clearly defined orogenic belts, towards each of which it acted as a rigid foreland. These belts are believed to represent the sites of ancient ocean basins. On the north-west the Caledonian belt extends through Scandinavia to Britain. In Scandinavia the south-eastern front of the belt is well preserved. Outward transport of nappes towards the Baltic Shield is conspicuous in many places with displacements of up to 100 km and more. Within the belt itself the rocks are contorted, metamorphosed, and intruded by granites. The north-western front of the belt is cut off by the Atlantic, except in the North-West Highlands, where the thrusts illustrated in Figure 31.6 splay out over another shield area, of which only a narrow strip now remains as land (Figure 31.4). This relic of the north-west foreland, together with the outer Hebrides, represents part of the Laurentian (North American–Greenland) Shield which was left behind during Mesozoic–Cenozoic opening of the North Atlantic when North America drifted away from Europe. The south-eastern front of the Caledonian belt is poorly defined in Britain. Much of it is hidden by later sediments, and where the older Palaeozoic rocks are exposed – as in Shropshire, England – the folding is open and broadly undulating.

The Caledonian mountains of Norway and Scotland present familiar present-day examples of ranges that owe their mountainous relief to the rejuvenation brought about by a renewal of uplift, following a long period of erosion, submergence, and burial. Equally remarkable is the fact that some stretches of the Caledonian belt appear to be still submerged and deeply buried. Between Norway and Britain the belt disappears beneath the North Sea. In view of the shallowness of the North Sea this may not seem to be particularly significant. The North Sea, however, has been a predominantly subsiding area since Early Carboniferous times. The Caledonian belt here floors an oil-bearing basin filled practically to the brim with sediments which, in the deepest known parts, reach a thickness of at least 7.5 km. It developed mainly as a 'failed' arm of a rift during opening of the Atlantic Ocean basin that divided the Caledonian Chain from its south-westerly continuation, the Appalachian Mountains of eastern North America.

While the final Caledonian movements were

Figure 31.5 Escarpment of Cambro-Ordovician dolomitic limestone overlooking the Precambrian rocks of the north-west foreland (Laurentian) of the Caledonian orogenic belt. Near Elphin, North-west Highlands of Scotland. (L.S. Paterson)

taking place towards the end of the Silurian or during the early Devonian, other orogenic basins were already beginning to develop and fill up with sediment:

(a) on the east along the site of the Urals, parts of which, particularly along the western flanks and in the Timan range, which branches off in the north, already had a Caledonian ancestry; and

(b) along the south across Europe from the promontories of south-west Ireland to north of the Sea of Azov. These became transformed into Hercynian (or Variscan) orogenic belts by compression and inversion during Carboniferous and early Permian times.

Most of the Uralian belt is continuously preserved, but the much wider Hercynian belt of the south has become broken into a series of isolated blocks or **massifs** (Figure 31.6), during Mesozoic extension associated with the opening of the North Atlantic, the North Sea and the Tethyan area. These include south-west Ireland, South Wales, and Cornwall and Devon; Brittany and the Central Plateau of France; the Ardennes, Vosges, and Black Forest (Figure 31.7); and the Harz and Bohemian Mountains. The depressed regions between are buried beneath later sediments, but here and there (e.g. in south-east England) the Hercynian foundation has been encountered in borings and mining operations. There is no doubt that the uplifted massifs referred to are parts of a highly complex belt that was originally continuous.

Certain parts, e.g. the massifs of Mt Blanc and the Aiguilles Rouges, after deep burial by Triassic and later sediments, rose again to become islands near the northern shores of the Tethys. These, together with parts of the Tethyan floor itself, were incorporated in the Alpine ranges which later completed the southern framework of Europe during Africa–Europe collision and **inversion** (i.e. uplift) of the Tethyan basins.

In most places the northern Hercynian front lies well to the north of the northern Alpine front, but the Carpathians were thrust forward beyond the Hercynian front, so that here the Russian Platform became the foreland. In Silesia (Poland) the structural relations between the mountains and their foreland have been made clear by mining operations. The broad coalfield of Silesia – belonging to a Hercynian depression in the older foreland – was partly overridden by the advancing nappes of the Carpathian arc. Nevertheless, the buried half of the coalfield has been located beneath the nappes by borings sunk through them into the coal seams beneath.

Just as the Alpine structures encroach on the older Hercynian belt, so in the British Isles (see Figure 31.3) the Hercynian structures in turn

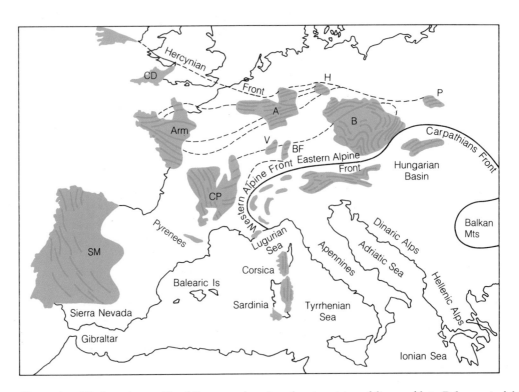

Figure 31.6 Hercynian (Variscan) massifs of Europe, showing dominant trend lines of late Palaeozoic folding. A, Ardennes; Arm, *Armorica*; B, Bohemia; BF, Black Forest; CD, Cornwall and Devon (*Cornubia*); CP, Central Plateau of France (Massif Central); H, Harz Mts (*Hercynia*); P, Polish Massif; SM, Spanish Meseta; V, Vosges.

Figure 31.7 The dissected plateau of the Black Forest viewed from the Feldberg. (*Dorien Leigh Limited*)

encroach on the still older Caledonian belt. From South Wales to the south-west shores of Ireland the Lower Palaeozoic rocks with their typical Caledonian structures disappear beneath the folded and overthrust Devonian and Carboniferous sediments which represent the northern Hercynian front. The two orogenic fronts, gradually converging across Europe, ultimately meet, and the northern Hercynian front begins to cross the Caledonian belt of south-west Britain. What is probably the completion of the crossing is found in the Appalachian Mountains (Figures 31.8, 31.9) on the other side of the Atlantic. Such crossings reflect the fact that the orogenic belts represent the sites of ocean basins that opened and closed at different times and on different trends.

31.3 THE APPALACHIAN MOUNTAINS

These mountains (Figure 31.8) appear to be a closely knit complex of two sets of orogenic belts (here referred to as 'Older' and 'Newer') which approximately correspond to the Caledonian and Hercynian belts of Europe. Because of the pioneer work of the Rogers Brothers and James Hall more than a century ago, the Appalachians were long regarded as the standard example of a **geosyncline** (see Section 30.1 and Section 31.4) which

developed into a unilateral system of fold-mountains, with the major structures directed towards the Laurentian Shield and its buried continuation beneath the Interior Lowlands (a region corresponding to the Russian Platform). This apparent simplicity, however, only reflects the fact that along the Atlantic seaboard the south-eastern side of the orogenic belt is everywhere concealed by the later sediments of the Coastal Plain (Figures 31.8 and 31.9). The foreland on the south-east is not seen, because the Atlantic now occupies the site where we should naturally look for it. Recognition of the south-eastern foreland depends on what rifted from ancestral North America (including north-west Scotland) (Figure 31.4) at the end of the Precambrian to form the basins eventually inverted in the Appalachian orogeny. The foreland is traditionally regarded as Europe and Africa but has recently been suggested to have been South America!

In the north the overthrust Older Appalachian front begins in Newfoundland. Continuing along the line of the St Lawrence as far as the hills opposite Quebec, it then turns southwards towards New York. During the Ordovician (the Taconic orogenic phase) and again during the Devonian (the Acadian phase) all this northern section was folded, invaded by granite, uplifted and overthrust in much the same way, though

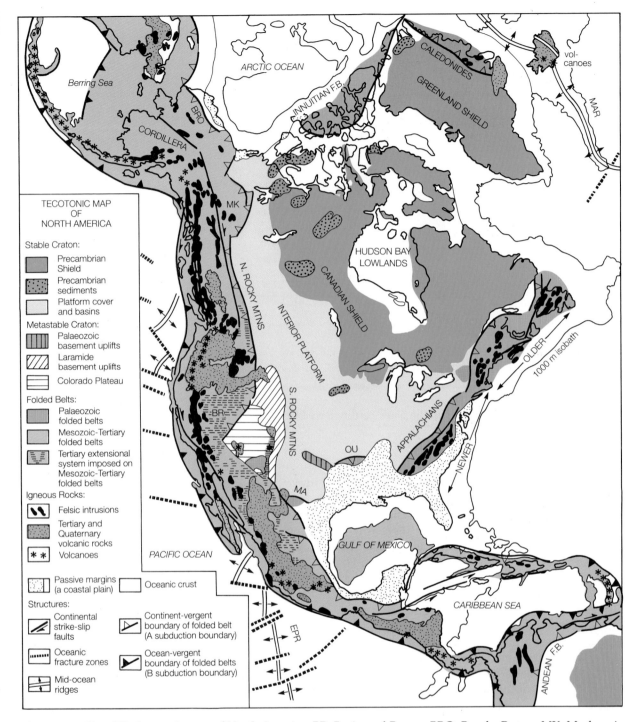

Figure 31.8 Simplified tectonic map of North America. BR, Basin and Range; BRO, Brooks Range; MK, Mackenzie Mountains; MA, Marathon uplift; OU, Ouachita Mountains; MAR, mid-Atlantic ridge; EPR, East Pacific Rise; FB, Fold belt. (*After A.W. Bally et al, 1989*)

with somewhat different timing, as the Caledonian belts of Scotland and East Greenland. South-west of New York the Caledonian part of the chain is now on the seaward side and here the old basement rocks, representing the floor of the geosyncline, have been further metamorphosed and intruded by granite, and uplifted (Figure 31.9a,b). Only a few infolded and highly metamorphosed remnants of the basinal sediments still

remain, except in the Carolina slate belt, where the structures are directed towards the hidden or vanished south-east foreland. Radiometric dating of the basement rocks shows that the floor of the Appalachian orogenic basin was an extension of the Grenville Province of the Laurentian Shield (about 1000 Ma). Where recrystallization has been complete, however, potassium–argon ages date the metamorphic event at about 470 Ma

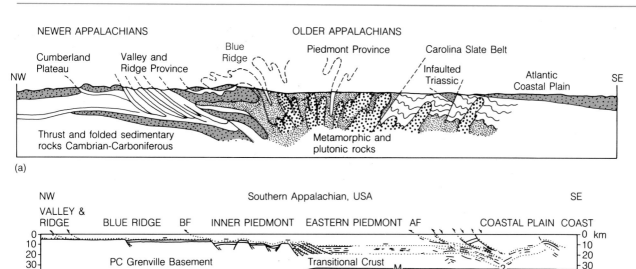

Figure 31.9 (a) Diagrammatic section across the composite orogenic belt of the southern Appalachians. Length of section, 560 km. (*After P.B. King, 1950*) (b) Appalachian crustal structure from seismic-reflection data. Note 'thin-skinned' tectonics with deformation largely confined to the supracrustal sedimentary wedge above a Precambrian basement. Abbreviations for faults: BF, Brevard fault; AF, Augusta fault; M, Mohorovič discontinuity.

corresponding to the Taconic orogenic phase.

On the inland side, between the Blue Ridge and the Appalachian Plateau (the Cumberland Plateau in the south), Palaeozoic sediments are largely preserved in a broad series of deep folds, broken and interrupted by thrusts directed towards the north-west. Still farther inland, in the Plateau, the folding gradually flattens out. In these 'Newer Appalachians' not only Lower Palaeozoic sediments are found, but also a far thicker sequence belonging to the Upper Palaeozoic. The whole assemblage was folded during the Alleghanian orogenic phase, which began in the Carboniferous and reached its climax in the Permian. The 'Newer Appalachians' thus correspond with the later phases of the Hercynian orogenesis. The Taconic, Acadian and Allaghanian phases together constitute the 'Appalachian orogenic revolution'.

Farther north, in the New England States, relics of strongly folded Carboniferous rocks mark the continuation of the 'Newer Appalachians.' But here they lie on the seaward side of the Caledonian front of the Appalachian system, whereas in the south the folded belt is on the inland side. The 'Newer' or Hercynian front appears on the coast near Boston, reaches the 'Older' or Caledonian front behind New York, and then, as E. B. Bailey puts it, 'steps clear of its Caledonian predecessor.' The crossing of the Caledonian belt by the outer front of the Hercynian belt, begun in Britain, is finally accomplished in America (Figures 31.3 and 31.8). As mentioned above, such crossings are now thought to represent the intersection of a younger ocean basin with an older one.

31.4 'GEOSYNCLINES' AND SEDIMENTATION

The concept of an orogenic basin or **geosyncline** was born in the Appalachians. In 1859 James Hall, as a result of his work as State Geologist of New York on the stratigraphy and structure of the northern Appalachians, recorded his great discovery that the folded Palaeozoic sediments of the mountain ranges are shallow-water marine types which reach a thickness of just over 12 km, and are everywhere ten to twenty times as thick as the unfolded strata of corresponding ages in the interior lowlands to the west.

The accumulation of so thick a sequence of sandstones, shales, and limestones clearly implies that the underlying floor of older rocks must have subsided by a like amount. The mountains were evidently preceded by long periods of downwarping during which sedimentation more or less kept pace with the depression of the crust. Such elongated belts of long-continued subsidence and sedimentation were called **geosynclinals** by Dana in 1873. Later the modified form (geosyncline) came into general use. With the advent of knowledge of the ocean basins following the Second World War, and especially with the understanding of plate tectonic theory in the 1960s and 1970s, the word geosyncline fell into disuse. It has been replaced by more actualistic models of basins of sedimentation in orogenic belts. None the less, there is a close similarity between the various types of geosyncline described in the literature, and modern continental margins and ocean basins, as we have seen in Chapter 30.

Hall and many of his followers thought that the weight of the ever-increasing load of sediments was itself sufficient to depress the crust and so automatically to provide room for still more sediments. Dana, however, could not accept this hypothesis. Instead, he proposed that the sediments accumulated where subsidence by earth movements had already made room for their reception. While enthusiastically supporting the principle that thick accumulation of sediments in a geosyncline is a necessary preliminary to orogenic mountain building, Dana clearly realized that Hall's discovery, momentous and abundantly confirmed as it was, failed to account for the folding of the strata and the uplift of the mountains. His sense of humour led him to declare that Hall had promoted 'a theory for the origin of mountains with the origin of mountains left out.' This famous comment has, of course, a double or ambiguous meaning, since the term 'origin' can refer either to the beginning of the process (the geosyncline) or to the finished end-product (the mountain range).

It is easy to show from the standpoint of isostasy that while yielding of the crust under a growing load of marine sediment undoubtedly occurs, the total effect is strictly limited. Suppose the initial depth of seawater to be c. 30 m (stage (a) in Figure 31.10) and that sediments of density 2.4 accumulate, and depress the crust isostatically, until the region becomes completely silted up (stage (b)). Let the maximum thickness of marine sediments so deposited be h metres. The crust is then depressed by $(h - 30)$ metres, and this must also be the thickness of the deep-seated sima, density 3.4, displaced below the base of the crust. At stage (b)

the weight of sediment added is proportional to $2.4 \times h$
the weight of water displaced is proportional to 1.0×30
the weight of sima displaced is proportional to $3.4 \times (h - 30)$

For isostasy to be maintained the weight lost must be equal to the weight gained. Thus, we have

$2.4\,h = 30 + 3.4\,h - 102$; whence $h = c.\ 73$ m.

For the c. 12 000 m of Appalachian sediments to accumulate under such conditions, the initial depth of water would have had to be over 5000 m. Taking 3.3 as the density of the displaced sima and 1.03 as the density of seawater makes no practical difference, the initial depth then being only a little less at c. 4900 m. Actually, however, the seas of the Appalachian orogenic basin were very shallow at the start as shown by the abundance of shore and deltaic deposits. Many sediment-filled tectonic basins illustrate the same significant phenomenon. Both in Europe and North America the Coal Measures provide some of the most familiar and convincing examples. In the South Wales coalfield the floor of the Upper Carboniferous sank at least 2500 m; and similar and greater thicknesses of coal measures have been recorded elsewhere. Yet all were kept filled with sediments which, like the coal seams themselves, were deposited never far above or below the sea-level of the time. It follows that sedimentary load cannot be the main cause for the depression of the floors of orogenic and related tectonic basins. It is primarily the independent downwarping of the crust that makes room for the sediments to accumulate. This is proved in many Alpine and other orogenic basins by the occurrence of sediments of deep-sea types here and there in the sequence. Evidently there were times when the floors were downwarped to depths that were beyond the reach of ordinary sediments except when transported and re-deposited by turbidity currents.

The subsidence mechanism is now recognized to be dominantly thermal (Chapter 28). Stretching and thinning of the continental crust in the initial stages of rifting to form an oceanic basement results in a denser column, but heating from below keeps initial subsidence to a minimum. Subsequent slow thermal cooling accounts for the tremendous thickness of shallow-water sediments. Phase changes in the underlying lithosphere may also play a role.

In every orogenic belt that has been mapped in detail it has been found that the floor of the parental basin was not a continuous, uniformly subsiding trough, but rather a string of elongated

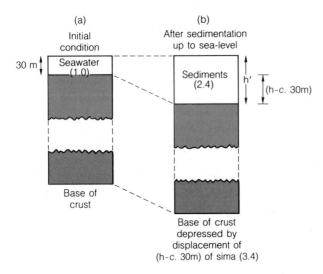

Figure 31.10 Diagram to illustrate the isostatic response of the Earth's crust to loading by sedimentation.

basins, which might be arc-shaped or straight, and in many examples arranged *en echelon*. Modern examples are the individual basins of the Mediterranean Sea, the marginal basins of the western Pacific (Figure 30.12), and the numerous troughs along the margins of the continents bordering the Atlantic Ocean (e.g. Figure 30.7).

Thus, the thicker accumulations of sediments and other infillings are separated by stretches in which accumulation during the same interval of time was limited by the depths then available. Moreover, there are generally two or more roughly parallel troughs with intervening barriers of various kinds, such as uplifted ridges of basement rocks or newly rising chains of fold-mountains. These may continue to rise until, becoming subject to erosion, they supply detritus to the adjoining depressions while the latter continue to deepen. Barriers which ascend from within a 'geosynclinal' system were referred to as **geanticlines**. Barriers of another kind may be built up in the form of volcanic arcs, the emergent cones and vents of which supply volcanic products to the surrounding slopes and depressions. Again, present-day examples are to be found in the bathymetric highs separating the basins of the western Pacific, Mediterranean and Atlantic margins.

Following a classification devised by H. Stille, geosynclines characterized by intermittent volcanic activity during their infilling were commonly referred to as **eugeosynclines**, while those with a paucity if volcanic products were distinguished as **miogeosynclines**. The Greek suffixes *eu* and *mio* are meant to indicate relatively high and low status from an igneous or mobility point of view. As exemplified by Figure 31.11 the two classes tend to occur side by side with a geanticline between. Attempts have been made to correlate certain assemblages of sediments with each of the two classes, based largely on the idea that the miogeosyncline adjoins a lowland platform, while the eugeosyncline is fed from a more actively

rising and more vigorously eroded borderland or volcanic arc. When these conditions are fulfilled, typical sediments of the 'eu' class are dark shales and poorly sorted sandstones (greywackes), often exhibiting graded bedding, and interpreted as deposits from turbidity currents. In the Alps monotonous thicknesses of such sediments, amounting to many thousands of metres, are referred to as **flysch**, and this term has been widely applied to similar deposits in other areas. The appearance of coarse materials in the sequence reflects a marked uplift of the source areas. Such sediments are therefore 'synorogenic' in the sense that sedimentation accompanied uplift of at least local tracts of these orogens.

Sediments of the 'mio' class characteristically include well-sorted sandstones, often seen as white- or pale-tinted quartzites; and limestones, some of which may be dolomitic. Related deposits also commonly occur in the depressions that develop in front of the rising mountains during the waning stages of an orogenic cycle, e.g. the Coal Measures in the foreland basins in front of the Hercynian and Appalachian ranges, and the **molasse** of the Swiss plain in front of the Alps.

Miogeosynclines are now regarded as former rifted continental margins such as those bordering the Atlantic Ocean, and eugeosynclines are believed to represent the deformed and inverted equivalents of small oceanic basins like the marginal basins (see Section 30.6) of the western Pacific such as the Sea of Japan and the Sea of Okhotsk.

The presence of rocks of oceanic affinities in Newfoundland (the Bay of Islands complex) and Scotland (the Ballantrae complex), for example, led to the idea that the Caledonides and 'Older' Appalachians represent the site of the ocean basin referred to as Iapetus (Section 3.8). Differences between the Cambro-Ordovician benthic fauna on the north-western and south-eastern margins of the belt and palaeomagnetic data suggest that

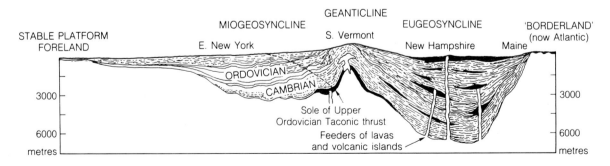

Figure 31.11 Traditional restored section of the Cambrian and Ordovician strata and volcanic rocks according to geosynclinal theory, as they were before the Taconic orogenic phase of the Older Appalachians of New England; illustrating the divisions of the Appalachian geosyncline from the foreland on the west to the 'borderland' (now the Atlantic) on the east. Length of section (425 km). (*After M. Kay, 1951*)

Iapetus was a deep and significant ocean basin, although perhaps not quite as wide as the present Atlantic Ocean. It was bordered in parts (for example Newfoundland and New England) by marginal basins and island-arc systems as in the present western Pacific. The entire Iapetus system was, however, closed and inverted during the Caledonian/Taconic and Acadian phases with the oceanic rocks marking the site of the suture between the two opposing continental margins, before opening of a complex series of basins that eventually gave rise to the Hercynian/Alleghenian orogen. Those basins, in turn, were closed and inverted to form the Hercynian mountain belt before the opening of the present-day Atlantic Ocean basin in the Mesozoic.

31.5 OROGENIC BELTS AND IGNEOUS ACTIVITY

The volcanic rocks interbedded with so-called 'eugeosynclinal' sediments are characteristically submarine lavas (Chapters 5 and 13), including pillow lavas (Figure 13.33 and Plate IXb) now seen in various stages of deformation and metamorphism (Figure 31.13). They are mostly of basic composition, with a high proportion of greenish minerals such as chlorite, serpentine, epidote, and hornblende. Ultrabasic rocks composed largely of serpentine (**serpentinites**) are often associated, especially in the earlier stages. In some localities, e.g. Turkey, there is evidence that these rocks were emplaced as submarine lavas. At the surface ultrabasic liquids would require a practically impossibly high temperature, but this difficulty does not arise when there is a considerable overhead pressure of water, as there may commonly

have been in certain stages of a major orogenic basin. However, whether these basic and ultrabasic rocks were extruded or intruded, they are collectively known as **green rocks** or **green schists** (according to their degree of alteration), and more generally as the **ophiolites**. The latter term (from the Greek, *ophis*, a serpent) was introduced by G. Steinmann, who drew attention in 1905 to the frequent association of **spilites** (altered basalts), serpentine rocks, and cherts in mountain systems of the Alpine type. Following E. B. Bailey, this assemblage became known to the geological world as the 'Steinmann trinity'. The presence of chert suggests that seawater in contact with newly erupted submarine lavas became enriched in SiO_2, which would be deposited as the temperature fell. When radiolarians (Figure 24.4) first appeared such eruptions would present them with highly favourable conditions for flourishing. But since chert is also found in association with the extremely old pillow lavas of Zimbabwe for example, it is clear that contributions from radiolarians *per se*, and perhaps from organic material generally, are not essential.

The term **ophiolite**, and particularly **ophiolite suite**, is now reserved for occurrences on land of the entire sequence of rocks believed to constitute the oceanic lithosphere. That is to say an ultramafic rock such as peridotite at the base, gabbro, sheeted dykes, and pillow lavas, overlain by bedded chert. The Deep Sea Drilling Project and its successor the Ocean Drilling Program (Section 24.9) have drilled as deep as gabbro, and ultramafic material has been dredged from escarpments beneath the oceans.

The volcanic rocks of orogenic basins are not invariably of basic composition. Acid varieties in

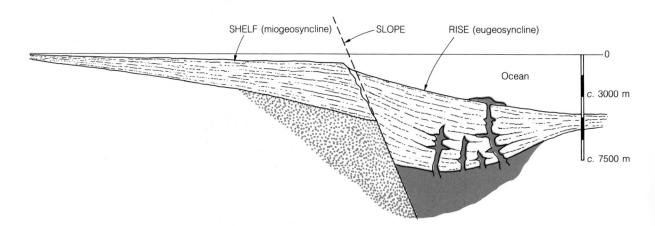

Figure 31.12 Hypothetical cross-section of a passive margin from Dietz (1963). Both Dietz (1963) and Drake *et al* (1959) attempted to interpret the entire geosyncline in terms of modern passive margins, assuming that the 'tectonic land' separating the mio- and eugeosynclines corresponded to the outer basement highs of modern passive margins, which appear as prominent basement ridges between the shelf and slope in seismic profiles. (*After G. Bond and M. Kominz, 1988*)

which the characteristic feldspar is albite (as in their more abundant basic associates) are also included in the spilitic suite. Moreover, ordinary andesites, accompanied by basalts on the one hand, and by dacites and rhyolites on the other, together with still greater volumes of ignimbrites, have made large contributions to the infillings of many major orogenic basins. Examples can be seen in the Cordilleran orogenic belts of North America. Others are to be found in the Caledonides of Wales, and the Appalachians and the Andes. The high volcanoes and Mesozoic granitic batholiths of the Andean Cordillera (see later) are the type example of subduction-related magmatism in a present-day orogenic belt.

What seems to have been the oldest and most mobile of all the Caledonian orogenic basins of Britain lay between the Highland Boundary Fault and the Moine Thrust belt (see Figure 31.4) and their continuations across Ireland, enclosing Donegal and Galway. The floor of this Highland geosyncline where exposed, consists of the metamorphosed sediments known as the Moine Series. Pegmatites traversing these rocks have a minimum radiometric age of 740 Ma and the Moine sediments may originally have been a connecting link between those of the Grenville Province in North America and rocks of similar age in southern Scandinavia. But, as we see them now, the Moines are largely the product of early Caledonian metamorphism, roughly approximating in age to the Taconic phase of the Appalachians.

It is customary to think of the Caledonian orogenic basin in the Highlands as having begun in very late Precambrian times with the deposition of a sedimentary series, now almost everywhere metamorphosed, known as the Dalradian. In Connemara Ordovician pillow lavas rest unconformably on typically metamorphosed Dalradian rocks, indicating the intervention of an orogenic phase that might correspond approximately with that known as the Taconic in the Appalachians.

Throughout the Caledonian belts that border much of the North Atlantic the Ordovician was the period in which volcanic activity was most intense. In Britain the earliest Ordovician volcano was Rhobell Fawr, east of Harlech in North Wales, followed soon afterwards by eruptions in southwest Wales and the south and western border of the Scottish Midland Valley. Activity became more widespread later, particularly around the Harlech Dome in Wales and in the English Lake District. In both regions it died out after bringing to the surface the last of the lavas and tuffs responsible for the rugged scenery of Snowdonia in North Wales (Figure 15.2) and of Borrowdale in the Lake District (Figure 3.2). During the Silurian there was no volcanic activity in Britain and only a little in the west of Ireland. Nevertheless, the Late Silurian and the Early Devonian saw the final phase of the Caledonian orogenesis. This reached its greatest intensity in what are now the Scottish Highlands where high-grade metamorphic rocks are now exposed and intruded by numerous gra-

Figure 31.13 Deformed pillow lava from the Svecofennian orogenic belt, Djupsund Sibbo, southern Finland. (*Eugene Wegmann*)

nitic stocks and batholiths. The last of these were associated with and followed by the great volcanic eruptions of the Lower Old Red Sandstone (e.g. Ben Nevis, Figure 31.14; and Glen Coe). At this time there was also andesitic volcanic activity within the Midland Valley tract, the products of which are still topographically conspicuous as the Sidlaw and Ochil Hills on the northern side, and as the Pentlands and their extensions to the outskirts of Edinburgh on the southern side. The Lower Old Red Sandstone volcanics were generated by north-west directed subductions of Iapetus near the time of final suturing.

In the crumpled rocks of the Moffat area to the south-east, thought to represent a fore-arc accretionary wedge, there are several Caledonian granitic masses, the largest being in Galloway, while Ireland has the Newry Complex and the larger but less well-known Leinster Granite south of Dublin. A few smaller occurrences are exposed in the Lake District (Figure 18.9) and the Isle of Man, but the only example with associated lavas of 'Old Red' age is the granitic core of the Cheviot Hills. The Welsh Caledonian basin, despite the great depth to which it was filled (c. 10 500 m), has only minor granitic intrusions to show; and most of these are

products of Ordovician activity, 'Old Red', lavas being unrepresented. This remarkable contrast with the activity to the north-west shows that even when the thickness of sediments and interbedded volcanic rocks exceeds 12 km, the rise of temperature and liberation of volatiles may fall short of the conditions necessary for andesitic vulcanism. Some additional source of heat and emanations is required to bring about the culminating manifestations of mobility and orogenesis, such as are seen in the Grampians and Northern Highlands, and in the corresponding parts of the Appalachians. Modern examples, in particular the Andes, indicate that subduction of oceanic lithosphere with its cover of hydrated pelagic sediments and trench turbidites is required to generate large batholithic bodies and andesitic volcanics (Sections 5.6 and 12.7).

Another example of this contrast is reviewed later in dealing with the North American Cordillera which was an Andean type margin until development of the San Andreas fault in the Cenozoic (Section 30.12). The tectonic, thermal, geochemical, and magmatic changes that accompany major crustal revolutions – those in which vast quantities of newly re-constituted granitic rocks

Figure 31.14 Ben Nevis viewed from Banavie, north-east of Fort William, Scotland. The summit (1343 m) is formed from a core of Lower Old Red Sandstone volcanic rocks about 600 m thick and a mile across; this is surrounded by a ring of fine-grained granite, which in turn is enclosed on three sides by coarser granite. (*Doris L. Reynolds*)

appear – are superbly illustrated by the Sierra Nevada and the Coast Ranges. In the Rockies (see later), however, immense blocks of Precambrian basement rocks rose up through many thousands of metres without going through the mobilization stages that would have generated 'new' granitic rocks at the levels where we now see them, except on a limited and localized scale. Table 31.1 gives a list of the radiometric ages of some of the uplifted basement rocks and shows how successfully they escaped the widespread **palingenesis** (to use Sederholm's term for the **rebirth** of an ancestral granite) that rejuvenated similar old rocks during the Nevadan orogenesis farther west.

It would appear that during the Nevadan orogenesis granites were generated by partial melting above a subduction zone, as in the Andean example. During the Laramide orogenesis, however, partial melting beneath the Rockies was limited, and only along fracture zones and shear planes did fluids migrate upwards and so made possible the generation of small stocks and intrusions of porphyry and the introduction of ore deposits. In the Great Basin (Section 29.4), relics of formerly widespread sheets of ignimbrite testify to a time when emanations were highly concentrated and unusually hot, so that molten and highly gas-charged materials from depth ascended through fissures to the surface and there escaped as incandescent spray. The Sierra Nevada, the Great Basin, and the Rockies thus exemplify three recognizably different types of orogenic environment.

Examples from the tectonically active regions of the world help us understand these differences. In the Andes, where the Nazca Plate (Section 28.15) is being segmented as it is being subducted beneath the South American plate, and continent, volca-

noes are present only above the 'normally' dipping segments of the downgoing lithosphere. These are called the northern, central, and southern volcanic zones. In the intervening zones the Nazca Plate is subducting at a low angle, so that it scrapes the base of the South American lithosphere and there is no room for a wedge of asthenospheric material, and hence there are no volcanoes there, and presumably no granites of Sierra Nevada type in depth. Instead there are prominent uplifts of Precambrian basement similar to the Laramide uplifts of North America. The earthquakes in these so-called flat slab zones are the result of east–west compression, and the uplift is driven by the compression.

With regard to the Great Basin, recent detailed structural studies and seismic traverses indicate that the volcanic activity there is taking place in a zone of extreme extension, amounting to several hundred per cent. The cessation appears to be related to the latest Mesozoic cessation of the subduction that resulted in the emplacement of the Sierra Nevada batholith. Seismic tomography (Section 25.10) indicates that the lower crust and mantle beneath the Great Basin are still hot today. Hence the three different types of orogenic environment identified in the American west can readily be explained in terms of plate tectonics.

31.6 THE DIVERSITY OF OROGENIC BELTS

For a long time geologists and geophysicists have sought a simple, unifying theory of orogenesis. However, the orogenic belts of the world differ widely in their structure, and history, and it is unlikely that such a theory will arise. The theory of

Table 31.1 Radiometric ages (in Ma) of Precambrian basement rocks uplifted during the Laramide and later orogenic phases

West		North		East	
		Little Rocky Mts	1800		
Little Belt Mts	2470				
Beartooth Mts	3120				
		Bighorn Range	2700		
Teton Range	2660			Black Hills	1650
		Wind River Range	2720	Laramide Range	2450
Wasatch Range	1580	Park Range	1660	Medicine Bow Mts	1720
Sheeprock Mts	1700	Sawatch Range	1400	Front Range	1700
Monument Uplift	1650				
		South			

plate tectonics has, however, at least provided a basic framework within which this diversity can be examined in order to understand the principal processes that operate in the formation of mountains.

The Western Alps, extensively studied by European geologists, represents a mountain belt formed as a result of the convergence of two continental masses. As Europe and Africa, traditionally regarded as forelands of the Alpine orogenic belt, moved eastwards relative to the Americas during the opening of the Atlantic Ocean, they underwent relative motion themselves. At times they diverged, as in the Late Jurassic when the Tethyan carbonate platform broke up and subsided to form small ocean basins between the Hercynian massifs; at times they moved laterally with respect to each other in either a sinistral or dextral sense. The main deformation, crustal thickening, and orogenic uplift resulted from periods of rapid convergence or so-called continental collision. Such collision is even more marked in the case of the Himalayas that resulted from direct collision of India with Asia as the former drifted rapidly north after the breakup of Gondwanaland (Section 29.3).

The Andes, on the other hand have been situated adjacent to the open Pacific Ocean since the mid-Palaeozoic, and that orogenic belt results from underthrusting of Pacific Ocean lithosphere beneath the South American continent as it moved west relative to Africa during the opening of the South Atlantic Ocean basin. No collision was involved in this case, but the North American cordillera, although also a continental margin orogenic belt, was involved in a number of collisional events that were critical in its development. In this case the collision was with island-arcs, oceanic plateaus and microcontinents located on the subducting Pacific Ocean lithosphere, rather than a full-scale continent–continent collision as in the case of the Western Alps or the Himalayas.

Finally, by way of contrast, the Transantarctic Mountains that extend for 4000 km from south of New Zealand to south of Africa, were uplifted in an extensional environment at the margin of a Cenozoic rift system. This orogenic belt did not involve plate convergence, let alone continental collision. Rather the mountain belt represents the uplift of a Cenozoic rift margin reactivating a long-standing zone of weakness in the lithosphere.

31.7 THE WESTERN ALPS

Since the discovery by Escher van der Linth in 1841 of gigantic recumbent folds and overthrusts in the Swiss Alps, many brilliant geologists have devoted their professional lives to the unravelling of tectonic structures which, for many years after they were first described, appeared to be quite incredible to those who had not actually traced them from peak to peak. By patient mapping, supplemented by underground observations made possible by an incomparable series of tunnels, many parts of the intricate pattern of the Western Alps have now been untangled. The surface and near surface observations by generations of geologists have now been supplemented by deep geophysical profiles (Figures 31.15, 31.16b) using several different techniques. It is generally accepted that the Alps were formed by the inversion of basins in the Tethys as Europe and the Adriatic Promontory of Africa collided during opening of the Atlantic basin to the west (Figure 31.15b). (Inversion is the reactivation of normal faults as reverse faults in compression. Hence the basin floor forms the highest thrust sheet.) The over-ridden European crust was subducted beneath the African crust. Hence the asymmetry of the belt referred to earlier.

Geologically the Alps are divided into the **Western Alps**, which curve in a broad arc from the Mediterranean to Lakes Constance and Como, and the **Eastern Alps**, which continue in a gentler curve towards the Danube (Figure 31.15a). Beyond Vienna the vast bow of the Carpathians begins, while on the southern side the Lombardy ranges of northern Italy swing round to join the Dinaric Alps. It is in the Western Alps that the nature of the general structure has been at least partially disclosed. The essential feature, portrayed in Figure 31.16 with some artistic licence, is the occurrence of a series of gigantic recumbent folds (Figures 8.4, 8.9) and nappes (Section 31.8), each of which has travelled forward for many miles towards the foreland, and in many places far across it. The Alpine rivers have cut deeply into the nappes, thus exposing the underlying rocks in many a steep-walled gorge (Figure 31.17). If the nappes were everywhere at the same level, only the outer parts of the structure would be exposed to view in this way, and even the tunnels would add little more. But as the nappes are traced along the trend of the ranges from south-west to north-east they are found to undulate up and down in an alternating succession of broad **culminations** and **depressions** (Figure 31.18). Where the depressions have carried the nappes downwards, the uppermost ones are still well-preserved in the mountain peaks. Where the culminations might have raised the upper nappes to levels now far above the highest peaks – that is to say, where they have already been swept away by erosion or removed by gravity gliding – the lower nappes come to the surface and are themselves cut through by the valleys. Thus, although no single section across the mountains provides more than part of the

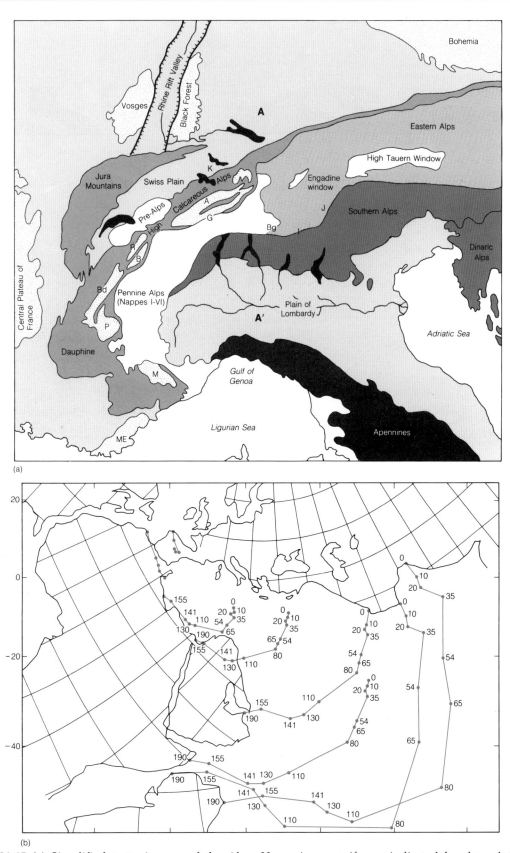

(a)

(b)

Figure 31.15 (a) Simplified tectonic map of the Alps. Hercynian massifs are indicated by close dotting. The relationships between the Southern, Eastern and Dinaric Alps are still obscure. A, Aar massif; B, Mt Blanc; Bd, Belledonne; Bg, Bergell Granite; G, St Gotthard massif; I, Insubric Line; J, Judicaria Line; K, Mythen Klippes; M, Mercantour massif; ME, Maures and Esterel massifs; P, Pelvoux massif; R, Aiguilles Rouges. (b) Relative motion history of Africa, Iberia, Arabia and India with respect to Eurasia from Early Jurassic to Present time. Position of continents at 190 Ma. Ages in Ma. Present latitudes and longitudes every 10° on continents. Interpolated flow-lines: continuous lines. (*Le Pichon, F. Bezerat and M.J. Roulet, 1988*)

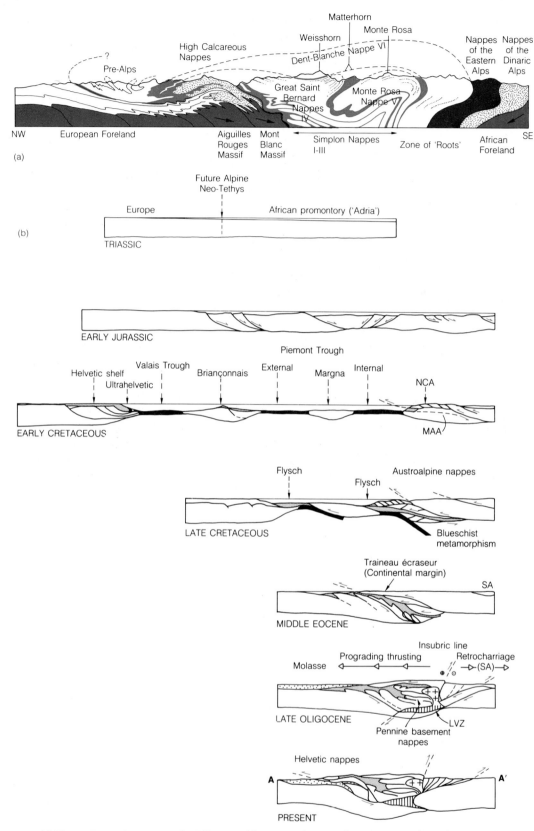

Figure 31.16 (a) Tectonic section across the Western Alps according to the concepts of Emile Argand and R. Staub, based on the idea that the great recumbent folds and nappes were driven from the Tethys geosyncline and its basement by the vice-like approach of the African and European forelands. (b) The evolution of the Swiss Alps along a line that goes roughly from A in the north to A' to the south as shown in Fig. 31.15(a): LVZ, low velocity zone; MAA, Middle Austroalpine nappes; NCA, Northern Calcareous Alps; SA, Southern Alps; white, continental crust; black, oceanic crust; fine stippling, shallow marine sediments; heavy stippling, deep marine clastic rocks, flysch and melange complexes; +, Bergell granite. (*A. M. Celâl, Şengör, 1990*)

Figure 31.17 Gorge of the Via Mala, cut through the Pennine nappes near their south-eastern exposures; looking upstream between Thusis and Zillis, Switzerland. (*F.N. Ashcroft*)

beneath a widespread cover of a different and locally higher series of nappes. Thus, because the evidence is largely hidden, the structure of the Eastern Alps is less well-known. Only here and there, as in the Engadine and the High Tauern (Figure 31.15a) has erosion removed the cover and so opened 'windows' in the surrounding framework through which the underlying structures are locally seen.

The chief subdivisions of the Western Alps as seen in plan, subdivisions representing successive zones which provide a first clue to the general structure, can be clearly detected from the air. Each has its distinctive topography, and each comes into view in turn during a flight from, say, Besançon to Milan. In order, these zones are as follows:

1. The **Jura Mountains**: a crescent-shaped assemblage of hills standing well in front of the Alps proper. The folds of the sedimentary cover rest on a faulted swell of Hercynian basement rocks, and extend from near Grenoble to just north of Zurich (Figure 31.15a). At each end there is one open anticline, but the number of folds increases greatly towards the middle, where there are many bundles of bold anticlines, some of which outline the actual hills (Figures 31.19 and 31.20). Unfolded tabular areas intervene between some of groups of folds, separated from the latter by thrusts directed towards the north-west. On the convex side of the arc the **folded Jura** pass into the broken tableland of the **tabular Jura**, which is a faulted sedimentary cover to the depressed parts of the Hercynian

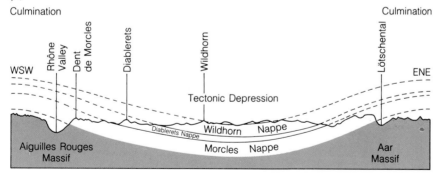

Figure 31.18 Longitudinal section showing the nappes of the High Calcareous Alps exposed in the tectonic depression between Aiguilles Rouges and the Aar culminations. The nappes advanced at right angles to the plane of the section in the direction away from the observer.

picture, the whole complicated structure can be visualized by taking a series of several sections in order across the successive culminations and depressions.

In the Eastern Alps, the Pennine and neighbouring zones of the Western Alps disappear

basement, extending from the Central Plateau of France on the one side to the Bohemian massif on the other. The folded Jura terminate just beyond the southern exposures of the Black Forest. The strata are largely Jurassic but range from Triassic to Mid-Cretaceous, with some Tertiary, mainly

Figure 31.19 Anticline in the Moutier Gorge, Jura Mts, Switzerland; an example of concordant morphology. (*Peter Christ*) *Left to right: Professor A. Buxtorf, Professor A. Holmes, Professor M. Reinhard.*

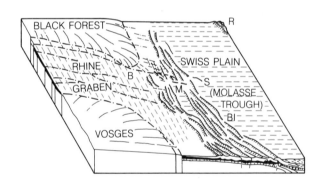

Figure 31.20 Block diagram (about 125 km square) showing the north-eastern part of the Jura Mts and their geological environment. B, Basle; M, Moutier; R, Rigi; S, Solothurn. (*J.H.F. Umbgrove*)

Miocene. Faulting began towards the end of the Oligocene, and after folding in the Late Miocene and Early Pliocene the region was more or less

levelled by erosion, only to be rejuvenated by renewed folding and uplift during the Pleistocene. It has long been thought that the advance of the Alps was responsible for pushing forward the Jura across the gap between the Hercynian massifs of the Central Plateau and the Vosges and Black Forest. Although the broad expanse of the Swiss Plain with its almost unfolded sediments makes this hypothesis seem untenable, a flat-lying detachment fault underlies the molasse (Figure 31.21). Although apparently the simplest of fold-mountains, the Jura have turned out to be the product of a long series of deformations, including faulting, folding, and shearing, and other responses of the sedimentary cover to differential movements, essentially vertical, of the crystalline basement.

2. The **Swiss Plain (Molasse Trough)**: a low-land area broadening to the north-east, underlain by mid-Tertiary sediments which occupy a fore-

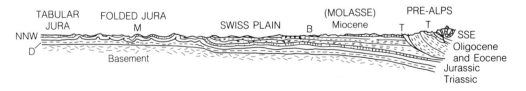

Figure 31.21 Section from the tabular Jura to the front of the Pre-Alps. B, Berne; D, décollement; M, Moutier; T, thrust faults. Length of section, 84 km. (*After J. Goguel*)

deep that developed in front of the rising Alps (Figure 31.22). Far to the south-east the front of the High Calcareous Alps appears in the distance like a great wall (Figure 31.22). South and east of Lake Geneva a remarkable patch of foothills 'without roots' protrudes on to the Plain. These are:

3. The **Pre-Alps**: consisting of isolated piles of nappes, much folded and sliced by minor thrusts, and extending in front of the Helvetian Alps from south of Lake Geneva to Lake Thun. They have travelled far from the region where their sediments were deposited and come to rest on the Swiss Plain, where the Molasse itself has been churned up by their forward movements. A nappe or pile of nappes detached by erosion or gravity gliding from the parental mass, of which it is now only a remnant, is known as a nappe-outlier or **klippe** (Figure 8.27b). The Pre-Alps are giant klippes but smaller examples occur at intervals as far as Vienna. Of these the most familiar are the Mythen (Figure 31.23). The Mesozoic strata of the Pre-Alps include types that are entirely different from anything seen in the Jura or the High Calcareous Alps; they are foreign both to their surroundings and to their foundations. Exactly where they were deposited has long been one of the unsolved

problems of Alpine geology. According to the interpretation presented in Figure 31.16 the uppermost of these rock-sheets were thought to be disconnected outliers of the most far-travelled of all the Alpine nappes. A similar assemblage of Triassic and Jurassic sediments, located south-east of the Pennine Alps, was tentatively suggested as the 'root' of the Pre-Alps.

4. The **High Calcareous or Helvetian Alps**: a high range of rugged mountains built up of several recumbent folds and nappes, notably the classical three named on Figure 31.18. The zone includes the Bernese Oberland with its snowfields and glaciers and its many familiar peaks (e.g. the Jungfrau). In south-east France the zone swings southwards through the Dauphinée country and then south-eastwards to reach the Mediterranean coast on both sides of Nice. The sediments were deposited over the Hercynian massifs and mostly behind them, in what was then a region of troughs and archipelagoes, continually changing in distribution during Triassic to Mid-Cretaceous times. Great thicknesses of flysch then accumulated, and as Africa began to impinge upon southern Europe, a succession of nappes was thrust northward. The higher members of this pile of nappes travelled the farthest from their source and their isolated

Figure 31.22 The Swiss Plain and the High Calcareous (Helvetian) Alps viewed from the Jura Mts near Solothurn (see Fig. 31.20 for locality).

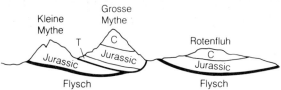

Figure 31.23 The Mythen and Rotenfluh overlooking Schwyz and Brunnen, north east of the great bend to the southern branch of Lake Lucerne, Switzerland. Klippes (outliers of nappes) composed of Mesozoic strata resting on flysch of mainly Eocene age. The main glide planes are shown by thick lines on the adjoining sketch; T, Triassic; C, Cretaceous. (*Robert Campbell*)

relics now form the long line of outliers known as klippes, and probably also the larger masses distinguished as the Pre-Alps.

5. The **Hercynian massifs**: an arcuate chain of long isolated blocks consisting largely of crystalline rocks having a Hercynian ancestry. These have been sheared by innumerable small thrusts, and except for the more massive granitic rocks (Figure 31.24) they fall a ready prey to the splintering action of frost (Figure 31.25). The skylines are in consequence characteristically jagged, as in the appropriately named Aiguilles Rouges. The latter massif, together with the adjoining massif of Mt Blanc (4810 m), emerges along the crest of a great culmination. To the north-east the Hercynian foundation disappears beneath the nappes of the High Calcareous Alps to emerge again in another culmination as the Aar and St Gotthard massifs (Figure 31.15 and 31.18). Here the Rhine and the Rhône have their sources. The Rhône flows to the west-southwest through a long trough-like valley which, after leaving the Hercynian massifs, follows the boundary between the High Calcareous Alps and the broad zone of the Pennine Alps. To

the south-west and round the great bend to the Mediterranean successive culminations bring up the Belledonne, Pelvoux, and Mercantour massifs.

Although all these crystalline massifs are uplifted blocks consisting mainly of wedges of rocks inherited from the Hercynian orogenic belt (cf. Figure 31.15), they were metamorphosed afresh several times during the Alpine orogeny. The recrystallized rock material is therefore of Late Cretaceous or Tertiary age. Within the heart of wedges that moved *en bloc* older minerals and structures have been preserved, while new ones were developing towards the margins, where shearing was concentrated. Striking confirmation of these phases of rejuvenation has recently been forthcoming by dating appropriate samples. In addition to the pre-Alpine relics, three generations of Alpine minerals, each representing an orogenic phase, have been recognized in the St Gotthard massif, and dated. The first produced elongated crystals of hornblende and biotite aligned north and south; these have retained sufficient of the argon generated within them to give minimum ages of 44–47 Ma. In the second

Figure 31.24 Mt Blanc (4810 m) the highest culmination of the Hercynian massifs; viewed from Mégève, looking west. (*Warren Redding*)

phase large crystals of hornblende and garnet grew across the earlier foliation; their minimum ages are 23–27 Ma. Large biotites belonging to migmatites of later formation give Rb/Sr ages of 15–17 Ma, corresponding to the major orogenic phase that occurred during the Miocene. It is of interest to notice that zircons and micas from relics of Hercynian migmatites in the St Gotthard and Aar massifs give radiometric ages ranging from 270–310 Ma, and that Hercynian biotite from the Aiguilles Rouges has a similar age, 286 Ma. Evidence that the Hercynian migmatites of the St Gotthard massif were in turn made out of still older rocks – Caledonian or even Precambrian – is provided by minerals from but little altered inclusions; zircons from the latter, for example, give ages of 485–650 Ma. Once thought to represent the autochthonous (unmoved) European foreland exposed in uplifts, the Hercynian massifs are now believed to have travelled northward on deeply buried shear surfaces generated during the African overriding of Europe (Figure 31.16b).

6. The **Pennine nappes**: an involved series of nappes, interpreted by Emile Argand and others as having been driven northward as gigantic recumbent folds having the boldly outlined structure depicted in Figure 31.16a. On this interpretation each of the major nappes has a core of Hercynian or older rocks, mainly gneisses, mig-

matites, and granitic rocks from the stretched continental floor of Tethys. The core is interfolded and partially enwrapped by lustrous schists (**schistes lustrées**) and crystalline limestones, representing the Mesozoic sediments and ophiolitic volcanics of the Tethys. The Pennine nappes are a tectonic 'sandwich' with the higher nappes being of Adriatic (i.e. African) origin, the lower ones of European origin, and between them nappes with oceanic affinities caught up as the 'filling' (Figure 31.16b). Out of these the Pennine Alps have been constructed. The region is one of lofty and heavily glaciated mountains, rising above the snowfields into pyramidal peaks, amongst which the Matterhorn (4482 m) is the most famous, although the less shapely Monte Rosa (4638 m) exceeds it in height. The summit levels decline eastwards, and beyond the margin of the overlapping Eastern Alps the Pennine nappes are unexposed, except possibly in the windows of the Lower Engadine and the High Tauern.

As numbered on Figure 31.16a the Pennine nappes were formerly described on the following lines:

I–III, The **Simplon Nappes**, folded into intricate convolutions as they were compressed by the higher nappes against the highly resistant obstruction of Mt Blanc. Despite their penetra-

Figure 31.25 Frost-splintered 'aiguilles' topography along a ridge of the Mt Blanc massif, with the shrunken Glacier de Tacul and a large glacial erratic in front. (*A.G. Kilchberg*)

tion by the Simplon Tunnel, they continue to present a difficult structural problem.

IV, The **Great St Bernard Nappe**, shown in the figure as riding over the Simplon nappes and bulging out backwards in response to the pressure exerted by

V, The **Monte Rosa Nappe**, plunging into the back of the Great St Bernard Nappe.

VI, The **Dent Blanche Nappe**, supposed to have travelled far forward over all the nappes in front, forming above them a widespread carapace, most of which was removed by erosion as the region gradually ascended.

There is now ample radiometric evidence that the 'Hercynian' cores of the Pennine nappes were granitized and largely turned into Alpine migmatites and intrusive granitic rocks during the late Cretaceous and early Tertiary orogenic phases. Biotite grown during the latest recrystallization dates from 18–23 Ma, corresponding to the Miocene movements and metamorphism. However, the greatest mobility was reached at an earlier date along the so-called 'zone of roots', where numerous bodies of granite and related rocks ascended diapirically through the mantling migmatites and schists (e.g. Adamello, Baveno, Bergell (Figure 31.15), and Bregaglia (Figure 31.26).

7. The '**Zone of Roots**', as shown on Figure 31.16, is situated on both sides of the Swiss–Italian frontier. On the northern side the Pennine nappes appear to turn vertically down and so to be 'rooted' in the ground, while on the other side the nappes of the Southern Alps appear to do likewise. The Pennine and Southern zones are distinct units, separated by a narrow belt of faults known as the **Insubric** or **Tonale** Line (Figures 31.15 and 31.16). Along this zone tectonic subsidence and subsequent uplift reached their maximum amplitude, so that today in the Ticino valleys we can see migmatites formed from schists that were originally Triassic sediments.

Figure 31.27 shows the smooth summit level – an easily recognizable erosion surface – of the Lepontine continuation of the Pennine Alps. This illustrates a feature that characterizes all the Alpine zones passed in review. Taken together the summit levels form a very broad arch, uplifted between the Molasse basin in the north-west and the basin of the River Po in the south-east, with superimposed minor undulations both along the arch and across it. Each of the major Tertiary orogenic phases, and particularly that of the Miocene, was followed by an interval of several million years during which uplift failed to keep pace with lowering of the surface by denudation, possibly aided by sub-crustal erosion. Towards the close of the Pliocene the Alps had been reduced to a region of low relief, the complex underlying structures being truncated by the ero-

Figure 31.26 Miocene granite of Bregaglia on the Swiss–Italian frontier; emplaced through the Dent Blanche nappe on the south side of the Bregaglia valley, south-west of St Moritz. (*J. Gaberell*)

Figure 31.27 The northern end of L. Maggiore with Locarno at the head and, just beyond, the Insubric Line showing as an east–west break through the mountains. The Southern Alps are seen in front and the Pennine (Lepontine) Alps of the Ticino behind. The summit level seen along the skyline represents a late Tertiary erosion plane which has since been uplifted. (*A. Pancaldi, Ascona, Italy*)

sion surface. Powerful upwarping of the latter then made possible the carving of the peaks and valleys that make up the Alpine scenery of today.

8. The **Southern Alps**: built mainly out of thick Permian ophiolites and Triassic limestones, which accumulated in a deep basin south of the Insubric

Line while the Pennine region was still a shallow submerged platform. The Pennine basin began, also with ophiolites, only in the late Triassic; the Helvetian geosyncline only during the Jurassic. Along the section we have followed, but from south-east to north-west, a spreading of activity is evident. In this case, however, the migration refers to the beginnings of basinal subsidence, and also to the times compression and deformation first set in. Traced eastwards the Southern Alps gradually widen and from the obscurity of this complex of highly metamorphosed rocks two other sets of structural zones emerge: the Eastern or Austrian Alps with their northward-directed nappes; and the ranges north of Trieste which pass into the Dinaric Alps, with nappes and folds directed towards the Adriatic.

31.8   THE NAPPE PROBLEM

The interpretation of Alpine structures in terms of nappes began in 1841, when Escher van der Linth (1807–72) in a short lecture-demonstration announced his discovery of 'colossal overthrusting' in the Helvetian zone of the Glarus district. He described the highly spectacular occurrence of **Verrucano** (a thick Permian formation of purplish-red conglomerates and sandstones) over light-coloured Mesozoic and Tertiary beds (Figure 31.28). As shown in Figure 31.29a he conceived of the structures in the simplest possibly way as two giant recumbent folds, one overturned towards the south, the other towards the north. Even so, a lateral displacement of the Verrucano amounting to at least 15 km was implied. At that time such an implication was sufficiently startling to make Escher exclaim: 'No one would believe me if I published my sections; they would put me in an asylum.' It was Albert Heim who published some of them in 1875, and Heim who, in a later book, supported a still more revolutionary interpretation of the Glarus structures, proposed by Marcel Bertrand in 1884. This combined the two opposite-facing folds into a single nappe directed from the south-southeast (Figure 31.29b). While removing a number of stratigraphical and structural difficulties, Bertrand's view required a nappe movement of at least 35 km, and this was widely regarded as mechanically impossible.

However by this time several equally 'impossible' overthrusts of gigantic proportions had been discovered in other mountain ranges. The celebrated Moine thrust (Figures 31.4 and 31.30) had been described by Calloway and Lapworth in 1883 and confirmed by the Geological Survey in 1884. By 1883 Törnebohm had recognized Caledonian overthrusts directed from the Scandinavian mountains over their foreland, the Baltic shield; and in 1888 he showed that there had been a lateral displacement of at least 100 km. By 1896 he felt justified in raising the total to 130 km. Mean-

Figure 31.28 The sole of the Glarus Nappe is seen at the base of the dark Verrucano which caps a ridge south-west of Elm in the east of Switzerland. The advancing nappe has sheared the underlying Jurassic limestone into a white marble-like mylonite which is here interfolded with Eocene flysch, as indicated in Fig. 31.29. A 30 m tunnel, eroded through a tongue of flysch, and known as St Martin's-loch, appears black in the picture (to the left of centre). (*P.G.H. Boswell*)

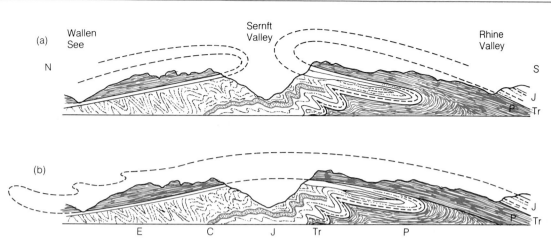

Figure 31.29 (a) Glarus 'double-fold' of Escher van der Linth; and (b) its re-interpretation as a nappe by M. Bertrand. P, Palaeozoic strata, including Verrucano; Tr, Triassic; J, Jurassic; C, Cretaceous; E, Eocene Flysch. Length of section, 40 km.

Figure 31.30 Outcrop of Moine Thrust-plane, Sutherland, showing Moine schist and an underlying wedge of Lewisian gneiss on white crystalline marble (Cambrian). (*British Geological Survey*)

while, after a long period of patient mapping by Swiss geologists, the second major Alpine discovery was announced by Schardt in 1893. This was an incontestable proof that the Pre-Alps (Figure 31.23) are outliers of a whole pile of far-travelled nappes. Doubters had only to 'go and see' to be converted. By the turn of the century the nappe theory was finding applications in most of the orogenic belts of the world as they came to be explored.

Nevertheless, the mechanism of nappe move-

ment long remained an unsolved riddle. Some nappes can be traced down-dip back to a narrow 'root' region, so-called, from which they appear to have been squeezed outwards and upwards by intense lateral compression between two supposedly rigid blocks, such as the Mt Blanc and Aiguilles Rouges massifs (Figure 31.31). As already indicated, this kind of mechanism is not only geologically 'awkward' and improbable, but it has to meet the objection that 'squeezing out' or pushing from the rear would merely shatter the

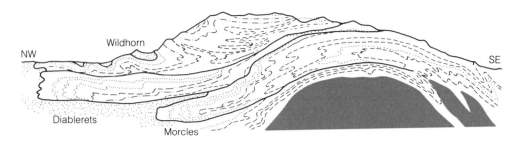

NW

Wildhorn

SE

Diablerets

Morcles

Figure 31.31 Section of the Helvetian nappes seen along the right-hand side of the Rhône valley. The crystalline basement is the south-east part of the Aiguilles Rouges massif. The Mt Blanc massif lies farther to the south-east and the nappes appear to be rooted in the depression between the two massifs. See Fig. 31.18 and Section 31.7. Length of section, 23 km. (*After Jean Goguel, translated by Hans E. Thalmann, 1962*).

sheets of rock instead of leaving them intact, unless they were many times thicker than the nappes actually observed. Moreover, on any hypothesis, including gravity gliding, rock friction appeared to be a fatal obstacle unless steep slopes of impossibly great length were available for the nappes to slide **down**. But many nappes appear to have travelled along upward-sloping surfaces, thus greatly magnifying the mechanical difficulties. Beds of wet clay or marl, especially if associated with evaporites, have been appealed to as lubricants, but the coefficient of friction still remains too high; in fact the presence of water tends to increase this coefficient rather than to reduce it. These considerations led M. King Hubbert and William W. Rubey to join forces in a determined effort to find a simple and adequate means of reducing the critical value of the shear stress required to produce large-scale sliding of overthrust blocks. Their quest has been remarkably successful.

On the old view, i.e. taking representative rocks and values of the coefficient of friction, they found that a layer 1 km thick could not be pushed from the rear along a horizontal surface for more than 8 km without becoming a chaotic mass of rubble. Even for a layer 5 km thick the thrust block could not travel 18 km and remain intact. For gravity gliding they find that the tilt of the surface would have to be 30° or steeper. These results are in practical agreement with those obtained by earlier critics of the nappe theory. But in these calculations the influence of the pressure of interstitial fluids has been neglected. Hubbert and Rubey found that it is this influence – a powerful buoyancy effect not to be confused with lubrication – that leads to an acceptable solution of the enigma.

At a given depth the hydrostatic pressure of interstitial water is the weight of a column of water per unit area extending from that depth to the surface. In the drilling of very deep oil-wells, however, the fluid pressure (which may be that of water, oil, or gas) is sometimes found to approach

or reach the entire weight per unit area of the overburden of saturated rock. This overburden pressure – the combined weight of solids and fluids – is technically distinguished as the **geostatic** or **lithostatic** pressure. Within 1 or 2 km of the surface the ratio of the fluid pressure to the lithostatic pressure averages about 0.465, corresponding to hydrostatic conditions. But at greater depths much higher values may be encountered, up to 0.9 in parts of the Gulf Coast, California, and Burma, and even reaching 0.95 in Iran and East Pakistan. In these circumstances a very powerful buoyancy effect comes into play, and the weight of a column of rock (per unit area at a given depth) may be largely supported by the fluid pressure. Entrapped water comes under increasing pressure as the overburden increases (e.g. with continuing 'geosynclinal' sedimentation) and especially when the latter includes relatively impermeable beds of shale and clay, which reduce the tendency to loss by leakage.

The column of rock may eventually be almost floating in its own interstitial fluid, and as this condition is approached the shear stress required to move the strata approaches zero. A small push from behind would then suffice to start nappe movement. Again taking a layer 5 km thick, the buoyancy effect due to hydrostatic pressure alone would extend the distance of horizontal travel without shattering to nearly 27 km. With a fluid-pressure ratio of 0.7 the distance extends to 41 km, and with a ratio of 0.9 it reaches 106 km. These are maximum theoretical figures, because once a nappe starts moving some fracturing and leakage is inevitable. Moreover, it is difficult to envisage horizontal travel in the sedimentary depths of a well-filled orogenic basin. Gliding down a gentle slope is simpler and the buoyancy effect makes this easily possible. The slope required for gravity gliding is reduced from 30° to 6.6° for a ratio of 0.8; to 3.3° when the ratio is 0.9; and to an angle as low as 1.6° where the ratio is 0.95.

Hubbert and Rubey described an amusing

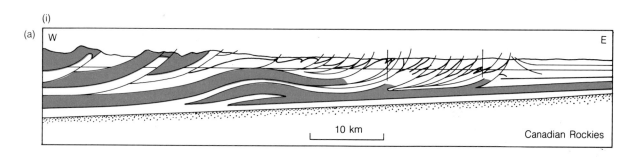

(a)

(i)

W   E

10 km

Canadian Rockies

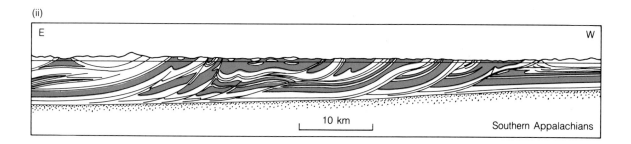

(ii)

E   W

10 km

Southern Appalachians

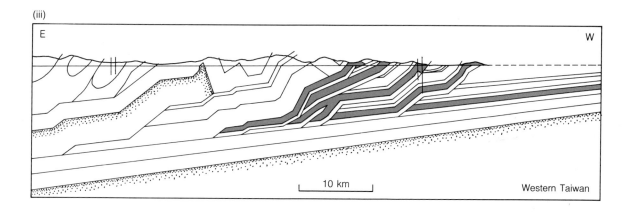

(iii)

E   W

10 km

Western Taiwan

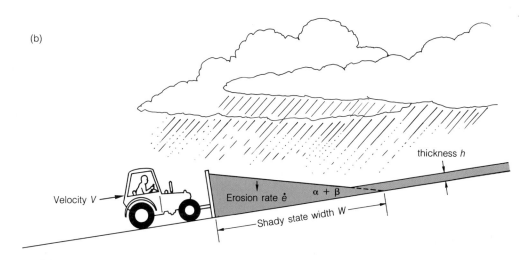

(b)

Velocity V

Erosion rate $\dot{e}$ $\alpha + \beta$

thickness h

Shady state width W

Figure 31.32 (a) Cross-sections of several foreland fold-and-thrust belts: (i) Canadian Rockies; (ii) southern Appalachians; and (iii) western Taiwan. (*D. Davis, J. Suppe and F.A. Dahlen*) (b) An eroding wedge attains a dynamic steady-state width given by $\dot{e}W = hV$, where $\dot{e}$ = rate of erosion, W = wedge width, h = thickness of sediment and V = velocity. (*F.A. Dahlen, 1990*)

experiment with a beer can which convincingly confirms their theory. The empty can with its open end upward is placed on a sheet of wet glass which can be tilted until the critical angle is reached at which the can is just able to slide down the wet surface. This angle is about 17°. Next, the can is chilled in a refrigerator and the experiment is repeated. The lowest angle of sliding is found to be the same as before. Finally, the sheet of wet glass is fixed with a slope of only 1°. The chilled can is again placed on the glass but this time with its open end downward. After a short time the can slides down the slope, but stops abruptly when it reaches the edge. The reason for this behaviour is that the cold imprisoned air warms up and expands, and so causes the pressure to increase inside the can. This in turn partly supports the weight of the can and the frictional resistance to sliding is very effectively reduced. When the can reaches the edge of the glass, the internal excess of pressure is released and the can stops sliding.

Active fold-thrust belts have recently been examined as accretionary wedges of material by F. A. Dahlen, D. M. Davis, and their colleagues. In this approach orogenic activity in localities such as the island of Taiwan, the site of ongoing collision between the Luzon island-arc on the Phillipine Sea Plate and the stable continental margin of China situated on the Eurasian Plate, are compared to the wedge of sand in front of a moving bulldozer (Figure 31.32b). The sediment or rock deforms until it develops a constant critical taper. This taper is dependent on the resistance to sliding along the décollement beneath the wedge. The greater the resistance, the higher the critical taper because the drag is the fundamental cause of the deformation within the wedge. On the other hand, an increase in the strength of the material 'bulldozed' decreases the critical taper. The state of stress in the wedge is always near Coulomb failure, since the process of deformation in the upper crust is by brittle failure.

The critical taper model resolves much of the debate concerning the Hubbert-Rubey model, although problems related to fluid flow and fluid pressure within accretionary wedges is a complex one. Hubbert and Rubey analysed the maximum possible length of individual thrust sheets and demonstrated that fluid pressure increases that value. They thus analysed the toe of a deforming wedge where flat-lying sediments are being accreted. They analysed the maximum length attainable without a taper, whereas there is no real limit to the length that can be attained with a taper. The Hubbert and Rubey idea led the way towards understanding of nappe and thrust-sheet emplacement. It is now apparent that the critical taper model, while so far applied to geologically simple situations, is appropriate to upper crustal mountain-building settings in general.

31.9 THE ANDES

The Andean Cordillera forms a continuous chain extending for nearly 10 000 km along the entire western margin of South America from the Caribbean Sea to the Scotia Sea (Figure 31.33). It reaches over 400 km in width and an elevation of approximately 7000 km. Cerro Aconcagua on the border of Argentina and Chile is the highest mountain in the western hemisphere. What particularly distinguishes the Andes as a mountain range is its vast number of very high volcanoes (Plate XVa). Due to the equatorial bulge of the Earth, the summit of Cotopaxi, a spectacular snow-capped volcano just south of the Equator near Quito, the capital of Ecuador, is actually farther from the centre of the Earth than the top of Mount Everest. The petrographic term 'andesite' refers to the most common lava erupted by the Andean volcanoes. Another feature of the Andes are enormous granitic batholiths. The Patagonian batholith, for example, extends along the Pacific margin of southern South America for over 1000 km and is almost 100 km wide. The average composition of these batholiths is approximately that of granodiorite, the plutonic equivalent of andesite. We are therefore undoubtedly seeing in the batholiths the deep-seated roots of former volcanic chains like those of today. These impressive quantities of volcanic and plutonic rocks typify the difference between the Alpine and Andean chains. Granitic plutons and intermediate volcanics are rare in the Western Alps. On the other hand ophiolitic rocks and high-grade metamorphic assemblages are rare in the Andes, while common in the Alps. The difference reflects subduction of thousands of kilometres of oceanic lithosphere beneath the Andes and the closure of many small ocean basins in formation of the Alps. The first geologist to study the Andes was Charles Darwin. In the first half of the 19th century he explored the fjords of Tierra del Fuego and Patagonia, made a traverse on horseback across the Main Cordillera of the Andes near Aconcagua, and experienced a major earthquake in the Chilean seaport of Concepcion. He noted the abundance of volcanic material and marvelled at the evidence for recent uplift and for rapid erosion.

The western margin of South America is not only remarkable for the presence of the Andean Cordillera. Offshore lies the deep Peru–Chile trench (Figures 31.33, 31.34 and Plate XVI), and the margin is the locus for earthquakes such as the one experienced by Darwin. The hypocentres of

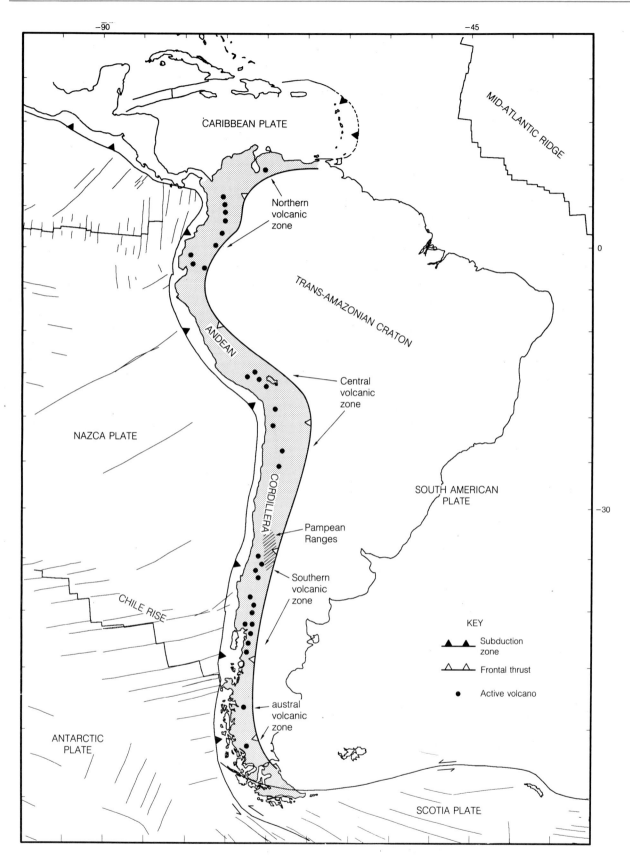

Figure 31.33 Tectonic setting for the Andean Cordillera.

the earthquakes descend eastward beneath the continent, reflecting the subduction of the lithosphere of the Pacific Ocean beneath the volcanoes (Figures 31.34 and 31.40). Along most of the Andean margin the Nazca Plate (named for a small town in southern Peru) is converging with South

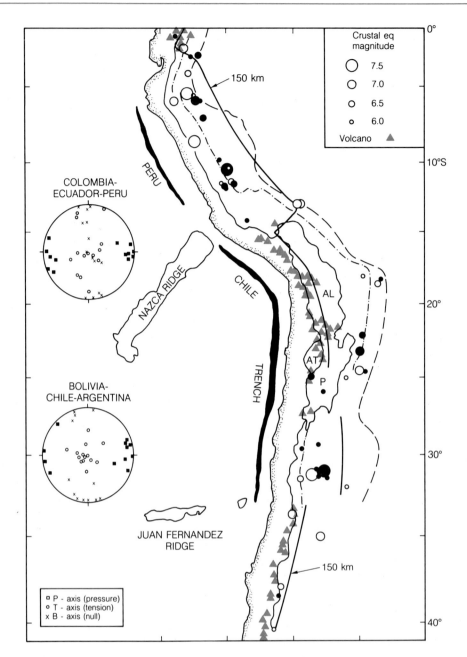

Figure 31.34 Map of western South America and the eastern Pacific ocean floor, showing distribution of large crustal earthquakes, Holocene volcanoes and geologic provinces, relative to the geometry of the subducted Nazca Plate (150 km contour on top of Benioff zone shown). Earthquakes shown have magnitudes greater than 5¾. Those shown with black circles are included in equal-area projections at left (lower hemisphere of common focal sphere) of P, T, and B axes of focal mechanism solutions. Geologic province boundaries are shown by dashed line (eastern limit of recognized Neogene deformation), dash-dotted line (boundary between foreland and hinterland) and light solid line (drainage divide along the Andean crest) that also encloses the Altiplano (AL), Puna (P), and Atacama basins (AT). The trench is mapped by the 6000 m isobath. (*T.E. Jordan et al, 1983*)

America at approximately 90 mm/year. South of the Chile Rise at 47°S, the Antarctic Plate is converging with the South American Plate at only 20 mm/year. The lower convergence, and hence subduction, rate is exemplified in the diminished height of the mountain range and a paucity of volcanoes to the south of the **triple junction** (Section 28.6).

Radiometric dating has shown that the great coastal batholiths of Peru and Chile are upper Mesozoic, predominantly Cretaceous. As the rocks forming these batholiths have the same composition as the andesites and related lavas of the active volcanoes, it is reasonable to deduce that subduction has been going on for at least the past 130 million years. That was the time when the South Atlantic Ocean began to form between South America and Africa, and it is understandable that convergence of the South American continent with the Pacific Ocean floor would have been a consequence, along the Andean margin, of separation from Africa. However, igneous rocks of the type being generated along the present Andean margin were typical of the Panthalassic (present Pacific) margin not only of South America, but also of Antarctica and Australia during the early Palaeozoic. It therefore, seems likely that subduction beneath the Andean margin was initiated after rifting of the margin from another continent in the late Precambrian. Throughout the Andes, however, Mesozoic strata rest with profound unconformity on a basement of older rocks, and just as the tectonic history of Europe can be considered in terms of Alpine and older events, so the Mesozoic and Cenozoic Andean history of South America can be distinguished from an older one that predated the break up of the Gondwanaland supercontinent. The so-called Andean cycle embraces all the geologic processes leading to the formation of the Andean Cordillera.

Although the processes that created the present-day Andes appear to have been active along the western margin of South America in the Palaeozoic and earliest Mesozoic before Gondwanaland fragmentation, there is one major difference. There is no evidence of major collisional events along the Pacific margin during the late Mesozoic–Cenozoic Andean orogenesis. Indeed remarkably little material has been added at the inner wall of the Peru–Chile trench given the amount of oceanic lithosphere and its sedimentary cover that must have been subducted during the past 130 million years. On the other hand the basement of the Andes from northern Chile south to Cape Horn is formed of material accreted to the Panthalassic margin of Gondwanaland during the Palaeozoic and early Mesozoic. Collisional events involving units as large as seamounts are well-documented, and Victor Ramos and his colleagues in the University of Buenos Aires believe that microcontinental collision may also have occurred.

After Darwin's pioneering study, understanding of the Andean Cordillera during the latter half of the 19th century and the first half of the 20th century was largely gained through the remarkable travels of continental European geologists. Scientists such as Albert Heim who had an intimate knowledge of the Western Alps were immediately struck by the absence of many of the rock types and structural styles characteristic of the Tethyan region. Where were the ophiolites, the thick flysch units, and the great overfolds and thrusts that they were familiar with in 'geosynclinal' settings? As mapping proceeded, it came to be appreciated that these are confined to the northernmost and southernmost parts of the Cordillera, while the central Andes, or 'Andes proper' are characterized by intermediate to silicic volcanic and volcaniclastic rocks, by subaerial sequences of sedimentary strata, and by comparatively open folding and block faulting. Jean Auboin and his French colleagues regarded the Central Andes as a 'liminal' mountain chain originating with sedimentation in basins literally 'on the lip' of the continent, as opposed to the more oceanic or 'geosynclinal' terminations to the north and south (Table 31.2).

Recent work has allowed us to develop a more unified view of the evolution of the Andean Cordillera. Uplift of the surface of Gondwanaland immediately prior to break up of the supercontinent was accompanied by widespread bimodal silicic and mafic magmatism and by extension of the continental crust. The uplift, magmatism and extension were all diachronous. The magmatism included the Karroo mafic flows and sills of southern Africa, the Ferrar dolerites and basalts of Antarctica, the Tasman dolerites of south-eastern Australia, and the Paraná basalts of South America, together with their less spectacular felsic equivalents. The pattern of extension related to these magmatic rocks, which led to the birth of the southern ocean basins that separate the formerly contiguous continental fragments of Gondwanaland, bears no obvious relationship to the former Panthalassic margin. Contemporaneous with this magmatism and extension, however, an extremely long composite basin came into existence along

Table 31.2 Characteristics of 'liminal' and 'geosynclinal' chains after J. Aubouin (1972)

	Liminal	Geosynclinal
Sedimentation	Marine and continental 'flysch' absent	Marine 'flysch' characteristic
Magmatism	Andesitic	Ophiolitic
Tectonism	Open, low amplitude folds	Nappe-like folds
Metamorphism	No high P/T assemblages	High P/T assemblages

the western margin of South America (Figure 31.35). Like the marginal basins of the western Pacific today, this basin was divided by old basement ridges into several segments with different characteristics. Notably the west Peruvian trough immediately south of the Equator, and the Andean basin of central Chile differ, as mentioned above, from the contemporaneous Rocas Verdes (Green Rocks) or Magellan basin in the Patagonian and Fuegian Andes, in being less oceanic in character. The Rocas Verdes basin has thick sequences of deep-marine turbidites and gabbro, sheeted dyke, pillow lava sequences. All the basins co-existed in the Late Jurassic to Early Cretaceous and were infilled mainly from the calc-alkaline Andean magmatic arc along the oceanic margin above the subducting plate or plate that formed the floor of the Pacific Ocean (Figure 31.36).

Approximately 100 m.y. ago, in the mid-Cretaceous, an abrupt change occurred. The sedimentary and volcanic strata of the basins were subjected to compression along the entire length of the margin. Folds and thrusts were formed that trend parallel to the length of the basins and the

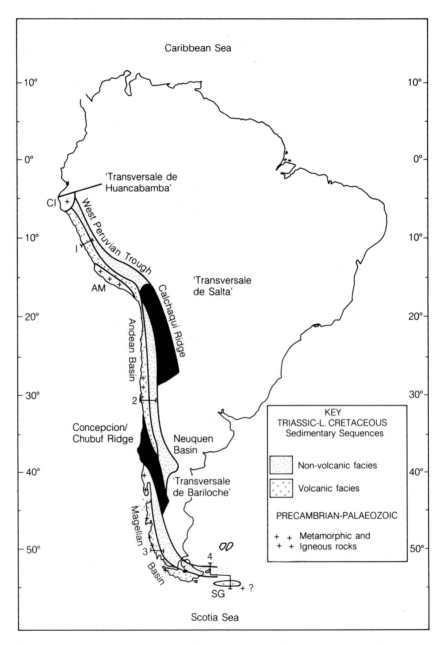

Figure 31.35 The Triassic-Lower Cretaceous sedimentary basins of the Andean Cordillera south of the 'Transversale de Huancabamba' at lat. 5°S. The significance of the ridges (shown in solid black) is discussed in the text. AM, Arequipa Massif; CI, Cerros de Illescas; SG, South Georgia microcontinent (palinspastically restored). (*I.W.D. Dalziel, 1981*) (1, 2, 3 and 4 are lines of sections in Figure 31.37.)

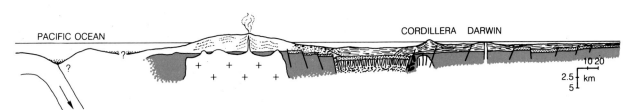

Figure 31.36 Interpretive cross-section of the southernmost Andes and South Georgia in the Early Cretaceous; diagram shows the maximum possible width of mafic crust based on present-day outcrop. (*I. W. D. Dalziel, 1981*)

margin, and the basins (presumably the weakest part of the continental lithosphere at the time) were inverted so that the deepest-seated parts were uplifted the most. In the Rocas Verdes basin, for example, the mafic floor was **obducted** (thrust up) onto the edge of the continental interior. The accompanying shearing was similar to the deformation associated with the emplacement of ophiolites in the western Alps. Deformation in the shallower epicontinental basins of the Central Andes was, however, less intense. This mid-Cretaceous event is variously known as the Patagonian, Mirano, or (following old European terminology for an event of the same age beneath the Harz Mountains of northern Germany) the sub-Hercynian orogeny. It represents the first tectonic thickening of the continental crust along the length of the Andean margin (Figure 31.37) and appears to have been remarkably contemporaneous given the enormous distance involved. In both Peru and Tierra del Fuego, for example, folded Aptian sedimentary strata are cut by plutons dated radiometrically between 90 and 105 Ma. The reason for the change from extension to compression, and for the contemporaneity along the length of the belt, probably lies in the fact that the South Atlantic Ocean began to open very rapidly in the mid-Cretaceous, thereby increasing the rate of convergence, and hence the compressive stress, along the Pacific margin of the South American continent (Figure 31.38).

It was immediately following this compressive event that the bulk of the granitic plutons that constitute the great batholiths of the Andes were emplaced. A simple explanation is that the increase in convergence rate not only increased the compression along the Pacific margin, but also increased the rate of magma production beneath the margin. It would have taken more time, however, for the granitic magma so generated to reach the surface than for the compressive stress to affect the rocks of the basins. Hence the widely observed field relations showing the batholiths generally cutting the folded, thrust and cleaved rocks.

A critical aspect of the mid-Cretaceous deformation with respect to the subsequent development of the Andes was the fact that the basin inversion

and crustal thickening along the Pacific margin of the continent at that time resulted in the formation of flexural downwarps (classically termed foredeeps) along the continental side of the embryonic mountain range. These in turn became the loci for the deposition of clastic sediment eroded from the rising mountains and zones of structural weakness in the lithosphere (Figure 31.39). Hence as the South Atlantic Ocean basin continued to expand, convergence, and hence subduction, continued along the Pacific margin, and deformation and calc-alkaline magmatism contributed to thickening of the continental crust throughout the Late Cretaceous and Cenozoic to the present day. Generally speaking the axis of magmatic activity and the locus of deformation has continued to move eastward away from the Pacific (Plate XVI). It takes less work to widen a mountain range than to continue lithospheric thickening and hence uplift.

At the present time, seismologic evidence indicates that eastward subduction of the Pacific Ocean floor beneath the continent is still going on (Figure 31.40). It is accompanied by westward underthrusting of the Andean belt by the South American fragment of the Gondwanaland craton to form a foreland fold-thrust belt. Hence the growth of the cordillera, both laterally and vertically, takes place in a compressive stress regime. Normal faulting is confined to high elevations and appears to be driven by the gravitational instability induced by the vertical relief. The crystalline core of the mountain belt therefore remains at depth.

As mentioned above, the mid-Cretaceous compression and basin-inversion event was essentially contemporaneous along the length of the Cordillera. Analysis of the present-day tectonic situation of the Andes reveals a different situation. The downgoing lithospheric slabs are segmented along tear faults at a high angle to the continental margin. Thus the active volcanoes occur above the more steeply-dipping segments of the Nazca Plate where there is a significant wedge of asthenospheric material between the oceanic lithosphere and the base of the continental lithosphere (Figure 31.34). Flat-dipping segments

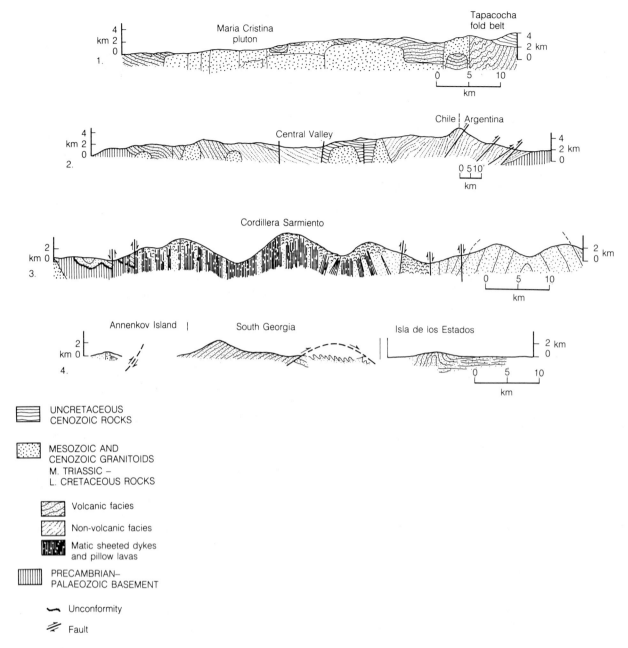

Figure 31.37 Simplified cross-sections across the West Peruvian Trough (Section 1), the Andean Basin (Section 2), the Northern Magellan Basin (Section 3), and the Southern Magellan Basin (Section 4). The lines of section are shown in Figure 31.35. Variation in the vertical exaggeration should be noted in comparing the sections.

of the oceanic plate result in high compressive stress being transmitted to the base of the continent and the uplift of basement highs comparable to the Wind River uplift and related features in the eastern part of the North American continent, (e.g. Pampean Ranges, Argentina, Fig. 31.33).

Although there are major faults striking parallel to the Andean Cordillera, there is no evidence of major displacement along its length as there is in the case of the North American Cordillera. The only evidence of this type of displacement comes from south of the Patagonian orocline where the

South Georgia microcontinent has been displaced approximately 1500 km to the east, and uplift of the deep-seated metamorphic core of the Cordillera to form the spectacular metamorphic complex, first observed by Charles Darwin (Figure 31.41), may have taken place in the Late Cretaceous due to relaxation of the far-field compressive stresses as the strike-slip regime developed between South America and Antarctica in the mid-Cretaceous. 'Core complexes' of this type are far more common in the North American Cordillera.

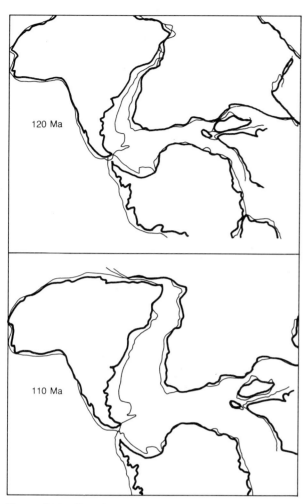

Figure 31.38 Development of the South Atlantic Ocean basin at 120 and 110 Ma.

31.10 THE NORTH AMERICAN CORDILLERA

The mountain ranges of the modern Cordillera of North America, which parallel the Pacific Ocean from Alaska to Mexico, are young features uplifted mainly in Tertiary times. Present-day tectonic activity reflects the continuing dynamic setting that characterizes the whole of the Pacific rim, although high relief and present-day volcanic and seismic activity record only the most recent events of this evolving mountain chain.

The geological record points to the existence of an ocean west of the continent since the earliest Phanerozoic. Throughout this long period of Earth history the continental margin has been modified by a variety of events, the most significant of which occurred in the Mesozoic when new **terranes** (amalgamated tectonic units, see later) were accreted to the continental margin.

1. The palaeocontinental margin
In late Precambrian time coarse clastics were deposited in an active rift system that extended along the length of the future Cordillera. It is uncertain whether this rifting occurred along the pre-existing western edge of the North American craton or developed within a larger continental framework. If there was a western continuation of the craton, it has yet to be identified; possible candidates include Australia, Siberia and Antarctica. By the Middle Cambrian a continental margin was well-established and a thick wedge of sediments had accumulated on the shelf and slope. Throughout most of the Palaeozoic the margin remained passive, and sediments continued to accumulate in a classical miogeoclinal setting. However, in the Ordovician and Silurian there are indications in the western distal sections of the miogeocline of arc-related volcanic activity and deformation. This early deformation culminated in the Late Devonian and was primarily restricted to the western edge of the miogeocline.

2. Terrane accretion
In the Early Jurassic, the region to the west of the North American continental margin was in some respects analogous to the modern western Pacific, with offshore arcs and intervening marginal basins. Westward drift of the North American plate accompanying the opening of the North Atlantic Ocean was opposed by south-easterly to north-easterly movement of the Pacific Ocean floor as it was obliquely subducted beneath the continental margin. Geological and palaeomagnetic data indicate that deformation of the margin in the Mesozoic was complex and included the sweeping of offshore island-arcs and marginal basins toward the North American margin where they became intensely deformed and accreted to the North American plate. The exact nature of these Mesozoic events is only partially understood, but there appear to have been several major arc systems and associated sedimentary basins that had been displaced northward prior to collision with, and accretion to, the North American plate margin. Such oblique convergence and ultimate collision is termed '**transpression**'. The individual 'terranes' are identified by local geographic names.

In general, the original positions and in some cases the ages of the terranes are uncertain. Some of the most easterly terranes are clearly displaced and deformed remnants of the North American palaeocontinental margin, but the more westerly terranes may include rocks from other areas of the present Pacific rim. The uncertainty surrounding the source areas for most of the terranes has led to the collective term '**suspect terranes**' (Figure 31.42a, see Figure 31.42b for other geological features).

The terranes are preserved because they were too buoyant to be passively subducted beneath

Figure 31.39 Upper Cretaceous turbidites in the foredeep of the southernmost Andes; note perecontemporaneous slump folds (*I. W. D. Dalziel*)

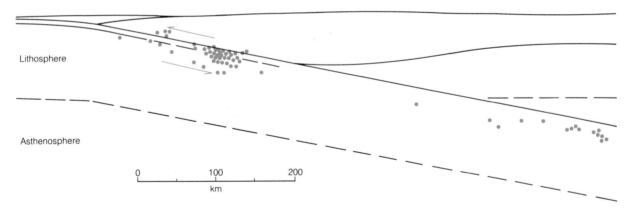

Lithosphere

Asthenosphere

0   100   200
km

Figure 31.40 Cross-section for seismicity through central Peru (*from Suárez et al, 1983*). The dip of the slab is inferred from the hypocentres of earthquakes.

the overriding North American plate. Instead the terranes were delaminated from their underlying lithosphere as it was subducted; the remaining buoyant slivers were obducted (thrust) onto the North American plate.

3. Crustal thickening

Terrane obduction onto the North American plate was accompanied by significant shortening and thickening of the western edge of the **miogeocline** (the prograding wedge of shallow water sediments located on a passive cratonic margin). A well documented example is the Early Jurassic obduction in the Canadian Cordillera (Figure 31.43), where slices of marginal basin and island-arc terranes were thrust northeastward onto the continental margin and then deformed into westerly directed folds and back thrusts as the edge of

Figure 31.41 Cordillera Darwin: a tectonically denuded core complex in the Andes of Tierra del Fuego. (*R. L. Brown*)

the North American plate yielded and thickened. Sediments that had been accumulating along the continental shelf and slope became deeply buried, were polydeformed and metamorphosed to high grade, and were intruded by granitic plutons.

The amalgamated terranes incorporated into the Canadian Cordillera during the Jurassic are collectively known as the **Intermontane Superterrane**. Transpression continued in the Cretaceous and additional terranes were accreted onto the continental margin. This new collage of accreted terranes has been called the **Insular Superterrane**. In Figure 31.44, the belts of the southern Canadian Cordillera are shown; how they are defined is explained in the caption.

Detailed geological studies have demonstrated that in the North American Cordillera the Mesozoic arrival and collision of terranes coincided with significant crustal thickening of the continental margin. However, in the Late Cretaceous and Palaeocene, the North American Cordillera experienced a climactic episode of crustal thickening with at least half of the total shortening in the southern Rocky Mountains occurring at this time. This period of crustal thickening (see Figure 31.43) does not coincide with arrival of new terranes;

rather, it appears to be related to a faster relative westward drift of the North American plate.

4. Tectonic denudation

During periods of rapid plate convergence and crustal thickening in the Mesozoic and early Tertiary, high relief was developed in the Cordillera as the thickened continental crust re-established isostatic equilibrium. The Himalayas and the Andes are contemporary examples of high-standing mountain belts in a tectonic setting of rapid active plate convergence, where crustal thickening and shortening are ongoing. However, in the highest parts of the ranges (the Tibetan plateau of the Himalayas and the Altiplano of the Andes), there are active normal faults indicative of extension and crustal thinning. This paradox is readily understood when the role of gravity in orogenic processes is considered. Mountain ranges stand high because of the buoyant nature of the thickened continental crust. However, this thick mass of continental crust can sustain only a limited load, beyond which it will yield due to its own weight. The upper part of the crust is brittle and strong relative to the plastic middle and lower crust. When gravitationally-induced vertical compressive stress exceeds the horizontal compressive

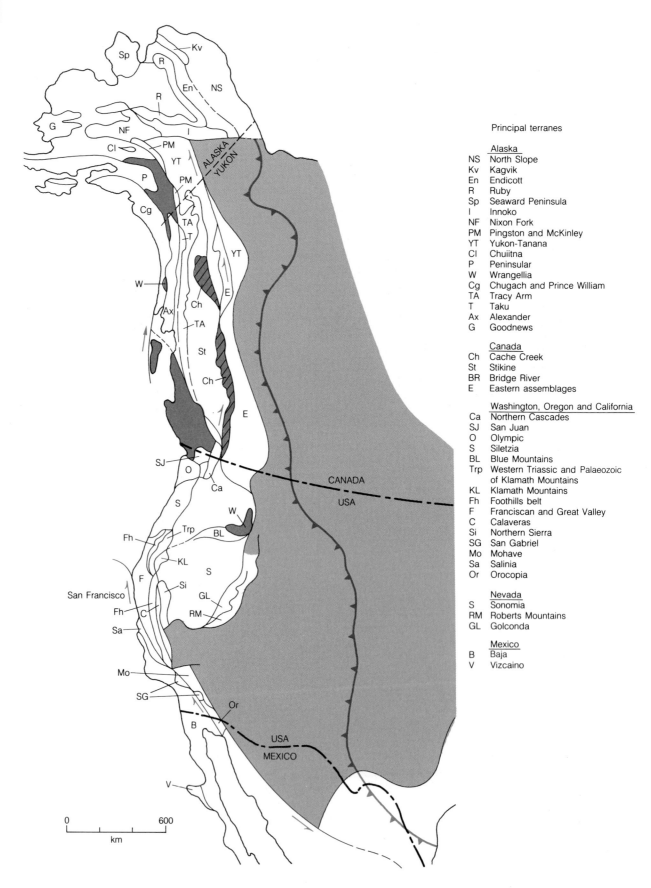

Principal terranes

Alaska
NS North Slope
Kv Kagvik
En Endicott
R Ruby
Sp Seaward Peninsula
I Innoko
NF Nixon Fork
PM Pingston and McKinley
YT Yukon-Tanana
Cl Chuiitna
P Peninsular
W Wrangellia
Cg Chugach and Prince William
TA Tracy Arm
T Taku
Ax Alexander
G Goodnews

Canada
Ch Cache Creek
St Stikine
BR Bridge River
E Eastern assemblages

Washington, Oregon and California
Ca Northern Cascades
SJ San Juan
O Olympic
S Siletzia
BL Blue Mountains
Trp Western Triassic and Palaeozoic
 of Klamath Mountains
KL Klamath Mountains
Fh Foothills belt
F Franciscan and Great Valley
C Calaveras
Si Northern Sierra
SG San Gabriel
Mo Mohave
Sa Salinia
Or Orocopia

Nevada
S Sonomia
RM Roberts Mountains
GL Golconda

Mexico
B Baja
V Vizcaino

Figure 31.42 (a) Generalized map of suspect terranes in western North America (From Keary and Vine 1990). Gray shaded area is underlain by North American cratonic basement; barbed line indicates eastern limit of Cordilleron deformation.

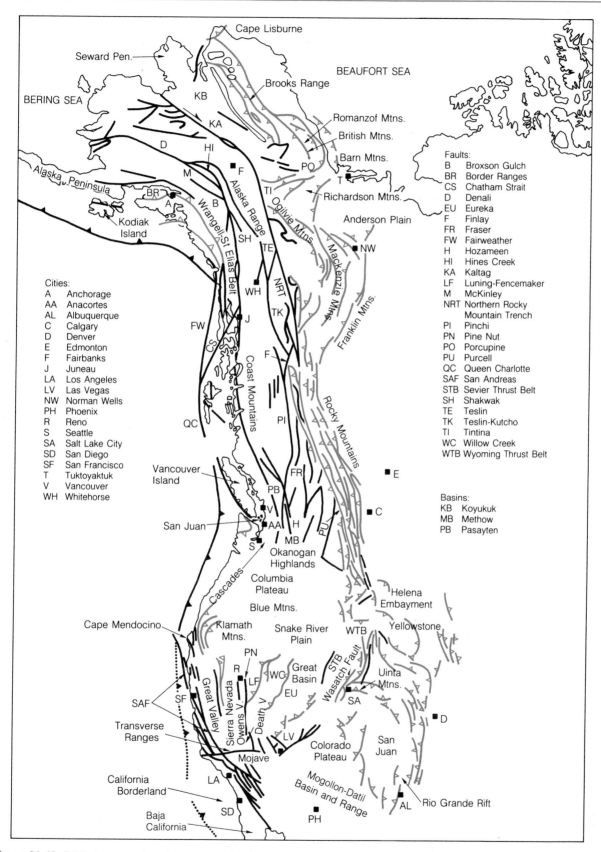

Figure 31.42 (b) Index map showing geographic locations and major faults of the North American Cordillera. (*After J. S. Oldow et al, 1989*)

stress induced by plate convergence, the ductile crust will tend to spread laterally. This lateral spreading of the middle and lower crust by distributed strain, or on discrete ductile shear zones,

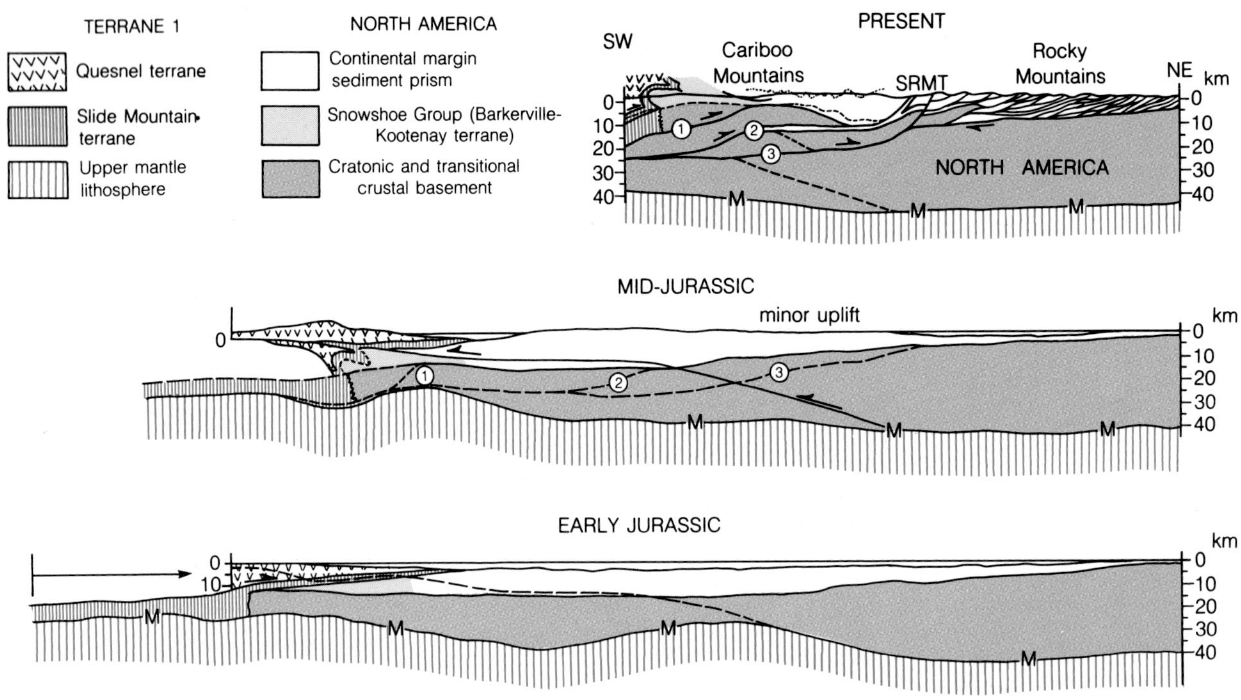

Figure 31.43 Model of terrane accretion, backthrusting and development of basement duplex in southern Canadian Cordillera. (*Modified from Murphy, 1987* and *Brown et al, 1986*) Terrane 1 is equivalent to part of the Intermontane Superterrane. SRMT – Southern Rocky Mountain Trench.

is accommodated in the brittle upper crust by normal faulting.

In regions of slow uplift the development of high relief may be prevented by erosion. However, in the uplifted hinterland of much of the North American Cordillera, field mapping and laboratory analysis have demonstrated that Tertiary uplift was rapid and that these regions became exposed largely as a result of extensional faulting rather than erosion. These tectonically denuded regions consist of metamorphic terranes that are bounded by normal faults carrying upper crustal rocks in their hanging walls. Uplift of the middle crustal rocks took place as the hanging walls of the normal faults moved down relative to the footwalls; as a result of this tectonic unloading the footwall rocks became arched and domed. These resulting structural culminations have been called 'metamorphic core complexes'. The normal faults commonly have very gentle dips of less than 30° and estimated displacement on the faults is on the order of tens of kilometres. In typical examples the middle crust was at depths of up to 15 km below the erosion surface prior to rapid uplift. This is determined by analysing the mineral assemblages in the exposed metamorphic rocks. The pressure at the time of peak metamorphic conditions can commonly be established, and geochronological techniques permit estimation of the age of the metamorphism and faulting.

The discovery that crustal extension in thickened orogenic belts can occur in regions of active plate convergence dictates that an understanding of orogeny must include the role of gravitationally-induced stresses, as well as the role of plate boundary stresses. In the North American Cordillera, extension and tectonic denudation occurred primarily in the Tertiary, although there is increasing evidence of earlier extensional events. In the southern Canadian Cordillera, the onset of Tertiary extension is closely linked to the culmination of crustal thickening. Thrust faulting in the Rocky Mountain Belt and in the Omineca Belt to the west (Figure 31.44) culminated in the Late Palaeocene. Low-angle normal faulting and ductile spreading within the southern Omineca Belt were underway by the latest Palaeocene. Locally it has been established that the time gap between the end of thrust faulting and the onset of extension is less than 2 Ma. On a regional scale it appears that thrust faulting was still active while normal faults developed at higher structural levels. These observations suggest that gravity spreading was driving extension in much the same manner as is currently observed in the High Andes and in the Tibetan Plateau of the Himalayas. However, in the case of the North American Cordillera, the Eocene appears to have been a time of reduced plate convergence compared with the Late Cretaceous and Palaeocene, and most of the extension

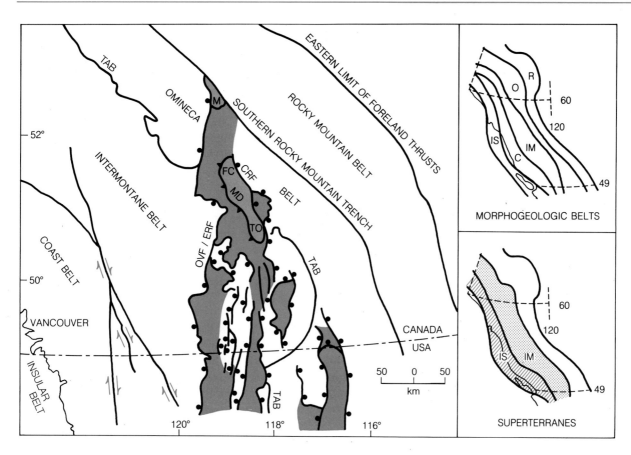

Figure 31.44 Tectonic map of southern Omineca Belt, Canadian Cordillera. (*Modified from Brown and Carr, 1990*) The upper inset locates the morphogeologic belts of the Cordillera. The Rocky Mountain Belt (R) is underlain by the North American craton and the western margin miogeocline; the Omineca Belt (O) is underlain by the Intermontane Superterrane overlapping the North American craton and miogeocline; the Intermontane Belt (IM) is underlain by the Intermontane Superterrane; the Coast Belt (C) is underlain by the Insular Superterrane overlapping the Intermontane Superterrane; and the Insular Belt (IS) is underlain by the Insular Superterrane. The lower inset locates the Intermontane (IM) and Insular (IS) superterranes. The Omineca Belt is bounded on the east by the southern Rocky Mountain Trench and on the west by the Okanagan Valley – Eagle River normal fault system (OVF/ERF) for the southern portion and terrane accretion boundary (TAB) for the northern portion. Intermontane Superterrane lies west of the terrane accretion boundary. Diagonal stripe pattern indicates the Shuswap complex. Basement rocks (dark grey pattern) are exposed in the Frenchman Cap (FC) and Thor-Odin (TO) culminations of the Monashee complex and in the Malton complex (M). The Monashee décollement (MD) is a Mesozoic-Palaeocene thrust fault; the OVF/ERF and the Columbia River fault (CRF) are Eocene normal faults.

occurred after the completion of crustal thickening.

In the Basin and Range Province of the southwestern United States (see Figure 31.42b for location), major extension was underway in the Miocene, considerably later than in the regions farther north. This activity continues today and has affected areas where there is little evidence of prior crustal thickening. This later Tertiary to Recent extension is most readily attributed directly to plate boundary stresses in a transpressional to transtensional setting.

5. Orogenic float

Surface geology together with reflection seismology support the view that the upper and part or all of the middle crustal rocks of the Cordillera have

been detached from the underlying lithosphere. It has long been recognized that the Rocky Mountain Belt is **'thin skinned'**, in that the thrust belt is underlain by a shear zone forming the sole thrust for the upper crustal faults; the crust below the sole thrust remained rigid while the overlying thrust sheets moved easterly (Figure 31.43). In contrast, the region to the west of the Rocky Mountain Belt has generally been considered 'thick skinned', with sections having large scale folding and thrusting involving the whole crust. Although ductile deformation has indeed affected the middle and lower crust of the hinterland, it has recently been demonstrated that the style of deformation in the deeper part of the crust is significantly different from that in higher levels and that this style change occurs within the middle crust well below the brittle-plastic transition zone. In the southern Canadian Cordillera the shear zone that forms the sole thrust beneath the Rocky Mountain Belt extends westward into the Omineca Belt, where it is recognized as a ductile shear zone separating middle and upper crustal rocks from middle and lower crust (MD, Figure 31.44). During the Mesozoic deformation the middle crust appears to have been weak relative to rocks above and below (Figure 31.45). This weak zone within the middle crust facilitated a decoupling of overlying crust from underlying crust. The crust above the zone of decoupling appears to float on the crust below. A similar decoupling model was proposed for the Alps by H. P. Laubscher in 1976 when he referred to the zone of nappes as the 'orogenic float'.

Detachment of upper and middle crustal rocks from the underlying crust characterized the compressional phase of the orogen; similarly, seismic data together with surface observations indicate a zone of detachment during Tertiary to Recent extension. Although in some instances low-angle normal faults extend as ductile shear zones into the lower crust and may offset the Moho, most of the major fault systems appear to flatten into the ductile middle crust. Below this middle crustal zone of detachment, extension has been accomplished by distributed strain rather than by development of clearly defined shear zones. Thus, strain partitioning that gave rise to the orogenic float appears to have been the style of crustal deformation during both compression and extension. Apparently, the boundary in the middle crust was a zone of weakness throughout the development of the orogen. It is interesting to note that this zone of detachment is generally characterized by the presence of large volumes of crustal melts in the form of granitic and pegmatitic sills. Formation of these igneous bodies would greatly weaken the crust and readily account for development of a detachment zone at this level in the crust.

6. The nature of the Moho

The North American Cordilleran assemblage (of attenuated, highly-deformed continental-margin rocks and the overthrust collage of accreted terranes) was variously extended and thinned from the Tertiary onwards. From British Columbia in

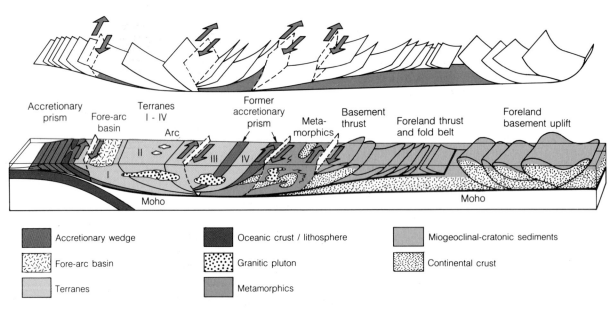

Figure 31.45 Conceptual diagram of a Cordilleran-type transpressional orogenic float. (*Modified from Oldow et al, 1989*) In order to simplify the diagram, syn- or post-orogenic extension is not considered here. The top diagram shows major decoupling systems stripped of the rock units involved in the deformation. From a deep decoupling zone at the base of the crust emanates a hierarchy of compressional decoupling levels within the basement, at the base of and within the sediments. Segments bound by strike-slip faults moved from out of the plane of the section into the plane of the section. From left to right: an accretionary wedge, involving deep-sea sediments and oceanic crust (ophiolites); several accreted terranes and former accretionary prisms (I–IV); a metamorphic core of dynamothermally reworked slices of transitional crust of the former North American passive margin; an interior basement thrust sheet (often a basement duplex); décollement foreland fold belt; and basement-involved foreland uplifts and basins. The last two zones would typically include foredeep sediments (for graphic simplicity not shown on this diagram) deposited on top of the miogeosynclinal-cratonic platform sequence. Units of the foreland belt shown on this diagram may be restored into the area now occupied by terranes transported from out of the plane of the section.

southern Canada to as far south as Mexico, the present crustal thickness beneath the extended hinterland is remarkably thin. The Moho is imaged in reflection and refraction seismic lines at a depth of 30–35 km. In reflection data, the boundary is usually sharp and clearly defined. Variations in crustal thickness and topography do not influence the depth of the Moho, which is characteristically flat and nearly horizontal. Reflections above the Moho in the lower crust are generally horizontal, and this is particularly apparent in the Basin and Range Province. These characteristics have led to the suggestion that the Moho beneath the extended terranes is a youthful feature that was established during crustal thinning. However, reflection seismic data from lines in southern British Columbia indicate that compressional shear zones are preserved in the lower crust and become horizontal adjacent to the Moho. There is no indication that these compressional features are truncated by the Moho, which would be expected if the boundary had formed during extension. It appears likely, therefore, from the seismic observations that the Moho was a zone of detachment during both the compressional and extensional events that formed the Cordillera. Limited geophysical data that give some indication of the strength of upper mantle rocks beneath the Moho indicate that the asthenosphere–lithosphere boundary lies at or near the Moho. If there was once a thicker lithosphere, it has somehow been removed from beneath the extended regions of the North American Cordillera.

7. Lithospheric root model

Rapid uplift of the Himalayan region has given rise to the highest mountains in the world, and this uplift has triggered crustal extension. There is not yet a consensus on the explanation for the Himalayan uplift, but serious consideration is being given to a model that invokes generation and mechanical detachment of a lithospheric root. The model is based on the assumption that during compression the lithosphere of the upper mantle is thickened together with the overlying continental crust, and thus a lithospheric root is generated in the upper mantle beneath the thickened crust of the orogen. The lithosphere–asthenosphere boundary is a thermal boundary layer within the mantle that defines the lower limit of rock able to sustain shear stresses. Because the temperature of the lithosphere is lower than that of the asthenosphere, the upper mantle rocks of the lithosphere have a higher density than the underlying asthenosphere. Theoretical modelling suggests that such a root would be mechanically unstable, and would tend to detach itself from the lithosphere and sink into the asthenosphere, where it would

heat up and gradually be incorporated into the asthenosphere. Detachment of such a lithospheric root would disturb the isostatic equilibrium, since the orogen would have been held down by the presence of the high density root. With detachment of this root rapid uplift would be expected until equilibrium were re-established. The model appears to satisfy the available observational data in the Himalayan orogen and may also be applicable in the North American orogen. Loss of lithosphere would explain the rapid uplift, and relaxation of the compressional stresses due to the downward pull of the root could also account for the onset of extension in the overlying crust. In the southern Canadian Cordillera, rapid uplift and extension also coincided with a change in plate boundary stresses that facilitated relaxation and crustal extension normal to the trend of the orogen (Figure 31.46).

8. Transpression

Major strike-slip displacements, which overlapped in time with episodes of crustal shortening, have significantly affected the North American Cordillera. This partitioning of strain into orogen-parallel and orogen-normal components has occurred in order to accommodate the oblique convergence of the plate boundaries. It is not yet clear how the orogen-parallel component is accommodated in the lower crust beneath the detachment zone. However, there is support from seismic reflection data for the notion that the nearly vertical strike-slip faults cutting the upper crust flatten out in the middle crust to merge with the detachment zone. Recognition of this style of deformation implies that orogen-parallel deformation must be accommodated by distributed strain in the lower crust. These observations clearly demonstrate that an understanding of the tectonic evolution of an orogenic belt requires a three-dimensional knowledge of its crustal structure. Attempts to draw balanced cross-sections based on the assumption of plane strain must be approached with this in mind.

An outstanding problem in the unravelling of the tectonics of the North American Cordillera is reconciliation of palaeomagnetic and geological data. Palaeomagnetic interpretations calling for up to 2000 km of latitudinal displacements of accreted terranes relative to the continental margin of North America are not supported by the known surface geology. Right-lateral strike-slip faults such as the Tintina fault in the Northern Rocky Mountain Trench may account for at least 500 km of displacement, but this decreases to significantly less in the south and has no recognized strike-slip component at the latitude of the Canada–USA border. There is no agreement at the moment as to

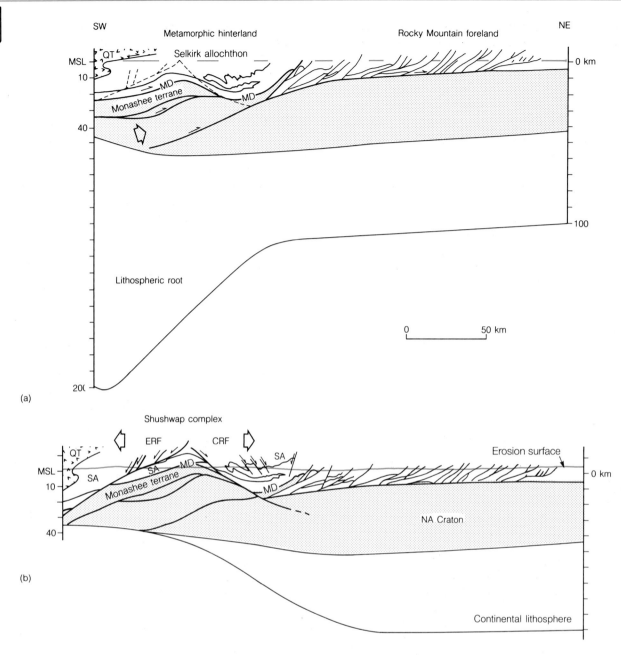

Figure 31.46 (a) Simplified cross-section at end of compressive phase in the southern Canadian Cordillera illustrates attached lithospheric root drawn to balance crustal shortening in Late Cretaceous and Palaeocene times. QT, Quesnel terrane of the Intermontane Superterrane; metamorphic hinterland, Omineca Belt; MD, Monashee décollement; MSL, mean sea-level. Rocky Mountain foreland, simplified from *Price and Mountjoy (1970)*; crustal duplex of metamorphic hinterland modified from *Brown et al (1986)* and *Journeay (1986)*; geometry of Selkirk allochthon simplified from *Brown and Lane (1988)*. (b) In the Late Palaeocene to Eocene the root is detached and the crustal welt is uplifted and denuded. ERF, Eagle River fault; CRF, Columbia River fault; SA, Selkirk allochthon; topography is present-day erosion level. Crustal section is modified after *Brown and Journeay (1987)*, as published in *Ranalli et al (1989)* and *Brown and Carr (1990)*.

how this strike-slip motion has been distributed. Other faults account for perhaps an additional 500 km of right-lateral displacement, but again there are major difficulties in attempting to join these faults or have them linked by some type of distributed strain. The answer in part may lie in suggested revisions of palaeomagnetic interpretations based on tilting estimates and palaeomag-

netic reference pole recalculations. Such revisions would reduce the amount of required displacements, so that the palaeomagnetic results would have error estimates overlapping the range of estimated displacements based on geological data. Also, further geological evidence of large orogen-parallel displacements may be forthcoming. Nevertheless, considerable work remains to be done before a clear picture of the origin and displacement history of these accreted terranes emerges.

31.11 THE TRANSANTARCTIC MOUNTAINS

When Captain Robert Falcon Scott and his companions first set out into the interior of the Antarctic continent in search of a route to the South Pole, they found their way barred by a formidable mountain barrier bordering the western side of the Ross Sea. Eventually a hazardous way was found through this barrier by traversing one of the heavily crevassed glaciers that flow through structural weaknesses in the mountains to form the ice-shelf that covers much of the Ross Sea embay-

ment of the Antarctic continent. It was such a route that Amundsen and Scott both took in their race to the Pole in 1911–1912. The Transantarctic Mountains, that are now known to cross the frozen continent from the Ross Sea south of New Zealand to the Weddell Sea south of Africa (Figure 31.47), constitute one of the Earth's most imposing ranges. Unlike the Alpine–Himalayan belt, the Andes and the North American Cordillera, however, they were not formed at a convergent plate boundary. Instead they flank a late-Mesozoic and Cenozoic intraplate and intracontinental rift system. They are 100–200 km wide and have the highest elevation (4–5 km), and the greatest relief (5–7 km) of any rift-flanking mountain chain. By way of comparison, the Wasatch Mountains bor-

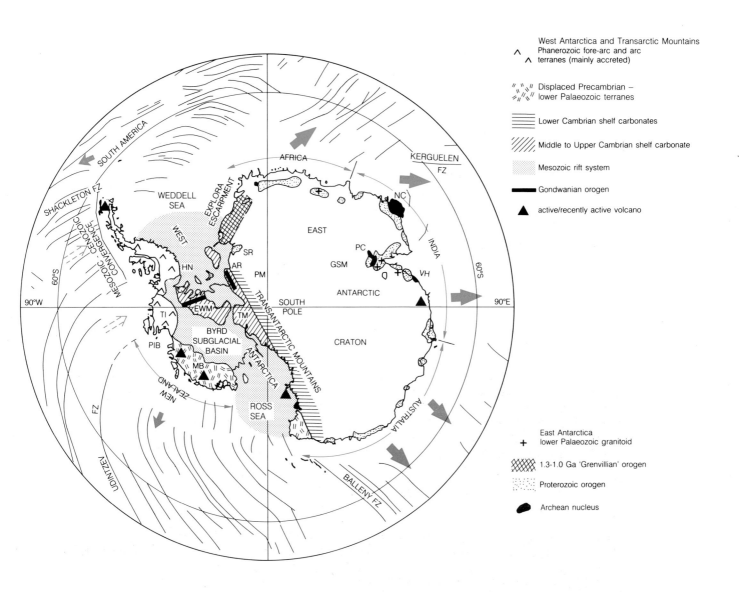

Figure 31.47 Simplified tectonic map of the Antarctic and surrounding ocean basins. Bold arrows show directions taken by the major fragments of the Gondwana supercontinent during fragmentation. (*I. W. D. Dalziel, 1992*)

dering the Basin and Range Province of North America, and the uplift along the margins of the Red Sea and Gulf of Suez are no more than 1000 to 2000 km high. Moreover, the 'half-wavelength' of these uplifts, that is to say the horizontal distance between the axis of maximum uplift and the adjacent topographic minimum, is generally on the order of 150 km while that of the Transantarctic Mountains is 400–500 km. The range has fascinated geologists since the days of the early explorers. The isolated cone of Mount Erebus at the base of the mountain front (Plate XVe) was erupting violently when Sir James Clark Ross first penetrated the ice-bound sea that bears his name. The geologists on the expeditions of Scott and Sir Ernest Shackleton, drew attention to the absence of internal deformation in the Transantarctic Mountains, in contrast to the Alps, and the absence of andesitic volcanics and granitic batholiths compared with the Andean and North American cordilleras. Instead they were found to have a basically simple structure and non-volcanic stratigraphy. A Precambrian and lower Palaeozoic basement, comprising igneous and metamorphic rocks, are unconformably overlain by flat-lying Devonian to Triassic sedimentary strata of the

Beacon Supergroup (Figure 31.48).

The Precambrian rocks are part of the Antarctic fragment of the Gondwanaland Precambrian craton from which a major continent, possibly North America, was rifted in the late Precambrian. The Palaeozoic rocks constitute the complex Ross orogen, an eroded mountain range that formed in a convergent regime along the craton margin continuous with the basement rocks of the Andes in South America. The Ross orogen trends parallel to the present-day Transantarctic Mountains. The basement was uplifted and eroded in the Silurian and Early Devonian to form the Kukri peneplain. The overlying Beacon Supergroup consists of fluvial, deltaic and shallow-marine sedimentary strata about 2.5 km thick. Both the basement and the cover are intruded by numerous dykes and transgressing sills of the Middle Jurassic Ferrar dolerites. The mafic magma also reached the surface to form the Kirkpatrick basalts, although not in the volumes associated with plateau lavas. There are a few felsic differentiates. The light-coloured sub-horizontal strata of the Beacon Supergroup (with their sills of black Ferrar dolerite (Figures 3.12, 31.48) exposed in ice-free valleys like the Grand Canyon of the Colorado River in

Figure 31.48 Transantarctic Mountains: flat-lying Palaeozoic to Lower Mesozoic strata of the Beacon Supergroup cut by Jurassic dolerites (dark coloured). (*I. W. D. Dalziel*)

North America, accentuate the fact that uplift has taken place by vertical crustal movements with limited differential tilting of fault blocks. The side of the mountains bordering the Ross Sea is step-faulted down to the coast. There are Late Pliocene or younger faults with up to 300 m of displacement along the Transantarctic Mountains front. The Kukri peneplain can be used as a reference surface for the uplift of the mountains because usually it lies parallel to the Beacon strata and the Ferrar dolerite sills. The axis of maximum uplift lies approximately 30 km inland from the coast. West of the axis of maximum uplift the peneplain dips gently to the west under the polar ice-cap, although the elevation of the mountains does not diminish as the thickness of the Beacon Supergroup increases in that direction (Figure 31.49). The glacier-filled transverse valleys appear, nonetheless, to have major structural significance. For example, the Byrd Glacier coincides with an offset in the coast, the Transantarctic Mountains front and associated gravity anomaly, and the locus of Cenozoic volcanic centres on the western side of the Ross Sea. The associated lineament continues to the northern coast of Marie Byrd Land along the Pacific margin.

Apatite fission track analysis has provided data on the timing and rates of uplift of the Transantarctic Mountains. Ages obtained by this method on the basement rocks of the Dry Valleys of South Victoria Land by P.G. Fitzgerald and colleagues show a strong correlation with elevation, and the shape of the age-elevation profile indicates a two-stage uplift history (Figure 31.50). The older part of the profile has a shallow gradient, while the part younger than c. 50–55 Ma, has a much steeper gradient. The shallow gradient is believed to represent partial annealing of the crystals following the thermal event associated with the intrusion of the Ferrar dolerites, a fossil 'partial annealing zone'. The pronounced increase in gradient at 50–55 Ma marks uplift of the Transantarctic Mountains. However, this was probably initiated at about 60 Ma, as the base of the annealing zone had

to be uplifted through the partial annealing zone.

The thickness of Palaeozoic granite, Beacon sedimentary strata, and diabase/basalt above the break in slope provides a minimum value for the thickness prior to uplift. When added to the elevation of the base of the uplifted annealing zone, Fitzgerald and his colleagues obtained a range of 4.8–5.3 km for minimum uplift since the Jurassic if they assumed an elevation of 500 m for the land surface following extrusion of the Kirkpatrick basalts. This in turn yields an estimated average uplift rate of 100 ± 5 m/Ma during the time interval 50 Ma to the Present. An alternative method of measuring the uplift based on the measured thermal gradient in the Dry Valleys and the present elevation of the base of the uplifted annealing zone yields 4.8 km, assuming the same value for the elevation of the Jurassic surface and a mean annual surface temperature of 0°C. The elevation estimate for the Kirkpatrick basalts seems to be reasonable given the fact that they were extruded in an alluvial flood-plain setting with ponds and streams. The distinctive shape of an apatite partial-annealing zone will not be preserved unless the pre-uplift denudation rate is less than 30 m/Ma. Hence a period of thermal and tectonic stability must have existed in the Late Cretaceous, before uplift began in the early Cenozoic.

A virtually identical uplift of c. 4.7–5.2 km was determined for most of North Victoria Land indicating, despite differences in the Palaeozoic tectonic history along the length of the orogen, a uniform Cenozoic uplift history for the Transantarctic Mountains in the area of the Ross embayment. This uplift and erosion rate is similar to the rate of 100–200 m/Ma determined for the Sinai Peninsula bordering the Red Sea rift, and somewhat lower than the rate of uplift of the Wasatch Mountains bordering the Basin and Range Province of North America (2–2.5km/Ma). Even though the figure of 100 m/Ma for uplift of the Transantarctic Mountains is an average over the past 50 Ma, and the rate may have been significantly higher at

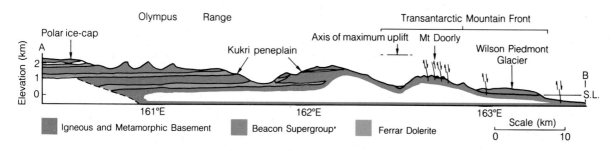

Figure 31.49 Geological cross-section of the Transantarctic Mountains through the Dry Valleys region showing the step faulted nature of the Transantarctic Mountain Front. Note the gentle westerly dip of the Kukri Peneplain and Beacon Supergroup strata, eventually resulting in their disappearance under the polar ice-cap. (*P.G. Fitzgerald et al, 1986*)

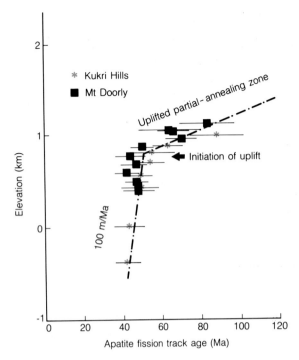

Figure 31.50 Apatite fission-track age versus elevation graph for two sampling profiles in South Victoria Land. The 'break in slope' of the graph at about 50 Ma represents the start of uplift of the Transantarctic Mountains in this area and the gradient of the lower slope represents the average uplift rate. The data from the two separate profiles plot almost perfectly together indicating that they lie in similar positions within the Transantarctic Mountain Front. (*After P.G. Fitzgerald et al, 1986*)

certain periods within that time interval, uplift rates in orogens at convergent plate boundaries are commonly an order of magnitude higher, for example 1000 m/Ma for Nanga Parbat in the Himalayas, 500–1000 m/Ma for Cordillera Darwin in the southernmost Andes, and 400–1000 m/Ma in the European Alps. The Southern Alps of New Zealand, along the obliquely-convergent Indian–Pacific plate boundary have uplift rates calculated to be as high as 10 000 m/Ma. Detailed study of the fission track lengths in samples from the Transantarctic Mountains bordering the Ross Sea has confirmed the indications from the dip of the Kukri peneplain that the amount of uplift dies out toward the west. Differential tilting of fault blocks is apparent from observation of the Kukri peneplain as well as from the fission track data, and the Rennick graben that cuts across north Victoria Land parallel to the Terror Rift in the Ross Sea, is a zone of differential movement directly related to the uplift of the mountains, because it marks a boundary between zones of greater (to the east) and lesser uplift.

Understanding the uplift of the Transantarctic Mountains means taking into account not merely the mountains themselves, but also the basins of the Ross embayment and the East Antarctic craton adjacent to the uplift zone. On the cratonic side of the mountains there is a subglacial basin 200–600 km wide that is set back 400–500 km from the mountains and marks the axis of a negative free-air gravity anomaly of 20 mgal or more, frequently referred to as the 'Transantarctic gravity anomaly'. Like the mountains themselves, the anomaly is most pronounced in the area adjacent to the Ross Sea. The basin is known as the Wilkes Subglacial Basin. It continues through the South Pole Basin and the Pensacola Basin to the Weddell Sea embayment. The gravity anomaly of −20 to −40 mgal has been interpreted as the result of 2–3 km of low-density sediments within the basin.

The Ross embayment is filled with an epicontinental sea partially covered by an ice-shelf approximately the size of Texas. The presence of calc-silicate gneiss at the base of a Deep Sea Drilling Project drill site there suggests that the basement of the Ross Sea is continental. The continental shelf in the embayment averages approximately 500 m deep, ranging from 200 m to 1100 m. It is divided by three north–south to northeast–southwest trending ridges into three sedimentary basins, the Eastern Basin, the Central Trough, and the Victoria Basin. Seismic reflection profiles show the basin infills to be deformed only by occasional normal faults that do not reach the surface. It is the structure of the 150 km wide Victoria Land Basin adjacent to the Transantarctic Mountain front that is particularly important from the point of view of understanding the orogenic processes (Figure 31.51). The basin is an asymmetric half-graben. It contains a 14 km thick sequence of sub-horizontal strata that are offset by normal faults. The faults, some of which reach the sea floor, show a progressive displacement with depth, and are therefore syndepositional. The syndepositional faulting has created a 50–60 km wide central rift, the Terror Rift, along the axis of which there are submerged volcanic centres. The rift line connects the active Mount Erebus volcano on Ross Island with Mount Melbourne on Cape Washington that has fumaroles, thus demonstrating the synchroneity of sedimentation, faulting, and volcanism. The Central Trough and the Eastern Basin are both shallower than the Victoria Land Basin. Neither has a central rift nor volcanic edifices, but both appear to have formed in an extensional regime. All three basins are marked by positive Bouguer gravity anomalies that extend for some distance over the Ross Ice Shelf.

The sedimentary sequence in the Victoria Land Basin has been divided from seismic work into seven acoustic units separated by discontinuous unconformities. Units 1 and 2 are rocks younger

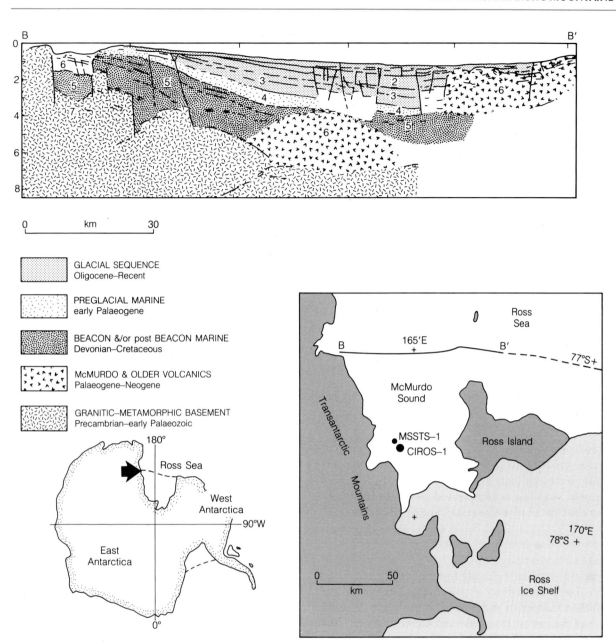

Figure 31.51 Cross-section across the western side of the Victoria Land Basin based on a sketch from multichannel seismic data presented in Cooper and Davey (1987). Numbers 1–7 refer to seismic units. (*After P.J. Barrett, 1989*)

than late Oligocene, the age of onset of glaciation in the Ross Sea; Late Oligocene and younger sedimentary rocks have been drilled in the Ross Sea. Units 3 and 4 include likely Palaeogene and older marine sediments; recycled Late Cretaceous foraminifera and Late Cretaceous and Tertiary palynomorphs suggest that strata of this age may have been exposed on the margins of the Ross Sea. Unit 5 is a high-velocity sedimentary unit of possible late Mesozoic to Palaeogene age filling in the lower 6–14 km of the basin. Unit 6 is interpreted as volcanics of possible Mesozoic to late Cenozoic age. Unit 7 is the acoustic basement of probable pre-Jurassic sedimentary, igneous and metamorphic rocks. The units are thickest along

the axis of the Terror Rift and thin onto its flanks. At the south end of the Victoria Land Basin a feature known as the Lee Arch fills the eastern part of the Terror Rift, the remaining part of the rift in the west being referred to as the Discovery Graben. The sedimentary section of the Victoria Land Basin as a whole presumably thins and ends at the northern end of the basin, but may extend south under the Ross Ice Shelf. The extent of the Terror Rift in the south is uncertain because active volcanism ends at Mount Erebus, although the Cenozoic volcanic zone can be traced as far as Mount Early at the Scott Glacier (87° 04'S).

Adjacent to the Transantarctic Mountains front, a zone of horsts, grabens, intrusives, and large-offset normal faults has been recognized beneath the western edge and flank of the basin. There are graben and arch structures in the Terror Rift, and normal faults beneath the eastern flanks of the Victoria Land Basin. Grabens and half-grabens occur on the Coulman High that bounds the basin in the north-east, and intrusives and associated volcanic structures are common in the southern part of the basin towards Ross Island. The crust at Ross Island is depressed approximately 1800 m by the volcanic load, and there is a 100 m high outer flexural bulge at a distance of 200 km from the centre of the mass. The flexural rigidity of the crust that can be calculated from this data is 10^{23} Nm, that is low for crust that may be approximately 500 Ma. This indicates that the thermal age of the crust in the western Ross Sea has been reset during the Cenozoic rifting and volcanism.

It is generally believed that there were two structural phases in the development of the Ross Sea embayment, an early rifting phase in which the down-faulting of the basement half-graben underlying the Victoria Land Basin occurred, and a late rifting phase in which the younger faults of the Transantarctic Mountains front, and the faults and volcanics of the Terror Rift, were formed. The Victoria Land Basin appears to have formed on top of the broad fluvial basin in which the Beacon strata were deposited. It is likely that uplift and regional warping accompanied the Jurassic Ferrar magmatism as elsewhere in Gondwanaland, and that the Victoria Land Basin may have originated as a graben at that time, as did the grabens of the Weddell Sea embayment.

Crustal extension during the early rift phase may have been related to the initiation of break up between Antarctica and Australia and New Zealand in the Cretaceous. Renewed volcanism in the southern part of the basin and initial uplift of the Transantarctic Mountains during the early Palaeogene started the late rift phase that has continued to the present day. This phase could have been triggered by changes in the plate configuration of the Antarctic–Australasian area during the Eocene that is marked by a 40° change in the magnetic anomaly pattern south of New Zealand.

T. Stern and U. ten Brink contend that the uplift of the Transantarctic Mountains is supported by a broad flexure of the East Antarctic lithosphere and are, therefore, regionally rather than locally compensated. They document the flexure of the lithosphere at the Transantarctic margin of the East Antarctic craton using the topography of the Transantarctic Mountains and the sub-ice topography of the Wilkes basin to the west, as determined by radioecho sounding. They then develop a flexural model using geological observations to constrain the uplift, half-wavelength of the flexure, and

regional dip of the Kukri peneplain. The apatite fission track studies discussed above indicate a geologically reasonable uplift of c. 5 km over the past 50 Ma. The half-wavelength is determined from the stacked topographic profiles as 450–500 km, the distance from the axis of maximum uplift of the Transantarctic Mountains to the deepest part of the Wilkes basin. A dip of 2–3° beneath the polar plateau, at 30–40 km inland from the coast, was adopted as the regional dip of the Kukri peneplain. It was found that the simplest model that can account for the uplift profile and the observed gravity anomalies across the Transantarctic Mountains is one where two lithospheric plates of vastly different effective thermal ages are juxtaposed and where no shear stresses are transferred across their common boundary (Figure 31.52). Clearly this could account for the fact that the uplift of the Transantarctic Mountains is much higher than that associated with the Red Sea, East African, and Basin and Range systems, where the rocks on either side of the bounding faults are of essentially the same age and crustal properties. It could also account for the fact that there is far less relief on the Marie Byrd Land side of the rift where the bounding fault system is wholly within the Panthalassic margin Phanerozoic accretionary

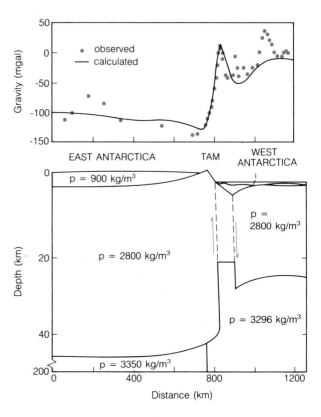

Figure 31.52 Gravity profile and crustal model of the Transantarctic Mountains. (*After T. Stern and U. ten Brink, 1989*)

wedge. The late Precambrian origin of the Transantarctic margin of the East Antarctic Shield as a major Atlantic-type rift system may be very important, with the rift geometry possibly controlling orogenic uplift of the Transantarctic Mountains 500 m.y. later.

The three principal uplift forces for the Transantarctic Mountains were found to be thermal uplift, erosion, and the Vening Meinez effect (whereby the footwall of a lithospheric fault experiences a positive buoyancy and the hanging wall a negative buoyancy after elastic failure along an inclined surface). The unusually pronounced flexural effect along the margin of the East Antarctic craton is ascribed to its high rigidity, 1×10^{25} Nm, corresponding to a lithospheric elastic thickness of 115 ± 10 km and an effective thermal age of about 600 Ma. This contrasts with 4×10^{22} Nm, 19 ± 5 km and 25 Ma for the lithosphere between the Ross embayment. The highly rigid East Antarctic lithosphere gives rise to a long (c. 500 km) flexural half-wavelength. Stern and ten Brink suggest that, in addition, the long-lived ice-cap has slowed and evenly distributed the erosional process. Finally, the Wilkes basin can be regarded as a flexurally-controlled depression in the lithosphere produced as a form of regional compensation for the uplift of the Transantarctic Mountains. Despite several suggestions over the years that the ice-cap may play a major role in lithospheric deformation, Stern and ten Brink's analysis indicates that this is not the case if they are correct in their proposal that the East and West Antarctic lithospheric plates are effectively uncoupled. Under these circumstances there will be differential movement between the two as ice-sheets wax and wane, but the movements will be small enough ($\pm$ c. 100 m) to constitute no more than noise on the signal of Transantarctic Mountains uplift. The effect may, however, have influenced sedimentation in the Ross Sea basins and be responsible for erosional unconformities.

The remaining questions regarding this hypothesis seem to be the geological ones that centre on how the thinner lithosphere of West Antarctica was created, how it was juxtaposed with the East Antarctic craton, and the cause of Cenozoic heating and stretching in the Ross embayment. The origin of the thinner West Antarctic lithosphere seems to lie in its origin as an accretionary and magmatic addition to the Pacific side of the craton during the Phanerozoic. Some extension, thinning and heating clearly accompanied diachronous fragmentation of the Gondwanaland supercontinent during the late Mesozoic, and it is probable that rifting in the Ross embayment was started at that time. Stern and ten Brink calculate that it will take about 70 m.y. for the associated density contrast that is necessary to generate thermal uplift to penetrate

50 km beneath the edge of the East Antarctic craton. The apatite fission track data from the Transantarctic Mountains are now believed to indicate that uplift was initiated at approximately 60 Ma. Volcanism in the West Antarctic Rift Zone goes back to approximately 40 Ma.

There are three possible explanations of the thermal and mechanical regime in the Cenozoic. First, the Ross embayment could be the site of a failed arm of rifting during the later stages of Gondwanaland fragmentation, comparable to the Benue Trough in West Africa. Second, the Antarctic plate may have overridden anomalously hot mantle. Third, the stationary position of the Antarctic plate during the Cenozoic may have been critical. There is no systematic trend to the numerous linear Cenozoic alkaline volcanic chains on the plate, and ridges have grown away from the continent as it became stationary over the south pole.

As an alternative type of hypothesis, Fitzgerald and his colleagues have suggested that the Transantarctic Mountains uplift is the result of a lithospheric low-angle simple shear zone of the type originally proposed by Wernicke (for the Basin and Range Province). According to this concept, failure of the lithosphere produces asymmetric uplift and subsidence on either side of a low-angle shear. In the Antarctic case the East Antarctic craton would be on the upper plate. The effect would be the opposite of the Vening–Meinesz model, where the uplifted craton margin is on the footwall and this may be testable by seismic experiments that are now planned. As pointed out by Stern and ten Brink, however, if reasonable estimates of the amount of extension in the Ross embayment are assumed, the simple shear mechanism would account for only 1 km rather than the observed 5 km of uplift in the upper plate.

31.12 FURTHER READING

Dalziel, I. W. D. (1992) Antarctica: a tale of two supercontinents? *Ann. Rev. Earth Planet. Sci.*, **20**, 501–26.

Dalziel, I. W. D. and Brown, R. L. (1989) Tectonic denudation of the Darwin metamorphic core complex in the Andes of Tierra del Fuego, Southern Chile. *Geology*, **17**, 699–703.

Keary, P. and Vine, F.J. (1990) *Global Tectonics*, Blackwell Scientific Publications, Oxford.

Oldow, J.S., Bally, A.W., Ave Lallemant, H.G. and Leeman, W.P. (1989) Phanerozoic evolution of the North American Cordillera; United States and Canada, in *The geology of North America – an Overview* (eds A.W. Bally and A.R. Palmer), Geological Society of America, Boulder, Colorado.

INDEX

Page numbers appearing in *italic* refer to figures and page numbers appearing in **bold** refer to tables.